Student Solutions Manual
for
MULTIVARIABLE CALCULUS
EIGHTH EDITION

DAN CLEGG
Palomar College

BARBARA FRANK
Cape Fear Community College

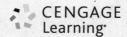
CENGAGE
Learning·

Australia · Brazil · Mexico · Singapore · United Kingdom · United States

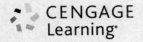
CENGAGE
Learning®

For product information and technology assistance, contact us at
Cengage Learning Customer & Sales Support,
1-800-354-9706

For permission to use material from this text or product, submit all requests online at
www.cengage.com/permissions
Further permissions questions can be emailed to
permissionrequest@cengage.com

ISBN: 978-1-305-27182-1

Cengage Learning
20 Channel Center Street
Boston, MA 02210
USA

Cengage Learning is a leading provider of customized learning solutions with employees residing in nearly 40 different countries and sales in more than 125 countries around the world. Find your local representative at: **www.cengage.com**.

Cengage Learning products are represented in Canada by Nelson Education, Ltd.

To learn more about Cengage Learning Solutions, visit **www.cengage.com**.

Purchase any of our products at your local college store or at our preferred online store **www.cengagebrain.com**.

Printed in the United States of America
Print Number: 02   Print Year: 2015

PREFACE

This *Student Solutions Manual* contains detailed solutions to selected exercises in the text *Multivariable Calculus,* Eighth Edition (Chapters 10–17 of *Calculus,* Eighth Edition, and *Calculus: Early Transcendentals,* Eighth Edition) by James Stewart. Specifically, it includes solutions to the odd-numbered exercises in each chapter section, review section, True-False Quiz, and Problems Plus section.

Because of differences between the regular version and the *Early Transcendentals* version of the text, some references are given in a dual format. In these cases, readers of the *Early Transcendentals* text should use the references denoted by "ET."

Each solution is presented in the context of the corresponding section of the text. In general, solutions to the initial exercises involving a new concept illustrate that concept in more detail; this knowledge is then utilized in subsequent solutions. Thus, while the intermediate steps of a solution are given, you may need to refer back to earlier exercises in the section or prior sections for addition-al explanation of the concepts involved. Note that, in many cases, different routes to an answer may exist which are equally valid; also, answers can be expressed in different but equivalent forms. Thus, the goal of this manual is not to give the definitive solution to each exercise, but rather to assist you as a student in understanding the concepts of the text and learning how to apply them to the chal-lenge of solving a problem.

We would like to thank James Stewart for entrusting us with the writing of this manual and offering suggestions, Gina Sanders for reviewing solutions for accuracy and style, and the staff of TECH-arts for typesetting and producing this manual as well as creating the illustrations. We also thank Richard Stratton, Neha Taleja, Samantha Lugtu, Stacy Green, and Terry Boyle of Cengage Learning, for their trust, assistance, and patience.

DAN CLEGG
Palomar College

BARBARA FRANK
Cape Fear Community College

ABBREVIATIONS AND SYMBOLS

CD	concave downward
CU	concave upward
D	the domain of f
FDT	First Derivative Test
HA	horizontal asymptote(s)
I	interval of convergence
IP	inflection point(s)
R	radius of convergence
VA	vertical asymptote(s)
$\overset{\text{CAS}}{=}$	indicates the use of a computer algebra system.
$\overset{\text{PR}}{\Rightarrow}$	indicates the use of the Product Rule.
$\overset{\text{QR}}{\Rightarrow}$	indicates the use of the Quotient Rule.
$\overset{\text{CR}}{\Rightarrow}$	indicates the use of the Chain Rule.
$\overset{\text{H}}{=}$	indicates the use of l'Hospital's Rule.
$\overset{j}{=}$	indicates the use of Formula j in the Table of Integrals in the back endpapers.
$\overset{s}{=}$	indicates the use of the substitution $\{u = \sin x, \, du = \cos x \, dx\}$.
$\overset{c}{=}$	indicates the use of the substitution $\{u = \cos x, \, du = -\sin x \, dx\}$.

CONTENTS

16   VECTOR CALCULUS   303

17   SECOND-ORDER DIFFERENTIAL EQUATIONS   345

APPENDIX   357

10 ☐ PARAMETRIC EQUATIONS AND POLAR COORDINATES

10.1 Curves Defined by Parametric Equations

1. $x = 1 - t^2$, $y = 2t - t^2$, $-1 \leq t \leq 2$

t	-1	0	1	2
x	0	1	0	-3
y	-3	0	1	0

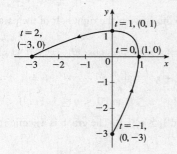

3. $x = t + \sin t$, $y = \cos t$, $-\pi \leq t \leq \pi$

t	$-\pi$	$-\pi/2$	0	$\pi/2$	π
x	$-\pi$	$-\pi/2 + 1$	0	$\pi/2 + 1$	π
y	-1	0	1	0	-1

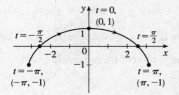

5. $x = 2t - 1$, $y = \frac{1}{2}t + 1$

(a)

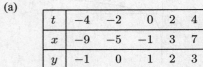

t	-4	-2	0	2	4
x	-9	-5	-1	3	7
y	-1	0	1	2	3

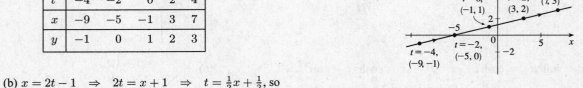

(b) $x = 2t - 1 \Rightarrow 2t = x + 1 \Rightarrow t = \frac{1}{2}x + \frac{1}{2}$, so

$$y = \frac{1}{2}t + 1 = \frac{1}{2}\left(\frac{1}{2}x + \frac{1}{2}\right) + 1 = \frac{1}{4}x + \frac{1}{4} + 1 \Rightarrow y = \frac{1}{4}x + \frac{5}{4}$$

7. $x = t^2 - 3$, $y = t + 2$, $-3 \leq t \leq 3$

(a)

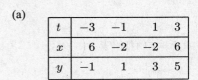

t	-3	-1	1	3
x	6	-2	-2	6
y	-1	1	3	5

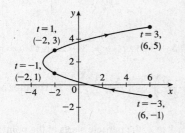

(b) $y = t + 2 \Rightarrow t = y - 2$, so

$$x = t^2 - 3 = (y - 2)^2 - 3 = y^2 - 4y + 4 - 3 \Rightarrow$$

$$x = y^2 - 4y + 1, \ -1 \leq y \leq 5$$

9. $x = \sqrt{t}$, $y = 1 - t$

(a)

t	0	1	2	3	4
x	0	1	1.414	1.732	2
y	1	0	-1	-2	-3

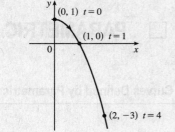

(b) $x = \sqrt{t} \;\Rightarrow\; t = x^2 \;\Rightarrow\; y = 1 - t = 1 - x^2$. Since $t \geq 0$, $x \geq 0$.

So the curve is the right half of the parabola $y = 1 - x^2$.

11. (a) $x = \sin \frac{1}{2}\theta$, $y = \cos \frac{1}{2}\theta$, $-\pi \leq \theta \leq \pi$.

$x^2 + y^2 = \sin^2 \frac{1}{2}\theta + \cos^2 \frac{1}{2}\theta = 1$. For $-\pi \leq \theta \leq 0$, we have

$-1 \leq x \leq 0$ and $0 \leq y \leq 1$. For $0 < \theta \leq \pi$, we have $0 < x \leq 1$

and $1 > y \geq 0$. The graph is a semicircle.

(b)

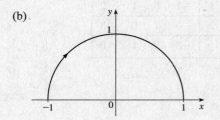

13. (a) $x = \sin t$, $y = \csc t$, $0 < t < \frac{\pi}{2}$. $y = \csc t = \dfrac{1}{\sin t} = \dfrac{1}{x}$.

For $0 < t < \frac{\pi}{2}$, we have $0 < x < 1$ and $y > 1$. Thus, the curve is

the portion of the hyperbola $y = 1/x$ with $y > 1$.

(b)

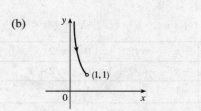

15. (a) $y = \ln t \;\Rightarrow\; t = e^y$, so $x = t^2 = (e^y)^2 = e^{2y}$.

(b)

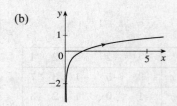

17. (a) $x = \sinh t$, $y = \cosh t \;\Rightarrow\; y^2 - x^2 = \cosh^2 t - \sinh^2 t = 1$.

Since $y = \cosh t \geq 1$, we have the upper branch of the hyperbola

$y^2 - x^2 = 1$.

(b)

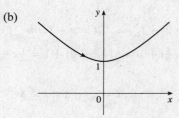

19. $x = 5 + 2\cos \pi t$, $y = 3 + 2\sin \pi t \;\Rightarrow\; \cos \pi t = \dfrac{x - 5}{2}$, $\sin \pi t = \dfrac{y - 3}{2}$. $\cos^2(\pi t) + \sin^2(\pi t) = 1 \;\Rightarrow$

$\left(\dfrac{x - 5}{2}\right)^2 + \left(\dfrac{y - 3}{2}\right)^2 = 1$. The motion of the particle takes place on a circle centered at $(5, 3)$ with a radius 2. As t goes

from 1 to 2, the particle starts at the point $(3, 3)$ and moves counterclockwise along the circle $\left(\dfrac{x - 5}{2}\right)^2 + \left(\dfrac{y - 3}{2}\right)^2 = 1$ to

$(7, 3)$ [one-half of a circle].

21. $x = 5\sin t$, $y = 2\cos t$ $\Rightarrow$ $\sin t = \dfrac{x}{5}$, $\cos t = \dfrac{y}{2}$. $\sin^2 t + \cos^2 t = 1$ $\Rightarrow$ $\left(\dfrac{x}{5}\right)^2 + \left(\dfrac{y}{2}\right)^2 = 1$. The motion of the

particle takes place on an ellipse centered at $(0,0)$. As t goes from $-\pi$ to 5π, the particle starts at the point $(0,-2)$ and moves

clockwise around the ellipse 3 times.

23. We must have $1 \le x \le 4$ and $2 \le y \le 3$. So the graph of the curve must be contained in the rectangle $[1,4]$ by $[2,3]$.

25. When $t = -1$, $(x,y) = (1,1)$. As t increases to 0, x and y both decrease to 0.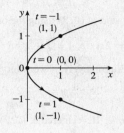

As t increases from 0 to 1, x increases from 0 to 1 and y decreases from 0 to

-1. As t increases beyond 1, x continues to increase and y continues to

decrease. For $t < -1$, x and y are both positive and decreasing. We could

achieve greater accuracy by estimating x- and y-values for selected values of t

from the given graphs and plotting the corresponding points.

27. When $t = -1$, $(x,y) = (0,1)$. As t increases to 0, x increases from 0 to 1 and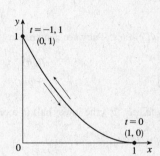

y decreases from 1 to 0. As t increases from 0 to 1, the curve is retraced in the

opposite direction with x decreasing from 1 to 0 and y increasing from 0 to 1.

We could achieve greater accuracy by estimating x- and y-values for selected

values of t from the given graphs and plotting the corresponding points.

29. Use $y = t$ and $x = t - 2\sin \pi t$ with a t-interval of $[-\pi, \pi]$.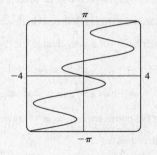

31. (a) $x = x_1 + (x_2 - x_1)t$, $y = y_1 + (y_2 - y_1)t$, $0 \le t \le 1$. Clearly the curve passes through $P_1(x_1, y_1)$ when $t = 0$ and

through $P_2(x_2, y_2)$ when $t = 1$. For $0 < t < 1$, x is strictly between x_1 and x_2 and y is strictly between y_1 and y_2. For

every value of t, x and y satisfy the relation $y - y_1 = \dfrac{y_2 - y_1}{x_2 - x_1}(x - x_1)$, which is the equation of the line through

$P_1(x_1, y_1)$ and $P_2(x_2, y_2)$.

Finally, any point (x,y) on that line satisfies $\dfrac{y - y_1}{y_2 - y_1} = \dfrac{x - x_1}{x_2 - x_1}$; if we call that common value t, then the given

parametric equations yield the point (x,y); and any (x,y) on the line between $P_1(x_1, y_1)$ and $P_2(x_2, y_2)$ yields a value of

t in $[0,1]$. So the given parametric equations exactly specify the line segment from $P_1(x_1, y_1)$ to $P_2(x_2, y_2)$.

(b) $x = -2 + [3 - (-2)]t = -2 + 5t$ and $y = 7 + (-1 - 7)t = 7 - 8t$ for $0 \le t \le 1$.

33. The circle $x^2 + (y-1)^2 = 4$ has center $(0, 1)$ and radius 2, so by Example 4 it can be represented by $x = 2\cos t$, $y = 1 + 2\sin t, 0 \le t \le 2\pi$. This representation gives us the circle with a counterclockwise orientation starting at $(2, 1)$.

(a) To get a clockwise orientation, we could change the equations to $x = 2\cos t, y = 1 - 2\sin t, 0 \le t \le 2\pi$.

(b) To get three times around in the counterclockwise direction, we use the original equations $x = 2\cos t, y = 1 + 2\sin t$ with the domain expanded to $0 \le t \le 6\pi$.

(c) To start at $(0, 3)$ using the original equations, we must have $x_1 = 0$; that is, $2\cos t = 0$. Hence, $t = \frac{\pi}{2}$. So we use
$$x = 2\cos t, y = 1 + 2\sin t, \frac{\pi}{2} \le t \le \frac{3\pi}{2}.$$

Alternatively, if we want t to start at 0, we could change the equations of the curve. For example, we could use
$$x = -2\sin t, y = 1 + 2\cos t, 0 \le t \le \pi.$$

35. *Big circle:* It's centered at $(2, 2)$ with a radius of 2, so by Example 4, parametric equations are
$$x = 2 + 2\cos t, \qquad y = 2 + 2\sin t, \qquad 0 \le t \le 2\pi$$

Small circles: They are centered at $(1, 3)$ and $(3, 3)$ with a radius of 0.1. By Example 4, parametric equations are

$$\begin{array}{llll} \textit{(left)} & x = 1 + 0.1\cos t, & y = 3 + 0.1\sin t, & 0 \le t \le 2\pi \\ \text{and} \quad \textit{(right)} & x = 3 + 0.1\cos t, & y = 3 + 0.1\sin t, & 0 \le t \le 2\pi \end{array}$$

Semicircle: It's the lower half of a circle centered at $(2, 2)$ with radius 1. By Example 4, parametric equations are
$$x = 2 + 1\cos t, \qquad y = 2 + 1\sin t, \qquad \pi \le t \le 2\pi$$

To get all four graphs on the same screen with a typical graphing calculator, we need to change the last t-interval to $[0, 2\pi]$ in order to match the others. We can do this by changing t to $0.5t$. This change gives us the upper half. There are several ways to get the lower half—one is to change the "+" to a "−" in the y-assignment, giving us
$$x = 2 + 1\cos(0.5t), \qquad y = 2 - 1\sin(0.5t), \qquad 0 \le t \le 2\pi$$

37. (a) $x = t^3 \;\Rightarrow\; t = x^{1/3}$, so $y = t^2 = x^{2/3}$.

We get the entire curve $y = x^{2/3}$ traversed in a left to right direction.

(b) $x = t^6 \;\Rightarrow\; t = x^{1/6}$, so $y = t^4 = x^{4/6} = x^{2/3}$.

Since $x = t^6 \ge 0$, we only get the right half of the curve $y = x^{2/3}$.

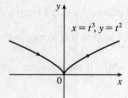

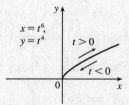

(c) $x = e^{-3t} = (e^{-t})^3$ [so $e^{-t} = x^{1/3}$],
$y = e^{-2t} = (e^{-t})^2 = (x^{1/3})^2 = x^{2/3}$.

If $t < 0$, then x and y are both larger than 1. If $t > 0$, then x and y are between 0 and 1. Since $x > 0$ and $y > 0$, the curve never quite reaches the origin.

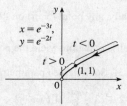

39. The case $\frac{\pi}{2} < \theta < \pi$ is illustrated. C has coordinates $(r\theta, r)$ as in Example 7,

and Q has coordinates $(r\theta, r + r\cos(\pi - \theta)) = (r\theta, r(1 - \cos\theta))$

[since $\cos(\pi - \alpha) = \cos\pi\cos\alpha + \sin\pi\sin\alpha = -\cos\alpha$], so P has

coordinates $(r\theta - r\sin(\pi - \theta), r(1 - \cos\theta)) = (r(\theta - \sin\theta), r(1 - \cos\theta))$

[since $\sin(\pi - \alpha) = \sin\pi\cos\alpha - \cos\pi\sin\alpha = \sin\alpha$]. Again we have the

parametric equations $x = r(\theta - \sin\theta)$, $y = r(1 - \cos\theta)$.

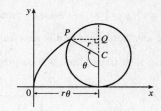

41. It is apparent that $x = |OQ|$ and $y = |QP| = |ST|$. From the diagram,

$x = |OQ| = a\cos\theta$ and $y = |ST| = b\sin\theta$. Thus, the parametric equations are

$x = a\cos\theta$ and $y = b\sin\theta$. To eliminate θ we rearrange: $\sin\theta = y/b$ $\Rightarrow$

$\sin^2\theta = (y/b)^2$ and $\cos\theta = x/a$ $\Rightarrow$ $\cos^2\theta = (x/a)^2$. Adding the two

equations: $\sin^2\theta + \cos^2\theta = 1 = x^2/a^2 + y^2/b^2$. Thus, we have an ellipse.

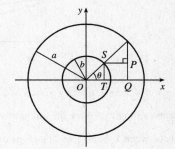

43. $C = (2a\cot\theta, 2a)$, so the x-coordinate of P is $x = 2a\cot\theta$. Let $B = (0, 2a)$.

Then $\angle OAB$ is a right angle and $\angle OBA = \theta$, so $|OA| = 2a\sin\theta$ and

$A = ((2a\sin\theta)\cos\theta, (2a\sin\theta)\sin\theta)$. Thus, the y-coordinate of P

is $y = 2a\sin^2\theta$.

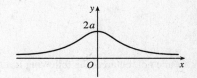

45. (a)

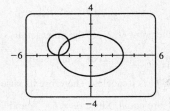

There are 2 points of intersection:

$(-3, 0)$ and approximately $(-2.1, 1.4)$.

(b) A collision point occurs when $x_1 = x_2$ and $y_1 = y_2$ for the same t. So solve the equations:

$$3\sin t = -3 + \cos t \quad \textbf{(1)}$$

$$2\cos t = 1 + \sin t \quad \textbf{(2)}$$

From **(2)**, $\sin t = 2\cos t - 1$. Substituting into **(1)**, we get $3(2\cos t - 1) = -3 + \cos t$ $\Rightarrow$ $5\cos t = 0$ $(\star)$ $\Rightarrow$

$\cos t = 0$ $\Rightarrow$ $t = \frac{\pi}{2}$ or $\frac{3\pi}{2}$. We check that $t = \frac{3\pi}{2}$ satisfies **(1)** and **(2)** but $t = \frac{\pi}{2}$ does not. So the only collision point

occurs when $t = \frac{3\pi}{2}$, and this gives the point $(-3, 0)$. [We could check our work by graphing x_1 and x_2 together as

functions of t and, on another plot, y_1 and y_2 as functions of t. If we do so, we see that the only value of t for which *both*

pairs of graphs intersect is $t = \frac{3\pi}{2}$.]

(c) The circle is centered at $(3, 1)$ instead of $(-3, 1)$. There are still 2 intersection points: $(3, 0)$ and $(2.1, 1.4)$, but there are

no collision points, since $(\star)$ in part (b) becomes $5\cos t = 6$ $\Rightarrow$ $\cos t = \frac{6}{5} > 1$.

47. $x = t^2, y = t^3 - ct$. We use a graphing device to produce the graphs for various values of c with $-\pi \le t \le \pi$. Note that all the members of the family are symmetric about the x-axis. For $c < 0$, the graph does not cross itself, but for $c = 0$ it has a cusp at $(0, 0)$ and for $c > 0$ the graph crosses itself at $x = c$, so the loop grows larger as c increases.

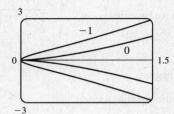 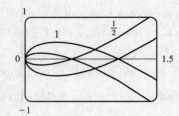

49. $x = t + a\cos t, y = t + a\sin t, a > 0$. From the first figure, we see that curves roughly follow the line $y = x$, and they start having loops when a is between 1.4 and 1.6. The loops increase in size as a increases.

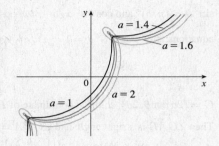

While not required, the following is a solution to determine the *exact* values for which the curve has a loop, that is, we seek the values of a for which there exist parameter values t and u such that $t < u$ and

$$(t + a\cos t, t + a\sin t) = (u + a\cos u, u + a\sin u).$$

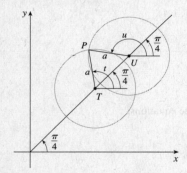

In the diagram at the left, T denotes the point (t, t), U the point (u, u), and P the point $(t + a\cos t, t + a\sin t) = (u + a\cos u, u + a\sin u)$. Since $\overline{PT} = \overline{PU} = a$, the triangle PTU is isosceles. Therefore its base angles, $\alpha = \angle PTU$ and $\beta = \angle PUT$ are equal. Since $\alpha = t - \frac{\pi}{4}$ and $\beta = 2\pi - \frac{3\pi}{4} - u = \frac{5\pi}{4} - u$, the relation $\alpha = \beta$ implies that

$$u + t = \frac{3\pi}{2} \text{ (1)}.$$

Since $\overline{TU} = \text{distance}((t, t), (u, u)) = \sqrt{2(u - t)^2} = \sqrt{2}(u - t)$, we see that

$$\cos\alpha = \frac{\frac{1}{2}\overline{TU}}{\overline{PT}} = \frac{(u - t)/\sqrt{2}}{a}, \text{ so } u - t = \sqrt{2}\,a\cos\alpha, \text{ that is,}$$

$u - t = \sqrt{2}\,a\cos\left(t - \frac{\pi}{4}\right)$ **(2)**. Now $\cos\left(t - \frac{\pi}{4}\right) = \sin\left[\frac{\pi}{2} - \left(t - \frac{\pi}{4}\right)\right] = \sin\left(\frac{3\pi}{4} - t\right)$,

so we can rewrite **(2)** as $u - t = \sqrt{2}\,a\sin\left(\frac{3\pi}{4} - t\right)$ **(2′)**. Subtracting **(2′)** from **(1)** and

dividing by 2, we obtain $t = \frac{3\pi}{4} - \frac{\sqrt{2}}{2}a\sin\left(\frac{3\pi}{4} - t\right)$, or $\frac{3\pi}{4} - t = \frac{a}{\sqrt{2}}\sin\left(\frac{3\pi}{4} - t\right)$ **(3)**.

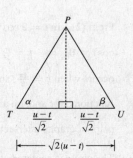

Since $a > 0$ and $t < u$, it follows from (2′) that $\sin\left(\frac{3\pi}{4} - t\right) > 0$. Thus from (3) we see that $t < \frac{3\pi}{4}$. [We have

implicitly assumed that $0 < t < \pi$ by the way we drew our diagram, but we lost no generality by doing so since replacing t

by $t + 2\pi$ merely increases x and y by 2π. The curve's basic shape repeats every time we change t by 2π.] Solving for a in

(3), we get $a = \dfrac{\sqrt{2}\left(\frac{3\pi}{4} - t\right)}{\sin\left(\frac{3\pi}{4} - t\right)}$. Write $z = \frac{3\pi}{4} - t$. Then $a = \dfrac{\sqrt{2}\,z}{\sin z}$, where $z > 0$. Now $\sin z < z$ for $z > 0$, so $a > \sqrt{2}$.

$\left[\text{As } z \to 0^+, \text{ that is, as } t \to \left(\frac{3\pi}{4}\right)^-, a \to \sqrt{2}\,\right].$

51. Note that all the Lissajous figures are symmetric about the x-axis. The parameters a and b simply stretch the graph in the

x- and y-directions respectively. For $a = b = n = 1$ the graph is simply a circle with radius 1. For $n = 2$ the graph crosses

itself at the origin and there are loops above and below the x-axis. In general, the figures have $n - 1$ points of intersection,

all of which are on the y-axis, and a total of n closed loops.

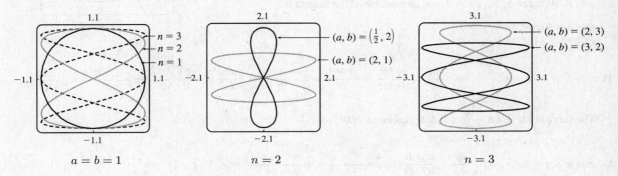

10.2   Calculus with Parametric Curves

1. $x = \dfrac{t}{1+t}$, $y = \sqrt{1+t}$ $\Rightarrow$ $\dfrac{dy}{dt} = \dfrac{1}{2}(1+t)^{-1/2} = \dfrac{1}{2\sqrt{1+t}}$, $\dfrac{dx}{dt} = \dfrac{(1+t)(1) - t(1)}{(1+t)^2} = \dfrac{1}{(1+t)^2}$, and

$\dfrac{dy}{dx} = \dfrac{dy/dt}{dx/dt} = \dfrac{1/(2\sqrt{1+t})}{1/(1+t)^2} = \dfrac{(1+t)^2}{2\sqrt{1+t}} = \dfrac{1}{2}(1+t)^{3/2}.$

3. $x = t^3 + 1$, $y = t^4 + t$; $t = -1$. $\dfrac{dy}{dt} = 4t^3 + 1$, $\dfrac{dx}{dt} = 3t^2$, and $\dfrac{dy}{dx} = \dfrac{dy/dt}{dx/dt} = \dfrac{4t^3 + 1}{3t^2}$. When $t = -1$, $(x, y) = (0, 0)$

and $dy/dx = -3/3 = -1$, so an equation of the tangent to the curve at the point corresponding to $t = -1$ is

$y - 0 = -1(x - 0)$, or $y = -x$.

5. $x = t\cos t$, $y = t\sin t$; $t = \pi$. $\dfrac{dy}{dt} = t\cos t + \sin t$, $\dfrac{dx}{dt} = t(-\sin t) + \cos t$, and $\dfrac{dy}{dx} = \dfrac{dy/dt}{dx/dt} = \dfrac{t\cos t + \sin t}{-t\sin t + \cos t}.$

When $t = \pi$, $(x, y) = (-\pi, 0)$ and $dy/dx = -\pi/(-1) = \pi$, so an equation of the tangent to the curve at the point

corresponding to $t = \pi$ is $y - 0 = \pi[x - (-\pi)]$, or $y = \pi x + \pi^2$.

7. (a) $x = 1 + \ln t$, $y = t^2 + 2$; $(1, 3)$. $\dfrac{dy}{dt} = 2t$, $\dfrac{dx}{dt} = \dfrac{1}{t}$, and $\dfrac{dy}{dx} = \dfrac{dy/dt}{dx/dt} = \dfrac{2t}{1/t} = 2t^2$. At $(1, 3)$,

$x = 1 + \ln t = 1 \;\Rightarrow\; \ln t = 0 \;\Rightarrow\; t = 1$ and $\dfrac{dy}{dx} = 2$, so an equation of the tangent is $y - 3 = 2(x - 1)$,

or $y = 2x + 1$.

(b) $x = 1 + \ln t \;\Rightarrow\; \ln t = x - 1 \;\Rightarrow\; t = e^{x-1}$, so $y = t^2 + 2 = (e^{x-1})^2 + 2 = e^{2x-2} + 2$, and $y' = e^{2x-2} \cdot 2$.

At $(1, 3)$, $y' = e^{2(1)-2} \cdot 2 = 2$, so an equation of the tangent is $y - 3 = 2(x - 1)$, or $y = 2x + 1$.

9. $x = t^2 - t$, $y = t^2 + t + 1$; $(0, 3)$. $\dfrac{dy}{dx} = \dfrac{dy/dt}{dx/dt} = \dfrac{2t + 1}{2t - 1}$. To find the

value of t corresponding to the point $(0, 3)$, solve $x = 0 \;\Rightarrow\;$

$t^2 - t = 0 \;\Rightarrow\; t(t - 1) = 0 \;\Rightarrow\; t = 0$ or $t = 1$. Only $t = 1$ gives

$y = 3$. With $t = 1$, $dy/dx = 3$, and an equation of the tangent is

$y - 3 = 3(x - 0)$, or $y = 3x + 3$.

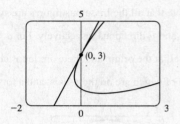

11. $x = t^2 + 1$, $y = t^2 + t \;\Rightarrow\; \dfrac{dy}{dx} = \dfrac{dy/dt}{dx/dt} = \dfrac{2t + 1}{2t} = 1 + \dfrac{1}{2t} \;\Rightarrow\; \dfrac{d^2y}{dx^2} = \dfrac{\dfrac{d}{dt}\left(\dfrac{dy}{dx}\right)}{dx/dt} = \dfrac{-1/(2t^2)}{2t} = -\dfrac{1}{4t^3}$.

The curve is CU when $\dfrac{d^2y}{dx^2} > 0$, that is, when $t < 0$.

13. $x = e^t$, $y = te^{-t} \;\Rightarrow\; \dfrac{dy}{dx} = \dfrac{dy/dt}{dx/dt} = \dfrac{-te^{-t} + e^{-t}}{e^t} = \dfrac{e^{-t}(1 - t)}{e^t} = e^{-2t}(1 - t) \;\Rightarrow\;$

$\dfrac{d^2y}{dx^2} = \dfrac{\dfrac{d}{dt}\left(\dfrac{dy}{dx}\right)}{dx/dt} = \dfrac{e^{-2t}(-1) + (1 - t)(-2e^{-2t})}{e^t} = \dfrac{e^{-2t}(-1 - 2 + 2t)}{e^t} = e^{-3t}(2t - 3)$. The curve is CU when

$\dfrac{d^2y}{dx^2} > 0$, that is, when $t > \dfrac{3}{2}$.

15. $x = t - \ln t$, $y = t + \ln t$ [note that $t > 0$] $\;\Rightarrow\; \dfrac{dy}{dx} = \dfrac{dy/dt}{dx/dt} = \dfrac{1 + 1/t}{1 - 1/t} = \dfrac{t + 1}{t - 1} \;\Rightarrow\;$

$\dfrac{d^2y}{dx^2} = \dfrac{\dfrac{d}{dt}\left(\dfrac{dy}{dx}\right)}{dx/dt} = \dfrac{\dfrac{(t - 1)(1) - (t + 1)(1)}{(t - 1)^2}}{(t - 1)/t} = \dfrac{-2t}{(t - 1)^3}$. The curve is CU when $\dfrac{d^2y}{dx^2} > 0$, that is, when $0 < t < 1$.

17. $x = t^3 - 3t$, $y = t^2 - 3$. $\dfrac{dy}{dt} = 2t$, so $\dfrac{dy}{dt} = 0 \;\Leftrightarrow\; t = 0 \;\Leftrightarrow\;$

$(x, y) = (0, -3)$. $\dfrac{dx}{dt} = 3t^2 - 3 = 3(t + 1)(t - 1)$, so $\dfrac{dx}{dt} = 0 \;\Leftrightarrow\;$

$t = -1$ or $1 \;\Leftrightarrow\; (x, y) = (2, -2)$ or $(-2, -2)$. The curve has a horizontal

tangent at $(0, -3)$ and vertical tangents at $(2, -2)$ and $(-2, -2)$.

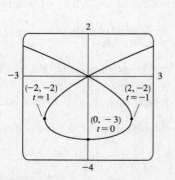

19. $x = \cos\theta$, $y = \cos 3\theta$. The whole curve is traced out for $0 \le \theta \le \pi$.

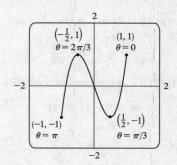

$\dfrac{dy}{d\theta} = -3\sin 3\theta$, so $\dfrac{dy}{d\theta} = 0 \Leftrightarrow \sin 3\theta = 0 \Leftrightarrow 3\theta = 0, \pi, 2\pi$, or $3\pi \Leftrightarrow$

$\theta = 0, \frac{\pi}{3}, \frac{2\pi}{3}$, or $\pi \Leftrightarrow (x,y) = (1,1), \left(\frac{1}{2}, -1\right), \left(-\frac{1}{2}, 1\right)$, or $(-1,-1)$.

$\dfrac{dx}{d\theta} = -\sin\theta$, so $\dfrac{dx}{d\theta} = 0 \Leftrightarrow \sin\theta = 0 \Leftrightarrow \theta = 0$ or $\pi \Leftrightarrow$

$(x,y) = (1,1)$ or $(-1,-1)$. Both $\dfrac{dy}{d\theta}$ and $\dfrac{dx}{d\theta}$ equal 0 when $\theta = 0$ and π.

To find the slope when $\theta = 0$, we find $\displaystyle\lim_{\theta\to 0} \dfrac{dy}{dx} = \lim_{\theta\to 0} \dfrac{-3\sin 3\theta}{-\sin\theta} \overset{H}{=} \lim_{\theta\to 0} \dfrac{-9\cos 3\theta}{-\cos\theta} = 9$, which is the same slope when $\theta = \pi$.

Thus, the curve has horizontal tangents at $\left(\frac{1}{2}, -1\right)$ and $\left(-\frac{1}{2}, 1\right)$, and there are no vertical tangents.

21. From the graph, it appears that the rightmost point on the curve $x = t - t^6$, $y = e^t$

is about $(0.6, 2)$. To find the exact coordinates, we find the value of t for which the

graph has a vertical tangent, that is, $0 = dx/dt = 1 - 6t^5 \Leftrightarrow t = 1/\sqrt[5]{6}$.

Hence, the rightmost point is

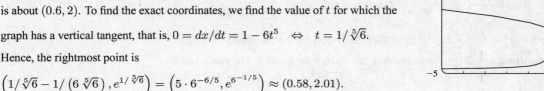

$$\left(1/\sqrt[5]{6} - 1/\left(6\sqrt[5]{6}\right), e^{1/\sqrt[5]{6}}\right) = \left(5 \cdot 6^{-6/5}, e^{6^{-1/5}}\right) \approx (0.58, 2.01).$$

23. We graph the curve $x = t^4 - 2t^3 - 2t^2$, $y = t^3 - t$ in the viewing rectangle $[-2, 1.1]$ by $[-0.5, 0.5]$. This rectangle

corresponds approximately to $t \in [-1, 0.8]$.

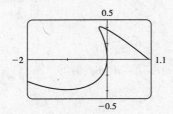
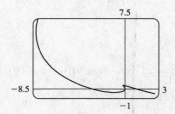

We estimate that the curve has horizontal tangents at about $(-1, -0.4)$ and $(-0.17, 0.39)$ and vertical tangents at

about $(0,0)$ and $(-0.19, 0.37)$. We calculate $\dfrac{dy}{dx} = \dfrac{dy/dt}{dx/dt} = \dfrac{3t^2 - 1}{4t^3 - 6t^2 - 4t}$. The horizontal tangents occur when

$dy/dt = 3t^2 - 1 = 0 \Leftrightarrow t = \pm\frac{1}{\sqrt{3}}$, so both horizontal tangents are shown in our graph. The vertical tangents occur when

$dx/dt = 2t(2t^2 - 3t - 2) = 0 \Leftrightarrow 2t(2t+1)(t-2) = 0 \Leftrightarrow t = 0, -\frac{1}{2}$ or 2. It seems that we have missed one vertical

tangent, and indeed if we plot the curve on the t-interval $[-1.2, 2.2]$ we see that there is another vertical tangent at $(-8, 6)$.

25. $x = \cos t$, $y = \sin t \cos t$. $dx/dt = -\sin t$,

$dy/dt = -\sin^2 t + \cos^2 t = \cos 2t$. $(x,y) = (0,0) \Leftrightarrow \cos t = 0 \Leftrightarrow t$ is

an odd multiple of $\frac{\pi}{2}$. When $t = \frac{\pi}{2}$, $dx/dt = -1$ and $dy/dt = -1$, so $dy/dx = 1$.

When $t = \frac{3\pi}{2}$, $dx/dt = 1$ and $dy/dt = -1$. So $dy/dx = -1$. Thus, $y = x$ and

$y = -x$ are both tangent to the curve at $(0,0)$.

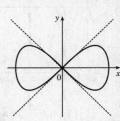

27. $x = r\theta - d\sin\theta, \; y = r - d\cos\theta.$

(a) $\dfrac{dx}{d\theta} = r - d\cos\theta, \; \dfrac{dy}{d\theta} = d\sin\theta$, so $\dfrac{dy}{dx} = \dfrac{d\sin\theta}{r - d\cos\theta}.$

(b) If $0 < d < r$, then $|d\cos\theta| \le d < r$, so $r - d\cos\theta \ge r - d > 0$. This shows that $dx/d\theta$ never vanishes,

so the trochoid can have no vertical tangent if $d < r$.

29. $x = 3t^2 + 1, \; y = t^3 - 1 \;\Rightarrow\; \dfrac{dy}{dx} = \dfrac{dy/dt}{dx/dt} = \dfrac{3t^2}{6t} = \dfrac{t}{2}$. The tangent line has slope $\dfrac{1}{2}$ when $\dfrac{t}{2} = \dfrac{1}{2} \;\Leftrightarrow\; t = 1$, so the

point is $(4, 0)$.

31. By symmetry of the ellipse about the x- and y-axes,

$$A = 4\int_0^a y\,dx = 4\int_{\pi/2}^0 b\sin\theta\,(-a\sin\theta)\,d\theta = 4ab\int_0^{\pi/2}\sin^2\theta\,d\theta = 4ab\int_0^{\pi/2}\tfrac{1}{2}(1 - \cos 2\theta)\,d\theta$$

$$= 2ab\big[\theta - \tfrac{1}{2}\sin 2\theta\big]_0^{\pi/2} = 2ab\big(\tfrac{\pi}{2}\big) = \pi ab$$

33. The curve $x = t^3 + 1, \, y = 2t - t^2 = t(2 - t)$ intersects the x-axis when $y = 0$, that

is, when $t = 0$ and $t = 2$. The corresponding values of x are 1 and 9. The shaded area

is given by

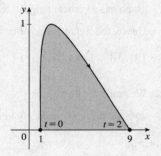

$$\int_{x=1}^{x=9}(y_T - y_B)\,dx = \int_{t=0}^{t=2}[y(t) - 0]\,x'(t)\,dt = \int_0^2(2t - t^2)(3t^2)\,dt$$

$$= 3\int_0^2(2t^3 - t^4)\,dt = 3\Big[\tfrac{1}{2}t^4 - \tfrac{1}{5}t^5\Big]_0^2 = 3\big(8 - \tfrac{32}{5}\big) = \tfrac{24}{5}$$

35. $x = r\theta - d\sin\theta, \; y = r - d\cos\theta.$

$$A = \int_0^{2\pi r} y\,dx = \int_0^{2\pi}(r - d\cos\theta)(r - d\cos\theta)\,d\theta = \int_0^{2\pi}(r^2 - 2dr\cos\theta + d^2\cos^2\theta)\,d\theta$$

$$= \big[r^2\theta - 2dr\sin\theta + \tfrac{1}{2}d^2\big(\theta + \tfrac{1}{2}\sin 2\theta\big)\big]_0^{2\pi} = 2\pi r^2 + \pi d^2$$

37. $x = t + e^{-t}, \; y = t - e^{-t}, \; 0 \le t \le 2.$ $dx/dt = 1 - e^{-t}$ and $dy/dt = 1 + e^{-t}$, so

$(dx/dt)^2 + (dy/dt)^2 = (1 - e^{-t})^2 + (1 + e^{-t})^2 = 1 - 2e^{-t} + e^{-2t} + 1 + 2e^{-t} + e^{-2t} = 2 + 2e^{-2t}.$

Thus, $L = \int_a^b \sqrt{(dx/dt)^2 + (dy/dt)^2}\,dt = \int_0^2 \sqrt{2 + 2e^{-2t}}\,dt \approx 3.1416.$

39. $x = t - 2\sin t, \; y = 1 - 2\cos t, \; 0 \le t \le 4\pi.$ $dx/dt = 1 - 2\cos t$ and $dy/dt = 2\sin t$, so

$(dx/dt)^2 + (dy/dt)^2 = (1 - 2\cos t)^2 + (2\sin t)^2 = 1 - 4\cos t + 4\cos^2 t + 4\sin^2 t = 5 - 4\cos t.$

Thus, $L = \int_a^b \sqrt{(dx/dt)^2 + (dy/dt)^2}\,dt = \int_0^{4\pi} \sqrt{5 - 4\cos t}\,dt \approx 26.7298.$

41. $x = 1 + 3t^2, \; y = 4 + 2t^3, \; 0 \le t \le 1.$ $dx/dt = 6t$ and $dy/dt = 6t^2$, so $(dx/dt)^2 + (dy/dt)^2 = 36t^2 + 36t^4.$

Thus, $L = \int_0^1 \sqrt{36t^2 + 36t^4}\,dt = \int_0^1 6t\sqrt{1 + t^2}\,dt = 6\int_1^2 \sqrt{u}\,\big(\tfrac{1}{2}\,du\big) \quad [u = 1 + t^2, du = 2t\,dt]$

$$= 3\Big[\tfrac{2}{3}u^{3/2}\Big]_1^2 = 2(2^{3/2} - 1) = 2(2\sqrt{2} - 1)$$

43. $x = t \sin t$, $y = t \cos t$, $0 \leq t \leq 1$. $\dfrac{dx}{dt} = t \cos t + \sin t$ and $\dfrac{dy}{dt} = -t \sin t + \cos t$, so

$$\left(\frac{dx}{dt}\right)^2 + \left(\frac{dy}{dt}\right)^2 = t^2 \cos^2 t + 2t \sin t \cos t + \sin^2 t + t^2 \sin^2 t - 2t \sin t \cos t + \cos^2 t$$

$$= t^2 (\cos^2 t + \sin^2 t) + \sin^2 t + \cos^2 t = t^2 + 1.$$

Thus, $L = \int_0^1 \sqrt{t^2 + 1}\, dt \overset{21}{=} \left[\frac{1}{2} t \sqrt{t^2 + 1} + \frac{1}{2} \ln\left(t + \sqrt{t^2 + 1}\,\right) \right]_0^1 = \frac{1}{2}\sqrt{2} + \frac{1}{2} \ln\left(1 + \sqrt{2}\,\right)$.

45.

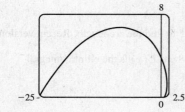

$x = e^t \cos t$, $y = e^t \sin t$, $0 \leq t \leq \pi$.

$$\left(\frac{dx}{dt}\right)^2 + \left(\frac{dy}{dt}\right)^2 = [e^t (\cos t - \sin t)]^2 + [e^t (\sin t + \cos t)]^2$$

$$= (e^t)^2 (\cos^2 t - 2 \cos t \sin t + \sin^2 t)$$

$$+ (e^t)^2 (\sin^2 t + 2 \sin t \cos t + \cos^2 t)$$

$$= e^{2t} (2 \cos^2 t + 2 \sin^2 t) = 2 e^{2t}$$

Thus, $L = \int_0^\pi \sqrt{2 e^{2t}}\, dt = \int_0^\pi \sqrt{2}\, e^t\, dt = \sqrt{2} \left[e^t \right]_0^\pi = \sqrt{2} \left(e^\pi - 1 \right)$.

47.

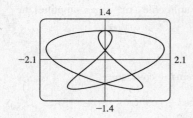

The figure shows the curve $x = \sin t + \sin 1.5t$, $y = \cos t$ for $0 \leq t \leq 4\pi$.

$dx/dt = \cos t + 1.5 \cos 1.5t$ and $dy/dt = -\sin t$, so

$(dx/dt)^2 + (dy/dt)^2 = \cos^2 t + 3 \cos t \cos 1.5t + 2.25 \cos^2 1.5t + \sin^2 t$.

Thus, $L = \int_0^{4\pi} \sqrt{1 + 3 \cos t \cos 1.5t + 2.25 \cos^2 1.5t}\, dt \approx 16.7102$.

49. $x = t - e^t$, $y = t + e^t$, $-6 \leq t \leq 6$.

$\left(\frac{dx}{dt}\right)^2 + \left(\frac{dy}{dt}\right)^2 = (1 - e^t)^2 + (1 + e^t)^2 = (1 - 2e^t + e^{2t}) + (1 + 2e^t + e^{2t}) = 2 + 2e^{2t}$, so $L = \int_{-6}^6 \sqrt{2 + 2e^{2t}}\, dt$.

Set $f(t) = \sqrt{2 + 2e^{2t}}$. Then by Simpson's Rule with $n = 6$ and $\Delta t = \frac{6 - (-6)}{6} = 2$, we get

$L \approx \frac{2}{3} [f(-6) + 4f(-4) + 2f(-2) + 4f(0) + 2f(2) + 4f(4) + f(6)] \approx 612.3053$.

51. $x = \sin^2 t$, $y = \cos^2 t$, $0 \leq t \leq 3\pi$.

$(dx/dt)^2 + (dy/dt)^2 = (2 \sin t \cos t)^2 + (-2 \cos t \sin t)^2 = 8 \sin^2 t \cos^2 t = 2 \sin^2 2t \quad \Rightarrow$

Distance $= \int_0^{3\pi} \sqrt{2}\, |\sin 2t|\, dt = 6\sqrt{2} \int_0^{\pi/2} \sin 2t\, dt$ [by symmetry] $= -3\sqrt{2} \left[\cos 2t \right]_0^{\pi/2} = -3\sqrt{2}(-1 - 1) = 6\sqrt{2}$.

The full curve is traversed as t goes from 0 to $\frac{\pi}{2}$, because the curve is the segment of $x + y = 1$ that lies in the first quadrant

(since $x, y \geq 0$), and this segment is completely traversed as t goes from 0 to $\frac{\pi}{2}$. Thus, $L = \int_0^{\pi/2} \sin 2t\, dt = \sqrt{2}$, as above.

53. $x = a \sin \theta$, $y = b \cos \theta$, $0 \leq \theta \leq 2\pi$.

$$\left(\frac{dx}{d\theta}\right)^2 + \left(\frac{dy}{d\theta}\right)^2 = (a \cos \theta)^2 + (-b \sin \theta)^2 = a^2 \cos^2 \theta + b^2 \sin^2 \theta = a^2 (1 - \sin^2 \theta) + b^2 \sin^2 \theta$$

$$= a^2 - (a^2 - b^2) \sin^2 \theta = a^2 - c^2 \sin^2 \theta = a^2 \left(1 - \frac{c^2}{a^2} \sin^2 \theta \right) = a^2 (1 - e^2 \sin^2 \theta)$$

So $L = 4 \int_0^{\pi/2} \sqrt{a^2 (1 - e^2 \sin^2 \theta)}\, d\theta$ [by symmetry] $= 4a \int_0^{\pi/2} \sqrt{1 - e^2 \sin^2 \theta}\, d\theta$.

55. (a) $x = 11\cos t - 4\cos(11t/2)$, $y = 11\sin t - 4\sin(11t/2)$.

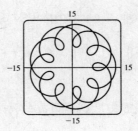

Notice that $0 \leq t \leq 2\pi$ does not give the complete curve because

$x(0) \neq x(2\pi)$. In fact, we must take $t \in [0, 4\pi]$ in order to obtain the

complete curve, since the first term in each of the parametric equations has

period 2π and the second has period $\frac{2\pi}{11/2} = \frac{4\pi}{11}$, and the least common

integer multiple of these two numbers is 4π.

(b) We use the CAS to find the derivatives dx/dt and dy/dt, and then use Theorem 5 to find the arc length. Recent versions

of Maple express the integral $\int_0^{4\pi} \sqrt{(dx/dt)^2 + (dy/dt)^2}\, dt$ as $88E(2\sqrt{2}\,i)$, where $E(x)$ is the elliptic integral

$\int_0^1 \dfrac{\sqrt{1 - x^2 t^2}}{\sqrt{1 - t^2}}\, dt$ and i is the imaginary number $\sqrt{-1}$.

Some earlier versions of Maple (as well as Mathematica) cannot do the integral exactly, so we use the command

`evalf(Int(sqrt(diff(x,t)^2+diff(y,t)^2),t=0..4*Pi));` to estimate the length, and find that the arc

length is approximately 294.03. Derive's `Para_arc_length` function in the utility file `Int_apps` simplifies the

integral to $11\int_0^{4\pi} \sqrt{-4\cos t\, \cos\left(\frac{11t}{2}\right) - 4\sin t\, \sin\left(\frac{11t}{2}\right) + 5}\, dt$.

57. $x = t\sin t$, $y = t\cos t$, $0 \leq t \leq \pi/2$. $dx/dt = t\cos t + \sin t$ and $dy/dt = -t\sin t + \cos t$, so

$(dx/dt)^2 + (dy/dt)^2 = t^2\cos^2 t + 2t\sin t\,\cos t + \sin^2 t + t^2\sin^2 t - 2t\sin t\,\cos t + \cos^2 t$

$= t^2(\cos^2 t + \sin^2 t) + \sin^2 t + \cos^2 t = t^2 + 1$

$S = \int 2\pi y\, ds = \int_0^{\pi/2} 2\pi t\cos t\sqrt{t^2 + 1}\, dt \approx 4.7394$.

59. $x = t + e^t$, $y = e^{-t}$, $0 \leq t \leq 1$.

$dx/dt = 1 + e^t$ and $dy/dt = -e^{-t}$, so $(dx/dt)^2 + (dy/dt)^2 = (1 + e^t)^2 + (-e^{-t})^2 = 1 + 2e^t + e^{2t} + e^{-2t}$.

$S = \int 2\pi y\, ds = \int_0^1 2\pi e^{-t}\sqrt{1 + 2e^t + e^{2t} + e^{-2t}}\, dt \approx 10.6705$.

61. $x = t^3$, $y = t^2$, $0 \leq t \leq 1$. $\left(\frac{dx}{dt}\right)^2 + \left(\frac{dy}{dt}\right)^2 = (3t^2)^2 + (2t)^2 = 9t^4 + 4t^2$.

$$S = \int_0^1 2\pi y\sqrt{\left(\frac{dx}{dt}\right)^2 + \left(\frac{dy}{dt}\right)^2}\, dt = \int_0^1 2\pi t^2\sqrt{9t^4 + 4t^2}\, dt = 2\pi\int_0^1 t^2\sqrt{t^2(9t^2 + 4)}\, dt$$

$$= 2\pi\int_4^{13}\left(\frac{u - 4}{9}\right)\sqrt{u}\left(\frac{1}{18}\,du\right)\quad \begin{bmatrix} u = 9t^2 + 4,\ t^2 = (u - 4)/9, \\ du = 18t\,dt,\ \text{so}\ t\,dt = \frac{1}{18}\,du \end{bmatrix}\quad = \frac{2\pi}{9\cdot 18}\int_4^{13}(u^{3/2} - 4u^{1/2})\, du$$

$$= \frac{\pi}{81}\left[\frac{2}{5}u^{5/2} - \frac{8}{3}u^{3/2}\right]_4^{13} = \frac{\pi}{81}\cdot\frac{2}{15}\left[3u^{5/2} - 20u^{3/2}\right]_4^{13}$$

$$= \frac{2\pi}{1215}\left[(3\cdot 13^2\sqrt{13} - 20\cdot 13\sqrt{13}) - (3\cdot 32 - 20\cdot 8)\right] = \frac{2\pi}{1215}\left(247\sqrt{13} + 64\right)$$

63. $x = a\cos^3\theta$, $y = a\sin^3\theta$, $0 \leq \theta \leq \frac{\pi}{2}$. $\left(\frac{dx}{d\theta}\right)^2 + \left(\frac{dy}{d\theta}\right)^2 = (-3a\cos^2\theta\,\sin\theta)^2 + (3a\sin^2\theta\,\cos\theta)^2 = 9a^2\sin^2\theta\,\cos^2\theta$.

$S = \int_0^{\pi/2} 2\pi\cdot a\sin^3\theta\cdot 3a\sin\theta\,\cos\theta\, d\theta = 6\pi a^2\int_0^{\pi/2}\sin^4\theta\,\cos\theta\, d\theta = \frac{6}{5}\pi a^2\left[\sin^5\theta\right]_0^{\pi/2} = \frac{6}{5}\pi a^2$

65. $x = 3t^2$, $y = 2t^3$, $0 \le t \le 5$ $\Rightarrow$ $\left(\frac{dx}{dt}\right)^2 + \left(\frac{dy}{dt}\right)^2 = (6t)^2 + (6t^2)^2 = 36t^2(1 + t^2)$ $\Rightarrow$

$$S = \int_0^5 2\pi x \sqrt{(dx/dt)^2 + (dy/dt)^2}\, dt = \int_0^5 2\pi (3t^2) 6t \sqrt{1 + t^2}\, dt = 18\pi \int_0^5 t^2 \sqrt{1 + t^2}\, 2t\, dt$$

$$= 18\pi \int_1^{26} (u - 1)\sqrt{u}\, du \quad \begin{bmatrix} u = 1 + t^2, \\ du = 2t\, dt \end{bmatrix} \quad = 18\pi \int_1^{26} (u^{3/2} - u^{1/2})\, du = 18\pi \left[\tfrac{2}{5} u^{5/2} - \tfrac{2}{3} u^{3/2} \right]_1^{26}$$

$$= 18\pi \left[\left(\tfrac{2}{5} \cdot 676\sqrt{26} - \tfrac{2}{3} \cdot 26\sqrt{26} \right) - \left(\tfrac{2}{5} - \tfrac{2}{3} \right) \right] = \tfrac{24}{5} \pi (949\sqrt{26} + 1)$$

67. If f' is continuous and $f'(t) \ne 0$ for $a \le t \le b$, then either $f'(t) > 0$ for all t in $[a, b]$ or $f'(t) < 0$ for all t in $[a, b]$. Thus, f

is monotonic (in fact, strictly increasing or strictly decreasing) on $[a, b]$. It follows that f has an inverse. Set $F = g \circ f^{-1}$,

that is, define F by $F(x) = g(f^{-1}(x))$. Then $x = f(t)$ $\Rightarrow$ $f^{-1}(x) = t$, so $y = g(t) = g(f^{-1}(x)) = F(x)$.

69. (a) $\phi = \tan^{-1}\left(\frac{dy}{dx}\right)$ $\Rightarrow$ $\frac{d\phi}{dt} = \frac{d}{dt} \tan^{-1}\left(\frac{dy}{dx}\right) = \frac{1}{1 + (dy/dx)^2}\left[\frac{d}{dt}\left(\frac{dy}{dx}\right)\right]$. But $\frac{dy}{dx} = \frac{dy/dt}{dx/dt} = \frac{\dot{y}}{\dot{x}}$ $\Rightarrow$

$\frac{d}{dt}\left(\frac{dy}{dx}\right) = \frac{d}{dt}\left(\frac{\dot{y}}{\dot{x}}\right) = \frac{\ddot{y}\dot{x} - \ddot{x}\dot{y}}{\dot{x}^2}$ $\Rightarrow$ $\frac{d\phi}{dt} = \frac{1}{1 + (\dot{y}/\dot{x})^2}\left(\frac{\ddot{y}\dot{x} - \ddot{x}\dot{y}}{\dot{x}^2}\right) = \frac{\dot{x}\ddot{y} - \ddot{x}\dot{y}}{\dot{x}^2 + \dot{y}^2}$. Using the Chain Rule, and the

fact that $s = \int_0^t \sqrt{\left(\frac{dx}{dt}\right)^2 + \left(\frac{dy}{dt}\right)^2}\, dt$ $\Rightarrow$ $\frac{ds}{dt} = \sqrt{\left(\frac{dx}{dt}\right)^2 + \left(\frac{dy}{dt}\right)^2} = (\dot{x}^2 + \dot{y}^2)^{1/2}$, we have that

$\frac{d\phi}{ds} = \frac{d\phi/dt}{ds/dt} = \left(\frac{\dot{x}\ddot{y} - \ddot{x}\dot{y}}{\dot{x}^2 + \dot{y}^2}\right) \frac{1}{(\dot{x}^2 + \dot{y}^2)^{1/2}} = \frac{\dot{x}\ddot{y} - \ddot{x}\dot{y}}{(\dot{x}^2 + \dot{y}^2)^{3/2}}$. So $\kappa = \left|\frac{d\phi}{ds}\right| = \left|\frac{\dot{x}\ddot{y} - \ddot{x}\dot{y}}{(\dot{x}^2 + \dot{y}^2)^{3/2}}\right| = \frac{|\dot{x}\ddot{y} - \ddot{x}\dot{y}|}{(\dot{x}^2 + \dot{y}^2)^{3/2}}$.

(b) $x = x$ and $y = f(x)$ $\Rightarrow$ $\dot{x} = 1, \ddot{x} = 0$ and $\dot{y} = \frac{dy}{dx}, \ddot{y} = \frac{d^2y}{dx^2}$.

So $\kappa = \frac{\left|1 \cdot (d^2y/dx^2) - 0 \cdot (dy/dx)\right|}{[1 + (dy/dx)^2]^{3/2}} = \frac{|d^2y/dx^2|}{[1 + (dy/dx)^2]^{3/2}}$.

71. $x = \theta - \sin\theta$ $\Rightarrow$ $\dot{x} = 1 - \cos\theta$ $\Rightarrow$ $\ddot{x} = \sin\theta$, and $y = 1 - \cos\theta$ $\Rightarrow$ $\dot{y} = \sin\theta$ $\Rightarrow$ $\ddot{y} = \cos\theta$. Therefore,

$\kappa = \frac{|\cos\theta - \cos^2\theta - \sin^2\theta|}{[(1 - \cos\theta)^2 + \sin^2\theta]^{3/2}} = \frac{|\cos\theta - (\cos^2\theta + \sin^2\theta)|}{(1 - 2\cos\theta + \cos^2\theta + \sin^2\theta)^{3/2}} = \frac{|\cos\theta - 1|}{(2 - 2\cos\theta)^{3/2}}$. The top of the arch is

characterized by a horizontal tangent, and from Example 2(b) in Section 10.2, the tangent is horizontal when $\theta = (2n - 1)\pi$,

so take $n = 1$ and substitute $\theta = \pi$ into the expression for κ: $\kappa = \frac{|\cos\pi - 1|}{(2 - 2\cos\pi)^{3/2}} = \frac{|-1 - 1|}{[2 - 2(-1)]^{3/2}} = \frac{1}{4}$.

73. The coordinates of T are $(r\cos\theta, r\sin\theta)$. Since TP was unwound from

arc TA, TP has length $r\theta$. Also $\angle PTQ = \angle PTR - \angle QTR = \frac{1}{2}\pi - \theta$,

so P has coordinates $x = r\cos\theta + r\theta\cos\left(\frac{1}{2}\pi - \theta\right) = r(\cos\theta + \theta\sin\theta)$,

$y = r\sin\theta - r\theta\sin\left(\frac{1}{2}\pi - \theta\right) = r(\sin\theta - \theta\cos\theta)$.

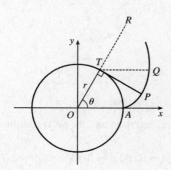

10.3 Polar Coordinates

1. (a) $\left(1, \frac{\pi}{4}\right)$

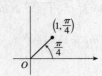

By adding 2π to $\frac{\pi}{4}$, we obtain the point $\left(1, \frac{9\pi}{4}\right)$, which satisfies the $r > 0$ requirement. The direction opposite $\frac{\pi}{4}$ is $\frac{5\pi}{4}$, so $\left(-1, \frac{5\pi}{4}\right)$ is a point that satisfies the $r < 0$ requirement.

(b) $\left(-2, \frac{3\pi}{2}\right)$

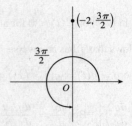

$\mathbf{r > 0:}$ $\left(-(-2), \frac{3\pi}{2} - \pi\right) = \left(2, \frac{\pi}{2}\right)$

$\mathbf{r < 0:}$ $\left(-2, \frac{3\pi}{2} + 2\pi\right) = \left(-2, \frac{7\pi}{2}\right)$

(c) $\left(3, -\frac{\pi}{3}\right)$

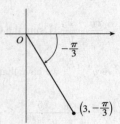

$\mathbf{r > 0:}$ $\left(3, -\frac{\pi}{3} + 2\pi\right) = \left(3, \frac{5\pi}{3}\right)$

$\mathbf{r < 0:}$ $\left(-3, -\frac{\pi}{3} + \pi\right) = \left(-3, \frac{2\pi}{3}\right)$

3. (a)

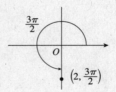

$x = 2\cos\frac{3\pi}{2} = 2(0) = 0$ and $y = 2\sin\frac{3\pi}{2} = 2(-1) = -2$ give us the Cartesian coordinates $(0, -2)$.

(b)

$x = \sqrt{2}\cos\frac{\pi}{4} = \sqrt{2}\left(\frac{1}{\sqrt{2}}\right) = 1$ and $y = \sqrt{2}\sin\frac{\pi}{4} = \sqrt{2}\left(\frac{1}{\sqrt{2}}\right) = 1$

give us the Cartesian coordinates $(1, 1)$.

(c)

$x = -1\cos\left(-\frac{\pi}{6}\right) = -1\left(\frac{\sqrt{3}}{2}\right) = -\frac{\sqrt{3}}{2}$ and

$y = -1\sin\left(-\frac{\pi}{6}\right) = -1\left(-\frac{1}{2}\right) = \frac{1}{2}$ give us the Cartesian

coordinates $\left(-\frac{\sqrt{3}}{2}, \frac{1}{2}\right)$.

5. (a) $x = -4$ and $y = 4$ $\Rightarrow$ $r = \sqrt{(-4)^2 + 4^2} = 4\sqrt{2}$ and $\tan\theta = \frac{4}{-4} = -1$ $[\theta = -\frac{\pi}{4} + n\pi]$. Since $(-4, 4)$ is in the second quadrant, the polar coordinates are (i) $\left(4\sqrt{2}, \frac{3\pi}{4}\right)$ and (ii) $\left(-4\sqrt{2}, \frac{7\pi}{4}\right)$.

(b) $x = 3$ and $y = 3\sqrt{3}$ $\Rightarrow$ $r = \sqrt{3^2 + \left(3\sqrt{3}\right)^2} = \sqrt{9 + 27} = 6$ and $\tan\theta = \frac{3\sqrt{3}}{3} = \sqrt{3}$ $[\theta = \frac{\pi}{3} + n\pi]$.

Since $\left(3, 3\sqrt{3}\right)$ is in the first quadrant, the polar coordinates are (i) $\left(6, \frac{\pi}{3}\right)$ and (ii) $\left(-6, \frac{4\pi}{3}\right)$.

7. $r \geq 1$.   The curve $r = 1$ represents a circle with center
O and radius 1. So $r \geq 1$ represents the region on or
outside the circle. Note that θ can take on any value.

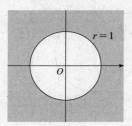

9. $r \geq 0$, $\pi/4 \leq \theta \leq 3\pi/4$.

$\theta = k$ represents a line through O.

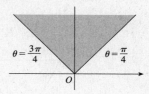

11. $2 < r < 3$, $\frac{5\pi}{3} \leq \theta \leq \frac{7\pi}{3}$

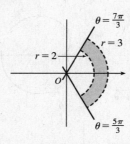

13. Converting the polar coordinates $\left(4, \frac{4\pi}{3}\right)$ and $\left(6, \frac{5\pi}{3}\right)$ to Cartesian coordinates gives us $\left(4\cos\frac{4\pi}{3}, 4\sin\frac{4\pi}{3}\right) = \left(-2, -2\sqrt{3}\right)$

and $\left(6\cos\frac{5\pi}{3}, 6\sin\frac{5\pi}{3}\right) = \left(3, -3\sqrt{3}\right)$. Now use the distance formula

$$d = \sqrt{(x_2 - x_1)^2 + (y_2 - y_1)^2} = \sqrt{[3 - (-2)]^2 + \left[-3\sqrt{3} - (-2\sqrt{3})\right]^2}$$

$$= \sqrt{5^2 + \left(-\sqrt{3}\right)^2} = \sqrt{25 + 3} = \sqrt{28} = 2\sqrt{7}$$

15. $r^2 = 5 \iff x^2 + y^2 = 5$, a circle of radius $\sqrt{5}$ centered at the origin.

17. $r = 5\cos\theta \implies r^2 = 5r\cos\theta \iff x^2 + y^2 = 5x \iff x^2 - 5x + \frac{25}{4} + y^2 = \frac{25}{4} \iff \left(x - \frac{5}{2}\right)^2 + y^2 = \frac{25}{4}$,

a circle of radius $\frac{5}{2}$ centered at $\left(\frac{5}{2}, 0\right)$. The first two equations are actually equivalent since $r^2 = 5r\cos\theta \implies$

$r(r - 5\cos\theta) = 0 \implies r = 0$ or $r = 5\cos\theta$. But $r = 5\cos\theta$ gives the point $r = 0$ (the pole) when $\theta = 0$. Thus, the

equation $r = 5\cos\theta$ is equivalent to the compound condition $(r = 0$ or $r = 5\cos\theta)$.

19. $r^2\cos 2\theta = 1 \iff r^2(\cos^2\theta - \sin^2\theta) = 1 \iff (r\cos\theta)^2 - (r\sin\theta)^2 = 1 \iff x^2 - y^2 = 1$, a hyperbola centered at

the origin with foci on the x-axis.

21. $y = 2 \iff r\sin\theta = 2 \iff r = \dfrac{2}{\sin\theta} \iff r = 2\csc\theta$

23. $y = 1 + 3x \iff r\sin\theta = 1 + 3r\cos\theta \iff r\sin\theta - 3r\cos\theta = 1 \iff r(\sin\theta - 3\cos\theta) = 1 \iff$

$r = \dfrac{1}{\sin\theta - 3\cos\theta}$

25. $x^2 + y^2 = 2cx \iff r^2 = 2cr\cos\theta \iff r^2 - 2cr\cos\theta = 0 \iff r(r - 2c\cos\theta) = 0 \iff r = 0$ or $r = 2c\cos\theta$.

$r = 0$ is included in $r = 2c\cos\theta$ when $\theta = \frac{\pi}{2} + n\pi$, so the curve is represented by the single equation $r = 2c\cos\theta$.

27. (a) The description leads immediately to the polar equation $\theta = \frac{\pi}{6}$, and the Cartesian equation $y = \tan\left(\frac{\pi}{6}\right)x = \frac{1}{\sqrt{3}}x$ is

slightly more difficult to derive.

(b) The easier description here is the Cartesian equation $x = 3$.

29. $r = -2\sin\theta$

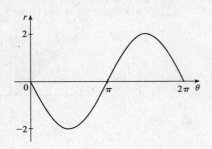 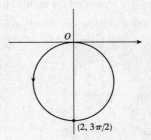

31. $r = 2(1 + \cos\theta)$

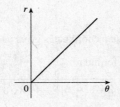 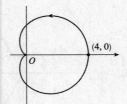

33. $r = \theta$, $\quad \theta \geq 0$

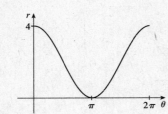 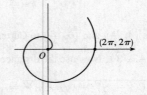

35. $r = 3\cos 3\theta$

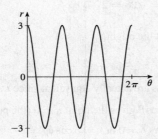 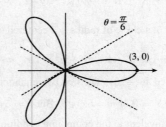

37. $r = 2\cos 4\theta$

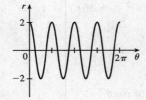 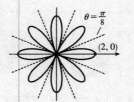

39. $r = 1 + 3\cos\theta$

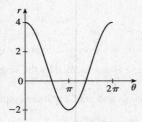 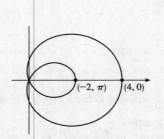

41. $r^2 = 9 \sin 2\theta$

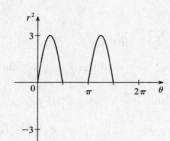

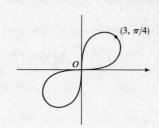

43. $r = 2 + \sin 3\theta$

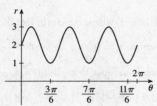

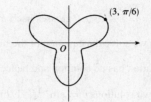

45. $r = \sin(\theta/2)$

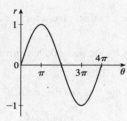

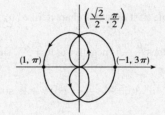

47. For $\theta = 0$, π, and 2π, r has its minimum value of about 0.5. For $\theta = \frac{\pi}{2}$ and $\frac{3\pi}{2}$, r attains its maximum value of 2.

We see that the graph has a similar shape for $0 \le \theta \le \pi$ and $\pi \le \theta \le 2\pi$.

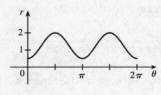

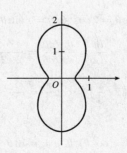

49. $x = r \cos \theta = (4 + 2 \sec \theta) \cos \theta = 4 \cos \theta + 2$. Now, $r \to \infty$ $\Rightarrow$

$(4 + 2 \sec \theta) \to \infty$ $\Rightarrow$ $\theta \to \left(\frac{\pi}{2}\right)^-$ or $\theta \to \left(\frac{3\pi}{2}\right)^+$ [since we need only

consider $0 \le \theta < 2\pi$], so $\lim\limits_{r \to \infty} x = \lim\limits_{\theta \to \pi/2^-} (4 \cos \theta + 2) = 2$. Also,

$r \to -\infty$ $\Rightarrow$ $(4 + 2 \sec \theta) \to -\infty$ $\Rightarrow$ $\theta \to \left(\frac{\pi}{2}\right)^+$ or $\theta \to \left(\frac{3\pi}{2}\right)^-$, so

$\lim\limits_{r \to -\infty} x = \lim\limits_{\theta \to \pi/2^+} (4 \cos \theta + 2) = 2$. Therefore, $\lim\limits_{r \to \pm\infty} x = 2$ $\Rightarrow$ $x = 2$ is a vertical asymptote.

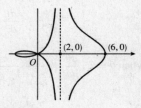

51. To show that $x = 1$ is an asymptote we must prove $\lim\limits_{r \to \pm\infty} x = 1$.

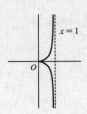

$x = (r)\cos\theta = (\sin\theta\,\tan\theta)\cos\theta = \sin^2\theta$. Now, $r \to \infty \;\Rightarrow\; \sin\theta\,\tan\theta \to \infty \;\Rightarrow$

$\theta \to \left(\frac{\pi}{2}\right)^-$, so $\lim\limits_{r \to \infty} x = \lim\limits_{\theta \to \pi/2^-} \sin^2\theta = 1$. Also, $r \to -\infty \;\Rightarrow\; \sin\theta\,\tan\theta \to -\infty \;\Rightarrow$

$\theta \to \left(\frac{\pi}{2}\right)^+$, so $\lim\limits_{r \to -\infty} x = \lim\limits_{\theta \to \pi/2^+} \sin^2\theta = 1$. Therefore, $\lim\limits_{r \to \pm\infty} x = 1 \;\Rightarrow\; x = 1$ is

a vertical asymptote. Also notice that $x = \sin^2\theta \ge 0$ for all θ, and $x = \sin^2\theta \le 1$ for all θ. And $x \ne 1$, since the curve is not

defined at odd multiples of $\frac{\pi}{2}$. Therefore, the curve lies entirely within the vertical strip $0 \le x < 1$.

53. (a) We see that the curve $r = 1 + c\sin\theta$ crosses itself at the origin, where $r = 0$ (in fact the inner loop corresponds to

negative r-values,) so we solve the equation of the limaçon for $r = 0 \;\Leftrightarrow\; c\sin\theta = -1 \;\Leftrightarrow\; \sin\theta = -1/c$. Now if

$|c| < 1$, then this equation has no solution and hence there is no inner loop. But if $c < -1$, then on the interval $(0, 2\pi)$

the equation has the two solutions $\theta = \sin^{-1}(-1/c)$ and $\theta = \pi - \sin^{-1}(-1/c)$, and if $c > 1$, the solutions are

$\theta = \pi + \sin^{-1}(1/c)$ and $\theta = 2\pi - \sin^{-1}(1/c)$. In each case, $r < 0$ for θ between the two solutions, indicating a loop.

(b) For $0 < c < 1$, the dimple (if it exists) is characterized by the fact that y has a local maximum at $\theta = \frac{3\pi}{2}$. So we

determine for what c-values $\dfrac{d^2y}{d\theta^2}$ is negative at $\theta = \frac{3\pi}{2}$, since by the Second Derivative Test this indicates a maximum:

$y = r\sin\theta = \sin\theta + c\sin^2\theta \;\Rightarrow\; \dfrac{dy}{d\theta} = \cos\theta + 2c\sin\theta\cos\theta = \cos\theta + c\sin 2\theta \;\Rightarrow\; \dfrac{d^2y}{d\theta^2} = -\sin\theta + 2c\cos 2\theta$.

At $\theta = \frac{3\pi}{2}$, this is equal to $-(-1) + 2c(-1) = 1 - 2c$, which is negative only for $c > \frac{1}{2}$. A similar argument shows that

for $-1 < c < 0$, y only has a local minimum at $\theta = \frac{\pi}{2}$ (indicating a dimple) for $c < -\frac{1}{2}$.

55. $r = 2\cos\theta \;\Rightarrow\; x = r\cos\theta = 2\cos^2\theta,\; y = r\sin\theta = 2\sin\theta\cos\theta = \sin 2\theta \;\Rightarrow$

$$\frac{dy}{dx} = \frac{dy/d\theta}{dx/d\theta} = \frac{2\cos 2\theta}{2 \cdot 2\cos\theta(-\sin\theta)} = \frac{\cos 2\theta}{-\sin 2\theta} = -\cot 2\theta$$

When $\theta = \frac{\pi}{3}$, $\dfrac{dy}{dx} = -\cot\left(2 \cdot \dfrac{\pi}{3}\right) = \cot\dfrac{\pi}{3} = \dfrac{1}{\sqrt{3}}$. [*Another method:* Use Equation 3.]

57. $r = 1/\theta \;\Rightarrow\; x = r\cos\theta = (\cos\theta)/\theta,\; y = r\sin\theta = (\sin\theta)/\theta \;\Rightarrow$

$$\frac{dy}{dx} = \frac{dy/d\theta}{dx/d\theta} = \frac{\sin\theta(-1/\theta^2) + (1/\theta)\cos\theta}{\cos\theta(-1/\theta^2) - (1/\theta)\sin\theta} \cdot \frac{\theta^2}{\theta^2} = \frac{-\sin\theta + \theta\cos\theta}{-\cos\theta - \theta\sin\theta}$$

When $\theta = \pi$, $\dfrac{dy}{dx} = \dfrac{-0 + \pi(-1)}{-(-1) - \pi(0)} = \dfrac{-\pi}{1} = -\pi$.

59. $r = \cos 2\theta \Rightarrow x = r \cos \theta = \cos 2\theta \cos \theta, \, y = r \sin \theta = \cos 2\theta \sin \theta \Rightarrow$

$$\frac{dy}{dx} = \frac{dy/d\theta}{dx/d\theta} = \frac{\cos 2\theta \cos \theta + \sin \theta \, (-2 \sin 2\theta)}{\cos 2\theta \, (-\sin \theta) + \cos \theta \, (-2 \sin 2\theta)}$$

When $\theta = \dfrac{\pi}{4}, \, \dfrac{dy}{dx} = \dfrac{0(\sqrt{2}/2) + (\sqrt{2}/2)(-2)}{0(-\sqrt{2}/2) + (\sqrt{2}/2)(-2)} = \dfrac{-\sqrt{2}}{-\sqrt{2}} = 1.$

61. $r = 3 \cos \theta \Rightarrow x = r \cos \theta = 3 \cos \theta \cos \theta, \, y = r \sin \theta = 3 \cos \theta \sin \theta \Rightarrow$

$\frac{dy}{d\theta} = -3 \sin^2 \theta + 3 \cos^2 \theta = 3 \cos 2\theta = 0 \Rightarrow 2\theta = \frac{\pi}{2}$ or $\frac{3\pi}{2} \Leftrightarrow \theta = \frac{\pi}{4}$ or $\frac{3\pi}{4}$.

So the tangent is horizontal at $\left(\frac{3}{\sqrt{2}}, \frac{\pi}{4}\right)$ and $\left(-\frac{3}{\sqrt{2}}, \frac{3\pi}{4}\right)$ $\left[\text{same as } \left(\frac{3}{\sqrt{2}}, -\frac{\pi}{4}\right)\right]$.

$\frac{dx}{d\theta} = -6 \sin \theta \cos \theta = -3 \sin 2\theta = 0 \Rightarrow 2\theta = 0$ or $\pi \Leftrightarrow \theta = 0$ or $\frac{\pi}{2}$. So the tangent is vertical at $(3, 0)$ and $\left(0, \frac{\pi}{2}\right)$.

63. $r = 1 + \cos \theta \Rightarrow x = r \cos \theta = \cos \theta \, (1 + \cos \theta), \, y = r \sin \theta = \sin \theta \, (1 + \cos \theta) \Rightarrow$

$\frac{dy}{d\theta} = (1 + \cos \theta) \cos \theta - \sin^2 \theta = 2 \cos^2 \theta + \cos \theta - 1 = (2 \cos \theta - 1)(\cos \theta + 1) = 0 \Rightarrow \cos \theta = \frac{1}{2}$ or $-1 \Rightarrow$

$\theta = \frac{\pi}{3}, \pi,$ or $\frac{5\pi}{3} \Rightarrow$ horizontal tangent at $\left(\frac{3}{2}, \frac{\pi}{3}\right), (0, \pi),$ and $\left(\frac{3}{2}, \frac{5\pi}{3}\right)$.

$\frac{dx}{d\theta} = -(1 + \cos \theta) \sin \theta - \cos \theta \sin \theta = -\sin \theta \, (1 + 2 \cos \theta) = 0 \Rightarrow \sin \theta = 0$ or $\cos \theta = -\frac{1}{2} \Rightarrow$

$\theta = 0, \pi, \frac{2\pi}{3},$ or $\frac{4\pi}{3} \Rightarrow$ vertical tangent at $(2, 0), \left(\frac{1}{2}, \frac{2\pi}{3}\right),$ and $\left(\frac{1}{2}, \frac{4\pi}{3}\right)$.

Note that the tangent is horizontal, not vertical when $\theta = \pi$, since $\lim\limits_{\theta \to \pi} \dfrac{dy/d\theta}{dx/d\theta} = 0$.

65. $r = a \sin \theta + b \cos \theta \Rightarrow r^2 = ar \sin \theta + br \cos \theta \Rightarrow x^2 + y^2 = ay + bx \Rightarrow$

$x^2 - bx + \left(\frac{1}{2}b\right)^2 + y^2 - ay + \left(\frac{1}{2}a\right)^2 = \left(\frac{1}{2}b\right)^2 + \left(\frac{1}{2}a\right)^2 \Rightarrow \left(x - \frac{1}{2}b\right)^2 + \left(y - \frac{1}{2}a\right)^2 = \frac{1}{4}(a^2 + b^2)$, and this is a circle

with center $\left(\frac{1}{2}b, \frac{1}{2}a\right)$ and radius $\frac{1}{2}\sqrt{a^2 + b^2}$.

67. $r = 1 + 2 \sin(\theta/2)$. The parameter interval is $[0, 4\pi]$.

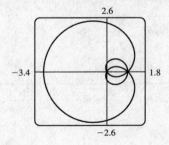

69. $r = e^{\sin \theta} - 2 \cos(4\theta)$.

The parameter interval is $[0, 2\pi]$.

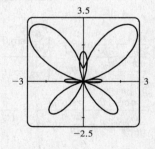

71. $r = 1 + \cos^{999} \theta$. The parameter interval is $[0, 2\pi]$.

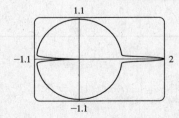

73. It appears that the graph of $r = 1 + \sin\left(\theta - \frac{\pi}{6}\right)$ is the same shape as the graph of $r = 1 + \sin\theta$, but rotated counterclockwise about the origin by $\frac{\pi}{6}$. Similarly, the graph of $r = 1 + \sin\left(\theta - \frac{\pi}{3}\right)$ is rotated by $\frac{\pi}{3}$. In general, the graph of $r = f(\theta - \alpha)$ is the same shape as that of $r = f(\theta)$, but rotated counterclockwise through α about the origin. That is, for any point (r_0, θ_0) on the curve $r = f(\theta)$, the point

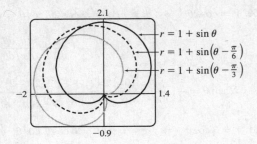

$(r_0, \theta_0 + \alpha)$ is on the curve $r = f(\theta - \alpha)$, since $r_0 = f(\theta_0) = f((\theta_0 + \alpha) - \alpha)$.

75. Consider curves with polar equation $r = 1 + c\cos\theta$, where c is a real number. If $c = 0$, we get a circle of radius 1 centered at the pole. For $0 < c \leq 0.5$, the curve gets slightly larger, moves right, and flattens out a bit on the left side. For $0.5 < c < 1$, the left side has a dimple shape. For $c = 1$, the dimple becomes a cusp. For $c > 1$, there is an internal loop. For $c \geq 0$, the rightmost point on the curve is $(1 + c, 0)$. For $c < 0$, the curves are reflections through the vertical axis of the curves with $c > 0$.

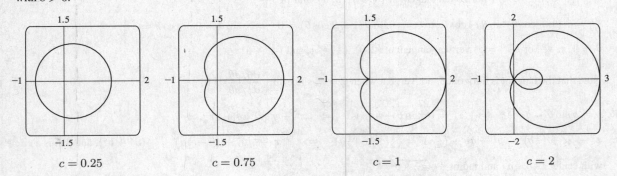

77. $\tan\psi = \tan(\phi - \theta) = \dfrac{\tan\phi - \tan\theta}{1 + \tan\phi\tan\theta} = \dfrac{\dfrac{dy}{dx} - \tan\theta}{1 + \dfrac{dy}{dx}\tan\theta} = \dfrac{\dfrac{dy/d\theta}{dx/d\theta} - \tan\theta}{1 + \dfrac{dy/d\theta}{dx/d\theta}\tan\theta}$

$= \dfrac{\dfrac{dy}{d\theta} - \dfrac{dx}{d\theta}\tan\theta}{\dfrac{dx}{d\theta} + \dfrac{dy}{d\theta}\tan\theta} = \dfrac{\left(\dfrac{dr}{d\theta}\sin\theta + r\cos\theta\right) - \tan\theta\left(\dfrac{dr}{d\theta}\cos\theta - r\sin\theta\right)}{\left(\dfrac{dr}{d\theta}\cos\theta - r\sin\theta\right) + \tan\theta\left(\dfrac{dr}{d\theta}\sin\theta + r\cos\theta\right)} = \dfrac{r\cos\theta + r\cdot\dfrac{\sin^2\theta}{\cos\theta}}{\dfrac{dr}{d\theta}\cos\theta + \dfrac{dr}{d\theta}\cdot\dfrac{\sin^2\theta}{\cos\theta}}$

$= \dfrac{r\cos^2\theta + r\sin^2\theta}{\dfrac{dr}{d\theta}\cos^2\theta + \dfrac{dr}{d\theta}\sin^2\theta} = \dfrac{r}{dr/d\theta}$

10.4 Areas and Lengths in Polar Coordinates

1. $r = e^{-\theta/4}$, $\pi/2 \leq \theta \leq \pi$.

$A = \displaystyle\int_{\pi/2}^{\pi} \tfrac{1}{2}r^2\,d\theta = \int_{\pi/2}^{\pi} \tfrac{1}{2}(e^{-\theta/4})^2\,d\theta = \int_{\pi/2}^{\pi} \tfrac{1}{2}e^{-\theta/2}\,d\theta = \tfrac{1}{2}\left[-2e^{-\theta/2}\right]_{\pi/2}^{\pi} = -1(e^{-\pi/2} - e^{-\pi/4}) = e^{-\pi/4} - e^{-\pi/2}$

3. $r = \sin\theta + \cos\theta$, $0 \le \theta \le \pi$.

$$A = \int_0^\pi \frac{1}{2} r^2\, d\theta = \int_0^\pi \frac{1}{2}(\sin\theta + \cos\theta)^2\, d\theta = \int_0^\pi \frac{1}{2}(\sin^2\theta + 2\sin\theta\,\cos\theta + \cos^2\theta)\, d\theta = \int_0^\pi \frac{1}{2}(1 + \sin 2\theta)\, d\theta$$

$$= \frac{1}{2}\Big[\theta - \frac{1}{2}\cos 2\theta\Big]_0^\pi = \frac{1}{2}\big[(\pi - \frac{1}{2}) - (0 - \frac{1}{2})\big] = \frac{\pi}{2}$$

5. $r^2 = \sin 2\theta$, $0 \le \theta \le \pi/2$.

$$A = \int_0^{\pi/2} \frac{1}{2} r^2\, d\theta = \int_0^{\pi/2} \frac{1}{2}\sin 2\theta\, d\theta = \Big[-\frac{1}{4}\cos 2\theta\Big]_0^{\pi/2} = -\frac{1}{4}(\cos\pi - \cos 0) = -\frac{1}{4}(-1 - 1) = \frac{1}{2}$$

7. $r = 4 + 3\sin\theta$, $-\frac{\pi}{2} \le \theta \le \frac{\pi}{2}$.

$$A = \int_{-\pi/2}^{\pi/2} \frac{1}{2}((4 + 3\sin\theta)^2\, d\theta = \frac{1}{2}\int_{-\pi/2}^{\pi/2}(16 + 24\sin\theta + 9\sin^2\theta)\, d\theta$$

$$= \frac{1}{2}\int_{-\pi/2}^{\pi/2}(16 + 9\sin^2\theta)\, d\theta \qquad \text{[by Theorem 4.5.6(b) [ET 5.5.7(b)]]}$$

$$= \frac{1}{2}\cdot 2\int_0^{\pi/2}\Big[16 + 9\cdot\frac{1}{2}(1 - \cos 2\theta)\Big]\, d\theta \qquad \text{[by Theorem 4.5.6(a) [ET 5.5.7(a)]]}$$

$$= \int_0^{\pi/2}\Big(\frac{41}{2} - \frac{9}{2}\cos 2\theta\Big)\, d\theta = \Big[\frac{41}{2}\theta - \frac{9}{4}\sin 2\theta\Big]_0^{\pi/2} = \Big(\frac{41\pi}{4} - 0\Big) - (0 - 0) = \frac{41\pi}{4}$$

9. The area is bounded by $r = 2\sin\theta$ for $\theta = 0$ to $\theta = \pi$.

$$A = \int_0^\pi \frac{1}{2} r^2\, d\theta = \frac{1}{2}\int_0^\pi (2\sin\theta)^2\, d\theta = \frac{1}{2}\int_0^\pi 4\sin^2\theta\, d\theta$$

$$= 2\int_0^\pi \frac{1}{2}(1 - \cos 2\theta)\, d\theta = \Big[\theta - \frac{1}{2}\sin 2\theta\Big]_0^\pi = \pi$$

Also, note that this is a circle with radius 1, so its area is $\pi(1)^2 = \pi$.

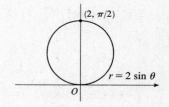

11. $A = \int_0^{2\pi} \frac{1}{2} r^2\, d\theta = \int_0^{2\pi} \frac{1}{2}(3 + 2\cos\theta)^2\, d\theta = \frac{1}{2}\int_0^{2\pi}(9 + 12\cos\theta + 4\cos^2\theta)\, d\theta$

$$= \frac{1}{2}\int_0^{2\pi}\Big[9 + 12\cos\theta + 4\cdot\frac{1}{2}(1 + \cos 2\theta)\Big]\, d\theta$$

$$= \frac{1}{2}\int_0^{2\pi}(11 + 12\cos\theta + 2\cos 2\theta)\, d\theta = \frac{1}{2}\Big[11\theta + 12\sin\theta + \sin 2\theta\Big]_0^{2\pi}$$

$$= \frac{1}{2}(22\pi) = 11\pi$$

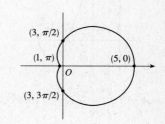

13. $A = \int_0^{2\pi} \frac{1}{2} r^2\, d\theta = \int_0^{2\pi} \frac{1}{2}(2 + \sin 4\theta)^2\, d\theta = \frac{1}{2}\int_0^{2\pi}(4 + 4\sin 4\theta + \sin^2 4\theta)\, d\theta$

$$= \frac{1}{2}\int_0^{2\pi}\Big[4 + 4\sin 4\theta + \frac{1}{2}(1 - \cos 8\theta)\Big]\, d\theta$$

$$= \frac{1}{2}\int_0^{2\pi}\Big(\frac{9}{2} + 4\sin 4\theta - \frac{1}{2}\cos 8\theta\Big)\, d\theta = \frac{1}{2}\Big[\frac{9}{2}\theta - \cos 4\theta - \frac{1}{16}\sin 8\theta\Big]_0^{2\pi}$$

$$= \frac{1}{2}[(9\pi - 1) - (-1)] = \frac{9}{2}\pi$$

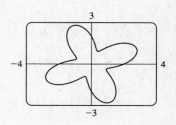

15. $A = \int_0^{2\pi} \frac{1}{2} r^2 \, d\theta = \int_0^{2\pi} \frac{1}{2} \left(\sqrt{1 + \cos^2 5\theta} \right)^2 d\theta$

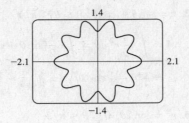

$\qquad = \frac{1}{2} \int_0^{2\pi} (1 + \cos^2 5\theta) \, d\theta = \frac{1}{2} \int_0^{2\pi} \left[1 + \frac{1}{2}(1 + \cos 10\theta) \right] d\theta$

$\qquad = \frac{1}{2} \left[\frac{3}{2}\theta + \frac{1}{20} \sin 10\theta \right]_0^{2\pi} = \frac{1}{2}(3\pi) = \frac{3}{2}\pi$

17. The curve passes through the pole when $r = 0 \;\Rightarrow\; 4\cos 3\theta = 0 \;\Rightarrow\; \cos 3\theta = 0 \;\Rightarrow\; 3\theta = \frac{\pi}{2} + \pi n \;\Rightarrow$

$\theta = \frac{\pi}{6} + \frac{\pi}{3} n$. The part of the shaded loop above the polar axis is traced out for

$\theta = 0$ to $\theta = \pi/6$, so we'll use $-\pi/6$ and $\pi/6$ as our limits of integration.

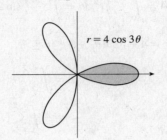

$r = 4\cos 3\theta$

$A = \int_{-\pi/6}^{\pi/6} \frac{1}{2}(4\cos 3\theta)^2 \, d\theta = 2 \int_0^{\pi/6} \frac{1}{2}(16 \cos^2 3\theta) \, d\theta$

$\qquad = 16 \int_0^{\pi/6} \frac{1}{2}(1 + \cos 6\theta) \, d\theta = 8 \left[\theta + \frac{1}{6} \sin 6\theta \right]_0^{\pi/6} = 8 \left(\frac{\pi}{6} \right) = \frac{4}{3}\pi$

19. $r = 0 \;\Rightarrow\; \sin 4\theta = 0 \;\Rightarrow\; 4\theta = \pi n \;\Rightarrow\; \theta = \frac{\pi}{4} n$.

$r = \sin 4\theta$

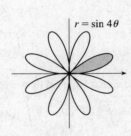

$A = \int_0^{\pi/4} \frac{1}{2}(\sin 4\theta)^2 \, d\theta = \frac{1}{2} \int_0^{\pi/4} \sin^2 4\theta \, d\theta = \frac{1}{2} \int_0^{\pi/4} \frac{1}{2}(1 - \cos 8\theta) \, d\theta$

$\qquad = \frac{1}{4} \left[\theta - \frac{1}{8} \sin 8\theta \right]_0^{\pi/4} = \frac{1}{4} \left(\frac{\pi}{4} \right) = \frac{1}{16} \pi$

21.

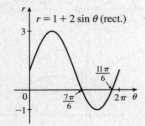

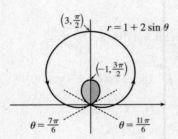

This is a limaçon, with inner loop traced

out between $\theta = \frac{7\pi}{6}$ and $\frac{11\pi}{6}$ [found by

solving $r = 0$].

$A = 2 \int_{7\pi/6}^{3\pi/2} \frac{1}{2}(1 + 2\sin\theta)^2 \, d\theta = \int_{7\pi/6}^{3\pi/2} \left(1 + 4\sin\theta + 4\sin^2\theta \right) d\theta = \int_{7\pi/6}^{3\pi/2} \left[1 + 4\sin\theta + 4 \cdot \frac{1}{2}(1 - \cos 2\theta) \right] d\theta$

$\qquad = \left[\theta - 4\cos\theta + 2\theta - \sin 2\theta \right]_{7\pi/6}^{3\pi/2} = \left(\frac{9\pi}{2} \right) - \left(\frac{7\pi}{2} + 2\sqrt{3} - \frac{\sqrt{3}}{2} \right) = \pi - \frac{3\sqrt{3}}{2}$

23. $4\sin\theta = 2 \;\Rightarrow\; \sin\theta = \frac{1}{2} \;\Rightarrow\; \theta = \frac{\pi}{6}$ or $\frac{5\pi}{6} \;\Rightarrow$

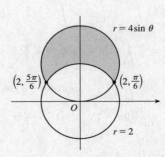

$A = \int_{\pi/6}^{5\pi/6} \frac{1}{2}[(4\sin\theta)^2 - 2^2] \, d\theta = 2 \int_{\pi/6}^{\pi/2} \frac{1}{2}(16\sin^2\theta - 4) \, d\theta$

$\qquad = \int_{\pi/6}^{\pi/2} \left[16 \cdot \frac{1}{2}(1 - \cos 2\theta) - 4 \right] d\theta = \int_{\pi/6}^{\pi/2} (4 - 8\cos 2\theta) \, d\theta$

$\qquad = \left[4\theta - 4\sin 2\theta \right]_{\pi/6}^{\pi/2} = (2\pi - 0) - \left(\frac{2\pi}{3} - 2\sqrt{3} \right) = \frac{4\pi}{3} + 2\sqrt{3}$

25. To find the area inside the leminiscate $r^2 = 8\cos 2\theta$ and outside the circle $r = 2$,

we first note that the two curves intersect when $r^2 = 8\cos 2\theta$ and $r = 2$,

that is, when $\cos 2\theta = \frac{1}{2}$. For $-\pi < \theta \le \pi$, $\cos 2\theta = \frac{1}{2}$ $\Leftrightarrow$ $2\theta = \pm\pi/3$

or $\pm 5\pi/3$ $\Leftrightarrow$ $\theta = \pm\pi/6$ or $\pm 5\pi/6$. The figure shows that the desired area

is 4 times the area between the curves from 0 to $\pi/6$. Thus,

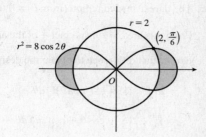

$$A = 4\int_0^{\pi/6}\left[\tfrac{1}{2}(8\cos 2\theta) - \tfrac{1}{2}(2)^2\right]d\theta = 8\int_0^{\pi/6}(2\cos 2\theta - 1)\,d\theta$$

$$= 8\left[\sin 2\theta - \theta\right]_0^{\pi/6} = 8\left(\sqrt{3}/2 - \pi/6\right) = 4\sqrt{3} - 4\pi/3$$

27. $3\cos\theta = 1 + \cos\theta$ $\Leftrightarrow$ $\cos\theta = \frac{1}{2}$ $\Rightarrow$ $\theta = \frac{\pi}{3}$ or $-\frac{\pi}{3}$.

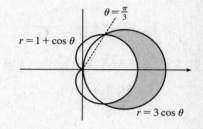

$$A = 2\int_0^{\pi/3}\tfrac{1}{2}\left[(3\cos\theta)^2 - (1+\cos\theta)^2\right]d\theta$$

$$= \int_0^{\pi/3}(8\cos^2\theta - 2\cos\theta - 1)\,d\theta = \int_0^{\pi/3}[4(1+\cos 2\theta) - 2\cos\theta - 1]\,d\theta$$

$$= \int_0^{\pi/3}(3 + 4\cos 2\theta - 2\cos\theta)\,d\theta = \left[3\theta + 2\sin 2\theta - 2\sin\theta\right]_0^{\pi/3}$$

$$= \pi + \sqrt{3} - \sqrt{3} = \pi$$

29. $3\sin\theta = 3\cos\theta$ $\Rightarrow$ $\dfrac{3\sin\theta}{3\cos\theta} = 1$ $\Rightarrow$ $\tan\theta = 1$ $\Rightarrow$ $\theta = \frac{\pi}{4}$ $\Rightarrow$

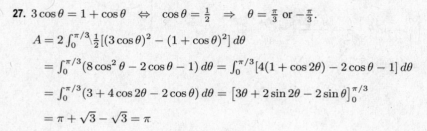

$$A = 2\int_0^{\pi/4}\tfrac{1}{2}(3\sin\theta)^2\,d\theta = \int_0^{\pi/4}9\sin^2\theta\,d\theta = \int_0^{\pi/4}9\cdot\tfrac{1}{2}(1-\cos 2\theta)\,d\theta$$

$$= \int_0^{\pi/4}\left(\tfrac{9}{2} - \tfrac{9}{2}\cos 2\theta\right)d\theta = \left[\tfrac{9}{2}\theta - \tfrac{9}{4}\sin 2\theta\right]_0^{\pi/4} = \left(\tfrac{9\pi}{8} - \tfrac{9}{4}\right) - (0 - 0)$$

$$= \tfrac{9\pi}{8} - \tfrac{9}{4}$$

31. $\sin 2\theta = \cos 2\theta$ $\Rightarrow$ $\dfrac{\sin 2\theta}{\cos 2\theta} = 1$ $\Rightarrow$ $\tan 2\theta = 1$ $\Rightarrow$ $2\theta = \frac{\pi}{4}$ $\Rightarrow$

$\theta = \frac{\pi}{8}$ $\Rightarrow$

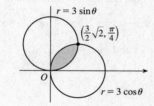

$$A = 8\cdot 2\int_0^{\pi/8}\tfrac{1}{2}\sin^2 2\theta\,d\theta = 8\int_0^{\pi/8}\tfrac{1}{2}(1-\cos 4\theta)\,d\theta$$

$$= 4\left[\theta - \tfrac{1}{4}\sin 4\theta\right]_0^{\pi/8} = 4\left(\tfrac{\pi}{8} - \tfrac{1}{4}\cdot 1\right) = \tfrac{\pi}{2} - 1$$

33. From the figure, we see that the shaded region is 4 times the shaded region

from $\theta = 0$ to $\theta = \pi/4$. $r^2 = 2\sin 2\theta$ and $r = 1$ $\Rightarrow$

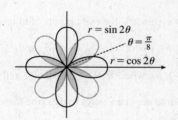
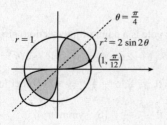

$2\sin 2\theta = 1^2$ $\Rightarrow$ $\sin 2\theta = \frac{1}{2}$ $\Rightarrow$ $2\theta = \frac{\pi}{6}$ $\Rightarrow$ $\theta = \frac{\pi}{12}$.

$$A = 4\int_0^{\pi/12}\tfrac{1}{2}(2\sin 2\theta)\,d\theta + 4\int_{\pi/12}^{\pi/4}\tfrac{1}{2}(1)^2\,d\theta$$

$$= \int_0^{\pi/12}4\sin 2\theta\,d\theta + \int_{\pi/12}^{\pi/4}2\,d\theta = \left[-2\cos 2\theta\right]_0^{\pi/12} + \left[2\theta\right]_{\pi/12}^{\pi/4}$$

$$= (-\sqrt{3} + 2) + \left(\tfrac{\pi}{2} - \tfrac{\pi}{6}\right) = -\sqrt{3} + 2 + \tfrac{\pi}{3}$$

35. The darker shaded region (from $\theta = 0$ to $\theta = 2\pi/3$) represents $\frac{1}{2}$ of the desired area plus $\frac{1}{2}$ of the area of the inner loop.

From this area, we'll subtract $\frac{1}{2}$ of the area of the inner loop (the lighter shaded region from $\theta = 2\pi/3$ to $\theta = \pi$), and then

double that difference to obtain the desired area.

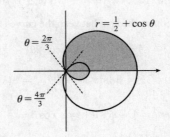

$$A = 2\left[\int_0^{2\pi/3} \frac{1}{2}\left(\frac{1}{2} + \cos\theta\right)^2 d\theta - \int_{2\pi/3}^{\pi} \frac{1}{2}\left(\frac{1}{2} + \cos\theta\right)^2 d\theta\right]$$

$$= \int_0^{2\pi/3} \left(\frac{1}{4} + \cos\theta + \cos^2\theta\right) d\theta - \int_{2\pi/3}^{\pi} \left(\frac{1}{4} + \cos\theta + \cos^2\theta\right) d\theta$$

$$= \int_0^{2\pi/3} \left[\frac{1}{4} + \cos\theta + \frac{1}{2}(1 + \cos 2\theta)\right] d\theta$$

$$\qquad - \int_{2\pi/3}^{\pi} \left[\frac{1}{4} + \cos\theta + \frac{1}{2}(1 + \cos 2\theta)\right] d\theta$$

$$= \left[\frac{\theta}{4} + \sin\theta + \frac{\theta}{2} + \frac{\sin 2\theta}{4}\right]_0^{2\pi/3} - \left[\frac{\theta}{4} + \sin\theta + \frac{\theta}{2} + \frac{\sin 2\theta}{4}\right]_{2\pi/3}^{\pi}$$

$$= \left(\frac{\pi}{6} + \frac{\sqrt{3}}{2} + \frac{\pi}{3} - \frac{\sqrt{3}}{8}\right) - \left(\frac{\pi}{4} + \frac{\pi}{2}\right) + \left(\frac{\pi}{6} + \frac{\sqrt{3}}{2} + \frac{\pi}{3} - \frac{\sqrt{3}}{8}\right)$$

$$= \frac{\pi}{4} + \frac{3}{4}\sqrt{3} = \frac{1}{4}\left(\pi + 3\sqrt{3}\right)$$

37. The pole is a point of intersection. $\sin\theta = 1 - \sin\theta \;\Rightarrow\; 2\sin\theta = 1 \;\Rightarrow$

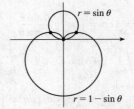

$\sin\theta = \frac{1}{2} \;\Rightarrow\; \theta = \frac{\pi}{6}$ or $\frac{5\pi}{6}$. So the other points of intersection are

$\left(\frac{1}{2}, \frac{\pi}{6}\right)$ and $\left(\frac{1}{2}, \frac{5\pi}{6}\right)$.

39. $2\sin 2\theta = 1 \;\Rightarrow\; \sin 2\theta = \frac{1}{2} \;\Rightarrow\; 2\theta = \frac{\pi}{6}, \frac{5\pi}{6}, \frac{13\pi}{6},$ or $\frac{17\pi}{6}$.

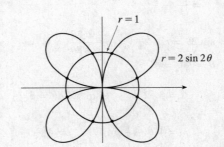

By symmetry, the eight points of intersection are given by

$(1, \theta)$, where $\theta = \frac{\pi}{12}, \frac{5\pi}{12}, \frac{13\pi}{12},$ and $\frac{17\pi}{12}$, and

$(-1, \theta)$, where $\theta = \frac{7\pi}{12}, \frac{11\pi}{12}, \frac{19\pi}{12},$ and $\frac{23\pi}{12}$.

[There are many ways to describe these points.]

41. The pole is a point of intersection. $\sin\theta = \sin 2\theta = 2\sin\theta\cos\theta \;\Leftrightarrow$

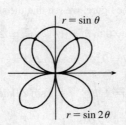

$\sin\theta(1 - 2\cos\theta) = 0 \;\Leftrightarrow\; \sin\theta = 0$ or $\cos\theta = \frac{1}{2} \;\Rightarrow$

$\theta = 0, \pi, \frac{\pi}{3},$ or $-\frac{\pi}{3} \;\Rightarrow$ the other intersection points are $\left(\frac{\sqrt{3}}{2}, \frac{\pi}{3}\right)$

and $\left(\frac{\sqrt{3}}{2}, \frac{2\pi}{3}\right)$ [by symmetry].

43.

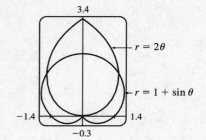

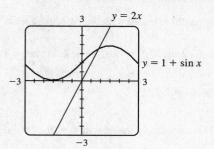

From the first graph, we see that the pole is one point of intersection. By zooming in or using the cursor, we find the θ-values of the intersection points to be $\alpha \approx 0.88786 \approx 0.89$ and $\pi - \alpha \approx 2.25$. (The first of these values may be more easily estimated by plotting $y = 1 + \sin x$ and $y = 2x$ in rectangular coordinates; see the second graph.) By symmetry, the total area contained is twice the area contained in the first quadrant, that is,

$$A = 2 \int_0^\alpha \tfrac{1}{2}(2\theta)^2 \, d\theta + 2 \int_\alpha^{\pi/2} \tfrac{1}{2}(1 + \sin\theta)^2 \, d\theta = \int_0^\alpha 4\theta^2 \, d\theta + \int_\alpha^{\pi/2} \left[1 + 2\sin\theta + \tfrac{1}{2}(1 - \cos 2\theta)\right] d\theta$$

$$= \left[\tfrac{4}{3}\theta^3\right]_0^\alpha + \left[\theta - 2\cos\theta + \left(\tfrac{1}{2}\theta - \tfrac{1}{4}\sin 2\theta\right)\right]_\alpha^{\pi/2} = \tfrac{4}{3}\alpha^3 + \left[\left(\tfrac{\pi}{2} + \tfrac{\pi}{4}\right) - \left(\alpha - 2\cos\alpha + \tfrac{1}{2}\alpha - \tfrac{1}{4}\sin 2\alpha\right)\right] \approx 3.4645$$

45. $L = \int_a^b \sqrt{r^2 + (dr/d\theta)^2} \, d\theta = \int_0^\pi \sqrt{(2\cos\theta)^2 + (-2\sin\theta)^2} \, d\theta$

$$= \int_0^\pi \sqrt{4(\cos^2\theta + \sin^2\theta)} \, d\theta = \int_0^\pi \sqrt{4} \, d\theta = \left[2\theta\right]_0^\pi = 2\pi$$

As a check, note that the curve is a circle of radius 1, so its circumference is $2\pi(1) = 2\pi$.

47. $L = \int_a^b \sqrt{r^2 + (dr/d\theta)^2} \, d\theta = \int_0^{2\pi} \sqrt{(\theta^2)^2 + (2\theta)^2} \, d\theta = \int_0^{2\pi} \sqrt{\theta^4 + 4\theta^2} \, d\theta$

$$= \int_0^{2\pi} \sqrt{\theta^2(\theta^2 + 4)} \, d\theta = \int_0^{2\pi} \theta\sqrt{\theta^2 + 4} \, d\theta$$

Now let $u = \theta^2 + 4$, so that $du = 2\theta \, d\theta$ $\left[\theta \, d\theta = \tfrac{1}{2} \, du\right]$ and

$$\int_0^{2\pi} \theta\sqrt{\theta^2 + 4} \, d\theta = \int_4^{4\pi^2 + 4} \tfrac{1}{2}\sqrt{u} \, du = \tfrac{1}{2} \cdot \tfrac{2}{3}\left[u^{3/2}\right]_4^{4(\pi^2 + 1)} = \tfrac{1}{3}\left[4^{3/2}(\pi^2 + 1)^{3/2} - 4^{3/2}\right] = \tfrac{8}{3}\left[(\pi^2 + 1)^{3/2} - 1\right]$$

49. The curve $r = \cos^4(\theta/4)$ is completely traced with $0 \le \theta \le 4\pi$.

$r^2 + (dr/d\theta)^2 = \left[\cos^4(\theta/4)\right]^2 + \left[4\cos^3(\theta/4) \cdot (-\sin(\theta/4)) \cdot \tfrac{1}{4}\right]^2$

$\qquad = \cos^8(\theta/4) + \cos^6(\theta/4)\sin^2(\theta/4)$

$\qquad = \cos^6(\theta/4)\left[\cos^2(\theta/4) + \sin^2(\theta/4)\right] = \cos^6(\theta/4)$

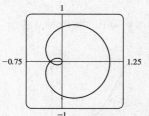

$L = \int_0^{4\pi} \sqrt{\cos^6(\theta/4)} \, d\theta = \int_0^{4\pi} \left|\cos^3(\theta/4)\right| \, d\theta$

$\quad = 2 \int_0^{2\pi} \cos^3(\theta/4) \, d\theta$ $\quad$ [since $\cos^3(\theta/4) \ge 0$ for $0 \le \theta \le 2\pi$] $\quad = 8 \int_0^{\pi/2} \cos^3 u \, du$ $\quad \left[u = \tfrac{1}{4}\theta\right]$

$\quad = 8 \int_0^{\pi/2} (1 - \sin^2 u)\cos u \, du = 8 \int_0^1 (1 - x^2) \, dx$ $\quad \begin{bmatrix} x = \sin u, \\ dx = \cos u \, du \end{bmatrix}$

$\quad = 8\left[x - \tfrac{1}{3}x^3\right]_0^1 = 8\left(1 - \tfrac{1}{3}\right) = \tfrac{16}{3}$

51. One loop of the curve $r = \cos 2\theta$ is traced with $-\pi/4 \le \theta \le \pi/4$.

$$r^2 + \left(\frac{dr}{d\theta}\right)^2 = \cos^2 2\theta + (-2\sin 2\theta)^2 = \cos^2 2\theta + 4\sin^2 2\theta = 1 + 3\sin^2 2\theta \quad \Rightarrow$$

$$L = \int_{-\pi/4}^{\pi/4} \sqrt{1 + 3\sin^2 2\theta}\, d\theta \approx 2.4221.$$

53. The curve $r = \sin(6\sin\theta)$ is completely traced with $0 \le \theta \le \pi$. $\quad r = \sin(6\sin\theta) \quad \Rightarrow$

$$\frac{dr}{d\theta} = \cos(6\sin\theta) \cdot 6\cos\theta, \text{ so } r^2 + \left(\frac{dr}{d\theta}\right)^2 = \sin^2(6\sin\theta) + 36\cos^2\theta\cos^2(6\sin\theta) \quad \Rightarrow$$

$$L = \int_0^\pi \sqrt{\sin^2(6\sin\theta) + 36\cos^2\theta\,\cos^2(6\sin\theta)}\, d\theta \approx 8.0091.$$

55. (a) From (10.2.6),

$$S = \int_a^b 2\pi y \sqrt{(dx/d\theta)^2 + (dy/d\theta)^2}\, d\theta$$

$$= \int_a^b 2\pi y \sqrt{r^2 + (dr/d\theta)^2}\, d\theta \qquad \text{[from the derivation of Equation 10.4.5]}$$

$$= \int_a^b 2\pi r\sin\theta \sqrt{r^2 + (dr/d\theta)^2}\, d\theta$$

(b) The curve $r^2 = \cos 2\theta$ goes through the pole when $\cos 2\theta = 0 \quad \Rightarrow$

$2\theta = \frac{\pi}{2} \quad \Rightarrow \quad \theta = \frac{\pi}{4}$. We'll rotate the curve from $\theta = 0$ to $\theta = \frac{\pi}{4}$ and double

this value to obtain the total surface area generated.

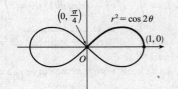

$$r^2 = \cos 2\theta \quad \Rightarrow \quad 2r\frac{dr}{d\theta} = -2\sin 2\theta \quad \Rightarrow \quad \left(\frac{dr}{d\theta}\right)^2 = \frac{\sin^2 2\theta}{r^2} = \frac{\sin^2 2\theta}{\cos 2\theta}.$$

$$S = 2\int_0^{\pi/4} 2\pi \sqrt{\cos 2\theta}\sin\theta \sqrt{\cos 2\theta + (\sin^2 2\theta)/\cos 2\theta}\, d\theta = 4\pi\int_0^{\pi/4} \sqrt{\cos 2\theta}\sin\theta \sqrt{\frac{\cos^2 2\theta + \sin^2 2\theta}{\cos 2\theta}}\, d\theta$$

$$= 4\pi\int_0^{\pi/4} \sqrt{\cos 2\theta}\sin\theta \frac{1}{\sqrt{\cos 2\theta}}\, d\theta = 4\pi\int_0^{\pi/4}\sin\theta\, d\theta = 4\pi\left[-\cos\theta\right]_0^{\pi/4} = -4\pi\left(\frac{\sqrt{2}}{2} - 1\right) = 2\pi\left(2 - \sqrt{2}\right)$$

10.5 Conic Sections

1. $x^2 = 6y$ and $x^2 = 4py \quad \Rightarrow \quad 4p = 6 \quad \Rightarrow \quad p = \frac{3}{2}$.

The vertex is $(0, 0)$, the focus is $\left(0, \frac{3}{2}\right)$, and the directrix

is $y = -\frac{3}{2}$.

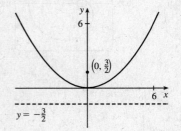

3. $2x = -y^2 \quad \Rightarrow \quad y^2 = -2x$. $4p = -2 \quad \Rightarrow \quad p = -\frac{1}{2}$.

The vertex is $(0, 0)$, the focus is $\left(-\frac{1}{2}, 0\right)$, and the

directrix is $x = \frac{1}{2}$.

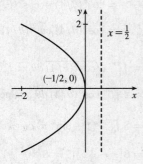

5. $(x+2)^2 = 8(y-3)$. $4p = 8$, so $p = 2$. The vertex is $(-2, 3)$, the focus is $(-2, 5)$, and the directrix is $y = 1$.

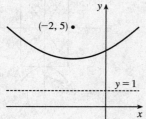

7. $y^2 + 6y + 2x + 1 = 0 \;\Leftrightarrow\; y^2 + 6y = -2x - 1$
$\Leftrightarrow\; y^2 + 6y + 9 = -2x + 8 \;\Leftrightarrow$
$(y+3)^2 = -2(x-4)$. $4p = -2$, so $p = -\frac{1}{2}$.
The vertex is $(4, -3)$, the focus is $\left(\frac{7}{2}, -3\right)$, and the directrix is $x = \frac{9}{2}$.

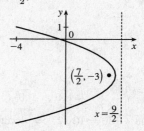

9. The equation has the form $y^2 = 4px$, where $p < 0$. Since the parabola passes through $(-1, 1)$, we have $1^2 = 4p(-1)$, so $4p = -1$ and an equation is $y^2 = -x$ or $x = -y^2$. $4p = -1$, so $p = -\frac{1}{4}$ and the focus is $\left(-\frac{1}{4}, 0\right)$ while the directrix is $x = \frac{1}{4}$.

11. $\dfrac{x^2}{2} + \dfrac{y^2}{4} = 1 \;\Rightarrow\; a = \sqrt{4} = 2, b = \sqrt{2}, c = \sqrt{a^2 - b^2} = \sqrt{4-2} = \sqrt{2}$. The ellipse is centered at $(0, 0)$, with vertices at $(0, \pm 2)$. The foci are $\left(0, \pm\sqrt{2}\right)$.

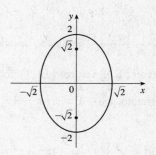

13. $x^2 + 9y^2 = 9 \;\Leftrightarrow\; \dfrac{x^2}{9} + \dfrac{y^2}{1} = 1 \;\Rightarrow\; a = \sqrt{9} = 3$,
$b = \sqrt{1} = 1, c = \sqrt{a^2 - b^2} = \sqrt{9-1} = \sqrt{8} = 2\sqrt{2}$.
The ellipse is centered at $(0, 0)$, with vertices $(\pm 3, 0)$.
The foci are $(\pm 2\sqrt{2}, 0)$.

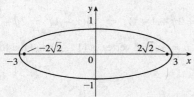

15. $9x^2 - 18x + 4y^2 = 27 \;\Leftrightarrow$
$9(x^2 - 2x + 1) + 4y^2 = 27 + 9 \;\Leftrightarrow$
$9(x-1)^2 + 4y^2 = 36 \;\Leftrightarrow\; \dfrac{(x-1)^2}{4} + \dfrac{y^2}{9} = 1 \;\Rightarrow$
$a = 3, b = 2, c = \sqrt{5} \;\Rightarrow\;$ center $(1, 0)$,
vertices $(1, \pm 3)$, foci $\left(1, \pm\sqrt{5}\right)$

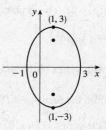

17. The center is $(0, 0)$, $a = 3$, and $b = 2$, so an equation is $\dfrac{x^2}{4} + \dfrac{y^2}{9} = 1$. $c = \sqrt{a^2 - b^2} = \sqrt{5}$, so the foci are $\left(0, \pm\sqrt{5}\right)$.

19. $\dfrac{y^2}{25} - \dfrac{x^2}{9} = 1 \Rightarrow a = 5, b = 3, c = \sqrt{25+9} = \sqrt{34} \Rightarrow$

center $(0,0)$, vertices $(0,\pm5)$, foci $\left(0, \pm\sqrt{34}\right)$, asymptotes $y = \pm\frac{5}{3}x$.

Note: It is helpful to draw a $2a$-by-$2b$ rectangle whose center is the center of

the hyperbola. The asymptotes are the extended diagonals of the rectangle.

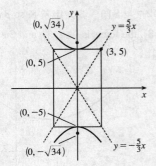

21. $x^2 - y^2 = 100 \Leftrightarrow \dfrac{x^2}{100} - \dfrac{y^2}{100} = 1 \Rightarrow a = b = 10,$

$c = \sqrt{100 + 100} = 10\sqrt{2} \Rightarrow$ center $(0,0)$, vertices $(\pm10, 0)$,

foci $\left(\pm10\sqrt{2}, 0\right)$, asymptotes $y = \pm\frac{10}{10}x = \pm x$

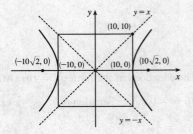

23. $x^2 - y^2 + 2y = 2 \Leftrightarrow x^2 - (y^2 - 2y + 1) = 2 - 1 \Leftrightarrow$

$\dfrac{x^2}{1} - \dfrac{(y-1)^2}{1} = 1 \Rightarrow a = b = 1, c = \sqrt{1+1} = \sqrt{2} \Rightarrow$

center $(0,1)$, vertices $(\pm1, 1)$, foci $\left(\pm\sqrt{2}, 1\right)$,

asymptotes $y - 1 = \pm\frac{1}{1}x = \pm x$.

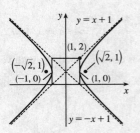

25. $4x^2 = y^2 + 4 \Leftrightarrow 4x^2 - y^2 = 4 \Leftrightarrow \dfrac{x^2}{1} - \dfrac{y^2}{4} = 1$. This is an equation of a *hyperbola* with vertices $(\pm1, 0)$.

The foci are at $\left(\pm\sqrt{1+4}, 0\right) = \left(\pm\sqrt{5}, 0\right)$.

27. $x^2 = 4y - 2y^2 \Leftrightarrow x^2 + 2y^2 - 4y = 0 \Leftrightarrow x^2 + 2(y^2 - 2y + 1) = 2 \Leftrightarrow x^2 + 2(y-1)^2 = 2 \Leftrightarrow$

$\dfrac{x^2}{2} + \dfrac{(y-1)^2}{1} = 1$. This is an equation of an *ellipse* with vertices at $\left(\pm\sqrt{2}, 1\right)$. The foci are at $\left(\pm\sqrt{2-1}, 1\right) = (\pm1, 1)$.

29. $3x^2 - 6x - 2y = 1 \Leftrightarrow 3x^2 - 6x = 2y + 1 \Leftrightarrow 3(x^2 - 2x + 1) = 2y + 1 + 3 \Leftrightarrow 3(x-1)^2 = 2y + 4 \Leftrightarrow$

$(x-1)^2 = \frac{2}{3}(y+2)$. This is an equation of a *parabola* with $4p = \frac{2}{3}$, so $p = \frac{1}{6}$. The vertex is $(1, -2)$ and the focus is

$\left(1, -2 + \frac{1}{6}\right) = \left(1, -\frac{11}{6}\right)$.

31. The parabola with vertex $(0,0)$ and focus $(1,0)$ opens to the right and has $p = 1$, so its equation is $y^2 = 4px$, or $y^2 = 4x$.

33. The distance from the focus $(-4, 0)$ to the directrix $x = 2$ is $2 - (-4) = 6$, so the distance from the focus to the vertex is

$\frac{1}{2}(6) = 3$ and the vertex is $(-1, 0)$. Since the focus is to the left of the vertex, $p = -3$. An equation is $y^2 = 4p(x+1) \Rightarrow$

$y^2 = -12(x+1)$.

35. The parabola with vertex $(3, -1)$ having a horizontal axis has equation $[y - (-1)]^2 = 4p(x - 3)$. Since it passes through

$(-15, 2)$, $(2 + 1)^2 = 4p(-15 - 3)$ $\Rightarrow$ $9 = 4p(-18)$ $\Rightarrow$ $4p = -\frac{1}{2}$. An equation is $(y + 1)^2 = -\frac{1}{2}(x - 3)$.

37. The ellipse with foci $(\pm 2, 0)$ and vertices $(\pm 5, 0)$ has center $(0, 0)$ and a horizontal major axis, with $a = 5$ and $c = 2$,

so $b^2 = a^2 - c^2 = 25 - 4 = 21$. An equation is $\dfrac{x^2}{25} + \dfrac{y^2}{21} = 1$.

39. Since the vertices are $(0, 0)$ and $(0, 8)$, the ellipse has center $(0, 4)$ with a vertical axis and $a = 4$. The foci at $(0, 2)$ and $(0, 6)$

are 2 units from the center, so $c = 2$ and $b = \sqrt{a^2 - c^2} = \sqrt{4^2 - 2^2} = \sqrt{12}$. An equation is $\dfrac{(x - 0)^2}{b^2} + \dfrac{(y - 4)^2}{a^2} = 1$ $\Rightarrow$

$\dfrac{x^2}{12} + \dfrac{(y - 4)^2}{16} = 1$.

41. An equation of an ellipse with center $(-1, 4)$ and vertex $(-1, 0)$ is $\dfrac{(x + 1)^2}{b^2} + \dfrac{(y - 4)^2}{4^2} = 1$. The focus $(-1, 6)$ is 2 units

from the center, so $c = 2$. Thus, $b^2 + 2^2 = 4^2$ $\Rightarrow$ $b^2 = 12$, and the equation is $\dfrac{(x + 1)^2}{12} + \dfrac{(y - 4)^2}{16} = 1$.

43. An equation of a hyperbola with vertices $(\pm 3, 0)$ is $\dfrac{x^2}{3^2} - \dfrac{y^2}{b^2} = 1$. Foci $(\pm 5, 0)$ $\Rightarrow$ $c = 5$ and $3^2 + b^2 = 5^2$ $\Rightarrow$

$b^2 = 25 - 9 = 16$, so the equation is $\dfrac{x^2}{9} - \dfrac{y^2}{16} = 1$.

45. The center of a hyperbola with vertices $(-3, -4)$ and $(-3, 6)$ is $(-3, 1)$, so $a = 5$ and an equation is

$\dfrac{(y - 1)^2}{5^2} - \dfrac{(x + 3)^2}{b^2} = 1$. Foci $(-3, -7)$ and $(-3, 9)$ $\Rightarrow$ $c = 8$, so $5^2 + b^2 = 8^2$ $\Rightarrow$ $b^2 = 64 - 25 = 39$ and the

equation is $\dfrac{(y - 1)^2}{25} - \dfrac{(x + 3)^2}{39} = 1$.

47. The center of a hyperbola with vertices $(\pm 3, 0)$ is $(0, 0)$, so $a = 3$ and an equation is $\dfrac{x^2}{3^2} - \dfrac{y^2}{b^2} = 1$.

Asymptotes $y = \pm 2x$ $\Rightarrow$ $\dfrac{b}{a} = 2$ $\Rightarrow$ $b = 2(3) = 6$ and the equation is $\dfrac{x^2}{9} - \dfrac{y^2}{36} = 1$.

49. In Figure 8, we see that the point on the ellipse closest to a focus is the closer vertex (which is a distance

$a - c$ from it) while the farthest point is the other vertex (at a distance of $a + c$). So for this lunar orbit,

$(a - c) + (a + c) = 2a = (1728 + 110) + (1728 + 314)$, or $a = 1940$; and $(a + c) - (a - c) = 2c = 314 - 110$,

or $c = 102$. Thus, $b^2 = a^2 - c^2 = 3{,}753{,}196$, and the equation is $\dfrac{x^2}{3{,}763{,}600} + \dfrac{y^2}{3{,}753{,}196} = 1$.

51. (a) Set up the coordinate system so that A is $(-200, 0)$ and B is $(200, 0)$.

$|PA| - |PB| = (1200)(980) = 1{,}176{,}000 \text{ ft} = \frac{2450}{11} \text{ mi} = 2a$ $\Rightarrow$ $a = \frac{1225}{11}$, and $c = 200$ so

$b^2 = c^2 - a^2 = \dfrac{3{,}339{,}375}{121}$ $\Rightarrow$ $\dfrac{121x^2}{1{,}500{,}625} - \dfrac{121y^2}{3{,}339{,}375} = 1$.

(b) Due north of B $\Rightarrow$ $x = 200$ $\Rightarrow$ $\dfrac{(121)(200)^2}{1{,}500{,}625} - \dfrac{121y^2}{3{,}339{,}375} = 1$ $\Rightarrow$ $y = \dfrac{133{,}575}{539} \approx 248$ mi

53. The function whose graph is the upper branch of this hyperbola is concave upward. The function is

$$y = f(x) = a\sqrt{1 + \frac{x^2}{b^2}} = \frac{a}{b}\sqrt{b^2 + x^2}, \text{ so } y' = \frac{a}{b}x(b^2 + x^2)^{-1/2} \text{ and}$$

$$y'' = \frac{a}{b}\left[(b^2 + x^2)^{-1/2} - x^2(b^2 + x^2)^{-3/2}\right] = ab(b^2 + x^2)^{-3/2} > 0 \text{ for all } x, \text{ and so } f \text{ is concave upward.}$$

55. (a) If $k > 16$, then $k - 16 > 0$, and $\dfrac{x^2}{k} + \dfrac{y^2}{k - 16} = 1$ is an *ellipse* since it is the sum of two squares on the left side.

(b) If $0 < k < 16$, then $k - 16 < 0$, and $\dfrac{x^2}{k} + \dfrac{y^2}{k - 16} = 1$ is a *hyperbola* since it is the difference of two squares on the

left side.

(c) If $k < 0$, then $k - 16 < 0$, and there is *no curve* since the left side is the sum of two negative terms, which cannot equal 1.

(d) In case (a), $a^2 = k$, $b^2 = k - 16$, and $c^2 = a^2 - b^2 = 16$, so the foci are at $(\pm 4, 0)$. In case (b), $k - 16 < 0$, so $a^2 = k$,

$b^2 = 16 - k$, and $c^2 = a^2 + b^2 = 16$, and so again the foci are at $(\pm 4, 0)$.

57. $x^2 = 4py$ $\Rightarrow$ $2x = 4py'$ $\Rightarrow$ $y' = \dfrac{x}{2p}$, so the tangent line at (x_0, y_0) is $y - \dfrac{x_0^2}{4p} = \dfrac{x_0}{2p}(x - x_0)$. This line passes

through the point $(a, -p)$ on the directrix, so $-p - \dfrac{x_0^2}{4p} = \dfrac{x_0}{2p}(a - x_0)$ $\Rightarrow$ $-4p^2 - x_0^2 = 2ax_0 - 2x_0^2$ $\Leftrightarrow$

$x_0^2 - 2ax_0 - 4p^2 = 0$ $\Leftrightarrow$ $x_0^2 - 2ax_0 + a^2 = a^2 + 4p^2$ $\Leftrightarrow$

$(x_0 - a)^2 = a^2 + 4p^2$ $\Leftrightarrow$ $x_0 = a \pm \sqrt{a^2 + 4p^2}$. The slopes of the tangent

lines at $x = a \pm \sqrt{a^2 + 4p^2}$ are $\dfrac{a \pm \sqrt{a^2 + 4p^2}}{2p}$, so the product of the two

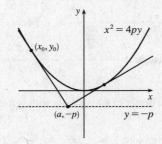

slopes is

$$\frac{a + \sqrt{a^2 + 4p^2}}{2p} \cdot \frac{a - \sqrt{a^2 + 4p^2}}{2p} = \frac{a^2 - (a^2 + 4p^2)}{4p^2} = \frac{-4p^2}{4p^2} = -1,$$

showing that the tangent lines are perpendicular.

59. $9x^2 + 4y^2 = 36$ $\Leftrightarrow$ $\dfrac{x^2}{4} + \dfrac{y^2}{9} = 1$. We use the parametrization $x = 2\cos t$, $y = 3\sin t$, $0 \le t \le 2\pi$. The circumference

is given by

$$L = \int_0^{2\pi} \sqrt{(dx/dt)^2 + (dy/dt)^2}\, dt = \int_0^{2\pi} \sqrt{(-2\sin t)^2 + (3\cos t)^2}\, dt$$

$$= \int_0^{2\pi} \sqrt{4\sin^2 t + 9\cos^2 t}\, dt = \int_0^{2\pi} \sqrt{4 + 5\cos^2 t}\, dt$$

Now use Simpson's Rule with $n = 8$, $\Delta t = \dfrac{2\pi - 0}{8} = \dfrac{\pi}{4}$, and $f(t) = \sqrt{4 + 5\cos^2 t}$ to get

$$L \approx S_8 = \tfrac{\pi/4}{3}\left[f(0) + 4f\left(\tfrac{\pi}{4}\right) + 2f\left(\tfrac{\pi}{2}\right) + 4f\left(\tfrac{3\pi}{4}\right) + 2f(\pi) + 4f\left(\tfrac{5\pi}{4}\right) + 2f\left(\tfrac{3\pi}{2}\right) + 4f\left(\tfrac{7\pi}{4}\right) + f(2\pi)\right] \approx 15.9.$$

61. $\dfrac{x^2}{a^2} - \dfrac{y^2}{b^2} = 1$ $\Rightarrow$ $\dfrac{y^2}{b^2} = \dfrac{x^2 - a^2}{a^2}$ $\Rightarrow$ $y = \pm\dfrac{b}{a}\sqrt{x^2 - a^2}$.

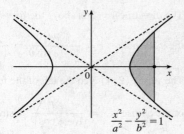

$$A = 2\int_a^c \dfrac{b}{a}\sqrt{x^2 - a^2}\,dx \overset{39}{=} \dfrac{2b}{a}\left[\dfrac{x}{2}\sqrt{x^2 - a^2} - \dfrac{a^2}{2}\ln\left|x + \sqrt{x^2 - a^2}\right|\right]_a^c$$

$$= \dfrac{b}{a}\left[c\sqrt{c^2 - a^2} - a^2\ln\left|c + \sqrt{c^2 - a^2}\right| + a^2\ln|a|\right]$$

Since $a^2 + b^2 = c^2$, $c^2 - a^2 = b^2$, and $\sqrt{c^2 - a^2} = b$.

$$= \dfrac{b}{a}\left[cb - a^2\ln(c + b) + a^2\ln a\right] = \dfrac{b}{a}\left[cb + a^2(\ln a - \ln(b + c))\right]$$

$$= b^2 c/a + ab\ln[a/(b + c)], \text{ where } c^2 = a^2 + b^2.$$

63. $9x^2 + 4y^2 = 36$ $\Leftrightarrow$ $\dfrac{x^2}{4} + \dfrac{y^2}{9} = 1$ $\Rightarrow$ $a = 3, b = 2$. By symmetry, $\overline{x} = 0$. By Example 2 in Section 7.3, the area of the

top half of the ellipse is $\frac{1}{2}(\pi ab) = 3\pi$. Solve $9x^2 + 4y^2 = 36$ for y to get an equation for the top half of the ellipse:

$9x^2 + 4y^2 = 36$ $\Leftrightarrow$ $4y^2 = 36 - 9x^2$ $\Leftrightarrow$ $y^2 = \frac{9}{4}(4 - x^2)$ $\Rightarrow$ $y = \frac{3}{2}\sqrt{4 - x^2}$. Now

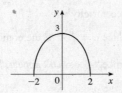

$$\overline{y} = \dfrac{1}{A}\int_a^b \dfrac{1}{2}[f(x)]^2\,dx = \dfrac{1}{3\pi}\int_{-2}^2 \dfrac{1}{2}\left(\dfrac{3}{2}\sqrt{4 - x^2}\right)^2\,dx = \dfrac{3}{8\pi}\int_{-2}^2 (4 - x^2)\,dx$$

$$= \dfrac{3}{8\pi}\cdot 2\int_0^2 (4 - x^2)\,dx = \dfrac{3}{4\pi}\left[4x - \dfrac{1}{3}x^3\right]_0^2 = \dfrac{3}{4\pi}\left(\dfrac{16}{3}\right) = \dfrac{4}{\pi}$$

so the centroid is $(0, 4/\pi)$.

65. Differentiating implicitly, $\dfrac{x^2}{a^2} + \dfrac{y^2}{b^2} = 1$ $\Rightarrow$ $\dfrac{2x}{a^2} + \dfrac{2yy'}{b^2} = 0$ $\Rightarrow$ $y' = -\dfrac{b^2 x}{a^2 y}$ $[y \neq 0]$. Thus, the slope of the tangent

line at P is $-\dfrac{b^2 x_1}{a^2 y_1}$. The slope of $F_1 P$ is $\dfrac{y_1}{x_1 + c}$ and of $F_2 P$ is $\dfrac{y_1}{x_1 - c}$. By the formula from Problems Plus, we have

$$\tan\alpha = \dfrac{\dfrac{y_1}{x_1 + c} + \dfrac{b^2 x_1}{a^2 y_1}}{1 - \dfrac{b^2 x_1 y_1}{a^2 y_1(x_1 + c)}} = \dfrac{a^2 y_1^2 + b^2 x_1(x_1 + c)}{a^2 y_1(x_1 + c) - b^2 x_1 y_1} = \dfrac{a^2 b^2 + b^2 c x_1}{c^2 x_1 y_1 + a^2 c y_1} \quad \left[\begin{array}{l}\text{using } b^2 x_1^2 + a^2 y_1^2 = a^2 b^2, \\ \text{and } a^2 - b^2 = c^2\end{array}\right]$$

$$= \dfrac{b^2(cx_1 + a^2)}{cy_1(cx_1 + a^2)} = \dfrac{b^2}{cy_1}$$

and $\quad \tan\beta = \dfrac{-\dfrac{b^2 x_1}{a^2 y_1} - \dfrac{y_1}{x_1 - c}}{1 - \dfrac{b^2 x_1 y_1}{a^2 y_1(x_1 - c)}} = \dfrac{-a^2 y_1^2 - b^2 x_1(x_1 - c)}{a^2 y_1(x_1 - c) - b^2 x_1 y_1} = \dfrac{-a^2 b^2 + b^2 c x_1}{c^2 x_1 y_1 - a^2 c y_1} = \dfrac{b^2(cx_1 - a^2)}{cy_1(cx_1 - a^2)} = \dfrac{b^2}{cy_1}$

Thus, $\alpha = \beta$.

10.6   Conic Sections in Polar Coordinates

1. The directrix $x = 4$ is to the right of the focus at the origin, so we use the form with "$+ e\cos\theta$" in the denominator.

(See Theorem 6 and Figure 2.) An equation of the ellipse is $r = \dfrac{ed}{1 + e\cos\theta} = \dfrac{\frac{1}{2}\cdot 4}{1 + \frac{1}{2}\cos\theta} = \dfrac{4}{2 + \cos\theta}$.

3. The directrix $y = 2$ is above the focus at the origin, so we use the form with "$+ e \sin \theta$" in the denominator. An equation of

the hyperbola is $r = \dfrac{ed}{1 + e \sin \theta} = \dfrac{1.5(2)}{1 + 1.5 \sin \theta} = \dfrac{6}{2 + 3 \sin \theta}$.

5. The vertex $(2, \pi)$ is to the left of the focus at the origin, so we use the form with "$-e \cos \theta$" in the denominator. An equation

of the ellipse is $r = \dfrac{ed}{1 - e \cos \theta}$. Using eccentricity $e = \dfrac{2}{3}$ with $\theta = \pi$ and $r = 2$, we get $2 = \dfrac{\frac{2}{3}d}{1 - \frac{2}{3}(-1)}$ $\Rightarrow$

$2 = \dfrac{2d}{5}$ $\Rightarrow$ $d = 5$, so we have $r = \dfrac{\frac{2}{3}(5)}{1 - \frac{2}{3} \cos \theta} = \dfrac{10}{3 - 2 \cos \theta}$.

7. The vertex $\left(3, \frac{\pi}{2}\right)$ is 3 units above the focus at the origin, so the directrix is 6 units above the focus ($d = 6$), and we use the

form "$+e \sin \theta$" in the denominator. $e = 1$ for a parabola, so an equation is $r = \dfrac{ed}{1 + e \sin \theta} = \dfrac{1(6)}{1 + 1 \sin \theta} = \dfrac{6}{1 + \sin \theta}$.

9. $r = \dfrac{4}{5 - 4 \sin \theta} \cdot \dfrac{1/5}{1/5} = \dfrac{4/5}{1 - \frac{4}{5} \sin \theta}$, where $e = \frac{4}{5}$ and $ed = \frac{4}{5}$ $\Rightarrow$ $d = 1$.

(a) Eccentricity $= e = \frac{4}{5}$

(b) Since $e = \frac{4}{5} < 1$, the conic is an ellipse.

(c) Since "$- e \sin \theta$" appears in the denominator, the directrix is below the focus

at the origin, $d = |Fl| = 1$, so an equation of the directrix is $y = -1$.

(d) The vertices are $\left(4, \frac{\pi}{2}\right)$ and $\left(\frac{4}{9}, \frac{3\pi}{2}\right)$.

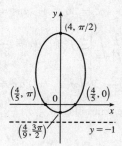

11. $r = \dfrac{2}{3 + 3 \sin \theta} \cdot \dfrac{1/3}{1/3} = \dfrac{2/3}{1 + 1 \sin \theta}$, where $e = 1$ and $ed = \frac{2}{3}$ $\Rightarrow$ $d = \frac{2}{3}$.

(a) Eccentricity $= e = 1$

(b) Since $e = 1$, the conic is a parabola.

(c) Since "$+ e \sin \theta$" appears in the denominator, the directrix is above the focus

at the origin. $d = |Fl| = \frac{2}{3}$, so an equation of the directrix is $y = \frac{2}{3}$.

(d) The vertex is at $\left(\frac{1}{3}, \frac{\pi}{2}\right)$, midway between the focus and directrix.

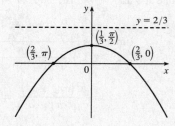

13. $r = \dfrac{9}{6 + 2 \cos \theta} \cdot \dfrac{1/6}{1/6} = \dfrac{3/2}{1 + \frac{1}{3} \cos \theta}$, where $e = \frac{1}{3}$ and $ed = \frac{3}{2}$ $\Rightarrow$ $d = \frac{9}{2}$.

(a) Eccentricity $= e = \frac{1}{3}$

(b) Since $e = \frac{1}{3} < 1$, the conic is an ellipse.

(c) Since "$+ e \cos \theta$" appears in the denominator, the directrix is to the right of

the focus at the origin. $d = |Fl| = \frac{9}{2}$, so an equation of the directrix is

$x = \frac{9}{2}$.

(d) The vertices are $\left(\frac{9}{8}, 0\right)$ and $\left(\frac{9}{4}, \pi\right)$, so the center is midway between them,

that is, $\left(\frac{9}{16}, \pi\right)$.

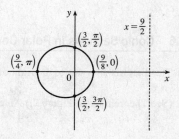

15. $r = \dfrac{3}{4 - 8\cos\theta} \cdot \dfrac{1/4}{1/4} = \dfrac{3/4}{1 - 2\cos\theta}$, where $e = 2$ and $ed = \frac{3}{4} \;\Rightarrow\; d = \frac{3}{8}$.

(a) Eccentricity $= e = 2$

(b) Since $e = 2 > 1$, the conic is a hyperbola.

(c) Since "$-e\cos\theta$" appears in the denominator, the directrix is to the left of

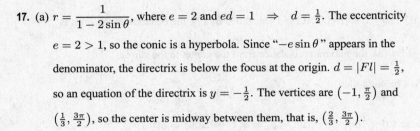

the focus at the origin. $d = |Fl| = \frac{3}{8}$, so an equation of the directrix is

$x = -\frac{3}{8}$.

(d) The vertices are $\left(-\frac{3}{4}, 0\right)$ and $\left(\frac{1}{4}, \pi\right)$, so the center is midway between them,

that is, $\left(\frac{1}{2}, \pi\right)$.

17. (a) $r = \dfrac{1}{1 - 2\sin\theta}$, where $e = 2$ and $ed = 1 \;\Rightarrow\; d = \frac{1}{2}$. The eccentricity

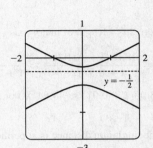

$e = 2 > 1$, so the conic is a hyperbola. Since "$-e\sin\theta$" appears in the

denominator, the directrix is below the focus at the origin. $d = |Fl| = \frac{1}{2}$,

so an equation of the directrix is $y = -\frac{1}{2}$. The vertices are $\left(-1, \frac{\pi}{2}\right)$ and

$\left(\frac{1}{3}, \frac{3\pi}{2}\right)$, so the center is midway between them, that is, $\left(\frac{2}{3}, \frac{3\pi}{2}\right)$.

(b) By the discussion that precedes Example 4, the equation

is $r = \dfrac{1}{1 - 2\sin\left(\theta - \frac{3\pi}{4}\right)}$.

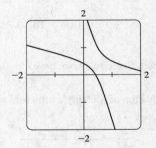

19. For $e < 1$ the curve is an ellipse. It is nearly circular when e is close to 0. As e

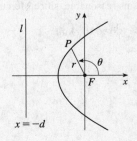

increases, the graph is stretched out to the right, and grows larger (that is, its

right-hand focus moves to the right while its left-hand focus remains at the

origin.) At $e = 1$, the curve becomes a parabola with focus at the origin.

21. $|PF| = e\,|Pl| \;\Rightarrow\; r = e[d - r\cos(\pi - \theta)] = e(d + r\cos\theta) \;\Rightarrow$

$r(1 - e\cos\theta) = ed \;\Rightarrow\; r = \dfrac{ed}{1 - e\cos\theta}$

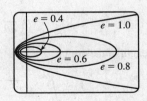

23. $|PF| = e|Pl| \Rightarrow r = e[d - r\sin(\theta - \pi)] = e(d + r\sin\theta) \Rightarrow$

$r(1 - e\sin\theta) = ed \Rightarrow r = \dfrac{ed}{1 - e\sin\theta}$

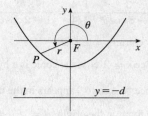

25. We are given $e = 0.093$ and $a = 2.28 \times 10^8$. By (7), we have

$$r = \frac{a(1 - e^2)}{1 + e\cos\theta} = \frac{2.28 \times 10^8[1 - (0.093)^2]}{1 + 0.093\cos\theta} \approx \frac{2.26 \times 10^8}{1 + 0.093\cos\theta}$$

27. Here $2a = $ length of major axis $= 36.18$ AU $\Rightarrow a = 18.09$ AU and $e = 0.97$. By (7), the equation of the orbit is

$r = \dfrac{18.09[1 - (0.97)^2]}{1 + 0.97\cos\theta} \approx \dfrac{1.07}{1 + 0.97\cos\theta}$. By (8), the maximum distance from the comet to the sun is

$18.09(1 + 0.97) \approx 35.64$ AU or about 3.314 billion miles.

29. The minimum distance is at perihelion, where $4.6 \times 10^7 = r = a(1 - e) = a(1 - 0.206) = a(0.794) \Rightarrow$

$a = 4.6 \times 10^7/0.794$. So the maximum distance, which is at aphelion, is

$r = a(1 + e) = (4.6 \times 10^7/0.794)(1.206) \approx 7.0 \times 10^7$ km.

31. From Exercise 29, we have $e = 0.206$ and $a(1 - e) = 4.6 \times 10^7$ km. Thus, $a = 4.6 \times 10^7/0.794$. From (7), we can write the

equation of Mercury's orbit as $r = a\dfrac{1 - e^2}{1 + e\cos\theta}$. So since

$\dfrac{dr}{d\theta} = \dfrac{a(1 - e^2)e\sin\theta}{(1 + e\cos\theta)^2} \Rightarrow$

$$r^2 + \left(\frac{dr}{d\theta}\right)^2 = \frac{a^2(1 - e^2)^2}{(1 + e\cos\theta)^2} + \frac{a^2(1 - e^2)^2 e^2 \sin^2\theta}{(1 + e\cos\theta)^4} = \frac{a^2(1 - e^2)^2}{(1 + e\cos\theta)^4}(1 + 2e\cos\theta + e^2)$$

the length of the orbit is

$$L = \int_0^{2\pi} \sqrt{r^2 + (dr/d\theta)^2}\, d\theta = a(1 - e^2)\int_0^{2\pi} \frac{\sqrt{1 + e^2 + 2e\cos\theta}}{(1 + e\cos\theta)^2}\, d\theta \approx 3.6 \times 10^8 \text{ km}$$

This seems reasonable, since Mercury's orbit is nearly circular, and the circumference of a circle of radius a

is $2\pi a \approx 3.6 \times 10^8$ km.

10   Review

TRUE-FALSE QUIZ

1. False.   Consider the curve defined by $x = f(t) = (t-1)^3$ and $y = g(t) = (t-1)^2$. Then $g'(t) = 2(t-1)$, so $g'(1) = 0$, but its graph has a *vertical* tangent when $t = 1$.   *Note:* The statement is true if $f'(1) \neq 0$ when $g'(1) = 0$.

3. False.   For example, if $f(t) = \cos t$ and $g(t) = \sin t$ for $0 \le t \le 4\pi$, then the curve is a circle of radius 1, hence its length is 2π, but $\int_0^{4\pi} \sqrt{[f'(t)]^2 + [g'(t)]^2}\, dt = \int_0^{4\pi} \sqrt{(-\sin t)^2 + (\cos t)^2}\, dt = \int_0^{4\pi} 1\, dt = 4\pi$, since as t increases from 0 to 4π, the circle is traversed twice.

5. True.   The curve $r = 1 - \sin 2\theta$ is unchanged if we rotate it through $180°$ about O because $1 - \sin 2(\theta + \pi) = 1 - \sin(2\theta + 2\pi) = 1 - \sin 2\theta$. So it's unchanged if we replace r by $-r$. (See the discussion after Example 8 in Section 10.3.) In other words, it's the same curve as $r = -(1 - \sin 2\theta) = \sin 2\theta - 1$.

7. False.   The first pair of equations gives the portion of the parabola $y = x^2$ with $x \ge 0$, whereas the second pair of equations traces out the whole parabola $y = x^2$.

9. True.   By rotating and translating the parabola, we can assume it has an equation of the form $y = cx^2$, where $c > 0$. The tangent at the point (a, ca^2) is the line $y - ca^2 = 2ca(x - a)$; i.e., $y = 2cax - ca^2$. This tangent meets the parabola at the points (x, cx^2) where $cx^2 = 2cax - ca^2$. This equation is equivalent to $x^2 = 2ax - a^2$ [since $c > 0$]. But $x^2 = 2ax - a^2 \Leftrightarrow x^2 - 2ax + a^2 = 0 \Leftrightarrow (x - a)^2 = 0 \Leftrightarrow x = a \Leftrightarrow (x, cx^2) = (a, ca^2)$. This shows that each tangent meets the parabola at exactly one point.

EXERCISES

1. $x = t^2 + 4t,\ y = 2 - t,\ -4 \le t \le 1.\ t = 2 - y$, so

$x = (2-y)^2 + 4(2-y) = 4 - 4y + y^2 + 8 - 4y = y^2 - 8y + 12 \Leftrightarrow$

$x + 4 = y^2 - 8y + 16 = (y-4)^2$. This is part of a parabola with vertex $(-4, 4)$, opening to the right.

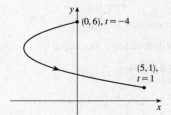

3. $y = \sec\theta = \dfrac{1}{\cos\theta} = \dfrac{1}{x}$. Since $0 \le \theta \le \pi/2,\ 0 < x \le 1$ and $y \ge 1$.

This is part of the hyperbola $y = 1/x$.

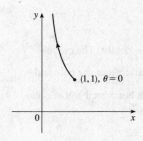

5. Three different sets of parametric equations for the curve $y = \sqrt{x}$ are

(i) $x = t, \;\; y = \sqrt{t}$

(ii) $x = t^4, \;\; y = t^2$

(iii) $x = \tan^2 t, \;\; y = \tan t, \;\; 0 \le t < \pi/2$

There are many other sets of equations that also give this curve.

7. (a)

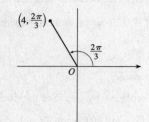

The Cartesian coordinates are $x = 4 \cos \frac{2\pi}{3} = 4\left(-\frac{1}{2}\right) = -2$ and

$y = 4 \sin \frac{2\pi}{3} = 4\left(\frac{\sqrt{3}}{2}\right) = 2\sqrt{3}$, that is, the point $\left(-2, 2\sqrt{3}\right)$.

(b) Given $x = -3$ and $y = 3$, we have $r = \sqrt{(-3)^2 + 3^2} = \sqrt{18} = 3\sqrt{2}$. Also, $\tan\theta = \frac{y}{x} \;\; \Rightarrow \;\; \tan\theta = \frac{3}{-3}$, and since

$(-3, 3)$ is in the second quadrant, $\theta = \frac{3\pi}{4}$. Thus, one set of polar coordinates for $(-3, 3)$ is $\left(3\sqrt{2}, \frac{3\pi}{4}\right)$, and two others are

$\left(3\sqrt{2}, \frac{11\pi}{4}\right)$ and $\left(-3\sqrt{2}, \frac{7\pi}{4}\right)$.

9. $r = 1 + \sin\theta$. This cardioid is symmetric about the $\theta = \pi/2$ axis.

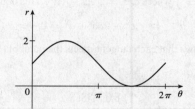

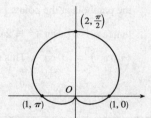

11. $r = \cos 3\theta$. This is a three-leaved rose. The curve is traced twice.

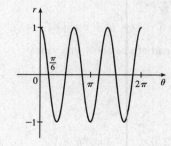

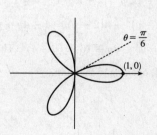

13. $r = 1 + \cos 2\theta$. The curve is symmetric about the pole and both the horizontal and vertical axes.

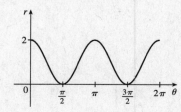

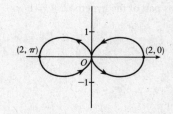

15. $r = \dfrac{3}{1 + 2\sin\theta}$ $\Rightarrow$ $e = 2 > 1$, so the conic is a hyperbola. $de = 3$ $\Rightarrow$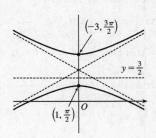

$d = \frac{3}{2}$ and the form "$+2\sin\theta$" imply that the directrix is above the focus at

the origin and has equation $y = \frac{3}{2}$. The vertices are $\left(1, \frac{\pi}{2}\right)$ and $\left(-3, \frac{3\pi}{2}\right)$.

17. $x + y = 2$ $\Leftrightarrow$ $r\cos\theta + r\sin\theta = 2$ $\Leftrightarrow$ $r(\cos\theta + \sin\theta) = 2$ $\Leftrightarrow$ $r = \dfrac{2}{\cos\theta + \sin\theta}$

19. $r = (\sin\theta)/\theta$. As $\theta \to \pm\infty$, $r \to 0$.

As $\theta \to 0$, $r \to 1$. In the first figure,

there are an infinite number of

x-intercepts at $x = \pi n$, n a nonzero

integer. These correspond to pole

points in the second figure.

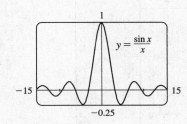

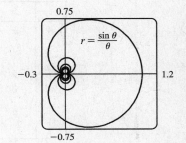

21. $x = \ln t$, $y = 1 + t^2$; $t = 1$. $\dfrac{dy}{dt} = 2t$ and $\dfrac{dx}{dt} = \dfrac{1}{t}$, so $\dfrac{dy}{dx} = \dfrac{dy/dt}{dx/dt} = \dfrac{2t}{1/t} = 2t^2$.

When $t = 1$, $(x, y) = (0, 2)$ and $dy/dx = 2$.

23. $r = e^{-\theta}$ $\Rightarrow$ $y = r\sin\theta = e^{-\theta}\sin\theta$ and $x = r\cos\theta = e^{-\theta}\cos\theta$ $\Rightarrow$

$\dfrac{dy}{dx} = \dfrac{dy/d\theta}{dx/d\theta} = \dfrac{\frac{dr}{d\theta}\sin\theta + r\cos\theta}{\frac{dr}{d\theta}\cos\theta - r\sin\theta} = \dfrac{-e^{-\theta}\sin\theta + e^{-\theta}\cos\theta}{-e^{-\theta}\cos\theta - e^{-\theta}\sin\theta} \cdot \dfrac{-e^{\theta}}{-e^{\theta}} = \dfrac{\sin\theta - \cos\theta}{\cos\theta + \sin\theta}$.

When $\theta = \pi$, $\dfrac{dy}{dx} = \dfrac{0 - (-1)}{-1 + 0} = \dfrac{1}{-1} = -1$.

25. $x = t + \sin t$, $y = t - \cos t$ $\Rightarrow$ $\dfrac{dy}{dx} = \dfrac{dy/dt}{dx/dt} = \dfrac{1 + \sin t}{1 + \cos t}$ $\Rightarrow$

$\dfrac{d^2y}{dx^2} = \dfrac{\dfrac{d}{dt}\left(\dfrac{dy}{dx}\right)}{dx/dt} = \dfrac{\dfrac{(1 + \cos t)\cos t - (1 + \sin t)(-\sin t)}{(1 + \cos t)^2}}{1 + \cos t} = \dfrac{\cos t + \cos^2 t + \sin t + \sin^2 t}{(1 + \cos t)^3} = \dfrac{1 + \cos t + \sin t}{(1 + \cos t)^3}$

27. We graph the curve $x = t^3 - 3t$, $y = t^2 + t + 1$ for $-2.2 \leq t \leq 1.2$.

By zooming in or using a cursor, we find that the lowest point is about

$(1.4, 0.75)$. To find the exact values, we find the t-value at which

$dy/dt = 2t + 1 = 0$ $\Leftrightarrow$ $t = -\frac{1}{2}$ $\Leftrightarrow$ $(x, y) = \left(\frac{11}{8}, \frac{3}{4}\right)$.

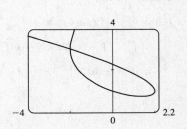

29. $x = 2a\cos t - a\cos 2t \Rightarrow \dfrac{dx}{dt} = -2a\sin t + 2a\sin 2t = 2a\sin t(2\cos t - 1) = 0 \Leftrightarrow$

$\sin t = 0$ or $\cos t = \frac{1}{2} \Rightarrow t = 0, \frac{\pi}{3}, \pi,$ or $\frac{5\pi}{3}$.

$y = 2a\sin t - a\sin 2t \Rightarrow \dfrac{dy}{dt} = 2a\cos t - 2a\cos 2t = 2a\big(1 + \cos t - 2\cos^2 t\big) = 2a(1 - \cos t)(1 + 2\cos t) = 0 \Rightarrow$

$t = 0, \frac{2\pi}{3},$ or $\frac{4\pi}{3}$.

Thus the graph has vertical tangents where $t = \frac{\pi}{3}, \pi$ and $\frac{5\pi}{3}$, and horizontal tangents where $t = \frac{2\pi}{3}$ and $\frac{4\pi}{3}$. To determine

what the slope is where $t = 0$, we use l'Hospital's Rule to evaluate $\displaystyle\lim_{t\to 0} \frac{dy/dt}{dx/dt} = 0$, so there is a horizontal tangent there.

t	x	y
0	a	0
$\frac{\pi}{3}$	$\frac{3}{2}a$	$\frac{\sqrt{3}}{2}a$
$\frac{2\pi}{3}$	$-\frac{1}{2}a$	$\frac{3\sqrt{3}}{2}a$
π	$-3a$	0
$\frac{4\pi}{3}$	$-\frac{1}{2}a$	$-\frac{3\sqrt{3}}{2}a$
$\frac{5\pi}{3}$	$\frac{3}{2}a$	$-\frac{\sqrt{3}}{2}a$

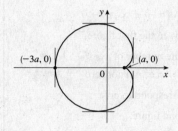

31. The curve $r^2 = 9\cos 5\theta$ has 10 "petals." For instance, for $-\frac{\pi}{10} \le \theta \le \frac{\pi}{10}$, there are two petals, one with $r > 0$ and one

with $r < 0$.

$$A = 10 \int_{-\pi/10}^{\pi/10} \tfrac{1}{2} r^2 \, d\theta = 5 \int_{-\pi/10}^{\pi/10} 9\cos 5\theta \, d\theta = 5 \cdot 9 \cdot 2 \int_0^{\pi/10} \cos 5\theta \, d\theta = 18\big[\sin 5\theta\big]_0^{\pi/10} = 18$$

33. The curves intersect when $4\cos\theta = 2 \Rightarrow \cos\theta = \frac{1}{2} \Rightarrow \theta = \pm\frac{\pi}{3}$

for $-\pi \le \theta \le \pi$. The points of intersection are $\big(2, \frac{\pi}{3}\big)$ and $\big(2, -\frac{\pi}{3}\big)$.

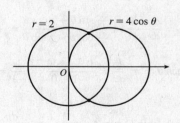

35. The curves intersect where $2\sin\theta = \sin\theta + \cos\theta \Rightarrow$

$\sin\theta = \cos\theta \Rightarrow \theta = \frac{\pi}{4}$, and also at the origin (at which $\theta = \frac{3\pi}{4}$

on the second curve).

$$A = \int_0^{\pi/4} \tfrac{1}{2}(2\sin\theta)^2 \, d\theta + \int_{\pi/4}^{3\pi/4} \tfrac{1}{2}(\sin\theta + \cos\theta)^2 \, d\theta$$

$$= \int_0^{\pi/4} (1 - \cos 2\theta) \, d\theta + \tfrac{1}{2}\int_{\pi/4}^{3\pi/4} (1 + \sin 2\theta) \, d\theta$$

$$= \big[\theta - \tfrac{1}{2}\sin 2\theta\big]_0^{\pi/4} + \big[\tfrac{1}{2}\theta - \tfrac{1}{4}\cos 2\theta\big]_{\pi/4}^{3\pi/4} = \tfrac{1}{2}(\pi - 1)$$

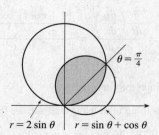

37. $x = 3t^2$, $y = 2t^3$.

$L = \int_0^2 \sqrt{(dx/dt)^2 + (dy/dt)^2}\, dt = \int_0^2 \sqrt{(6t)^2 + (6t^2)^2}\, dt = \int_0^2 \sqrt{36t^2 + 36t^4}\, dt = \int_0^2 \sqrt{36t^2}\sqrt{1 + t^2}\, dt$

$= \int_0^2 6\,|t|\,\sqrt{1 + t^2}\, dt = 6\int_0^2 t\,\sqrt{1 + t^2}\, dt = 6\int_1^5 u^{1/2}\left(\tfrac{1}{2}du\right)\qquad \left[u = 1 + t^2,\, du = 2t\,dt\right]$

$= 6\cdot\tfrac{1}{2}\cdot\tfrac{2}{3}\Big[u^{3/2}\Big]_1^5 = 2(5^{3/2} - 1) = 2\big(5\sqrt{5} - 1\big)$

39. $L = \int_\pi^{2\pi} \sqrt{r^2 + (dr/d\theta)^2}\, d\theta = \int_\pi^{2\pi} \sqrt{(1/\theta)^2 + (-1/\theta^2)^2}\, d\theta = \int_\pi^{2\pi} \dfrac{\sqrt{\theta^2 + 1}}{\theta^2}\, d\theta$

$\overset{24}{=} \left[-\dfrac{\sqrt{\theta^2 + 1}}{\theta} + \ln\!\left(\theta + \sqrt{\theta^2 + 1}\right)\right]_\pi^{2\pi} = \dfrac{\sqrt{\pi^2 + 1}}{\pi} - \dfrac{\sqrt{4\pi^2 + 1}}{2\pi} + \ln\!\left(\dfrac{2\pi + \sqrt{4\pi^2 + 1}}{\pi + \sqrt{\pi^2 + 1}}\right)$

$= \dfrac{2\sqrt{\pi^2 + 1} - \sqrt{4\pi^2 + 1}}{2\pi} + \ln\!\left(\dfrac{2\pi + \sqrt{4\pi^2 + 1}}{\pi + \sqrt{\pi^2 + 1}}\right)$

41. $x = 4\sqrt{t}$, $y = \dfrac{t^3}{3} + \dfrac{1}{2t^2}$, $1 \le t \le 4\ \Rightarrow$

$S = \int_1^4 2\pi y\,\sqrt{(dx/dt)^2 + (dy/dt)^2}\, dt = \int_1^4 2\pi\big(\tfrac{1}{3}t^3 + \tfrac{1}{2}t^{-2}\big)\sqrt{\big(2/\sqrt{t}\big)^2 + (t^2 - t^{-3})^2}\, dt$

$= 2\pi\int_1^4 \big(\tfrac{1}{3}t^3 + \tfrac{1}{2}t^{-2}\big)\sqrt{(t^2 + t^{-3})^2}\, dt = 2\pi\int_1^4 \big(\tfrac{1}{3}t^5 + \tfrac{5}{6} + \tfrac{1}{2}t^{-5}\big)\, dt = 2\pi\Big[\tfrac{1}{18}t^6 + \tfrac{5}{6}t - \tfrac{1}{8}t^{-4}\Big]_1^4 = \dfrac{471{,}295}{1024}\pi$

43. For all c except -1, the curve is asymptotic to the line $x = 1$. For $c < -1$, the curve bulges to the right near $y = 0$. As c increases, the bulge becomes smaller, until at $c = -1$ the curve is the straight line $x = 1$. As c continues to increase, the curve bulges to the left, until at $c = 0$ there is a cusp at the origin. For $c > 0$, there is a loop to the left of the origin, whose size and roundness increase as c increases. Note that the x-intercept of the curve is always $-c$.

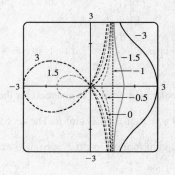

45. $\dfrac{x^2}{9} + \dfrac{y^2}{8} = 1$ is an ellipse with center $(0, 0)$.

$a = 3, b = 2\sqrt{2}, c = 1\ \Rightarrow$
foci $(\pm 1, 0)$, vertices $(\pm 3, 0)$.

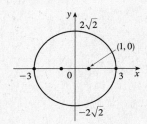

47. $6y^2 + x - 36y + 55 = 0\ \Leftrightarrow$
$6(y^2 - 6y + 9) = -(x + 1)\ \Leftrightarrow$
$(y - 3)^2 = -\tfrac{1}{6}(x + 1)$, a parabola with vertex $(-1, 3)$,
opening to the left, $p = -\tfrac{1}{24}\ \Rightarrow$ focus $\left(-\tfrac{25}{24}, 3\right)$ and
directrix $x = -\tfrac{23}{24}$.

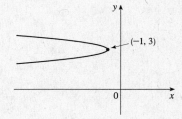

49. The ellipse with foci $(\pm 4, 0)$ and vertices $(\pm 5, 0)$ has center $(0, 0)$ and a horizontal major axis, with $a = 5$ and $c = 4$,

so $b^2 = a^2 - c^2 = 5^2 - 4^2 = 9$. An equation is $\dfrac{x^2}{25} + \dfrac{y^2}{9} = 1$.

51. The center of a hyperbola with foci $(0, \pm 4)$ is $(0, 0)$, so $c = 4$ and an equation is $\dfrac{y^2}{a^2} - \dfrac{x^2}{b^2} = 1$.

The asymptote $y = 3x$ has slope 3, so $\dfrac{a}{b} = \dfrac{3}{1} \;\Rightarrow\; a = 3b$ and $a^2 + b^2 = c^2 \;\Rightarrow\; (3b)^2 + b^2 = 4^2 \;\Rightarrow\;$

$10b^2 = 16 \;\Rightarrow\; b^2 = \frac{8}{5}$ and so $a^2 = 16 - \frac{8}{5} = \frac{72}{5}$. Thus, an equation is $\dfrac{y^2}{72/5} - \dfrac{x^2}{8/5} = 1$, or $\dfrac{5y^2}{72} - \dfrac{5x^2}{8} = 1$.

53. $x^2 + y = 100 \;\Leftrightarrow\; x^2 = -(y - 100)$ has its vertex at $(0, 100)$, so one of the vertices of the ellipse is $(0, 100)$. Another

form of the equation of a parabola is $x^2 = 4p(y - 100)$ so $4p(y - 100) = -(y - 100) \;\Rightarrow\; 4p = -1 \;\Rightarrow\; p = -\frac{1}{4}$.

Therefore the shared focus is found at $\left(0, \frac{399}{4}\right)$ so $2c = \frac{399}{4} - 0 \;\Rightarrow\; c = \frac{399}{8}$ and the center of the ellipse is $\left(0, \frac{399}{8}\right)$. So

$a = 100 - \frac{399}{8} = \frac{401}{8}$ and $b^2 = a^2 - c^2 = \dfrac{401^2 - 399^2}{8^2} = 25$. So the equation of the ellipse is $\dfrac{x^2}{b^2} + \dfrac{\left(y - \frac{399}{8}\right)^2}{a^2} = 1 \;\Rightarrow\;$

$\dfrac{x^2}{25} + \dfrac{\left(y - \frac{399}{8}\right)^2}{\left(\frac{401}{8}\right)^2} = 1$, or $\dfrac{x^2}{25} + \dfrac{(8y - 399)^2}{160{,}801} = 1$.

55. Directrix $x = 4 \;\Rightarrow\; d = 4$, so $e = \frac{1}{3} \;\Rightarrow\; r = \dfrac{ed}{1 + e\cos\theta} = \dfrac{4}{3 + \cos\theta}$.

57. In polar coordinates, an equation for the circle is $r = 2a\sin\theta$. Thus, the coordinates of Q are $x = r\cos\theta = 2a\sin\theta\,\cos\theta$

and $y = r\sin\theta = 2a\sin^2\theta$. The coordinates of R are $x = 2a\cot\theta$ and $y = 2a$. Since P is the midpoint of QR, we use the

midpoint formula to get $x = a(\sin\theta\,\cos\theta + \cot\theta)$ and $y = a(1 + \sin^2\theta)$.

PROBLEMS PLUS

1. See the figure. The circle with center $(-1, 0)$ and radius $\sqrt{2}$ has equation

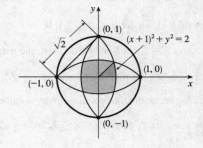

$(x+1)^2 + y^2 = 2$ and describes the circular arc from $(0, -1)$ to $(0, 1)$.
Converting the equation to polar coordinates gives us

$(r\cos\theta + 1)^2 + (r\sin\theta)^2 = 2 \quad \Rightarrow$

$r^2\cos^2\theta + 2r\cos\theta + 1 + r^2\sin^2\theta = 2 \quad \Rightarrow$

$r^2(\cos^2\theta + \sin^2\theta) + 2r\cos\theta = 1 \quad \Rightarrow \quad r^2 + 2r\cos\theta = 1$. Using the

quadratic formula to solve for r gives us

$$r = \frac{-2\cos\theta \pm \sqrt{4\cos^2\theta + 4}}{2} = -\cos\theta + \sqrt{\cos^2\theta + 1} \text{ for } r > 0.$$

The darkest shaded region is $\frac{1}{8}$ of the entire shaded region A, so $\frac{1}{8}A = \int_0^{\pi/4} \frac{1}{2}r^2\,d\theta = \frac{1}{2}\int_0^{\pi/4}(1 - 2r\cos\theta)\,d\theta \quad \Rightarrow$

$$\frac{1}{4}A = \int_0^{\pi/4}\left[1 - 2\cos\theta\left(-\cos\theta + \sqrt{\cos^2\theta + 1}\right)\right]d\theta = \int_0^{\pi/4}\left(1 + 2\cos^2\theta - 2\cos\theta\sqrt{\cos^2\theta + 1}\right)d\theta$$

$$= \int_0^{\pi/4}\left[1 + 2\cdot\frac{1}{2}(1 + \cos 2\theta) - 2\cos\theta\sqrt{(1 - \sin^2\theta) + 1}\right]d\theta$$

$$= \int_0^{\pi/4}(2 + \cos 2\theta)\,d\theta - 2\int_0^{\pi/4}\cos\theta\sqrt{2 - \sin^2\theta}\,d\theta$$

$$= \left[2\theta + \frac{1}{2}\sin 2\theta\right]_0^{\pi/4} - 2\int_0^{1/\sqrt{2}}\sqrt{2 - u^2}\,du \qquad \begin{bmatrix} u = \sin\theta, \\ du = \cos\theta\,d\theta \end{bmatrix}$$

$$= \left(\frac{\pi}{2} + \frac{1}{2}\right) - (0 + 0) - 2\left[\frac{u}{2}\sqrt{2 - u^2} + \sin^{-1}\frac{u}{\sqrt{2}}\right]_0^{1/\sqrt{2}} \qquad \begin{bmatrix} \text{Formula 30,} \\ a = \sqrt{2} \end{bmatrix}$$

$$= \frac{\pi}{2} + \frac{1}{2} - 2\left(\frac{1}{2\sqrt{2}}\cdot\frac{\sqrt{3}}{\sqrt{2}} + \frac{\pi}{6}\right) = \frac{\pi}{2} + \frac{1}{2} - \frac{1}{2}\sqrt{3} - \frac{\pi}{3} = \frac{\pi}{6} + \frac{1}{2} - \frac{1}{2}\sqrt{3}.$$

Thus, $A = 4\left(\dfrac{\pi}{6} + \dfrac{1}{2} - \dfrac{1}{2}\sqrt{3}\right) = \dfrac{2\pi}{3} + 2 - 2\sqrt{3}$.

3. In terms of x and y, we have $x = r\cos\theta = (1 + c\sin\theta)\cos\theta = \cos\theta + c\sin\theta\cos\theta = \cos\theta + \frac{1}{2}c\sin 2\theta$ and

$y = r\sin\theta = (1 + c\sin\theta)\sin\theta = \sin\theta + c\sin^2\theta$. Now $-1 \le \sin\theta \le 1 \quad \Rightarrow \quad -1 \le \sin\theta + c\sin^2\theta \le 1 + c \le 2$, so

$-1 \le y \le 2$. Furthermore, $y = 2$ when $c = 1$ and $\theta = \frac{\pi}{2}$, while $y = -1$ for $c = 0$ and $\theta = \frac{3\pi}{2}$. Therefore, we need a viewing

rectangle with $-1 \le y \le 2$.

To find the x-values, look at the equation $x = \cos\theta + \frac{1}{2}c\sin 2\theta$ and use the fact that $\sin 2\theta \ge 0$ for $0 \le \theta \le \frac{\pi}{2}$ and

$\sin 2\theta \le 0$ for $-\frac{\pi}{2} \le \theta \le 0$. [Because $r = 1 + c\sin\theta$ is symmetric about the y-axis, we only need to consider

[continued]

$-\frac{\pi}{2} \le \theta \le \frac{\pi}{2}$.] So for $-\frac{\pi}{2} \le \theta \le 0$, x has a maximum value when $c = 0$ and then $x = \cos\theta$ has a maximum value

of 1 at $\theta = 0$. Thus, the maximum value of x must occur on $\left[0, \frac{\pi}{2}\right]$ with $c = 1$. Then $x = \cos\theta + \frac{1}{2}\sin 2\theta$ $\Rightarrow$

$\frac{dx}{d\theta} = -\sin\theta + \cos 2\theta = -\sin\theta + 1 - 2\sin^2\theta$ $\Rightarrow$ $\frac{dx}{d\theta} = -(2\sin\theta - 1)(\sin\theta + 1) = 0$ when $\sin\theta = -1$ or $\frac{1}{2}$

[but $\sin\theta \ne -1$ for $0 \le \theta \le \frac{\pi}{2}$]. If $\sin\theta = \frac{1}{2}$, then $\theta = \frac{\pi}{6}$ and

$x = \cos\frac{\pi}{6} + \frac{1}{2}\sin\frac{\pi}{3} = \frac{3}{4}\sqrt{3}$. Thus, the maximum value of x is $\frac{3}{4}\sqrt{3}$, and,

by symmetry, the minimum value is $-\frac{3}{4}\sqrt{3}$. Therefore, the smallest

viewing rectangle that contains every member of the family of polar curves

$r = 1 + c\sin\theta$, where $0 \le c \le 1$, is $\left[-\frac{3}{4}\sqrt{3}, \frac{3}{4}\sqrt{3}\right] \times [-1, 2]$.

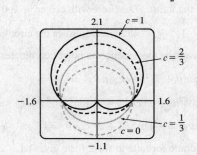

5. Without loss of generality, assume the hyperbola has equation $\frac{x^2}{a^2} - \frac{y^2}{b^2} = 1$. Use implicit differentiation to get

$\frac{2x}{a^2} - \frac{2y\,y'}{b^2} = 0$, so $y' = \frac{b^2 x}{a^2 y}$. The tangent line at the point (c, d) on the hyperbola has equation $y - d = \frac{b^2 c}{a^2 d}(x - c)$.

The tangent line intersects the asymptote $y = \frac{b}{a}x$ when $\frac{b}{a}x - d = \frac{b^2 c}{a^2 d}(x - c)$ $\Rightarrow$ $abdx - a^2 d^2 = b^2 cx - b^2 c^2$ $\Rightarrow$

$abdx - b^2 cx = a^2 d^2 - b^2 c^2$ $\Rightarrow$ $x = \frac{a^2 d^2 - b^2 c^2}{b(ad - bc)} = \frac{ad + bc}{b}$ and the y-value is $\frac{b}{a}\frac{ad + bc}{b} = \frac{ad + bc}{a}$.

Similarly, the tangent line intersects $y = -\frac{b}{a}x$ at $\left(\frac{bc - ad}{b}, \frac{ad - bc}{a}\right)$. The midpoint of these intersection points is

$\left(\frac{1}{2}\left(\frac{ad + bc}{b} + \frac{bc - ad}{b}\right), \frac{1}{2}\left(\frac{ad + bc}{a} + \frac{ad - bc}{a}\right)\right) = \left(\frac{1}{2}\frac{2bc}{b}, \frac{1}{2}\frac{2ad}{a}\right) = (c, d)$, the point of tangency.

Note: If $y = 0$, then at $(\pm a, 0)$, the tangent line is $x = \pm a$, and the points of intersection are clearly equidistant from the point

of tangency.

11 □ INFINITE SEQUENCES AND SERIES

11.1 Sequences

1. (a) A sequence is an ordered list of numbers. It can also be defined as a function whose domain is the set of positive integers.

(b) The terms a_n approach 8 as n becomes large. In fact, we can make a_n as close to 8 as we like by taking n sufficiently large.

(c) The terms a_n become large as n becomes large. In fact, we can make a_n as large as we like by taking n sufficiently large.

3. $a_n = \dfrac{2^n}{2n+1}$, so the sequence is $\left\{\dfrac{2^1}{2(1)+1}, \dfrac{2^2}{2(2)+1}, \dfrac{2^3}{2(3)+1}, \dfrac{2^4}{2(4)+1}, \dfrac{2^5}{2(5)+1}, \ldots\right\} = \left\{\dfrac{2}{3}, \dfrac{4}{5}, \dfrac{8}{7}, \dfrac{16}{9}, \dfrac{32}{11}, \ldots\right\}$.

5. $a_n = \dfrac{(-1)^{n-1}}{5^n}$, so the sequence is $\left\{\dfrac{1}{5^1}, \dfrac{-1}{5^2}, \dfrac{1}{5^3}, \dfrac{-1}{5^4}, \dfrac{1}{5^5}, \ldots\right\} = \left\{\dfrac{1}{5}, -\dfrac{1}{25}, \dfrac{1}{125}, -\dfrac{1}{625}, \dfrac{1}{3125}, \ldots\right\}$.

7. $a_n = \dfrac{1}{(n+1)!}$, so the sequence is $\left\{\dfrac{1}{2!}, \dfrac{1}{3!}, \dfrac{1}{4!}, \dfrac{1}{5!}, \dfrac{1}{6!}, \ldots\right\} = \left\{\dfrac{1}{2}, \dfrac{1}{6}, \dfrac{1}{24}, \dfrac{1}{120}, \dfrac{1}{720}, \ldots\right\}$.

9. $a_1 = 1$, $a_{n+1} = 5a_n - 3$. Each term is defined in terms of the preceding term. $a_2 = 5a_1 - 3 = 5(1) - 3 = 2$.
$a_3 = 5a_2 - 3 = 5(2) - 3 = 7$. $a_4 = 5a_3 - 3 = 5(7) - 3 = 32$. $a_5 = 5a_4 - 3 = 5(32) - 3 = 157$.
The sequence is $\{1, 2, 7, 32, 157, \ldots\}$.

11. $a_1 = 2$, $a_{n+1} = \dfrac{a_n}{1+a_n}$. $a_2 = \dfrac{a_1}{1+a_1} = \dfrac{2}{1+2} = \dfrac{2}{3}$. $a_3 = \dfrac{a_2}{1+a_2} = \dfrac{2/3}{1+2/3} = \dfrac{2}{5}$. $a_4 = \dfrac{a_3}{1+a_3} = \dfrac{2/5}{1+2/5} = \dfrac{2}{7}$.
$a_5 = \dfrac{a_4}{1+a_4} = \dfrac{2/7}{1+2/7} = \dfrac{2}{9}$. The sequence is $\left\{2, \dfrac{2}{3}, \dfrac{2}{5}, \dfrac{2}{7}, \dfrac{2}{9}, \ldots\right\}$.

13. $\left\{\dfrac{1}{2}, \dfrac{1}{4}, \dfrac{1}{6}, \dfrac{1}{8}, \dfrac{1}{10}, \ldots\right\}$. The denominator is two times the number of the term, n, so $a_n = \dfrac{1}{2n}$.

15. $\left\{-3, 2, -\dfrac{4}{3}, \dfrac{8}{9}, -\dfrac{16}{27}, \ldots\right\}$. The first term is -3 and each term is $-\dfrac{2}{3}$ times the preceding one, so $a_n = -3\left(-\dfrac{2}{3}\right)^{n-1}$.

17. $\left\{\dfrac{1}{2}, -\dfrac{4}{3}, \dfrac{9}{4}, -\dfrac{16}{5}, \dfrac{25}{6}, \ldots\right\}$. The numerator of the nth term is n^2 and its denominator is $n+1$. Including the alternating signs, we get $a_n = (-1)^{n+1}\dfrac{n^2}{n+1}$.

19.

n	$a_n = \dfrac{3n}{1 + 6n}$
1	0.4286
2	0.4615
3	0.4737
4	0.4800
5	0.4839
6	0.4865
7	0.4884
8	0.4898
9	0.4909
10	0.4918

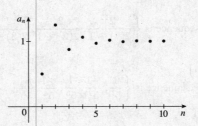

It appears that $\lim\limits_{n \to \infty} a_n = 0.5$.

$$\lim_{n \to \infty} \frac{3n}{1 + 6n} = \lim_{n \to \infty} \frac{(3n)/n}{(1 + 6n)/n} = \lim_{n \to \infty} \frac{3}{1/n + 6} = \frac{3}{6} = \frac{1}{2}$$

21.

n	$a_n = 1 + \left(-\frac{1}{2}\right)^n$
1	0.5000
2	1.2500
3	0.8750
4	1.0625
5	0.9688
6	1.0156
7	0.9922
8	1.0039
9	0.9980
10	1.0010

It appears that $\lim\limits_{n \to \infty} a_n = 1$.

$$\lim_{n \to \infty} \left(1 + \left(-\tfrac{1}{2}\right)^n\right) = \lim_{n \to \infty} 1 + \lim_{n \to \infty} \left(-\tfrac{1}{2}\right)^n = 1 + 0 = 1 \text{ since}$$

$$\lim_{n \to \infty} \left(-\tfrac{1}{2}\right)^n = 0 \text{ by (9)}.$$

23. $a_n = \dfrac{3 + 5n^2}{n + n^2} = \dfrac{(3 + 5n^2)/n^2}{(n + n^2)/n^2} = \dfrac{5 + 3/n^2}{1 + 1/n}$, so $a_n \to \dfrac{5 + 0}{1 + 0} = 5$ as $n \to \infty$. Converges

25. $a_n = \dfrac{n^4}{n^3 - 2n} = \dfrac{n^4/n^3}{(n^3 - 2n)/n^3} = \dfrac{n}{1 - 2/n^2}$, so $a_n \to \infty$ as $n \to \infty$ since $\lim\limits_{n \to \infty} n = \infty$ and

$\lim\limits_{n \to \infty} \left(1 - \dfrac{2}{n^2}\right) = 1 - 0 = 1$. Diverges

27. $a_n = 3^n 7^{-n} = \dfrac{3^n}{7^n} = \left(\dfrac{3}{7}\right)^n$, so $\lim\limits_{n \to \infty} a_n = 0$ by (9) with $r = \dfrac{3}{7}$. Converges

29. Because the natural exponential function is continuous at 0, Theorem 7 enables us to write

$\lim\limits_{n \to \infty} a_n = \lim\limits_{n \to \infty} e^{-1/\sqrt{n}} = e^{\lim\limits_{n \to \infty} (-1/\sqrt{n})} = e^0 = 1$. Converges

31. $a_n = \sqrt{\dfrac{1 + 4n^2}{1 + n^2}} = \sqrt{\dfrac{(1 + 4n^2)/n^2}{(1 + n^2)/n^2}} = \sqrt{\dfrac{(1/n^2) + 4}{(1/n^2) + 1}} \to \sqrt{4} = 2$ as $n \to \infty$ since $\lim\limits_{n \to \infty} (1/n^2) = 0$. Converges

33. $a_n = \dfrac{n^2}{\sqrt{n^3 + 4n}} = \dfrac{n^2/\sqrt{n^3}}{\sqrt{n^3 + 4n}/\sqrt{n^3}} = \dfrac{\sqrt{n}}{\sqrt{1 + 4/n^2}}$, so $a_n \to \infty$ as $n \to \infty$ since $\lim\limits_{n\to\infty} \sqrt{n} = \infty$ and

$\lim\limits_{n\to\infty} \sqrt{1 + 4/n^2} = 1$.   Diverges

35. $\lim\limits_{n\to\infty} |a_n| = \lim\limits_{n\to\infty} \left| \dfrac{(-1)^n}{2\sqrt{n}} \right| = \dfrac{1}{2} \lim\limits_{n\to\infty} \dfrac{1}{n^{1/2}} = \dfrac{1}{2}(0) = 0$, so $\lim\limits_{n\to\infty} a_n = 0$ by (6).   Converges

37. $a_n = \dfrac{(2n-1)!}{(2n+1)!} = \dfrac{(2n-1)!}{(2n+1)(2n)(2n-1)!} = \dfrac{1}{(2n+1)(2n)} \to 0$ as $n \to \infty$.   Converges

39. $a_n = \sin n$. This sequence diverges since the terms don't approach any particular real number as $n \to \infty$. The terms take on

values between -1 and 1.   Diverges

41. $a_n = n^2 e^{-n} = \dfrac{n^2}{e^n}$. Since $\lim\limits_{x\to\infty} \dfrac{x^2}{e^x} \overset{\text{H}}{=} \lim\limits_{x\to\infty} \dfrac{2x}{e^x} \overset{\text{H}}{=} \lim\limits_{x\to\infty} \dfrac{2}{e^x} = 0$, it follows from Theorem 3 that $\lim\limits_{n\to\infty} a_n = 0$.   Converges

43. $0 \le \dfrac{\cos^2 n}{2^n} \le \dfrac{1}{2^n}$ [since $0 \le \cos^2 n \le 1$],   so since $\lim\limits_{n\to\infty} \dfrac{1}{2^n} = 0$, $\left\{ \dfrac{\cos^2 n}{2^n} \right\}$ converges to 0 by the Squeeze Theorem.

45. $a_n = n\sin(1/n) = \dfrac{\sin(1/n)}{1/n}$.   Since $\lim\limits_{x\to\infty} \dfrac{\sin(1/x)}{1/x} = \lim\limits_{t\to 0^+} \dfrac{\sin t}{t}$ [where $t = 1/x$] $= 1$, it follows from Theorem 3

that $\{a_n\}$ converges to 1.

47. $y = \left(1 + \dfrac{2}{x}\right)^x \Rightarrow \ln y = x \ln\left(1 + \dfrac{2}{x}\right)$, so

$\lim\limits_{x\to\infty} \ln y = \lim\limits_{x\to\infty} \dfrac{\ln(1 + 2/x)}{1/x} \overset{\text{H}}{=} \lim\limits_{x\to\infty} \dfrac{\left(\dfrac{1}{1 + 2/x}\right)\left(-\dfrac{2}{x^2}\right)}{-1/x^2} = \lim\limits_{x\to\infty} \dfrac{2}{1 + 2/x} = 2 \Rightarrow$

$\lim\limits_{x\to\infty} \left(1 + \dfrac{2}{x}\right)^x = \lim\limits_{x\to\infty} e^{\ln y} = e^2$, so by Theorem 3, $\lim\limits_{n\to\infty} \left(1 + \dfrac{2}{n}\right)^n = e^2$.   Converges

49. $a_n = \ln(2n^2 + 1) - \ln(n^2 + 1) = \ln\left(\dfrac{2n^2 + 1}{n^2 + 1}\right) = \ln\left(\dfrac{2 + 1/n^2}{1 + 1/n^2}\right) \to \ln 2$ as $n \to \infty$.   Converges

51. $a_n = \arctan(\ln n)$. Let $f(x) = \arctan(\ln x)$. Then $\lim\limits_{x\to\infty} f(x) = \dfrac{\pi}{2}$ since $\ln x \to \infty$ as $x \to \infty$ and $\arctan$ is continuous.

Thus, $\lim\limits_{n\to\infty} a_n = \lim\limits_{n\to\infty} f(n) = \dfrac{\pi}{2}$.   Converges

53. $\{0, 1, 0, 0, 1, 0, 0, 0, 1, \ldots\}$ diverges since the sequence takes on only two values, 0 and 1, and never stays arbitrarily close to

either one (or any other value) for n sufficiently large.

55. $a_n = \dfrac{n!}{2^n} = \dfrac{1}{2} \cdot \dfrac{2}{2} \cdot \dfrac{3}{2} \cdot \ldots \cdot \dfrac{(n-1)}{2} \cdot \dfrac{n}{2} \ge \dfrac{1}{2} \cdot \dfrac{n}{2}$ [for $n > 1$] $= \dfrac{n}{4} \to \infty$ as $n \to \infty$, so $\{a_n\}$ diverges.

57.

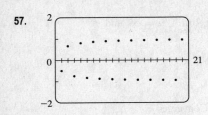

From the graph, it appears that the sequence $\{a_n\} = \left\{ (-1)^n \dfrac{n}{n+1} \right\}$ is

divergent, since it oscillates between 1 and -1 (approximately). To prove this,

suppose that $\{a_n\}$ converges to L. If $b_n = \dfrac{n}{n+1}$, then $\{b_n\}$ converges to 1,

and $\lim\limits_{n\to\infty} \dfrac{a_n}{b_n} = \dfrac{L}{1} = L$. But $\dfrac{a_n}{b_n} = (-1)^n$, so $\lim\limits_{n\to\infty} \dfrac{a_n}{b_n}$ does not exist. This

contradiction shows that $\{a_n\}$ diverges.

59.

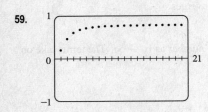

From the graph, it appears that the sequence converges to a number between

0.7 and 0.8.

$$a_n = \arctan\left(\frac{n^2}{n^2 + 4} \right) = \arctan\left(\frac{n^2/n^2}{(n^2+4)/n^2} \right) = \arctan\left(\frac{1}{1 + 4/n^2} \right) \to$$

$\arctan 1 = \dfrac{\pi}{4}$ $[\approx 0.785]$ as $n \to \infty$.

61.

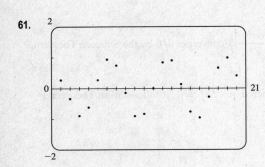

From the graph, it appears that the sequence $\{a_n\} = \left\{ \dfrac{n^2 \cos n}{1 + n^2} \right\}$ is

divergent, since it oscillates between 1 and -1 (approximately). To

prove this, suppose that $\{a_n\}$ converges to L. If $b_n = \dfrac{n^2}{1 + n^2}$, then

$\{b_n\}$ converges to 1, and $\lim\limits_{n\to\infty} \dfrac{a_n}{b_n} = \dfrac{L}{1} = L$. But $\dfrac{a_n}{b_n} = \cos n$, so

$\lim\limits_{n\to\infty} \dfrac{a_n}{b_n}$ does not exist. This contradiction shows that $\{a_n\}$ diverges.

63.

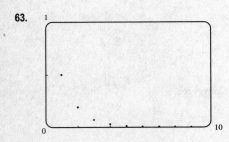

From the graph, it appears that the sequence approaches 0.

$$0 < a_n = \frac{1 \cdot 3 \cdot 5 \cdot \cdots \cdot (2n-1)}{(2n)^n} = \frac{1}{2n} \cdot \frac{3}{2n} \cdot \frac{5}{2n} \cdot \cdots \cdot \frac{2n-1}{2n}$$

$$\leq \frac{1}{2n} \cdot (1) \cdot (1) \cdot \cdots \cdot (1) = \frac{1}{2n} \to 0 \text{ as } n \to \infty$$

So by the Squeeze Theorem, $\left\{ \dfrac{1 \cdot 3 \cdot 5 \cdot \cdots \cdot (2n-1)}{(2n)^n} \right\}$ converges to 0.

65. (a) $a_n = 1000(1.06)^n \Rightarrow a_1 = 1060,\ a_2 = 1123.60,\ a_3 = 1191.02,\ a_4 = 1262.48,$ and $a_5 = 1338.23$.

(b) $\lim\limits_{n\to\infty} a_n = 1000 \lim\limits_{n\to\infty} (1.06)^n$, so the sequence diverges by (9) with $r = 1.06 > 1$.

67. (a) We are given that the initial population is 5000, so $P_0 = 5000$. The number of catfish increases by 8% per month and is

decreased by 300 per month, so $P_1 = P_0 + 8\%P_0 - 300 = 1.08P_0 - 300,\ P_2 = 1.08P_1 - 300$, and so on. Thus,

$P_n = 1.08P_{n-1} - 300$.

(b) Using the recursive formula with $P_0 = 5000$, we get $P_1 = 5100,\ P_2 = 5208,\ P_3 = 5325$ (rounding any portion of a

catfish), $P_4 = 5451,\ P_5 = 5587$, and $P_6 = 5734$, which is the number of catfish in the pond after six months.

69. If $|r| \geq 1$, then $\{r^n\}$ diverges by (9), so $\{nr^n\}$ diverges also, since $|nr^n| = n|r^n| \geq |r^n|$. If $|r| < 1$ then

$$\lim_{x \to \infty} xr^x = \lim_{x \to \infty} \frac{x}{r^{-x}} \overset{\text{H}}{=} \lim_{x \to \infty} \frac{1}{(-\ln r)\, r^{-x}} = \lim_{x \to \infty} \frac{r^x}{-\ln r} = 0, \text{ so } \lim_{n \to \infty} nr^n = 0, \text{ and hence } \{nr^n\} \text{ converges}$$

whenever $|r| < 1$.

71. Since $\{a_n\}$ is a decreasing sequence, $a_n > a_{n+1}$ for all $n \geq 1$. Because all of its terms lie between 5 and 8, $\{a_n\}$ is a bounded sequence. By the Monotonic Sequence Theorem, $\{a_n\}$ is convergent; that is, $\{a_n\}$ has a limit L. L must be less than 8 since $\{a_n\}$ is decreasing, so $5 \leq L < 8$.

73. $a_n = \dfrac{1}{2n+3}$ is decreasing since $a_{n+1} = \dfrac{1}{2(n+1)+3} = \dfrac{1}{2n+5} < \dfrac{1}{2n+3} = a_n$ for each $n \geq 1$. The sequence is

bounded since $0 < a_n \leq \frac{1}{5}$ for all $n \geq 1$. Note that $a_1 = \frac{1}{5}$.

75. The terms of $a_n = n(-1)^n$ alternate in sign, so the sequence is not monotonic. The first five terms are $-1, 2, -3, 4,$ and -5.

Since $\lim\limits_{n \to \infty} |a_n| = \lim\limits_{n \to \infty} n = \infty$, the sequence is not bounded.

77. $a_n = 3 - 2ne^{-n}$. Let $f(x) = 3 - 2xe^{-x}$. Then $f'(x) = 0 - 2[x(-e^{-x}) + e^{-x}] = 2e^{-x}(x - 1)$, which is positive for

$x > 1$, so f is increasing on $(1, \infty)$. It follows that the sequence $\{a_n\} = \{f(n)\}$ is increasing. The sequence is bounded

below by $a_1 = 3 - 2e^{-1} \approx 2.26$ and above by 3, so the sequence is bounded.

79. For $\left\{\sqrt{2},\, \sqrt{2\sqrt{2}},\, \sqrt{2\sqrt{2\sqrt{2}}},\, \dots\right\}$, $a_1 = 2^{1/2}$, $a_2 = 2^{3/4}$, $a_3 = 2^{7/8}$, ..., so $a_n = 2^{(2^n - 1)/2^n} = 2^{1 - (1/2^n)}$.

$$\lim_{n \to \infty} a_n = \lim_{n \to \infty} 2^{1 - (1/2^n)} = 2^1 = 2.$$

Alternate solution: Let $L = \lim\limits_{n \to \infty} a_n$. (We could show the limit exists by showing that $\{a_n\}$ is bounded and increasing.)

Then L must satisfy $L = \sqrt{2 \cdot L}$ $\Rightarrow$ $L^2 = 2L$ $\Rightarrow$ $L(L - 2) = 0$. $L \neq 0$ since the sequence increases, so $L = 2$.

81. $a_1 = 1$, $a_{n+1} = 3 - \dfrac{1}{a_n}$.   We show by induction that $\{a_n\}$ is increasing and bounded above by 3. Let P_n be the proposition

that $a_{n+1} > a_n$ and $0 < a_n < 3$. Clearly P_1 is true. Assume that P_n is true. Then $a_{n+1} > a_n$ $\Rightarrow$ $\dfrac{1}{a_{n+1}} < \dfrac{1}{a_n}$ $\Rightarrow$

$-\dfrac{1}{a_{n+1}} > -\dfrac{1}{a_n}$. Now $a_{n+2} = 3 - \dfrac{1}{a_{n+1}} > 3 - \dfrac{1}{a_n} = a_{n+1}$ $\Leftrightarrow$ P_{n+1}. This proves that $\{a_n\}$ is increasing and bounded

above by 3, so $1 = a_1 < a_n < 3$, that is, $\{a_n\}$ is bounded, and hence convergent by the Monotonic Sequence Theorem.

If $L = \lim\limits_{n \to \infty} a_n$, then $\lim\limits_{n \to \infty} a_{n+1} = L$ also, so L must satisfy $L = 3 - 1/L$ $\Rightarrow$ $L^2 - 3L + 1 = 0$ $\Rightarrow$ $L = \frac{3 \pm \sqrt{5}}{2}$.

But $L > 1$, so $L = \frac{3 + \sqrt{5}}{2}$.

83. (a) Let a_n be the number of rabbit pairs in the nth month. Clearly $a_1 = 1 = a_2$. In the nth month, each pair that is

2 or more months old (that is, a_{n-2} pairs) will produce a new pair to add to the a_{n-1} pairs already present. Thus,

$a_n = a_{n-1} + a_{n-2}$, so that $\{a_n\} = \{f_n\}$, the Fibonacci sequence.

(b) $a_n = \dfrac{f_{n+1}}{f_n}$ $\Rightarrow$ $a_{n-1} = \dfrac{f_n}{f_{n-1}} = \dfrac{f_{n-1} + f_{n-2}}{f_{n-1}} = 1 + \dfrac{f_{n-2}}{f_{n-1}} = 1 + \dfrac{1}{f_{n-1}/f_{n-2}} = 1 + \dfrac{1}{a_{n-2}}$. If $L = \lim\limits_{n\to\infty} a_n$,

then $L = \lim\limits_{n\to\infty} a_{n-1}$ and $L = \lim\limits_{n\to\infty} a_{n-2}$, so L must satisfy $L = 1 + \dfrac{1}{L}$ $\Rightarrow$ $L^2 - L - 1 = 0$ $\Rightarrow$ $L = \frac{1+\sqrt{5}}{2}$

[since L must be positive].

85. (a)

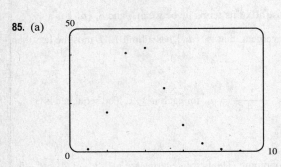

From the graph, it appears that the sequence $\left\{\dfrac{n^5}{n!}\right\}$

converges to 0, that is, $\lim\limits_{n\to\infty} \dfrac{n^5}{n!} = 0$.

(b)

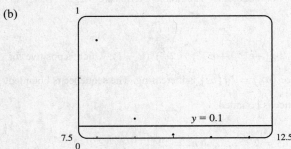 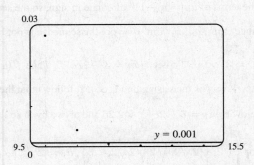

From the first graph, it seems that the smallest possible value of N corresponding to $\varepsilon = 0.1$ is 9, since $n^5/n! < 0.1$

whenever $n \geq 10$, but $9^5/9! > 0.1$. From the second graph, it seems that for $\varepsilon = 0.001$, the smallest possible value for N

is 11 since $n^5/n! < 0.001$ whenever $n \geq 12$.

87. Theorem 6: If $\lim\limits_{n\to\infty} |a_n| = 0$ then $\lim\limits_{n\to\infty} -|a_n| = 0$, and since $-|a_n| \leq a_n \leq |a_n|$, we have that $\lim\limits_{n\to\infty} a_n = 0$ by the

Squeeze Theorem.

89. To Prove: If $\lim\limits_{n\to\infty} a_n = 0$ and $\{b_n\}$ is bounded, then $\lim\limits_{n\to\infty} (a_n b_n) = 0$.

Proof: Since $\{b_n\}$ is bounded, there is a positive number M such that $|b_n| \leq M$ and hence, $|a_n|\,|b_n| \leq |a_n|\,M$ for

all $n \geq 1$. Let $\varepsilon > 0$ be given. Since $\lim\limits_{n\to\infty} a_n = 0$, there is an integer N such that $|a_n - 0| < \dfrac{\varepsilon}{M}$ if $n > N$. Then

$|a_n b_n - 0| = |a_n b_n| = |a_n|\,|b_n| \leq |a_n|\,M = |a_n - 0|\,M < \dfrac{\varepsilon}{M} \cdot M = \varepsilon$ for all $n > N$. Since ε was arbitrary,

$\lim\limits_{n\to\infty} (a_n b_n) = 0$.

91. (a) First we show that $a > a_1 > b_1 > b$.

$a_1 - b_1 = \frac{a+b}{2} - \sqrt{ab} = \frac{1}{2}\left(a - 2\sqrt{ab} + b\right) = \frac{1}{2}\left(\sqrt{a} - \sqrt{b}\right)^2 > 0$ [since $a > b$] $\Rightarrow$ $a_1 > b_1$. Also

$a - a_1 = a - \frac{1}{2}(a+b) = \frac{1}{2}(a-b) > 0$ and $b - b_1 = b - \sqrt{ab} = \sqrt{b}\left(\sqrt{b} - \sqrt{a}\right) < 0$, so $a > a_1 > b_1 > b$. In the same way we can show that $a_1 > a_2 > b_2 > b_1$ and so the given assertion is true for $n = 1$. Suppose it is true for $n = k$, that is, $a_k > a_{k+1} > b_{k+1} > b_k$. Then

$$a_{k+2} - b_{k+2} = \tfrac{1}{2}(a_{k+1} + b_{k+1}) - \sqrt{a_{k+1}b_{k+1}} = \tfrac{1}{2}\left(a_{k+1} - 2\sqrt{a_{k+1}b_{k+1}} + b_{k+1}\right) = \tfrac{1}{2}\left(\sqrt{a_{k+1}} - \sqrt{b_{k+1}}\right)^2 > 0,$$

$$a_{k+1} - a_{k+2} = a_{k+1} - \tfrac{1}{2}(a_{k+1} + b_{k+1}) = \tfrac{1}{2}(a_{k+1} - b_{k+1}) > 0, \text{ and}$$

$$b_{k+1} - b_{k+2} = b_{k+1} - \sqrt{a_{k+1}b_{k+1}} = \sqrt{b_{k+1}}\left(\sqrt{b_{k+1}} - \sqrt{a_{k+1}}\right) < 0 \quad \Rightarrow \quad a_{k+1} > a_{k+2} > b_{k+2} > b_{k+1},$$

so the assertion is true for $n = k + 1$. Thus, it is true for all n by mathematical induction.

(b) From part (a) we have $a > a_n > a_{n+1} > b_{n+1} > b_n > b$, which shows that both sequences, $\{a_n\}$ and $\{b_n\}$, are monotonic and bounded. So they are both convergent by the Monotonic Sequence Theorem.

(c) Let $\lim\limits_{n\to\infty} a_n = \alpha$ and $\lim\limits_{n\to\infty} b_n = \beta$. Then $\lim\limits_{n\to\infty} a_{n+1} = \lim\limits_{n\to\infty} \dfrac{a_n + b_n}{2} \ \Rightarrow \ \alpha = \dfrac{\alpha + \beta}{2} \ \Rightarrow$

$2\alpha = \alpha + \beta \ \Rightarrow \ \alpha = \beta$.

93. (a) Suppose $\{p_n\}$ converges to p. Then $p_{n+1} = \dfrac{bp_n}{a + p_n} \ \Rightarrow \ \lim\limits_{n\to\infty} p_{n+1} = \dfrac{b\lim\limits_{n\to\infty} p_n}{a + \lim\limits_{n\to\infty} p_n} \ \Rightarrow \ p = \dfrac{bp}{a + p} \ \Rightarrow$

$p^2 + ap = bp \ \Rightarrow \ p(p + a - b) = 0 \ \Rightarrow \ p = 0 \text{ or } p = b - a.$

(b) $p_{n+1} = \dfrac{bp_n}{a + p_n} = \dfrac{\left(\dfrac{b}{a}\right)p_n}{1 + \dfrac{p_n}{a}} < \left(\dfrac{b}{a}\right)p_n$ since $1 + \dfrac{p_n}{a} > 1$.

(c) By part (b), $p_1 < \left(\dfrac{b}{a}\right)p_0$, $p_2 < \left(\dfrac{b}{a}\right)p_1 < \left(\dfrac{b}{a}\right)^2 p_0$, $p_3 < \left(\dfrac{b}{a}\right)p_2 < \left(\dfrac{b}{a}\right)^3 p_0$, etc. In general, $p_n < \left(\dfrac{b}{a}\right)^n p_0$,

so $\lim\limits_{n\to\infty} p_n \le \lim\limits_{n\to\infty} \left(\dfrac{b}{a}\right)^n \cdot p_0 = 0$ since $b < a$. $\left[\text{By (9), } \lim\limits_{n\to\infty} r^n = 0 \text{ if } -1 < r < 1. \text{ Here } r = \dfrac{b}{a} \in (0, 1).\right]$

(d) Let $a < b$. We first show, by induction, that if $p_0 < b - a$, then $p_n < b - a$ and $p_{n+1} > p_n$.

For $n = 0$, we have $p_1 - p_0 = \dfrac{bp_0}{a + p_0} - p_0 = \dfrac{p_0(b - a - p_0)}{a + p_0} > 0$ since $p_0 < b - a$. So $p_1 > p_0$.

Now we suppose the assertion is true for $n = k$, that is, $p_k < b - a$ and $p_{k+1} > p_k$. Then

$$b - a - p_{k+1} = b - a - \dfrac{bp_k}{a + p_k} = \dfrac{a(b - a) + bp_k - ap_k - bp_k}{a + p_k} = \dfrac{a(b - a - p_k)}{a + p_k} > 0 \text{ because } p_k < b - a. \text{ So}$$

$p_{k+1} < b - a$. And $p_{k+2} - p_{k+1} = \dfrac{bp_{k+1}}{a + p_{k+1}} - p_{k+1} = \dfrac{p_{k+1}(b - a - p_{k+1})}{a + p_{k+1}} > 0$ since $p_{k+1} < b - a$. Therefore,

$p_{k+2} > p_{k+1}$. Thus, the assertion is true for $n = k + 1$. It is therefore true for all n by mathematical induction.

A similar proof by induction shows that if $p_0 > b - a$, then $p_n > b - a$ and $\{p_n\}$ is decreasing.

In either case the sequence $\{p_n\}$ is bounded and monotonic, so it is convergent by the Monotonic Sequence Theorem. It then follows from part (a) that $\lim\limits_{n\to\infty} p_n = b - a$.

11.2 Series

1. (a) A sequence is an ordered list of numbers whereas a series is the *sum* of a list of numbers.

(b) A series is convergent if the sequence of partial sums is a convergent sequence. A series is divergent if it is not convergent.

3. $\displaystyle\sum_{n=1}^{\infty} a_n = \lim_{n\to\infty} s_n = \lim_{n\to\infty} [2 - 3(0.8)^n] = \lim_{n\to\infty} 2 - 3\lim_{n\to\infty} (0.8)^n = 2 - 3(0) = 2$

5. For $\displaystyle\sum_{n=1}^{\infty} \frac{1}{n^4 + n^2}$, $a_n = \dfrac{1}{n^4 + n^2}$. $s_1 = a_1 = \dfrac{1}{1^4 + 1^2} = \dfrac{1}{2} = 0.5$, $s_2 = s_1 + a_2 = \dfrac{1}{2} + \dfrac{1}{16 + 4} = 0.55$,

$s_3 = s_2 + a_3 \approx 0.5611$, $s_4 = s_3 + a_4 \approx 0.5648$, $s_5 = s_4 + a_5 \approx 0.5663$, $s_6 = s_5 + a_6 \approx 0.5671$,

$s_7 = s_6 + a_7 \approx 0.5675$, and $s_8 = s_7 + a_8 \approx 0.5677$. It appears that the series is convergent.

7. For $\displaystyle\sum_{n=1}^{\infty} \sin n$, $a_n = \sin n$. $s_1 = a_1 = \sin 1 \approx 0.8415$, $s_2 = s_1 + a_2 \approx 1.7508$,

$s_3 = s_2 + a_3 \approx 1.8919$, $s_4 = s_3 + a_4 \approx 1.1351$, $s_5 = s_4 + a_5 \approx 0.1762$, $s_6 = s_5 + a_6 \approx -0.1033$,

$s_7 = s_6 + a_7 \approx 0.5537$, and $s_8 = s_7 + a_8 \approx 1.5431$. It appears that the series is divergent.

9.

n	s_n
1	-2.40000
2	-1.92000
3	-2.01600
4	-1.99680
5	-2.00064
6	-1.99987
7	-2.00003
8	-1.99999
9	-2.00000
10	-2.00000

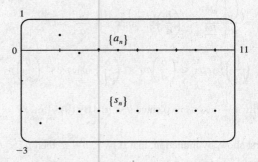

From the graph and the table, it seems that the series converges to -2. In fact, it is a geometric

series with $a = -2.4$ and $r = -\frac{1}{5}$, so its sum is $\displaystyle\sum_{n=1}^{\infty} \frac{12}{(-5)^n} = \frac{-2.4}{1 - \left(-\frac{1}{5}\right)} = \frac{-2.4}{1.2} = -2$.

Note that the dot corresponding to $n = 1$ is part of both $\{a_n\}$ and $\{s_n\}$.

TI-86 Note: To graph $\{a_n\}$ and $\{s_n\}$, set your calculator to Param mode and DrawDot mode. (DrawDot is under

GRAPH, MORE, FORMT (F3).) Now under E(t) = make the assignments: xt1=t, yt1=12/(-5)^t, xt2=t,

yt2=sum seq(yt1,t,1,t,1). (sum and seq are under LIST, OPS (F5), MORE.) Under WIND use

1,10,1,0,10,1,-3,1,1 to obtain a graph similar to the one above. Then use TRACE (F4) to see the values.

11.

n	s_n
1	0.44721
2	1.15432
3	1.98637
4	2.88080
5	3.80927
6	4.75796
7	5.71948
8	6.68962
9	7.66581
10	8.64639

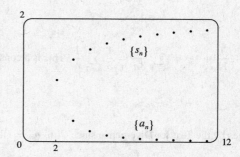

The series $\sum_{n=1}^{\infty} \dfrac{n}{\sqrt{n^2+4}}$ diverges, since its terms do not approach 0.

13.

n	s_n
2	1.00000
3	1.33333
4	1.50000
5	1.60000
6	1.66667
7	1.71429
8	1.75000
9	1.77778
10	1.80000
11	1.81818

From the graph and the table, we see that the terms are getting smaller and may approach 0, and that the series may approach a number near 2. Using partial fractions, we have

$$\sum_{n=2}^{k} \frac{2}{n^2-n} = \sum_{n=2}^{k}\left(\frac{2}{n-1} - \frac{2}{n}\right)$$

$$= \left(\frac{2}{1} - \frac{2}{2}\right) + \left(\frac{2}{2} - \frac{2}{3}\right) + \left(\frac{2}{3} - \frac{2}{4}\right)$$

$$+ \cdots + \left(\frac{2}{k-2} - \frac{2}{k-1}\right) + \left(\frac{2}{k-1} - \frac{2}{k}\right)$$

$$= 2 - \frac{2}{k}$$

As $k \to \infty$, $2 - \dfrac{2}{k} \to 2$, so $\sum_{n=2}^{\infty} \dfrac{2}{n^2-n} = 2$.

15. (a) $\lim\limits_{n\to\infty} a_n = \lim\limits_{n\to\infty} \dfrac{2n}{3n+1} = \dfrac{2}{3}$, so the *sequence* $\{a_n\}$ is convergent by (11.1.1).

(b) Since $\lim\limits_{n\to\infty} a_n = \frac{2}{3} \neq 0$, the *series* $\sum_{n=1}^{\infty} a_n$ is divergent by the Test for Divergence.

17. $3 - 4 + \frac{16}{3} - \frac{64}{9} + \cdots$ is a geometric series with ratio $r = -\frac{4}{3}$. Since $|r| = \frac{4}{3} > 1$, the series diverges.

19. $10 - 2 + 0.4 - 0.08 + \cdots$ is a geometric series with ratio $-\frac{2}{10} = -\frac{1}{5}$. Since $|r| = \frac{1}{5} < 1$, the series converges to

$$\frac{a}{1-r} = \frac{10}{1-(-1/5)} = \frac{10}{6/5} = \frac{50}{6} = \frac{25}{3}.$$

21. $\sum_{n=1}^{\infty} 12\,(0.73)^{n-1}$ is a geometric series with first term $a = 12$ and ratio $r = 0.73$. Since $|r| = 0.73 < 1$, the series converges

to $\dfrac{a}{1-r} = \dfrac{12}{1-0.73} = \dfrac{12}{0.27} = \dfrac{12(100)}{27} = \dfrac{400}{9}$.

23. $\sum_{n=1}^{\infty} \dfrac{(-3)^{n-1}}{4^n} = \dfrac{1}{4} \sum_{n=1}^{\infty} \left(-\dfrac{3}{4}\right)^{n-1}$. The latter series is geometric with $a = 1$ and ratio $r = -\frac{3}{4}$. Since $|r| = \frac{3}{4} < 1$, it

converges to $\dfrac{1}{1-(-3/4)} = \frac{4}{7}$. Thus, the given series converges to $\left(\frac{1}{4}\right)\left(\frac{4}{7}\right) = \frac{1}{7}$.

25. $\sum_{n=1}^{\infty} \dfrac{e^{2n}}{6^{n-1}} = \sum_{n=1}^{\infty} \dfrac{(e^2)^n}{6^n 6^{-1}} = 6 \sum_{n=1}^{\infty} \left(\dfrac{e^2}{6}\right)^n$ is a geometric series with ratio $r = \dfrac{e^2}{6}$. Since $|r| = \dfrac{e^2}{6}[\approx 1.23] > 1$, the series

diverges.

27. $\dfrac{1}{3} + \dfrac{1}{6} + \dfrac{1}{9} + \dfrac{1}{12} + \dfrac{1}{15} + \cdots = \sum_{n=1}^{\infty} \dfrac{1}{3n} = \dfrac{1}{3} \sum_{n=1}^{\infty} \dfrac{1}{n}$. This is a constant multiple of the divergent harmonic series, so

it diverges.

29. $\sum_{n=1}^{\infty} \dfrac{2+n}{1-2n}$ diverges by the Test for Divergence since $\lim_{n\to\infty} a_n = \lim_{n\to\infty} \dfrac{2+n}{1-2n} = \lim_{n\to\infty} \dfrac{2/n+1}{1/n-2} = -\dfrac{1}{2} \neq 0$.

31. $\sum_{n=1}^{\infty} 3^{n+1}4^{-n} = \sum_{n=1}^{\infty} \dfrac{3^n \cdot 3^1}{4^n} = 3 \sum_{n=1}^{\infty} \left(\dfrac{3}{4}\right)^n$. The latter series is geometric with $a = \dfrac{3}{4}$ and ratio $r = \dfrac{3}{4}$. Since $|r| = \dfrac{3}{4} < 1$,

it converges to $\dfrac{3/4}{1-3/4} = 3$. Thus, the given series converges to $3(3) = 9$.

33. $\sum_{n=1}^{\infty} \dfrac{1}{4+e^{-n}}$ diverges by the Test for Divergence since $\lim_{n\to\infty} \dfrac{1}{4+e^{-n}} = \dfrac{1}{4+0} = \dfrac{1}{4} \neq 0$.

35. $\sum_{k=1}^{\infty} (\sin 100)^k$ is a geometric series with first term $a = \sin 100\ [\approx -0.506]$ and ratio $r = \sin 100$. Since $|r| < 1$, the series

converges to $\dfrac{\sin 100}{1 - \sin 100} \approx -0.336$.

37. $\sum_{n=1}^{\infty} \ln\left(\dfrac{n^2+1}{2n^2+1}\right)$ diverges by the Test for Divergence since

$\lim_{n\to\infty} a_n = \lim_{n\to\infty} \ln\left(\dfrac{n^2+1}{2n^2+1}\right) = \ln\left(\lim_{n\to\infty} \dfrac{n^2+1}{2n^2+1}\right) = \ln\frac{1}{2} \neq 0$.

39. $\sum_{n=1}^{\infty} \arctan n$ diverges by the Test for Divergence since $\lim_{n\to\infty} a_n = \lim_{n\to\infty} \arctan n = \dfrac{\pi}{2} \neq 0$.

41. $\displaystyle\sum_{n=1}^{\infty} \frac{1}{e^n} = \sum_{n=1}^{\infty} \left(\frac{1}{e}\right)^n$ is a geometric series with first term $a = \dfrac{1}{e}$ and ratio $r = \dfrac{1}{e}$. Since $|r| = \dfrac{1}{e} < 1$, the series converges

to $\dfrac{1/e}{1 - 1/e} = \dfrac{1/e}{1 - 1/e} \cdot \dfrac{e}{e} = \dfrac{1}{e - 1}$. By Example 8, $\displaystyle\sum_{n=1}^{\infty} \frac{1}{n(n+1)} = 1$. Thus, by Theorem 8(ii),

$$\sum_{n=1}^{\infty} \left(\frac{1}{e^n} + \frac{1}{n(n+1)}\right) = \sum_{n=1}^{\infty} \frac{1}{e^n} + \sum_{n=1}^{\infty} \frac{1}{n(n+1)} = \frac{1}{e-1} + 1 = \frac{1}{e-1} + \frac{e-1}{e-1} = \frac{e}{e-1}.$$

43. Using partial fractions, the partial sums of the series $\displaystyle\sum_{n=2}^{\infty} \frac{2}{n^2 - 1}$ are

$$s_n = \sum_{i=2}^{n} \frac{2}{(i-1)(i+1)} = \sum_{i=2}^{n} \left(\frac{1}{i-1} - \frac{1}{i+1}\right)$$

$$= \left(1 - \frac{1}{3}\right) + \left(\frac{1}{2} - \frac{1}{4}\right) + \left(\frac{1}{3} - \frac{1}{5}\right) + \cdots + \left(\frac{1}{n-3} - \frac{1}{n-1}\right) + \left(\frac{1}{n-2} - \frac{1}{n}\right)$$

This sum is a telescoping series and $s_n = 1 + \dfrac{1}{2} - \dfrac{1}{n-1} - \dfrac{1}{n}$.

Thus, $\displaystyle\sum_{n=2}^{\infty} \frac{2}{n^2 - 1} = \lim_{n\to\infty} s_n = \lim_{n\to\infty}\left(1 + \frac{1}{2} - \frac{1}{n-1} - \frac{1}{n}\right) = \frac{3}{2}$.

45. For the series $\displaystyle\sum_{n=1}^{\infty} \frac{3}{n(n+3)}$, $s_n = \displaystyle\sum_{i=1}^{n} \frac{3}{i(i+3)} = \sum_{i=1}^{n}\left(\frac{1}{i} - \frac{1}{i+3}\right)$ [using partial fractions]. The latter sum is

$$\left(1 - \tfrac{1}{4}\right) + \left(\tfrac{1}{2} - \tfrac{1}{5}\right) + \left(\tfrac{1}{3} - \tfrac{1}{6}\right) + \left(\tfrac{1}{4} - \tfrac{1}{7}\right) + \cdots + \left(\tfrac{1}{n-3} - \tfrac{1}{n}\right) + \left(\tfrac{1}{n-2} - \tfrac{1}{n+1}\right) + \left(\tfrac{1}{n-1} - \tfrac{1}{n+2}\right) + \left(\tfrac{1}{n} - \tfrac{1}{n+3}\right)$$

$$= 1 + \tfrac{1}{2} + \tfrac{1}{3} - \tfrac{1}{n+1} - \tfrac{1}{n+2} - \tfrac{1}{n+3} \quad \text{[telescoping series]}$$

Thus, $\displaystyle\sum_{n=1}^{\infty} \frac{3}{n(n+3)} = \lim_{n\to\infty} s_n = \lim_{n\to\infty}\left(1 + \tfrac{1}{2} + \tfrac{1}{3} - \tfrac{1}{n+1} - \tfrac{1}{n+2} - \tfrac{1}{n+3}\right) = 1 + \tfrac{1}{2} + \tfrac{1}{3} = \tfrac{11}{6}.$ Converges

47. For the series $\displaystyle\sum_{n=1}^{\infty} \left(e^{1/n} - e^{1/(n+1)}\right)$,

$$s_n = \sum_{i=1}^{n}\left(e^{1/i} - e^{1/(i+1)}\right) = \left(e^1 - e^{1/2}\right) + \left(e^{1/2} - e^{1/3}\right) + \cdots + \left(e^{1/n} - e^{1/(n+1)}\right) = e - e^{1/(n+1)}$$

[telescoping series]

Thus, $\displaystyle\sum_{n=1}^{\infty} \left(e^{1/n} - e^{1/(n+1)}\right) = \lim_{n\to\infty} s_n = \lim_{n\to\infty}\left(e - e^{1/(n+1)}\right) = e - e^0 = e - 1.$ Converges

49. (a) Many people would guess that $x < 1$, but note that x consists of an infinite number of 9s.

(b) $x = 0.99999\ldots = \dfrac{9}{10} + \dfrac{9}{100} + \dfrac{9}{1000} + \dfrac{9}{10,000} + \cdots = \displaystyle\sum_{n=1}^{\infty} \frac{9}{10^n}$, which is a geometric series with $a_1 = 0.9$ and

$r = 0.1$. Its sum is $\dfrac{0.9}{1 - 0.1} = \dfrac{0.9}{0.9} = 1$, that is, $x = 1$.

(c) The number 1 has two decimal representations, $1.00000\ldots$ and $0.99999\ldots$.

(d) Except for 0, all rational numbers that have a terminating decimal representation can be written in more than one way. For example, 0.5 can be written as $0.49999\ldots$ as well as $0.50000\ldots$.

51. $0.\overline{8} = \dfrac{8}{10} + \dfrac{8}{10^2} + \cdots$ is a geometric series with $a = \dfrac{8}{10}$ and $r = \dfrac{1}{10}$. It converges to $\dfrac{a}{1-r} = \dfrac{8/10}{1-1/10} = \dfrac{8}{9}$.

53. $2.\overline{516} = 2 + \dfrac{516}{10^3} + \dfrac{516}{10^6} + \cdots$. Now $\dfrac{516}{10^3} + \dfrac{516}{10^6} + \cdots$ is a geometric series with $a = \dfrac{516}{10^3}$ and $r = \dfrac{1}{10^3}$. It converges to

$\dfrac{a}{1-r} = \dfrac{516/10^3}{1-1/10^3} = \dfrac{516/10^3}{999/10^3} = \dfrac{516}{999}$. Thus, $2.\overline{516} = 2 + \dfrac{516}{999} = \dfrac{2514}{999} = \dfrac{838}{333}$.

55. $1.234\overline{567} = 1.234 + \dfrac{567}{10^6} + \dfrac{567}{10^9} + \cdots$. Now $\dfrac{567}{10^6} + \dfrac{567}{10^9} + \cdots$ is a geometric series with $a = \dfrac{567}{10^6}$ and

$r = \dfrac{1}{10^3}$. It converges to $\dfrac{a}{1-r} = \dfrac{567/10^6}{1-1/10^3} = \dfrac{567/10^6}{999/10^3} = \dfrac{567}{999,000} = \dfrac{21}{37,000}$. Thus,

$1.234\overline{567} = 1.234 + \dfrac{21}{37,000} = \dfrac{1234}{1000} + \dfrac{21}{37,000} = \dfrac{45,658}{37,000} + \dfrac{21}{37,000} = \dfrac{45,679}{37,000}$.

57. $\displaystyle\sum_{n=1}^{\infty} (-5)^n x^n = \sum_{n=1}^{\infty} (-5x)^n$ is a geometric series with $r = -5x$, so the series converges $\Leftrightarrow |r| < 1 \Leftrightarrow$

$|-5x| < 1 \Leftrightarrow |x| < \frac{1}{5}$, that is, $-\frac{1}{5} < x < \frac{1}{5}$. In that case, the sum of the series is $\dfrac{a}{1-r} = \dfrac{-5x}{1-(-5x)} = \dfrac{-5x}{1+5x}$.

59. $\displaystyle\sum_{n=0}^{\infty} \dfrac{(x-2)^n}{3^n} = \sum_{n=0}^{\infty} \left(\dfrac{x-2}{3}\right)^n$ is a geometric series with $r = \dfrac{x-2}{3}$, so the series converges $\Leftrightarrow |r| < 1 \Leftrightarrow$

$\left|\dfrac{x-2}{3}\right| < 1 \Leftrightarrow -1 < \dfrac{x-2}{3} < 1 \Leftrightarrow -3 < x-2 < 3 \Leftrightarrow -1 < x < 5$. In that case, the sum of the series is

$\dfrac{a}{1-r} = \dfrac{1}{1-\dfrac{x-2}{3}} = \dfrac{1}{\dfrac{3-(x-2)}{3}} = \dfrac{3}{5-x}$.

61. $\displaystyle\sum_{n=0}^{\infty} \dfrac{2^n}{x^n} = \sum_{n=0}^{\infty} \left(\dfrac{2}{x}\right)^n$ is a geometric series with $r = \dfrac{2}{x}$, so the series converges $\Leftrightarrow |r| < 1 \Leftrightarrow \left|\dfrac{2}{x}\right| < 1 \Leftrightarrow$

$2 < |x| \Leftrightarrow x > 2$ or $x < -2$. In that case, the sum of the series is $\dfrac{a}{1-r} = \dfrac{1}{1-2/x} = \dfrac{x}{x-2}$.

63. $\displaystyle\sum_{n=0}^{\infty} e^{nx} = \sum_{n=0}^{\infty} (e^x)^n$ is a geometric series with $r = e^x$, so the series converges $\Leftrightarrow |r| < 1 \Leftrightarrow |e^x| < 1 \Leftrightarrow$

$-1 < e^x < 1 \Leftrightarrow 0 < e^x < 1 \Leftrightarrow x < 0$. In that case, the sum of the series is $\dfrac{a}{1-r} = \dfrac{1}{1-e^x}$.

65. After defining f, We use `convert(f,parfrac);` in Maple, `Apart` in Mathematica, or `Expand Rational` and

`Simplify` in Derive to find that the general term is $\dfrac{3n^2 + 3n + 1}{(n^2+n)^3} = \dfrac{1}{n^3} - \dfrac{1}{(n+1)^3}$. So the nth partial sum is

$$s_n = \sum_{k=1}^{n} \left(\dfrac{1}{k^3} - \dfrac{1}{(k+1)^3}\right) = \left(1 - \dfrac{1}{2^3}\right) + \left(\dfrac{1}{2^3} - \dfrac{1}{3^3}\right) + \cdots + \left(\dfrac{1}{n^3} - \dfrac{1}{(n+1)^3}\right) = 1 - \dfrac{1}{(n+1)^3}$$

The series converges to $\lim\limits_{n\to\infty} s_n = 1$. This can be confirmed by directly computing the sum using

`sum(f,n=1..infinity);` (in Maple), `Sum[f,{n,1,Infinity}]` (in Mathematica), or `Calculus Sum`

(from 1 to ∞) and `Simplify` (in Derive).

67. For $n = 1$, $a_1 = 0$ since $s_1 = 0$. For $n > 1$,

$$a_n = s_n - s_{n-1} = \frac{n-1}{n+1} - \frac{(n-1)-1}{(n-1)+1} = \frac{(n-1)n - (n+1)(n-2)}{(n+1)n} = \frac{2}{n(n+1)}$$

Also, $\displaystyle\sum_{n=1}^{\infty} a_n = \lim_{n\to\infty} s_n = \lim_{n\to\infty} \frac{1 - 1/n}{1 + 1/n} = 1$.

69. (a) The quantity of the drug in the body after the first tablet is 100 mg. After the second tablet, there is 100 mg plus 20% of

the first 100-mg tablet; that is, $100 + 0.20(100) = 120$ mg. After the third tablet, the quantity is $100 + 0.20(120)$ or,

equivalently, $100 + 100(0.20) + 100(0.20)^2$. Either expression gives us 124 mg.

(b) From part (a), we see that $Q_{n+1} = 100 + 0.20\, Q_n$.

(c) $Q_n = 100 + 100(0.20)^1 + 100(0.20)^2 + \cdots + 100(0.20)^{n-1}$

$\displaystyle = \sum_{i=1}^{n} 100(0.20)^{i-1}$ [geometric with $a = 100$ and $r = 0.20$].

The quantity of the antibiotic that remains in the body in the long run is $\displaystyle\lim_{n\to\infty} Q_n = \frac{100}{1 - 0.20} = \frac{100}{4/5} = 125$ mg.

71. (a) The quantity of the drug in the body after the first tablet is 150 mg. After the second tablet, there is 150 mg plus 5%

of the first 150-mg tablet, that is, $[150 + 150(0.05)]$ mg. After the third tablet, the quantity is

$[150 + 150(0.05) + 150(0.05)^2] = 157.875$ mg. After n tablets, the quantity (in mg) is

$150 + 150(0.05) + \cdots + 150(0.05)^{n-1}$. We can use Formula 3 to write this as $\dfrac{150(1 - 0.05^n)}{1 - 0.05} = \dfrac{3000}{19}(1 - 0.05^n)$.

(b) The number of milligrams remaining in the body in the long run is $\displaystyle\lim_{n\to\infty} \left[\frac{3000}{19}(1 - 0.05^n) \right] = \frac{3000}{19}(1 - 0) \approx 157.895$,

only 0.02 mg more than the amount after 3 tablets.

73. (a) The first step in the chain occurs when the local government spends D dollars. The people who receive it spend a

fraction c of those D dollars, that is, Dc dollars. Those who receive the Dc dollars spend a fraction c of it, that is,

Dc^2 dollars. Continuing in this way, we see that the total spending after n transactions is

$S_n = D + Dc + Dc^2 + \cdots + Dc^{n-1} = \dfrac{D(1 - c^n)}{1 - c}$ by (3).

(b) $\displaystyle\lim_{n\to\infty} S_n = \lim_{n\to\infty} \frac{D(1 - c^n)}{1 - c} = \frac{D}{1 - c} \lim_{n\to\infty} (1 - c^n) = \frac{D}{1 - c}$ $\left[$since $0 < c < 1$ $\Rightarrow$ $\displaystyle\lim_{n\to\infty} c^n = 0 \right]$

$= \dfrac{D}{s}$ [since $c + s = 1$] $= kD$ [since $k = 1/s$].

If $c = 0.8$, then $s = 1 - c = 0.2$ and the multiplier is $k = 1/s = 5$.

75. $\displaystyle\sum_{n=2}^{\infty} (1 + c)^{-n}$ is a geometric series with $a = (1 + c)^{-2}$ and $r = (1 + c)^{-1}$, so the series converges when

$\left| (1 + c)^{-1} \right| < 1$ $\Leftrightarrow$ $|1 + c| > 1$ $\Leftrightarrow$ $1 + c > 1$ or $1 + c < -1$ $\Leftrightarrow$ $c > 0$ or $c < -2$. We calculate the sum of the series and set it equal to 2: $\dfrac{(1+c)^{-2}}{1-(1+c)^{-1}} = 2 \iff \left(\dfrac{1}{1+c}\right)^2 = 2 - 2\left(\dfrac{1}{1+c}\right) \iff 1 = 2(1+c)^2 - 2(1+c) \iff$

$2c^2 + 2c - 1 = 0 \iff c = \dfrac{-2 \pm \sqrt{12}}{4} = \dfrac{\pm\sqrt{3}-1}{2}$. However, the negative root is inadmissible because $-2 < \dfrac{-\sqrt{3}-1}{2} < 0$.

So $c = \dfrac{\sqrt{3}-1}{2}$.

77. $e^{s_n} = e^{1+\frac{1}{2}+\frac{1}{3}+\cdots+\frac{1}{n}} = e^1 e^{1/2} e^{1/3} \cdots e^{1/n} > (1+1)\left(1+\tfrac{1}{2}\right)\left(1+\tfrac{1}{3}\right)\cdots\left(1+\tfrac{1}{n}\right)$ $\qquad [e^x > 1+x]$

$\qquad = \dfrac{2}{1}\dfrac{3}{2}\dfrac{4}{3}\cdots\dfrac{n+1}{n} = n+1$

Thus, $e^{s_n} > n+1$ and $\lim\limits_{n\to\infty} e^{s_n} = \infty$. Since $\{s_n\}$ is increasing, $\lim\limits_{n\to\infty} s_n = \infty$, implying that the harmonic series is

divergent.

79. Let d_n be the diameter of C_n. We draw lines from the centers of the C_i to

the center of D (or C), and using the Pythagorean Theorem, we can write

$1^2 + \left(1 - \tfrac{1}{2}d_1\right)^2 = \left(1 + \tfrac{1}{2}d_1\right)^2 \iff$

$1 = \left(1 + \tfrac{1}{2}d_1\right)^2 - \left(1 - \tfrac{1}{2}d_1\right)^2 = 2d_1$ [difference of squares] $\Rightarrow d_1 = \tfrac{1}{2}$.

Similarly,

$1 = \left(1 + \tfrac{1}{2}d_2\right)^2 - \left(1 - d_1 - \tfrac{1}{2}d_2\right)^2 = 2d_2 + 2d_1 - d_1^2 - d_1 d_2$

$\quad = (2 - d_1)(d_1 + d_2) \iff$

$d_2 = \dfrac{1}{2-d_1} - d_1 = \dfrac{(1-d_1)^2}{2-d_1}, 1 = \left(1 + \tfrac{1}{2}d_3\right)^2 - \left(1 - d_1 - d_2 - \tfrac{1}{2}d_3\right)^2 \iff d_3 = \dfrac{[1-(d_1+d_2)]^2}{2-(d_1+d_2)}$, and in general,

$d_{n+1} = \dfrac{\left(1 - \sum_{i=1}^n d_i\right)^2}{2 - \sum_{i=1}^n d_i}$. If we actually calculate d_2 and d_3 from the formulas above, we find that they are $\dfrac{1}{6} = \dfrac{1}{2\cdot 3}$ and

$\dfrac{1}{12} = \dfrac{1}{3\cdot 4}$ respectively, so we suspect that in general, $d_n = \dfrac{1}{n(n+1)}$. To prove this, we use induction: Assume that for all

$k \le n$, $d_k = \dfrac{1}{k(k+1)} = \dfrac{1}{k} - \dfrac{1}{k+1}$. Then $\sum\limits_{i=1}^n d_i = 1 - \dfrac{1}{n+1} = \dfrac{n}{n+1}$ [telescoping sum]. Substituting this into our

formula for d_{n+1}, we get $d_{n+1} = \dfrac{\left[1 - \dfrac{n}{n+1}\right]^2}{2 - \left(\dfrac{n}{n+1}\right)} = \dfrac{\dfrac{1}{(n+1)^2}}{\dfrac{n+2}{n+1}} = \dfrac{1}{(n+1)(n+2)}$, and the induction is complete.

Now, we observe that the partial sums $\sum_{i=1}^n d_i$ of the diameters of the circles approach 1 as $n \to \infty$; that is,

$\sum\limits_{n=1}^{\infty} a_n = \sum\limits_{n=1}^{\infty} \dfrac{1}{n(n+1)} = 1$, which is what we wanted to prove.

81. The series $1 - 1 + 1 - 1 + 1 - 1 + \cdots$ diverges (geometric series with $r = -1$) so we cannot say that

$0 = 1 - 1 + 1 - 1 + 1 - 1 + \cdots$.

83. $\sum\limits_{n=1}^{\infty} ca_n = \lim\limits_{n\to\infty} \sum\limits_{i=1}^n ca_i = \lim\limits_{n\to\infty} c\sum\limits_{i=1}^n a_i = c\lim\limits_{n\to\infty} \sum\limits_{i=1}^n a_i = c\sum\limits_{n=1}^{\infty} a_n$, which exists by hypothesis.

85. Suppose on the contrary that $\sum(a_n + b_n)$ converges. Then $\sum(a_n + b_n)$ and $\sum a_n$ are convergent series. So by Theorem 8(iii), $\sum[(a_n + b_n) - a_n]$ would also be convergent. But $\sum[(a_n + b_n) - a_n] = \sum b_n$, a contradiction, since $\sum b_n$ is given to be divergent.

87. The partial sums $\{s_n\}$ form an increasing sequence, since $s_n - s_{n-1} = a_n > 0$ for all n. Also, the sequence $\{s_n\}$ is bounded since $s_n \leq 1000$ for all n. So by the Monotonic Sequence Theorem, the sequence of partial sums converges, that is, the series $\sum a_n$ is convergent.

89. (a) At the first step, only the interval $\left(\frac{1}{3}, \frac{2}{3}\right)$ (length $\frac{1}{3}$) is removed. At the second step, we remove the intervals $\left(\frac{1}{9}, \frac{2}{9}\right)$ and $\left(\frac{7}{9}, \frac{8}{9}\right)$, which have a total length of $2 \cdot \left(\frac{1}{3}\right)^2$. At the third step, we remove 2^2 intervals, each of length $\left(\frac{1}{3}\right)^3$. In general, at the nth step we remove 2^{n-1} intervals, each of length $\left(\frac{1}{3}\right)^n$, for a length of $2^{n-1} \cdot \left(\frac{1}{3}\right)^n = \frac{1}{3}\left(\frac{2}{3}\right)^{n-1}$. Thus, the total length of all removed intervals is $\sum_{n=1}^{\infty} \frac{1}{3}\left(\frac{2}{3}\right)^{n-1} = \frac{1/3}{1 - 2/3} = 1$ [geometric series with $a = \frac{1}{3}$ and $r = \frac{2}{3}$]. Notice that at the nth step, the leftmost interval that is removed is $\left(\left(\frac{1}{3}\right)^n, \left(\frac{2}{3}\right)^n\right)$, so we never remove 0, and 0 is in the Cantor set. Also, the rightmost interval removed is $\left(1 - \left(\frac{2}{3}\right)^n, 1 - \left(\frac{1}{3}\right)^n\right)$, so 1 is never removed. Some other numbers in the Cantor set are $\frac{1}{3}, \frac{2}{3}, \frac{1}{9}, \frac{2}{9}, \frac{7}{9}$, and $\frac{8}{9}$.

(b) The area removed at the first step is $\frac{1}{9}$; at the second step, $8 \cdot \left(\frac{1}{9}\right)^2$; at the third step, $(8)^2 \cdot \left(\frac{1}{9}\right)^3$. In general, the area removed at the nth step is $(8)^{n-1}\left(\frac{1}{9}\right)^n = \frac{1}{9}\left(\frac{8}{9}\right)^{n-1}$, so the total area of all removed squares is

$$\sum_{n=1}^{\infty} \frac{1}{9}\left(\frac{8}{9}\right)^{n-1} = \frac{1/9}{1 - 8/9} = 1.$$

91. (a) For $\sum_{n=1}^{\infty} \frac{n}{(n+1)!}$, $s_1 = \frac{1}{1 \cdot 2} = \frac{1}{2}$, $s_2 = \frac{1}{2} + \frac{2}{1 \cdot 2 \cdot 3} = \frac{5}{6}$, $s_3 = \frac{5}{6} + \frac{3}{1 \cdot 2 \cdot 3 \cdot 4} = \frac{23}{24}$,

$s_4 = \frac{23}{24} + \frac{4}{1 \cdot 2 \cdot 3 \cdot 4 \cdot 5} = \frac{119}{120}$. The denominators are $(n+1)!$, so a guess would be $s_n = \frac{(n+1)! - 1}{(n+1)!}$.

(b) For $n = 1$, $s_1 = \frac{1}{2} = \frac{2! - 1}{2!}$, so the formula holds for $n = 1$. Assume $s_k = \frac{(k+1)! - 1}{(k+1)!}$. Then

$$s_{k+1} = \frac{(k+1)! - 1}{(k+1)!} + \frac{k+1}{(k+2)!} = \frac{(k+1)! - 1}{(k+1)!} + \frac{k+1}{(k+1)!(k+2)} = \frac{(k+2)! - (k+2) + k + 1}{(k+2)!}$$

$$= \frac{(k+2)! - 1}{(k+2)!}$$

Thus, the formula is true for $n = k + 1$. So by induction, the guess is correct.

(c) $\lim_{n \to \infty} s_n = \lim_{n \to \infty} \frac{(n+1)! - 1}{(n+1)!} = \lim_{n \to \infty} \left[1 - \frac{1}{(n+1)!}\right] = 1$ and so $\sum_{n=1}^{\infty} \frac{n}{(n+1)!} = 1$.

11.3 The Integral Test and Estimates of Sums

1. The picture shows that $a_2 = \dfrac{1}{2^{1.3}} < \displaystyle\int_1^2 \dfrac{1}{x^{1.3}}\, dx$,

$a_3 = \dfrac{1}{3^{1.3}} < \displaystyle\int_2^3 \dfrac{1}{x^{1.3}}\, dx$, and so on, so $\displaystyle\sum_{n=2}^{\infty} \dfrac{1}{n^{1.3}} < \displaystyle\int_1^{\infty} \dfrac{1}{x^{1.3}}\, dx$. The

integral converges by (7.8.2) with $p = 1.3 > 1$, so the series converges.

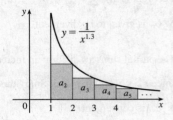

3. The function $f(x) = x^{-3}$ is continuous, positive, and decreasing on $[1, \infty)$, so the Integral Test applies.

$$\int_1^{\infty} x^{-3}\, dx = \lim_{t \to \infty} \int_1^t x^{-3}\, dx = \lim_{t \to \infty} \left[\dfrac{x^{-2}}{-2}\right]_1^t = \lim_{t \to \infty} \left(-\dfrac{1}{2t^2} + \dfrac{1}{2}\right) = \dfrac{1}{2}.$$

Since this improper integral is convergent, the series $\displaystyle\sum_{n=1}^{\infty} n^{-3}$ is also convergent by the Integral Test.

5. The function $f(x) = \dfrac{2}{5x - 1}$ is continuous, positive, and decreasing on $[1, \infty)$, so the Integral Test applies.

$$\int_1^{\infty} \dfrac{2}{5x - 1}\, dx = \lim_{t \to \infty} \int_1^t \dfrac{2}{5x - 1}\, dx = \lim_{t \to \infty} \left[\dfrac{2}{5} \ln(5x - 1)\right]_1^t = \lim_{t \to \infty} \left[\dfrac{2}{5} \ln(5t - 1) - \dfrac{2}{5} \ln 4\right] = \infty.$$

Since this improper integral is divergent, the series $\displaystyle\sum_{n=1}^{\infty} \dfrac{2}{5n - 1}$ is also divergent by the Integral Test.

7. The function $f(x) = \dfrac{x}{x^2 + 1}$ is continuous, positive, and decreasing on $[1, \infty)$, so the Integral Test applies.

$$\int_1^{\infty} \dfrac{x}{x^2 + 1}\, dx = \lim_{t \to \infty} \int_1^t \dfrac{x}{x^2 + 1}\, dx = \lim_{t \to \infty} \left[\dfrac{1}{2} \ln(x^2 + 1)\right]_1^t = \dfrac{1}{2} \lim_{t \to \infty} [\ln(t^2 + 1) - \ln 2] = \infty.$$ Since this improper

integral is divergent, the series $\displaystyle\sum_{n=1}^{\infty} \dfrac{n}{n^2 + 1}$ is also divergent by the Integral Test.

9. $\displaystyle\sum_{n=1}^{\infty} \dfrac{1}{n^{\sqrt{2}}}$ is a p-series with $p = \sqrt{2} > 1$, so it converges by (1).

11. $1 + \dfrac{1}{8} + \dfrac{1}{27} + \dfrac{1}{64} + \dfrac{1}{125} + \cdots = \displaystyle\sum_{n=1}^{\infty} \dfrac{1}{n^3}$. This is a p-series with $p = 3 > 1$, so it converges by (1).

13. $\dfrac{1}{3} + \dfrac{1}{7} + \dfrac{1}{11} + \dfrac{1}{15} + \dfrac{1}{19} + \cdots = \displaystyle\sum_{n=1}^{\infty} \dfrac{1}{4n - 1}$. The function $f(x) = \dfrac{1}{4x - 1}$ is continuous, positive, and decreasing on

$[1, \infty)$, so the Integral Test applies.

$$\int_1^{\infty} \dfrac{1}{4x - 1}\, dx = \lim_{t \to \infty} \int_1^t \dfrac{1}{4x - 1}\, dx = \lim_{t \to \infty} \left[\tfrac{1}{4} \ln(4x - 1)\right]_1^t = \lim_{t \to \infty} \left[\tfrac{1}{4} \ln(4t - 1) - \tfrac{1}{4} \ln 3\right] = \infty,$$ so the series

$\displaystyle\sum_{n=1}^{\infty} \dfrac{1}{4n - 1}$ diverges.

15. $\sum\limits_{n=1}^{\infty} \dfrac{\sqrt{n}+4}{n^2} = \sum\limits_{n=1}^{\infty}\left(\dfrac{\sqrt{n}}{n^2}+\dfrac{4}{n^2}\right) = \sum\limits_{n=1}^{\infty}\dfrac{1}{n^{3/2}}+\sum\limits_{n=1}^{\infty}\dfrac{4}{n^2}.$ $\sum\limits_{n=1}^{\infty}\dfrac{1}{n^{3/2}}$ is a convergent p-series with $p=\frac{3}{2}>1$.

$\sum\limits_{n=1}^{\infty}\dfrac{4}{n^2} = 4\sum\limits_{n=1}^{\infty}\dfrac{1}{n^2}$ is a constant multiple of a convergent p-series with $p=2>1$, so it converges. The sum of two

convergent series is convergent, so the original series is convergent.

17. The function $f(x)=\dfrac{1}{x^2+4}$ is continuous, positive, and decreasing on $[1,\infty)$, so we can apply the Integral Test.

$$\int_1^{\infty}\frac{1}{x^2+4}\,dx = \lim_{t\to\infty}\int_1^t \frac{1}{x^2+4}\,dx = \lim_{t\to\infty}\left[\frac{1}{2}\tan^{-1}\frac{x}{2}\right]_1^t = \frac{1}{2}\lim_{t\to\infty}\left[\tan^{-1}\left(\frac{t}{2}\right)-\tan^{-1}\left(\frac{1}{2}\right)\right]$$

$$= \frac{1}{2}\left[\frac{\pi}{2}-\tan^{-1}\left(\frac{1}{2}\right)\right]$$

Therefore, the series $\sum\limits_{n=1}^{\infty}\dfrac{1}{n^2+4}$ converges.

19. The function $f(x)=\dfrac{x^3}{x^4+4}$ is continuous and positive on $[2,\infty)$, and is also decreasing since

$$f'(x)=\frac{(x^4+4)(3x^2)-x^3(4x^3)}{(x^4+4)^2}=\frac{12x^2-x^6}{(x^4+4)^2}=\frac{x^2(12-x^4)}{(x^4+4)^2}<0 \text{ for } x>\sqrt[4]{12}\approx 1.86, \text{ so we can use the}$$

Integral Test on $[2,\infty)$.

$$\int_2^{\infty}\frac{x^3}{x^4+4}\,dx=\lim_{t\to\infty}\int_2^t\frac{x^3}{x^4+4}\,dx=\lim_{t\to\infty}\left[\tfrac{1}{4}\ln(x^4+4)\right]_2^t=\lim_{t\to\infty}\left[\tfrac{1}{4}\ln(t^4+4)-\tfrac{1}{4}\ln 20\right]=\infty, \text{ so the series}$$

$\sum\limits_{n=2}^{\infty}\dfrac{n^3}{n^4+4}$ diverges, and it follows that $\sum\limits_{n=1}^{\infty}\dfrac{n^3}{n^4+4}$ diverges as well.

21. $f(x)=\dfrac{1}{x\ln x}$ is continuous and positive on $[2,\infty)$, and also decreasing since $f'(x)=-\dfrac{1+\ln x}{x^2(\ln x)^2}<0$ for $x>2$, so we can

use the Integral Test. $\int_2^{\infty}\dfrac{1}{x\ln x}\,dx=\lim_{t\to\infty}\left[\ln(\ln x)\right]_2^t=\lim_{t\to\infty}\left[\ln(\ln t)-\ln(\ln 2)\right]=\infty$, so the series $\sum\limits_{n=2}^{\infty}\dfrac{1}{n\ln n}$ diverges.

23. The function $f(x)=xe^{-x}=\dfrac{x}{e^x}$ is continuous and positive on $[1,\infty)$, and also decreasing since

$$f'(x)=\frac{e^x\cdot 1-xe^x}{(e^x)^2}=\frac{e^x(1-x)}{(e^x)^2}=\frac{1-x}{e^x}<0 \text{ for } x>1 \text{ [and } f(1)>f(2)], \text{ so we can use the Integral Test on } [1,\infty).$$

$$\int_1^{\infty}xe^{-x}\,dx=\lim_{t\to\infty}\int_1^t xe^{-x}\,dx=\lim_{t\to\infty}\left(\left[-xe^{-x}\right]_1^t+\int_1^t e^{-x}\,dx\right) \quad \left[\begin{array}{c}\text{by parts with}\\ u=x,\,dv=e^{-x}\,dx\end{array}\right]$$

$$=\lim_{t\to\infty}\left(-te^{-t}+e^{-1}+\left[-e^{-x}\right]_1^t\right)=\lim_{t\to\infty}\left(-\frac{t}{e^t}+\frac{1}{e}-\frac{1}{e^t}+\frac{1}{e}\right)$$

$$\overset{\text{H}}{=}\lim_{t\to\infty}\left(-\frac{1}{e^t}+\frac{1}{e}-0+\frac{1}{e}\right)=\frac{2}{e},$$

so the series $\sum\limits_{k=1}^{\infty}ke^{-k}$ converges.

25. The function $f(x) = \dfrac{1}{x^2 + x^3} = \dfrac{1}{x^2} - \dfrac{1}{x} + \dfrac{1}{x+1}$ [by partial fractions] is continuous, positive and decreasing on $[1, \infty)$,

so the Integral Test applies.

$$\int_1^\infty f(x)\,dx = \lim_{t \to \infty} \int_1^t \left(\frac{1}{x^2} - \frac{1}{x} + \frac{1}{x+1} \right) dx = \lim_{t \to \infty} \left[-\frac{1}{x} - \ln x + \ln(x+1) \right]_1^t$$

$$= \lim_{t \to \infty} \left[-\frac{1}{t} + \ln \frac{t+1}{t} + 1 - \ln 2 \right] = 0 + 0 + 1 - \ln 2$$

The integral converges, so the series $\displaystyle\sum_{n=1}^\infty \frac{1}{n^2 + n^3}$ converges.

27. The function $f(x) = \dfrac{\cos \pi x}{\sqrt{x}}$ is neither positive nor decreasing on $[1, \infty)$, so the hypotheses of the Integral Test are not

satisfied for the series $\displaystyle\sum_{n=1}^\infty \frac{\cos \pi n}{\sqrt{n}}$.

29. We have already shown (in Exercise 21) that when $p = 1$ the series $\displaystyle\sum_{n=2}^\infty \frac{1}{n(\ln n)^p}$ diverges, so assume that $p \neq 1$.

$f(x) = \dfrac{1}{x(\ln x)^p}$ is continuous and positive on $[2, \infty)$, and $f'(x) = -\dfrac{p + \ln x}{x^2 (\ln x)^{p+1}} < 0$ if $x > e^{-p}$, so that f is eventually

decreasing and we can use the Integral Test.

$$\int_2^\infty \frac{1}{x(\ln x)^p}\,dx = \lim_{t \to \infty} \left[\frac{(\ln x)^{1-p}}{1-p} \right]_2^t \quad \text{[for } p \neq 1\text{]} = \lim_{t \to \infty} \left[\frac{(\ln t)^{1-p}}{1-p} - \frac{(\ln 2)^{1-p}}{1-p} \right]$$

This limit exists whenever $1 - p < 0 \iff p > 1$, so the series converges for $p > 1$.

31. Clearly the series cannot converge if $p \geq -\frac{1}{2}$, because then $\displaystyle\lim_{n \to \infty} n(1 + n^2)^p \neq 0$. So assume $p < -\frac{1}{2}$. Then

$f(x) = x(1 + x^2)^p$ is continuous, positive, and eventually decreasing on $[1, \infty)$, and we can use the Integral Test.

$$\int_1^\infty x(1 + x^2)^p\,dx = \lim_{t \to \infty} \left[\frac{1}{2} \cdot \frac{(1 + x^2)^{p+1}}{p+1} \right]_1^t = \frac{1}{2(p+1)} \lim_{t \to \infty} [(1 + t^2)^{p+1} - 2^{p+1}].$$

This limit exists and is finite $\iff p + 1 < 0 \iff p < -1$, so the series $\displaystyle\sum_{n=1}^\infty n(1 + n^2)^p$ converges whenever $p < -1$.

33. Since this is a p-series with $p = x$, $\zeta(x)$ is defined when $x > 1$. Unless specified otherwise, the domain of a function f is the

set of real numbers x such that the expression for $f(x)$ makes sense and defines a real number. So, in the case of a series, it's

the set of real numbers x such that the series is convergent.

35. (a) $\displaystyle\sum_{n=1}^\infty \left(\frac{3}{n} \right)^4 = \sum_{n=1}^\infty \frac{81}{n^4} = 81 \sum_{n=1}^\infty \frac{1}{n^4} = 81 \left(\frac{\pi^4}{90} \right) = \frac{9\pi^4}{10}$

(b) $\displaystyle\sum_{k=5}^\infty \frac{1}{(k-2)^4} = \frac{1}{3^4} + \frac{1}{4^4} + \frac{1}{5^4} + \cdots = \sum_{k=3}^\infty \frac{1}{k^4} = \frac{\pi^4}{90} - \left(\frac{1}{1^4} + \frac{1}{2^4} \right)$ [subtract a_1 and a_2] $= \frac{\pi^4}{90} - \frac{17}{16}$

37. (a) $f(x) = \dfrac{1}{x^2}$ is positive and continuous and $f'(x) = -\dfrac{2}{x^3}$ is negative for $x > 0$, and so the Integral Test applies.

$$\sum_{n=1}^{\infty} \frac{1}{n^2} \approx s_{10} = \frac{1}{1^2} + \frac{1}{2^2} + \frac{1}{3^2} + \cdots + \frac{1}{10^2} \approx 1.549768.$$

$$R_{10} \leq \int_{10}^{\infty} \frac{1}{x^2}\, dx = \lim_{t \to \infty} \left[\frac{-1}{x}\right]_{10}^{t} = \lim_{t \to \infty} \left(-\frac{1}{t} + \frac{1}{10}\right) = \frac{1}{10}, \text{ so the error is at most } 0.1.$$

(b) $s_{10} + \displaystyle\int_{11}^{\infty} \frac{1}{x^2}\, dx \leq s \leq s_{10} + \int_{10}^{\infty} \frac{1}{x^2}\, dx \;\Rightarrow\; s_{10} + \frac{1}{11} \leq s \leq s_{10} + \frac{1}{10} \;\Rightarrow$

$1.549768 + 0.090909 = 1.640677 \leq s \leq 1.549768 + 0.1 = 1.649768$, so we get $s \approx 1.64522$ (the average of 1.640677

and 1.649768) with error ≤ 0.005 (the maximum of $1.649768 - 1.64522$ and $1.64522 - 1.640677$, rounded up).

(c) The estimate in part (b) is $s \approx 1.64522$ with error ≤ 0.005. The exact value given in Exercise 34 is $\pi^2/6 \approx 1.644934$.

The difference is less than 0.0003.

(d) $R_n \leq \displaystyle\int_{n}^{\infty} \frac{1}{x^2}\, dx = \frac{1}{n}$. So $R_n < 0.001$ if $\dfrac{1}{n} < \dfrac{1}{1000} \;\Leftrightarrow\; n > 1000$.

39. $f(x) = 1/(2x+1)^6$ is continuous, positive, and decreasing on $[1, \infty)$, so the Integral Test applies. Using (2),

$$R_n \leq \int_{n}^{\infty} (2x+1)^{-6}\, dx = \lim_{t \to \infty} \left[\frac{-1}{10(2x+1)^5}\right]_{n}^{t} = \frac{1}{10(2n+1)^5}. \text{ To be correct to five decimal places, we want}$$

$$\frac{1}{10(2n+1)^5} \leq \frac{5}{10^6} \;\Leftrightarrow\; (2n+1)^5 \geq 20{,}000 \;\Leftrightarrow\; n \geq \tfrac{1}{2}\left(\sqrt[5]{20{,}000} - 1\right) \approx 3.12, \text{ so use } n = 4.$$

$$s_4 = \sum_{n=1}^{4} \frac{1}{(2n+1)^6} = \frac{1}{3^6} + \frac{1}{5^6} + \frac{1}{7^6} + \frac{1}{9^6} \approx 0.001\,446 \approx 0.00145.$$

41. $\displaystyle\sum_{n=1}^{\infty} n^{-1.001} = \sum_{n=1}^{\infty} \frac{1}{n^{1.001}}$ is a convergent p-series with $p = 1.001 > 1$. Using (2), we get

$$R_n \leq \int_{n}^{\infty} x^{-1.001}\, dx = \lim_{t \to \infty} \left[\frac{x^{-0.001}}{-0.001}\right]_{n}^{t} = -1000 \lim_{t \to \infty} \left[\frac{1}{x^{0.001}}\right]_{n}^{t} = -1000\left(-\frac{1}{n^{0.001}}\right) = \frac{1000}{n^{0.001}}. \text{ We want}$$

$$R_n < 0.000\,000\,005 \;\Leftrightarrow\; \frac{1000}{n^{0.001}} < 5 \times 10^{-9} \;\Leftrightarrow\; n^{0.001} > \frac{1000}{5 \times 10^{-9}} \;\Leftrightarrow$$

$$n > \left(2 \times 10^{11}\right)^{1000} = 2^{1000} \times 10^{11{,}000} \approx 1.07 \times 10^{301} \times 10^{11{,}000} = 1.07 \times 10^{11{,}301}.$$

43. (a) From the figure, $a_2 + a_3 + \cdots + a_n \leq \int_1^n f(x)\, dx$, so with

$$f(x) = \frac{1}{x}, \; \frac{1}{2} + \frac{1}{3} + \frac{1}{4} + \cdots + \frac{1}{n} \leq \int_1^n \frac{1}{x}\, dx = \ln n.$$

Thus, $s_n = 1 + \dfrac{1}{2} + \dfrac{1}{3} + \dfrac{1}{4} + \cdots + \dfrac{1}{n} \leq 1 + \ln n.$

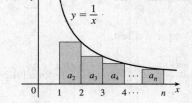

(b) By part (a), $s_{10^6} \leq 1 + \ln 10^6 \approx 14.82 < 15$ and

$s_{10^9} \leq 1 + \ln 10^9 \approx 21.72 < 22.$

45. $b^{\ln n} = \left(e^{\ln b}\right)^{\ln n} = \left(e^{\ln n}\right)^{\ln b} = n^{\ln b} = \dfrac{1}{n^{-\ln b}}$. This is a p-series, which converges for all b such that $-\ln b > 1$ ⇔

$\ln b < -1$ ⇔ $b < e^{-1}$ ⇔ $b < 1/e$ [with $b > 0$].

11.4 The Comparison Tests

1. (a) We cannot say anything about $\sum a_n$. If $a_n > b_n$ for all n and $\sum b_n$ is convergent, then $\sum a_n$ could be convergent or

divergent. (See the note after Example 2.)

(b) If $a_n < b_n$ for all n, then $\sum a_n$ is convergent. [This is part (i) of the Comparison Test.]

3. $\dfrac{1}{n^3 + 8} < \dfrac{1}{n^3}$ for all $n \geq 1$, so $\displaystyle\sum_{n=1}^{\infty} \dfrac{1}{n^3 + 8}$ converges by comparison with $\displaystyle\sum_{n=1}^{\infty} \dfrac{1}{n^3}$, which converges because it is a p-series

with $p = 3 > 1$.

5. $\dfrac{n+1}{n\sqrt{n}} > \dfrac{n}{n\sqrt{n}} = \dfrac{1}{\sqrt{n}}$ for all $n \geq 1$, so $\displaystyle\sum_{n=1}^{\infty} \dfrac{n+1}{n\sqrt{n}}$ diverges by comparison with $\displaystyle\sum_{n=1}^{\infty} \dfrac{1}{\sqrt{n}}$, which diverges because it is a

p-series with $p = \frac{1}{2} \leq 1$.

7. $\dfrac{9^n}{3 + 10^n} < \dfrac{9^n}{10^n} = \left(\dfrac{9}{10}\right)^n$ for all $n \geq 1$. $\displaystyle\sum_{n=1}^{\infty} \left(\frac{9}{10}\right)^n$ is a convergent geometric series $\left(|r| = \frac{9}{10} < 1\right)$, so $\displaystyle\sum_{n=1}^{\infty} \dfrac{9^n}{3 + 10^n}$

converges by the Comparison Test.

9. $\dfrac{\ln k}{k} > \dfrac{1}{k}$ for all k ≥ 3 [since $\ln k > 1$ for $k \geq 3$], so $\displaystyle\sum_{k=3}^{\infty} \dfrac{\ln k}{k}$ diverges by comparison with $\displaystyle\sum_{k=3}^{\infty} \dfrac{1}{k}$, which diverges because it

is a p-series with $p = 1 \leq 1$ (the harmonic series). Thus, $\displaystyle\sum_{k=1}^{\infty} \dfrac{\ln k}{k}$ diverges since a finite number of terms doesn't affect the

convergence or divergence of a series.

11. $\dfrac{\sqrt[3]{k}}{\sqrt{k^3 + 4k + 3}} < \dfrac{\sqrt[3]{k}}{\sqrt{k^3}} = \dfrac{k^{1/3}}{k^{3/2}} = \dfrac{1}{k^{7/6}}$ for all $k \geq 1$, so $\displaystyle\sum_{k=1}^{\infty} \dfrac{\sqrt[3]{k}}{\sqrt{k^3 + 4k + 3}}$ converges by comparison with $\displaystyle\sum_{k=1}^{\infty} \dfrac{1}{k^{7/6}}$,

which converges because it is a p-series with $p = \frac{7}{6} > 1$.

13. $\dfrac{1 + \cos n}{e^n} < \dfrac{2}{e^n}$ for all $n \geq 1$. $\displaystyle\sum_{n=1}^{\infty} \dfrac{2}{e^n}$ is a convergent geometric series $\left(|r| = \frac{1}{e} < 1\right)$, so $\displaystyle\sum_{n=1}^{\infty} \dfrac{1 + \cos n}{e^n}$ converges by the

Comparison Test.

15. $\dfrac{4^{n+1}}{3^n - 2} > \dfrac{4 \cdot 4^n}{3^n} = 4\left(\dfrac{4}{3}\right)^n$ for all $n \geq 1$. $\displaystyle\sum_{n=1}^{\infty} 4\left(\frac{4}{3}\right)^n = 4\displaystyle\sum_{n=1}^{\infty} \left(\frac{4}{3}\right)^n$ is a divergent geometric series $\left(|r| = \frac{4}{3} > 1\right)$, so

$\displaystyle\sum_{n=1}^{\infty} \dfrac{4^{n+1}}{3^n - 2}$ diverges by the Comparison Test.

17. Use the Limit Comparison Test with $a_n = \dfrac{1}{\sqrt{n^2 + 1}}$ and $b_n = \dfrac{1}{n}$:

$$\lim_{n \to \infty} \frac{a_n}{b_n} = \lim_{n \to \infty} \frac{n}{\sqrt{n^2 + 1}} = \lim_{n \to \infty} \frac{1}{\sqrt{1 + (1/n^2)}} = 1 > 0. \text{ Since the harmonic series } \sum_{n=1}^{\infty} \frac{1}{n} \text{ diverges, so does}$$

$$\sum_{n=1}^{\infty} \frac{1}{\sqrt{n^2 + 1}}.$$

19. Use the Limit Comparison Test with $a_n = \dfrac{n+1}{n^3 + n}$ and $b_n = \dfrac{1}{n^2}$:

$$\lim_{n \to \infty} \frac{a_n}{b_n} = \lim_{n \to \infty} \frac{(n+1)n^2}{n(n^2 + 1)} = \lim_{n \to \infty} \frac{n^2 + n}{n^2 + 1} = \lim_{n \to \infty} \frac{1 + 1/n}{1 + 1/n^2} = 1 > 0. \text{ Since } \sum_{n=1}^{\infty} \frac{1}{n^2} \text{ is a convergent } p\text{-series}$$

$[p = 2 > 1]$, the series $\sum_{n=1}^{\infty} \dfrac{n+1}{n^3 + n}$ also converges.

21. Use the Limit Comparison Test with $a_n = \dfrac{\sqrt{1 + n}}{2 + n}$ and $b_n = \dfrac{1}{\sqrt{n}}$:

$$\lim_{n \to \infty} \frac{a_n}{b_n} = \lim_{n \to \infty} \frac{\sqrt{1 + n}\sqrt{n}}{2 + n} = \lim_{n \to \infty} \frac{\sqrt{n^2 + n}/\sqrt{n^2}}{(2 + n)/n} = \lim_{n \to \infty} \frac{\sqrt{1 + 1/n}}{2/n + 1} = 1 > 0. \text{ Since } \sum_{n=1}^{\infty} \frac{1}{\sqrt{n}} \text{ is a divergent } p\text{-series}$$

$[\, p = \tfrac{1}{2} \le 1]$, the series $\sum_{n=1}^{\infty} \dfrac{\sqrt{1 + n}}{2 + n}$ also diverges.

23. Use the Limit Comparison Test with $a_n = \dfrac{5 + 2n}{(1 + n^2)^2}$ and $b_n = \dfrac{1}{n^3}$:

$$\lim_{n \to \infty} \frac{a_n}{b_n} = \lim_{n \to \infty} \frac{n^3(5 + 2n)}{(1 + n^2)^2} = \lim_{n \to \infty} \frac{5n^3 + 2n^4}{(1 + n^2)^2} \cdot \frac{1/n^4}{1/(n^2)^2} = \lim_{n \to \infty} \frac{\frac{5}{n} + 2}{\left(\frac{1}{n^2} + 1\right)^2} = 2 > 0. \text{ Since } \sum_{n=1}^{\infty} \frac{1}{n^3} \text{ is a convergent}$$

p-series $[p = 3 > 1]$, the series $\sum_{n=1}^{\infty} \dfrac{5 + 2n}{(1 + n^2)^2}$ also converges.

25. $\dfrac{e^n + 1}{ne^n + 1} \ge \dfrac{e^n + 1}{ne^n + n} = \dfrac{e^n + 1}{n(e^n + 1)} = \dfrac{1}{n}$ for $n \ge 1$, so the series $\sum_{n=1}^{\infty} \dfrac{e^n + 1}{ne^n + 1}$ diverges by comparison with the divergent

harmonic series $\sum_{n=1}^{\infty} \dfrac{1}{n}$.   *Or:* Use the Limit Comparison Test with $a_n = \dfrac{e^n + 1}{ne^n + 1}$ and $b_n = \dfrac{1}{n}$.

27. Use the Limit Comparison Test with $a_n = \left(1 + \dfrac{1}{n}\right)^2 e^{-n}$ and $b_n = e^{-n}$: $\displaystyle\lim_{n \to \infty} \frac{a_n}{b_n} = \lim_{n \to \infty} \left(1 + \frac{1}{n}\right)^2 = 1 > 0.$ Since

$\sum_{n=1}^{\infty} e^{-n} = \sum_{n=1}^{\infty} \dfrac{1}{e^n}$ is a convergent geometric series $\left[|r| = \tfrac{1}{e} < 1 \right]$, the series $\sum_{n=1}^{\infty} \left(1 + \dfrac{1}{n}\right)^2 e^{-n}$ also converges.

29. Clearly $n! = n(n-1)(n-2)\cdots(3)(2) \ge 2 \cdot 2 \cdot 2 \cdot \cdots \cdot 2 \cdot 2 = 2^{n-1}$, so $\dfrac{1}{n!} \le \dfrac{1}{2^{n-1}}$. $\sum_{n=1}^{\infty} \dfrac{1}{2^{n-1}}$ is a convergent geometric

series $\left[|r| = \tfrac{1}{2} < 1 \right]$, so $\sum_{n=1}^{\infty} \dfrac{1}{n!}$ converges by the Comparison Test.

31. Use the Limit Comparison Test with $a_n = \sin\left(\dfrac{1}{n}\right)$ and $b_n = \dfrac{1}{n}$. Then $\sum a_n$ and $\sum b_n$ are series with positive terms and

$$\lim_{n\to\infty} \frac{a_n}{b_n} = \lim_{n\to\infty} \frac{\sin(1/n)}{1/n} = \lim_{\theta\to 0} \frac{\sin\theta}{\theta} = 1 > 0. \quad \text{Since } \sum_{n=1}^{\infty} b_n \text{ is the divergent harmonic series,}$$

$\sum_{n=1}^{\infty} \sin(1/n)$ also diverges.   [Note that we could also use l'Hospital's Rule to evaluate the limit:

$$\lim_{x\to\infty} \frac{\sin(1/x)}{1/x} \overset{\text{H}}{=} \lim_{x\to\infty} \frac{\cos(1/x)\cdot(-1/x^2)}{-1/x^2} = \lim_{x\to\infty} \cos\frac{1}{x} = \cos 0 = 1.]$$

33. $\displaystyle\sum_{n=1}^{10} \frac{1}{5+n^5} = \frac{1}{5+1^5} + \frac{1}{5+2^5} + \frac{1}{5+3^5} + \cdots + \frac{1}{5+10^5} \approx 0.19926.$ Now $\dfrac{1}{5+n^5} < \dfrac{1}{n^5}$, so the error is

$$R_{10} \le T_{10} \le \int_{10}^{\infty} \frac{1}{x^5}\,dx = \lim_{t\to\infty} \int_{10}^{t} x^{-5}\,dx = \lim_{t\to\infty} \left[\frac{-1}{4x^4}\right]_{10}^{t} = \lim_{t\to\infty}\left(\frac{-1}{4t^4} + \frac{1}{40{,}000}\right) = \frac{1}{40{,}000} = 0.000\,025.$$

35. $\displaystyle\sum_{n=1}^{10} 5^{-n}\cos^2 n = \frac{\cos^2 1}{5} + \frac{\cos^2 2}{5^2} + \frac{\cos^2 3}{5^3} + \cdots + \frac{\cos^2 10}{5^{10}} \approx 0.07393.$ Now $\dfrac{\cos^2 n}{5^n} \le \dfrac{1}{5^n}$, so the error is

$$R_{10} \le T_{10} \le \int_{10}^{\infty} \frac{1}{5^x}\,dx = \lim_{t\to\infty}\int_{10}^{t} 5^{-x}\,dx = \lim_{t\to\infty}\left[-\frac{5^{-x}}{\ln 5}\right]_{10}^{t} = \lim_{t\to\infty}\left(-\frac{5^{-t}}{\ln 5} + \frac{5^{-10}}{\ln 5}\right) = \frac{1}{5^{10}\ln 5} < 6.4\times 10^{-8}.$$

37. Since $\dfrac{d_n}{10^n} \le \dfrac{9}{10^n}$ for each n, and since $\displaystyle\sum_{n=1}^{\infty} \frac{9}{10^n}$ is a convergent geometric series $\left(|r| = \frac{1}{10} < 1\right)$, $0.d_1 d_2 d_3 \ldots = \displaystyle\sum_{n=1}^{\infty} \frac{d_n}{10^n}$

will always converge by the Comparison Test.

39. Since $\sum a_n$ converges, $\lim_{n\to\infty} a_n = 0$, so there exists N such that $|a_n - 0| < 1$ for all $n > N$ $\Rightarrow$ $0 \le a_n < 1$ for

all $n > N$ $\Rightarrow$ $0 \le a_n^2 \le a_n$. Since $\sum a_n$ converges, so does $\sum a_n^2$ by the Comparison Test.

41. (a) Since $\lim_{n\to\infty} \dfrac{a_n}{b_n} = \infty$, there is an integer N such that $\dfrac{a_n}{b_n} > 1$ whenever $n > N$. (Take $M = 1$ in Definition 11.1.5.)

   Then $a_n > b_n$ whenever $n > N$ and since $\sum b_n$ is divergent, $\sum a_n$ is also divergent by the Comparison Test.

   (b) (i) If $a_n = \dfrac{1}{\ln n}$ and $b_n = \dfrac{1}{n}$ for $n \ge 2$, then $\lim_{n\to\infty}\dfrac{a_n}{b_n} = \lim_{n\to\infty}\dfrac{n}{\ln n} = \lim_{x\to\infty}\dfrac{x}{\ln x} \overset{\text{H}}{=} \lim_{x\to\infty}\dfrac{1}{1/x} = \lim_{x\to\infty} x = \infty$,

   so by part (a), $\displaystyle\sum_{n=2}^{\infty} \frac{1}{\ln n}$ is divergent.

   (ii) If $a_n = \dfrac{\ln n}{n}$ and $b_n = \dfrac{1}{n}$, then $\displaystyle\sum_{n=1}^{\infty} b_n$ is the divergent harmonic series and $\lim_{n\to\infty}\dfrac{a_n}{b_n} = \lim_{n\to\infty} \ln n = \lim_{x\to\infty} \ln x = \infty$,

   so $\displaystyle\sum_{n=1}^{\infty} a_n$ diverges by part (a).

43. $\lim_{n\to\infty} na_n = \lim_{n\to\infty} \dfrac{a_n}{1/n}$, so we apply the Limit Comparison Test with $b_n = \dfrac{1}{n}$. Since $\lim_{n\to\infty} na_n > 0$ we know that either both

series converge or both series diverge, and we also know that $\displaystyle\sum_{n=1}^{\infty} \frac{1}{n}$ diverges [p-series with $p = 1$]. Therefore, $\sum a_n$ must be

divergent.

45. Yes. Since $\sum a_n$ is a convergent series with positive terms, $\lim\limits_{n\to\infty} a_n = 0$ by Theorem 11.2.6, and $\sum b_n = \sum \sin(a_n)$ is a

series with positive terms (for large enough n). We have $\lim\limits_{n\to\infty} \dfrac{b_n}{a_n} = \lim\limits_{n\to\infty} \dfrac{\sin(a_n)}{a_n} = 1 > 0$ by Theorem 2.4.2

[ET Theorem 3.3.2]. Thus, $\sum b_n$ is also convergent by the Limit Comparison Test.

11.5 Alternating Series

1. (a) An alternating series is a series whose terms are alternately positive and negative.

(b) An alternating series $\sum\limits_{n=1}^{\infty} a_n = \sum\limits_{n=1}^{\infty} (-1)^{n-1} b_n$, where $b_n = |a_n|$, converges if $0 < b_{n+1} \le b_n$ for all n and $\lim\limits_{n\to\infty} b_n = 0$.

(This is the Alternating Series Test.)

(c) The error involved in using the partial sum s_n as an approximation to the total sum s is the remainder $R_n = s - s_n$ and the

size of the error is smaller than b_{n+1}; that is, $|R_n| \le b_{n+1}$. (This is the Alternating Series Estimation Theorem.)

3. $-\dfrac{2}{5} + \dfrac{4}{6} - \dfrac{6}{7} + \dfrac{8}{8} - \dfrac{10}{9} + \cdots = \sum\limits_{n=1}^{\infty} (-1)^n \dfrac{2n}{n+4}$. Now $\lim\limits_{n\to\infty} b_n = \lim\limits_{n\to\infty} \dfrac{2n}{n+4} = \lim\limits_{n\to\infty} \dfrac{2}{1+4/n} = \dfrac{2}{1} \ne 0$. Since

$\lim\limits_{n\to\infty} a_n \ne 0$ (in fact the limit does not exist), the series diverges by the Test for Divergence.

5. $\sum\limits_{n=1}^{\infty} a_n = \sum\limits_{n=1}^{\infty} \dfrac{(-1)^{n-1}}{3+5n} = \sum\limits_{n=1}^{\infty} (-1)^{n-1} b_n$. Now $b_n = \dfrac{1}{3+5n} > 0$, $\{b_n\}$ is decreasing, and $\lim\limits_{n\to\infty} b_n = 0$, so the series

converges by the Alternating Series Test.

7. $\sum\limits_{n=1}^{\infty} a_n = \sum\limits_{n=1}^{\infty} (-1)^n \dfrac{3n-1}{2n+1} = \sum\limits_{n=1}^{\infty} (-1)^n b_n$. Now $\lim\limits_{n\to\infty} b_n = \lim\limits_{n\to\infty} \dfrac{3-1/n}{2+1/n} = \dfrac{3}{2} \ne 0$. Since $\lim\limits_{n\to\infty} a_n \ne 0$

(in fact the limit does not exist), the series diverges by the Test for Divergence.

9. $\sum\limits_{n=1}^{\infty} a_n = \sum\limits_{n=1}^{\infty} (-1)^n e^{-n} = \sum\limits_{n=1}^{\infty} (-1)^n b_n$. Now $b_n = \dfrac{1}{e^n} > 0$, $\{b_n\}$ is decreasing, and $\lim\limits_{n\to\infty} b_n = 0$, so the series converges

by the Alternating Series Test.

11. $b_n = \dfrac{n^2}{n^3+4} > 0$ for $n \ge 1$. $\{b_n\}$ is decreasing for $n \ge 2$ since

$\left(\dfrac{x^2}{x^3+4}\right)' = \dfrac{(x^3+4)(2x) - x^2(3x^2)}{(x^3+4)^2} = \dfrac{x(2x^3+8-3x^3)}{(x^3+4)^2} = \dfrac{x(8-x^3)}{(x^3+4)^2} < 0$ for $x > 2$. Also,

$\lim\limits_{n\to\infty} b_n = \lim\limits_{n\to\infty} \dfrac{1/n}{1+4/n^3} = 0$. Thus, the series $\sum\limits_{n=1}^{\infty} (-1)^{n+1} \dfrac{n^2}{n^3+4}$ converges by the Alternating Series Test.

13. $\lim\limits_{n\to\infty} b_n = \lim\limits_{n\to\infty} e^{2/n} = e^0 = 1$, so $\lim\limits_{n\to\infty} (-1)^{n-1} e^{2/n}$ does not exist. Thus, the series $\sum\limits_{n=1}^{\infty} (-1)^{n-1} e^{2/n}$ diverges by the

Test for Divergence.

15. $a_n = \dfrac{\sin\left(n + \frac{1}{2}\right)\pi}{1+\sqrt{n}} = \dfrac{(-1)^n}{1+\sqrt{n}}$. Now $b_n = \dfrac{1}{1+\sqrt{n}} > 0$ for $n \ge 0$, $\{b_n\}$ is decreasing, and $\lim\limits_{n\to\infty} b_n = 0$, so the series

$\sum\limits_{n=0}^{\infty} \dfrac{\sin\left(n + \frac{1}{2}\right)\pi}{1+\sqrt{n}}$ converges by the Alternating Series Test.

17. $\sum_{n=1}^{\infty} (-1)^n \sin\left(\frac{\pi}{n}\right)$. $b_n = \sin\left(\frac{\pi}{n}\right) > 0$ for $n \geq 2$ and $\sin\left(\frac{\pi}{n}\right) \geq \sin\left(\frac{\pi}{n+1}\right)$, and $\lim_{n\to\infty} \sin\left(\frac{\pi}{n}\right) = \sin 0 = 0$, so the

series converges by the Alternating Series Test.

19. $\dfrac{n^n}{n!} = \dfrac{n \cdot n \cdot \cdots \cdot n}{1 \cdot 2 \cdot \cdots \cdot n} \geq n \implies \lim_{n\to\infty} \dfrac{n^n}{n!} = \infty \implies \lim_{n\to\infty} \dfrac{(-1)^n n^n}{n!}$ does not exist. So the series $\sum_{n=1}^{\infty} (-1)^n \dfrac{n^n}{n!}$ diverges

by the Test for Divergence.

21.

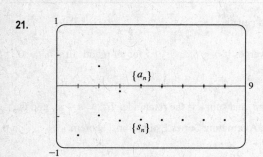

The graph gives us an estimate for the sum of the series

$$\sum_{n=1}^{\infty} \frac{(-0.8)^n}{n!} \text{ of } -0.55.$$

$$b_8 = \frac{(0.8)^n}{8!} \approx 0.000\,004, \text{ so}$$

$$\sum_{n=1}^{\infty} \frac{(-0.8)^n}{n!} \approx s_7 = \sum_{n=1}^{7} \frac{(-0.8)^n}{n!}$$

$$\approx -0.8 + 0.32 - 0.085\overline{3} + 0.0170\overline{6} - 0.002\,731 + 0.000\,364 - 0.000\,042 \approx -0.5507$$

Adding b_8 to s_7 does not change the fourth decimal place of s_7, so the sum of the series, correct to four decimal places,

is -0.5507.

23. The series $\sum_{n=1}^{\infty} \dfrac{(-1)^{n+1}}{n^6}$ satisfies (i) of the Alternating Series Test because $\dfrac{1}{(n+1)^6} < \dfrac{1}{n^6}$ and (ii) $\lim_{n\to\infty} \dfrac{1}{n^6} = 0$, so the

series is convergent. Now $b_5 = \dfrac{1}{5^6} = 0.000064 > 0.00005$ and $b_6 = \dfrac{1}{6^6} \approx 0.00002 < 0.00005$, so by the Alternating Series

Estimation Theorem, $n = 5$. (That is, since the 6th term is less than the desired error, we need to add the first 5 terms to get the

sum to the desired accuracy.)

25. The series $\sum_{n=1}^{\infty} \dfrac{(-1)^{n-1}}{n^2 2^n}$ satisfies (i) of the Alternating Series Test because $\dfrac{1}{(n+1)^2 2^{n+1}} < \dfrac{1}{n^2 2^n}$ and (ii) $\lim_{n\to\infty} \dfrac{1}{n^2 2^n} = 0$,

so the series is convergent. Now $b_5 = \dfrac{1}{5^2 2^5} = 0.00125 > 0.0005$ and $b_6 = \dfrac{1}{6^2 2^6} \approx 0.0004 < 0.0005$, so by the Alternating

Series Estimation Theorem, $n = 5$. (That is, since the 6th term is less than the desired error, we need to add the first 5 terms to

get the sum to the desired accuracy.)

27. $b_4 = \dfrac{1}{8!} = \dfrac{1}{40{,}320} \approx 0.000\,025$, so

$$\sum_{n=1}^{\infty} \frac{(-1)^n}{(2n)!} \approx s_3 = \sum_{n=1}^{3} \frac{(-1)^n}{(2n)!} = -\frac{1}{2} + \frac{1}{24} - \frac{1}{720} \approx -0.459\,722$$

Adding b_4 to s_3 does not change the fourth decimal place of s_3, so by the Alternating Series Estimation Theorem, the sum of

the series, correct to four decimal places, is -0.4597.

29. $\sum_{n=1}^{\infty} (-1)^n n e^{-2n} \approx s_5 = -\dfrac{1}{e^2} + \dfrac{2}{e^4} - \dfrac{3}{e^6} + \dfrac{4}{e^8} - \dfrac{5}{e^{10}} \approx -0.105\,025$. Adding $b_6 = 6/e^{12} \approx 0.000\,037$ to s_5 does not

change the fourth decimal place of s_5, so by the Alternating Series Estimation Theorem, the sum of the series, correct to four

decimal places, is -0.1050.

31. $\sum_{n=1}^{\infty} \dfrac{(-1)^{n-1}}{n} = 1 - \dfrac{1}{2} + \dfrac{1}{3} - \dfrac{1}{4} + \cdots + \dfrac{1}{49} - \dfrac{1}{50} + \dfrac{1}{51} - \dfrac{1}{52} + \cdots$. The 50th partial sum of this series is an

underestimate, since $\sum_{n=1}^{\infty} \dfrac{(-1)^{n-1}}{n} = s_{50} + \left(\dfrac{1}{51} - \dfrac{1}{52}\right) + \left(\dfrac{1}{53} - \dfrac{1}{54}\right) + \cdots$, and the terms in parentheses are all positive.

The result can be seen geometrically in Figure 1.

33. Clearly $b_n = \dfrac{1}{n+p}$ is decreasing and eventually positive and $\lim_{n \to \infty} b_n = 0$ for any p. So the series $\sum_{n=1}^{\infty} \dfrac{(-1)^n}{n+p}$ converges (by

the Alternating Series Test) for any p for which every b_n is defined, that is, $n + p \neq 0$ for $n \geq 1$, or p is not a negative integer.

35. $\sum b_{2n} = \sum 1/(2n)^2$ clearly converges (by comparison with the p-series for $p = 2$). So suppose that $\sum (-1)^{n-1} b_n$

converges. Then by Theorem 11.2.8(ii), so does $\sum \left[(-1)^{n-1} b_n + b_n\right] = 2\left(1 + \dfrac{1}{3} + \dfrac{1}{5} + \cdots\right) = 2\sum \dfrac{1}{2n-1}$. But this

diverges by comparison with the harmonic series, a contradiction. Therefore, $\sum (-1)^{n-1} b_n$ must diverge. The Alternating

Series Test does not apply since $\{b_n\}$ is not decreasing.

11.6 Absolute Convergence and the Ratio and Root Tests

1. (a) Since $\lim_{n \to \infty} \left| \dfrac{a_{n+1}}{a_n} \right| = 8 > 1$, part (b) of the Ratio Test tells us that the series $\sum a_n$ is divergent.

(b) Since $\lim_{n \to \infty} \left| \dfrac{a_{n+1}}{a_n} \right| = 0.8 < 1$, part (a) of the Ratio Test tells us that the series $\sum a_n$ is absolutely convergent (and

therefore convergent).

(c) Since $\lim_{n \to \infty} \left| \dfrac{a_{n+1}}{a_n} \right| = 1$, the Ratio Test fails and the series $\sum a_n$ might converge or it might diverge.

3. $b_n = \dfrac{1}{5n+1} > 0$ for $n \geq 0$, $\{b_n\}$ is decreasing for $n \geq 0$, and $\lim_{n \to \infty} b_n = 0$, so $\sum_{n=0}^{\infty} \dfrac{(-1)^n}{5n+1}$ converges by the Alternating

Series Test. To determine absolute convergence, choose $a_n = \dfrac{1}{n}$ to get

$\lim_{n \to \infty} \dfrac{a_n}{b_n} = \lim_{n \to \infty} \dfrac{1/n}{1/(5n+1)} = \lim_{n \to \infty} \dfrac{5n+1}{n} = 5 > 0$, so $\sum_{n=1}^{\infty} \dfrac{1}{5n+1}$ diverges by the Limit Comparison Test with the

harmonic series. Thus, the series $\sum_{n=0}^{\infty} \dfrac{(-1)^n}{5n+1}$ is conditionally convergent.

5. $0 < \left| \dfrac{\sin n}{2^n} \right| < \dfrac{1}{2^n}$ for $n \geq 1$ and $\sum_{n=1}^{\infty} \dfrac{1}{2^n}$ is a convergent geometric series ($r = \frac{1}{2} < 1$), so $\sum_{n=1}^{\infty} \left| \dfrac{\sin n}{2^n} \right|$ converges by

comparison and the series $\sum_{n=1}^{\infty} \dfrac{\sin n}{2^n}$ is absolutely convergent.

7. $\lim\limits_{n\to\infty}\left|\dfrac{a_{n+1}}{a_n}\right| = \lim\limits_{n\to\infty}\left|\dfrac{n+1}{5^{n+1}}\cdot\dfrac{5^n}{n}\right| = \lim\limits_{n\to\infty}\left|\dfrac{1}{5}\cdot\dfrac{n+1}{n}\right| = \dfrac{1}{5}\lim\limits_{n\to\infty}\dfrac{1+1/n}{1} = \dfrac{1}{5}(1) = \dfrac{1}{5} < 1$, so the series $\sum\limits_{n=1}^{\infty}\dfrac{n}{5^n}$ is

absolutely convergent by the Ratio Test.

9. $\lim\limits_{n\to\infty}\left|\dfrac{a_{n+1}}{a_n}\right| = \lim\limits_{n\to\infty}\left|\dfrac{(-1)^n 3^{n+1}}{2^{n+1}(n+1)^3}\cdot\dfrac{2^n n^3}{(-1)^{n-1} 3^n}\right| = \lim\limits_{n\to\infty}\left|\left(-\dfrac{3}{2}\right)\dfrac{n^3}{(n+1)^3}\right| = \dfrac{3}{2}\lim\limits_{n\to\infty}\dfrac{1}{(1+1/n)^3} = \dfrac{3}{2}(1) = \dfrac{3}{2} > 1$,

so the series $\sum\limits_{n=1}^{\infty}(-1)^{n-1}\dfrac{3^n}{2^n n^3}$ is divergent by the Ratio Test.

11. $\lim\limits_{k\to\infty}\left|\dfrac{a_{k+1}}{a_k}\right| = \lim\limits_{k\to\infty}\left|\dfrac{1}{(k+1)!}\cdot\dfrac{k!}{1}\right| = \lim\limits_{k\to\infty}\dfrac{1}{k+1} = 0 < 1$, so the series $\sum\limits_{k=1}^{\infty}\dfrac{1}{k!}$ is absolutely convergent by the Ratio Test.

Since the terms of this series are positive, absolute convergence is the same as convergence.

13. $\lim\limits_{n\to\infty}\left|\dfrac{a_{n+1}}{a_n}\right| = \lim\limits_{n\to\infty}\left[\dfrac{10^{n+1}}{(n+2)\,4^{2n+3}}\cdot\dfrac{(n+1)\,4^{2n+1}}{10^n}\right] = \lim\limits_{n\to\infty}\left(\dfrac{10}{4^2}\cdot\dfrac{n+1}{n+2}\right) = \dfrac{5}{8} < 1$, so the series $\sum\limits_{n=1}^{\infty}\dfrac{10^n}{(n+1)4^{2n+1}}$

is absolutely convergent by the Ratio Test. Since the terms of this series are positive, absolute convergence is the same as

convergence.

15. $\lim\limits_{n\to\infty}\left|\dfrac{a_{n+1}}{a_n}\right| = \lim\limits_{n\to\infty}\left|\dfrac{(n+1)\pi^{n+1}}{(-3)^n}\cdot\dfrac{(-3)^{n-1}}{n\pi^n}\right| = \lim\limits_{n\to\infty}\left|\dfrac{\pi}{-3}\cdot\dfrac{n+1}{n}\right| = \dfrac{\pi}{3}\lim\limits_{n\to\infty}\dfrac{1+1/n}{1} = \dfrac{\pi}{3}(1) = \dfrac{\pi}{3} > 1$, so the

series $\sum\limits_{n=1}^{\infty}\dfrac{n\pi^n}{(-3)^{n-1}}$ diverges by the Ratio Test. *Or:* Since $\lim\limits_{n\to\infty}|a_n| = \infty$, the series diverges by the Test for Divergence.

17. $\lim\limits_{n\to\infty}\left|\dfrac{a_{n+1}}{a_n}\right| = \lim\limits_{n\to\infty}\left|\dfrac{\cos[(n+1)\pi/3]}{(n+1)!}\cdot\dfrac{n!}{\cos(n\pi/3)}\right| = \lim\limits_{n\to\infty}\left|\dfrac{\cos[(n+1)\pi/3]}{(n+1)\cos(n\pi/3)}\right| = \lim\limits_{n\to\infty}\dfrac{c}{n+1} = 0 < 1$ (where

$0 < c \le 2$ for all positive integers n), so the series $\sum\limits_{n=1}^{\infty}\dfrac{\cos(n\pi/3)}{n!}$ is absolutely convergent by the Ratio Test.

19. $\lim\limits_{n\to\infty}\left|\dfrac{a_{n+1}}{a_n}\right| = \lim\limits_{n\to\infty}\left|\dfrac{(n+1)^{100}100^{n+1}}{(n+1)!}\cdot\dfrac{n!}{n^{100}100^n}\right| = \lim\limits_{n\to\infty}\dfrac{100}{n+1}\left(\dfrac{n+1}{n}\right)^{100} = \lim\limits_{n\to\infty}\dfrac{100}{n+1}\left(1+\dfrac{1}{n}\right)^{100}$

$= 0\cdot 1 = 0 < 1$

so the series $\sum\limits_{n=1}^{\infty}\dfrac{n^{100}100^n}{n!}$ is absolutely convergent by the Ratio Test.

21. $\lim\limits_{n\to\infty}\left|\dfrac{a_{n+1}}{a_n}\right| = \lim\limits_{n\to\infty}\left|\dfrac{(-1)^n(n+1)!}{1\cdot 3\cdot 5\cdot\,\cdots\,\cdot(2n-1)(2n+1)}\cdot\dfrac{1\cdot 3\cdot 5\cdot\,\cdots\,\cdot(2n-1)}{(-1)^{n-1}n!}\right| = \lim\limits_{n\to\infty}\dfrac{n+1}{2n+1}$

$= \lim\limits_{n\to\infty}\dfrac{1+1/n}{2+1/n} = \dfrac{1}{2} < 1,$

so the series $1 - \dfrac{2!}{1\cdot 3} + \dfrac{3!}{1\cdot 3\cdot 5} - \dfrac{4!}{1\cdot 3\cdot 5\cdot 7} + \cdots + (-1)^{n-1}\dfrac{n!}{1\cdot 3\cdot 5\cdot\,\cdots\,\cdot(2n-1)} + \cdots$ is absolutely convergent by

the Ratio Test.

23. $\lim\limits_{n\to\infty}\left|\dfrac{a_{n+1}}{a_n}\right| = \lim\limits_{n\to\infty}\left|\dfrac{2\cdot 4\cdot 6\cdot\,\cdots\,\cdot(2n)(2n+2)}{(n+1)!}\cdot\dfrac{n!}{2\cdot 4\cdot 6\cdot\,\cdots\,\cdot(2n)}\right| = \lim\limits_{n\to\infty}\dfrac{2n+2}{n+1} = \lim\limits_{n\to\infty}\dfrac{2(n+1)}{n+1} = 2 > 1$, so

the series $\sum\limits_{n=1}^{\infty}\dfrac{2\cdot 4\cdot 6\cdot\,\cdots\,\cdot(2n)}{n!}$ diverges by the Ratio Test.

25. $\lim\limits_{n\to\infty} \sqrt[n]{|a_n|} = \lim\limits_{n\to\infty} \dfrac{n^2+1}{2n^2+1} = \lim\limits_{n\to\infty} \dfrac{1+1/n^2}{2+1/n^2} = \dfrac{1}{2} < 1$, so the series $\sum\limits_{n=1}^{\infty} \left(\dfrac{n^2+1}{2n^2+1}\right)^n$ is absolutely convergent by the

Root Test.

27. $\lim\limits_{n\to\infty} \sqrt[n]{|a_n|} = \lim\limits_{n\to\infty} \sqrt[n]{\left|\dfrac{(-1)^{n-1}}{(\ln n)^n}\right|} = \lim\limits_{n\to\infty} \dfrac{1}{\ln n} = 0 < 1$, so the series $\sum\limits_{n=2}^{\infty} \dfrac{(-1)^{n-1}}{(\ln n)^n}$ is absolutely convergent by the Root

Test.

29. $\lim\limits_{n\to\infty} \sqrt[n]{|a_n|} = \lim\limits_{n\to\infty} \sqrt[n]{\left(1+\dfrac{1}{n}\right)^{n^2}} = \lim\limits_{n\to\infty} \left(1+\dfrac{1}{n}\right)^n = e > 1$ [by Equation 6.4.9 (or 6.4*.9) [ET 3.6.6]], so the series

$\sum\limits_{n=1}^{\infty} \left(1+\dfrac{1}{n}\right)^{n^2}$ diverges by the Root Test.

31. $\sum\limits_{n=2}^{\infty} \dfrac{(-1)^n}{\ln n}$ converges by the Alternating Series Test since $\lim\limits_{n\to\infty} \dfrac{1}{\ln n} = 0$ and $\left\{\dfrac{1}{\ln n}\right\}$ is decreasing. Now $\ln n < n$, so

$\dfrac{1}{\ln n} > \dfrac{1}{n}$, and since $\sum\limits_{n=2}^{\infty} \dfrac{1}{n}$ is the divergent (partial) harmonic series, $\sum\limits_{n=2}^{\infty} \dfrac{1}{\ln n}$ diverges by the Comparison Test. Thus,

$\sum\limits_{n=2}^{\infty} \dfrac{(-1)^n}{\ln n}$ is conditionally convergent.

33. $\lim\limits_{n\to\infty} \left|\dfrac{a_{n+1}}{a_n}\right| = \lim\limits_{n\to\infty} \left|\dfrac{(-9)^{n+1}}{(n+1)10^{n+2}} \cdot \dfrac{n10^{n+1}}{(-9)^n}\right| = \lim\limits_{n\to\infty} \left|\dfrac{(-9)n}{10(n+1)}\right| = \dfrac{9}{10} \lim\limits_{n\to\infty} \dfrac{1}{1+1/n} = \dfrac{9}{10}(1) = \dfrac{9}{10} < 1$, so the

series $\sum\limits_{n=1}^{\infty} \dfrac{(-9)^n}{n10^{n+1}}$ is absolutely convergent by the Ratio Test.

35. $\lim\limits_{n\to\infty} \sqrt[n]{|a_n|} = \lim\limits_{n\to\infty} \sqrt[n]{\left|\left(\dfrac{n}{\ln n}\right)^n\right|} = \lim\limits_{n\to\infty} \dfrac{n}{\ln n} = \lim\limits_{x\to\infty} \dfrac{x}{\ln x} \overset{\text{H}}{=} \lim\limits_{x\to\infty} \dfrac{1}{1/x} = \lim\limits_{x\to\infty} x = \infty$, so the series $\sum\limits_{n=2}^{\infty} \left(\dfrac{n}{\ln n}\right)^n$

diverges by the Root Test.

37. $\left|\dfrac{(-1)^n \arctan n}{n^2}\right| < \dfrac{\pi/2}{n^2}$, so since $\sum\limits_{n=1}^{\infty} \dfrac{\pi/2}{n^2} = \dfrac{\pi}{2} \sum\limits_{n=1}^{\infty} \dfrac{1}{n^2}$ converges $(p=2 > 1)$, the given series $\sum\limits_{n=1}^{\infty} \dfrac{(-1)^n \arctan n}{n^2}$

converges absolutely by the Comparison Test.

39. By the recursive definition, $\lim\limits_{n\to\infty} \left|\dfrac{a_{n+1}}{a_n}\right| = \lim\limits_{n\to\infty} \left|\dfrac{5n+1}{4n+3}\right| = \dfrac{5}{4} > 1$, so the series diverges by the Ratio Test.

41. The series $\sum\limits_{n=1}^{\infty} \dfrac{b_n^n \cos n\pi}{n} = \sum\limits_{n=1}^{\infty} (-1)^n \dfrac{b_n^n}{n}$, where $b_n > 0$ for $n \geq 1$ and $\lim\limits_{n\to\infty} b_n = \dfrac{1}{2}$.

$\lim\limits_{n\to\infty} \left|\dfrac{a_{n+1}}{a_n}\right| = \lim\limits_{n\to\infty} \left|\dfrac{(-1)^{n+1} b_n^{n+1}}{n+1} \cdot \dfrac{n}{(-1)^n b_n^n}\right| = \lim\limits_{n\to\infty} b_n \dfrac{n}{n+1} = \dfrac{1}{2}(1) = \dfrac{1}{2} < 1$, so the series $\sum\limits_{n=1}^{\infty} \dfrac{b_n^n \cos n\pi}{n}$ is

absolutely convergent by the Ratio Test.

43. (a) $\lim\limits_{n\to\infty} \left|\dfrac{1/(n+1)^3}{1/n^3}\right| = \lim\limits_{n\to\infty} \dfrac{n^3}{(n+1)^3} = \lim\limits_{n\to\infty} \dfrac{1}{(1+1/n)^3} = 1.$ Inconclusive

(b) $\lim\limits_{n\to\infty} \left|\dfrac{(n+1)}{2^{n+1}} \cdot \dfrac{2^n}{n}\right| = \lim\limits_{n\to\infty} \dfrac{n+1}{2n} = \lim\limits_{n\to\infty} \left(\dfrac{1}{2} + \dfrac{1}{2n}\right) = \dfrac{1}{2}.$ Conclusive (convergent)

(c) $\lim\limits_{n\to\infty} \left| \dfrac{(-3)^n}{\sqrt{n+1}} \cdot \dfrac{\sqrt{n}}{(-3)^{n-1}} \right| = 3 \lim\limits_{n\to\infty} \sqrt{\dfrac{n}{n+1}} = 3 \lim\limits_{n\to\infty} \sqrt{\dfrac{1}{1+1/n}} = 3.$ Conclusive (divergent)

(d) $\lim\limits_{n\to\infty} \left| \dfrac{\sqrt{n+1}}{1+(n+1)^2} \cdot \dfrac{1+n^2}{\sqrt{n}} \right| = \lim\limits_{n\to\infty} \left[\sqrt{1+\dfrac{1}{n}} \cdot \dfrac{1/n^2+1}{1/n^2+(1+1/n)^2} \right] = 1.$ Inconclusive

45. (a) $\lim\limits_{n\to\infty} \left| \dfrac{a_{n+1}}{a_n} \right| = \lim\limits_{n\to\infty} \left| \dfrac{x^{n+1}}{(n+1)!} \cdot \dfrac{n!}{x^n} \right| = \lim\limits_{n\to\infty} \left| \dfrac{x}{n+1} \right| = |x| \lim\limits_{n\to\infty} \dfrac{1}{n+1} = |x| \cdot 0 = 0 < 1,$ so by the Ratio Test the

series $\sum\limits_{n=0}^{\infty} \dfrac{x^n}{n!}$ converges for all x.

(b) Since the series of part (a) always converges, we must have $\lim\limits_{n\to\infty} \dfrac{x^n}{n!} = 0$ by Theorem 11.2.6.

47. (a) $s_5 = \sum\limits_{n=1}^{5} \dfrac{1}{n2^n} = \dfrac{1}{2} + \dfrac{1}{8} + \dfrac{1}{24} + \dfrac{1}{64} + \dfrac{1}{160} = \dfrac{661}{960} \approx 0.68854.$ Now the ratios

$r_n = \dfrac{a_{n+1}}{a_n} = \dfrac{n2^n}{(n+1)2^{n+1}} = \dfrac{n}{2(n+1)}$ form an increasing sequence, since

$r_{n+1} - r_n = \dfrac{n+1}{2(n+2)} - \dfrac{n}{2(n+1)} = \dfrac{(n+1)^2 - n(n+2)}{2(n+1)(n+2)} = \dfrac{1}{2(n+1)(n+2)} > 0.$ So by Exercise 46(b), the error

in using s_5 is $R_5 \le \dfrac{a_6}{1 - \lim\limits_{n\to\infty} r_n} = \dfrac{1/(6 \cdot 2^6)}{1 - 1/2} = \dfrac{1}{192} \approx 0.00521.$

(b) The error in using s_n as an approximation to the sum is $R_n = \dfrac{a_{n+1}}{1 - \frac{1}{2}} = \dfrac{2}{(n+1)2^{n+1}}.$ We want $R_n < 0.00005 \iff$

$\dfrac{1}{(n+1)2^n} < 0.00005 \iff (n+1)2^n > 20{,}000.$ To find such an n we can use trial and error or a graph. We calculate

$(11+1)2^{11} = 24{,}576,$ so $s_{11} = \sum\limits_{n=1}^{11} \dfrac{1}{n2^n} \approx 0.693109$ is within 0.00005 of the actual sum.

49. (i) Following the hint, we get that $|a_n| < r^n$ for $n \ge N$, and so since the geometric series $\sum_{n=1}^{\infty} r^n$ converges $[0 < r < 1]$,

the series $\sum_{n=N}^{\infty} |a_n|$ converges as well by the Comparison Test, and hence so does $\sum_{n=1}^{\infty} |a_n|$, so $\sum_{n=1}^{\infty} a_n$ is absolutely

convergent.

(ii) If $\lim\limits_{n\to\infty} \sqrt[n]{|a_n|} = L > 1$, then there is an integer N such that $\sqrt[n]{|a_n|} > 1$ for all $n \ge N$, so $|a_n| > 1$ for $n \ge N$. Thus,

$\lim\limits_{n\to\infty} a_n \ne 0$, so $\sum_{n=1}^{\infty} a_n$ diverges by the Test for Divergence.

(iii) Consider $\sum\limits_{n=1}^{\infty} \dfrac{1}{n}$ [diverges] and $\sum\limits_{n=1}^{\infty} \dfrac{1}{n^2}$ [converges]. For each sum, $\lim\limits_{n\to\infty} \sqrt[n]{|a_n|} = 1$, so the Root Test is inconclusive.

51. (a) Since $\sum a_n$ is absolutely convergent, and since $|a_n^+| \le |a_n|$ and $|a_n^-| \le |a_n|$ (because a_n^+ and a_n^- each equal

either a_n or 0), we conclude by the Comparison Test that both $\sum a_n^+$ and $\sum a_n^-$ must be absolutely convergent.

Or: Use Theorem 11.2.8.

(b) We will show by contradiction that both $\sum a_n^+$ and $\sum a_n^-$ must diverge. For suppose that $\sum a_n^+$ converged. Then so

would $\sum (a_n^+ - \frac{1}{2}a_n)$ by Theorem 11.2.8. But $\sum (a_n^+ - \frac{1}{2}a_n) = \sum \left[\frac{1}{2}(a_n + |a_n|) - \frac{1}{2}a_n \right] = \frac{1}{2} \sum |a_n|,$ which

diverges because $\sum a_n$ is only conditionally convergent. Hence, $\sum a_n^+$ can't converge. Similarly, neither can $\sum a_n^-$.

53. Suppose that $\sum a_n$ is conditionally convergent.

(a) $\sum n^2 a_n$ is divergent: Suppose $\sum n^2 a_n$ converges. Then $\lim_{n\to\infty} n^2 a_n = 0$ by Theorem 6 in Section 11.2, so there is an

integer $N > 0$ such that $n > N \Rightarrow n^2 |a_n| < 1$. For $n > N$, we have $|a_n| < \dfrac{1}{n^2}$, so $\sum_{n>N} |a_n|$ converges by

comparison with the convergent p-series $\sum_{n>N} \dfrac{1}{n^2}$. In other words, $\sum a_n$ converges absolutely, contradicting the

assumption that $\sum a_n$ is conditionally convergent. This contradiction shows that $\sum n^2 a_n$ diverges.

Remark: The same argument shows that $\sum n^p a_n$ diverges for any $p > 1$.

(b) $\sum_{n=2}^{\infty} \dfrac{(-1)^n}{n \ln n}$ is conditionally convergent. It converges by the Alternating Series Test, but does not converge absolutely

$\left[\text{by the Integral Test, since the function } f(x) = \dfrac{1}{x \ln x} \text{ is continuous, positive, and decreasing on } [2, \infty) \text{ and}\right.$

$\left.\displaystyle\int_2^{\infty} \dfrac{dx}{x \ln x} = \lim_{t\to\infty} \int_2^t \dfrac{dx}{x \ln x} = \lim_{t\to\infty} \Big[\ln(\ln x)\Big]_2^t = \infty\right].$ Setting $a_n = \dfrac{(-1)^n}{n \ln n}$ for $n \geq 2$, we find that

$\displaystyle\sum_{n=2}^{\infty} n a_n = \sum_{n=2}^{\infty} \dfrac{(-1)^n}{\ln n}$ converges by the Alternating Series Test.

It is easy to find conditionally convergent series $\sum a_n$ such that $\sum n a_n$ diverges. Two examples are $\displaystyle\sum_{n=1}^{\infty} \dfrac{(-1)^{n-1}}{n}$ and

$\displaystyle\sum_{n=1}^{\infty} \dfrac{(-1)^{n-1}}{\sqrt{n}}$, both of which converge by the Alternating Series Test and fail to converge absolutely because $\sum |a_n|$ is a

p-series with $p \leq 1$. In both cases, $\sum n a_n$ diverges by the Test for Divergence.

11.7 Strategy for Testing Series

1. Use the Limit Comparison Test with $a_n = \dfrac{n^2 - 1}{n^3 + 1}$ and $b_n = \dfrac{1}{n}$:

$\lim_{n\to\infty} \dfrac{a_n}{b_n} = \lim_{n\to\infty} \dfrac{(n^2 - 1)n}{n^3 + 1} = \lim_{n\to\infty} \dfrac{n^3 - n}{n^3 + 1} = \lim_{n\to\infty} \dfrac{1 - 1/n^2}{1 + 1/n^3} = 1 > 0.$ Since $\displaystyle\sum_{n=1}^{\infty} \dfrac{1}{n}$ is the divergent harmonic series, the

series $\displaystyle\sum_{n=1}^{\infty} \dfrac{n^2 - 1}{n^3 + 1}$ also diverges.

3. $\displaystyle\sum_{n=1}^{\infty} (-1)^n \dfrac{n^2 - 1}{n^3 + 1} = \sum_{n=1}^{\infty} (-1)^n b_n.$ Now $b_n = \dfrac{n^2 - 1}{n^3 + 1} > 0$ for $n \geq 2$, $\{b_n\}$ is decreasing for $n \geq 2$, and $\lim_{n\to\infty} b_n = 0$, so

the series $\displaystyle\sum_{n=1}^{\infty} (-1)^n \dfrac{n^2 - 1}{n^3 + 1}$ converges by the Alternating Series Test. By Exercise 1, $\displaystyle\sum_{n=1}^{\infty} \dfrac{n^2 - 1}{n^3 + 1}$ diverges, so the series

$\displaystyle\sum_{n=1}^{\infty} (-1)^n \dfrac{n^2 - 1}{n^3 + 1}$ is conditionally convergent.

5. $\lim_{x\to\infty} \dfrac{e^x}{x^2} \overset{\text{H}}{=} \lim_{x\to\infty} \dfrac{e^x}{2x} \overset{\text{H}}{=} \lim_{x\to\infty} \dfrac{e^x}{2} = \infty,$ so $\lim_{n\to\infty} \dfrac{e^n}{n^2} = \infty.$ Thus, the series $\displaystyle\sum_{n=1}^{\infty} \dfrac{e^n}{n^2}$ diverges by the Test for Divergence.

7. Let $f(x) = \dfrac{1}{x \sqrt{\ln x}}$. Then f is positive, continuous, and decreasing on $[2, \infty)$, so we can apply the Integral Test.

Since $\displaystyle\int \dfrac{1}{x\sqrt{\ln x}}\, dx \begin{bmatrix} u = \ln x, \\ du = dx/x \end{bmatrix} = \int u^{-1/2}\, du = 2u^{1/2} + C = 2\sqrt{\ln x} + C$, we find

$\displaystyle\int_2^\infty \dfrac{dx}{x\sqrt{\ln x}} = \lim_{t\to\infty} \int_2^t \dfrac{dx}{x\sqrt{\ln x}} = \lim_{t\to\infty} \left[2\sqrt{\ln x}\right]_2^t = \lim_{t\to\infty}\left(2\sqrt{\ln t} - 2\sqrt{\ln 2}\right) = \infty$. Since the integral diverges, the

given series $\displaystyle\sum_{n=2}^\infty \dfrac{1}{n\sqrt{\ln n}}$ diverges.

9. $\displaystyle\lim_{n\to\infty}\left|\dfrac{a_{n+1}}{a_n}\right| = \lim_{n\to\infty}\left|\dfrac{\pi^{2n+2}}{(2n+2)!} \cdot \dfrac{(2n)!}{\pi^{2n}}\right| = \lim_{n\to\infty}\dfrac{\pi^2}{(2n+2)(2n+1)} = 0 < 1$, so the series $\displaystyle\sum_{n=0}^\infty (-1)^n \dfrac{\pi^{2n}}{(2n)!}$ is absolutely

convergent (and therefore convergent) by the Ratio Test.

11. $\displaystyle\sum_{n=1}^\infty \left(\dfrac{1}{n^3} + \dfrac{1}{3^n}\right) = \sum_{n=1}^\infty \dfrac{1}{n^3} + \sum_{n=1}^\infty \left(\dfrac{1}{3}\right)^n$. The first series converges since it is a p-series with $p = 3 > 1$ and the second

series converges since it is geometric with $|r| = \frac{1}{3} < 1$. The sum of two convergent series is convergent.

13. $\displaystyle\lim_{n\to\infty}\left|\dfrac{a_{n+1}}{a_n}\right| = \lim_{n\to\infty}\left|\dfrac{3^{n+1}(n+1)^2}{(n+1)!} \cdot \dfrac{n!}{3^n n^2}\right| = \lim_{n\to\infty}\dfrac{3(n+1)^2}{(n+1)n^2} = 3\lim_{n\to\infty}\dfrac{n+1}{n^2} = 0 < 1$, so the series $\displaystyle\sum_{n=1}^\infty \dfrac{3^n n^2}{n!}$

converges by the Ratio Test.

15. $a_k = \dfrac{2^{k-1}3^{k+1}}{k^k} = \dfrac{2^k 2^{-1} 3^k 3^1}{k^k} = \dfrac{3}{2}\left(\dfrac{2\cdot 3}{k}\right)^k$. By the Root Test, $\displaystyle\lim_{k\to\infty}\sqrt[k]{\left(\dfrac{6}{k}\right)^k} = \lim_{k\to\infty}\dfrac{6}{k} = 0 < 1$, so the series

$\displaystyle\sum_{k=1}^\infty \left(\dfrac{6}{k}\right)^k$ converges. It follows from Theorem 8(i) in Section 11.2 that the given series, $\displaystyle\sum_{k=1}^\infty \dfrac{2^{k-1}3^{k+1}}{k^k} = \sum_{k=1}^\infty \dfrac{3}{2}\left(\dfrac{6}{k}\right)^k$,

also converges.

17. $\displaystyle\lim_{n\to\infty}\left|\dfrac{a_{n+1}}{a_n}\right| = \lim_{n\to\infty}\left|\dfrac{1\cdot 3\cdot 5\cdot\cdots\cdot(2n-1)(2n+1)}{2\cdot 5\cdot 8\cdot\cdots\cdot(3n-1)(3n+2)} \cdot \dfrac{2\cdot 5\cdot 8\cdot\cdots\cdot(3n-1)}{1\cdot 3\cdot 5\cdot\cdots\cdot(2n-1)}\right| = \lim_{n\to\infty}\dfrac{2n+1}{3n+2}$

$= \displaystyle\lim_{n\to\infty}\dfrac{2 + 1/n}{3 + 2/n} = \dfrac{2}{3} < 1$,

so the series $\displaystyle\sum_{n=1}^\infty \dfrac{1\cdot 3\cdot 5\cdot\cdots\cdot(2n-1)}{2\cdot 5\cdot 8\cdot\cdots\cdot(3n-1)}$ converges by the Ratio Test.

19. Let $f(x) = \dfrac{\ln x}{\sqrt{x}}$. Then $f'(x) = \dfrac{2 - \ln x}{2x^{3/2}} < 0$ when $\ln x > 2$ or $x > e^2$, so $\dfrac{\ln n}{\sqrt{n}}$ is decreasing for $n > e^2$.

By l'Hospital's Rule, $\displaystyle\lim_{n\to\infty}\dfrac{\ln n}{\sqrt{n}} = \lim_{n\to\infty}\dfrac{1/n}{1/\left(2\sqrt{n}\right)} = \lim_{n\to\infty}\dfrac{2}{\sqrt{n}} = 0$, so the series $\displaystyle\sum_{n=1}^\infty (-1)^n \dfrac{\ln n}{\sqrt{n}}$ converges by the

Alternating Series Test.

21. $\displaystyle\lim_{n\to\infty}|a_n| = \lim_{n\to\infty}\left|(-1)^n \cos(1/n^2)\right| = \lim_{n\to\infty}\left|\cos(1/n^2)\right| = \cos 0 = 1$, so the series $\displaystyle\sum_{n=1}^\infty (-1)^n \cos(1/n^2)$ diverges by the

Test for Divergence.

23. Using the Limit Comparison Test with $a_n = \tan\left(\dfrac{1}{n}\right)$ and $b_n = \dfrac{1}{n}$, we have

$$\lim_{n \to \infty} \frac{a_n}{b_n} = \lim_{n \to \infty} \frac{\tan(1/n)}{1/n} = \lim_{x \to \infty} \frac{\tan(1/x)}{1/x} \stackrel{\text{H}}{=} \lim_{x \to \infty} \frac{\sec^2(1/x) \cdot (-1/x^2)}{-1/x^2} = \lim_{x \to \infty} \sec^2(1/x) = 1^2 = 1 > 0. \text{ Since}$$

$\displaystyle\sum_{n=1}^{\infty} b_n$ is the divergent harmonic series, $\displaystyle\sum_{n=1}^{\infty} a_n$ is also divergent.

25. Use the Ratio Test. $\displaystyle\lim_{n \to \infty} \left| \frac{a_{n+1}}{a_n} \right| = \lim_{n \to \infty} \left| \frac{(n+1)!}{e^{(n+1)^2}} \cdot \frac{e^{n^2}}{n!} \right| = \lim_{n \to \infty} \frac{(n+1)n! \cdot e^{n^2}}{e^{n^2+2n+1}n!} = \lim_{n \to \infty} \frac{n+1}{e^{2n+1}} = 0 < 1$, so $\displaystyle\sum_{n=1}^{\infty} \frac{n!}{e^{n^2}}$

converges.

27. $\displaystyle\int_{2}^{\infty} \frac{\ln x}{x^2}\,dx = \lim_{t \to \infty} \left[-\frac{\ln x}{x} - \frac{1}{x} \right]_{1}^{t}$ [using integration by parts] $\stackrel{\text{H}}{=} 1$. So $\displaystyle\sum_{n=1}^{\infty} \frac{\ln n}{n^2}$ converges by the Integral Test, and since

$\dfrac{k \ln k}{(k+1)^3} < \dfrac{k \ln k}{k^3} = \dfrac{\ln k}{k^2}$, the given series $\displaystyle\sum_{k=1}^{\infty} \frac{k \ln k}{(k+1)^3}$ converges by the Comparison Test.

29. $\displaystyle\sum_{n=1}^{\infty} a_n = \sum_{n=1}^{\infty} (-1)^n \frac{1}{\cosh n} = \sum_{n=1}^{\infty} (-1)^n b_n$. Now $b_n = \dfrac{1}{\cosh n} > 0$, $\{b_n\}$ is decreasing, and $\displaystyle\lim_{n \to \infty} b_n = 0$, so the series

converges by the Alternating Series Test.

Or: Write $\dfrac{1}{\cosh n} = \dfrac{2}{e^n + e^{-n}} < \dfrac{2}{e^n}$ and $\displaystyle\sum_{n=1}^{\infty} \frac{1}{e^n}$ is a convergent geometric series, so $\displaystyle\sum_{n=1}^{\infty} \frac{1}{\cosh n}$ is convergent by the

Comparison Test. So $\displaystyle\sum_{n=1}^{\infty} (-1)^n \frac{1}{\cosh n}$ is absolutely convergent and therefore convergent.

31. $\displaystyle\lim_{k \to \infty} a_k = \lim_{k \to \infty} \frac{5^k}{3^k + 4^k} = $ [divide by 4^k] $\displaystyle\lim_{k \to \infty} \frac{(5/4)^k}{(3/4)^k + 1} = \infty$ since $\displaystyle\lim_{k \to \infty} \left(\frac{3}{4}\right)^k = 0$ and $\displaystyle\lim_{k \to \infty} \left(\frac{5}{4}\right)^k = \infty$.

Thus, $\displaystyle\sum_{k=1}^{\infty} \frac{5^k}{3^k + 4^k}$ diverges by the Test for Divergence.

33. $\displaystyle\lim_{n \to \infty} \sqrt[n]{|a_n|} = \lim_{n \to \infty} \left(\frac{n}{n+1}\right)^{n^2/n} = \lim_{n \to \infty} \frac{1}{[(n+1)/n]^n} = \frac{1}{\displaystyle\lim_{n \to \infty} (1 + 1/n)^n} = \frac{1}{e} < 1$, so the series $\displaystyle\sum_{n=1}^{\infty} \left(\frac{n}{n+1}\right)^{n^2}$

converges by the Root Test.

35. $a_n = \dfrac{1}{n^{1+1/n}} = \dfrac{1}{n \cdot n^{1/n}}$, so let $b_n = \dfrac{1}{n}$ and use the Limit Comparison Test. $\displaystyle\lim_{n \to \infty} \frac{a_n}{b_n} = \lim_{n \to \infty} \frac{1}{n^{1/n}} = 1 > 0$

(see Exercise 6.8.63 [ET 4.4.63]), so the series $\displaystyle\sum_{n=1}^{\infty} \frac{1}{n^{1+1/n}}$ diverges by comparison with the divergent harmonic series.

37. $\displaystyle\lim_{n \to \infty} \sqrt[n]{|a_n|} = \lim_{n \to \infty} (2^{1/n} - 1) = 1 - 1 = 0 < 1$, so the series $\displaystyle\sum_{n=1}^{\infty} \left(\sqrt[n]{2} - 1\right)^n$ converges by the Root Test.

11.8 Power Series

1. A power series is a series of the form $\sum_{n=0}^{\infty} c_n x^n = c_0 + c_1 x + c_2 x^2 + c_3 x^3 + \cdots$, where x is a variable and the c_n's are

constants called the coefficients of the series.

More generally, a series of the form $\sum_{n=0}^{\infty} c_n (x-a)^n = c_0 + c_1(x-a) + c_2(x-a)^2 + \cdots$ is called a power series in

$(x-a)$ or a power series centered at a or a power series about a, where a is a constant.

3. If $a_n = (-1)^n n x^n$, then

$$\lim_{n \to \infty} \left| \frac{a_{n+1}}{a_n} \right| = \lim_{n \to \infty} \left| \frac{(-1)^{n+1}(n+1)x^{n+1}}{(-1)^n n x^n} \right| = \lim_{n \to \infty} \left| (-1)\frac{n+1}{n} x \right| = \lim_{n \to \infty} \left[\left(1 + \frac{1}{n} \right) |x| \right] = |x|. \text{ By the Ratio Test, the}$$

series $\displaystyle\sum_{n=1}^{\infty} (-1)^n n x^n$ converges when $|x| < 1$, so the radius of convergence $R = 1$. Now we'll check the endpoints, that is,

$x = \pm 1$. Both series $\displaystyle\sum_{n=1}^{\infty} (-1)^n n (\pm 1)^n = \sum_{n=1}^{\infty} (\mp 1)^n n$ diverge by the Test for Divergence since $\displaystyle\lim_{n \to \infty} |(\mp 1)^n n| = \infty$. Thus,

the interval of convergence is $I = (-1, 1)$.

5. If $a_n = \dfrac{x^n}{2n-1}$, then $\displaystyle\lim_{n \to \infty} \left| \frac{a_{n+1}}{a_n} \right| = \lim_{n \to \infty} \left| \frac{x^{n+1}}{2n+1} \cdot \frac{2n-1}{x^n} \right| = \lim_{n \to \infty} \left(\frac{2n-1}{2n+1} |x| \right) = \lim_{n \to \infty} \left(\frac{2-1/n}{2+1/n} |x| \right) = |x|.$ By

the Ratio Test, the series $\displaystyle\sum_{n=1}^{\infty} \frac{x^n}{2n-1}$ converges when $|x| < 1$, so $R = 1$. When $x = 1$, the series $\displaystyle\sum_{n=1}^{\infty} \frac{1}{2n-1}$ diverges by

comparison with $\displaystyle\sum_{n=1}^{\infty} \frac{1}{2n}$ since $\dfrac{1}{2n-1} > \dfrac{1}{2n}$ and $\dfrac{1}{2} \displaystyle\sum_{n=1}^{\infty} \frac{1}{n}$ diverges since it is a constant multiple of the harmonic series.

When $x = -1$, the series $\displaystyle\sum_{n=1}^{\infty} \frac{(-1)^n}{2n-1}$ converges by the Alternating Series Test. Thus, the interval of convergence is $[-1, 1)$.

7. If $a_n = \dfrac{x^n}{n!}$, then $\displaystyle\lim_{n \to \infty} \left| \frac{a_{n+1}}{a_n} \right| = \lim_{n \to \infty} \left| \frac{x^{n+1}}{(n+1)!} \cdot \frac{n!}{x^n} \right| = \lim_{n \to \infty} \left| \frac{x}{n+1} \right| = |x| \lim_{n \to \infty} \frac{1}{n+1} = |x| \cdot 0 = 0 < 1$ for *all* real x.

So, by the Ratio Test, $R = \infty$ and $I = (-\infty, \infty)$.

9. If $a_n = \dfrac{x^n}{n^4 \, 4^n}$, then

$$\lim_{n \to \infty} \left| \frac{a_{n+1}}{a_n} \right| = \lim_{n \to \infty} \left| \frac{x^{n+1}}{(n+1)^4 \, 4^{n+1}} \cdot \frac{n^4 \, 4^n}{x^n} \right| = \lim_{n \to \infty} \left| \frac{n^4}{(n+1)^4} \cdot \frac{x}{4} \right| = \lim_{n \to \infty} \left(\frac{n}{n+1} \right)^4 \frac{|x|}{4} = 1^4 \cdot \frac{|x|}{4} = \frac{|x|}{4}. \text{ By the}$$

Ratio Test, the series $\displaystyle\sum_{n=1}^{\infty} \frac{x^n}{n^4 \, 4^n}$ converges when $\dfrac{|x|}{4} < 1 \iff |x| < 4$, so $R = 4$. When $x = 4$, the series $\displaystyle\sum_{n=1}^{\infty} \frac{1}{n^4}$

converges since it is a p-series ($p = 4 > 1$). When $x = -4$, the series $\displaystyle\sum_{n=1}^{\infty} \frac{(-1)^n}{n^4}$ converges by the Alternating Series Test.

Thus, the interval of convergence is $[-4, 4]$.

11. If $a_n = \dfrac{(-1)^n \, 4^n}{\sqrt{n}} x^n$, then $\displaystyle\lim_{n \to \infty} \left| \frac{a_{n+1}}{a_n} \right| = \lim_{n \to \infty} \left| \frac{(-1)^{n+1} \, 4^{n+1} \, x^{n+1}}{\sqrt{n+1}} \cdot \frac{\sqrt{n}}{(-1)^n \, 4^n \, x^n} \right| = \lim_{n \to \infty} \sqrt{\frac{n}{n+1}} \cdot 4 |x| = 4 |x|.$

By the Ratio Test, the series $\displaystyle\sum_{n=1}^{\infty} \frac{(-1)^n \, 4^n}{\sqrt{n}} x^n$ converges when $4 |x| < 1 \iff |x| < \frac{1}{4}$, so $R = \frac{1}{4}$. When $x = \frac{1}{4}$, the series

$\displaystyle\sum_{n=1}^{\infty} \frac{(-1)^n}{\sqrt{n}}$ converges by the Alternating Series Test. When $x = -\frac{1}{4}$, the series $\displaystyle\sum_{n=1}^{\infty} \frac{1}{\sqrt{n}}$ diverges since it is a p-series

($p = \frac{1}{2} \le 1$). Thus, the interval of convergence is $\left(-\frac{1}{4}, \frac{1}{4} \right]$.

13. If $a_n = \dfrac{n}{2^n(n^2+1)}\, x^n$, then

$$\lim_{n\to\infty}\left|\frac{a_{n+1}}{a_n}\right| = \lim_{n\to\infty}\left|\frac{(n+1)x^{n+1}}{2^{n+1}(n^2+2n+2)}\cdot\frac{2^n(n^2+1)}{n\,x^n}\right| = \lim_{n\to\infty}\frac{n^3+n^2+n+1}{n^3+2n^2+2n}\cdot\frac{|x|}{2}$$

$$= \lim_{n\to\infty}\frac{1+1/n+1/n^2+1/n^3}{1+2/n+2/n^2}\cdot\frac{|x|}{2} = \frac{|x|}{2}$$

By the Ratio Test, the series $\displaystyle\sum_{n=1}^{\infty}\frac{n}{2^n(n^2+1)}\,x^n$ converges when $\dfrac{|x|}{2}<1 \;\Leftrightarrow\; |x|<2$, so $R=2$. When $x=2$, the series

$\displaystyle\sum_{n=1}^{\infty}\frac{n}{n^2+1}$ diverges by the Limit Comparison Test with $b_n=\dfrac{1}{n}$. When $x=-2$, the series $\displaystyle\sum_{n=1}^{\infty}\frac{(-1)^n n}{n^2+1}$ converges by the

Alternating Series Test. Thus, the interval of convergence is $[-2,2)$.

15. If $a_n = \dfrac{(x-2)^n}{n^2+1}$, then $\displaystyle\lim_{n\to\infty}\left|\frac{a_{n+1}}{a_n}\right| = \lim_{n\to\infty}\left|\frac{(x-2)^{n+1}}{(n+1)^2+1}\cdot\frac{n^2+1}{(x-2)^n}\right| = |x-2|\lim_{n\to\infty}\frac{n^2+1}{(n+1)^2+1} = |x-2|$. By the

Ratio Test, the series $\displaystyle\sum_{n=0}^{\infty}\frac{(x-2)^n}{n^2+1}$ converges when $|x-2|<1$ $[R=1]$ $\Leftrightarrow$ $-1<x-2<1$ $\Leftrightarrow$ $1<x<3$. When

$x=1$, the series $\displaystyle\sum_{n=0}^{\infty}(-1)^n\frac{1}{n^2+1}$ converges by the Alternating Series Test; when $x=3$, the series $\displaystyle\sum_{n=0}^{\infty}\frac{1}{n^2+1}$ converges by

comparison with the p-series $\displaystyle\sum_{n=1}^{\infty}\frac{1}{n^2}$ $[p=2>1]$. Thus, the interval of convergence is $I=[1,3]$.

17. If $a_n = \dfrac{(x+2)^n}{2^n \ln n}$, then $\displaystyle\lim_{n\to\infty}\left|\frac{(x+2)^{n+1}}{2^{n+1}\ln(n+1)}\cdot\frac{2^n \ln n}{(x+2)^n}\right| = \lim_{n\to\infty}\frac{\ln n}{\ln(n+1)}\cdot\frac{|x+2|}{2} = \frac{|x+2|}{2}$ since

$$\lim_{n\to\infty}\frac{\ln n}{\ln(n+1)} = \lim_{x\to\infty}\frac{\ln x}{\ln(x+1)} \overset{\text{H}}{=} \lim_{x\to\infty}\frac{1/x}{1/(x+1)} = \lim_{x\to\infty}\frac{x+1}{x} = \lim_{x\to\infty}\left(1+\frac{1}{x}\right) = 1.$$ By the Ratio Test, the series

$\displaystyle\sum_{n=2}^{\infty}\frac{(x+2)^n}{2^n \ln n}$ converges when $\dfrac{|x+2|}{2}<1 \;\Leftrightarrow\; |x+2|<2$ $[R=2]$ $\Leftrightarrow$ $-2<x+2<2$ $\Leftrightarrow$ $-4<x<0$.

When $x=-4$, the series $\displaystyle\sum_{n=2}^{\infty}\frac{(-1)^n}{\ln n}$ converges by the Alternating Series Test. When $x=0$, the series $\displaystyle\sum_{n=2}^{\infty}\frac{1}{\ln n}$ diverges by

the Limit Comparison Test with $b_n=\dfrac{1}{n}$ (or by comparison with the harmonic series). Thus, the interval of convergence is

$[-4,0)$.

19. If $a_n = \dfrac{(x-2)^n}{n^n}$, then $\displaystyle\lim_{n\to\infty}\sqrt[n]{|a_n|} = \lim_{n\to\infty}\frac{|x-2|}{n} = 0$, so the series converges for all x (by the Root Test).

$R=\infty$ and $I=(-\infty,\infty)$.

21. $a_n = \dfrac{n}{b^n}(x-a)^n$, where $b>0$.

$$\lim_{n\to\infty}\left|\frac{a_{n+1}}{a_n}\right| = \lim_{n\to\infty}\frac{(n+1)|x-a|^{n+1}}{b^{n+1}}\cdot\frac{b^n}{n|x-a|^n} = \lim_{n\to\infty}\left(1+\frac{1}{n}\right)\frac{|x-a|}{b} = \frac{|x-a|}{b}.$$

By the Ratio Test, the series converges when $\dfrac{|x-a|}{b}<1 \;\Leftrightarrow\; |x-a|<b$ $[\text{so } R=b]$ $\Leftrightarrow$ $-b<x-a<b$ $\Leftrightarrow$

$a-b<x<a+b$. When $|x-a|=b$, $\displaystyle\lim_{n\to\infty}|a_n| = \lim_{n\to\infty}n = \infty$, so the series diverges. Thus, $I=(a-b,a+b)$.

23. If $a_n = n! \, (2x - 1)^n$, then $\displaystyle\lim_{n\to\infty} \left| \dfrac{a_{n+1}}{a_n} \right| = \lim_{n\to\infty} \left| \dfrac{(n+1)! \, (2x-1)^{n+1}}{n!(2x-1)^n} \right| = \lim_{n\to\infty} (n+1) \, |2x-1| \to \infty$ as $n \to \infty$

for all $x \neq \frac{1}{2}$. Since the series diverges for all $x \neq \frac{1}{2}$, $R = 0$ and $I = \left\{ \frac{1}{2} \right\}$.

25. If $a_n = \dfrac{(5x-4)^n}{n^3}$, then

$$\lim_{n\to\infty} \left| \frac{a_{n+1}}{a_n} \right| = \lim_{n\to\infty} \left| \frac{(5x-4)^{n+1}}{(n+1)^3} \cdot \frac{n^3}{(5x-4)^n} \right| = \lim_{n\to\infty} |5x-4| \left(\frac{n}{n+1} \right)^3 = \lim_{n\to\infty} |5x-4| \left(\frac{1}{1+1/n} \right)^3$$

$$= |5x-4| \cdot 1 = |5x-4|$$

By the Ratio Test, $\displaystyle\sum_{n=1}^{\infty} \dfrac{(5x-4)^n}{n^3}$ converges when $|5x-4| < 1 \iff |x - \frac{4}{5}| < \frac{1}{5} \iff -\frac{1}{5} < x - \frac{4}{5} < \frac{1}{5} \iff$

$\frac{3}{5} < x < 1$, so $R = \frac{1}{5}$. When $x = 1$, the series $\displaystyle\sum_{n=1}^{\infty} \dfrac{1}{n^3}$ is a convergent p-series ($p = 3 > 1$). When $x = \frac{3}{5}$, the series

$\displaystyle\sum_{n=1}^{\infty} \dfrac{(-1)^n}{n^3}$ converges by the Alternating Series Test. Thus, the interval of convergence is $I = \left[\frac{3}{5}, 1 \right]$.

27. If $a_n = \dfrac{x^n}{1 \cdot 3 \cdot 5 \cdot \, \cdots \, \cdot (2n-1)}$, then

$$\lim_{n\to\infty} \left| \frac{a_{n+1}}{a_n} \right| = \lim_{n\to\infty} \left| \frac{x^{n+1}}{1 \cdot 3 \cdot 5 \cdot \, \cdots \, \cdot (2n-1)(2n+1)} \cdot \frac{1 \cdot 3 \cdot 5 \cdot \, \cdots \, \cdot (2n-1)}{x^n} \right| = \lim_{n\to\infty} \frac{|x|}{2n+1} = 0 < 1. \text{ Thus, by}$$

the Ratio Test, the series $\displaystyle\sum_{n=1}^{\infty} \dfrac{x^n}{1 \cdot 3 \cdot 5 \cdot \, \cdots \, \cdot (2n-1)}$ converges for *all* real x and we have $R = \infty$ and $I = (-\infty, \infty)$.

29. (a) We are given that the power series $\sum_{n=0}^{\infty} c_n x^n$ is convergent for $x = 4$. So by Theorem 4, it must converge for at least

$-4 < x \leq 4$. In particular, it converges when $x = -2$; that is, $\sum_{n=0}^{\infty} c_n (-2)^n$ is convergent.

(b) It does not follow that $\sum_{n=0}^{\infty} c_n (-4)^n$ is necessarily convergent. [See the comments after Theorem 4 about convergence at

the endpoint of an interval. An example is $c_n = (-1)^n / (n4^n)$.]

31. If $a_n = \dfrac{(n!)^k}{(kn)!} x^n$, then

$$\lim_{n\to\infty} \left| \frac{a_{n+1}}{a_n} \right| = \lim_{n\to\infty} \frac{[(n+1)!]^k \, (kn)!}{(n!)^k \, [k(n+1)]!} |x| = \lim_{n\to\infty} \frac{(n+1)^k}{(kn+k)(kn+k-1) \cdots (kn+2)(kn+1)} |x|$$

$$= \lim_{n\to\infty} \left[\frac{(n+1)}{(kn+1)} \frac{(n+1)}{(kn+2)} \cdots \frac{(n+1)}{(kn+k)} \right] |x|$$

$$= \lim_{n\to\infty} \left[\frac{n+1}{kn+1} \right] \lim_{n\to\infty} \left[\frac{n+1}{kn+2} \right] \cdots \lim_{n\to\infty} \left[\frac{n+1}{kn+k} \right] |x|$$

$$= \left(\frac{1}{k} \right)^k |x| < 1 \iff |x| < k^k \text{ for convergence, and the radius of convergence is } R = k^k.$$

33. No. If a power series is centered at a, its interval of convergence is symmetric about a. If a power series has an infinite radius

of convergence, then its interval of convergence must be $(-\infty, \infty)$, not $[0, \infty)$.

35. (a) If $a_n = \dfrac{(-1)^n \, x^{2n+1}}{n!(n+1)! \, 2^{2n+1}}$, then

$$\lim_{n\to\infty} \left| \frac{a_{n+1}}{a_n} \right| = \lim_{n\to\infty} \left| \frac{x^{2n+3}}{(n+1)!(n+2)! \, 2^{2n+3}} \cdot \frac{n!(n+1)! \, 2^{2n+1}}{x^{2n+1}} \right| = \left(\frac{x}{2} \right)^2 \lim_{n\to\infty} \frac{1}{(n+1)(n+2)} = 0 \text{ for all } x.$$

So $J_1(x)$ converges for all x and its domain is $(-\infty, \infty)$.

(b), (c) The initial terms of $J_1(x)$ up to $n = 5$ are $a_0 = \dfrac{x}{2}$,

$a_1 = -\dfrac{x^3}{16}, a_2 = \dfrac{x^5}{384}, a_3 = -\dfrac{x^7}{18{,}432}, a_4 = \dfrac{x^9}{1{,}474{,}560}$,

and $a_5 = -\dfrac{x^{11}}{176{,}947{,}200}$. The partial sums seem to

approximate $J_1(x)$ well near the origin, but as $|x|$ increases,

we need to take a large number of terms to get a good

approximation.

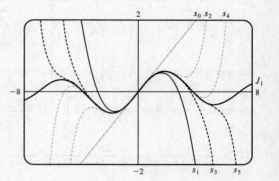

37. $s_{2n-1} = 1 + 2x + x^2 + 2x^3 + x^4 + 2x^5 + \cdots + x^{2n-2} + 2x^{2n-1}$

$\qquad = 1(1+2x) + x^2(1+2x) + x^4(1+2x) + \cdots + x^{2n-2}(1+2x) = (1+2x)(1+x^2+x^4+\cdots+x^{2n-2})$

$\qquad = (1+2x)\dfrac{1-x^{2n}}{1-x^2}$ [by (11.2.3) with $r = x^2$] $\;\to\; \dfrac{1+2x}{1-x^2}$ as $n \to \infty$ by (11.2.4), when $|x| < 1$.

Also $s_{2n} = s_{2n-1} + x^{2n} \to \dfrac{1+2x}{1-x^2}$ since $x^{2n} \to 0$ for $|x| < 1$. Therefore, $s_n \to \dfrac{1+2x}{1-x^2}$ since s_{2n} and s_{2n-1} both

approach $\dfrac{1+2x}{1-x^2}$ as $n \to \infty$. Thus, the interval of convergence is $(-1, 1)$ and $f(x) = \dfrac{1+2x}{1-x^2}$.

39. We use the Root Test on the series $\sum c_n x^n$. We need $\displaystyle\lim_{n\to\infty} \sqrt[n]{|c_n x^n|} = |x| \lim_{n\to\infty} \sqrt[n]{|c_n|} = c|x| < 1$ for convergence, or

$|x| < 1/c$, so $R = 1/c$.

41. For $2 < x < 3$, $\sum c_n x^n$ diverges and $\sum d_n x^n$ converges. By Exercise 11.2.85, $\sum (c_n + d_n) x^n$ diverges. Since both series

converge for $|x| < 2$, the radius of convergence of $\sum (c_n + d_n) x^n$ is 2.

11.9   Representations of Functions as Power Series

1. If $f(x) = \displaystyle\sum_{n=0}^{\infty} c_n x^n$ has radius of convergence 10, then $f'(x) = \displaystyle\sum_{n=1}^{\infty} n c_n x^{n-1}$ also has radius of convergence 10 by

Theorem 2.

3. Our goal is to write the function in the form $\dfrac{1}{1-r}$, and then use Equation 1 to represent the function as a sum of a power

series. $f(x) = \dfrac{1}{1+x} = \dfrac{1}{1-(-x)} = \displaystyle\sum_{n=0}^{\infty} (-x)^n = \displaystyle\sum_{n=0}^{\infty} (-1)^n x^n$ with $|-x| < 1 \;\Leftrightarrow\; |x| < 1$, so $R = 1$ and $I = (-1, 1)$.

5. $f(x) = \dfrac{2}{3-x} = \dfrac{2}{3}\left(\dfrac{1}{1-x/3} \right) = \dfrac{2}{3} \displaystyle\sum_{n=0}^{\infty} \left(\dfrac{x}{3} \right)^n$ or, equivalently, $2 \displaystyle\sum_{n=0}^{\infty} \dfrac{1}{3^{n+1}} x^n$. The series converges when $\left| \dfrac{x}{3} \right| < 1$,

that is, when $|x| < 3$, so $R = 3$ and $I = (-3, 3)$.

7. $f(x) = \dfrac{x^2}{x^4 + 16} = \dfrac{x^2}{16}\left(\dfrac{1}{1 + x^4/16}\right) = \dfrac{x^2}{16}\left(\dfrac{1}{1 - [-(x/2)^4]}\right) = \dfrac{x^2}{16}\sum_{n=0}^{\infty}\left[-\left(\dfrac{x}{2}\right)^4\right]^n$ or, equivalently, $\sum_{n=0}^{\infty}\dfrac{(-1)^n\,x^{4n+2}}{2^{4n+4}}$.

The series converges when $\left|-\left(\dfrac{x}{2}\right)^4\right| < 1 \;\Rightarrow\; \left|\dfrac{x}{2}\right| < 1 \;\Rightarrow\; |x| < 2$, so $R = 2$ and $I = (-2, 2)$.

9. $f(x) = \dfrac{x-1}{x+2} = \dfrac{x+2-3}{x+2} = 1 - \dfrac{3}{x+2} = 1 - \dfrac{3/2}{x/2+1} = 1 - \dfrac{3}{2}\cdot\dfrac{1}{1-(-x/2)}$

$= 1 - \dfrac{3}{2}\sum_{n=0}^{\infty}\left(-\dfrac{x}{2}\right)^n = 1 - \dfrac{3}{2} - \dfrac{3}{2}\sum_{n=1}^{\infty}\left(-\dfrac{x}{2}\right)^n = -\dfrac{1}{2} - \sum_{n=1}^{\infty}\dfrac{(-1)^n\,3x^n}{2^{n+1}}$.

The geometric series $\sum_{n=0}^{\infty}\left(-\dfrac{x}{2}\right)^n$ converges when $\left|-\dfrac{x}{2}\right| < 1 \;\Leftrightarrow\; |x| < 2$, so $R = 2$ and $I = (-2, 2)$.

Alternatively, you could write $f(x) = 1 - 3\left(\dfrac{1}{x+2}\right)$ and use the series for $\dfrac{1}{x+2}$ found in Example 2.

11. $f(x) = \dfrac{2x-4}{x^2-4x+3} = \dfrac{2x-4}{(x-1)(x-3)} = \dfrac{A}{x-1} + \dfrac{B}{x-3} \;\Rightarrow\; 2x-4 = A(x-3) + B(x-1)$. Let $x = 1$ to get

$-2 = -2A \;\Leftrightarrow\; A = 1$ and $x = 3$ to get $2 = 2B \;\Leftrightarrow\; B = 1$. Thus,

$\dfrac{2x-4}{x^2-4x+3} = \dfrac{1}{x-1} + \dfrac{1}{x-3} = \dfrac{-1}{1-x} + \dfrac{1}{-3}\left[\dfrac{1}{1-(x/3)}\right] = -\sum_{n=0}^{\infty}x^n - \dfrac{1}{3}\sum_{n=0}^{\infty}\left(\dfrac{x}{3}\right)^n = \sum_{n=0}^{\infty}\left(-1 - \dfrac{1}{3^{n+1}}\right)x^n$.

We represented f as the sum of two geometric series; the first converges for $x \in (-1, 1)$ and the second converges for

$x \in (-3, 3)$. Thus, the sum converges for $x \in (-1, 1) = I$.

13. (a) $f(x) = \dfrac{1}{(1+x)^2} = \dfrac{d}{dx}\left(\dfrac{-1}{1+x}\right) = -\dfrac{d}{dx}\left[\sum_{n=0}^{\infty}(-1)^n x^n\right]$ [from Exercise 3]

$= \sum_{n=1}^{\infty}(-1)^{n+1}n x^{n-1}$ [from Theorem 2(i)] $= \sum_{n=0}^{\infty}(-1)^n(n+1)x^n$ with $R = 1$.

In the last step, note that we *decreased* the initial value of the summation variable n by 1, and then *increased* each

occurrence of n in the term by 1 [also note that $(-1)^{n+2} = (-1)^n$].

(b) $f(x) = \dfrac{1}{(1+x)^3} = -\dfrac{1}{2}\dfrac{d}{dx}\left[\dfrac{1}{(1+x)^2}\right] = -\dfrac{1}{2}\dfrac{d}{dx}\left[\sum_{n=0}^{\infty}(-1)^n(n+1)x^n\right]$ [from part (a)]

$= -\dfrac{1}{2}\sum_{n=1}^{\infty}(-1)^n(n+1)n x^{n-1} = \dfrac{1}{2}\sum_{n=0}^{\infty}(-1)^n(n+2)(n+1)x^n$ with $R = 1$.

(c) $f(x) = \dfrac{x^2}{(1+x)^3} = x^2 \cdot \dfrac{1}{(1+x)^3} = x^2 \cdot \dfrac{1}{2}\sum_{n=0}^{\infty}(-1)^n(n+2)(n+1)x^n$ [from part (b)]

$= \dfrac{1}{2}\sum_{n=0}^{\infty}(-1)^n(n+2)(n+1)x^{n+2}$

To write the power series with x^n rather than x^{n+2}, we will *decrease* each occurrence of n in the term by 2 and *increase*

the initial value of the summation variable by 2. This gives us $\dfrac{1}{2}\sum_{n=2}^{\infty}(-1)^n(n)(n-1)x^n$ with $R = 1$.

15. $f(x) = \ln(5-x) = -\displaystyle\int\dfrac{dx}{5-x} = -\dfrac{1}{5}\int\dfrac{dx}{1-x/5} = -\dfrac{1}{5}\int\left[\sum_{n=0}^{\infty}\left(\dfrac{x}{5}\right)^n\right]dx = C - \dfrac{1}{5}\sum_{n=0}^{\infty}\dfrac{x^{n+1}}{5^n(n+1)} = C - \sum_{n=1}^{\infty}\dfrac{x^n}{n5^n}$

Putting $x = 0$, we get $C = \ln 5$.   The series converges for $|x/5| < 1 \;\Leftrightarrow\; |x| < 5$, so $R = 5$.

17. We know that $\dfrac{1}{1+4x} = \dfrac{1}{1-(-4x)} = \displaystyle\sum_{n=0}^{\infty} (-4x)^n$. Differentiating, we get

$$\frac{-4}{(1+4x)^2} = \sum_{n=1}^{\infty} (-4)^n n x^{n-1} = \sum_{n=0}^{\infty} (-4)^{n+1}(n+1)x^n, \text{ so}$$

$$f(x) = \frac{x}{(1+4x)^2} = \frac{-x}{4} \cdot \frac{-4}{(1+4x)^2} = \frac{-x}{4} \sum_{n=0}^{\infty} (-4)^{n+1}(n+1)x^n = \sum_{n=0}^{\infty} (-1)^n 4^n (n+1)x^{n+1}$$

for $|-4x| < 1 \quad\Leftrightarrow\quad |x| < \frac{1}{4}$, so $R = \frac{1}{4}$.

19. By Example 5, $\dfrac{1}{(1-x)^2} = \displaystyle\sum_{n=0}^{\infty} (n+1)x^n$. Thus,

$$f(x) = \frac{1+x}{(1-x)^2} = \frac{1}{(1-x)^2} + \frac{x}{(1-x)^2} = \sum_{n=0}^{\infty} (n+1)x^n + \sum_{n=0}^{\infty} (n+1)x^{n+1}$$

$$= \sum_{n=0}^{\infty} (n+1)x^n + \sum_{n=1}^{\infty} n x^n \qquad \text{[make the starting values equal]}$$

$$= 1 + \sum_{n=1}^{\infty} [(n+1) + n]x^n = 1 + \sum_{n=1}^{\infty} (2n+1)x^n = \sum_{n=0}^{\infty} (2n+1)x^n \text{ with } R = 1.$$

21. $f(x) = \dfrac{x^2}{x^2+1} = x^2 \left(\dfrac{1}{1-(-x^2)} \right) = x^2 \displaystyle\sum_{n=0}^{\infty} (-x^2)^n = \sum_{n=0}^{\infty} (-1)^n x^{2n+2}$. This series converges when $|-x^2| < 1 \quad\Leftrightarrow$

$x^2 < 1 \quad\Leftrightarrow\quad |x| < 1$, so $R = 1$. The partial sums are $s_1 = x^2$,

$s_2 = s_1 - x^4$, $s_3 = s_2 + x^6$, $s_4 = s_3 - x^8$, $s_5 = s_4 + x^{10}, \ldots$.

Note that s_1 corresponds to the first term of the infinite sum,

regardless of the value of the summation variable and the value of the

exponent. As n increases, $s_n(x)$ approximates f better on the

interval of convergence, which is $(-1, 1)$.

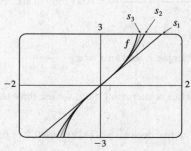

23. $f(x) = \ln\left(\dfrac{1+x}{1-x}\right) = \ln(1+x) - \ln(1-x) = \displaystyle\int \frac{dx}{1+x} + \int \frac{dx}{1-x} = \int \frac{dx}{1-(-x)} + \int \frac{dx}{1-x}$

$$= \int \left[\sum_{n=0}^{\infty} (-1)^n x^n + \sum_{n=0}^{\infty} x^n \right] dx = \int [(1 - x + x^2 - x^3 + x^4 - \cdots) + (1 + x + x^2 + x^3 + x^4 + \cdots)]\,dx$$

$$= \int (2 + 2x^2 + 2x^4 + \cdots)\,dx = \int \sum_{n=0}^{\infty} 2x^{2n}\,dx = C + \sum_{n=0}^{\infty} \frac{2x^{2n+1}}{2n+1}$$

But $f(0) = \ln\frac{1}{1} = 0$, so $C = 0$ and we have $f(x) = \displaystyle\sum_{n=0}^{\infty} \frac{2x^{2n+1}}{2n+1}$ with $R = 1$. If $x = \pm 1$, then $f(x) = \pm 2 \displaystyle\sum_{n=0}^{\infty} \frac{1}{2n+1}$,

which both diverge by the Limit Comparison Test with $b_n = \dfrac{1}{n}$.

The partial sums are $s_1 = \dfrac{2x}{1}$, $s_2 = s_1 + \dfrac{2x^3}{3}$, $s_3 = s_2 + \dfrac{2x^5}{5}, \ldots$.

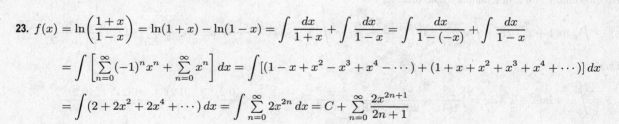

As n increases, $s_n(x)$ approximates f better on the interval of

convergence, which is $(-1, 1)$.

25. $\dfrac{t}{1-t^8} = t \cdot \dfrac{1}{1-t^8} = t \displaystyle\sum_{n=0}^{\infty} (t^8)^n = \sum_{n=0}^{\infty} t^{8n+1}$ $\Rightarrow$ $\displaystyle\int \dfrac{t}{1-t^8}\,dt = C + \sum_{n=0}^{\infty} \dfrac{t^{8n+2}}{8n+2}$. The series for $\dfrac{1}{1-t^8}$ converges

when $\left|t^8\right| < 1$ $\Leftrightarrow$ $|t| < 1$, so $R = 1$ for that series and also the series for $t/(1-t^8)$. By Theorem 2, the series for

$\displaystyle\int \dfrac{t}{1-t^8}\,dt$ also has $R = 1$.

27. From Example 6, $\ln(1+x) = \displaystyle\sum_{n=1}^{\infty} (-1)^{n-1}\dfrac{x^n}{n}$ for $|x| < 1$, so $x^2 \ln(1+x) = \displaystyle\sum_{n=1}^{\infty} (-1)^{n-1}\dfrac{x^{n+2}}{n}$ and

$\displaystyle\int x^2 \ln(1+x)\,dx = C + \sum_{n=1}^{\infty} (-1)^{n-1}\dfrac{x^{n+3}}{n(n+3)}$. $R = 1$ for the series for $\ln(1+x)$, so $R = 1$ for the series representing

$x^2 \ln(1+x)$ as well. By Theorem 2, the series for $\displaystyle\int x^2 \ln(1+x)\,dx$ also has $R = 1$.

29. $\dfrac{x}{1+x^3} = x\left[\dfrac{1}{1-(-x^3)}\right] = x \displaystyle\sum_{n=0}^{\infty} (-x^3)^n = \sum_{n=0}^{\infty} (-1)^n x^{3n+1}$ $\Rightarrow$

$\displaystyle\int \dfrac{x}{1+x^3}\,dx = \int \sum_{n=0}^{\infty} (-1)^n x^{3n+1}\,dx = C + \sum_{n=0}^{\infty} (-1)^n \dfrac{x^{3n+2}}{3n+2}$. Thus,

$I = \displaystyle\int_0^{0.3} \dfrac{x}{1+x^3}\,dx = \left[\dfrac{x^2}{2} - \dfrac{x^5}{5} + \dfrac{x^8}{8} - \dfrac{x^{11}}{11} + \cdots\right]_0^{0.3} = \dfrac{(0.3)^2}{2} - \dfrac{(0.3)^5}{5} + \dfrac{(0.3)^8}{8} - \dfrac{(0.3)^{11}}{11} + \cdots$.

The series is alternating, so if we use the first three terms, the error is at most $(0.3)^{11}/11 \approx 1.6 \times 10^{-7}$. So

$I \approx (0.3)^2/2 - (0.3)^5/5 + (0.3)^8/8 \approx 0.044\,522$ to six decimal places.

31. We substitute x^2 for x in Example 6, and find that

$$\int x\ln(1+x^2)\,dx = \int x \sum_{n=1}^{\infty} (-1)^{n-1}\dfrac{(x^2)^n}{n}\,dx = \int \sum_{n=1}^{\infty} (-1)^{n-1}\dfrac{x^{2n+1}}{n}\,dx = C + \sum_{n=1}^{\infty} (-1)^{n-1}\dfrac{x^{2n+2}}{n(2n+2)}$$

Thus,

$$I \approx \int_0^{0.2} x\ln(1+x^2)\,dx = \left[\dfrac{x^4}{1(4)} - \dfrac{x^6}{2(6)} + \dfrac{x^8}{3(8)} - \dfrac{x^{10}}{4(10)} + \cdots\right]_0^{0.2} = \dfrac{(0.2)^4}{4} - \dfrac{(0.2)^6}{12} + \dfrac{(0.2)^8}{24} - \dfrac{(0.2)^{10}}{40} + \cdots$$

The series is alternating, so if we use two terms, the error is at most $(0.2)^8/24 \approx 1.1 \times 10^{-7}$. So

$I \approx \dfrac{(0.2)^4}{4} - \dfrac{(0.2)^6}{12} \approx 0.000\,395$ to six decimal places.

33. By Example 7, $\arctan x = x - \dfrac{x^3}{3} + \dfrac{x^5}{5} - \dfrac{x^7}{7} + \cdots$, so $\arctan 0.2 = 0.2 - \dfrac{(0.2)^3}{3} + \dfrac{(0.2)^5}{5} - \dfrac{(0.2)^7}{7} + \cdots$.

The series is alternating, so if we use three terms, the error is at most $\dfrac{(0.2)^7}{7} \approx 0.000\,002$.

Thus, to five decimal places, $\arctan 0.2 \approx 0.2 - \dfrac{(0.2)^3}{3} + \dfrac{(0.2)^5}{5} \approx 0.197\,40$.

35. (a) $J_0(x) = \sum_{n=0}^{\infty} \dfrac{(-1)^n\, x^{2n}}{2^{2n}(n!)^2}$, $J_0'(x) = \sum_{n=1}^{\infty} \dfrac{(-1)^n\, 2nx^{2n-1}}{2^{2n}(n!)^2}$, and $J_0''(x) = \sum_{n=1}^{\infty} \dfrac{(-1)^n\, 2n(2n-1)x^{2n-2}}{2^{2n}(n!)^2}$, so

$$x^2 J_0''(x) + x J_0'(x) + x^2 J_0(x) = \sum_{n=1}^{\infty} \dfrac{(-1)^n\, 2n(2n-1)x^{2n}}{2^{2n}(n!)^2} + \sum_{n=1}^{\infty} \dfrac{(-1)^n\, 2nx^{2n}}{2^{2n}(n!)^2} + \sum_{n=0}^{\infty} \dfrac{(-1)^n\, x^{2n+2}}{2^{2n}(n!)^2}$$

$$= \sum_{n=1}^{\infty} \dfrac{(-1)^n\, 2n(2n-1)x^{2n}}{2^{2n}(n!)^2} + \sum_{n=1}^{\infty} \dfrac{(-1)^n\, 2nx^{2n}}{2^{2n}(n!)^2} + \sum_{n=1}^{\infty} \dfrac{(-1)^{n-1}\, x^{2n}}{2^{2n-2}\,[(n-1)!]^2}$$

$$= \sum_{n=1}^{\infty} \dfrac{(-1)^n\, 2n(2n-1)x^{2n}}{2^{2n}(n!)^2} + \sum_{n=1}^{\infty} \dfrac{(-1)^n\, 2nx^{2n}}{2^{2n}(n!)^2} + \sum_{n=1}^{\infty} \dfrac{(-1)^n(-1)^{-1}2^2 n^2 x^{2n}}{2^{2n}(n!)^2}$$

$$= \sum_{n=1}^{\infty} (-1)^n \left[\dfrac{2n(2n-1) + 2n - 2^2 n^2}{2^{2n}(n!)^2} \right] x^{2n}$$

$$= \sum_{n=1}^{\infty} (-1)^n \left[\dfrac{4n^2 - 2n + 2n - 4n^2}{2^{2n}(n!)^2} \right] x^{2n} = 0$$

(b) $\displaystyle\int_0^1 J_0(x)\,dx = \int_0^1 \left[\sum_{n=0}^{\infty} \dfrac{(-1)^n\, x^{2n}}{2^{2n}(n!)^2} \right] dx = \int_0^1 \left(1 - \dfrac{x^2}{4} + \dfrac{x^4}{64} - \dfrac{x^6}{2304} + \cdots \right) dx$

$$= \left[x - \dfrac{x^3}{3\cdot 4} + \dfrac{x^5}{5\cdot 64} - \dfrac{x^7}{7\cdot 2304} + \cdots \right]_0^1 = 1 - \dfrac{1}{12} + \dfrac{1}{320} - \dfrac{1}{16{,}128} + \cdots$$

Since $\frac{1}{16{,}128} \approx 0.000062$, it follows from The Alternating Series Estimation Theorem that, correct to three decimal places,

$\int_0^1 J_0(x)\,dx \approx 1 - \frac{1}{12} + \frac{1}{320} \approx 0.920$.

37. (a) $f(x) = \sum_{n=0}^{\infty} \dfrac{x^n}{n!} \;\Rightarrow\; f'(x) = \sum_{n=1}^{\infty} \dfrac{nx^{n-1}}{n!} = \sum_{n=1}^{\infty} \dfrac{x^{n-1}}{(n-1)!} = \sum_{n=0}^{\infty} \dfrac{x^n}{n!} = f(x)$

(b) By Theorem 9.4.2, the only solution to the differential equation $df(x)/dx = f(x)$ is $f(x) = Ke^x$, but $f(0) = 1$, so $K = 1$ and $f(x) = e^x$.

Or: We could solve the equation $df(x)/dx = f(x)$ as a separable differential equation.

39. If $a_n = \dfrac{x^n}{n^2}$, then by the Ratio Test, $\lim\limits_{n\to\infty} \left| \dfrac{a_{n+1}}{a_n} \right| = \lim\limits_{n\to\infty} \left| \dfrac{x^{n+1}}{(n+1)^2} \cdot \dfrac{n^2}{x^n} \right| = |x| \lim\limits_{n\to\infty} \left(\dfrac{n}{n+1} \right)^2 = |x| < 1$ for

convergence, so $R = 1$. When $x = \pm 1$, $\sum_{n=1}^{\infty} \left| \dfrac{x^n}{n^2} \right| = \sum_{n=1}^{\infty} \dfrac{1}{n^2}$ which is a convergent p-series ($p = 2 > 1$), so the interval of

convergence for f is $[-1, 1]$. By Theorem 2, the radii of convergence of f' and f'' are both 1, so we need only check the

endpoints. $f(x) = \sum_{n=1}^{\infty} \dfrac{x^n}{n^2} \;\Rightarrow\; f'(x) = \sum_{n=1}^{\infty} \dfrac{nx^{n-1}}{n^2} = \sum_{n=0}^{\infty} \dfrac{x^n}{n+1}$, and this series diverges for $x = 1$ (harmonic series)

and converges for $x = -1$ (Alternating Series Test), so the interval of convergence is $[-1, 1)$. $f''(x) = \sum_{n=1}^{\infty} \dfrac{nx^{n-1}}{n+1}$ diverges

at both 1 and -1 (Test for Divergence) since $\lim\limits_{n\to\infty} \dfrac{n}{n+1} = 1 \neq 0$, so its interval of convergence is $(-1, 1)$.

41. By Example 7, $\tan^{-1} x = \sum\limits_{n=0}^{\infty} (-1)^n \dfrac{x^{2n+1}}{2n+1}$ for $|x| < 1$. In particular, for $x = \dfrac{1}{\sqrt{3}}$, we

have $\dfrac{\pi}{6} = \tan^{-1}\left(\dfrac{1}{\sqrt{3}}\right) = \sum\limits_{n=0}^{\infty} (-1)^n \dfrac{(1/\sqrt{3})^{2n+1}}{2n+1} = \sum\limits_{n=0}^{\infty} (-1)^n \left(\dfrac{1}{3}\right)^n \dfrac{1}{\sqrt{3}} \dfrac{1}{2n+1}$, so

$\pi = \dfrac{6}{\sqrt{3}} \sum\limits_{n=0}^{\infty} \dfrac{(-1)^n}{(2n+1)3^n} = 2\sqrt{3} \sum\limits_{n=0}^{\infty} \dfrac{(-1)^n}{(2n+1)3^n}$.

11.10 Taylor and Maclaurin Series

1. Using Theorem 5 with $\sum\limits_{n=0}^{\infty} b_n (x-5)^n$, $b_n = \dfrac{f^{(n)}(a)}{n!}$, so $b_8 = \dfrac{f^{(8)}(5)}{8!}$.

3. Since $f^{(n)}(0) = (n+1)!$, Equation 7 gives the Maclaurin series

$\sum\limits_{n=0}^{\infty} \dfrac{f^{(n)}(0)}{n!} x^n = \sum\limits_{n=0}^{\infty} \dfrac{(n+1)!}{n!} x^n = \sum\limits_{n=0}^{\infty} (n+1)x^n$. Applying the Ratio Test with $a_n = (n+1)x^n$ gives us

$\lim\limits_{n\to\infty} \left| \dfrac{a_{n+1}}{a_n} \right| = \lim\limits_{n\to\infty} \left| \dfrac{(n+2)x^{n+1}}{(n+1)x^n} \right| = |x| \lim\limits_{n\to\infty} \dfrac{n+2}{n+1} = |x| \cdot 1 = |x|$. For convergence, we must have $|x| < 1$, so the

radius of convergence $R = 1$.

5.

n	$f^{(n)}(x)$	$f^{(n)}(0)$
0	xe^x	0
1	$(x+1)e^x$	1
2	$(x+2)e^x$	2
3	$(x+3)e^x$	3
4	$(x+4)e^x$	4

Using Equation 6 with $n = 0$ to 4 and $a = 0$, we get

$\sum\limits_{n=0}^{4} \dfrac{f^{(n)}(0)}{n!} (x-0)^n = \dfrac{0}{0!} x^0 + \dfrac{1}{1!} x^1 + \dfrac{2}{2!} x^2 + \dfrac{3}{3!} x^3 + \dfrac{4}{4!} x^4$

$= x + x^2 + \tfrac{1}{2}x^3 + \tfrac{1}{6}x^4$

7.

n	$f^{(n)}(x)$	$f^{(n)}(8)$
0	$\sqrt[3]{x}$	2
1	$\dfrac{1}{3x^{2/3}}$	$\dfrac{1}{12}$
2	$-\dfrac{2}{9x^{5/3}}$	$-\dfrac{2}{288}$
3	$\dfrac{10}{27x^{8/3}}$	$\dfrac{10}{6912}$

$\sum\limits_{n=0}^{3} \dfrac{f^{(n)}(8)}{n!} (x-8)^n = \dfrac{2}{0!} (x-8)^0 + \dfrac{\frac{1}{12}}{1!} (x-8)^1$

$\qquad\qquad - \dfrac{\frac{2}{288}}{2!} (x-8)^2 + \dfrac{\frac{10}{6912}}{3!} (x-8)^3$

$= 2 + \tfrac{1}{12}(x-8) - \tfrac{1}{288}(x-8)^2 + \tfrac{5}{20,736}(x-8)^3$

9.

n	$f^{(n)}(x)$	$f^{(n)}(\pi/6)$
0	$\sin x$	$1/2$
1	$\cos x$	$\sqrt{3}/2$
2	$-\sin x$	$-1/2$
3	$-\cos x$	$-\sqrt{3}/2$

$$\sum_{n=0}^{3} \frac{f^{(n)}(\pi/6)}{n!}\left(x-\frac{\pi}{6}\right)^n = \frac{1/2}{0!}\left(x-\frac{\pi}{6}\right)^0 + \frac{\sqrt{3}/2}{1!}\left(x-\frac{\pi}{6}\right)^1 - \frac{1/2}{2!}\left(x-\frac{\pi}{6}\right)^2 - \frac{\sqrt{3}/2}{3!}\left(x-\frac{\pi}{6}\right)^3$$

$$= \frac{1}{2} + \frac{\sqrt{3}}{2}\left(x-\frac{\pi}{6}\right) - \frac{1}{4}\left(x-\frac{\pi}{6}\right)^2 - \frac{\sqrt{3}}{12}\left(x-\frac{\pi}{6}\right)^3$$

11.

n	$f^{(n)}(x)$	$f^{(n)}(0)$
0	$(1-x)^{-2}$	1
1	$2(1-x)^{-3}$	2
2	$6(1-x)^{-4}$	6
3	$24(1-x)^{-5}$	24
4	$120(1-x)^{-6}$	120
$\vdots$	$\vdots$	$\vdots$

$$(1-x)^{-2} = f(0) + f'(0)x + \frac{f''(0)}{2!}x^2 + \frac{f'''(0)}{3!}x^3 + \frac{f^{(4)}(0)}{4!}x^4 + \cdots$$

$$= 1 + 2x + \frac{6}{2}x^2 + \frac{24}{6}x^3 + \frac{120}{24}x^4 + \cdots$$

$$= 1 + 2x + 3x^2 + 4x^3 + 5x^4 + \cdots = \sum_{n=0}^{\infty}(n+1)x^n$$

$$\lim_{n\to\infty}\left|\frac{a_{n+1}}{a_n}\right| = \lim_{n\to\infty}\left|\frac{(n+2)x^{n+1}}{(n+1)x^n}\right| = |x|\lim_{n\to\infty}\frac{n+2}{n+1} = |x|\,(1) = |x| < 1$$

for convergence, so $R = 1$.

13.

n	$f^{(n)}(x)$	$f^{(n)}(0)$
0	$\cos x$	1
1	$-\sin x$	0
2	$-\cos x$	-1
3	$\sin x$	0
4	$\cos x$	1
$\vdots$	$\vdots$	$\vdots$

$$\cos x = f(0) + f'(0)x + \frac{f''(0)}{2!}x^2 + \frac{f'''(0)}{3!}x^3 + \frac{f^{(4)}(0)}{4!}x^4 + \cdots$$

$$= 1 - \frac{1}{2!}x^2 + \frac{1}{4!}x^4 - \cdots$$

$$= \sum_{n=0}^{\infty}(-1)^n\frac{x^{2n}}{(2n)!} \qquad \text{[Agrees with (16).]}$$

$$\lim_{n\to\infty}\left|\frac{a_{n+1}}{a_n}\right| = \lim_{n\to\infty}\left|\frac{x^{2n+2}}{(2n+2)!}\cdot\frac{(2n)!}{x^{2n}}\right| = \lim_{n\to\infty}\frac{x^2}{(2n+2)(2n+1)} = 0 < 1$$

for all x, so $R = \infty$.

15.

n	$f^{(n)}(x)$	$f^{(n)}(0)$
0	2^x	1
1	$2^x(\ln 2)$	$\ln 2$
2	$2^x(\ln 2)^2$	$(\ln 2)^2$
3	$2^x(\ln 2)^3$	$(\ln 2)^3$
4	$2^x(\ln 2)^4$	$(\ln 2)^4$
$\vdots$	$\vdots$	$\vdots$

$$2^x = \sum_{n=0}^{\infty}\frac{f^{(n)}(0)}{n!}x^n = \sum_{n=0}^{\infty}\frac{(\ln 2)^n}{n!}x^n.$$

$$\lim_{n\to\infty}\left|\frac{a_{n+1}}{a_n}\right| = \lim_{n\to\infty}\left|\frac{(\ln 2)^{n+1}x^{n+1}}{(n+1)!}\cdot\frac{n!}{(\ln 2)^n x^n}\right|$$

$$= \lim_{n\to\infty}\frac{(\ln 2)\,|x|}{n+1} = 0 < 1 \quad \text{for all } x, \text{ so } R = \infty.$$

17.

n	$f^{(n)}(x)$	$f^{(n)}(0)$
0	$\sinh x$	0
1	$\cosh x$	1
2	$\sinh x$	0
3	$\cosh x$	1
4	$\sinh x$	0
⋮	⋮	⋮

$$f^{(n)}(0) = \begin{cases} 0 & \text{if } n \text{ is even} \\ 1 & \text{if } n \text{ is odd} \end{cases} \quad \text{so } \sinh x = \sum_{n=0}^{\infty} \frac{x^{2n+1}}{(2n+1)!}.$$

Use the Ratio Test to find R. If $a_n = \dfrac{x^{2n+1}}{(2n+1)!}$, then

$$\lim_{n\to\infty} \left| \frac{a_{n+1}}{a_n} \right| = \lim_{n\to\infty} \left| \frac{x^{2n+3}}{(2n+3)!} \cdot \frac{(2n+1)!}{x^{2n+1}} \right| = x^2 \cdot \lim_{n\to\infty} \frac{1}{(2n+3)(2n+2)}$$

$$= 0 < 1 \quad \text{for all } x, \text{ so } R = \infty.$$

19.

n	$f^{(n)}(x)$	$f^{(n)}(2)$
0	$x^5 + 2x^3 + x$	50
1	$5x^4 + 6x^2 + 1$	105
2	$20x^3 + 12x$	184
3	$60x^2 + 12$	252
4	$120x$	240
5	120	120
6	0	0
7	0	0
⋮	⋮	⋮

$f^{(n)}(x) = 0$ for $n \geq 6$, so f has a finite expansion about $a = 2$.

$$f(x) = x^5 + 2x^3 + x = \sum_{n=0}^{5} \frac{f^{(n)}(2)}{n!}(x-2)^n$$

$$= \frac{50}{0!}(x-2)^0 + \frac{105}{1!}(x-2)^1 + \frac{184}{2!}(x-2)^2 + \frac{252}{3!}(x-2)^3$$

$$+ \frac{240}{4!}(x-2)^4 + \frac{120}{5!}(x-2)^5$$

$$= 50 + 105(x-2) + 92(x-2)^2 + 42(x-2)^3$$

$$+ 10(x-2)^4 + (x-2)^5$$

A finite series converges for all x, so $R = \infty$.

21.

n	$f^{(n)}(x)$	$f^{(n)}(2)$
0	$\ln x$	$\ln 2$
1	$1/x$	$1/2$
2	$-1/x^2$	$-1/2^2$
3	$2/x^3$	$2/2^3$
4	$-6/x^4$	$-6/2^4$
5	$24/x^5$	$24/2^5$
⋮	⋮	⋮

$$f(x) = \ln x = \sum_{n=0}^{\infty} \frac{f^{(n)}(2)}{n!}(x-2)^n$$

$$= \frac{\ln 2}{0!}(x-2)^0 + \frac{1}{1!\,2^1}(x-2)^1 + \frac{-1}{2!\,2^2}(x-2)^2 + \frac{2}{3!\,2^3}(x-2)^3$$

$$+ \frac{-6}{4!\,2^4}(x-2)^4 + \frac{24}{5!\,2^5}(x-2)^5 + \cdots$$

$$= \ln 2 + \sum_{n=1}^{\infty} (-1)^{n+1} \frac{(n-1)!}{n!\,2^n}(x-2)^n$$

$$= \ln 2 + \sum_{n=1}^{\infty} (-1)^{n+1} \frac{1}{n\,2^n}(x-2)^n$$

$$\lim_{n\to\infty} \left| \frac{a_{n+1}}{a_n} \right| = \lim_{n\to\infty} \left| \frac{(-1)^{n+2}(x-2)^{n+1}}{(n+1)\,2^{n+1}} \cdot \frac{n\,2^n}{(-1)^{n+1}(x-2)^n} \right| = \lim_{n\to\infty} \left| \frac{(-1)(x-2)n}{(n+1)2} \right| = \lim_{n\to\infty} \left(\frac{n}{n+1} \right) \frac{|x-2|}{2}$$

$$= \frac{|x-2|}{2} < 1 \quad \text{for convergence, so } |x-2| < 2 \text{ and } R = 2.$$

23.

n	$f^{(n)}(x)$	$f^{(n)}(3)$
0	e^{2x}	e^6
1	$2e^{2x}$	$2e^6$
2	$2^2 e^{2x}$	$4e^6$
3	$2^3 e^{2x}$	$8e^6$
4	$2^4 e^{2x}$	$16e^6$
⋮	⋮	⋮

$$f(x) = e^{2x} = \sum_{n=0}^{\infty} \frac{f^{(n)}(3)}{n!}(x-3)^n$$

$$= \frac{e^6}{0!}(x-3)^0 + \frac{2e^6}{1!}(x-3)^1 + \frac{4e^6}{2!}(x-3)^2$$

$$+ \frac{8e^6}{3!}(x-3)^3 + \frac{16e^6}{4!}(x-3)^4 + \cdots$$

$$= \sum_{n=0}^{\infty} \frac{2^n e^6}{n!}(x-3)^n$$

$$\lim_{n\to\infty} \left| \frac{a_{n+1}}{a_n} \right| = \lim_{n\to\infty} \left| \frac{2^{n+1} e^6 (x-3)^{n+1}}{(n+1)!} \cdot \frac{n!}{2^n e^6 (x-3)^n} \right| = \lim_{n\to\infty} \frac{2|x-3|}{n+1} = 0 < 1 \quad \text{for all } x, \text{ so } R = \infty.$$

25.

n	$f^{(n)}(x)$	$f^{(n)}(\pi)$
0	$\sin x$	0
1	$\cos x$	-1
2	$-\sin x$	0
3	$-\cos x$	1
4	$\sin x$	0
5	$\cos x$	-1
6	$-\sin x$	0
7	$-\cos x$	1
⋮	⋮	⋮

$$f(x) = \sin x = \sum_{n=0}^{\infty} \frac{f^{(n)}(\pi)}{n!}(x-\pi)^n$$

$$= \frac{-1}{1!}(x-\pi)^1 + \frac{1}{3!}(x-\pi)^3 + \frac{-1}{5!}(x-\pi)^5 + \frac{1}{7!}(x-\pi)^7 + \cdots$$

$$= \sum_{n=0}^{\infty} \frac{(-1)^{n+1}}{(2n+1)!}(x-\pi)^{2n+1}$$

$$\lim_{n\to\infty} \left| \frac{a_{n+1}}{a_n} \right| = \lim_{n\to\infty} \left| \frac{(-1)^{n+2}(x-\pi)^{2n+3}}{(2n+3)!} \cdot \frac{(2n+1)!}{(-1)^{n+1}(x-\pi)^{2n+1}} \right|$$

$$= \lim_{n\to\infty} \frac{(x-\pi)^2}{(2n+3)(2n+2)} = 0 < 1 \quad \text{for all } x, \text{ so } R = \infty.$$

27. If $f(x) = \cos x$, then $f^{(n+1)}(x) = \pm \sin x$ or $\pm \cos x$. In each case, $\left| f^{(n+1)}(x) \right| \leq 1$, so by Formula 9 with $a = 0$ and

$M = 1$, $|R_n(x)| \leq \dfrac{1}{(n+1)!} |x|^{n+1}$. Thus, $|R_n(x)| \to 0$ as $n \to \infty$ by Equation 10. So $\displaystyle\lim_{n\to\infty} R_n(x) = 0$ and, by Theorem

8, the series in Exercise 13 represents $\cos x$ for all x.

29. If $f(x) = \sinh x$, then for all n, $f^{(n+1)}(x) = \cosh x$ or $\sinh x$. Since $|\sinh x| < |\cosh x| = \cosh x$ for all x, we have

$\left| f^{(n+1)}(x) \right| \leq \cosh x$ for all n. If d is any positive number and $|x| \leq d$, then $\left| f^{(n+1)}(x) \right| \leq \cosh x \leq \cosh d$, so by

Formula 9 with $a = 0$ and $M = \cosh d$, we have $|R_n(x)| \leq \dfrac{\cosh d}{(n+1)!} |x|^{n+1}$. It follows that $|R_n(x)| \to 0$ as $n \to \infty$ for

$|x| \leq d$ (by Equation 10). But d was an arbitrary positive number. So by Theorem 8, the series represents $\sinh x$ for all x.

31. $\sqrt[4]{1-x} = [1 + (-x)]^{1/4} = \displaystyle\sum_{n=0}^{\infty} \binom{1/4}{n} (-x)^n = 1 + \frac{1}{4}(-x) + \frac{\frac{1}{4}\left(-\frac{3}{4}\right)}{2!}(-x)^2 + \frac{\frac{1}{4}\left(-\frac{3}{4}\right)\left(-\frac{7}{4}\right)}{3!}(-x)^3 + \cdots$

$\qquad = 1 - \frac{1}{4}x + \displaystyle\sum_{n=2}^{\infty} \frac{(-1)^{n-1}(-1)^n \cdot [3 \cdot 7 \cdot \cdots \cdot (4n-5)]}{4^n \cdot n!} x^n$

$\qquad = 1 - \frac{1}{4}x - \displaystyle\sum_{n=2}^{\infty} \frac{3 \cdot 7 \cdot \cdots \cdot (4n-5)}{4^n \cdot n!} x^n$

and $|-x| < 1 \iff |x| < 1$, so $R = 1$.

33. $\dfrac{1}{(2+x)^3} = \dfrac{1}{[2(1+x/2)]^3} = \dfrac{1}{8}\left(1 + \dfrac{x}{2}\right)^{-3} = \dfrac{1}{8} \displaystyle\sum_{n=0}^{\infty} \binom{-3}{n} \left(\dfrac{x}{2}\right)^n$. The binomial coefficient is

$$\binom{-3}{n} = \frac{(-3)(-4)(-5) \cdot \cdots \cdot (-3-n+1)}{n!} = \frac{(-3)(-4)(-5) \cdot \cdots \cdot [-(n+2)]}{n!}$$

$$= \frac{(-1)^n \cdot 2 \cdot 3 \cdot 4 \cdot 5 \cdot \cdots \cdot (n+1)(n+2)}{2 \cdot n!} = \frac{(-1)^n (n+1)(n+2)}{2}$$

Thus, $\dfrac{1}{(2+x)^3} = \dfrac{1}{8} \displaystyle\sum_{n=0}^{\infty} \frac{(-1)^n (n+1)(n+2)}{2} \frac{x^n}{2^n} = \displaystyle\sum_{n=0}^{\infty} \frac{(-1)^n (n+1)(n+2)x^n}{2^{n+4}}$ for $\left|\dfrac{x}{2}\right| < 1 \iff |x| < 2$, so $R = 2$.

35. $\arctan x = \displaystyle\sum_{n=0}^{\infty} (-1)^n \frac{x^{2n+1}}{2n+1}$, so $f(x) = \arctan(x^2) = \displaystyle\sum_{n=0}^{\infty} (-1)^n \frac{(x^2)^{2n+1}}{2n+1} = \displaystyle\sum_{n=0}^{\infty} (-1)^n \frac{1}{2n+1} x^{4n+2}$, $R = 1$.

37. $\cos x = \displaystyle\sum_{n=0}^{\infty} (-1)^n \frac{x^{2n}}{(2n)!} \implies \cos 2x = \displaystyle\sum_{n=0}^{\infty} (-1)^n \frac{(2x)^{2n}}{(2n)!} = \displaystyle\sum_{n=0}^{\infty} (-1)^n \frac{2^{2n} x^{2n}}{(2n)!}$, so

$f(x) = x\cos 2x = \displaystyle\sum_{n=0}^{\infty} (-1)^n \frac{2^{2n}}{(2n)!} x^{2n+1}$, $R = \infty$.

39. $\cos x = \displaystyle\sum_{n=0}^{\infty} (-1)^n \frac{x^{2n}}{(2n)!} \implies \cos\left(\frac{1}{2}x^2\right) = \displaystyle\sum_{n=0}^{\infty} (-1)^n \frac{\left(\frac{1}{2}x^2\right)^{2n}}{(2n)!} = \displaystyle\sum_{n=0}^{\infty} (-1)^n \frac{x^{4n}}{2^{2n}(2n)!}$, so

$f(x) = x\cos\left(\frac{1}{2}x^2\right) = \displaystyle\sum_{n=0}^{\infty} (-1)^n \frac{1}{2^{2n}(2n)!} x^{4n+1}$, $R = \infty$.

41. We must write the binomial in the form $(1+$ expression$)$, so we'll factor out a 4.

$$\frac{x}{\sqrt{4+x^2}} = \frac{x}{\sqrt{4(1+x^2/4)}} = \frac{x}{2\sqrt{1+x^2/4}} = \frac{x}{2}\left(1 + \frac{x^2}{4}\right)^{-1/2} = \frac{x}{2} \displaystyle\sum_{n=0}^{\infty} \binom{-\frac{1}{2}}{n} \left(\frac{x^2}{4}\right)^n$$

$$= \frac{x}{2}\left[1 + \left(-\tfrac{1}{2}\right)\frac{x^2}{4} + \frac{\left(-\frac{1}{2}\right)\left(-\frac{3}{2}\right)}{2!}\left(\frac{x^2}{4}\right)^2 + \frac{\left(-\frac{1}{2}\right)\left(-\frac{3}{2}\right)\left(-\frac{5}{2}\right)}{3!}\left(\frac{x^2}{4}\right)^3 + \cdots\right]$$

$$= \frac{x}{2} + \frac{x}{2} \displaystyle\sum_{n=1}^{\infty} (-1)^n \frac{1 \cdot 3 \cdot 5 \cdot \cdots \cdot (2n-1)}{2^n \cdot 4^n \cdot n!} x^{2n}$$

$$= \frac{x}{2} + \displaystyle\sum_{n=1}^{\infty} (-1)^n \frac{1 \cdot 3 \cdot 5 \cdot \cdots \cdot (2n-1)}{n! \, 2^{3n+1}} x^{2n+1} \text{ and } \frac{x^2}{4} < 1 \iff \frac{|x|}{2} < 1 \iff |x| < 2, \text{ so } R = 2.$$

43. $\sin^2 x = \frac{1}{2}(1 - \cos 2x) = \frac{1}{2}\left[1 - \sum_{n=0}^{\infty} \frac{(-1)^n (2x)^{2n}}{(2n)!}\right] = \frac{1}{2}\left[1 - 1 - \sum_{n=1}^{\infty} \frac{(-1)^n (2x)^{2n}}{(2n)!}\right] = \sum_{n=1}^{\infty} \frac{(-1)^{n+1} 2^{2n-1} x^{2n}}{(2n)!}$,

$R = \infty$

45. $\cos x \stackrel{(16)}{=} \sum_{n=0}^{\infty} (-1)^n \frac{x^{2n}}{(2n)!} \quad \Rightarrow$

$$f(x) = \cos(x^2) = \sum_{n=0}^{\infty} \frac{(-1)^n (x^2)^{2n}}{(2n)!} = \sum_{n=0}^{\infty} \frac{(-1)^n x^{4n}}{(2n)!}$$

$$= 1 - \frac{1}{2}x^4 + \frac{1}{24}x^8 - \frac{1}{720}x^{12} + \cdots$$

The series for $\cos x$ converges for all x, so the same is true of the series for $f(x)$, that is, $R = \infty$. Notice that, as n increases, $T_n(x)$ becomes a better approximation to $f(x)$.

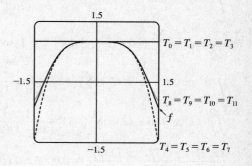

47. $e^x \stackrel{(11)}{=} \sum_{n=0}^{\infty} \frac{x^n}{n!}$, so $e^{-x} = \sum_{n=0}^{\infty} \frac{(-x)^n}{n!} = \sum_{n=0}^{\infty} (-1)^n \frac{x^n}{n!}$, so

$$f(x) = xe^{-x} = \sum_{n=0}^{\infty} (-1)^n \frac{1}{n!} x^{n+1}$$

$$= x - x^2 + \frac{1}{2}x^3 - \frac{1}{6}x^4 + \frac{1}{24}x^5 - \frac{1}{120}x^6 + \cdots$$

$$= \sum_{n=1}^{\infty} (-1)^{n-1} \frac{x^n}{(n-1)!}$$

The series for e^x converges for all x, so the same is true of the series for $f(x)$; that is, $R = \infty$. From the graphs of f and the first few Taylor polynomials, we see that $T_n(x)$ provides a closer fit to $f(x)$ near 0 as n increases.

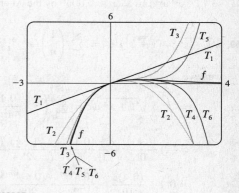

49. $5° = 5°\left(\frac{\pi}{180°}\right) = \frac{\pi}{36}$ radians and $\cos x = \sum_{n=0}^{\infty} (-1)^n \frac{x^{2n}}{(2n)!} = 1 - \frac{x^2}{2!} + \frac{x^4}{4!} - \frac{x^6}{6!} + \cdots$, so

$\cos\frac{\pi}{36} = 1 - \frac{(\pi/36)^2}{2!} + \frac{(\pi/36)^4}{4!} - \frac{(\pi/36)^6}{6!} + \cdots$. Now $1 - \frac{(\pi/36)^2}{2!} \approx 0.99619$ and adding $\frac{(\pi/36)^4}{4!} \approx 2.4 \times 10^{-6}$

does not affect the fifth decimal place, so $\cos 5° \approx 0.99619$ by the Alternating Series Estimation Theorem.

51. (a) $1/\sqrt{1 - x^2} = \left[1 + (-x^2)\right]^{-1/2} = 1 + \left(-\frac{1}{2}\right)(-x^2) + \frac{\left(-\frac{1}{2}\right)\left(-\frac{3}{2}\right)}{2!}(-x^2)^2 + \frac{\left(-\frac{1}{2}\right)\left(-\frac{3}{2}\right)\left(-\frac{5}{2}\right)}{3!}(-x^2)^3 + \cdots$

$$= 1 + \sum_{n=1}^{\infty} \frac{1 \cdot 3 \cdot 5 \cdots \cdot (2n-1)}{2^n \cdot n!} x^{2n}$$

(b) $\sin^{-1} x = \int \frac{1}{\sqrt{1-x^2}}\, dx = C + x + \sum_{n=1}^{\infty} \frac{1 \cdot 3 \cdot 5 \cdots \cdot (2n-1)}{(2n+1)2^n \cdot n!} x^{2n+1}$

$$= x + \sum_{n=1}^{\infty} \frac{1 \cdot 3 \cdot 5 \cdots \cdot (2n-1)}{(2n+1)2^n \cdot n!} x^{2n+1} \quad \text{since } 0 = \sin^{-1} 0 = C.$$

53. $\sqrt{1 + x^3} = (1 + x^3)^{1/2} = \sum_{n=0}^{\infty} \binom{\frac{1}{2}}{n} (x^3)^n = \sum_{n=0}^{\infty} \binom{\frac{1}{2}}{n} x^{3n} \quad \Rightarrow \quad \int \sqrt{1 + x^3}\, dx = C + \sum_{n=0}^{\infty} \binom{\frac{1}{2}}{n} \frac{x^{3n+1}}{3n+1}$,

with $R = 1$.

55. $\cos x \overset{(16)}{=} \sum\limits_{n=0}^{\infty} (-1)^n \dfrac{x^{2n}}{(2n)!}$ $\Rightarrow$ $\cos x - 1 = \sum\limits_{n=1}^{\infty} (-1)^n \dfrac{x^{2n}}{(2n)!}$ $\Rightarrow$ $\dfrac{\cos x - 1}{x} = \sum\limits_{n=1}^{\infty} (-1)^n \dfrac{x^{2n-1}}{(2n)!}$ $\Rightarrow$

$\displaystyle\int \dfrac{\cos x - 1}{x}\, dx = C + \sum\limits_{n=1}^{\infty} (-1)^n \dfrac{x^{2n}}{2n \cdot (2n)!}$, with $R = \infty$.

57. $\arctan x = \sum\limits_{n=0}^{\infty} (-1)^n \dfrac{x^{2n+1}}{2n+1}$ for $|x| < 1$, so $x^3 \arctan x = \sum\limits_{n=0}^{\infty} (-1)^n \dfrac{x^{2n+4}}{2n+1}$ for $|x| < 1$ and

$\displaystyle\int x^3 \arctan x\, dx = C + \sum\limits_{n=0}^{\infty} (-1)^n \dfrac{x^{2n+5}}{(2n+1)(2n+5)}$. Since $\frac{1}{2} < 1$, we have

$\displaystyle\int_0^{1/2} x^3 \arctan x\, dx = \sum\limits_{n=0}^{\infty} (-1)^n \dfrac{(1/2)^{2n+5}}{(2n+1)(2n+5)} = \dfrac{(1/2)^5}{1 \cdot 5} - \dfrac{(1/2)^7}{3 \cdot 7} + \dfrac{(1/2)^9}{5 \cdot 9} - \dfrac{(1/2)^{11}}{7 \cdot 11} + \cdots$. Now

$\dfrac{(1/2)^5}{1 \cdot 5} - \dfrac{(1/2)^7}{3 \cdot 7} + \dfrac{(1/2)^9}{5 \cdot 9} \approx 0.0059$ and subtracting $\dfrac{(1/2)^{11}}{7 \cdot 11} \approx 6.3 \times 10^{-6}$ does not affect the fourth decimal place,

so $\int_0^{1/2} x^3 \arctan x\, dx \approx 0.0059$ by the Alternating Series Estimation Theorem.

59. $\sqrt{1 + x^4} = (1 + x^4)^{1/2} = \sum\limits_{n=0}^{\infty} \binom{\frac{1}{2}}{n} (x^4)^n$, so $\displaystyle\int \sqrt{1 + x^4}\, dx = C + \sum\limits_{n=0}^{\infty} \binom{\frac{1}{2}}{n} \dfrac{x^{4n+1}}{4n+1}$ and hence, since $0.4 < 1$,

we have

$$I = \int_0^{0.4} \sqrt{1 + x^4}\, dx = \sum\limits_{n=0}^{\infty} \binom{\frac{1}{2}}{n} \dfrac{(0.4)^{4n+1}}{4n+1}$$

$$= (1)\dfrac{(0.4)^1}{0!} + \dfrac{\frac{1}{2}}{1!} \dfrac{(0.4)^5}{5} + \dfrac{\frac{1}{2}\left(-\frac{1}{2}\right)}{2!}\dfrac{(0.4)^9}{9} + \dfrac{\frac{1}{2}\left(-\frac{1}{2}\right)\left(-\frac{3}{2}\right)}{3!}\dfrac{(0.4)^{13}}{13} + \dfrac{\frac{1}{2}\left(-\frac{1}{2}\right)\left(-\frac{3}{2}\right)\left(-\frac{5}{2}\right)}{4!}\dfrac{(0.4)^{17}}{17} + \cdots$$

$$= 0.4 + \dfrac{(0.4)^5}{10} - \dfrac{(0.4)^9}{72} + \dfrac{(0.4)^{13}}{208} - \dfrac{5(0.4)^{17}}{2176} + \cdots$$

Now $\dfrac{(0.4)^9}{72} \approx 3.6 \times 10^{-6} < 5 \times 10^{-6}$, so by the Alternating Series Estimation Theorem, $I \approx 0.4 + \dfrac{(0.4)^5}{10} \approx 0.40102$

(correct to five decimal places).

61. $\lim\limits_{x \to 0} \dfrac{x - \ln(1 + x)}{x^2} = \lim\limits_{x \to 0} \dfrac{x - \left(x - \frac{1}{2}x^2 + \frac{1}{3}x^3 - \frac{1}{4}x^4 + \frac{1}{5}x^5 - \cdots\right)}{x^2} = \lim\limits_{x \to 0} \dfrac{\frac{1}{2}x^2 - \frac{1}{3}x^3 + \frac{1}{4}x^4 - \frac{1}{5}x^5 + \cdots}{x^2}$

$\qquad = \lim\limits_{x \to 0}\left(\frac{1}{2} - \frac{1}{3}x + \frac{1}{4}x^2 - \frac{1}{5}x^3 + \cdots\right) = \frac{1}{2}$

since power series are continuous functions.

63. $\lim\limits_{x \to 0} \dfrac{\sin x - x + \frac{1}{6}x^3}{x^5} = \lim\limits_{x \to 0} \dfrac{\left(x - \frac{1}{3!}x^3 + \frac{1}{5!}x^5 - \frac{1}{7!}x^7 + \cdots\right) - x + \frac{1}{6}x^3}{x^5}$

$\qquad = \lim\limits_{x \to 0} \dfrac{\frac{1}{5!}x^5 - \frac{1}{7!}x^7 + \cdots}{x^5} = \lim\limits_{x \to 0}\left(\dfrac{1}{5!} - \dfrac{x^2}{7!} + \dfrac{x^4}{9!} - \cdots\right) = \dfrac{1}{5!} = \dfrac{1}{120}$

since power series are continuous functions.

65. $\lim\limits_{x \to 0} \dfrac{x^3 - 3x + 3\tan^{-1} x}{x^5} = \lim\limits_{x \to 0} \dfrac{x^3 - 3x + 3\left(x - \frac{1}{3}x^3 + \frac{1}{5}x^5 - \frac{1}{7}x^7 + \cdots\right)}{x^5}$

$\qquad = \lim\limits_{x \to 0} \dfrac{x^3 - 3x + 3x - x^3 + \frac{3}{5}x^5 - \frac{3}{7}x^7 + \cdots}{x^5} = \lim\limits_{x \to 0} \dfrac{\frac{3}{5}x^5 - \frac{3}{7}x^7 + \cdots}{x^5}$

$\qquad = \lim\limits_{x \to 0}\left(\frac{3}{5} - \frac{3}{7}x^2 + \cdots\right) = \frac{3}{5}$ since power series are continuous functions.

67. From Equation 11, we have $e^{-x^2} = 1 - \dfrac{x^2}{1!} + \dfrac{x^4}{2!} - \dfrac{x^6}{3!} + \cdots$ and we know that $\cos x = 1 - \dfrac{x^2}{2!} + \dfrac{x^4}{4!} - \cdots$ from

Equation 16. Therefore, $e^{-x^2}\cos x = \left(1 - x^2 + \frac{1}{2}x^4 - \cdots\right)\left(1 - \frac{1}{2}x^2 + \frac{1}{24}x^4 - \cdots\right)$. Writing only the terms with

degree ≤ 4, we get $e^{-x^2}\cos x = 1 - \frac{1}{2}x^2 + \frac{1}{24}x^4 - x^2 + \frac{1}{2}x^4 + \frac{1}{2}x^4 + \cdots = 1 - \frac{3}{2}x^2 + \frac{25}{24}x^4 + \cdots$.

69. $\dfrac{x}{\sin x} \overset{(15)}{=} \dfrac{x}{x - \frac{1}{6}x^3 + \frac{1}{120}x^5 - \cdots}.$

$$
\begin{array}{r}
1 + \frac{1}{6}x^2 + \frac{7}{360}x^4 + \cdots \\[4pt]
x - \frac{1}{6}x^3 + \frac{1}{120}x^5 - \cdots\ \overline{\smash{\big)}\ x } \\[4pt]
\underline{x - \frac{1}{6}x^3 + \frac{1}{120}x^5 - \cdots} \\[4pt]
\frac{1}{6}x^3 - \frac{1}{120}x^5 + \cdots \\[4pt]
\underline{\frac{1}{6}x^3 - \frac{1}{36}x^5 + \cdots} \\[4pt]
\frac{7}{360}x^5 + \cdots \\[4pt]
\underline{\frac{7}{360}x^5 + \cdots} \\[4pt]
\cdots
\end{array}
$$

From the long division above, $\dfrac{x}{\sin x} = 1 + \frac{1}{6}x^2 + \frac{7}{360}x^4 + \cdots$.

71. $y = (\arctan x)^2 = \left(x - \frac{1}{3}x^3 + \frac{1}{5}x^5 - \frac{1}{7}x^7 + \cdots\right)\left(x - \frac{1}{3}x^3 + \frac{1}{5}x^5 - \frac{1}{7}x^7 + \cdots\right)$. Writing only the terms with

degree ≤ 6, we get $(\arctan x)^2 = x^2 - \frac{1}{3}x^4 + \frac{1}{5}x^6 - \frac{1}{3}x^4 + \frac{1}{9}x^6 + \frac{1}{5}x^6 + \cdots = x^2 - \frac{2}{3}x^4 + \frac{23}{45}x^6 + \cdots$.

73. $\displaystyle\sum_{n=0}^{\infty}(-1)^n\frac{x^{4n}}{n!} = \sum_{n=0}^{\infty}\frac{\left(-x^4\right)^n}{n!} = e^{-x^4}$, by (11).

75. $\displaystyle\sum_{n=1}^{\infty}(-1)^{n-1}\frac{3^n}{n5^n} = \sum_{n=1}^{\infty}(-1)^{n-1}\frac{(3/5)^n}{n} = \ln\left(1 + \frac{3}{5}\right)$ [from Table 1] $= \ln\frac{8}{5}$

77. $\displaystyle\sum_{n=0}^{\infty}\frac{(-1)^n\,\pi^{2n+1}}{4^{2n+1}(2n+1)!} = \sum_{n=0}^{\infty}\frac{(-1)^n\left(\frac{\pi}{4}\right)^{2n+1}}{(2n+1)!} = \sin\frac{\pi}{4} = \frac{1}{\sqrt{2}}$, by (15).

79. $3 + \dfrac{9}{2!} + \dfrac{27}{3!} + \dfrac{81}{4!} + \cdots = \dfrac{3^1}{1!} + \dfrac{3^2}{2!} + \dfrac{3^3}{3!} + \dfrac{3^4}{4!} + \cdots = \displaystyle\sum_{n=1}^{\infty}\frac{3^n}{n!} = \sum_{n=0}^{\infty}\frac{3^n}{n!} - 1 = e^3 - 1$, by (11).

81. If p is an nth-degree polynomial, then $p^{(i)}(x) = 0$ for $i > n$, so its Taylor series at a is $p(x) = \displaystyle\sum_{i=0}^{n}\frac{p^{(i)}(a)}{i!}(x - a)^i$.

Put $x - a = 1$, so that $x = a + 1$. Then $p(a + 1) = \displaystyle\sum_{i=0}^{n}\frac{p^{(i)}(a)}{i!}$.

This is true for any a, so replace a by x: $p(x + 1) = \displaystyle\sum_{i=0}^{n}\frac{p^{(i)}(x)}{i!}$

83. Assume that $|f'''(x)| \le M$, so $f'''(x) \le M$ for $a \le x \le a + d$. Now $\int_a^x f'''(t)\,dt \le \int_a^x M\,dt \quad\Rightarrow$

$f''(x) - f''(a) \le M(x - a) \quad\Rightarrow\quad f''(x) \le f''(a) + M(x - a)$. Thus, $\int_a^x f''(t)\,dt \le \int_a^x [f''(a) + M(t - a)]\,dt \quad\Rightarrow$

$f'(x) - f'(a) \le f''(a)(x - a) + \frac{1}{2}M(x - a)^2 \quad\Rightarrow\quad f'(x) \le f'(a) + f''(a)(x - a) + \frac{1}{2}M(x - a)^2 \quad\Rightarrow$

$\int_a^x f'(t)\,dt \le \int_a^x \left[f'(a) + f''(a)(t - a) + \frac{1}{2}M(t - a)^2 \right] dt \quad\Rightarrow$

$f(x) - f(a) \le f'(a)(x - a) + \frac{1}{2}f''(a)(x - a)^2 + \frac{1}{6}M(x - a)^3$. So

$f(x) - f(a) - f'(a)(x - a) - \frac{1}{2}f''(a)(x - a)^2 \le \frac{1}{6}M(x - a)^3$. But

$R_2(x) = f(x) - T_2(x) = f(x) - f(a) - f'(a)(x - a) - \frac{1}{2}f''(a)(x - a)^2$, so $R_2(x) \le \frac{1}{6}M(x - a)^3$.

A similar argument using $f'''(x) \ge -M$ shows that $R_2(x) \ge -\frac{1}{6}M(x - a)^3$. So $|R_2(x_2)| \le \frac{1}{6}M|x - a|^3$.

Although we have assumed that $x > a$, a similar calculation shows that this inequality is also true if $x < a$.

85. (a) $g(x) = \sum\limits_{n=0}^{\infty} \binom{k}{n} x^n \quad\Rightarrow\quad g'(x) = \sum\limits_{n=1}^{\infty} \binom{k}{n} nx^{n-1}$, so

$(1 + x)g'(x) = (1 + x) \sum\limits_{n=1}^{\infty} \binom{k}{n} nx^{n-1} = \sum\limits_{n=1}^{\infty} \binom{k}{n} nx^{n-1} + \sum\limits_{n=1}^{\infty} \binom{k}{n} nx^n$

$\displaystyle = \sum\limits_{n=0}^{\infty} \binom{k}{n+1} (n+1)x^n + \sum\limits_{n=0}^{\infty} \binom{k}{n} nx^n \qquad \left[\begin{array}{l} \text{Replace } n \text{ with } n + 1 \\ \text{in the first series} \end{array} \right]$

$\displaystyle = \sum\limits_{n=0}^{\infty} (n+1)\frac{k(k-1)(k-2)\cdots(k-n+1)(k-n)}{(n+1)!} x^n + \sum\limits_{n=0}^{\infty} \left[(n)\frac{k(k-1)(k-2)\cdots(k-n+1)}{n!} \right] x^n$

$\displaystyle = \sum\limits_{n=0}^{\infty} \frac{(n+1)k(k-1)(k-2)\cdots(k-n+1)}{(n+1)!} [(k-n) + n]\, x^n$

$\displaystyle = k \sum\limits_{n=0}^{\infty} \frac{k(k-1)(k-2)\cdots(k-n+1)}{n!} x^n = k \sum\limits_{n=0}^{\infty} \binom{k}{n} x^n = kg(x)$

Thus, $g'(x) = \dfrac{kg(x)}{1 + x}$.

(b) $h(x) = (1 + x)^{-k} g(x) \quad\Rightarrow$

$\qquad\qquad h'(x) = -k(1+x)^{-k-1}g(x) + (1+x)^{-k}g'(x) \qquad\qquad \text{[Product Rule]}$

$\qquad\qquad\qquad = -k(1+x)^{-k-1}g(x) + (1+x)^{-k}\dfrac{kg(x)}{1+x} \qquad\qquad \text{[from part (a)]}$

$\qquad\qquad\qquad = -k(1+x)^{-k-1}g(x) + k(1+x)^{-k-1}g(x) = 0$

(c) From part (b) we see that $h(x)$ must be constant for $x \in (-1, 1)$, so $h(x) = h(0) = 1$ for $x \in (-1, 1)$.

Thus, $h(x) = 1 = (1 + x)^{-k} g(x) \quad\Leftrightarrow\quad g(x) = (1 + x)^k$ for $x \in (-1, 1)$.

11.11 Applications of Taylor Polynomials

1. (a)

n	$f^{(n)}(x)$	$f^{(n)}(0)$	$T_n(x)$
0	$\sin x$	0	0
1	$\cos x$	1	x
2	$-\sin x$	0	x
3	$-\cos x$	-1	$x - \frac{1}{6}x^3$
4	$\sin x$	0	$x - \frac{1}{6}x^3$
5	$\cos x$	1	$x - \frac{1}{6}x^3 + \frac{1}{120}x^5$

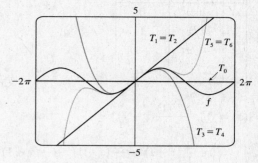

$$\textit{Note: } T_n(x) = \sum_{k=0}^{n} \frac{f^{(k)}(0)}{k!}x^k$$

(b)

x	f	$T_0(x)$	$T_1(x) = T_2(x)$	$T_3(x) = T_4(x)$	$T_5(x)$
$\frac{\pi}{4}$	0.7071	0	0.7854	0.7047	0.7071
$\frac{\pi}{2}$	1	0	1.5708	0.9248	1.0045
π	0	0	3.1416	-2.0261	0.5240

(c) As n increases, $T_n(x)$ is a good approximation to $f(x)$ on a larger and larger interval.

3.

n	$f^{(n)}(x)$	$f^{(n)}(1)$
0	e^x	e
1	e^x	e
2	e^x	e
3	e^x	e

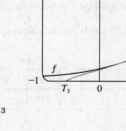

$$\begin{aligned}
T_3(x) &= \sum_{n=0}^{3} \frac{f^{(n)}(1)}{n!}(x-1)^n \\
&= \frac{e}{0!}(x-1)^0 + \frac{e}{1!}(x-1)^1 + \frac{e}{2!}(x-1)^2 + \frac{e}{3!}(x-1)^3 \\
&= e + e(x-1) + \tfrac{1}{2}e(x-1)^2 + \tfrac{1}{6}e(x-1)^3
\end{aligned}$$

5.

n	$f^{(n)}(x)$	$f^{(n)}(\pi/2)$
0	$\cos x$	0
1	$-\sin x$	-1
2	$-\cos x$	0
3	$\sin x$	1

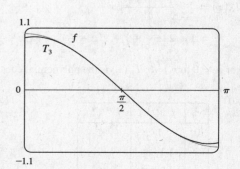

$$\begin{aligned}
T_3(x) &= \sum_{n=0}^{3} \frac{f^{(n)}(\pi/2)}{n!}\left(x - \tfrac{\pi}{2}\right)^n \\
&= -\left(x - \tfrac{\pi}{2}\right) + \tfrac{1}{6}\left(x - \tfrac{\pi}{2}\right)^3
\end{aligned}$$

7.

n	$f^{(n)}(x)$	$f^{(n)}(1)$
0	$\ln x$	0
1	$1/x$	1
2	$-1/x^2$	-1
3	$2/x^3$	2

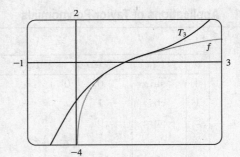

$$T_3(x) = \sum_{n=0}^{3} \frac{f^{(n)}(1)}{n!}(x-1)^n$$

$$= 0 + \frac{1}{1!}(x-1) + \frac{-1}{2!}(x-1)^2 + \frac{2}{3!}(x-1)^3$$

$$= (x-1) - \tfrac{1}{2}(x-1)^2 + \tfrac{1}{3}(x-1)^3$$

9.

n	$f^{(n)}(x)$	$f^{(n)}(0)$
0	xe^{-2x}	0
1	$(1-2x)e^{-2x}$	1
2	$4(x-1)e^{-2x}$	-4
3	$4(3-2x)e^{-2x}$	12

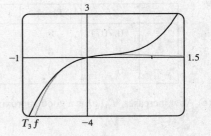

$$T_3(x) = \sum_{n=0}^{3} \frac{f^{(n)}(0)}{n!}x^n = \frac{0}{1}\cdot 1 + \frac{1}{1}x^1 + \frac{-4}{2}x^2 + \frac{12}{6}x^3 = x - 2x^2 + 2x^3$$

11. You may be able to simply find the Taylor polynomials for

$f(x) = \cot x$ using your CAS. We will list the values of $f^{(n)}(\pi/4)$

for $n = 0$ to $n = 5$.

n	0	1	2	3	4	5
$f^{(n)}(\pi/4)$	1	-2	4	-16	80	-512

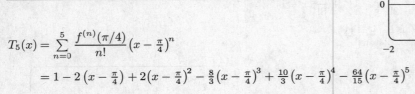

$$T_5(x) = \sum_{n=0}^{5} \frac{f^{(n)}(\pi/4)}{n!}\left(x - \tfrac{\pi}{4}\right)^n$$

$$= 1 - 2\left(x - \tfrac{\pi}{4}\right) + 2\left(x - \tfrac{\pi}{4}\right)^2 - \tfrac{8}{3}\left(x - \tfrac{\pi}{4}\right)^3 + \tfrac{10}{3}\left(x - \tfrac{\pi}{4}\right)^4 - \tfrac{64}{15}\left(x - \tfrac{\pi}{4}\right)^5$$

For $n = 2$ to $n = 5$, $T_n(x)$ is the polynomial consisting of all the terms up to and including the $\left(x - \tfrac{\pi}{4}\right)^n$ term.

13. (a)

n	$f^{(n)}(x)$	$f^{(n)}(1)$
0	$1/x$	1
1	$-1/x^2$	-1
2	$2/x^3$	2
3	$-6/x^4$	

$f(x) = 1/x \approx T_2(x)$

$$= \frac{1}{0!}(x-1)^0 - \frac{1}{1!}(x-1)^1 + \frac{2}{2!}(x-1)^2$$

$$= 1 - (x-1) + (x-1)^2$$

(b) $|R_2(x)| \le \dfrac{M}{3!}\,|x-1|^3$, where $|f'''(x)| \le M$. Now $0.7 \le x \le 1.3 \;\Rightarrow\; |x-1| \le 0.3 \;\Rightarrow\; |x-1|^3 \le 0.027$.

Since $|f'''(x)|$ is decreasing on $[0.7, 1.3]$, we can take $M = |f'''(0.7)| = 6/(0.7)^4$, so

$$|R_2(x)| \le \frac{6/(0.7)^4}{6}(0.027) = 0.112\,453\,1.$$

(c)

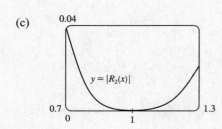

From the graph of $|R_2(x)| = \left|\dfrac{1}{x} - T_2(x)\right|$, it seems that the error is less than

$0.038\,571$ on $[0.7, 1.3]$.

15.

n	$f^{(n)}(x)$	$f^{(n)}(1)$
0	$x^{2/3}$	1
1	$\frac{2}{3}x^{-1/3}$	$\frac{2}{3}$
2	$-\frac{2}{9}x^{-4/3}$	$-\frac{2}{9}$
3	$\frac{8}{27}x^{-7/3}$	$\frac{8}{27}$
4	$-\frac{56}{81}x^{-10/3}$	

(a) $f(x) = x^{2/3} \approx T_3(x) = 1 + \frac{2}{3}(x-1) - \dfrac{2/9}{2!}(x-1)^2 + \dfrac{8/27}{3!}(x-1)^3$

$\qquad = 1 + \frac{2}{3}(x-1) - \frac{1}{9}(x-1)^2 + \frac{4}{81}(x-1)^3$

(b) $|R_3(x)| \le \dfrac{M}{4!}\,|x-1|^4$, where $\left|f^{(4)}(x)\right| \le M$. Now $0.8 \le x \le 1.2 \;\Rightarrow$

$|x-1| \le 0.2 \;\Rightarrow\; |x-1|^4 \le 0.0016$. Since $\left|f^{(4)}(x)\right|$ is decreasing

on $[0.8, 1.2]$, we can take $M = \left|f^{(4)}(0.8)\right| = \frac{56}{81}(0.8)^{-10/3}$, so

$$|R_3(x)| \le \frac{\frac{56}{81}(0.8)^{-10/3}}{24}(0.0016) \approx 0.000\,096\,97.$$

(c)

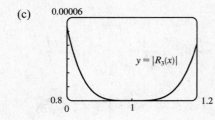

From the graph of $|R_3(x)| = \left|x^{2/3} - T_3(x)\right|$, it seems that the

error is less than $0.000\,053\,3$ on $[0.8, 1.2]$.

17.

n	$f^{(n)}(x)$	$f^{(n)}(0)$
0	$\sec x$	1
1	$\sec x \tan x$	0
2	$\sec x\,(2\sec^2 x - 1)$	1
3	$\sec x \tan x\,(6\sec^2 x - 1)$	

(a) $f(x) = \sec x \approx T_2(x) = 1 + \frac{1}{2}x^2$

(b) $|R_2(x)| \le \dfrac{M}{3!}\,|x|^3$, where $\left|f^{(3)}(x)\right| \le M$. Now $-0.2 \le x \le 0.2 \;\Rightarrow\; |x| \le 0.2 \;\Rightarrow\; |x|^3 \le (0.2)^3$.

$f^{(3)}(x)$ is an odd function and it is increasing on $[0, 0.2]$ since $\sec x$ and $\tan x$ are increasing on $[0, 0.2]$,

so $\left|f^{(3)}(x)\right| \le f^{(3)}(0.2) \approx 1.085\,158\,892$. Thus, $|R_2(x)| \le \dfrac{f^{(3)}(0.2)}{3!}(0.2)^3 \approx 0.001\,447$.

(c)

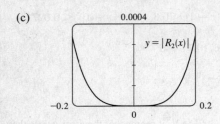

From the graph of $|R_2(x)| = |\sec x - T_2(x)|$, it seems that the error is less than $0.000\,339$ on $[-0.2, 0.2]$.

19.

n	$f^{(n)}(x)$	$f^{(n)}(0)$
0	e^{x^2}	1
1	$e^{x^2}(2x)$	0
2	$e^{x^2}(2 + 4x^2)$	2
3	$e^{x^2}(12x + 8x^3)$	0
4	$e^{x^2}(12 + 48x^2 + 16x^4)$	

(a) $f(x) = e^{x^2} \approx T_3(x) = 1 + \dfrac{2}{2!}x^2 = 1 + x^2$

(b) $|R_3(x)| \le \dfrac{M}{4!}|x|^4$, where $\left|f^{(4)}(x)\right| \le M$. Now $0 \le x \le 0.1 \;\Rightarrow$

$x^4 \le (0.1)^4$, and letting $x = 0.1$ gives

$$|R_3(x)| \le \frac{e^{0.01}(12 + 0.48 + 0.0016)}{24}(0.1)^4 \approx 0.00006.$$

(c)

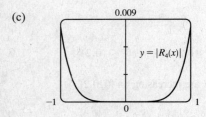

From the graph of $|R_3(x)| = \left|e^{x^2} - T_3(x)\right|$, it appears that the error is less than $0.000\,051$ on $[0, 0.1]$.

21.

n	$f^{(n)}(x)$	$f^{(n)}(0)$
0	$x \sin x$	0
1	$\sin x + x \cos x$	0
2	$2 \cos x - x \sin x$	2
3	$-3 \sin x - x \cos x$	0
4	$-4 \cos x + x \sin x$	-4
5	$5 \sin x + x \cos x$	

(a) $f(x) = x \sin x \approx T_4(x) = \dfrac{2}{2!}(x - 0)^2 + \dfrac{-4}{4!}(x - 0)^4 = x^2 - \dfrac{1}{6}x^4$

(b) $|R_4(x)| \le \dfrac{M}{5!}|x|^5$, where $\left|f^{(5)}(x)\right| \le M$. Now $-1 \le x \le 1 \;\Rightarrow$

$|x| \le 1$, and a graph of $f^{(5)}(x)$ shows that $\left|f^{(5)}(x)\right| \le 5$ for $-1 \le x \le 1$.

Thus, we can take $M = 5$ and get $|R_4(x)| \le \dfrac{5}{5!} \cdot 1^5 = \dfrac{1}{24} = 0.041\overline{6}$.

(c)

From the graph of $|R_4(x)| = |x \sin x - T_4(x)|$, it seems that the error is less than 0.0082 on $[-1, 1]$.

23. From Exercise 5, $\cos x = -\left(x - \frac{\pi}{2}\right) + \frac{1}{6}\left(x - \frac{\pi}{2}\right)^3 + R_3(x)$, where $|R_3(x)| \leq \frac{M}{4!}\left|x - \frac{\pi}{2}\right|^4$ with

$\left|f^{(4)}(x)\right| = |\cos x| \leq M = 1$. Now $x = 80° = (90° - 10°) = \left(\frac{\pi}{2} - \frac{\pi}{18}\right) = \frac{4\pi}{9}$ radians, so the error is

$\left|R_3\left(\frac{4\pi}{9}\right)\right| \leq \frac{1}{24}\left(\frac{\pi}{18}\right)^4 \approx 0.000\,039$, which means our estimate would *not* be accurate to five decimal places. However,

$T_3 = T_4$, so we can use $\left|R_4\left(\frac{4\pi}{9}\right)\right| \leq \frac{1}{120}\left(\frac{\pi}{18}\right)^5 \approx 0.000\,001$. Therefore, to five decimal places,

$\cos 80° \approx -\left(-\frac{\pi}{18}\right) + \frac{1}{6}\left(-\frac{\pi}{18}\right)^3 \approx 0.17365$.

25. All derivatives of e^x are e^x, so $|R_n(x)| \leq \dfrac{e^x}{(n+1)!}|x|^{n+1}$, where $0 < x < 0.1$. Letting $x = 0.1$,

$R_n(0.1) \leq \dfrac{e^{0.1}}{(n+1)!}(0.1)^{n+1} < 0.00001$, and by trial and error we find that $n = 3$ satisfies this inequality since

$R_3(0.1) < 0.0000046$. Thus, by adding the four terms of the Maclaurin series for e^x corresponding to $n = 0, 1, 2,$ and 3,

we can estimate $e^{0.1}$ to within 0.00001. (In fact, this sum is $1.1051\overline{6}$ and $e^{0.1} \approx 1.10517$.)

27. $\sin x = x - \dfrac{1}{3!}x^3 + \dfrac{1}{5!}x^5 - \cdots$. By the Alternating Series

Estimation Theorem, the error in the approximation

$\sin x = x - \dfrac{1}{3!}x^3$ is less than $\left|\dfrac{1}{5!}x^5\right| < 0.01 \iff$

$\left|x^5\right| < 120(0.01) \iff |x| < (1.2)^{1/5} \approx 1.037$. The curves

$y = x - \frac{1}{6}x^3$ and $y = \sin x - 0.01$ intersect at $x \approx 1.043$, so

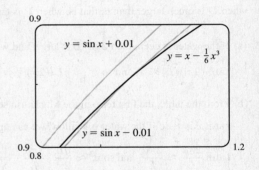

the graph confirms our estimate. Since both the sine function

and the given approximation are odd functions, we need to check the estimate only for $x > 0$. Thus, the desired range of

values for x is $-1.037 < x < 1.037$.

29. $\arctan x = x - \dfrac{x^3}{3} + \dfrac{x^5}{5} - \dfrac{x^7}{7} + \cdots$. By the Alternating Series

Estimation Theorem, the error is less than $\left|-\frac{1}{7}x^7\right| < 0.05 \iff$

$\left|x^7\right| < 0.35 \iff |x| < (0.35)^{1/7} \approx 0.8607$. The curves

$y = x - \frac{1}{3}x^3 + \frac{1}{5}x^5$ and $y = \arctan x + 0.05$ intersect at

$x \approx 0.9245$, so the graph confirms our estimate. Since both the

arctangent function and the given approximation are odd functions,

we need to check the estimate only for $x > 0$. Thus, the desired

range of values for x is $-0.86 < x < 0.86$.

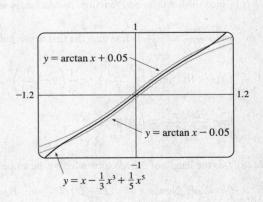

31. Let $s(t)$ be the position function of the car, and for convenience set $s(0) = 0$. The velocity of the car is $v(t) = s'(t)$ and the

acceleration is $a(t) = s''(t)$, so the second degree Taylor polynomial is $T_2(t) = s(0) + v(0)t + \dfrac{a(0)}{2}t^2 = 20t + t^2$. We

estimate the distance traveled during the next second to be $s(1) \approx T_2(1) = 20 + 1 = 21$ m. The function $T_2(t)$ would not be

accurate over a full minute, since the car could not possibly maintain an acceleration of 2 m/s^2 for that long (if it did, its final

speed would be 140 m/s $\approx$ 313 mi/h!).

33. $E = \dfrac{q}{D^2} - \dfrac{q}{(D+d)^2} = \dfrac{q}{D^2} - \dfrac{q}{D^2(1+d/D)^2} = \dfrac{q}{D^2}\left[1 - \left(1 + \dfrac{d}{D}\right)^{-2}\right].$

We use the Binomial Series to expand $(1 + d/D)^{-2}$:

$$E = \dfrac{q}{D^2}\left[1 - \left(1 - 2\left(\dfrac{d}{D}\right) + \dfrac{2\cdot 3}{2!}\left(\dfrac{d}{D}\right)^2 - \dfrac{2\cdot 3\cdot 4}{3!}\left(\dfrac{d}{D}\right)^3 + \cdots\right)\right] = \dfrac{q}{D^2}\left[2\left(\dfrac{d}{D}\right) - 3\left(\dfrac{d}{D}\right)^2 + 4\left(\dfrac{d}{D}\right)^3 - \cdots\right]$$

$$\approx \dfrac{q}{D^2}\cdot 2\left(\dfrac{d}{D}\right) = 2qd\cdot\dfrac{1}{D^3}$$

when D is much larger than d; that is, when P is far away from the dipole.

35. (a) If the water is deep, then $2\pi d/L$ is large, and we know that $\tanh x \to 1$ as $x \to \infty$. So we can approximate

$\tanh(2\pi d/L) \approx 1$, and so $v^2 \approx gL/(2\pi) \iff v \approx \sqrt{gL/(2\pi)}$.

(b) From the table, the first term in the Maclaurin series of

$\tanh x$ is x, so if the water is shallow, we can approximate

$\tanh\dfrac{2\pi d}{L} \approx \dfrac{2\pi d}{L}$, and so $v^2 \approx \dfrac{gL}{2\pi}\cdot\dfrac{2\pi d}{L} \iff v \approx \sqrt{gd}$.

n	$f^{(n)}(x)$	$f^{(n)}(0)$
0	$\tanh x$	0
1	$\operatorname{sech}^2 x$	1
2	$-2\operatorname{sech}^2 x \tanh x$	0
3	$2\operatorname{sech}^2 x\,(3\tanh^2 x - 1)$	-2

(c) Since $\tanh x$ is an odd function, its Maclaurin series is alternating, so the error in the approximation

$\tanh\dfrac{2\pi d}{L} \approx \dfrac{2\pi d}{L}$ is less than the first neglected term, which is $\dfrac{|f'''(0)|}{3!}\left(\dfrac{2\pi d}{L}\right)^3 = \dfrac{1}{3}\left(\dfrac{2\pi d}{L}\right)^3.$

If $L > 10d$, then $\dfrac{1}{3}\left(\dfrac{2\pi d}{L}\right)^3 < \dfrac{1}{3}\left(2\pi\cdot\dfrac{1}{10}\right)^3 = \dfrac{\pi^3}{375}$, so the error in the approximation $v^2 = gd$ is less

than $\dfrac{gL}{2\pi}\cdot\dfrac{\pi^3}{375} \approx 0.0132gL.$

37. (a) L is the length of the arc subtended by the angle θ, so $L = R\theta \Rightarrow$

$\theta = L/R$. Now $\sec\theta = (R+C)/R \Rightarrow R\sec\theta = R + C \Rightarrow$

$C = R\sec\theta - R = R\sec(L/R) - R.$

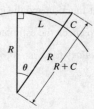

(b) First we'll find a Taylor polynomial $T_4(x)$ for $f(x) = \sec x$ at $x = 0$.

n	$f^{(n)}(x)$	$f^{(n)}(0)$
0	$\sec x$	1
1	$\sec x \tan x$	0
2	$\sec x(2\tan^2 x + 1)$	1
3	$\sec x \tan x(6\tan^2 x + 5)$	0
4	$\sec x(24\tan^4 x + 28\tan^2 x + 5)$	5

Thus, $f(x) = \sec x \approx T_4(x) = 1 + \frac{1}{2!}(x-0)^2 + \frac{5}{4!}(x-0)^4 = 1 + \frac{1}{2}x^2 + \frac{5}{24}x^4$. By part (a),

$$C \approx R\left[1 + \frac{1}{2}\left(\frac{L}{R}\right)^2 + \frac{5}{24}\left(\frac{L}{R}\right)^4\right] - R = R + \frac{1}{2}R \cdot \frac{L^2}{R^2} + \frac{5}{24}R \cdot \frac{L^4}{R^4} - R = \frac{L^2}{2R} + \frac{5L^4}{24R^3}.$$

(c) Taking $L = 100$ km and $R = 6370$ km, the formula in part (a) says that

$$C = R\sec(L/R) - R = 6370\sec(100/6370) - 6370 \approx 0.785\,009\,965\,44 \text{ km}.$$

The formula in part (b) says that $C \approx \dfrac{L^2}{2R} + \dfrac{5L^4}{24R^3} = \dfrac{100^2}{2 \cdot 6370} + \dfrac{5 \cdot 100^4}{24 \cdot 6370^3} \approx 0.785\,009\,957\,36$ km.

The difference between these two results is only $0.000\,000\,008\,08$ km, or $0.000\,008\,08$ m!

39. Using $f(x) = T_n(x) + R_n(x)$ with $n = 1$ and $x = r$, we have $f(r) = T_1(r) + R_1(r)$, where T_1 is the first-degree Taylor polynomial of f at a. Because $a = x_n$, $f(r) = f(x_n) + f'(x_n)(r - x_n) + R_1(r)$. But r is a root of f, so $f(r) = 0$ and we have $0 = f(x_n) + f'(x_n)(r - x_n) + R_1(r)$. Taking the first two terms to the left side gives us

$f'(x_n)(x_n - r) - f(x_n) = R_1(r)$. Dividing by $f'(x_n)$, we get $x_n - r - \dfrac{f(x_n)}{f'(x_n)} = \dfrac{R_1(r)}{f'(x_n)}$. By the formula for Newton's

method, the left side of the preceding equation is $x_{n+1} - r$, so $|x_{n+1} - r| = \left|\dfrac{R_1(r)}{f'(x_n)}\right|$. Taylor's Inequality gives us

$|R_1(r)| \leq \dfrac{|f''(r)|}{2!}|r - x_n|^2$. Combining this inequality with the facts $|f''(x)| \leq M$ and $|f'(x)| \geq K$ gives us

$|x_{n+1} - r| \leq \dfrac{M}{2K}|x_n - r|^2$.

11   Review

TRUE-FALSE QUIZ

---

1. False.   See Note 2 after Theorem 11.2.6.

3. True.   If $\lim\limits_{n\to\infty} a_n = L$, then as $n \to \infty$, $2n + 1 \to \infty$, so $a_{2n+1} \to L$.

5. False.   For example, take $c_n = (-1)^n/(n6^n)$.

7. False, since $\lim\limits_{n\to\infty}\left|\dfrac{a_{n+1}}{a_n}\right| = \lim\limits_{n\to\infty}\left|\dfrac{1}{(n+1)^3} \cdot \dfrac{n^3}{1}\right| = \lim\limits_{n\to\infty}\left|\dfrac{n^3}{(n+1)^3} \cdot \dfrac{1/n^3}{1/n^3}\right| = \lim\limits_{n\to\infty}\dfrac{1}{(1+1/n)^3} = 1$.

9. False. See the note after Example 11.4.2.

11. True. See (9) in Section 11.1.

13. True. By Theorem 11.10.5 the coefficient of x^3 is $\dfrac{f'''(0)}{3!} = \dfrac{1}{3}$ $\Rightarrow$ $f'''(0) = 2$.

Or: Use Theorem 11.9.2 to differentiate f three times.

15. False. For example, let $a_n = b_n = (-1)^n$. Then $\{a_n\}$ and $\{b_n\}$ are divergent, but $a_n b_n = 1$, so $\{a_n b_n\}$ is convergent.

17. True by Theorem 11.6.3. $\left[\sum (-1)^n \, a_n \text{ is absolutely convergent and hence convergent.} \right]$

19. True. $0.99999\ldots = 0.9 + 0.9(0.1)^1 + 0.9(0.1)^2 + 0.9(0.1)^3 + \cdots = \displaystyle\sum_{n=1}^{\infty} (0.9)(0.1)^{n-1} = \dfrac{0.9}{1-0.1} = 1$ by the formula

for the sum of a geometric series $[S = a_1/(1-r)]$ with ratio r satisfying $|r| < 1$.

21. True. A finite number of terms doesn't affect convergence or divergence of a series.

EXERCISES

1. $\left\{ \dfrac{2+n^3}{1+2n^3} \right\}$ converges since $\displaystyle\lim_{n\to\infty} \dfrac{2+n^3}{1+2n^3} = \lim_{n\to\infty} \dfrac{2/n^3 + 1}{1/n^3 + 2} = \dfrac{1}{2}$.

3. $\displaystyle\lim_{n\to\infty} a_n = \lim_{n\to\infty} \dfrac{n^3}{1+n^2} = \lim_{n\to\infty} \dfrac{n}{1/n^2 + 1} = \infty$, so the sequence diverges.

5. $|a_n| = \left| \dfrac{n \sin n}{n^2+1} \right| \le \dfrac{n}{n^2+1} < \dfrac{1}{n}$, so $|a_n| \to 0$ as $n \to \infty$. Thus, $\displaystyle\lim_{n\to\infty} a_n = 0$. The sequence $\{a_n\}$ is convergent.

7. $\left\{ \left(1 + \dfrac{3}{n} \right)^{4n} \right\}$ is convergent. Let $y = \left(1 + \dfrac{3}{x} \right)^{4x}$. Then

$\displaystyle\lim_{x\to\infty} \ln y = \lim_{x\to\infty} 4x \ln(1+3/x) = \lim_{x\to\infty} \dfrac{\ln(1+3/x)}{1/(4x)} \overset{\text{H}}{=} \lim_{x\to\infty} \dfrac{\dfrac{1}{1+3/x}\left(-\dfrac{3}{x^2}\right)}{-1/(4x^2)} = \lim_{x\to\infty} \dfrac{12}{1+3/x} = 12$, so

$\displaystyle\lim_{x\to\infty} y = \lim_{n\to\infty} \left(1 + \dfrac{3}{n} \right)^{4n} = e^{12}$.

9. We use induction, hypothesizing that $a_{n-1} < a_n < 2$. Note first that $1 < a_2 = \frac{1}{3}(1+4) = \frac{5}{3} < 2$, so the hypothesis holds

for $n = 2$. Now assume that $a_{k-1} < a_k < 2$. Then $a_k = \frac{1}{3}(a_{k-1}+4) < \frac{1}{3}(a_k+4) < \frac{1}{3}(2+4) = 2$. So $a_k < a_{k+1} < 2$,

and the induction is complete. To find the limit of the sequence, we note that $L = \displaystyle\lim_{n\to\infty} a_n = \lim_{n\to\infty} a_{n+1}$ $\Rightarrow$

$L = \frac{1}{3}(L+4)$ $\Rightarrow$ $L = 2$.

11. $\dfrac{n}{n^3+1} < \dfrac{n}{n^3} = \dfrac{1}{n^2}$, so $\displaystyle\sum_{n=1}^{\infty} \dfrac{n}{n^3+1}$ converges by the Comparison Test with the convergent p-series $\displaystyle\sum_{n=1}^{\infty} \dfrac{1}{n^2}$ $[p = 2 > 1]$.

13. $\lim\limits_{n\to\infty} \left| \dfrac{a_{n+1}}{a_n} \right| = \lim\limits_{n\to\infty} \left[\dfrac{(n+1)^3}{5^{n+1}} \cdot \dfrac{5^n}{n^3} \right] = \lim\limits_{n\to\infty} \left(1 + \dfrac{1}{n} \right)^3 \cdot \dfrac{1}{5} = \dfrac{1}{5} < 1$, so $\sum\limits_{n=1}^{\infty} \dfrac{n^3}{5^n}$ converges by the Ratio Test.

15. Let $f(x) = \dfrac{1}{x\sqrt{\ln x}}$. Then f is continuous, positive, and decreasing on $[2, \infty)$, so the Integral Test applies.

$$\int_2^{\infty} f(x)\,dx = \lim\limits_{t\to\infty} \int_2^t \dfrac{1}{x\sqrt{\ln x}}\,dx \quad \left[u = \ln x,\, du = \dfrac{1}{x}\,dx \right] = \lim\limits_{t\to\infty} \int_{\ln 2}^{\ln t} u^{-1/2}\,du = \lim\limits_{t\to\infty} \left[2\sqrt{u} \right]_{\ln 2}^{\ln t}$$

$$= \lim\limits_{t\to\infty} \left(2\sqrt{\ln t} - 2\sqrt{\ln 2} \right) = \infty,$$

so the series $\sum\limits_{n=2}^{\infty} \dfrac{1}{n\sqrt{\ln n}}$ diverges.

17. $|a_n| = \left| \dfrac{\cos 3n}{1 + (1.2)^n} \right| \le \dfrac{1}{1 + (1.2)^n} < \dfrac{1}{(1.2)^n} = \left(\dfrac{5}{6} \right)^n$, so $\sum\limits_{n=1}^{\infty} |a_n|$ converges by comparison with the convergent geometric

series $\sum\limits_{n=1}^{\infty} \left(\dfrac{5}{6} \right)^n$ $\left[r = \dfrac{5}{6} < 1 \right]$. It follows that $\sum\limits_{n=1}^{\infty} a_n$ converges (by Theorem 11.6.3).

19. $\lim\limits_{n\to\infty} \left| \dfrac{a_{n+1}}{a_n} \right| = \lim\limits_{n\to\infty} \dfrac{1 \cdot 3 \cdot 5 \cdot \cdots \cdot (2n-1)(2n+1)}{5^{n+1}\,(n+1)!} \cdot \dfrac{5^n\,n!}{1 \cdot 3 \cdot 5 \cdot \cdots \cdot (2n-1)} = \lim\limits_{n\to\infty} \dfrac{2n+1}{5(n+1)} = \dfrac{2}{5} < 1$, so the series

converges by the Ratio Test.

21. $b_n = \dfrac{\sqrt{n}}{n+1} > 0$, $\{b_n\}$ is decreasing, and $\lim\limits_{n\to\infty} b_n = 0$, so the series $\sum\limits_{n=1}^{\infty} (-1)^{n-1} \dfrac{\sqrt{n}}{n+1}$ converges by the Alternating

Series Test.

23. Consider the series of absolute values: $\sum\limits_{n=1}^{\infty} n^{-1/3}$ is a p-series with $p = \dfrac{1}{3} \le 1$ and is therefore divergent. But if we apply the

Alternating Series Test, we see that $b_n = \dfrac{1}{\sqrt[3]{n}} > 0$, $\{b_n\}$ is decreasing, and $\lim\limits_{n\to\infty} b_n = 0$, so the series $\sum\limits_{n=1}^{\infty} (-1)^{n-1}\, n^{-1/3}$

converges. Thus, $\sum\limits_{n=1}^{\infty} (-1)^{n-1}\, n^{-1/3}$ is conditionally convergent.

25. $\left| \dfrac{a_{n+1}}{a_n} \right| = \left| \dfrac{(-1)^{n+1}(n+2)3^{n+1}}{2^{2n+3}} \cdot \dfrac{2^{2n+1}}{(-1)^n(n+1)3^n} \right| = \dfrac{n+2}{n+1} \cdot \dfrac{3}{4} = \dfrac{1 + (2/n)}{1 + (1/n)} \cdot \dfrac{3}{4} \to \dfrac{3}{4} < 1$ as $n \to \infty$, so by the Ratio

Test, $\sum\limits_{n=1}^{\infty} \dfrac{(-1)^n(n+1)3^n}{2^{2n+1}}$ is absolutely convergent.

27. $\sum\limits_{n=1}^{\infty} \dfrac{(-3)^{n-1}}{2^{3n}} = \sum\limits_{n=1}^{\infty} \dfrac{(-3)^{n-1}}{(2^3)^n} = \sum\limits_{n=1}^{\infty} \dfrac{(-3)^{n-1}}{8^n} = \dfrac{1}{8} \sum\limits_{n=1}^{\infty} \dfrac{(-3)^{n-1}}{8^{n-1}} = \dfrac{1}{8} \sum\limits_{n=1}^{\infty} \left(-\dfrac{3}{8} \right)^{n-1} = \dfrac{1}{8} \left(\dfrac{1}{1 - (-3/8)} \right)$

$$= \dfrac{1}{8} \cdot \dfrac{8}{11} = \dfrac{1}{11}$$

29. $\displaystyle\sum_{n=1}^{\infty} [\tan^{-1}(n+1) - \tan^{-1} n] = \lim_{n\to\infty} s_n$

$$= \lim_{n\to\infty} [(\tan^{-1} 2 - \tan^{-1} 1) + (\tan^{-1} 3 - \tan^{-1} 2) + \cdots + (\tan^{-1}(n+1) - \tan^{-1} n)]$$

$$= \lim_{n\to\infty} [\tan^{-1}(n+1) - \tan^{-1} 1] = \tfrac{\pi}{2} - \tfrac{\pi}{4} = \tfrac{\pi}{4}$$

31. $1 - e + \dfrac{e^2}{2!} - \dfrac{e^3}{3!} + \dfrac{e^4}{4!} - \cdots = \displaystyle\sum_{n=0}^{\infty} (-1)^n \dfrac{e^n}{n!} = \sum_{n=0}^{\infty} \dfrac{(-e)^n}{n!} = e^{-e}$ since $e^x = \displaystyle\sum_{n=0}^{\infty} \dfrac{x^n}{n!}$ for all x.

33. $\cosh x = \dfrac{1}{2}(e^x + e^{-x}) = \dfrac{1}{2}\left(\displaystyle\sum_{n=0}^{\infty} \dfrac{x^n}{n!} + \sum_{n=0}^{\infty} \dfrac{(-x)^n}{n!} \right)$

$$= \dfrac{1}{2}\left[\left(1 + x + \dfrac{x^2}{2!} + \dfrac{x^3}{3!} + \dfrac{x^4}{4!} + \cdots \right) + \left(1 - x + \dfrac{x^2}{2!} - \dfrac{x^3}{3!} + \dfrac{x^4}{4!} - \cdots \right) \right]$$

$$= \dfrac{1}{2}\left(2 + 2 \cdot \dfrac{x^2}{2!} + 2 \cdot \dfrac{x^4}{4!} + \cdots \right) = 1 + \dfrac{1}{2}x^2 + \sum_{n=2}^{\infty} \dfrac{x^{2n}}{(2n)!} \geq 1 + \dfrac{1}{2}x^2 \quad \text{for all } x$$

35. $\displaystyle\sum_{n=1}^{\infty} \dfrac{(-1)^{n+1}}{n^5} = 1 - \dfrac{1}{32} + \dfrac{1}{243} - \dfrac{1}{1024} + \dfrac{1}{3125} - \dfrac{1}{7776} + \dfrac{1}{16,807} - \dfrac{1}{32,768} + \cdots$.

Since $b_8 = \dfrac{1}{8^5} = \dfrac{1}{32,768} < 0.000031$, $\displaystyle\sum_{n=1}^{\infty} \dfrac{(-1)^{n+1}}{n^5} \approx \sum_{n=1}^{7} \dfrac{(-1)^{n+1}}{n^5} \approx 0.9721$.

37. $\displaystyle\sum_{n=1}^{\infty} \dfrac{1}{2 + 5^n} \approx \sum_{n=1}^{8} \dfrac{1}{2 + 5^n} \approx 0.18976224$. To estimate the error, note that $\dfrac{1}{2 + 5^n} < \dfrac{1}{5^n}$, so the remainder term is

$$R_8 = \sum_{n=9}^{\infty} \dfrac{1}{2 + 5^n} < \sum_{n=9}^{\infty} \dfrac{1}{5^n} = \dfrac{1/5^9}{1 - 1/5} = 6.4 \times 10^{-7} \ \left[\text{geometric series with } a = \tfrac{1}{5^9} \text{ and } r = \tfrac{1}{5}\right].$$

39. Use the Limit Comparison Test. $\displaystyle\lim_{n\to\infty} \left| \dfrac{\left(\frac{n+1}{n} \right) a_n}{a_n} \right| = \lim_{n\to\infty} \dfrac{n+1}{n} = \lim_{n\to\infty} \left(1 + \dfrac{1}{n} \right) = 1 > 0.$

Since $\sum |a_n|$ is convergent, so is $\sum \left| \left(\dfrac{n+1}{n} \right) a_n \right|$, by the Limit Comparison Test.

41. $\displaystyle\lim_{n\to\infty} \left| \dfrac{a_{n+1}}{a_n} \right| = \lim_{n\to\infty} \left[\dfrac{|x+2|^{n+1}}{(n+1)\,4^{n+1}} \cdot \dfrac{n\,4^n}{|x+2|^n} \right] = \lim_{n\to\infty} \left[\dfrac{n}{n+1} \dfrac{|x+2|}{4} \right] = \dfrac{|x+2|}{4} < 1 \ \Leftrightarrow \ |x+2| < 4,$ so $R = 4.$

$|x+2| < 4 \ \Leftrightarrow \ -4 < x + 2 < 4 \ \Leftrightarrow \ -6 < x < 2.$ If $x = -6$, then the series $\displaystyle\sum_{n=1}^{\infty} \dfrac{(x+2)^n}{n\,4^n}$ becomes

$\displaystyle\sum_{n=1}^{\infty} \dfrac{(-4)^n}{n4^n} = \sum_{n=1}^{\infty} \dfrac{(-1)^n}{n}$, the alternating harmonic series, which converges by the Alternating Series Test. When $x = 2$, the

series becomes the harmonic series $\displaystyle\sum_{n=1}^{\infty} \dfrac{1}{n}$, which diverges. Thus, $I = [-6, 2).$

43. $\lim\limits_{n\to\infty}\left|\dfrac{a_{n+1}}{a_n}\right| = \lim\limits_{n\to\infty}\left|\dfrac{2^{n+1}(x-3)^{n+1}}{\sqrt{n+4}}\cdot\dfrac{\sqrt{n+3}}{2^n(x-3)^n}\right| = 2\,|x-3|\lim\limits_{n\to\infty}\sqrt{\dfrac{n+3}{n+4}} = 2\,|x-3| < 1 \;\Leftrightarrow\; |x-3| < \tfrac{1}{2},$

so $R = \tfrac{1}{2}$. $|x-3| < \tfrac{1}{2} \;\Leftrightarrow\; -\tfrac{1}{2} < x - 3 < \tfrac{1}{2} \;\Leftrightarrow\; \tfrac{5}{2} < x < \tfrac{7}{2}$. For $x = \tfrac{7}{2}$, the series $\sum\limits_{n=1}^{\infty}\dfrac{2^n(x-3)^n}{\sqrt{n+3}}$ becomes

$\sum\limits_{n=0}^{\infty}\dfrac{1}{\sqrt{n+3}} = \sum\limits_{n=3}^{\infty}\dfrac{1}{n^{1/2}}$, which diverges $\left[p = \tfrac{1}{2} \le 1\right]$, but for $x = \tfrac{5}{2}$, we get $\sum\limits_{n=0}^{\infty}\dfrac{(-1)^n}{\sqrt{n+3}}$, which is a convergent

alternating series, so $I = \left[\tfrac{5}{2}, \tfrac{7}{2}\right)$.

45.

n	$f^{(n)}(x)$	$f^{(n)}\!\left(\tfrac{\pi}{6}\right)$
0	$\sin x$	$\tfrac{1}{2}$
1	$\cos x$	$\tfrac{\sqrt{3}}{2}$
2	$-\sin x$	$-\tfrac{1}{2}$
3	$-\cos x$	$-\tfrac{\sqrt{3}}{2}$
4	$\sin x$	$\tfrac{1}{2}$
$\vdots$	$\vdots$	$\vdots$

$\sin x = f\!\left(\tfrac{\pi}{6}\right) + f'\!\left(\tfrac{\pi}{6}\right)\!\left(x - \tfrac{\pi}{6}\right) + \dfrac{f''\!\left(\tfrac{\pi}{6}\right)}{2!}\left(x - \tfrac{\pi}{6}\right)^2 + \dfrac{f^{(3)}\!\left(\tfrac{\pi}{6}\right)}{3!}\left(x - \tfrac{\pi}{6}\right)^3 + \dfrac{f^{(4)}\!\left(\tfrac{\pi}{6}\right)}{4!}\left(x - \tfrac{\pi}{6}\right)^4 + \cdots$

$\quad = \dfrac{1}{2}\left[1 - \dfrac{1}{2!}\left(x - \tfrac{\pi}{6}\right)^2 + \dfrac{1}{4!}\left(x - \tfrac{\pi}{6}\right)^4 - \cdots\right] + \dfrac{\sqrt{3}}{2}\left[\left(x - \tfrac{\pi}{6}\right) - \dfrac{1}{3!}\left(x - \tfrac{\pi}{6}\right)^3 + \cdots\right]$

$\quad = \dfrac{1}{2}\sum\limits_{n=0}^{\infty}(-1)^n\dfrac{1}{(2n)!}\left(x - \tfrac{\pi}{6}\right)^{2n} + \dfrac{\sqrt{3}}{2}\sum\limits_{n=0}^{\infty}(-1)^n\dfrac{1}{(2n+1)!}\left(x - \tfrac{\pi}{6}\right)^{2n+1}$

47. $\dfrac{1}{1+x} = \dfrac{1}{1-(-x)} = \sum\limits_{n=0}^{\infty}(-x)^n = \sum\limits_{n=0}^{\infty}(-1)^n x^n$ for $|x| < 1 \;\Rightarrow\; \dfrac{x^2}{1+x} = \sum\limits_{n=0}^{\infty}(-1)^n x^{n+2}$ with $R = 1$.

49. $\displaystyle\int\dfrac{1}{4-x}\,dx = -\ln(4-x) + C$ and

$\displaystyle\int\dfrac{1}{4-x}\,dx = \dfrac{1}{4}\int\dfrac{1}{1-x/4}\,dx = \dfrac{1}{4}\int\sum\limits_{n=0}^{\infty}\left(\dfrac{x}{4}\right)^n dx = \dfrac{1}{4}\int\sum\limits_{n=0}^{\infty}\dfrac{x^n}{4^n}\,dx = \dfrac{1}{4}\sum\limits_{n=0}^{\infty}\dfrac{x^{n+1}}{4^n(n+1)} + C$. So

$\ln(4-x) = -\dfrac{1}{4}\sum\limits_{n=0}^{\infty}\dfrac{x^{n+1}}{4^n(n+1)} + C = -\sum\limits_{n=0}^{\infty}\dfrac{x^{n+1}}{4^{n+1}(n+1)} + C = -\sum\limits_{n=1}^{\infty}\dfrac{x^n}{n4^n} + C$. Putting $x = 0$, we get $C = \ln 4$.

Thus, $f(x) = \ln(4-x) = \ln 4 - \sum\limits_{n=1}^{\infty}\dfrac{x^n}{n4^n}$. The series converges for $|x/4| < 1 \;\Leftrightarrow\; |x| < 4$, so $R = 4$.

Another solution:

$$\ln(4-x) = \ln[4(1 - x/4)] = \ln 4 + \ln(1 - x/4) = \ln 4 + \ln[1 + (-x/4)]$$

$$= \ln 4 + \sum\limits_{n=1}^{\infty}(-1)^{n+1}\dfrac{(-x/4)^n}{n} \quad\text{[from Table 1 in Section 11.10]}$$

$$= \ln 4 + \sum\limits_{n=1}^{\infty}(-1)^{2n+1}\dfrac{x^n}{n4^n} = \ln 4 - \sum\limits_{n=1}^{\infty}\dfrac{x^n}{n4^n}.$$

51. $\sin x = \sum\limits_{n=0}^{\infty}\dfrac{(-1)^n x^{2n+1}}{(2n+1)!} \;\Rightarrow\; \sin(x^4) = \sum\limits_{n=0}^{\infty}\dfrac{(-1)^n (x^4)^{2n+1}}{(2n+1)!} = \sum\limits_{n=0}^{\infty}\dfrac{(-1)^n x^{8n+4}}{(2n+1)!}$ for all x, so the radius of

convergence is ∞.

53. $f(x) = \dfrac{1}{\sqrt[4]{16-x}} = \dfrac{1}{\sqrt[4]{16(1-x/16)}} = \dfrac{1}{\sqrt[4]{16}\,(1-\frac{1}{16}x)^{1/4}} = \frac{1}{2}\left(1-\frac{1}{16}x\right)^{-1/4}$

$\qquad = \dfrac{1}{2}\left[1 + \left(-\dfrac{1}{4}\right)\left(-\dfrac{x}{16}\right) + \dfrac{\left(-\frac{1}{4}\right)\left(-\frac{5}{4}\right)}{2!}\left(-\dfrac{x}{16}\right)^2 + \dfrac{\left(-\frac{1}{4}\right)\left(-\frac{5}{4}\right)\left(-\frac{9}{4}\right)}{3!}\left(-\dfrac{x}{16}\right)^3 + \cdots\right]$

$\qquad = \dfrac{1}{2} + \displaystyle\sum_{n=1}^{\infty} \dfrac{1\cdot 5\cdot 9\cdot\,\cdots\,\cdot(4n-3)}{2\cdot 4^n\cdot n!\cdot 16^n}\,x^n = \dfrac{1}{2} + \displaystyle\sum_{n=1}^{\infty} \dfrac{1\cdot 5\cdot 9\cdot\,\cdots\,\cdot(4n-3)}{2^{6n+1}\,n!}\,x^n$

for $\left|-\dfrac{x}{16}\right| < 1 \quad\Leftrightarrow\quad |x| < 16$, so $R = 16$.

55. $e^x = \displaystyle\sum_{n=0}^{\infty}\dfrac{x^n}{n!}$, so $\dfrac{e^x}{x} = \dfrac{1}{x}\displaystyle\sum_{n=0}^{\infty}\dfrac{x^n}{n!} = \displaystyle\sum_{n=0}^{\infty}\dfrac{x^{n-1}}{n!} = x^{-1} + \displaystyle\sum_{n=1}^{\infty}\dfrac{x^{n-1}}{n!} = \dfrac{1}{x} + \displaystyle\sum_{n=1}^{\infty}\dfrac{x^{n-1}}{n!}$ and

$\qquad \displaystyle\int\dfrac{e^x}{x}\,dx = C + \ln|x| + \displaystyle\sum_{n=1}^{\infty}\dfrac{x^n}{n\cdot n!}$.

57. (a)

n	$f^{(n)}(x)$	$f^{(n)}(1)$
0	$x^{1/2}$	1
1	$\frac{1}{2}x^{-1/2}$	$\frac{1}{2}$
2	$-\frac{1}{4}x^{-3/2}$	$-\frac{1}{4}$
3	$\frac{3}{8}x^{-5/2}$	$\frac{3}{8}$
4	$-\frac{15}{16}x^{-7/2}$	$-\frac{15}{16}$
$\vdots$	$\vdots$	$\vdots$

$\sqrt{x} \approx T_3(x) = 1 + \dfrac{1/2}{1!}(x-1) - \dfrac{1/4}{2!}(x-1)^2 + \dfrac{3/8}{3!}(x-1)^3$

$\qquad\qquad = 1 + \frac{1}{2}(x-1) - \frac{1}{8}(x-1)^2 + \frac{1}{16}(x-1)^3$

(b)

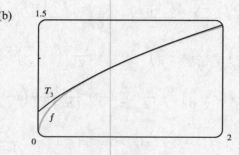

(c) $|R_3(x)| \le \dfrac{M}{4!}|x-1|^4$, where $\left|f^{(4)}(x)\right| \le M$ with $f^{(4)}(x) = -\frac{15}{16}x^{-7/2}$. Now $0.9 \le x \le 1.1 \quad\Rightarrow$

$\qquad -0.1 \le x - 1 \le 0.1 \quad\Rightarrow\quad (x-1)^4 \le (0.1)^4$, and letting $x = 0.9$ gives $M = \dfrac{15}{16(0.9)^{7/2}}$, so

$\qquad |R_3(x)| \le \dfrac{15}{16(0.9)^{7/2}\,4!}\,(0.1)^4 \approx 0.000\,005\,648 \approx 0.000\,006 = 6 \times 10^{-6}$.

(d)

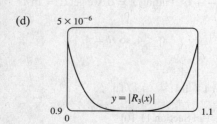

From the graph of $|R_3(x)| = |\sqrt{x} - T_3(x)|$, it appears that

the error is less than 5×10^{-6} on $[0.9, 1.1]$.

59. $\sin x = \displaystyle\sum_{n=0}^{\infty}(-1)^n\dfrac{x^{2n+1}}{(2n+1)!} = x - \dfrac{x^3}{3!} + \dfrac{x^5}{5!} - \dfrac{x^7}{7!} + \cdots$, so $\sin x - x = -\dfrac{x^3}{3!} + \dfrac{x^5}{5!} - \dfrac{x^7}{7!} + \cdots$ and

$\qquad \dfrac{\sin x - x}{x^3} = -\dfrac{1}{3!} + \dfrac{x^2}{5!} - \dfrac{x^4}{7!} + \cdots$. Thus, $\displaystyle\lim_{x\to 0}\dfrac{\sin x - x}{x^3} = \lim_{x\to 0}\left(-\dfrac{1}{6} + \dfrac{x^2}{120} - \dfrac{x^4}{5040} + \cdots\right) = -\dfrac{1}{6}$.

61. $f(x) = \sum\limits_{n=0}^{\infty} c_n\, x^n \quad\Rightarrow\quad f(-x) = \sum\limits_{n=0}^{\infty} c_n(-x)^n = \sum\limits_{n=0}^{\infty} (-1)^n c_n\, x^n$

(a) If f is an odd function, then $f(-x) = -f(x) \quad\Rightarrow\quad \sum\limits_{n=0}^{\infty} (-1)^n c_n x^n = \sum\limits_{n=0}^{\infty} -c_n x^n$. The coefficients of any power series

are uniquely determined (by Theorem 11.10.5), so $(-1)^n c_n = -c_n$.

If n is even, then $(-1)^n = 1$, so $c_n = -c_n \quad\Rightarrow\quad 2c_n = 0 \quad\Rightarrow\quad c_n = 0$. Thus, all even coefficients are 0, that is,
$c_0 = c_2 = c_4 = \cdots = 0$.

(b) If f is even, then $f(-x) = f(x) \quad\Rightarrow\quad \sum\limits_{n=0}^{\infty} (-1)^n c_n\, x^n = \sum\limits_{n=0}^{\infty} c_n\, x^n \quad\Rightarrow\quad (-1)^n c_n = c_n$.

If n is odd, then $(-1)^n = -1$, so $-c_n = c_n \quad\Rightarrow\quad 2c_n = 0 \quad\Rightarrow\quad c_n = 0$. Thus, all odd coefficients are 0,
that is, $c_1 = c_3 = c_5 = \cdots = 0$.

□ PROBLEMS PLUS

1. It would be far too much work to compute 15 derivatives of f. The key idea is to remember that $f^{(n)}(0)$ occurs in the

coefficient of x^n in the Maclaurin series of f. We start with the Maclaurin series for sin: $\sin x = x - \dfrac{x^3}{3!} + \dfrac{x^5}{5!} - \cdots$.

Then $\sin(x^3) = x^3 - \dfrac{x^9}{3!} + \dfrac{x^{15}}{5!} - \cdots$, and so the coefficient of x^{15} is $\dfrac{f^{(15)}(0)}{15!} = \dfrac{1}{5!}$. Therefore,

$$f^{(15)}(0) = \frac{15!}{5!} = 6 \cdot 7 \cdot 8 \cdot 9 \cdot 10 \cdot 11 \cdot 12 \cdot 13 \cdot 14 \cdot 15 = 10{,}897{,}286{,}400.$$

3. (a) From Formula 14a in Appendix D, with $x = y = \theta$, we get $\tan 2\theta = \dfrac{2\tan\theta}{1 - \tan^2\theta}$, so $\cot 2\theta = \dfrac{1 - \tan^2\theta}{2\tan\theta}$ $\Rightarrow$

$2\cot 2\theta = \dfrac{1 - \tan^2\theta}{\tan\theta} = \cot\theta - \tan\theta$. Replacing θ by $\tfrac{1}{2}x$, we get $2\cot x = \cot\tfrac{1}{2}x - \tan\tfrac{1}{2}x$, or

$\tan\tfrac{1}{2}x = \cot\tfrac{1}{2}x - 2\cot x$.

(b) From part (a) with $\dfrac{x}{2^{n-1}}$ in place of x, $\tan\dfrac{x}{2^n} = \cot\dfrac{x}{2^n} - 2\cot\dfrac{x}{2^{n-1}}$, so the nth partial sum of $\displaystyle\sum_{n=1}^{\infty} \dfrac{1}{2^n}\tan\dfrac{x}{2^n}$ is

$$
\begin{aligned}
s_n &= \frac{\tan(x/2)}{2} + \frac{\tan(x/4)}{4} + \frac{\tan(x/8)}{8} + \cdots + \frac{\tan(x/2^n)}{2^n} \\
&= \left[\frac{\cot(x/2)}{2} - \cot x\right] + \left[\frac{\cot(x/4)}{4} - \frac{\cot(x/2)}{2}\right] + \left[\frac{\cot(x/8)}{8} - \frac{\cot(x/4)}{4}\right] + \cdots \\
&\quad + \left[\frac{\cot(x/2^n)}{2^n} - \frac{\cot(x/2^{n-1})}{2^{n-1}}\right] = -\cot x + \frac{\cot(x/2^n)}{2^n} \quad \text{[telescoping sum]}
\end{aligned}
$$

Now $\dfrac{\cot(x/2^n)}{2^n} = \dfrac{\cos(x/2^n)}{2^n\sin(x/2^n)} = \dfrac{\cos(x/2^n)}{x} \cdot \dfrac{x/2^n}{\sin(x/2^n)} \to \dfrac{1}{x} \cdot 1 = \dfrac{1}{x}$ as $n \to \infty$ since $x/2^n \to 0$

for $x \neq 0$. Therefore, if $x \neq 0$ and $x \neq k\pi$ where k is any integer, then

$$\sum_{n=1}^{\infty} \frac{1}{2^n}\tan\frac{x}{2^n} = \lim_{n\to\infty} s_n = \lim_{n\to\infty}\left(-\cot x + \frac{1}{2^n}\cot\frac{x}{2^n}\right) = -\cot x + \frac{1}{x}$$

If $x = 0$, then all terms in the series are 0, so the sum is 0.

5. (a) At each stage, each side is replaced by four shorter sides, each of length

$\tfrac{1}{3}$ of the side length at the preceding stage. Writing s_0 and ℓ_0 for the

number of sides and the length of the side of the initial triangle, we

generate the table at right. In general, we have $s_n = 3 \cdot 4^n$ and

$s_0 = 3$	$\ell_0 = 1$
$s_1 = 3 \cdot 4$	$\ell_1 = 1/3$
$s_2 = 3 \cdot 4^2$	$\ell_2 = 1/3^2$
$s_3 = 3 \cdot 4^3$	$\ell_3 = 1/3^3$
$\vdots$	$\vdots$

$\ell_n = \left(\tfrac{1}{3}\right)^n$, so the length of the perimeter at the nth stage of construction

is $p_n = s_n\ell_n = 3 \cdot 4^n \cdot \left(\tfrac{1}{3}\right)^n = 3 \cdot \left(\tfrac{4}{3}\right)^n$.

(b) $p_n = \dfrac{4^n}{3^{n-1}} = 4\left(\dfrac{4}{3}\right)^{n-1}$. Since $\tfrac{4}{3} > 1$, $p_n \to \infty$ as $n \to \infty$.

(c) The area of each of the small triangles added at a given stage is one-ninth of the area of the triangle added at the preceding stage. Let a be the area of the original triangle. Then the area a_n of each of the small triangles added at stage n is

$$a_n = a \cdot \frac{1}{9^n} = \frac{a}{9^n}.$$ Since a small triangle is added to each side at every stage, it follows that the total area A_n added to the

figure at the nth stage is $A_n = s_{n-1} \cdot a_n = 3 \cdot 4^{n-1} \cdot \frac{a}{9^n} = a \cdot \frac{4^{n-1}}{3^{2n-1}}.$ Then the total area enclosed by the snowflake

curve is $A = a + A_1 + A_2 + A_3 + \cdots = a + a \cdot \frac{1}{3} + a \cdot \frac{4}{3^3} + a \cdot \frac{4^2}{3^5} + a \cdot \frac{4^3}{3^7} + \cdots.$ After the first term, this is a

geometric series with common ratio $\frac{4}{9}$, so $A = a + \frac{a/3}{1 - \frac{4}{9}} = a + \frac{a}{3} \cdot \frac{9}{5} = \frac{8a}{5}.$ But the area of the original equilateral

triangle with side 1 is $a = \frac{1}{2} \cdot 1 \cdot \sin \frac{\pi}{3} = \frac{\sqrt{3}}{4}.$ So the area enclosed by the snowflake curve is $\frac{8}{5} \cdot \frac{\sqrt{3}}{4} = \frac{2\sqrt{3}}{5}.$

7. (a) Let $a = \arctan x$ and $b = \arctan y$. Then, from Formula 14b in Appendix D,

$$\tan(a - b) = \frac{\tan a - \tan b}{1 + \tan a \tan b} = \frac{\tan(\arctan x) - \tan(\arctan y)}{1 + \tan(\arctan x)\tan(\arctan y)} = \frac{x - y}{1 + xy}$$

Now $\arctan x - \arctan y = a - b = \arctan(\tan(a - b)) = \arctan \frac{x - y}{1 + xy}$ since $-\frac{\pi}{2} < a - b < \frac{\pi}{2}.$

(b) From part (a) we have

$$\arctan \frac{120}{119} - \arctan \frac{1}{239} = \arctan \frac{\frac{120}{119} - \frac{1}{239}}{1 + \frac{120}{119} \cdot \frac{1}{239}} = \arctan \frac{\frac{28,561}{28,441}}{\frac{28,561}{28,441}} = \arctan 1 = \frac{\pi}{4}$$

(c) Replacing y by $-y$ in the formula of part (a), we get $\arctan x + \arctan y = \arctan \frac{x + y}{1 - xy}.$ So

$$4 \arctan \frac{1}{5} = 2\left(\arctan \frac{1}{5} + \arctan \frac{1}{5}\right) = 2 \arctan \frac{\frac{1}{5} + \frac{1}{5}}{1 - \frac{1}{5} \cdot \frac{1}{5}} = 2 \arctan \frac{5}{12} = \arctan \frac{5}{12} + \arctan \frac{5}{12}$$

$$= \arctan \frac{\frac{5}{12} + \frac{5}{12}}{1 - \frac{5}{12} \cdot \frac{5}{12}} = \arctan \frac{120}{119}$$

Thus, from part (b), we have $4 \arctan \frac{1}{5} - \arctan \frac{1}{239} = \arctan \frac{120}{119} - \arctan \frac{1}{239} = \frac{\pi}{4}.$

(d) From Example 11.9.7 we have $\arctan x = x - \frac{x^3}{3} + \frac{x^5}{5} - \frac{x^7}{7} + \frac{x^9}{9} - \frac{x^{11}}{11} + \cdots,$ so

$$\arctan \frac{1}{5} = \frac{1}{5} - \frac{1}{3 \cdot 5^3} + \frac{1}{5 \cdot 5^5} - \frac{1}{7 \cdot 5^7} + \frac{1}{9 \cdot 5^9} - \frac{1}{11 \cdot 5^{11}} + \cdots$$

This is an alternating series and the size of the terms decreases to 0, so by the Alternating Series Estimation Theorem, the sum lies between s_5 and s_6, that is, $0.197395560 < \arctan \frac{1}{5} < 0.197395562.$

(e) From the series in part (d) we get $\arctan \frac{1}{239} = \frac{1}{239} - \frac{1}{3 \cdot 239^3} + \frac{1}{5 \cdot 239^5} - \cdots.$ The third term is less than

2.6×10^{-13}, so by the Alternating Series Estimation Theorem, we have, to nine decimal places,

$\arctan \frac{1}{239} \approx s_2 \approx 0.004184076.$ Thus, $0.004184075 < \arctan \frac{1}{239} < 0.004184077.$

(f) From part (c) we have $\pi = 16\arctan\frac{1}{5} - 4\arctan\frac{1}{239}$, so from parts (d) and (e) we have

$$16(0.197395560) - 4(0.004184077) < \pi < 16(0.197395562) - 4(0.004184075) \quad \Rightarrow$$

$3.141592652 < \pi < 3.141592692$. So, to 7 decimal places, $\pi \approx 3.1415927$.

9. We want $\arctan\left(\dfrac{2}{n^2}\right)$ to equal $\arctan\dfrac{x-y}{1+xy}$. Note that $1 + xy = n^2 \quad \Leftrightarrow \quad xy = n^2 - 1 = (n+1)(n-1)$, so if we

let $x = n + 1$ and $y = n - 1$, then $x - y = 2$ and $xy \neq -1$. Thus, from Problem 7(a),

$$\arctan\left(\frac{2}{n^2}\right) = \arctan\frac{x-y}{1+xy} = \arctan x - \arctan y = \arctan(n+1) - \arctan(n-1). \text{ Therefore,}$$

$$\sum_{n=1}^{k} \arctan\left(\frac{2}{n^2}\right) = \sum_{n=1}^{k} [\arctan(n+1) - \arctan(n-1)]$$

$$= \sum_{n=1}^{k} [\arctan(n+1) - \arctan n + \arctan n - \arctan(n-1)]$$

$$= \sum_{n=1}^{k} [\arctan(n+1) - \arctan n] + \sum_{n=1}^{k} [\arctan n - \arctan(n-1)]$$

$$= [\arctan(k+1) - \arctan 1] + [\arctan k - \arctan 0] \quad \text{[since both sums are telescoping]}$$

$$= \arctan(k+1) - \tfrac{\pi}{4} + \arctan k - 0$$

Now $\displaystyle\sum_{n=1}^{k} \arctan\left(\frac{2}{n^2}\right) = \lim_{k \to 0} \sum_{n=1}^{k} \arctan\left(\frac{2}{n^2}\right) = \lim_{k \to \infty} \left[\arctan(k+1) - \frac{\pi}{4} + \arctan k\right] = \frac{\pi}{2} - \frac{\pi}{4} + \frac{\pi}{2} = \frac{3\pi}{4}$.

Note: For all $n \geq 1$, $0 \leq \arctan(n-1) < \arctan(n+1) < \frac{\pi}{2}$, so $-\frac{\pi}{2} < \arctan(n+1) - \arctan(n-1) < \frac{\pi}{2}$, and the

identity in Problem 7(a) holds.

11. We start with the geometric series $\displaystyle\sum_{n=0}^{\infty} x^n = \frac{1}{1-x}$, $|x| < 1$, and differentiate:

$$\sum_{n=1}^{\infty} nx^{n-1} = \frac{d}{dx}\left(\sum_{n=0}^{\infty} x^n\right) = \frac{d}{dx}\left(\frac{1}{1-x}\right) = \frac{1}{(1-x)^2} \text{ for } |x| < 1 \quad \Rightarrow \quad \sum_{n=1}^{\infty} nx^n = x\sum_{n=1}^{\infty} nx^{n-1} = \frac{x}{(1-x)^2}$$

for $|x| < 1$. Differentiate again:

$$\sum_{n=1}^{\infty} n^2 x^{n-1} = \frac{d}{dx}\frac{x}{(1-x)^2} = \frac{(1-x)^2 - x \cdot 2(1-x)(-1)}{(1-x)^4} = \frac{x+1}{(1-x)^3} \quad \Rightarrow \quad \sum_{n=1}^{\infty} n^2 x^n = \frac{x^2 + x}{(1-x)^3} \quad \Rightarrow$$

$$\sum_{n=1}^{\infty} n^3 x^{n-1} = \frac{d}{dx}\frac{x^2 + x}{(1-x)^3} = \frac{(1-x)^3(2x+1) - (x^2+x)3(1-x)^2(-1)}{(1-x)^6} = \frac{x^2 + 4x + 1}{(1-x)^4} \quad \Rightarrow$$

$$\sum_{n=1}^{\infty} n^3 x^n = \frac{x^3 + 4x^2 + x}{(1-x)^4}, \ |x| < 1. \text{ The radius of convergence is } 1 \text{ because that is the radius of convergence for the}$$

geometric series we started with. If $x = \pm 1$, the series is $\sum n^3(\pm 1)^n$, which diverges by the Test For Divergence, so the

interval of convergence is $(-1, 1)$.

13. $\ln\left(1 - \dfrac{1}{n^2}\right) = \ln\left(\dfrac{n^2 - 1}{n^2}\right) = \ln\dfrac{(n+1)(n-1)}{n^2} = \ln[(n+1)(n-1)] - \ln n^2$

$$= \ln(n+1) + \ln(n-1) - 2\ln n = \ln(n-1) - \ln n - \ln n + \ln(n+1)$$

$$= \ln\dfrac{n-1}{n} - [\ln n - \ln(n+1)] = \ln\dfrac{n-1}{n} - \ln\dfrac{n}{n+1}.$$

Let $s_k = \displaystyle\sum_{n=2}^{k} \ln\left(1 - \dfrac{1}{n^2}\right) = \sum_{n=2}^{k}\left(\ln\dfrac{n-1}{n} - \ln\dfrac{n}{n+1}\right)$ for $k \geq 2$. Then

$$s_k = \left(\ln\dfrac{1}{2} - \ln\dfrac{2}{3}\right) + \left(\ln\dfrac{2}{3} - \ln\dfrac{3}{4}\right) + \cdots + \left(\ln\dfrac{k-1}{k} - \ln\dfrac{k}{k+1}\right) = \ln\dfrac{1}{2} - \ln\dfrac{k}{k+1}, \text{ so}$$

$$\sum_{n=2}^{\infty} \ln\left(1 - \dfrac{1}{n^2}\right) = \lim_{k\to\infty} s_k = \lim_{k\to\infty}\left(\ln\dfrac{1}{2} - \ln\dfrac{k}{k+1}\right) = \ln\dfrac{1}{2} - \ln 1 = \ln 1 - \ln 2 - \ln 1 = -\ln 2 \ \left(\text{or } \ln\tfrac{1}{2}\right).$$

15. If L is the length of a side of the equilateral triangle, then the area is $A = \frac{1}{2}L \cdot \frac{\sqrt{3}}{2}L = \frac{\sqrt{3}}{4}L^2$ and so $L^2 = \frac{4}{\sqrt{3}}A$.

Let r be the radius of one of the circles. When there are n rows of circles, the figure shows that

$$L = \sqrt{3}\,r + r + (n-2)(2r) + r + \sqrt{3}\,r = r\big(2n - 2 + 2\sqrt{3}\big), \text{ so } r = \dfrac{L}{2\big(n + \sqrt{3} - 1\big)}.$$

The number of circles is $1 + 2 + \cdots + n = \dfrac{n(n+1)}{2}$, and so the total area of the circles is

$$A_n = \dfrac{n(n+1)}{2}\pi r^2 = \dfrac{n(n+1)}{2}\,\pi\,\dfrac{L^2}{4\big(n + \sqrt{3} - 1\big)^2}$$

$$= \dfrac{n(n+1)}{2}\,\pi\,\dfrac{4A/\sqrt{3}}{4\big(n + \sqrt{3} - 1\big)^2} = \dfrac{n(n+1)}{\big(n + \sqrt{3} - 1\big)^2}\,\dfrac{\pi A}{2\sqrt{3}} \quad\Rightarrow$$

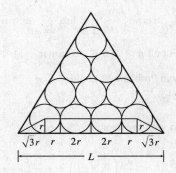

$$\dfrac{A_n}{A} = \dfrac{n(n+1)}{\big(n + \sqrt{3} - 1\big)^2}\,\dfrac{\pi}{2\sqrt{3}}$$

$$= \dfrac{1 + 1/n}{\big[1 + (\sqrt{3} - 1)/n\big]^2}\,\dfrac{\pi}{2\sqrt{3}} \to \dfrac{\pi}{2\sqrt{3}} \text{ as } n \to \infty$$

17. (a)

The x-intercepts of the curve occur where $\sin x = 0 \iff x = n\pi$, n an integer. So using the formula for disks (and either a CAS or $\sin^2 x = \frac{1}{2}(1 - \cos 2x)$ and Formula 99 to evaluate the integral), the volume of the nth bead is

$$V_n = \pi\int_{(n-1)\pi}^{n\pi}\big(e^{-x/10}\sin x\big)^2\,dx = \pi\int_{(n-1)\pi}^{n\pi} e^{-x/5}\sin^2 x\,dx$$

$$= \dfrac{250\pi}{101}\big(e^{-(n-1)\pi/5} - e^{-n\pi/5}\big)$$

(b) The total volume is

$$\pi\int_0^\infty e^{-x/5}\sin^2 x\,dx = \sum_{n=1}^{\infty} V_n = \dfrac{250\pi}{101}\sum_{n=1}^{\infty}\big[e^{-(n-1)\pi/5} - e^{-n\pi/5}\big] = \dfrac{250\pi}{101} \quad \text{[telescoping sum]}.$$

[continued]

Another method: If the volume in part (a) has been written as $V_n = \frac{250\pi}{101} e^{-n\pi/5} (e^{\pi/5} - 1)$, then we recognize $\sum\limits_{n=1}^{\infty} V_n$

as a geometric series with $a = \frac{250\pi}{101} (1 - e^{-\pi/5})$ and $r = e^{-\pi/5}$.

19. By Table 1 in Section 11.10, $\tan^{-1} x = \sum\limits_{n=0}^{\infty} (-1)^n \frac{x^{2n+1}}{2n+1}$ for $|x| < 1$. In particular, for $x = \dfrac{1}{\sqrt{3}}$, we

have $\dfrac{\pi}{6} = \tan^{-1}\left(\dfrac{1}{\sqrt{3}}\right) = \sum\limits_{n=0}^{\infty} (-1)^n \dfrac{(1/\sqrt{3})^{2n+1}}{2n+1} = \sum\limits_{n=0}^{\infty} (-1)^n \left(\dfrac{1}{3}\right)^n \dfrac{1}{\sqrt{3}} \dfrac{1}{2n+1}$, so

$\pi = \dfrac{6}{\sqrt{3}} \sum\limits_{n=0}^{\infty} \dfrac{(-1)^n}{(2n+1)3^n} = 2\sqrt{3} \sum\limits_{n=0}^{\infty} \dfrac{(-1)^n}{(2n+1)3^n} = 2\sqrt{3} \left(1 + \sum\limits_{n=1}^{\infty} \dfrac{(-1)^n}{(2n+1)3^n}\right) \;\Rightarrow\; \sum\limits_{n=1}^{\infty} \dfrac{(-1)^n}{(2n+1)3^n} = \dfrac{\pi}{2\sqrt{3}} - 1.$

21. Let $f(x)$ denote the left-hand side of the equation $1 + \dfrac{x}{2!} + \dfrac{x^2}{4!} + \dfrac{x^3}{6!} + \dfrac{x^4}{8!} + \cdots = 0$. If $x \geq 0$, then $f(x) \geq 1$ and there are

no solutions of the equation. Note that $f(-x^2) = 1 - \dfrac{x^2}{2!} + \dfrac{x^4}{4!} - \dfrac{x^6}{6!} + \dfrac{x^8}{8!} - \cdots = \cos x$. The solutions of $\cos x = 0$ for

$x < 0$ are given by $x = \dfrac{\pi}{2} - \pi k$, where k is a positive integer. Thus, the solutions of $f(x) = 0$ are $x = -\left(\dfrac{\pi}{2} - \pi k\right)^2$, where

k is a positive integer.

23. Call the series S. We group the terms according to the number of digits in their denominators:

$$S = \underbrace{\left(\tfrac{1}{1} + \tfrac{1}{2} + \cdots + \tfrac{1}{8} + \tfrac{1}{9}\right)}_{g_1} + \underbrace{\left(\tfrac{1}{11} + \cdots + \tfrac{1}{99}\right)}_{g_2} + \underbrace{\left(\tfrac{1}{111} + \cdots + \tfrac{1}{999}\right)}_{g_3} + \cdots$$

Now in the group g_n, since we have 9 choices for each of the n digits in the denominator, there are 9^n terms.

Furthermore, each term in g_n is less than $\frac{1}{10^{n-1}}$ [except for the first term in g_1]. So $g_n < 9^n \cdot \frac{1}{10^{n-1}} = 9\left(\frac{9}{10}\right)^{n-1}$.

Now $\sum\limits_{n=1}^{\infty} 9\left(\frac{9}{10}\right)^{n-1}$ is a geometric series with $a = 9$ and $r = \frac{9}{10} < 1$. Therefore, by the Comparison Test,

$S = \sum\limits_{n=1}^{\infty} g_n < \sum\limits_{n=1}^{\infty} 9\left(\frac{9}{10}\right)^{n-1} = \frac{9}{1 - 9/10} = 90.$

25. $u = 1 + \dfrac{x^3}{3!} + \dfrac{x^6}{6!} + \dfrac{x^9}{9!} + \cdots,\; v = x + \dfrac{x^4}{4!} + \dfrac{x^7}{7!} + \dfrac{x^{10}}{10!} + \cdots,\; w = \dfrac{x^2}{2!} + \dfrac{x^5}{5!} + \dfrac{x^8}{8!} + \cdots.$

Use the Ratio Test to show that the series for u, v, and w have positive radii of convergence (∞ in each case), so

Theorem 11.9.2 applies, and hence, we may differentiate each of these series:

$$\frac{du}{dx} = \frac{3x^2}{3!} + \frac{6x^5}{6!} + \frac{9x^8}{9!} + \cdots = \frac{x^2}{2!} + \frac{x^5}{5!} + \frac{x^8}{8!} + \cdots = w$$

Similarly, $\dfrac{dv}{dx} = 1 + \dfrac{x^3}{3!} + \dfrac{x^6}{6!} + \dfrac{x^9}{9!} + \cdots = u$, and $\dfrac{dw}{dx} = x + \dfrac{x^4}{4!} + \dfrac{x^7}{7!} + \dfrac{x^{10}}{10!} + \cdots = v$.

So $u' = w$, $v' = u$, and $w' = v$. Now differentiate the left-hand side of the desired equation:

$$\frac{d}{dx}(u^3 + v^3 + w^3 - 3uvw) = 3u^2u' + 3v^2v' + 3w^2w' - 3(u'vw + uv'w + uvw')$$

$$= 3u^2w + 3v^2u + 3w^2v - 3(vw^2 + u^2w + uv^2) = 0 \quad \Rightarrow$$

$u^3 + v^3 + w^3 - 3uvw = C$. To find the value of the constant C, we put $x = 0$ in the last equation and get

$1^3 + 0^3 + 0^3 - 3(1 \cdot 0 \cdot 0) = C \quad \Rightarrow \quad C = 1$, so $u^3 + v^3 + w^3 - 3uvw = 1$.

12 □ VECTORS AND THE GEOMETRY OF SPACE

12.1 Three-Dimensional Coordinate Systems

1. We start at the origin, which has coordinates $(0, 0, 0)$. First we move 4 units along the positive x-axis, affecting only the x-coordinate, bringing us to the point $(4, 0, 0)$. We then move 3 units straight downward, in the negative z-direction. Thus only the z-coordinate is affected, and we arrive at $(4, 0, -3)$.

3. The distance from a point to the yz-plane is the absolute value of the x-coordinate of the point. $C(2, 4, 6)$ has the x-coordinate with the smallest absolute value, so C is the point closest to the yz-plane. $A(-4, 0, -1)$ must lie in the xz-plane since the distance from A to the xz-plane, given by the y-coordinate of A, is 0.

5. In $\mathbb{R}^2$, the equation $x = 4$ represents a line parallel to the y-axis and 4 units to the right of it. In $\mathbb{R}^3$, the equation $x = 4$ represents the set $\{(x, y, z) \mid x = 4\}$, the set of all points whose x-coordinate is 4. This is the vertical plane that is parallel to the yz-plane and 4 units in front of it.

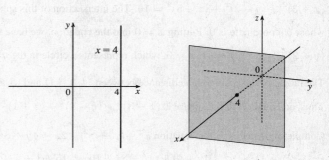

7. The equation $x + y = 2$ represents the set of all points in $\mathbb{R}^3$ whose x- and y-coordinates have a sum of 2, or equivalently where $y = 2 - x$. This is the set $\{(x, 2 - x, z) \mid x \in \mathbb{R}, z \in \mathbb{R}\}$ which is a vertical plane that intersects the xy-plane in the line $y = 2 - x$, $z = 0$.

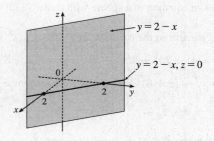

9. We can find the lengths of the sides of the triangle by using the distance formula between pairs of vertices:

$$|PQ| = \sqrt{(7 - 3)^2 + [0 - (-2)]^2 + [1 - (-3)]^2} = \sqrt{16 + 4 + 16} = 6$$

$$|QR| = \sqrt{(1 - 7)^2 + (2 - 0)^2 + (1 - 1)^2} = \sqrt{36 + 4 + 0} = \sqrt{40} = 2\sqrt{10}$$

$$|RP| = \sqrt{(3 - 1)^2 + (-2 - 2)^2 + (-3 - 1)^2} = \sqrt{4 + 16 + 16} = 6$$

The longest side is QR, but the Pythagorean Theorem is not satisfied: $|PQ|^2 + |RP|^2 \neq |QR|^2$. Thus PQR is not a right triangle. PQR is isosceles, as two sides have the same length.

11. (a) First we find the distances between points:

$$|AB| = \sqrt{(3-2)^2 + (7-4)^2 + (-2-2)^2} = \sqrt{26}$$

$$|BC| = \sqrt{(1-3)^2 + (3-7)^2 + [3-(-2)]^2} = \sqrt{45} = 3\sqrt{5}$$

$$|AC| = \sqrt{(1-2)^2 + (3-4)^2 + (3-2)^2} = \sqrt{3}$$

In order for the points to lie on a straight line, the sum of the two shortest distances must be equal to the longest distance.

Since $\sqrt{26} + \sqrt{3} \neq 3\sqrt{5}$, the three points do not lie on a straight line.

(b) First we find the distances between points:

$$|DE| = \sqrt{(1-0)^2 + [-2-(-5)]^2 + (4-5)^2} = \sqrt{11}$$

$$|EF| = \sqrt{(3-1)^2 + [4-(-2)]^2 + (2-4)^2} = \sqrt{44} = 2\sqrt{11}$$

$$|DF| = \sqrt{(3-0)^2 + [4-(-5)]^2 + (2-5)^2} = \sqrt{99} = 3\sqrt{11}$$

Since $|DE| + |EF| = |DF|$, the three points lie on a straight line.

13. An equation of the sphere with center $(-3, 2, 5)$ and radius 4 is $[x-(-3)]^2 + (y-2)^2 + (z-5)^2 = 4^2$ or

$(x+3)^2 + (y-2)^2 + (z-5)^2 = 16$. The intersection of this sphere with the yz-plane is the set of points on the sphere

whose x-coordinate is 0. Putting $x = 0$ into the equation, we have $9 + (y-2)^2 + (z-5)^2 = 16$, $x = 0$ or

$(y-2)^2 + (z-5)^2 = 7$, $x = 0$, which represents a circle in the yz-plane with center $(0, 2, 5)$ and radius $\sqrt{7}$.

15. The radius of the sphere is the distance between $(4, 3, -1)$ and $(3, 8, 1)$: $r = \sqrt{(3-4)^2 + (8-3)^2 + [1-(-1)]^2} = \sqrt{30}$.

Thus, an equation of the sphere is $(x-3)^2 + (y-8)^2 + (z-1)^2 = 30$.

17. Completing squares in the equation $x^2 + y^2 + z^2 - 2x - 4y + 8z = 15$ gives

$(x^2 - 2x + 1) + (y^2 - 4y + 4) + (z^2 + 8z + 16) = 15 + 1 + 4 + 16 \quad \Rightarrow \quad (x-1)^2 + (y-2)^2 + (z+4)^2 = 36$, which we

recognize as an equation of a sphere with center $(1, 2, -4)$ and radius 6.

19. Completing squares in the equation $2x^2 - 8x + 2y^2 + 2z^2 + 24z = 1$ gives

$2(x^2 - 4x + 4) + 2y^2 + 2(z^2 + 12z + 36) = 1 + 8 + 72 \quad \Rightarrow \quad 2(x-2)^2 + 2y^2 + 2(z+6)^2 = 81 \quad \Rightarrow$

$(x-2)^2 + y^2 + (z+6)^2 = \frac{81}{2}$, which we recognize as an equation of a sphere with center $(2, 0, -6)$ and

radius $\sqrt{\frac{81}{2}} = 9/\sqrt{2}$.

21. (a) If the midpoint of the line segment from $P_1(x_1, y_1, z_1)$ to $P_2(x_2, y_2, z_2)$ is $Q = \left(\dfrac{x_1 + x_2}{2}, \dfrac{y_1 + y_2}{2}, \dfrac{z_1 + z_2}{2} \right)$,

then the distances $|P_1Q|$ and $|QP_2|$ are equal, and each is half of $|P_1P_2|$. We verify that this is the case:

$$|P_1P_2| = \sqrt{(x_2 - x_1)^2 + (y_2 - y_1)^2 + (z_2 - z_1)^2}$$

$$|P_1Q| = \sqrt{\left[\tfrac{1}{2}(x_1 + x_2) - x_1\right]^2 + \left[\tfrac{1}{2}(y_1 + y_2) - y_1\right]^2 + \left[\tfrac{1}{2}(z_1 + z_2) - z_1\right]^2}$$

$$= \sqrt{\left(\tfrac{1}{2}x_2 - \tfrac{1}{2}x_1\right)^2 + \left(\tfrac{1}{2}y_2 - \tfrac{1}{2}y_1\right)^2 + \left(\tfrac{1}{2}z_2 - \tfrac{1}{2}z_1\right)^2}$$

$$= \sqrt{\left(\tfrac{1}{2}\right)^2 \left[(x_2 - x_1)^2 + (y_2 - y_1)^2 + (z_2 - z_1)^2\right]} = \tfrac{1}{2}\sqrt{(x_2 - x_1)^2 + (y_2 - y_1)^2 + (z_2 - z_1)^2}$$

$$= \tfrac{1}{2}|P_1P_2|$$

$$|QP_2| = \sqrt{\left[x_2 - \tfrac{1}{2}(x_1 + x_2)\right]^2 + \left[y_2 - \tfrac{1}{2}(y_1 + y_2)\right]^2 + \left[z_2 - \tfrac{1}{2}(z_1 + z_2)\right]^2}$$

$$= \sqrt{\left(\tfrac{1}{2}x_2 - \tfrac{1}{2}x_1\right)^2 + \left(\tfrac{1}{2}y_2 - \tfrac{1}{2}y_1\right)^2 + \left(\tfrac{1}{2}z_2 - \tfrac{1}{2}z_1\right)^2} = \sqrt{\left(\tfrac{1}{2}\right)^2\left[(x_2 - x_1)^2 + (y_2 - y_1)^2 + (z_2 - z_1)^2\right]}$$

$$= \tfrac{1}{2}\sqrt{(x_2 - x_1)^2 + (y_2 - y_1)^2 + (z_2 - z_1)^2} = \tfrac{1}{2}|P_1 P_2|$$

So Q is indeed the midpoint of $P_1 P_2$.

(b) By part (a), the midpoints of sides AB, BC and CA are $P_1\left(-\tfrac{1}{2}, 1, 4\right)$, $P_2\left(1, \tfrac{1}{2}, 5\right)$ and $P_3\left(\tfrac{5}{2}, \tfrac{3}{2}, 4\right)$. Then the lengths of the medians are:

$$|AP_2| = \sqrt{0^2 + \left(\tfrac{1}{2} - 2\right)^2 + (5 - 3)^2} = \sqrt{\tfrac{9}{4} + 4} = \sqrt{\tfrac{25}{4}} = \tfrac{5}{2}$$

$$|BP_3| = \sqrt{\left(\tfrac{5}{2} + 2\right)^2 + \left(\tfrac{3}{2}\right)^2 + (4 - 5)^2} = \sqrt{\tfrac{81}{4} + \tfrac{9}{4} + 1} = \sqrt{\tfrac{94}{4}} = \tfrac{1}{2}\sqrt{94}$$

$$|CP_1| = \sqrt{\left(-\tfrac{1}{2} - 4\right)^2 + (1 - 1)^2 + (4 - 5)^2} = \sqrt{\tfrac{81}{4} + 1} = \tfrac{1}{2}\sqrt{85}$$

23. (a) Since the sphere touches the xy-plane, its radius is the distance from its center, $(2, -3, 6)$, to the xy-plane, namely 6.

Therefore $r = 6$ and an equation of the sphere is $(x - 2)^2 + (y + 3)^2 + (z - 6)^2 = 6^2 = 36$.

(b) The radius of this sphere is the distance from its center $(2, -3, 6)$ to the yz-plane, which is 2. Therefore, an equation is

$(x - 2)^2 + (y + 3)^2 + (z - 6)^2 = 4$.

(c) Here the radius is the distance from the center $(2, -3, 6)$ to the xz-plane, which is 3. Therefore, an equation is

$(x - 2)^2 + (y + 3)^2 + (z - 6)^2 = 9$.

25. The equation $x = 5$ represents a plane parallel to the yz-plane and 5 units in front of it.

27. The inequality $y < 8$ represents a half-space consisting of all points to the left of the plane $y = 8$.

29. The inequality $0 \le z \le 6$ represents all points on or between the horizontal planes $z = 0$ (the xy-plane) and $z = 6$.

31. Because $z = -1$, all points in the region must lie in the horizontal plane $z = -1$. In addition, $x^2 + y^2 = 4$, so the region consists of all points that lie on a circle with radius 2 and center on the z-axis that is contained in the plane $z = -1$.

33. The equation $x^2 + y^2 + z^2 = 4$ is equivalent to $\sqrt{x^2 + y^2 + z^2} = 2$, so the region consists of those points whose distance from the origin is 2. This is the set of all points on a sphere with radius 2 and center $(0, 0, 0)$.

35. The inequalities $1 \le x^2 + y^2 + z^2 \le 5$ are equivalent to $1 \le \sqrt{x^2 + y^2 + z^2} \le \sqrt{5}$, so the region consists of those points whose distance from the origin is at least 1 and at most $\sqrt{5}$. This is the set of all points on or between spheres with radii 1 and $\sqrt{5}$ and centers $(0, 0, 0)$.

37. Here $x^2 + z^2 \le 9$ or equivalently $\sqrt{x^2 + z^2} \le 3$ which describes the set of all points in $\mathbb{R}^3$ whose distance from the y-axis is at most 3. Thus the inequality represents the region consisting of all points on or inside a circular cylinder of radius 3 with axis the y-axis.

39. This describes all points whose x-coordinate is between 0 and 5, that is, $0 < x < 5$.

41. This describes a region all of whose points have a distance to the origin which is greater than r, but smaller than R. So inequalities describing the region are $r < \sqrt{x^2 + y^2 + z^2} < R$, or $r^2 < x^2 + y^2 + z^2 < R^2$.

43. (a) To find the x- and y-coordinates of the point P, we project it onto L_2

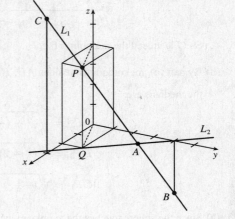

and project the resulting point Q onto the x- and y-axes. To find the z-coordinate, we project P onto either the xz-plane or the yz-plane (using our knowledge of its x- or y-coordinate) and then project the resulting point onto the z-axis. (Or, we could draw a line parallel to QO from P to the z-axis.) The coordinates of P are $(2, 1, 4)$.

(b) A is the intersection of L_1 and L_2, B is directly below the y-intercept of L_2, and C is directly above the x-intercept of L_2.

45. We need to find a set of points $\{P(x, y, z) \mid |AP| = |BP|\}$.

$$\sqrt{(x+1)^2 + (y-5)^2 + (z-3)^2} = \sqrt{(x-6)^2 + (y-2)^2 + (z+2)^2} \Rightarrow$$

$$(x+1)^2 + (y-5) + (z-3)^2 = (x-6)^2 + (y-2)^2 + (z+2)^2 \Rightarrow$$

$$x^2 + 2x + 1 + y^2 - 10y + 25 + z^2 - 6z + 9 = x^2 - 12x + 36 + y^2 - 4y + 4 + z^2 + 4z + 4 \Rightarrow 14x - 6y - 10z = 9.$$

Thus the set of points is a plane perpendicular to the line segment joining A and B (since this plane must contain the perpendicular bisector of the line segment AB).

47. The sphere $x^2 + y^2 + z^2 = 4$ has center $(0, 0, 0)$ and radius 2. Completing squares in $x^2 - 4x + y^2 - 4y + z^2 - 4z = -11$

gives $(x^2 - 4x + 4) + (y^2 - 4y + 4) + (z^2 - 4z + 4) = -11 + 4 + 4 + 4 \Rightarrow (x-2)^2 + (y-2)^2 + (z-2)^2 = 1$,

so this is the sphere with center $(2, 2, 2)$ and radius 1. The (shortest) distance between the spheres is measured along the line segment connecting their centers. The distance between $(0, 0, 0)$ and $(2, 2, 2)$ is

$\sqrt{(2-0)^2 + (2-0)^2 + (2-0)^2} = \sqrt{12} = 2\sqrt{3}$, and subtracting the radius of each circle, the distance between the spheres is $2\sqrt{3} - 2 - 1 = 2\sqrt{3} - 3$.

12.2 Vectors

1. (a) The cost of a theater ticket is a scalar, because it has only magnitude.

(b) The current in a river is a vector, because it has both magnitude (the speed of the current) and direction at any given location.

(c) If we assume that the initial path is linear, the initial flight path from Houston to Dallas is a vector, because it has both magnitude (distance) and direction.

(d) The population of the world is a scalar, because it has only magnitude.

3. Vectors are equal when they share the same length and direction (but not necessarily location). Using the symmetry of the parallelogram as a guide, we see that $\overrightarrow{AB} = \overrightarrow{DC}$, $\overrightarrow{DA} = \overrightarrow{CB}$, $\overrightarrow{DE} = \overrightarrow{EB}$, and $\overrightarrow{EA} = \overrightarrow{CE}$.

5. (a)

(b)

(c)

(d)

(e)

(f)

7. Because the tail of $\mathbf{d}$ is the midpoint of QR we have $\overrightarrow{QR} = 2\mathbf{d}$, and by the Triangle Law,

$\mathbf{a} + 2\mathbf{d} = \mathbf{b} \quad \Rightarrow \quad 2\mathbf{d} = \mathbf{b} - \mathbf{a} \quad \Rightarrow \quad \mathbf{d} = \frac{1}{2}(\mathbf{b} - \mathbf{a}) = \frac{1}{2}\mathbf{b} - \frac{1}{2}\mathbf{a}$. Again by the Triangle Law we have $\mathbf{c} + \mathbf{d} = \mathbf{b}$ so

$\mathbf{c} = \mathbf{b} - \mathbf{d} = \mathbf{b} - \left(\frac{1}{2}\mathbf{b} - \frac{1}{2}\mathbf{a}\right) = \frac{1}{2}\mathbf{a} + \frac{1}{2}\mathbf{b}$.

9. $\mathbf{a} = \langle 1 - (-2), 2 - 1 \rangle = \langle 3, 1 \rangle$

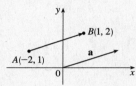

11. $\mathbf{a} = \langle 2 - 3, 3 - (-1) \rangle = \langle -1, 4 \rangle$

13. $\mathbf{a} = \langle 2 - 0, 3 - 3, -1 - 1 \rangle = \langle 2, 0, -2 \rangle$

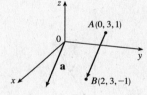

15. $\langle -1, 4 \rangle + \langle 6, -2 \rangle = \langle -1 + 6, 4 + (-2) \rangle = \langle 5, 2 \rangle$

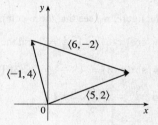

17. $\langle 3, 0, 1 \rangle + \langle 0, 8, 0 \rangle = \langle 3 + 0, 0 + 8, 1 + 0 \rangle$
$\qquad = \langle 3, 8, 1 \rangle$

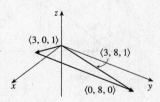

19. $\mathbf{a} + \mathbf{b} = \langle -3, 4 \rangle + \langle 9, -1 \rangle = \langle -3 + 9, 4 + (-1) \rangle = \langle 6, 3 \rangle$

$4\mathbf{a} + 2\mathbf{b} = 4\langle -3, 4 \rangle + 2\langle 9, -1 \rangle = \langle -12, 16 \rangle + \langle 18, -2 \rangle = \langle 6, 14 \rangle$

$|\mathbf{a}| = \sqrt{(-3)^2 + 4^2} = \sqrt{25} = 5$

$|\mathbf{a} - \mathbf{b}| = |\langle -3 - 9, 4 - (-1) \rangle| = |\langle -12, 5 \rangle| = \sqrt{(-12)^2 + 5^2} = \sqrt{169} = 13$

21. $\mathbf{a} + \mathbf{b} = (4\mathbf{i} - 3\mathbf{j} + 2\mathbf{k}) + (2\mathbf{i} - 4\mathbf{k}) = 6\mathbf{i} - 3\mathbf{j} - 2\mathbf{k}$

$4\mathbf{a} + 2\mathbf{b} = 4(4\mathbf{i} - 3\mathbf{j} + 2\mathbf{k}) + 2(2\mathbf{i} - 4\mathbf{k}) = 16\mathbf{i} - 12\mathbf{j} + 8\mathbf{k} + 4\mathbf{i} - 8\mathbf{k} = 20\mathbf{i} - 12\mathbf{j}$

$|\mathbf{a}| = \sqrt{4^2 + (-3)^2 + 2^2} = \sqrt{29}$

$|\mathbf{a} - \mathbf{b}| = |(4\mathbf{i} - 3\mathbf{j} + 2\mathbf{k}) - (2\mathbf{i} - 4\mathbf{k})| = |2\mathbf{i} - 3\mathbf{j} + 6\mathbf{k}| = \sqrt{2^2 + (-3)^2 + 6^2} = \sqrt{49} = 7$

23. The vector $\langle 6, -2 \rangle$ has length $|\langle 6, -2 \rangle| = \sqrt{6^2 + (-2)^2} = \sqrt{40} = 2\sqrt{10}$, so by Equation 4 the unit vector with the same

direction is $\dfrac{1}{2\sqrt{10}} \langle 6, -2 \rangle = \left\langle \dfrac{3}{\sqrt{10}}, -\dfrac{1}{\sqrt{10}} \right\rangle$.

25. The vector $8\mathbf{i} - \mathbf{j} + 4\mathbf{k}$ has length $|8\mathbf{i} - \mathbf{j} + 4\mathbf{k}| = \sqrt{8^2 + (-1)^2 + 4^2} = \sqrt{81} = 9$, so by Equation 4 the unit vector with

the same direction is $\frac{1}{9}(8\mathbf{i} - \mathbf{j} + 4\mathbf{k}) = \frac{8}{9}\mathbf{i} - \frac{1}{9}\mathbf{j} + \frac{4}{9}\mathbf{k}$.

27.

From the figure, we see that $\tan\theta = \dfrac{\sqrt{3}}{1} = \sqrt{3} \quad \Rightarrow \quad \theta = 60°$.

29. From the figure, we see that the x-component of $\mathbf{v}$ is

$v_1 = |\mathbf{v}|\cos(\pi/3) = 4 \cdot \frac{1}{2} = 2$ and the y-component is

$v_2 = |\mathbf{v}|\sin(\pi/3) = 4 \cdot \frac{\sqrt{3}}{2} = 2\sqrt{3}$. Thus

$\mathbf{v} = \langle v_1, v_2 \rangle = \langle 2, 2\sqrt{3} \rangle$.

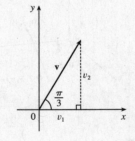

31. The velocity vector $\mathbf{v}$ makes an angle of $40°$ with the horizontal and

has magnitude equal to the speed at which the football was thrown.

From the figure, we see that the horizontal component of $\mathbf{v}$ is

$|\mathbf{v}|\cos 40° = 60\cos 40° \approx 45.96$ ft/s and the vertical component

is $|\mathbf{v}|\sin 40° = 60\sin 40° \approx 38.57$ ft/s.

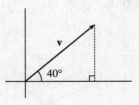

33. The given force vectors can be expressed in terms of their horizontal and vertical components as $-300\mathbf{i}$ and

$200\cos 60°\,\mathbf{i} + 200\sin 60°\,\mathbf{j} = 200(\frac{1}{2})\,\mathbf{i} + 200\left(\frac{\sqrt{3}}{2}\right)\mathbf{j} = 100\mathbf{i} + 100\sqrt{3}\,\mathbf{j}$. The resultant force $\mathbf{F}$ is the sum of

these two vectors: $\mathbf{F} = (-300 + 100)\,\mathbf{i} + (0 + 100\sqrt{3})\,\mathbf{j} = -200\mathbf{i} + 100\sqrt{3}\,\mathbf{j}$. Then we have

$|\mathbf{F}| \approx \sqrt{(-200)^2 + (100\sqrt{3})^2} = \sqrt{70{,}000} = 100\sqrt{7} \approx 264.6$ N. Let θ be the angle $\mathbf{F}$ makes with the positive x-axis. Then $\tan \theta = \dfrac{100\sqrt{3}}{-200} = -\dfrac{\sqrt{3}}{2}$ and the terminal point of $\mathbf{F}$ lies in the second quadrant, so

$$\theta = \tan^{-1}\left(-\frac{\sqrt{3}}{2}\right) + 180° \approx -40.9° + 180° = 139.1°.$$

35. With respect to the water's surface, the woman's velocity is the vector sum of the velocity of the ship with respect to the water, and the woman's velocity with respect to the ship. If we let north be the positive y-direction, then

$\mathbf{v} = \langle 0, 22 \rangle + \langle -3, 0 \rangle = \langle -3, 22 \rangle$. The woman's speed is $|\mathbf{v}| = \sqrt{9 + 484} \approx 22.2$ mi/h. The vector $\mathbf{v}$ makes an angle θ

with the east, where $\theta = \tan^{-1}\left(\frac{22}{-3}\right) \approx 98°$. Therefore, the woman's direction is about $\mathrm{N}(98 - 90)°\mathrm{W} = \mathrm{N}8°\mathrm{W}$.

37. Call the two tension vectors $\mathbf{T}_2$ and $\mathbf{T}_3$, corresponding to the ropes of length 2 m and 3 m. In terms of vertical and horizontal components,

$$\mathbf{T}_2 = -|\mathbf{T}_2|\cos 50°\mathbf{i} + |\mathbf{T}_2|\sin 50°\mathbf{j} \quad \mathbf{(1)} \qquad \text{and} \qquad \mathbf{T}_3 = |\mathbf{T}_3|\cos 38°\mathbf{i} + |\mathbf{T}_3|\sin 38°\mathbf{j} \quad \mathbf{(2)}$$

The resultant of these forces, $\mathbf{T}_2 + \mathbf{T}_3$, counterbalances the weight of the hoist (which is $-350\,\mathbf{j}$), so

$\mathbf{T}_2 + \mathbf{T}_3 = 350\,\mathbf{j} \quad \Rightarrow$

$(-|\mathbf{T}_2|\cos 50° + |\mathbf{T}_3|\cos 38°)\,\mathbf{i} + (|\mathbf{T}_2|\sin 50° + |\mathbf{T}_3|\sin 38°)\,\mathbf{j} = 350\,\mathbf{j}$. Equating components, we have

$-|\mathbf{T}_2|\cos 50° + |\mathbf{T}_3|\cos 38° = 0 \quad \Rightarrow \quad |\mathbf{T}_2| = |\mathbf{T}_3|\dfrac{\cos 38°}{\cos 50°}$ and

$|\mathbf{T}_2|\sin 50° + |\mathbf{T}_3|\sin 38° = 350$. Substituting the first equation into the second gives

$|\mathbf{T}_3|\dfrac{\cos 38°}{\cos 50°}\sin 50° + |\mathbf{T}_3|\sin 38° = 350 \quad \Rightarrow \quad |\mathbf{T}_3|(\cos 38° \tan 50° + \sin 38°) = 350$, so the magnitudes of the

tensions are $|\mathbf{T}_3| = \dfrac{350}{\cos 38° \tan 50° + \sin 38°} \approx 225.11$ N and $|\mathbf{T}_2| = |\mathbf{T}_3|\dfrac{\cos 38°}{\cos 50°} \approx 275.97$ N. Finally, from $\mathbf{(1)}$ and $\mathbf{(2)}$,

the tension vectors are $\mathbf{T}_2 \approx -177.39\,\mathbf{i} + 211.41\,\mathbf{j}$ and $\mathbf{T}_3 \approx 177.39\,\mathbf{i} + 138.59\,\mathbf{j}$.

39. (a) Set up coordinate axes so that the boatman is at the origin, the canal is bordered by the y-axis and the line $x = 3$, and the current flows in the negative y-direction. The boatman wants to reach the point $(3, 2)$. Let θ be the angle, measured from the positive y-axis, in the direction he should steer. (See the figure.)

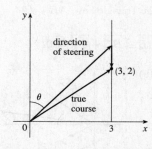

In still water, the boat has velocity $\mathbf{v}_b = \langle 13\sin\theta, 13\cos\theta \rangle$ and the velocity of the current is $\mathbf{v}_c \langle 0, -3.5 \rangle$, so the true path of the boat is determined by the velocity vector $\mathbf{v} = \mathbf{v}_b + \mathbf{v}_c = \langle 13\sin\theta, 13\cos\theta - 3.5 \rangle$. Let t be the time (in hours) after the boat departs; then the position of the boat at time t is given by $t\mathbf{v}$ and the boat crosses the canal when

$$t\mathbf{v} = \langle 13\sin\theta, 13\cos\theta - 3.5 \rangle\, t = \langle 3, 2 \rangle. \text{ Thus } 13(\sin\theta)t = 3 \quad \Rightarrow \quad t = \frac{3}{13\sin\theta} \text{ and } (13\cos\theta - 3.5)\,t = 2.$$

[continued]

Substituting gives $(13\cos\theta - 3.5)\dfrac{3}{13\sin\theta} = 2 \quad \Rightarrow \quad 39\cos\theta - 10.5 = 26\sin\theta$ **(1)**. Squaring both sides, we have

$$1521\cos^2\theta - 819\cos\theta + 110.25 = 676\sin^2\theta = 676\left(1 - \cos^2\theta\right)$$

$$2197\cos^2\theta - 819\cos\theta - 565.75 = 0$$

The quadratic formula gives

$$\cos\theta = \frac{819 \pm \sqrt{(-819)^2 - 4(2197)(-565.75)}}{2(2197)}$$

$$= \frac{819 \pm \sqrt{5,642,572}}{4394} \approx 0.72699 \text{ or } -0.35421$$

The acute value for θ is approximately $\cos^{-1}(0.72699) \approx 43.4°$. Thus the boatman should steer in the direction that is 43.4° from the bank, toward upstream.

Alternate solution: We could solve **(1)** graphically by plotting $y = 39\cos\theta - 10.5$ and $y = 26\sin\theta$ on a graphing device and finding the approximate intersection point $(0.757, 17.85)$. Thus $\theta \approx 0.757$ radians or equivalently 43.4°.

(b) From part (a) we know the trip is completed when $t = \dfrac{3}{13\sin\theta}$. But $\theta \approx 43.4°$, so the time required is approximately $\dfrac{3}{13\sin 43.4°} \approx 0.336$ hours or 20.2 minutes.

41. The slope of the tangent line to the graph of $y = x^2$ at the point $(2, 4)$ is

$$\left.\frac{dy}{dx}\right|_{x=2} = 2x\,\bigg|_{x=2} = 4$$

and a parallel vector is $\mathbf{i} + 4\mathbf{j}$ which has length $|\mathbf{i} + 4\mathbf{j}| = \sqrt{1^2 + 4^2} = \sqrt{17}$, so unit vectors parallel to the tangent line are $\pm\frac{1}{\sqrt{17}}(\mathbf{i} + 4\mathbf{j})$.

43. By the Triangle Law, $\overrightarrow{AB} + \overrightarrow{BC} = \overrightarrow{AC}$. Then $\overrightarrow{AB} + \overrightarrow{BC} + \overrightarrow{CA} = \overrightarrow{AC} + \overrightarrow{CA}$, but $\overrightarrow{AC} + \overrightarrow{CA} = \overrightarrow{AC} + \left(-\overrightarrow{AC}\right) = \mathbf{0}$.

So $\overrightarrow{AB} + \overrightarrow{BC} + \overrightarrow{CA} = \mathbf{0}$.

45. (a), (b)

(c) From the sketch, we estimate that $s \approx 1.3$ and $t \approx 1.6$.

(d) $\mathbf{c} = s\mathbf{a} + t\mathbf{b} \quad \Leftrightarrow \quad 7 = 3s + 2t$ and $1 = 2s - t$.

Solving these equations gives $s = \frac{9}{7}$ and $t = \frac{11}{7}$.

47. $|\mathbf{r} - \mathbf{r}_0|$ is the distance between the points (x, y, z) and (x_0, y_0, z_0), so the set of points is a sphere with radius 1 and center (x_0, y_0, z_0).

Alternate method: $|\mathbf{r} - \mathbf{r}_0| = 1 \quad \Leftrightarrow \quad \sqrt{(x - x_0)^2 + (y - y_0)^2 + (z - z_0)^2} = 1 \quad \Leftrightarrow$

$(x - x_0)^2 + (y - y_0)^2 + (z - z_0)^2 = 1$, which is the equation of a sphere with radius 1 and center (x_0, y_0, z_0).

49. $\mathbf{a} + (\mathbf{b} + \mathbf{c}) = \langle a_1, a_2 \rangle + (\langle b_1, b_2 \rangle + \langle c_1, c_2 \rangle) = \langle a_1, a_2 \rangle + \langle b_1 + c_1, b_2 + c_2 \rangle$

$$= \langle a_1 + b_1 + c_1, a_2 + b_2 + c_2 \rangle = \langle (a_1 + b_1) + c_1, (a_2 + b_2) + c_2 \rangle$$

$$= \langle a_1 + b_1, a_2 + b_2 \rangle + \langle c_1, c_2 \rangle = (\langle a_1, a_2 \rangle + \langle b_1, b_2 \rangle) + \langle c_1, c_2 \rangle$$

$$= (\mathbf{a} + \mathbf{b}) + \mathbf{c}$$

51. Consider triangle ABC, where D and E are the midpoints of AB and BC. We know that $\overrightarrow{AB} + \overrightarrow{BC} = \overrightarrow{AC}$ **(1)** and

$\overrightarrow{DB} + \overrightarrow{BE} = \overrightarrow{DE}$ **(2)**. However, $\overrightarrow{DB} = \frac{1}{2}\overrightarrow{AB}$, and $\overrightarrow{BE} = \frac{1}{2}\overrightarrow{BC}$. Substituting these expressions for $\overrightarrow{DB}$ and $\overrightarrow{BE}$ into

(2) gives $\frac{1}{2}\overrightarrow{AB} + \frac{1}{2}\overrightarrow{BC} = \overrightarrow{DE}$. Comparing this with **(1)** gives $\overrightarrow{DE} = \frac{1}{2}\overrightarrow{AC}$. Therefore $\overrightarrow{AC}$ and $\overrightarrow{DE}$ are parallel and

$\left| \overrightarrow{DE} \right| = \frac{1}{2} \left| \overrightarrow{AC} \right|$.

12.3 The Dot Product

1. (a) $\mathbf{a} \cdot \mathbf{b}$ is a scalar, and the dot product is defined only for vectors, so $(\mathbf{a} \cdot \mathbf{b}) \cdot \mathbf{c}$ has no meaning.

(b) $(\mathbf{a} \cdot \mathbf{b}) \mathbf{c}$ is a scalar multiple of a vector, so it does have meaning.

(c) Both $|\mathbf{a}|$ and $\mathbf{b} \cdot \mathbf{c}$ are scalars, so $|\mathbf{a}| (\mathbf{b} \cdot \mathbf{c})$ is an ordinary product of real numbers, and has meaning.

(d) Both $\mathbf{a}$ and $\mathbf{b} + \mathbf{c}$ are vectors, so the dot product $\mathbf{a} \cdot (\mathbf{b} + \mathbf{c})$ has meaning.

(e) $\mathbf{a} \cdot \mathbf{b}$ is a scalar, but $\mathbf{c}$ is a vector, and so the two quantities cannot be added and $\mathbf{a} \cdot \mathbf{b} + \mathbf{c}$ has no meaning.

(f) $|\mathbf{a}|$ is a scalar, and the dot product is defined only for vectors, so $|\mathbf{a}| \cdot (\mathbf{b} + \mathbf{c})$ has no meaning.

3. $\mathbf{a} \cdot \mathbf{b} = \langle 1.5, 0.4 \rangle \cdot \langle -4, 6 \rangle = (1.5)(-4) + (0.4)(6) = -6 + 2.4 = -3.6$

5. $\mathbf{a} \cdot \mathbf{b} = \langle 4, 1, \frac{1}{4} \rangle \cdot \langle 6, -3, -8 \rangle = (4)(6) + (1)(-3) + \left(\frac{1}{4} \right)(-8) = 19$

7. $\mathbf{a} \cdot \mathbf{b} = (2\mathbf{i} + \mathbf{j}) \cdot (\mathbf{i} - \mathbf{j} + \mathbf{k}) = (2)(1) + (1)(-1) + (0)(1) = 1$

9. By Theorem 3, $\mathbf{a} \cdot \mathbf{b} = |\mathbf{a}| \, |\mathbf{b}| \cos \theta = (7)(4) \cos 30° = 28 \left(\frac{\sqrt{3}}{2} \right) = 14 \sqrt{3} \approx 24.25$.

11. $\mathbf{u}$, $\mathbf{v}$, and $\mathbf{w}$ are all unit vectors, so the triangle is an equilateral triangle. Thus the angle between $\mathbf{u}$ and $\mathbf{v}$ is 60° and

$\mathbf{u} \cdot \mathbf{v} = |\mathbf{u}| \, |\mathbf{v}| \cos 60° = (1)(1)\left(\frac{1}{2} \right) = \frac{1}{2}$. If $\mathbf{w}$ is moved so it has the same initial point as $\mathbf{u}$, we can see that the angle

between them is 120° and we have $\mathbf{u} \cdot \mathbf{w} = |\mathbf{u}| \, |\mathbf{w}| \cos 120° = (1)(1)\left(-\frac{1}{2} \right) = -\frac{1}{2}$.

13. (a) $\mathbf{i} \cdot \mathbf{j} = \langle 1, 0, 0 \rangle \cdot \langle 0, 1, 0 \rangle = (1)(0) + (0)(1) + (0)(0) = 0$. Similarly, $\mathbf{j} \cdot \mathbf{k} = (0)(0) + (1)(0) + (0)(1) = 0$ and

$\mathbf{k} \cdot \mathbf{i} = (0)(1) + (0)(0) + (1)(0) = 0$.

Another method: Because $\mathbf{i}$, $\mathbf{j}$, and $\mathbf{k}$ are mutually perpendicular, the cosine factor in each dot product (see Theorem 3)

is $\cos \frac{\pi}{2} = 0$.

(b) By Property 1 of the dot product, $\mathbf{i} \cdot \mathbf{i} = |\mathbf{i}|^2 = 1^2 = 1$ since $\mathbf{i}$ is a unit vector. Similarly, $\mathbf{j} \cdot \mathbf{j} = |\mathbf{j}|^2 = 1$ and

$\mathbf{k} \cdot \mathbf{k} = |\mathbf{k}|^2 = 1$.

15. $|\mathbf{a}| = \sqrt{4^2 + 3^2} = 5$, $|\mathbf{b}| = \sqrt{2^2 + (-1)^2} = \sqrt{5}$, and $\mathbf{a} \cdot \mathbf{b} = (4)(2) + (3)(-1) = 5$. From Corollary 6, we have

$$\cos\theta = \frac{\mathbf{a} \cdot \mathbf{b}}{|\mathbf{a}|\,|\mathbf{b}|} = \frac{5}{5 \cdot \sqrt{5}} = \frac{1}{\sqrt{5}}. \text{ So the angle between } \mathbf{a} \text{ and } \mathbf{b} \text{ is } \theta = \cos^{-1}\left(\frac{1}{\sqrt{5}}\right) \approx 63°.$$

17. $|\mathbf{a}| = \sqrt{1^2 + (-4)^2 + 1^2} = \sqrt{18} = 3\sqrt{2}$, $|\mathbf{b}| = \sqrt{0^2 + 2^2 + (-2)^2} = \sqrt{8} = 2\sqrt{2}$, and

$\mathbf{a} \cdot \mathbf{b} = (1)(0) + (-4)(2) + (1)(-2) = -10$. From Corollary 6, we have $\cos\theta = \dfrac{\mathbf{a} \cdot \mathbf{b}}{|\mathbf{a}|\,|\mathbf{b}|} = \dfrac{-10}{3\sqrt{2} \cdot 2\sqrt{2}} = -\dfrac{10}{12} = -\dfrac{5}{6}$ and

the angle between $\mathbf{a}$ and $\mathbf{b}$ is $\theta = \cos^{-1}\left(-\frac{5}{6}\right) \approx 146°$.

19. $|\mathbf{a}| = \sqrt{4^2 + (-3)^2 + 1^2} = \sqrt{26}$, $|\mathbf{b}| = \sqrt{2^2 + 0^2 + (-1)^2} = \sqrt{5}$, and $\mathbf{a} \cdot \mathbf{b} = (4)(2) + (-3)(0) + (1)(-1) = 7$.

Then $\cos\theta = \dfrac{\mathbf{a} \cdot \mathbf{b}}{|\mathbf{a}|\,|\mathbf{b}|} = \dfrac{7}{\sqrt{26} \cdot \sqrt{5}} = \dfrac{7}{\sqrt{130}}$ and $\theta = \cos^{-1}\left(\dfrac{7}{\sqrt{130}}\right) \approx 52°$.

21. Let p, q, and r be the angles at vertices P, Q, and R respectively.

Then p is the angle between vectors $\overrightarrow{PQ}$ and $\overrightarrow{PR}$, q is the angle

between vectors $\overrightarrow{QP}$ and $\overrightarrow{QR}$, and r is the angle between vectors

$\overrightarrow{RP}$ and $\overrightarrow{RQ}$.

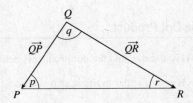

Thus $\cos p = \dfrac{\overrightarrow{PQ} \cdot \overrightarrow{PR}}{\left|\overrightarrow{PQ}\right|\left|\overrightarrow{PR}\right|} = \dfrac{\langle -2, 3\rangle \cdot \langle 1, 4\rangle}{\sqrt{(-2)^2 + 3^2}\sqrt{1^2 + 4^2}} = \dfrac{-2 + 12}{\sqrt{13}\sqrt{17}} = \dfrac{10}{\sqrt{221}}$ and $p = \cos^{-1}\left(\dfrac{10}{\sqrt{221}}\right) \approx 48°$. Similarly,

$\cos q = \dfrac{\overrightarrow{QP} \cdot \overrightarrow{QR}}{\left|\overrightarrow{QP}\right|\left|\overrightarrow{QR}\right|} = \dfrac{\langle 2, -3\rangle \cdot \langle 3, 1\rangle}{\sqrt{4 + 9}\sqrt{9 + 1}} = \dfrac{6 - 3}{\sqrt{13}\sqrt{10}} = \dfrac{3}{\sqrt{130}}$ so $q = \cos^{-1}\left(\dfrac{3}{\sqrt{130}}\right) \approx 75°$ and

$r \approx 180° - (48° + 75°) = 57°$.

Alternate solution: Apply the Law of Cosines three times as follows: $\cos p = \dfrac{\left|\overrightarrow{QR}\right|^2 - \left|\overrightarrow{PQ}\right|^2 - \left|\overrightarrow{PR}\right|^2}{2\left|\overrightarrow{PQ}\right|\left|\overrightarrow{PR}\right|}$,

$\cos q = \dfrac{\left|\overrightarrow{PR}\right|^2 - \left|\overrightarrow{PQ}\right|^2 - \left|\overrightarrow{QR}\right|^2}{2\left|\overrightarrow{PQ}\right|\left|\overrightarrow{QR}\right|}$, and $\cos r = \dfrac{\left|\overrightarrow{PQ}\right|^2 - \left|\overrightarrow{PR}\right|^2 - \left|\overrightarrow{QR}\right|^2}{2\left|\overrightarrow{PR}\right|\left|\overrightarrow{QR}\right|}$.

23. (a) $\mathbf{a} \cdot \mathbf{b} = (9)(-2) + (3)(6) = 0$, so $\mathbf{a}$ and $\mathbf{b}$ are orthogonal (and not parallel).

(b) $\mathbf{a} \cdot \mathbf{b} = (4)(3) + (5)(-1) + (-2)(5) = -3 \neq 0$, so $\mathbf{a}$ and $\mathbf{b}$ are not orthogonal. Also, since $\mathbf{a}$ is not a scalar multiple

of $\mathbf{b}$, $\mathbf{a}$ and $\mathbf{b}$ are not parallel.

(c) $\mathbf{a} \cdot \mathbf{b} = (-8)(6) + (12)(-9) + (4)(-3) = -168 \neq 0$, so $\mathbf{a}$ and $\mathbf{b}$ are not orthogonal. Because $\mathbf{a} = -\frac{4}{3}\mathbf{b}$, $\mathbf{a}$ and $\mathbf{b}$ are

parallel.

(d) $\mathbf{a} \cdot \mathbf{b} = (3)(5) + (-1)(9) + (3)(-2) = 0$, so $\mathbf{a}$ and $\mathbf{b}$ are orthogonal (and not parallel).

25. $\overrightarrow{QP} = \langle -1, -3, 2\rangle$, $\overrightarrow{QR} = \langle 4, -2, -1\rangle$, and $\overrightarrow{QP} \cdot \overrightarrow{QR} = -4 + 6 - 2 = 0$. Thus $\overrightarrow{QP}$ and $\overrightarrow{QR}$ are orthogonal, so the angle of

the triangle at vertex Q is a right angle.

27. Let $\mathbf{a} = a_1\,\mathbf{i} + a_2\,\mathbf{j} + a_3\,\mathbf{k}$ be a vector orthogonal to both $\mathbf{i} + \mathbf{j}$ and $\mathbf{i} + \mathbf{k}$. Then $\mathbf{a} \cdot (\mathbf{i} + \mathbf{j}) = 0 \quad \Leftrightarrow \quad a_1 + a_2 = 0$ and

$\mathbf{a} \cdot (\mathbf{i} + \mathbf{k}) = 0 \quad \Leftrightarrow \quad a_1 + a_3 = 0$, so $a_1 = -a_2 = -a_3$. Furthermore $\mathbf{a}$ is to be a unit vector, so $1 = a_1^2 + a_2^2 + a_3^2 = 3a_1^2$

implies $a_1 = \pm\frac{1}{\sqrt{3}}$. Thus $\mathbf{a} = \frac{1}{\sqrt{3}}\,\mathbf{i} - \frac{1}{\sqrt{3}}\,\mathbf{j} - \frac{1}{\sqrt{3}}\,\mathbf{k}$ and $\mathbf{a} = -\frac{1}{\sqrt{3}}\,\mathbf{i} + \frac{1}{\sqrt{3}}\,\mathbf{j} + \frac{1}{\sqrt{3}}\,\mathbf{k}$ are two such unit vectors.

29. The line $2x - y = 3 \quad \Leftrightarrow \quad y = 2x - 3$ has slope 2, so a vector parallel to the line is $\mathbf{a} = \langle 1, 2 \rangle$. The line $3x + y = 7 \quad \Leftrightarrow$

$y = -3x + 7$ has slope -3, so a vector parallel to the line is $\mathbf{b} = \langle 1, -3 \rangle$. The angle between the lines is the same as the

angle θ between the vectors. Here we have $\mathbf{a} \cdot \mathbf{b} = (1)(1) + (2)(-3) = -5$, $|\mathbf{a}| = \sqrt{1^2 + 2^2} = \sqrt{5}$, and

$|\mathbf{b}| = \sqrt{1^2 + (-3)^2} = \sqrt{10}$, so $\cos\theta = \dfrac{\mathbf{a} \cdot \mathbf{b}}{|\mathbf{a}|\,|\mathbf{b}|} = \dfrac{-5}{\sqrt{5} \cdot \sqrt{10}} = \dfrac{-5}{5\sqrt{2}} = -\dfrac{1}{\sqrt{2}}$ or $-\dfrac{\sqrt{2}}{2}$. Thus $\theta = 135°$, and the

acute angle between the lines is $180° - 135° = 45°$.

31. The curves $y = x^2$ and $y = x^3$ meet when $x^2 = x^3 \quad \Leftrightarrow \quad x^3 - x^2 = 0 \quad \Leftrightarrow \quad x^2(x - 1) = 0 \quad \Leftrightarrow \quad x = 0, x = 1$. We have

$\dfrac{d}{dx}x^2 = 2x$ and $\dfrac{d}{dx}x^3 = 3x^2$, so the tangent lines of both curves have slope 0 at $x = 0$. Thus the angle between the curves is

$0°$ at the point $(0, 0)$. For $x = 1$, $\dfrac{d}{dx}x^2 \Big|_{x=1} = 2$ and $\dfrac{d}{dx}x^3 \Big|_{x=1} = 3$ so the tangent lines at the point $(1, 1)$ have slopes 2 and

3. Vectors parallel to the tangent lines are $\langle 1, 2 \rangle$ and $\langle 1, 3 \rangle$, and the angle θ between them is given by

$$\cos\theta = \frac{\langle 1, 2 \rangle \cdot \langle 1, 3 \rangle}{|\langle 1, 2 \rangle|\,|\langle 1, 3 \rangle|} = \frac{1 + 6}{\sqrt{5}\,\sqrt{10}} = \frac{7}{5\sqrt{2}}$$

Thus $\theta = \cos^{-1}\left(\dfrac{7}{5\sqrt{2}}\right) \approx 8.1°$.

33. Since $|\langle 2, 1, 2 \rangle| = \sqrt{4 + 1 + 4} = \sqrt{9} = 3$, using Equations 8 and 9 we have $\cos\alpha = \frac{2}{3}$, $\cos\beta = \frac{1}{3}$, and $\cos\gamma = \frac{2}{3}$. The

direction angles are given by $\alpha = \cos^{-1}\left(\frac{2}{3}\right) \approx 48°$, $\beta = \cos^{-1}\left(\frac{1}{3}\right) \approx 71°$, and $\gamma = \cos^{-1}\left(\frac{2}{3}\right) = 48°$.

35. Since $|\,\mathbf{i} - 2\,\mathbf{j} - 3\,\mathbf{k}| = \sqrt{1 + 4 + 9} = \sqrt{14}$, Equations 8 and 9 give $\cos\alpha = \frac{1}{\sqrt{14}}$, $\cos\beta = \frac{-2}{\sqrt{14}}$, and $\cos\gamma = \frac{-3}{\sqrt{14}}$, while

$\alpha = \cos^{-1}\left(\frac{1}{\sqrt{14}}\right) \approx 74°$, $\beta = \cos^{-1}\left(-\frac{2}{\sqrt{14}}\right) \approx 122°$, and $\gamma = \cos^{-1}\left(-\frac{3}{\sqrt{14}}\right) \approx 143°$.

37. $|\langle c, c, c \rangle| = \sqrt{c^2 + c^2 + c^2} = \sqrt{3}\,c$ [since $c > 0$], so $\cos\alpha = \cos\beta = \cos\gamma = \dfrac{c}{\sqrt{3}\,c} = \dfrac{1}{\sqrt{3}}$ and

$\alpha = \beta = \gamma = \cos^{-1}\left(\frac{1}{\sqrt{3}}\right) \approx 55°$.

39. $|\mathbf{a}| = \sqrt{(-5)^2 + 12^2} = \sqrt{169} = 13$. The scalar projection of $\mathbf{b}$ onto $\mathbf{a}$ is $\text{comp}_{\mathbf{a}}\,\mathbf{b} = \dfrac{\mathbf{a} \cdot \mathbf{b}}{|\mathbf{a}|} = \dfrac{-5 \cdot 4 + 12 \cdot 6}{13} = 4$ and the

vector projection of $\mathbf{b}$ onto $\mathbf{a}$ is $\text{proj}_{\mathbf{a}}\,\mathbf{b} = \left(\dfrac{\mathbf{a} \cdot \mathbf{b}}{|\mathbf{a}|}\right)\dfrac{\mathbf{a}}{|\mathbf{a}|} = 4 \cdot \frac{1}{13}\langle -5, 12 \rangle = \left\langle -\frac{20}{13}, \frac{48}{13} \right\rangle$.

41. $|\mathbf{a}| = \sqrt{4^2 + 7^2 + (-4)^2} = \sqrt{81} = 9$ so the scalar projection of $\mathbf{b}$ onto $\mathbf{a}$ is

$\text{comp}_{\mathbf{a}}\mathbf{b} = \dfrac{\mathbf{a} \cdot \mathbf{b}}{|\mathbf{a}|} = \dfrac{(4)(3) + (7)(-1) + (-4)(1)}{9} = \dfrac{1}{9}$. The vector projection of $\mathbf{b}$ onto $\mathbf{a}$ is

$\text{proj}_{\mathbf{a}}\mathbf{b} = \left(\dfrac{\mathbf{a} \cdot \mathbf{b}}{|\mathbf{a}|}\right)\dfrac{\mathbf{a}}{|\mathbf{a}|} = \frac{1}{9} \cdot \frac{1}{9}\langle 4, 7, -4 \rangle = \frac{1}{81}\langle 4, 7, -4 \rangle = \left\langle \frac{4}{81}, \frac{7}{81}, -\frac{4}{81} \right\rangle$.

43. $|\mathbf{a}| = \sqrt{9+9+1} = \sqrt{19}$ so the scalar projection of $\mathbf{b}$ onto $\mathbf{a}$ is $\text{comp}_{\mathbf{a}}\,\mathbf{b} = \dfrac{\mathbf{a}\cdot\mathbf{b}}{|\mathbf{a}|} = \dfrac{6-12-1}{\sqrt{19}} = -\dfrac{7}{\sqrt{19}}$ while the vector

projection of $\mathbf{b}$ onto $\mathbf{a}$ is $\text{proj}_{\mathbf{a}}\,\mathbf{b} = -\dfrac{7}{\sqrt{19}}\dfrac{\mathbf{a}}{|\mathbf{a}|} = -\dfrac{7}{\sqrt{19}}\cdot\dfrac{1}{\sqrt{19}}(3\mathbf{i}-3\mathbf{j}+\mathbf{k}) = -\dfrac{7}{19}(3\mathbf{i}-3\mathbf{j}+\mathbf{k}) = -\dfrac{21}{19}\mathbf{i} + \dfrac{21}{19}\mathbf{j} - \dfrac{7}{19}\mathbf{k}.$

45. $(\text{orth}_{\mathbf{a}}\,\mathbf{b})\cdot\mathbf{a} = (\mathbf{b} - \text{proj}_{\mathbf{a}}\,\mathbf{b})\cdot\mathbf{a} = \mathbf{b}\cdot\mathbf{a} - (\text{proj}_{\mathbf{a}}\,\mathbf{b})\cdot\mathbf{a} = \mathbf{b}\cdot\mathbf{a} - \dfrac{\mathbf{a}\cdot\mathbf{b}}{|\mathbf{a}|^2}\,\mathbf{a}\cdot\mathbf{a} = \mathbf{b}\cdot\mathbf{a} - \dfrac{\mathbf{a}\cdot\mathbf{b}}{|\mathbf{a}|^2}\,|\mathbf{a}|^2 = \mathbf{b}\cdot\mathbf{a} - \mathbf{a}\cdot\mathbf{b} = 0.$

So they are orthogonal by (7).

47. $\text{comp}_{\mathbf{a}}\,\mathbf{b} = \dfrac{\mathbf{a}\cdot\mathbf{b}}{|\mathbf{a}|} = 2 \;\Leftrightarrow\; \mathbf{a}\cdot\mathbf{b} = 2\,|\mathbf{a}| = 2\sqrt{10}.$ If $\mathbf{b} = \langle b_1, b_2, b_3 \rangle$, then we need $3b_1 + 0b_2 - 1b_3 = 2\sqrt{10}.$

One possible solution is obtained by taking $b_1 = 0,\ b_2 = 0,\ b_3 = -2\sqrt{10}.$ In general, $\mathbf{b} = \langle s, t, 3s - 2\sqrt{10}\,\rangle,\ s, t \in \mathbb{R}.$

49. The displacement vector is $\mathbf{D} = (6-0)\,\mathbf{i} + (12-10)\,\mathbf{j} + (20-8)\,\mathbf{k} = 6\mathbf{i} + 2\mathbf{j} + 12\mathbf{k}$ so, by Equation 12, the work done is

$W = \mathbf{F}\cdot\mathbf{D} = (8\mathbf{i} - 6\mathbf{j} + 9\mathbf{k})\cdot(6\mathbf{i} + 2\mathbf{j} + 12\mathbf{k}) = 48 - 12 + 108 = 144$ joules.

51. Here $|\mathbf{D}| = 80$ ft, $|\mathbf{F}| = 30$ lb, and $\theta = 40°.$ Thus

$W = \mathbf{F}\cdot\mathbf{D} = |\mathbf{F}|\,|\mathbf{D}|\cos\theta = (30)(80)\cos 40° = 2400\cos 40° \approx 1839$ ft-lb.

53. First note that $\mathbf{n} = \langle a, b \rangle$ is perpendicular to the line, because if $Q_1 = (a_1, b_1)$ and $Q_2 = (a_2, b_2)$ lie on the line, then

$\mathbf{n}\cdot\overrightarrow{Q_1Q_2} = aa_2 - aa_1 + bb_2 - bb_1 = 0,$ since $aa_2 + bb_2 = -c = aa_1 + bb_1$ from the equation of the line.

Let $P_2 = (x_2, y_2)$ lie on the line. Then the distance from P_1 to the line is the absolute value of the scalar projection

of $\overrightarrow{P_1P_2}$ onto $\mathbf{n}.$ $\quad \text{comp}_{\mathbf{n}}\left(\overrightarrow{P_1P_2}\right) = \dfrac{|\mathbf{n}\cdot\langle x_2 - x_1, y_2 - y_1\rangle|}{|\mathbf{n}|} = \dfrac{|ax_2 - ax_1 + by_2 - by_1|}{\sqrt{a^2 + b^2}} = \dfrac{|ax_1 + by_1 + c|}{\sqrt{a^2 + b^2}}$

since $ax_2 + by_2 = -c.$ The required distance is $\dfrac{|(3)(-2) + (-4)(3) + 5|}{\sqrt{3^2 + (-4)^2}} = \dfrac{13}{5}.$

55. For convenience, consider the unit cube positioned so that its back left corner is at the origin, and its edges lie along the

coordinate axes. The diagonal of the cube that begins at the origin and ends at $(1, 1, 1)$ has vector representation $\langle 1, 1, 1 \rangle.$

The angle θ between this vector and the vector of the edge which also begins at the origin and runs along the x-axis [that is,

$\langle 1, 0, 0 \rangle$] is given by $\cos\theta = \dfrac{\langle 1, 1, 1 \rangle \cdot \langle 1, 0, 0 \rangle}{|\langle 1, 1, 1 \rangle|\,|\langle 1, 0, 0 \rangle|} = \dfrac{1}{\sqrt{3}} \;\Rightarrow\; \theta = \cos^{-1}\!\left(\dfrac{1}{\sqrt{3}}\right) \approx 55°.$

57. Consider the H—C—H combination consisting of the sole carbon atom and the two hydrogen atoms that are at $(1, 0, 0)$ and

$(0, 1, 0)$ (or any H—C—H combination, for that matter). Vector representations of the line segments emanating from the

carbon atom and extending to these two hydrogen atoms are $\langle 1 - \frac{1}{2}, 0 - \frac{1}{2}, 0 - \frac{1}{2} \rangle = \langle \frac{1}{2}, -\frac{1}{2}, -\frac{1}{2} \rangle$ and

$\langle 0 - \frac{1}{2}, 1 - \frac{1}{2}, 0 - \frac{1}{2} \rangle = \langle -\frac{1}{2}, \frac{1}{2}, -\frac{1}{2} \rangle.$ The bond angle, $\theta,$ is therefore given by

$\cos\theta = \dfrac{\langle \frac{1}{2}, -\frac{1}{2}, -\frac{1}{2} \rangle \cdot \langle -\frac{1}{2}, \frac{1}{2}, -\frac{1}{2} \rangle}{|\langle \frac{1}{2}, -\frac{1}{2}, -\frac{1}{2} \rangle|\,|\langle -\frac{1}{2}, \frac{1}{2}, -\frac{1}{2} \rangle|} = \dfrac{-\frac{1}{4} - \frac{1}{4} + \frac{1}{4}}{\sqrt{\frac{3}{4}}\sqrt{\frac{3}{4}}} = -\dfrac{1}{3} \;\Rightarrow\; \theta = \cos^{-1}\!\left(-\tfrac{1}{3}\right) \approx 109.5°.$

59. Let $\mathbf{a} = \langle a_1, a_2, a_3 \rangle$ and $\mathbf{b} = \langle b_1, b_2, b_3 \rangle$.

Property 2: $\mathbf{a} \cdot \mathbf{b} = \langle a_1, a_2, a_3 \rangle \cdot \langle b_1, b_2, b_3 \rangle = a_1 b_1 + a_2 b_2 + a_3 b_3$

$$= b_1 a_1 + b_2 a_2 + b_3 a_3 = \langle b_1, b_2, b_3 \rangle \cdot \langle a_1, a_2, a_3 \rangle = \mathbf{b} \cdot \mathbf{a}$$

Property 4: $(c\,\mathbf{a}) \cdot \mathbf{b} = \langle ca_1, ca_2, ca_3 \rangle \cdot \langle b_1, b_2, b_3 \rangle = (ca_1)b_1 + (ca_2)b_2 + (ca_3)b_3$

$$= c\,(a_1 b_1 + a_2 b_2 + a_3 b_3) = c\,(\mathbf{a} \cdot \mathbf{b}) = a_1(cb_1) + a_2(cb_2) + a_3(cb_3)$$

$$= \langle a_1, a_2, a_3 \rangle \cdot \langle cb_1, cb_2, cb_3 \rangle = \mathbf{a} \cdot (c\,\mathbf{b})$$

Property 5: $\mathbf{0} \cdot \mathbf{a} = \langle 0, 0, 0 \rangle \cdot \langle a_1, a_2, a_3 \rangle = (0)(a_1) + (0)(a_2) + (0)(a_3) = 0$

61. $|\mathbf{a} \cdot \mathbf{b}| = \big|\, |\mathbf{a}|\,|\mathbf{b}| \cos \theta \,\big| = |\mathbf{a}|\,|\mathbf{b}|\,|\cos \theta|$. Since $|\cos \theta| \le 1$, $|\mathbf{a} \cdot \mathbf{b}| = |\mathbf{a}|\,|\mathbf{b}|\,|\cos \theta| \le |\mathbf{a}|\,|\mathbf{b}|$.

Note: We have equality in the case of $\cos \theta = \pm 1$, so $\theta = 0$ or $\theta = \pi$, thus equality when $\mathbf{a}$ and $\mathbf{b}$ are parallel.

63. (a)

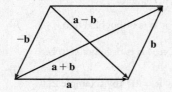

The Parallelogram Law states that the sum of the squares of the lengths of the diagonals of a parallelogram equals the sum of the squares of its (four) sides.

(b) $|\mathbf{a} + \mathbf{b}|^2 = (\mathbf{a} + \mathbf{b}) \cdot (\mathbf{a} + \mathbf{b}) = |\mathbf{a}|^2 + 2(\mathbf{a} \cdot \mathbf{b}) + |\mathbf{b}|^2$ and $|\mathbf{a} - \mathbf{b}|^2 = (\mathbf{a} - \mathbf{b}) \cdot (\mathbf{a} - \mathbf{b}) = |\mathbf{a}|^2 - 2(\mathbf{a} \cdot \mathbf{b}) + |\mathbf{b}|^2$.

Adding these two equations gives $|\mathbf{a} + \mathbf{b}|^2 + |\mathbf{a} - \mathbf{b}|^2 = 2\,|\mathbf{a}|^2 + 2\,|\mathbf{b}|^2$.

65.
$$\mathrm{proj}_{\mathbf{a}}\,\mathbf{b} \cdot \mathrm{proj}_{\mathbf{b}}\,\mathbf{a} = \frac{\mathbf{a} \cdot \mathbf{b}}{|\mathbf{a}|^2}\,\mathbf{a} \cdot \frac{\mathbf{b} \cdot \mathbf{a}}{|\mathbf{b}|^2}\,\mathbf{b} = \frac{\mathbf{a} \cdot \mathbf{b}}{|\mathbf{a}|^2} \cdot \frac{\mathbf{b} \cdot \mathbf{a}}{|\mathbf{b}|^2}\,(\mathbf{a} \cdot \mathbf{b}) \qquad \text{by Property 4 of the dot product}$$

$$= \frac{(\mathbf{a} \cdot \mathbf{b})^2}{|\mathbf{a}|^2\,|\mathbf{b}|^2}\,(\mathbf{a} \cdot \mathbf{b}) = \left(\frac{\mathbf{a} \cdot \mathbf{b}}{|\mathbf{a}|\,|\mathbf{b}|} \right)^2 (\mathbf{a} \cdot \mathbf{b}) \qquad \text{by Property 2}$$

$$= (\cos \theta)^2\,(\mathbf{a} \cdot \mathbf{b}) = (\mathbf{a} \cdot \mathbf{b}) \cos^2 \theta \qquad \text{by Corollary 6}$$

12.4   The Cross Product

1. $\mathbf{a} \times \mathbf{b} = \begin{vmatrix} \mathbf{i} & \mathbf{j} & \mathbf{k} \\ 2 & 3 & 0 \\ 1 & 0 & 5 \end{vmatrix} = \begin{vmatrix} 3 & 0 \\ 0 & 5 \end{vmatrix} \mathbf{i} - \begin{vmatrix} 2 & 0 \\ 1 & 5 \end{vmatrix} \mathbf{j} + \begin{vmatrix} 2 & 3 \\ 1 & 0 \end{vmatrix} \mathbf{k}$

$$= (15 - 0)\,\mathbf{i} - (10 - 0)\,\mathbf{j} + (0 - 3)\,\mathbf{k} = 15\,\mathbf{i} - 10\,\mathbf{j} - 3\,\mathbf{k}$$

Now $(\mathbf{a} \times \mathbf{b}) \cdot \mathbf{a} = \langle 15, -10, -3 \rangle \cdot \langle 2, 3, 0 \rangle = 30 - 30 + 0 = 0$ and

$(\mathbf{a} \times \mathbf{b}) \cdot \mathbf{b} = \langle 15, -10, -3 \rangle \cdot \langle 1, 0, 5 \rangle = 15 + 0 - 15 = 0$, so $\mathbf{a} \times \mathbf{b}$ is orthogonal to both $\mathbf{a}$ and $\mathbf{b}$.

3. $\mathbf{a} \times \mathbf{b} = \begin{vmatrix} \mathbf{i} & \mathbf{j} & \mathbf{k} \\ 0 & 2 & -4 \\ -1 & 3 & 1 \end{vmatrix} = \begin{vmatrix} 2 & -4 \\ 3 & 1 \end{vmatrix} \mathbf{i} - \begin{vmatrix} 0 & -4 \\ -1 & 1 \end{vmatrix} \mathbf{j} + \begin{vmatrix} 0 & 2 \\ -1 & 3 \end{vmatrix} \mathbf{k}$

$$= [2 - (-12)]\,\mathbf{i} - (0 - 4)\,\mathbf{j} + [0 - (-2)]\,\mathbf{k} = 14\,\mathbf{i} + 4\,\mathbf{j} + 2\,\mathbf{k}$$

Since $(\mathbf{a} \times \mathbf{b}) \cdot \mathbf{a} = (14\,\mathbf{i} + 4\,\mathbf{j} + 2\,\mathbf{k}) \cdot (2\,\mathbf{j} - 4\,\mathbf{k}) = 0 + 8 - 8 = 0$, $\mathbf{a} \times \mathbf{b}$ is orthogonal to $\mathbf{a}$.

Since $(\mathbf{a} \times \mathbf{b}) \cdot \mathbf{b} = (14\,\mathbf{i} + 4\,\mathbf{j} + 2\,\mathbf{k}) \cdot (-\mathbf{i} + 3\,\mathbf{j} + \mathbf{k}) = -14 + 12 + 2 = 0$, $\mathbf{a} \times \mathbf{b}$ is orthogonal to $\mathbf{b}$.

5. $\mathbf{a} \times \mathbf{b} = \begin{vmatrix} \mathbf{i} & \mathbf{j} & \mathbf{k} \\ \frac{1}{2} & \frac{1}{3} & \frac{1}{4} \\ 1 & 2 & -3 \end{vmatrix} = \begin{vmatrix} \frac{1}{3} & \frac{1}{4} \\ 2 & -3 \end{vmatrix} \mathbf{i} - \begin{vmatrix} \frac{1}{2} & \frac{1}{4} \\ 1 & -3 \end{vmatrix} \mathbf{j} + \begin{vmatrix} \frac{1}{2} & \frac{1}{3} \\ 1 & 2 \end{vmatrix} \mathbf{k}$

$= \left(-1 - \frac{1}{2}\right) \mathbf{i} - \left(-\frac{3}{2} - \frac{1}{4}\right) \mathbf{j} + \left(1 - \frac{1}{3}\right) \mathbf{k} = -\frac{3}{2} \mathbf{i} + \frac{7}{4} \mathbf{j} + \frac{2}{3} \mathbf{k}$

Since $(\mathbf{a} \times \mathbf{b}) \cdot \mathbf{a} = \left(-\frac{3}{2} \mathbf{i} + \frac{7}{4} \mathbf{j} + \frac{2}{3} \mathbf{k}\right) \cdot \left(\frac{1}{2} \mathbf{i} + \frac{1}{3} \mathbf{j} + \frac{1}{4} \mathbf{k}\right) = -\frac{3}{4} + \frac{7}{12} + \frac{1}{6} = 0$, $\mathbf{a} \times \mathbf{b}$ is orthogonal to $\mathbf{a}$.

Since $(\mathbf{a} \times \mathbf{b}) \cdot \mathbf{b} = \left(-\frac{3}{2} \mathbf{i} + \frac{7}{4} \mathbf{j} + \frac{2}{3} \mathbf{k}\right) \cdot (\mathbf{i} + 2\mathbf{j} - 3\mathbf{k}) = -\frac{3}{2} + \frac{7}{2} - 2 = 0$, $\mathbf{a} \times \mathbf{b}$ is orthogonal to $\mathbf{b}$.

7. $\mathbf{a} \times \mathbf{b} = \begin{vmatrix} \mathbf{i} & \mathbf{j} & \mathbf{k} \\ t & 1 & 1/t \\ t^2 & t^2 & 1 \end{vmatrix} = \begin{vmatrix} 1 & 1/t \\ t^2 & 1 \end{vmatrix} \mathbf{i} - \begin{vmatrix} t & 1/t \\ t^2 & 1 \end{vmatrix} \mathbf{j} + \begin{vmatrix} t & 1 \\ t^2 & t^2 \end{vmatrix} \mathbf{k}$

$= (1 - t) \mathbf{i} - (t - t) \mathbf{j} + (t^3 - t^2) \mathbf{k} = (1 - t) \mathbf{i} + (t^3 - t^2) \mathbf{k}$

Since $(\mathbf{a} \times \mathbf{b}) \cdot \mathbf{a} = \langle 1 - t, 0, t^3 - t^2 \rangle \cdot \langle t, 1, 1/t \rangle = t - t^2 + 0 + t^2 - t = 0$, $\mathbf{a} \times \mathbf{b}$ is orthogonal to $\mathbf{a}$.

Since $(\mathbf{a} \times \mathbf{b}) \cdot \mathbf{b} = \langle 1 - t, 0, t^3 - t^2 \rangle \cdot \langle t^2, t^2, 1 \rangle = t^2 - t^3 + 0 + t^3 - t^2 = 0$, $\mathbf{a} \times \mathbf{b}$ is orthogonal to $\mathbf{b}$.

9. According to the discussion following Example 4, $\mathbf{i} \times \mathbf{j} = \mathbf{k}$, so $(\mathbf{i} \times \mathbf{j}) \times \mathbf{k} = \mathbf{k} \times \mathbf{k} = \mathbf{0}$ [by Example 2].

11. $(\mathbf{j} - \mathbf{k}) \times (\mathbf{k} - \mathbf{i}) = (\mathbf{j} - \mathbf{k}) \times \mathbf{k} + (\mathbf{j} - \mathbf{k}) \times (-\mathbf{i})$ by Property 3 of the cross product

$= \mathbf{j} \times \mathbf{k} + (-\mathbf{k}) \times \mathbf{k} + \mathbf{j} \times (-\mathbf{i}) + (-\mathbf{k}) \times (-\mathbf{i})$ by Property 4

$= (\mathbf{j} \times \mathbf{k}) + (-1)(\mathbf{k} \times \mathbf{k}) + (-1)(\mathbf{j} \times \mathbf{i}) + (-1)^2(\mathbf{k} \times \mathbf{i})$ by Property 2

$= \mathbf{i} + (-1)\,\mathbf{0} + (-1)(-\mathbf{k}) + \mathbf{j} = \mathbf{i} + \mathbf{j} + \mathbf{k}$ by Example 2 and
                                                                              the discussion following Example 4

13. (a) Since $\mathbf{b} \times \mathbf{c}$ is a vector, the dot product $\mathbf{a} \cdot (\mathbf{b} \times \mathbf{c})$ is meaningful and is a scalar.

(b) $\mathbf{b} \cdot \mathbf{c}$ is a scalar, so $\mathbf{a} \times (\mathbf{b} \cdot \mathbf{c})$ is meaningless, as the cross product is defined only for two *vectors*.

(c) Since $\mathbf{b} \times \mathbf{c}$ is a vector, the cross product $\mathbf{a} \times (\mathbf{b} \times \mathbf{c})$ is meaningful and results in another vector.

(d) $\mathbf{b} \cdot \mathbf{c}$ is a scalar, so the dot product $\mathbf{a} \cdot (\mathbf{b} \cdot \mathbf{c})$ is meaningless, as the dot product is defined only for two vectors.

(e) Since $(\mathbf{a} \cdot \mathbf{b})$ and $(\mathbf{c} \cdot \mathbf{d})$ are both scalars, the cross product $(\mathbf{a} \cdot \mathbf{b}) \times (\mathbf{c} \cdot \mathbf{d})$ is meaningless.

(f) $\mathbf{a} \times \mathbf{b}$ and $\mathbf{c} \times \mathbf{d}$ are both vectors, so the dot product $(\mathbf{a} \times \mathbf{b}) \cdot (\mathbf{c} \times \mathbf{d})$ is meaningful and is a scalar.

15. If we sketch $\mathbf{u}$ and $\mathbf{v}$ starting from the same initial point, we see that the

angle between them is $60°$. Using Theorem 9, we have

$|\mathbf{u} \times \mathbf{v}| = |\mathbf{u}|\,|\mathbf{v}| \sin\theta = (12)(16)\sin 60° = 192 \cdot \dfrac{\sqrt{3}}{2} = 96\sqrt{3}.$

By the right-hand rule, $\mathbf{u} \times \mathbf{v}$ is directed into the page.

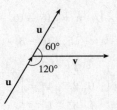

17. $\mathbf{a} \times \mathbf{b} = \begin{vmatrix} \mathbf{i} & \mathbf{j} & \mathbf{k} \\ 2 & -1 & 3 \\ 4 & 2 & 1 \end{vmatrix} = \begin{vmatrix} -1 & 3 \\ 2 & 1 \end{vmatrix} \mathbf{i} - \begin{vmatrix} 2 & 3 \\ 4 & 1 \end{vmatrix} \mathbf{j} + \begin{vmatrix} 2 & -1 \\ 4 & 2 \end{vmatrix} \mathbf{k} = (-1 - 6)\,\mathbf{i} - (2 - 12)\,\mathbf{j} + [4 - (-4)]\,\mathbf{k} = -7\,\mathbf{i} + 10\,\mathbf{j} + 8\,\mathbf{k}$

$$\mathbf{b}\times\mathbf{a}=\begin{vmatrix}\mathbf{i}&\mathbf{j}&\mathbf{k}\\4&2&1\\2&-1&3\end{vmatrix}=\begin{vmatrix}2&1\\-1&3\end{vmatrix}\mathbf{i}-\begin{vmatrix}4&1\\2&3\end{vmatrix}\mathbf{j}+\begin{vmatrix}4&2\\2&-1\end{vmatrix}\mathbf{k}=[6-(-1)]\,\mathbf{i}-(12-2)\,\mathbf{j}+(-4-4)\,\mathbf{k}=7\,\mathbf{i}-10\,\mathbf{j}-8\,\mathbf{k}$$

Notice $\mathbf{a}\times\mathbf{b}=-\mathbf{b}\times\mathbf{a}$ here, as we know is always true by Property 1 of the cross product.

19. By Theorem 8, the cross product of two vectors is orthogonal to both vectors. So we calculate

$$\langle 3,2,1\rangle\times\langle -1,1,0\rangle=\begin{vmatrix}\mathbf{i}&\mathbf{j}&\mathbf{k}\\3&2&1\\-1&1&0\end{vmatrix}=\begin{vmatrix}2&1\\1&0\end{vmatrix}\mathbf{i}-\begin{vmatrix}3&1\\-1&0\end{vmatrix}\mathbf{j}+\begin{vmatrix}3&2\\-1&1\end{vmatrix}\mathbf{k}=-\mathbf{i}-\mathbf{j}+5\,\mathbf{k}.$$

So two unit vectors orthogonal to both given vectors are $\pm\dfrac{\langle -1,-1,5\rangle}{\sqrt{1+1+25}}=\pm\dfrac{\langle -1,-1,5\rangle}{3\sqrt{3}}$, that is, $\left\langle -\dfrac{1}{3\sqrt{3}},-\dfrac{1}{3\sqrt{3}},\dfrac{5}{3\sqrt{3}}\right\rangle$

and $\left\langle \dfrac{1}{3\sqrt{3}},\dfrac{1}{3\sqrt{3}},-\dfrac{5}{3\sqrt{3}}\right\rangle$.

21. Let $\mathbf{a}=\langle a_1,a_2,a_3\rangle$. Then

$$\mathbf{0}\times\mathbf{a}=\begin{vmatrix}\mathbf{i}&\mathbf{j}&\mathbf{k}\\0&0&0\\a_1&a_2&a_3\end{vmatrix}=\begin{vmatrix}0&0\\a_2&a_3\end{vmatrix}\mathbf{i}-\begin{vmatrix}0&0\\a_1&a_3\end{vmatrix}\mathbf{j}+\begin{vmatrix}0&0\\a_1&a_2\end{vmatrix}\mathbf{k}=\mathbf{0},$$

$$\mathbf{a}\times\mathbf{0}=\begin{vmatrix}\mathbf{i}&\mathbf{j}&\mathbf{k}\\a_1&a_2&a_3\\0&0&0\end{vmatrix}=\begin{vmatrix}a_2&a_3\\0&0\end{vmatrix}\mathbf{i}-\begin{vmatrix}a_1&a_3\\0&0\end{vmatrix}\mathbf{j}+\begin{vmatrix}a_1&a_2\\0&0\end{vmatrix}\mathbf{k}=\mathbf{0}.$$

23. $\mathbf{a}\times\mathbf{b}=\langle a_2b_3-a_3b_2,a_3b_1-a_1b_3,a_1b_2-a_2b_1\rangle$

$$=\langle(-1)(b_2a_3-b_3a_2),(-1)(b_3a_1-b_1a_3),(-1)(b_1a_2-b_2a_1)\rangle$$

$$=-\langle b_2a_3-b_3a_2,b_3a_1-b_1a_3,b_1a_2-b_2a_1\rangle=-\mathbf{b}\times\mathbf{a}$$

25. $\mathbf{a}\times(\mathbf{b}+\mathbf{c})=\mathbf{a}\times\langle b_1+c_1,b_2+c_2,b_3+c_3\rangle$

$$=\langle a_2(b_3+c_3)-a_3(b_2+c_2),a_3(b_1+c_1)-a_1(b_3+c_3),a_1(b_2+c_2)-a_2(b_1+c_1)\rangle$$

$$=\langle a_2b_3+a_2c_3-a_3b_2-a_3c_2,a_3b_1+a_3c_1-a_1b_3-a_1c_3,a_1b_2+a_1c_2-a_2b_1-a_2c_1\rangle$$

$$=\langle(a_2b_3-a_3b_2)+(a_2c_3-a_3c_2),(a_3b_1-a_1b_3)+(a_3c_1-a_1c_3),(a_1b_2-a_2b_1)+(a_1c_2-a_2c_1)\rangle$$

$$=\langle a_2b_3-a_3b_2,a_3b_1-a_1b_3,a_1b_2-a_2b_1\rangle+\langle a_2c_3-a_3c_2,a_3c_1-a_1c_3,a_1c_2-a_2c_1\rangle$$

$$=(\mathbf{a}\times\mathbf{b})+(\mathbf{a}\times\mathbf{c})$$

27. By plotting the vertices, we can see that the parallelogram is determined

by the vectors $\overrightarrow{AB}=\langle 2,3\rangle$ and $\overrightarrow{AD}=\langle 6,-1\rangle$. We know that the area

of the parallelogram determined by two vectors is equal to the length of

the cross product of these vectors. In order to compute the cross product,

we consider the vector $\overrightarrow{AB}$ as the three-dimensional vector $\langle 2,3,0\rangle$

(and similarly for $\overrightarrow{AD}$), and then the area of parallelogram $ABCD$ is

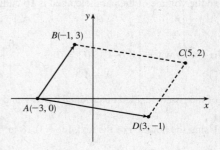

$$\left|\overrightarrow{AB}\times\overrightarrow{AD}\right|=\left\|\begin{vmatrix}\mathbf{i}&\mathbf{j}&\mathbf{k}\\2&3&0\\6&-1&0\end{vmatrix}\right\|=\left|(0-0)\,\mathbf{i}-(0-0)\,\mathbf{j}+(-2-18)\,\mathbf{k}\right|=\left|-20\,\mathbf{k}\right|=20$$

29. (a) Because the plane through P, Q, and R contains the vectors $\overrightarrow{PQ}$ and $\overrightarrow{PR}$, a vector orthogonal to both of these vectors

(such as their cross product) is also orthogonal to the plane. Here $\overrightarrow{PQ} = \langle -3, 1, 2 \rangle$ and $\overrightarrow{PR} = \langle 3, 2, 4 \rangle$, so

$$\overrightarrow{PQ} \times \overrightarrow{PR} = \langle (1)(4) - (2)(2), (2)(3) - (-3)(4), (-3)(2) - (1)(3) \rangle = \langle 0, 18, -9 \rangle$$

Therefore, $\langle 0, 18, -9 \rangle$ (or any nonzero scalar multiple thereof, such as $\langle 0, 2, -1 \rangle$) is orthogonal to the plane through P, Q,

and R.

(b) Note that the area of the triangle determined by P, Q, and R is equal to half of the area of the

parallelogram determined by the three points. From part (a), the area of the parallelogram is

$\left| \overrightarrow{PQ} \times \overrightarrow{PR} \right| = |\langle 0, 18, -9 \rangle| = \sqrt{0 + 324 + 81} = \sqrt{405} = 9\sqrt{5}$, so the area of the triangle is $\frac{1}{2} \cdot 9\sqrt{5} = \frac{9}{2}\sqrt{5}$.

31. (a) $\overrightarrow{PQ} = \langle 4, 3, -2 \rangle$ and $\overrightarrow{PR} = \langle 5, 5, 1 \rangle$, so a vector orthogonal to the plane through P, Q, and R is

$\overrightarrow{PQ} \times \overrightarrow{PR} = \langle (3)(1) - (-2)(5), (-2)(5) - (4)(1), (4)(5) - (3)(5) \rangle = \langle 13, -14, 5 \rangle$ [or any scalar mutiple thereof].

(b) The area of the parallelogram determined by $\overrightarrow{PQ}$ and $\overrightarrow{PR}$ is

$\left| \overrightarrow{PQ} \times \overrightarrow{PR} \right| = |\langle 13, -14, 5 \rangle| = \sqrt{13^2 + (-14)^2 + 5^2} = \sqrt{390}$, so the area of triangle PQR is $\frac{1}{2}\sqrt{390}$.

33. By Equation 14, the volume of the parallelepiped determined by $\mathbf{a}$, $\mathbf{b}$, and $\mathbf{c}$ is the magnitude of their scalar triple product,

which is $\mathbf{a} \cdot (\mathbf{b} \times \mathbf{c}) = \begin{vmatrix} 1 & 2 & 3 \\ -1 & 1 & 2 \\ 2 & 1 & 4 \end{vmatrix} = 1 \begin{vmatrix} 1 & 2 \\ 1 & 4 \end{vmatrix} - 2 \begin{vmatrix} -1 & 2 \\ 2 & 4 \end{vmatrix} + 3 \begin{vmatrix} -1 & 1 \\ 2 & 1 \end{vmatrix} = 1(4 - 2) - 2(-4 - 4) + 3(-1 - 2) = 9.$

Thus the volume of the parallelepiped is 9 cubic units.

35. $\mathbf{a} = \overrightarrow{PQ} = \langle 4, 2, 2 \rangle$, $\mathbf{b} = \overrightarrow{PR} = \langle 3, 3, -1 \rangle$, and $\mathbf{c} = \overrightarrow{PS} = \langle 5, 5, 1 \rangle$.

$\mathbf{a} \cdot (\mathbf{b} \times \mathbf{c}) = \begin{vmatrix} 4 & 2 & 2 \\ 3 & 3 & -1 \\ 5 & 5 & 1 \end{vmatrix} = 4 \begin{vmatrix} 3 & -1 \\ 5 & 1 \end{vmatrix} - 2 \begin{vmatrix} 3 & -1 \\ 5 & 1 \end{vmatrix} + 2 \begin{vmatrix} 3 & 3 \\ 5 & 5 \end{vmatrix} = 32 - 16 + 0 = 16,$

so the volume of the parallelepiped is 16 cubic units.

37. $\mathbf{u} \cdot (\mathbf{v} \times \mathbf{w}) = \begin{vmatrix} 1 & 5 & -2 \\ 3 & -1 & 0 \\ 5 & 9 & -4 \end{vmatrix} = 1 \begin{vmatrix} -1 & 0 \\ 9 & -4 \end{vmatrix} - 5 \begin{vmatrix} 3 & 0 \\ 5 & -4 \end{vmatrix} + (-2) \begin{vmatrix} 3 & -1 \\ 5 & 9 \end{vmatrix} = 4 + 60 - 64 = 0$, which says that the volume

of the parallelepiped determined by $\mathbf{u}$, $\mathbf{v}$ and $\mathbf{w}$ is 0, and thus these three vectors are coplanar.

39. Using the notation of the text, $|\mathbf{r}| = 0.18$ m, $|\mathbf{F}| = 60$ N, and the angle between $\mathbf{r}$ and $\mathbf{F}$ is $\theta = 70° + 10° = 80°$.

(Move $\mathbf{F}$ so that both vectors start from the same point.) Then the magnitude of the torque is

$|\boldsymbol{\tau}| = |\mathbf{r} \times \mathbf{F}| = |\mathbf{r}| \, |\mathbf{F}| \sin \theta = (0.18)(60) \sin 80° = 10.8 \sin 80° \approx 10.6$ N·m.

41. Using the notation of the text, $\mathbf{r} = \langle 0, 0.3, 0 \rangle$ (measuring in meters) and $\mathbf{F}$ has direction $\langle 0, 3, -4 \rangle$. The angle θ between them

can be determined by $\cos\theta = \dfrac{\langle 0, 0.3, 0 \rangle \cdot \langle 0, 3, -4 \rangle}{|\langle 0, 0.3, 0 \rangle| \, |\langle 0, 3, -4 \rangle|} \;\Rightarrow\; \cos\theta = \dfrac{0.9}{(0.3)(5)} \;\Rightarrow\; \cos\theta = 0.6 \;\Rightarrow$

$\theta = \cos^{-1}(0.6) \approx 53.1°$. Then $|\boldsymbol{\tau}| = |\mathbf{r}| \, |\mathbf{F}| \sin\theta \;\Rightarrow\; 100 \approx 0.3 \, |\mathbf{F}| \sin 53.1° \;\Rightarrow\; |\mathbf{F}| \approx \dfrac{100}{0.3 \sin 53.1°} \approx 417 \text{ N}.$

43. From Theorem 9 we have $|\mathbf{a} \times \mathbf{b}| = |\mathbf{a}| \, |\mathbf{b}| \sin\theta$, where θ is the angle between $\mathbf{a}$ and $\mathbf{b}$, and from Theorem 12.3.3 we have

$\mathbf{a} \cdot \mathbf{b} = |\mathbf{a}| \, |\mathbf{b}| \cos\theta \;\Rightarrow\; |\mathbf{a}| \, |\mathbf{b}| = \dfrac{\mathbf{a} \cdot \mathbf{b}}{\cos\theta}$. Substituting the second equation into the first gives $|\mathbf{a} \times \mathbf{b}| = \dfrac{\mathbf{a} \cdot \mathbf{b}}{\cos\theta} \sin\theta$, so

$\dfrac{|\mathbf{a} \times \mathbf{b}|}{\mathbf{a} \cdot \mathbf{b}} = \tan\theta$. Here $|\mathbf{a} \times \mathbf{b}| = |\langle 1, 2, 2 \rangle| = \sqrt{1 + 4 + 4} = 3$, so $\tan\theta = \dfrac{|\mathbf{a} \times \mathbf{b}|}{\mathbf{a} \cdot \mathbf{b}} = \dfrac{3}{\sqrt{3}} = \sqrt{3} \;\Rightarrow\; \theta = 60°.$

45. (a)

The distance between a point and a line is the length of the perpendicular

from the point to the line, here $\left| \overrightarrow{PS} \right| = d$. But referring to triangle PQS,

$d = \left| \overrightarrow{PS} \right| = \left| \overrightarrow{QP} \right| \sin\theta = |\mathbf{b}| \sin\theta$. But θ is the angle between $\overrightarrow{QP} = \mathbf{b}$

and $\overrightarrow{QR} = \mathbf{a}$. Thus by Theorem 9, $\sin\theta = \dfrac{|\mathbf{a} \times \mathbf{b}|}{|\mathbf{a}| \, |\mathbf{b}|}$

and so $d = |\mathbf{b}| \sin\theta = \dfrac{|\mathbf{b}| \, |\mathbf{a} \times \mathbf{b}|}{|\mathbf{a}| \, |\mathbf{b}|} = \dfrac{|\mathbf{a} \times \mathbf{b}|}{|\mathbf{a}|}.$

(b) $\mathbf{a} = \overrightarrow{QR} = \langle -1, -2, -1 \rangle$ and $\mathbf{b} = \overrightarrow{QP} = \langle 1, -5, -7 \rangle$. Then

$\mathbf{a} \times \mathbf{b} = \langle (-2)(-7) - (-1)(-5), (-1)(1) - (-1)(-7), (-1)(-5) - (-2)(1) \rangle = \langle 9, -8, 7 \rangle.$

Thus the distance is $d = \dfrac{|\mathbf{a} \times \mathbf{b}|}{|\mathbf{a}|} = \dfrac{1}{\sqrt{6}} \sqrt{81 + 64 + 49} = \sqrt{\dfrac{194}{6}} = \sqrt{\dfrac{97}{3}}.$

47. From Theorem 9 we have $|\mathbf{a} \times \mathbf{b}| = |\mathbf{a}| \, |\mathbf{b}| \sin\theta$ so

$$|\mathbf{a} \times \mathbf{b}|^2 = |\mathbf{a}|^2 \, |\mathbf{b}|^2 \sin^2\theta = |\mathbf{a}|^2 \, |\mathbf{b}|^2 \left(1 - \cos^2\theta \right)$$
$$= |\mathbf{a}|^2 \, |\mathbf{b}|^2 - (|\mathbf{a}| \, |\mathbf{b}| \cos\theta)^2 = |\mathbf{a}|^2 \, |\mathbf{b}|^2 - (\mathbf{a} \cdot \mathbf{b})^2$$

by Theorem 12.3.3.

49. $(\mathbf{a} - \mathbf{b}) \times (\mathbf{a} + \mathbf{b}) = (\mathbf{a} - \mathbf{b}) \times \mathbf{a} + (\mathbf{a} - \mathbf{b}) \times \mathbf{b}$ by Property 3 of the cross product

$= \mathbf{a} \times \mathbf{a} + (-\mathbf{b}) \times \mathbf{a} + \mathbf{a} \times \mathbf{b} + (-\mathbf{b}) \times \mathbf{b}$ by Property 4

$= (\mathbf{a} \times \mathbf{a}) - (\mathbf{b} \times \mathbf{a}) + (\mathbf{a} \times \mathbf{b}) - (\mathbf{b} \times \mathbf{b})$ by Property 2 (with $c = -1$)

$= \mathbf{0} - (\mathbf{b} \times \mathbf{a}) + (\mathbf{a} \times \mathbf{b}) - \mathbf{0}$ by Example 2

$= (\mathbf{a} \times \mathbf{b}) + (\mathbf{a} \times \mathbf{b})$ by Property 1

$= 2(\mathbf{a} \times \mathbf{b})$

51. $\mathbf{a} \times (\mathbf{b} \times \mathbf{c}) + \mathbf{b} \times (\mathbf{c} \times \mathbf{a}) + \mathbf{c} \times (\mathbf{a} \times \mathbf{b})$

$= [(\mathbf{a} \cdot \mathbf{c})\mathbf{b} - (\mathbf{a} \cdot \mathbf{b})\mathbf{c}] + [(\mathbf{b} \cdot \mathbf{a})\mathbf{c} - (\mathbf{b} \cdot \mathbf{c})\mathbf{a}] + [(\mathbf{c} \cdot \mathbf{b})\mathbf{a} - (\mathbf{c} \cdot \mathbf{a})\mathbf{b}]$ by Exercise 50

$= (\mathbf{a} \cdot \mathbf{c})\mathbf{b} - (\mathbf{a} \cdot \mathbf{b})\mathbf{c} + (\mathbf{a} \cdot \mathbf{b})\mathbf{c} - (\mathbf{b} \cdot \mathbf{c})\mathbf{a} + (\mathbf{b} \cdot \mathbf{c})\mathbf{a} - (\mathbf{a} \cdot \mathbf{c})\mathbf{b} = \mathbf{0}$

53. (a) No. If $\mathbf{a} \cdot \mathbf{b} = \mathbf{a} \cdot \mathbf{c}$, then $\mathbf{a} \cdot (\mathbf{b} - \mathbf{c}) = 0$, so $\mathbf{a}$ is perpendicular to $\mathbf{b} - \mathbf{c}$, which can happen if $\mathbf{b} \neq \mathbf{c}$. For example,

    let $\mathbf{a} = \langle 1, 1, 1 \rangle$, $\mathbf{b} = \langle 1, 0, 0 \rangle$ and $\mathbf{c} = \langle 0, 1, 0 \rangle$.

    (b) No. If $\mathbf{a} \times \mathbf{b} = \mathbf{a} \times \mathbf{c}$ then $\mathbf{a} \times (\mathbf{b} - \mathbf{c}) = \mathbf{0}$, which implies that $\mathbf{a}$ is parallel to $\mathbf{b} - \mathbf{c}$, which of course can happen

    if $\mathbf{b} \neq \mathbf{c}$.

    (c) Yes. Since $\mathbf{a} \cdot \mathbf{c} = \mathbf{a} \cdot \mathbf{b}$, $\mathbf{a}$ is perpendicular to $\mathbf{b} - \mathbf{c}$, by part (a). From part (b), $\mathbf{a}$ is also parallel to $\mathbf{b} - \mathbf{c}$. Thus since

    $\mathbf{a} \neq \mathbf{0}$ but is both parallel and perpendicular to $\mathbf{b} - \mathbf{c}$, we have $\mathbf{b} - \mathbf{c} = \mathbf{0}$, so $\mathbf{b} = \mathbf{c}$.

12.5 Equations of Lines and Planes

1. (a) True; each of the first two lines has a direction vector parallel to the direction vector of the third line, so these vectors are

    each scalar multiples of the third direction vector. Then the first two direction vectors are also scalar multiples of each

    other, so these vectors, and hence the two lines, are parallel.

    (b) False; for example, the x- and y-axes are both perpendicular to the z-axis, yet the x- and y-axes are not parallel.

    (c) True; each of the first two planes has a normal vector parallel to the normal vector of the third plane, so these two normal

    vectors are parallel to each other and the planes are parallel.

    (d) False; for example, the xy- and yz-planes are not parallel, yet they are both perpendicular to the xz-plane.

    (e) False; the x- and y-axes are not parallel, yet they are both parallel to the plane $z = 1$.

    (f) True; if each line is perpendicular to a plane, then the lines' direction vectors are both parallel to a normal vector for the

    plane. Thus, the direction vectors are parallel to each other and the lines are parallel.

    (g) False; the planes $y = 1$ and $z = 1$ are not parallel, yet they are both parallel to the x-axis.

    (h) True; if each plane is perpendicular to a line, then any normal vector for each plane is parallel to a direction vector for the

    line. Thus, the normal vectors are parallel to each other and the planes are parallel.

    (i) True; see Figure 9 and the accompanying discussion.

    (j) False; they can be skew, as in Example 3.

    (k) True. Consider any normal vector for the plane and any direction vector for the line. If the normal vector is perpendicular

    to the direction vector, the line and plane are parallel. Otherwise, the vectors meet at an angle θ, $0° \leq \theta < 90°$, and the

    line will intersect the plane at an angle $90° - \theta$.

3. For this line, we have $\mathbf{r}_0 = 2\mathbf{i} + 2.4\mathbf{j} + 3.5\mathbf{k}$ and $\mathbf{v} = 3\mathbf{i} + 2\mathbf{j} - \mathbf{k}$, so a vector equation is

    $\mathbf{r} = \mathbf{r}_0 + t\mathbf{v} = (2\mathbf{i} + 2.4\mathbf{j} + 3.5\mathbf{k}) + t(3\mathbf{i} + 2\mathbf{j} - \mathbf{k}) = (2 + 3t)\mathbf{i} + (2.4 + 2t)\mathbf{j} + (3.5 - t)\mathbf{k}$ and parametric equations are

    $x = 2 + 3t$, $y = 2.4 + 2t$, $z = 3.5 - t$.

5. A line perpendicular to the given plane has the same direction as a normal vector to the plane, such as

    $\mathbf{n} = \langle 1, 3, 1 \rangle$. So $\mathbf{r}_0 = \mathbf{i} + 6\mathbf{k}$, and we can take $\mathbf{v} = \mathbf{i} + 3\mathbf{j} + \mathbf{k}$. Then a vector equation is

    $\mathbf{r} = (\mathbf{i} + 6\mathbf{k}) + t(\mathbf{i} + 3\mathbf{j} + \mathbf{k}) = (1 + t)\mathbf{i} + 3t\mathbf{j} + (6 + t)\mathbf{k}$, and parametric equations are $x = 1 + t$, $y = 3t$, $z = 6 + t$.

7. The vector $\mathbf{v} = \langle 2 - 0, 1 - \frac{1}{2}, -3 - 1 \rangle = \langle 2, \frac{1}{2}, -4 \rangle$ is parallel to the line. Letting $P_0 = (2, 1, -3)$, parametric equations

are $x = 2 + 2t$, $y = 1 + \frac{1}{2}t$, $z = -3 - 4t$, while symmetric equations are $\dfrac{x-2}{2} = \dfrac{y-1}{1/2} = \dfrac{z+3}{-4}$ or

$\dfrac{x-2}{2} = 2y - 2 = \dfrac{z+3}{-4}$.

9. $\mathbf{v} = \langle 3 - (-8), -2 - 1, 4 - 4 \rangle = \langle 11, -3, 0 \rangle$, and letting $P_0 = (-8, 1, 4)$, parametric equations are $x = -8 + 11t$,

$y = 1 - 3t$, $z = 4 + 0t = 4$, while symmetric equations are $\dfrac{x+8}{11} = \dfrac{y-1}{-3}$, $z = 4$. Notice here that the direction number

$c = 0$, so rather than writing $\dfrac{z-4}{0}$ in the symmetric equation we must write the equation $z = 4$ separately.

11. The given line $\dfrac{x}{2} = \dfrac{y}{3} = \dfrac{z+1}{1}$ has direction $\mathbf{v} = \langle 2, 3, 1 \rangle$. Taking $(-6, 2, 3)$ as P_0, parametric equations are $x = -6 + 2t$,

$y = 2 + 3t$, $z = 3 + t$ and symmetric equations are $\dfrac{x+6}{2} = \dfrac{y-2}{3} = z - 3$.

13. Direction vectors of the lines are $\mathbf{v}_1 = \langle -2 - (-4), 0 - (-6), -3 - 1 \rangle = \langle 2, 6, -4 \rangle$ and

$\mathbf{v}_2 = \langle 5 - 10, 3 - 18, 14 - 4 \rangle = \langle -5, -15, 10 \rangle$, and since $\mathbf{v}_2 = -\frac{5}{2}\mathbf{v}_1$, the direction vectors and thus the lines are parallel.

15. (a) The line passes through the point $(1, -5, 6)$ and a direction vector for the line is $\langle -1, 2, -3 \rangle$, so symmetric equations for

the line are $\dfrac{x-1}{-1} = \dfrac{y+5}{2} = \dfrac{z-6}{-3}$.

(b) The line intersects the xy-plane when $z = 0$, so we need $\dfrac{x-1}{-1} = \dfrac{y+5}{2} = \dfrac{0-6}{-3}$ or $\dfrac{x-1}{-1} = 2$ $\Rightarrow$ $x = -1$,

$\dfrac{y+5}{2} = 2$ $\Rightarrow$ $y = -1$. Thus the point of intersection with the xy-plane is $(-1, -1, 0)$. Similarly for the yz-plane,

we need $x = 0$ $\Rightarrow$ $1 = \dfrac{y+5}{2} = \dfrac{z-6}{-3}$ $\Rightarrow$ $y = -3$, $z = 3$. Thus the line intersects the yz-plane at $(0, -3, 3)$. For

the xz-plane, we need $y = 0$ $\Rightarrow$ $\dfrac{x-1}{-1} = \dfrac{5}{2} = \dfrac{z-6}{-3}$ $\Rightarrow$ $x = -\frac{3}{2}$, $z = -\frac{3}{2}$. So the line intersects the xz-plane

at $\left(-\frac{3}{2}, 0, -\frac{3}{2}\right)$.

17. From Equation 4, the line segment from $\mathbf{r}_0 = 6\mathbf{i} - \mathbf{j} + 9\mathbf{k}$ to $\mathbf{r}_1 = 7\mathbf{i} + 6\mathbf{j}$ has vector equation

$\mathbf{r}(t) = (1 - t)\mathbf{r}_0 + t\mathbf{r}_1 = (1 - t)(6\mathbf{i} - \mathbf{j} + 9\mathbf{k}) + t(7\mathbf{i} + 6\mathbf{j})$

$= (6\mathbf{i} - \mathbf{j} + 9\mathbf{k}) - t(6\mathbf{i} - \mathbf{j} + 9\mathbf{k}) + t(7\mathbf{i} + 6\mathbf{j})$

$= (6\mathbf{i} - \mathbf{j} + 9\mathbf{k}) + t(\mathbf{i} + 7\mathbf{j} - 9\mathbf{k})$, $\quad 0 \le t \le 1$.

19. Since the direction vectors $\langle 2, -1, 3 \rangle$ and $\langle 4, -2, 5 \rangle$ are not scalar multiples of each other, the lines aren't parallel. For the

lines to intersect, we must be able to find one value of t and one value of s that produce the same point from the respective

parametric equations. Thus we need to satisfy the following three equations: $3 + 2t = 1 + 4s$, $4 - t = 3 - 2s$,

$1 + 3t = 4 + 5s$. Solving the last two equations we get $t = 1$, $s = 0$ and checking, we see that these values don't satisfy the

first equation. Thus the lines aren't parallel and don't intersect, so they must be skew lines.

21. Since the direction vectors $\langle 1, -2, -3 \rangle$ and $\langle 1, 3, -7 \rangle$ aren't scalar multiples of each other, the lines aren't parallel. Parametric equations of the lines are L_1: $x = 2 + t$, $y = 3 - 2t$, $z = 1 - 3t$ and L_2: $x = 3 + s$, $y = -4 + 3s$, $z = 2 - 7s$. Thus, for the lines to intersect, the three equations $2 + t = 3 + s$, $3 - 2t = -4 + 3s$, and $1 - 3t = 2 - 7s$ must be satisfied simultaneously. Solving the first two equations gives $t = 2$, $s = 1$ and checking, we see that these values do satisfy the third equation, so the lines intersect when $t = 2$ and $s = 1$, that is, at the point $(4, -1, -5)$.

23. Since the plane is perpendicular to the vector $\langle 1, -2, 5 \rangle$, we can take $\langle 1, -2, 5 \rangle$ as a normal vector to the plane. $(0, 0, 0)$ is a point on the plane, so setting $a = 1$, $b = -2$, $c = 5$ and $x_0 = 0$, $y_0 = 0$, $z_0 = 0$ in Equation 7 gives $1(x - 0) + (-2)(y - 0) + 5(z - 0) = 0$ or $x - 2y + 5z = 0$ as an equation of the plane.

25. $\mathbf{i} + 4\mathbf{j} + \mathbf{k} = \langle 1, 4, 1 \rangle$ is a normal vector to the plane and $\left(-1, \frac{1}{2}, 3\right)$ is a point on the plane, so setting $a = 1$, $b = 4$, $c = 1$, $x_0 = -1$, $y_0 = \frac{1}{2}$, $z_0 = 3$ in Equation 7 gives $1[x - (-1)] + 4\left(y - \frac{1}{2}\right) + 1(z - 3) = 0$ or $x + 4y + z = 4$ as an equation of the plane.

27. Since the two planes are parallel, they will have the same normal vectors. So we can take $\mathbf{n} = \langle 5, -1, -1 \rangle$, and an equation of the plane is $5(x - 1) - 1[y - (-1)] - 1[z - (-1)] = 0$ or $5x - y - z = 7$.

29. Since the two planes are parallel, they will have the same normal vectors. So we can take $\mathbf{n} = \langle 1, 1, 1 \rangle$, and an equation of the plane is $1(x - 1) + 1\left(y - \frac{1}{2}\right) + 1\left(z - \frac{1}{3}\right) = 0$ or $x + y + z = \frac{11}{6}$ or $6x + 6y + 6z = 11$.

31. The vector from $(0, 1, 1)$ to $(1, 0, 1)$, namely $\mathbf{a} = \langle 1 - 0, 0 - 1, 1 - 1 \rangle = \langle 1, -1, 0 \rangle$, and the vector from $(0, 1, 1)$ to $(1, 1, 0)$, $\mathbf{b} = \langle 1 - 0, 1 - 1, 0 - 1 \rangle = \langle 1, 0, -1 \rangle$, both lie in the plane, so $\mathbf{a} \times \mathbf{b}$ is a normal vector to the plane. Thus, we can take $\mathbf{n} = \mathbf{a} \times \mathbf{b} = \langle (-1)(-1) - (0)(0), (0)(1) - (1)(-1), (1)(0) - (-1)(1) \rangle = \langle 1, 1, 1 \rangle$. If P_0 is the point $(0, 1, 1)$, an equation of the plane is $1(x - 0) + 1(y - 1) + 1(z - 1) = 0$ or $x + y + z = 2$.

33. Here the vectors $\mathbf{a} = \langle 3 - 2, -8 - 1, 6 - 2 \rangle = \langle 1, -9, 4 \rangle$ and $\mathbf{b} = \langle -2 - 2, -3 - 1, 1 - 2 \rangle = \langle -4, -4, -1 \rangle$ lie in the plane, so a normal vector to the plane is $\mathbf{n} = \mathbf{a} \times \mathbf{b} = \langle 9 + 16, -16 + 1, -4 - 36 \rangle = \langle 25, -15, -40 \rangle$ and an equation of the plane is $25(x - 2) - 15(y - 1) - 40(z - 2) = 0$ or $25x - 15y - 40z = -45$ or $5x - 3y - 8z = -9$.

35. If we first find two nonparallel vectors in the plane, their cross product will be a normal vector to the plane. Since the given line lies in the plane, its direction vector $\mathbf{a} = \langle -1, 2, -3 \rangle$ is one vector in the plane. We can verify that the given point $(3, 5, -1)$ does not lie on this line, so to find another nonparallel vector $\mathbf{b}$ which lies in the plane, we can pick any point on the line and find a vector connecting the points. If we put $t = 0$, we see that $(4, -1, 0)$ is on the line, so $\mathbf{b} = \langle 4 - 3, -1 - 5, 0 - (-1) \rangle = \langle 1, -6, 1 \rangle$ and $\mathbf{n} = \mathbf{a} \times \mathbf{b} = \langle 2 - 18, -3 + 1, 6 - 2 \rangle = \langle -16, -2, 4 \rangle$. Thus, an equation of the plane is $-16(x - 3) - 2(y - 5) + 4[z - (-1)] = 0$ or $-16x - 2y + 4z = -62$ or $8x + y - 2z = 31$.

37. Normal vectors for the given planes are $\mathbf{n}_1 = \langle 1, 2, 3 \rangle$ and $\mathbf{n}_2 = \langle 2, -1, 1 \rangle$. A direction vector, then, for the line of intersection is $\mathbf{a} = \mathbf{n}_1 \times \mathbf{n}_2 = \langle 2 + 3, 6 - 1, -1 - 4 \rangle = \langle 5, 5, -5 \rangle$, and $\mathbf{a}$ is parallel to the desired plane. Another vector

parallel to the plane is the vector connecting any point on the line of intersection to the given point $(3, 1, 4)$ in the plane.

Setting $z = 0$, the equations of the planes reduce to $x + 2y = 1$ and $2x - y = -3$ with simultaneous solution $x = -1$ and

$y = 1$. So a point on the line is $(-1, 1, 0)$ and another vector parallel to the plane is $\mathbf{b} = \langle 3 - (-1), 1 - 1, 4 - 0 \rangle = \langle 4, 0, 4 \rangle$.

Then a normal vector to the plane is $\mathbf{n} = \mathbf{a} \times \mathbf{b} = \langle 20 - 0, -20 - 20, 0 - 20 \rangle = \langle 20, -40, -20 \rangle$. Equivalently, we can take

$\langle 1, -2, -1 \rangle$ as a normal vector, and an equation of the plane is $1(x - 3) - 2(y - 1) - 1(z - 4) = 0$ or $x - 2y - z = -3$.

39. If a plane is perpendicular to two other planes, its normal vector is perpendicular to the normal vectors of the other two planes.

Thus $\langle 2, 1, -2 \rangle \times \langle 1, 0, 3 \rangle = \langle 3 - 0, -2 - 6, 0 - 1 \rangle = \langle 3, -8, -1 \rangle$ is a normal vector to the desired plane. The point

$(1, 5, 1)$ lies on the plane, so an equation is $3(x - 1) - 8(y - 5) - (z - 1) = 0$ or $3x - 8y - z = -38$.

41. To find the x-intercept we set $y = z = 0$ in the equation $2x + 5y + z = 10$

and obtain $2x = 10 \quad \Rightarrow \quad x = 5$ so the x-intercept is $(5, 0, 0)$. When

$x = z = 0$ we get $5y = 10 \quad \Rightarrow \quad y = 2$, so the y-intercept is $(0, 2, 0)$.

Setting $x = y = 0$ gives $z = 10$, so the z-intercept is $(0, 0, 10)$ and we

graph the portion of the plane that lies in the first octant.

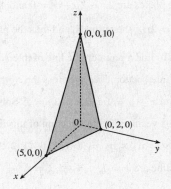

43. Setting $y = z = 0$ in the equation $6x - 3y + 4z = 6$ gives $6x = 6 \quad \Rightarrow$

$x = 1$, when $x = z = 0$ we have $-3y = 6 \quad \Rightarrow \quad y = -2$, and $x = y = 0$

implies $4z = 6 \quad \Rightarrow \quad z = \frac{3}{2}$, so the intercepts are $(1, 0, 0)$, $(0, -2, 0)$, and

$(0, 0, \frac{3}{2})$. The figure shows the portion of the plane cut off by the coordinate

planes.

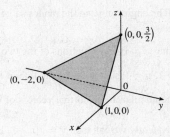

45. Substitute the parametric equations of the line into the equation of the plane: $x + 2y - z = 7 \quad \Rightarrow$

$(2 - 2t) + 2(3t) - (1 + t) = 7 \quad \Rightarrow \quad 3t + 1 = 7 \quad \Rightarrow \quad t = 2$. Therefore, the point of intersection of the line and the plane is

given by $x = 2 - 2(2) = -2$, $y = 3(2) = 6$, and $z = 1 + 2 = 3$, that is, the point $(-2, 6, 3)$.

47. Parametric equations for the line are $x = \frac{1}{5}t$, $y = 2t$, $z = t - 2$ and substitution into the equation of the plane gives

$10\left(\frac{1}{5}t\right) - 7(2t) + 3(t - 2) + 24 = 0 \quad \Rightarrow \quad -9t + 18 = 0 \quad \Rightarrow \quad t = 2$. Thus $x = \frac{1}{5}(2) = \frac{2}{5}$, $y = 2(2) = 4$, $z = 2 - 2 = 0$

and the point of intersection is $\left(\frac{2}{5}, 4, 0\right)$.

49. Setting $x = 0$, we see that $(0, 1, 0)$ satisfies the equations of both planes, so that they do in fact have a line of intersection.

$\mathbf{v} = \mathbf{n}_1 \times \mathbf{n}_2 = \langle 1, 1, 1 \rangle \times \langle 1, 0, 1 \rangle = \langle 1, 0, -1 \rangle$ is the direction of this line. Therefore, direction numbers of the intersecting

line are $1, 0, -1$.

51. Normal vectors for the planes are $\mathbf{n}_1 = \langle 1, 4, -3 \rangle$ and $\mathbf{n}_2 = \langle -3, 6, 7 \rangle$. The normals aren't parallel (they are not scalar multiples of each other), so neither are the planes. But $\mathbf{n}_1 \cdot \mathbf{n}_2 = -3 + 24 - 21 = 0$, so the normals, and thus the planes, are perpendicular.

53. Normal vectors for the planes are $\mathbf{n}_1 = \langle 1, 2, -1 \rangle$ and $\mathbf{n}_2 = \langle 2, -2, 1 \rangle$. The normals are not parallel (they are not scalar multiples of each other), so neither are the planes. Furthermore, $\mathbf{n}_1 \cdot \mathbf{n}_2 = 2 - 4 - 1 = -3 \neq 0$, so the planes aren't perpendicular. The angle between the planes is the same as the angle between the normals, given by

$$\cos\theta = \frac{\mathbf{n}_1 \cdot \mathbf{n}_2}{|\mathbf{n}_1|\,|\mathbf{n}_2|} = \frac{-3}{\sqrt{6}\,\sqrt{9}} = -\frac{1}{\sqrt{6}} \quad \Rightarrow \quad \theta = \cos^{-1}\left(-\frac{1}{\sqrt{6}}\right) \approx 114.1°.$$

55. The planes are $2x - 3y - z = 0$ and $4x - 6y - 2z = 3$ with normal vectors $\mathbf{n}_1 = \langle 2, -3, -1 \rangle$ and $\mathbf{n}_2 = \langle 4, -6, -2 \rangle$. Since $\mathbf{n}_2 = 2\mathbf{n}_1$, the normals, and thus the planes, are parallel.

57. (a) To find a point on the line of intersection, set one of the variables equal to a constant, say $z = 0$. (This will fail if the line of intersection does not cross the xy-plane; in that case, try setting x or y equal to 0.) The equations of the two planes reduce to $x + y = 1$ and $x + 2y = 1$. Solving these two equations gives $x = 1$, $y = 0$. Thus a point on the line is $(1, 0, 0)$. A vector $\mathbf{v}$ in the direction of this intersecting line is perpendicular to the normal vectors of both planes, so we can take $\mathbf{v} = \mathbf{n}_1 \times \mathbf{n}_2 = \langle 1, 1, 1 \rangle \times \langle 1, 2, 2 \rangle = \langle 2 - 2, 1 - 2, 2 - 1 \rangle = \langle 0, -1, 1 \rangle$. By Equations 2, parametric equations for the line are $x = 1$, $y = -t$, $z = t$.

(b) The angle between the planes satisfies $\cos\theta = \dfrac{\mathbf{n}_1 \cdot \mathbf{n}_2}{|\mathbf{n}_1|\,|\mathbf{n}_2|} = \dfrac{1 + 2 + 2}{\sqrt{3}\,\sqrt{9}} = \dfrac{5}{3\sqrt{3}}$. Therefore $\theta = \cos^{-1}\left(\dfrac{5}{3\sqrt{3}}\right) \approx 15.8°$.

59. Setting $z = 0$, the equations of the two planes become $5x - 2y = 1$ and $4x + y = 6$. Solving these two equations gives $x = 1$, $y = 2$ so a point on the line of intersection is $(1, 2, 0)$. A vector $\mathbf{v}$ in the direction of this intersecting line is perpendicular to the normal vectors of both planes. So we can use $\mathbf{v} = \mathbf{n}_1 \times \mathbf{n}_2 = \langle 5, -2, -2 \rangle \times \langle 4, 1, 1 \rangle = \langle 0, -13, 13 \rangle$ or equivalently we can take $\mathbf{v} = \langle 0, -1, 1 \rangle$, and symmetric equations for the line are $x = 1$, $\dfrac{y - 2}{-1} = \dfrac{z}{1}$ or $x = 1$, $y - 2 = -z$.

61. The distance from a point (x, y, z) to $(1, 0, -2)$ is $d_1 = \sqrt{(x - 1)^2 + y^2 + (z + 2)^2}$ and the distance from (x, y, z) to $(3, 4, 0)$ is $d_2 = \sqrt{(x - 3)^2 + (y - 4)^2 + z^2}$. The plane consists of all points (x, y, z) where $d_1 = d_2 \quad \Rightarrow \quad d_1^2 = d_2^2 \quad \Leftrightarrow$

$(x - 1)^2 + y^2 + (z + 2)^2 = (x - 3)^2 + (y - 4)^2 + z^2 \quad \Leftrightarrow$

$x^2 - 2x + y^2 + z^2 + 4z + 5 = x^2 - 6x + y^2 - 8y + z^2 + 25 \quad \Leftrightarrow \quad 4x + 8y + 4z = 20$ so an equation for the plane is $4x + 8y + 4z = 20$ or equivalently $x + 2y + z = 5$.

Alternatively, you can argue that the segment joining points $(1, 0, -2)$ and $(3, 4, 0)$ is perpendicular to the plane and the plane includes the midpoint of the segment.

63. The plane contains the points $(a, 0, 0)$, $(0, b, 0)$ and $(0, 0, c)$. Thus the vectors $\mathbf{a} = \langle -a, b, 0 \rangle$ and $\mathbf{b} = \langle -a, 0, c \rangle$ lie in the plane, and $\mathbf{n} = \mathbf{a} \times \mathbf{b} = \langle bc - 0, 0 + ac, 0 + ab \rangle = \langle bc, ac, ab \rangle$ is a normal vector to the plane. The equation of the plane is therefore $bcx + acy + abz = abc + 0 + 0$ or $bcx + acy + abz = abc$. Notice that if $a \neq 0$, $b \neq 0$ and $c \neq 0$ then we can

rewrite the equation as $\dfrac{x}{a} + \dfrac{y}{b} + \dfrac{z}{c} = 1$. This is a good equation to remember!

65. Two vectors which are perpendicular to the required line are the normal of the given plane, $\langle 1, 1, 1 \rangle$, and a direction vector for

the given line, $\langle 1, -1, 2 \rangle$. So a direction vector for the required line is $\langle 1, 1, 1 \rangle \times \langle 1, -1, 2 \rangle = \langle 3, -1, -2 \rangle$. Thus L is given

by $\langle x, y, z \rangle = \langle 0, 1, 2 \rangle + t\langle 3, -1, -2 \rangle$, or in parametric form, $x = 3t$, $y = 1 - t$, $z = 2 - 2t$.

67. Let P_i have normal vector $\mathbf{n}_i$. Then $\mathbf{n}_1 = \langle 3, 6, -3 \rangle$, $\mathbf{n}_2 = \langle 4, -12, 8 \rangle$, $\mathbf{n}_3 = \langle 3, -9, 6 \rangle$, $\mathbf{n}_4 = \langle 1, 2, -1 \rangle$. Now $\mathbf{n}_1 = 3\mathbf{n}_4$,

so $\mathbf{n}_1$ and $\mathbf{n}_4$ are parallel, and hence P_1 and P_4 are parallel; similarly P_2 and P_3 are parallel because $\mathbf{n}_2 = \frac{4}{3}\mathbf{n}_3$. However, $\mathbf{n}_1$

and $\mathbf{n}_2$ are not parallel (so not all four planes are parallel). Notice that the point $(2, 0, 0)$ lies on both P_1 and P_4, so these two

planes are identical. The point $\left(\frac{5}{4}, 0, 0\right)$ lies on P_2 but not on P_3, so these are different planes.

69. Let $Q = (1, 3, 4)$ and $R = (2, 1, 1)$, points on the line corresponding to $t = 0$ and $t = 1$. Let

$P = (4, 1, -2)$. Then $\mathbf{a} = \overrightarrow{QR} = \langle 1, -2, -3 \rangle$, $\mathbf{b} = \overrightarrow{QP} = \langle 3, -2, -6 \rangle$. The distance is

$$d = \frac{|\mathbf{a} \times \mathbf{b}|}{|\mathbf{a}|} = \frac{|\langle 1, -2, -3 \rangle \times \langle 3, -2, -6 \rangle|}{|\langle 1, -2, -3 \rangle|} = \frac{|\langle 6, -3, 4 \rangle|}{|\langle 1, -2, -3 \rangle|} = \frac{\sqrt{6^2 + (-3)^2 + 4^2}}{\sqrt{1^2 + (-2)^2 + (-3)^2}} = \frac{\sqrt{61}}{\sqrt{14}} = \sqrt{\frac{61}{14}}.$$

71. By Equation 9, the distance is $D = \dfrac{|ax_1 + by_1 + cz_1 + d|}{\sqrt{a^2 + b^2 + c^2}} = \dfrac{|3(1) + 2(-2) + 6(4) - 5|}{\sqrt{3^2 + 2^2 + 6^2}} = \dfrac{|18|}{\sqrt{49}} = \dfrac{18}{7}$.

73. Put $y = z = 0$ in the equation of the first plane to get the point $(2, 0, 0)$ on the plane. Because the planes are parallel, the

distance D between them is the distance from $(2, 0, 0)$ to the second plane. By Equation 9,

$$D = \frac{|4(2) - 6(0) + 2(0) - 3|}{\sqrt{4^2 + (-6)^2 + (2)^2}} = \frac{5}{\sqrt{56}} = \frac{5}{2\sqrt{14}} \text{ or } \frac{5\sqrt{14}}{28}.$$

75. The distance between two parallel planes is the same as the distance between a point on one of the planes and the other plane.

Let $P_0 = (x_0, y_0, z_0)$ be a point on the plane given by $ax + by + cz + d_1 = 0$. Then $ax_0 + by_0 + cz_0 + d_1 = 0$ and the

distance between P_0 and the plane given by $ax + by + cz + d_2 = 0$ is, from Equation 9,

$$D = \frac{|ax_0 + by_0 + cz_0 + d_2|}{\sqrt{a^2 + b^2 + c^2}} = \frac{|-d_1 + d_2|}{\sqrt{a^2 + b^2 + c^2}} = \frac{|d_1 - d_2|}{\sqrt{a^2 + b^2 + c^2}}.$$

77. $L_1: x = y = z \Rightarrow x = y$ **(1)**. $L_2: x + 1 = y/2 = z/3 \Rightarrow x + 1 = y/2$ **(2)**. The solution of **(1)** and **(2)** is

$x = y = -2$. However, when $x = -2$, $x = z \Rightarrow z = -2$, but $x - 1 = z/3 \Rightarrow z = -3$, a contradiction. Hence the

lines do not intersect. For L_1, $\mathbf{v}_1 = \langle 1, 1, 1 \rangle$, and for L_2, $\mathbf{v}_2 = \langle 1, 2, 3 \rangle$, so the lines are not parallel. Thus the lines are skew

lines. If two lines are skew, they can be viewed as lying in two parallel planes and so the distance between the skew lines

would be the same as the distance between these parallel planes. The common normal vector to the planes must be

perpendicular to both $\langle 1, 1, 1 \rangle$ and $\langle 1, 2, 3 \rangle$, the direction vectors of the two lines. So set

$\mathbf{n} = \langle 1, 1, 1 \rangle \times \langle 1, 2, 3 \rangle = \langle 3 - 2, -3 + 1, 2 - 1 \rangle = \langle 1, -2, 1 \rangle$. From above, we know that $(-2, -2, -2)$ and $(-2, -2, -3)$

are points of L_1 and L_2 respectively. So in the notation of Equation 8, $1(-2) - 2(-2) + 1(-2) + d_1 = 0 \Rightarrow d_1 = 0$ and $1(-2) - 2(-2) + 1(-3) + d_2 = 0 \Rightarrow d_2 = 1$.

By Exercise 75, the distance between these two skew lines is $D = \dfrac{|0 - 1|}{\sqrt{1 + 4 + 1}} = \dfrac{1}{\sqrt{6}}$.

Alternate solution (without reference to planes): A vector which is perpendicular to both of the lines is $\mathbf{n} = \langle 1, 1, 1 \rangle \times \langle 1, 2, 3 \rangle = \langle 1, -2, 1 \rangle$. Pick any point on each of the lines, say $(-2, -2, -2)$ and $(-2, -2, -3)$, and form the vector $\mathbf{b} = \langle 0, 0, 1 \rangle$ connecting the two points. The distance between the two skew lines is the absolute value of the scalar projection of $\mathbf{b}$ along $\mathbf{n}$, that is, $D = \dfrac{|\mathbf{n} \cdot \mathbf{b}|}{|\mathbf{n}|} = \dfrac{|1 \cdot 0 - 2 \cdot 0 + 1 \cdot 1|}{\sqrt{1 + 4 + 1}} = \dfrac{1}{\sqrt{6}}$.

79. A direction vector for L_1 is $\mathbf{v}_1 = \langle 2, 0, -1 \rangle$ and a direction vector for L_2 is $\mathbf{v}_2 = \langle 3, 2, 2 \rangle$. These vectors are not parallel so neither are the lines. Parametric equations for the lines are $L_1\colon x = 2t,\ y = 0,\ z = -t$, and $L_2\colon x = 1 + 3s,\ y = -1 + 2s$, $z = 1 + 2s$. No values of t and s satisfy these equations simultaneously, so the lines don't intersect and hence are skew. We can view the lines as lying in two parallel planes; a common normal vector to the planes is $\mathbf{n} = \mathbf{v}_1 \times \mathbf{v}_2 = \langle 2, -7, 4 \rangle$. Line L_1 passes through the origin, so $(0, 0, 0)$ lies on one of the planes, and $(1, -1, 1)$ is a point on L_2 and therefore on the other plane. Equations of the planes then are $2x - 7y + 4z = 0$ and $2x - 7y + 4z - 13 = 0$, and by Exercise 75, the distance between the two skew lines is $D = \dfrac{|0 - (-13)|}{\sqrt{4 + 49 + 16}} = \dfrac{13}{\sqrt{69}}$.

Alternate solution (without reference to planes): Direction vectors of the two lines are $\mathbf{v}_1 = \langle 2, 0, -1 \rangle$ and $\mathbf{v}_2 = \langle 3, 2, 2 \rangle$. Then $\mathbf{n} = \mathbf{v}_1 \times \mathbf{v}_2 = \langle 2, -7, 4 \rangle$ is perpendicular to both lines. Pick any point on each of the lines, say $(0, 0, 0)$ and $(1, -1, 1)$, and form the vector $\mathbf{b} = \langle 1, -1, 1 \rangle$ connecting the two points. Then the distance between the two skew lines is the absolute value of the scalar projection of $\mathbf{b}$ along $\mathbf{n}$, that is, $D = \dfrac{|\mathbf{n} \cdot \mathbf{b}|}{|\mathbf{n}|} = \dfrac{|2 + 7 + 4|}{\sqrt{4 + 49 + 16}} = \dfrac{13}{\sqrt{69}}$.

81. (a) A direction vector from tank A to tank B is $\langle 765 - 325,\ 675 - 810,\ 599 - 561 \rangle = \langle 440, -135, 38 \rangle$. Taking tank A's position $(325, 810, 561)$ as the initial point, parametric equations for the line of sight are $x = 325 + 440t$, $y = 810 - 135t,\ z = 561 + 38t$ for $0 \le t \le 1$.

(b) We divide the line of sight into 5 equal segments, corresponding to $\Delta t = 0.2$, and compute the elevation from the z-component of the parametric equations in part (a):

t	$z = 561 + 38t$	terrain elevation
0	561.0	
0.2	568.6	549
0.4	576.2	566
0.6	583.8	586
0.8	591.4	589
1.0	599.0	

Since the terrain is higher than the line of sight when $t = 0.6$, the tanks can't see each other.

83. If $a \neq 0$, then $ax + by + cz + d = 0 \Rightarrow a(x + d/a) + b(y - 0) + c(z - 0) = 0$ which by (7) is the scalar equation of the plane through the point $(-d/a, 0, 0)$ with normal vector $\langle a, b, c \rangle$. Similarly, if $b \neq 0$ (or if $c \neq 0$) the equation of the plane can be rewritten as $a(x - 0) + b(y + d/b) + c(z - 0) = 0$ [or as $a(x - 0) + b(y - 0) + c(z + d/c) = 0$] which by (7) is the scalar equation of a plane through the point $(0, -d/b, 0)$ [or the point $(0, 0, -d/c)$] with normal vector $\langle a, b, c \rangle$.

12.6 Cylinders and Quadric Surfaces

1. (a) In $\mathbb{R}^2$, the equation $y = x^2$ represents a parabola.

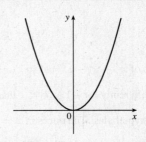

(b) In $\mathbb{R}^3$, the equation $y = x^2$ doesn't involve z, so any horizontal plane with equation $z = k$ intersects the graph in a curve with equation $y = x^2$. Thus, the surface is a parabolic cylinder, made up of infinitely many shifted copies of the same parabola. The rulings are parallel to the z-axis.

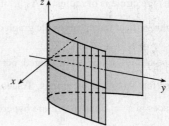

(c) In $\mathbb{R}^3$, the equation $z = y^2$ also represents a parabolic cylinder. Since x doesn't appear, the graph is formed by moving the parabola $z = y^2$ in the direction of the x-axis. Thus, the rulings of the cylinder are parallel to the x-axis.

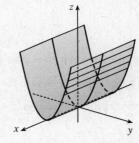

3. Since y is missing from the equation, the vertical traces $x^2 + z^2 = 1$, $y = k$, are copies of the same circle in the plane $y = k$. Thus the surface $x^2 + z^2 = 1$ is a circular cylinder with rulings parallel to the y-axis.

5. Since x is missing, each vertical trace $z = 1 - y^2$, $x = k$, is a copy of the same parabola in the plane $x = k$. Thus the surface $z = 1 - y^2$ is a parabolic cylinder with rulings parallel to the x-axis.

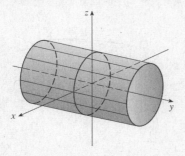

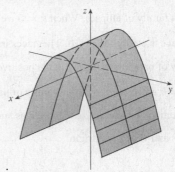

7. Since z is missing, each horizontal trace $xy = 1$,
$z = k$, is a copy of the same hyperbola in the plane
$z = k$. Thus the surface $xy = 1$ is a hyperbolic
cylinder with rulings parallel to the z-axis.

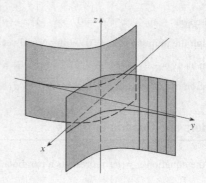

9. (a) The traces of $x^2 + y^2 - z^2 = 1$ in $x = k$ are $y^2 - z^2 = 1 - k^2$, a family of hyperbolas. (Note that the hyperbolas are
oriented differently for $-1 < k < 1$ than for $k < -1$ or $k > 1$.) The traces in $y = k$ are $x^2 - z^2 = 1 - k^2$, a similar
family of hyperbolas. The traces in $z = k$ are $x^2 + y^2 = 1 + k^2$, a family of circles. For $k = 0$, the trace in the
xy-plane, the circle is of radius 1. As $|k|$ increases, so does the radius of the circle. This behavior, combined with the
hyperbolic vertical traces, gives the graph of the hyperboloid of one sheet in Table 1.

(b) The shape of the surface is unchanged, but the hyperboloid is
rotated so that its axis is the y-axis. Traces in $y = k$ are circles,
while traces in $x = k$ and $z = k$ are hyperbolas.

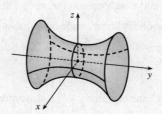

(c) Completing the square in y gives $x^2 + (y + 1)^2 - z^2 = 1$. The
surface is a hyperboloid identical to the one in part (a) but shifted
one unit in the negative y-direction.

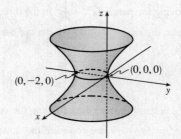

11. For $x = y^2 + 4z^2$, the traces in $x = k$ are $y^2 + 4z^2 = k$. When $k > 0$ we
have a family of ellipses. When $k = 0$ we have just a point at the origin, and
the trace is empty for $k < 0$. The traces in $y = k$ are $x = 4z^2 + k^2$, a
family of parabolas opening in the positive x-direction. Similarly, the traces
in $z = k$ are $x = y^2 + 4k^2$, a family of parabolas opening in the positive
x-direction. We recognize the graph as an elliptic paraboloid with axis the
x-axis and vertex the origin.

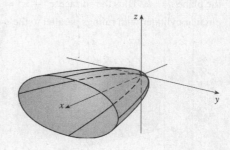

13. $x^2 = 4y^2 + z^2$. The traces in $x = k$ are the ellipses $4y^2 + z^2 = k^2$. The

traces in $y = k$ are $x^2 - z^2 = 4k^2$, hyperbolas for $k \neq 0$ and two

intersecting lines if $k = 0$. Similarly, the traces in $z = k$ are

$x^2 - 4y^2 = k^2$, hyperbolas for $k \neq 0$ and two intersecting lines if $k = 0$.

We recognize the graph as an elliptic cone with axis the x-axis and vertex

the origin.

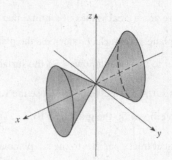

15. $9y^2 + 4z^2 = x^2 + 36$. The traces in $x = k$ are $9y^2 + 4z^2 = k^2 + 36$, a

family of ellipses. The traces in $y = k$ are $4z^2 - x^2 = 9(4 - k^2)$, a family

of hyperbolas for $|k| \neq 2$ and two intersecting lines when $|k| = 2$. (Note

that the hyperbolas are oriented differently for $|k| < 2$ than for $|k| > 2$.)

The traces in $z = k$ are $9y^2 - x^2 = 4(9 - k^2)$, a family of hyperbolas

when $|k| \neq 3$ (oriented differently for $|k| < 3$ than for $|k| > 3$) and two

intersecting lines when $|k| = 3$. We recognize the graph as a hyperboloid of

one sheet with axis the x-axis.

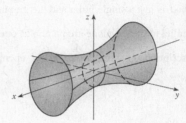

17. $\dfrac{x^2}{9} + \dfrac{y^2}{25} + \dfrac{z^2}{4} = 1$. The traces in $x = k$ are $\dfrac{y^2}{25} + \dfrac{z^2}{4} = 1 - \dfrac{k^2}{9}$, a family

of ellipses for $|k| < 3$. (The traces are a single point for $|k| = 3$ and are

empty for $|k| > 3$.) The traces in $y = k$ are the ellipses

$\dfrac{x^2}{9} + \dfrac{z^2}{4} = 1 - \dfrac{k^2}{25}$, $|k| < 5$, and the traces in $z = k$ are the ellipses

$\dfrac{x^2}{9} + \dfrac{y^2}{25} = 1 - \dfrac{k^2}{4}$, $|k| < 2$. The surface is an ellipsoid centered at the

origin with intercepts $x = \pm 3$, $y = \pm 5$, $z = \pm 2$.

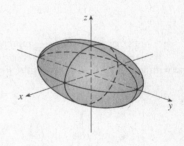

19. $y = z^2 - x^2$. The traces in $x = k$ are the parabolas $y = z^2 - k^2$, opening

in the positive y-direction. The traces in $y = k$ are $k = z^2 - x^2$, two

intersecting lines when $k = 0$ and a family of hyperbolas for $k \neq 0$ (note

that the hyperbolas are oriented differently for $k > 0$ than for $k < 0$). The

traces in $z = k$ are the parabolas $y = k^2 - x^2$ which open in the negative

y-direction. Thus the surface is a hyperbolic paraboloid centered at $(0, 0, 0)$.

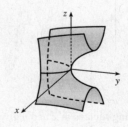

21. This is the equation of an ellipsoid: $x^2 + 4y^2 + 9z^2 = x^2 + \dfrac{y^2}{(1/2)^2} + \dfrac{z^2}{(1/3)^2} = 1$, with x-intercepts ± 1, y-intercepts $\pm\frac{1}{2}$

and z-intercepts $\pm\frac{1}{3}$. So the major axis is the x-axis and the only possible graph is VII.

23. This is the equation of a hyperboloid of one sheet, with $a = b = c = 1$. Since the coefficient of y^2 is negative, the axis of the

hyperboloid is the y-axis, hence the correct graph is II.

25. There are no real values of x and z that satisfy this equation for $y < 0$, so this surface does not extend to the left of the xz-plane. The surface intersects the plane $y = k > 0$ in an ellipse. Notice that y occurs to the first power whereas x and z occur to the second power. So the surface is an elliptic paraboloid with axis the y-axis. Its graph is VI.

27. This surface is a cylinder because the variable y is missing from the equation. The intersection of the surface and the xz-plane is an ellipse. So the graph is VIII.

29. Vertical traces parallel to the xz-plane are circles centered at the origin whose radii increase as y decreases. (The trace in $y = 1$ is just a single point and the graph suggests that traces in $y = k$ are empty for $k > 1$.) The traces in vertical planes parallel to the yz-plane are parabolas opening to the left that shift to the left as $|x|$ increases. One surface that fits this description is a circular paraboloid, opening to the left, with vertex $(0, 1, 0)$.

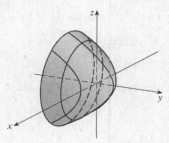

31. $y^2 = x^2 + \frac{1}{9}z^2$ or $y^2 = x^2 + \dfrac{z^2}{9}$ represents an elliptic cone with vertex $(0, 0, 0)$ and axis the y-axis.

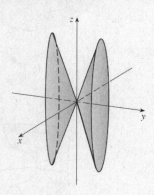

33. $x^2 + 2y - 2z^2 = 0$ or $2y = 2z^2 - x^2$ or $y = z^2 - \dfrac{x^2}{2}$ represents a hyperbolic paraboloid with center $(0, 0, 0)$.

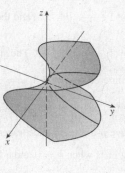

35. Completing squares in x and y gives

$\left(x^2 - 2x + 1\right) + \left(y^2 - 6y + 9\right) - z = 0 \iff$

$(x - 1)^2 + (y - 3)^2 - z = 0$ or $z = (x - 1)^2 + (y - 3)^2$, a circular paraboloid opening upward with vertex $(1, 3, 0)$ and axis the vertical line $x = 1, y = 3$.

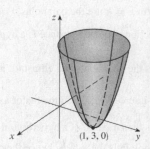

37. Completing squares in x and z gives

$$\left(x^2 - 4x + 4\right) - y^2 + \left(z^2 - 2z + 1\right) = 0 + 4 + 1 \quad \Leftrightarrow$$

$$(x - 2)^2 - y^2 + (z - 1)^2 = 5 \text{ or } \frac{(x-2)^2}{5} - \frac{y^2}{5} + \frac{(z-1)^2}{5} = 1, \text{ a}$$

hyperboloid of one sheet with center $(2, 0, 1)$ and axis the horizontal line

$x = 2, z = 1$.

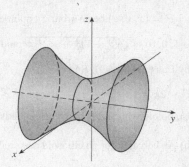

39. Solving the equation for z we get $z = \pm\sqrt{1 + 4x^2 + y^2}$, so we plot separately $z = \sqrt{1 + 4x^2 + y^2}$ and

$z = -\sqrt{1 + 4x^2 + y^2}$.

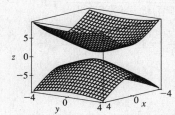 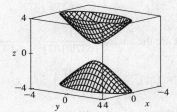

To restrict the z-range as in the second graph, we can use the option `view=-4..4` in Maple's `plot3d` command, or

`PlotRange->{-4,4}` in Mathematica's `Plot3D` command.

41. Solving the equation for z we get $z = \pm\sqrt{4x^2 + y^2}$, so we plot separately $z = \sqrt{4x^2 + y^2}$ and $z = -\sqrt{4x^2 + y^2}$.

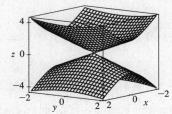 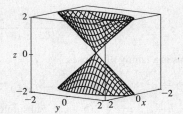

43.

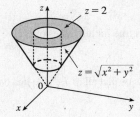

45. The curve $y = \sqrt{x}$ is equivalent to $x = y^2$, $y \geq 0$. Rotating the curve about

the x-axis creates a circular paraboloid with vertex at the origin, axis the

x-axis, opening in the positive x-direction. The trace in the xy-plane is

$x = y^2$, $z = 0$, and the trace in the xz-plane is a parabola of the same

shape: $x = z^2$, $y = 0$. An equation for the surface is $x = y^2 + z^2$.

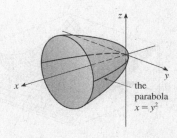

47. Let $P = (x, y, z)$ be an arbitrary point equidistant from $(-1, 0, 0)$ and the plane $x = 1$. Then the distance from P to

$(-1, 0, 0)$ is $\sqrt{(x+1)^2 + y^2 + z^2}$ and the distance from P to the plane $x = 1$ is $|x - 1| / \sqrt{1^2} = |x - 1|$

(by Equation 12.5.9). So $|x - 1| = \sqrt{(x+1)^2 + y^2 + z^2}$ $\Leftrightarrow$ $(x - 1)^2 = (x + 1)^2 + y^2 + z^2$ $\Leftrightarrow$

$x^2 - 2x + 1 = x^2 + 2x + 1 + y^2 + z^2$ $\Leftrightarrow$ $-4x = y^2 + z^2$. Thus the collection of all such points P is a circular

paraboloid with vertex at the origin, axis the x-axis, which opens in the negative x-direction.

49. (a) An equation for an ellipsoid centered at the origin with intercepts $x = \pm a$, $y = \pm b$, and $z = \pm c$ is $\dfrac{x^2}{a^2} + \dfrac{y^2}{b^2} + \dfrac{z^2}{c^2} = 1$.

Here the poles of the model intersect the z-axis at $z = \pm 6356.523$ and the equator intersects the x- and y-axes at

$x = \pm 6378.137$, $y = \pm 6378.137$, so an equation is

$$\frac{x^2}{(6378.137)^2} + \frac{y^2}{(6378.137)^2} + \frac{z^2}{(6356.523)^2} = 1$$

(b) Traces in $z = k$ are the circles $\dfrac{x^2}{(6378.137)^2} + \dfrac{y^2}{(6378.137)^2} = 1 - \dfrac{k^2}{(6356.523)^2}$ $\Leftrightarrow$

$x^2 + y^2 = (6378.137)^2 - \left(\dfrac{6378.137}{6356.523}\right)^2 k^2$.

(c) To identify the traces in $y = mx$ we substitute $y = mx$ into the equation of the ellipsoid:

$$\frac{x^2}{(6378.137)^2} + \frac{(mx)^2}{(6378.137)^2} + \frac{z^2}{(6356.523)^2} = 1$$

$$\frac{(1 + m^2)x^2}{(6378.137)^2} + \frac{z^2}{(6356.523)^2} = 1$$

$$\frac{x^2}{(6378.137)^2/(1 + m^2)} + \frac{z^2}{(6356.523)^2} = 1$$

As expected, this is a family of ellipses.

51. If (a, b, c) satisfies $z = y^2 - x^2$, then $c = b^2 - a^2$. L_1: $x = a + t$, $y = b + t$, $z = c + 2(b - a)t$,

L_2: $x = a + t$, $y = b - t$, $z = c - 2(b + a)t$. Substitute the parametric equations of L_1 into the equation of the hyperbolic

paraboloid in order to find the points of intersection: $z = y^2 - x^2$ $\Rightarrow$

$c + 2(b - a)t = (b + t)^2 - (a + t)^2 = b^2 - a^2 + 2(b - a)t$ $\Rightarrow$ $c = b^2 - a^2$. As this is true for all values of t,

L_1 lies on $z = y^2 - x^2$. Performing similar operations with L_2 gives: $z = y^2 - x^2$ $\Rightarrow$

$c - 2(b + a)t = (b - t)^2 - (a + t)^2 = b^2 - a^2 - 2(b + a)t$ $\Rightarrow$ $c = b^2 - a^2$. This tells us that all of L_2 also lies on

$z = y^2 - x^2$.

53.

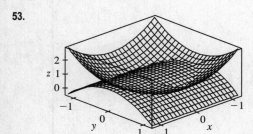

The curve of intersection looks like a bent ellipse. The projection

of this curve onto the xy-plane is the set of points $(x, y, 0)$ which

satisfy $x^2 + y^2 = 1 - y^2$ $\Leftrightarrow$ $x^2 + 2y^2 = 1$ $\Leftrightarrow$

$x^2 + \dfrac{y^2}{(1/\sqrt{2})^2} = 1$. This is an equation of an ellipse.

12 Review

TRUE-FALSE QUIZ

1. This is false, as the dot product of two vectors is a scalar, not a vector.

3. False. For example, if $\mathbf{u} = \mathbf{i}$ and $\mathbf{v} = \mathbf{j}$ then $|\mathbf{u} \cdot \mathbf{v}| = |0| = 0$ but $|\mathbf{u}|\,|\mathbf{v}| = 1 \cdot 1 = 1$. In fact, by Theorem 12.3.3, $|\mathbf{u} \cdot \mathbf{v}| = \big||\mathbf{u}|\,|\mathbf{v}| \cos\theta\big|$.

5. True, by Theorem 12.3.2, property 2.

7. True. If θ is the angle between $\mathbf{u}$ and $\mathbf{v}$, then by Theorem 12.4.9, $|\mathbf{u} \times \mathbf{v}| = |\mathbf{u}|\,|\mathbf{v}| \sin\theta = |\mathbf{v}|\,|\mathbf{u}| \sin\theta = |\mathbf{v} \times \mathbf{u}|$.
   (Or, by Theorem 12.4.11, $|\mathbf{u} \times \mathbf{v}| = |-\mathbf{v} \times \mathbf{u}| = |-1|\,|\mathbf{v} \times \mathbf{u}| = |\mathbf{v} \times \mathbf{u}|$.)

9. Theorem 12.4.11, property 2 tells us that this is true.

11. This is true by Theorem 12.4.11, property 5.

13. This is true because $\mathbf{u} \times \mathbf{v}$ is orthogonal to $\mathbf{u}$ (see Theorem 12.4.8), and the dot product of two orthogonal vectors is 0.

15. This is false. A normal vector to the plane is $\mathbf{n} = \langle 6, -2, 4\rangle$. Because $\langle 3, -1, 2\rangle = \frac{1}{2}\mathbf{n}$, the vector is parallel to $\mathbf{n}$ and hence perpendicular to the plane.

17. This is false. In $\mathbb{R}^2$, $x^2 + y^2 = 1$ represents a circle, but $\{(x, y, z) \mid x^2 + y^2 = 1\}$ represents a *three-dimensional surface*, namely, a circular cylinder with axis the z-axis.

19. False. For example, $\mathbf{i} \cdot \mathbf{j} = 0$ but $\mathbf{i} \neq \mathbf{0}$ and $\mathbf{j} \neq \mathbf{0}$.

21. This is true. If $\mathbf{u}$ and $\mathbf{v}$ are both nonzero, then by (7) in Section 12.3, $\mathbf{u} \cdot \mathbf{v} = 0$ implies that $\mathbf{u}$ and $\mathbf{v}$ are orthogonal. But $\mathbf{u} \times \mathbf{v} = \mathbf{0}$ implies that $\mathbf{u}$ and $\mathbf{v}$ are parallel (see Corollary 12.4.10). Two nonzero vectors can't be both parallel and orthogonal, so at least one of $\mathbf{u}$, $\mathbf{v}$ must be $\mathbf{0}$.

EXERCISES

1. (a) The radius of the sphere is the distance between the points $(-1, 2, 1)$ and $(6, -2, 3)$, namely,
   $\sqrt{[6 - (-1)]^2 + (-2 - 2)^2 + (3 - 1)^2} = \sqrt{69}$. By the formula for an equation of a sphere (see page 835 [ET 795]), an equation of the sphere with center $(-1, 2, 1)$ and radius $\sqrt{69}$ is $(x + 1)^2 + (y - 2)^2 + (z - 1)^2 = 69$.

   (b) The intersection of this sphere with the yz-plane is the set of points on the sphere whose x-coordinate is 0. Putting $x = 0$ into the equation, we have $(y - 2)^2 + (z - 1)^2 = 68, x = 0$ which represents a circle in the yz-plane with center $(0, 2, 1)$ and radius $\sqrt{68}$.

   (c) Completing squares gives $(x - 4)^2 + (y + 1)^2 + (z + 3)^2 = -1 + 16 + 1 + 9 = 25$. Thus the sphere is centered at $(4, -1, -3)$ and has radius 5.

3. $\mathbf{u} \cdot \mathbf{v} = |\mathbf{u}|\,|\mathbf{v}| \cos 45° = (2)(3)\frac{\sqrt{2}}{2} = 3\sqrt{2}$. $|\mathbf{u} \times \mathbf{v}| = |\mathbf{u}|\,|\mathbf{v}| \sin 45° = (2)(3)\frac{\sqrt{2}}{2} = 3\sqrt{2}$.
   By the right-hand rule, $\mathbf{u} \times \mathbf{v}$ is directed out of the page.

5. For the two vectors to be orthogonal, we need $\langle 3, 2, x \rangle \cdot \langle 2x, 4, x \rangle = 0 \quad \Leftrightarrow \quad (3)(2x) + (2)(4) + (x)(x) = 0 \quad \Leftrightarrow$
$x^2 + 6x + 8 = 0 \quad \Leftrightarrow \quad (x+2)(x+4) = 0 \quad \Leftrightarrow \quad x = -2$ or $x = -4$.

7. (a) $(\mathbf{u} \times \mathbf{v}) \cdot \mathbf{w} = \mathbf{u} \cdot (\mathbf{v} \times \mathbf{w}) = 2$

(b) $\mathbf{u} \cdot (\mathbf{w} \times \mathbf{v}) = \mathbf{u} \cdot [-(\mathbf{v} \times \mathbf{w})] = -\mathbf{u} \cdot (\mathbf{v} \times \mathbf{w}) = -2$

(c) $\mathbf{v} \cdot (\mathbf{u} \times \mathbf{w}) = (\mathbf{v} \times \mathbf{u}) \cdot \mathbf{w} = -(\mathbf{u} \times \mathbf{v}) \cdot \mathbf{w} = -2$

(d) $(\mathbf{u} \times \mathbf{v}) \cdot \mathbf{v} = \mathbf{u} \cdot (\mathbf{v} \times \mathbf{v}) = \mathbf{u} \cdot \mathbf{0} = 0$

9. For simplicity, consider a unit cube positioned with its back left corner at the origin. Vector representations of the diagonals joining the points $(0, 0, 0)$ to $(1, 1, 1)$ and $(1, 0, 0)$ to $(0, 1, 1)$ are $\langle 1, 1, 1 \rangle$ and $\langle -1, 1, 1 \rangle$. Let θ be the angle between these two vectors. $\langle 1, 1, 1 \rangle \cdot \langle -1, 1, 1 \rangle = -1 + 1 + 1 = 1 = |\langle 1, 1, 1 \rangle| \, |\langle -1, 1, 1 \rangle| \cos \theta = 3 \cos \theta \quad \Rightarrow \quad \cos \theta = \frac{1}{3} \quad \Rightarrow$
$\theta = \cos^{-1}\left(\frac{1}{3}\right) \approx 71°$.

11. $\overrightarrow{AB} = \langle 1, 0, -1 \rangle$, $\overrightarrow{AC} = \langle 0, 4, 3 \rangle$, so

(a) a vector perpendicular to the plane is $\overrightarrow{AB} \times \overrightarrow{AC} = \langle 0 + 4, -(3 + 0), 4 - 0 \rangle = \langle 4, -3, 4 \rangle$.

(b) $\frac{1}{2} \left| \overrightarrow{AB} \times \overrightarrow{AC} \right| = \frac{1}{2}\sqrt{16 + 9 + 16} = \frac{\sqrt{41}}{2}$.

13. Let F_1 be the magnitude of the force directed $20°$ away from the direction of shore, and let F_2 be the magnitude of the other force. Separating these forces into components parallel to the direction of the resultant force and perpendicular to it gives

$F_1 \cos 20° + F_2 \cos 30° = 255$ **(1)**, and $F_1 \sin 20° - F_2 \sin 30° = 0 \quad \Rightarrow \quad F_1 = F_2 \dfrac{\sin 30°}{\sin 20°}$ **(2)**. Substituting **(2)**

into **(1)** gives $F_2(\sin 30° \cot 20° + \cos 30°) = 255 \quad \Rightarrow \quad F_2 \approx 114$ N. Substituting this into **(2)** gives $F_1 \approx 166$ N.

15. The line has direction $\mathbf{v} = \langle -3, 2, 3 \rangle$. Letting $P_0 = (4, -1, 2)$, parametric equations are
$x = 4 - 3t, \ y = -1 + 2t, \ z = 2 + 3t$.

17. A direction vector for the line is a normal vector for the plane, $\mathbf{n} = \langle 2, -1, 5 \rangle$, and parametric equations for the line are
$x = -2 + 2t, \ y = 2 - t, \ z = 4 + 5t$.

19. Here the vectors $\mathbf{a} = \langle 4 - 3, 0 - (-1), 2 - 1 \rangle = \langle 1, 1, 1 \rangle$ and $\mathbf{b} = \langle 6 - 3, 3 - (-1), 1 - 1 \rangle = \langle 3, 4, 0 \rangle$ lie in the plane,
so $\mathbf{n} = \mathbf{a} \times \mathbf{b} = \langle -4, 3, 1 \rangle$ is a normal vector to the plane and an equation of the plane is
$-4(x - 3) + 3(y - (-1)) + 1(z - 1) = 0$ or $-4x + 3y + z = -14$.

21. Substitution of the parametric equations into the equation of the plane gives $2x - y + z = 2(2 - t) - (1 + 3t) + 4t = 2 \quad \Rightarrow$
$-t + 3 = 2 \quad \Rightarrow \quad t = 1$. When $t = 1$, the parametric equations give $x = 2 - 1 = 1, \ y = 1 + 3 = 4$ and $z = 4$. Therefore,
the point of intersection is $(1, 4, 4)$.

23. Since the direction vectors $\langle 2, 3, 4 \rangle$ and $\langle 6, -1, 2 \rangle$ aren't parallel, neither are the lines. For the lines to intersect, the three
equations $1 + 2t = -1 + 6s, \ 2 + 3t = 3 - s, \ 3 + 4t = -5 + 2s$ must be satisfied simultaneously. Solving the first two
equations gives $t = \frac{1}{5}, \ s = \frac{2}{5}$ and checking we see these values don't satisfy the third equation. Thus the lines aren't parallel
and they don't intersect, so they must be skew.

25. $\mathbf{n}_1 = \langle 1, 0, -1 \rangle$ and $\mathbf{n}_2 = \langle 0, 1, 2 \rangle$. Setting $z = 0$, it is easy to see that $(1, 3, 0)$ is a point on the line of intersection of

$x - z = 1$ and $y + 2z = 3$. The direction of this line is $\mathbf{v}_1 = \mathbf{n}_1 \times \mathbf{n}_2 = \langle 1, -2, 1 \rangle$. A second vector parallel to the desired

plane is $\mathbf{v}_2 = \langle 1, 1, -2 \rangle$, since it is perpendicular to $x + y - 2z = 1$. Therefore, the normal of the plane in question is

$\mathbf{n} = \mathbf{v}_1 \times \mathbf{v}_2 = \langle 4 - 1, 1 + 2, 1 + 2 \rangle = 3 \langle 1, 1, 1 \rangle$. Taking $(x_0, y_0, z_0) = (1, 3, 0)$, the equation we are looking for is

$(x - 1) + (y - 3) + z = 0 \quad \Leftrightarrow \quad x + y + z = 4$.

27. By Exercise 12.5.75, $D = \dfrac{|-2 - (-24)|}{\sqrt{3^2 + 1^2 + (-4)^2}} = \dfrac{22}{\sqrt{26}}$.

29. The equation $x = z$ represents a plane perpendicular to the xz-plane and intersecting the xz-plane in the line $x = z$, $y = 0$.

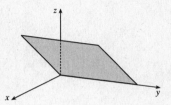

31. The equation $x^2 = y^2 + 4z^2$ represents a (right elliptical) cone with vertex at the origin and axis the x-axis.

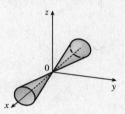

33. An equivalent equation is $-x^2 + \dfrac{y^2}{4} - z^2 = 1$, a hyperboloid of two sheets with axis the y-axis. For $|y| > 2$, traces parallel to the xz-plane are circles.

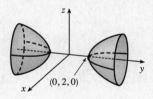

35. Completing the square in y gives

$$4x^2 + 4(y - 1)^2 + z^2 = 4 \text{ or } x^2 + (y - 1)^2 + \dfrac{z^2}{4} = 1,$$

an ellipsoid centered at $(0, 1, 0)$.

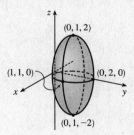

37. $4x^2 + y^2 = 16 \quad \Leftrightarrow \quad \dfrac{x^2}{4} + \dfrac{y^2}{16} = 1$. The equation of the ellipsoid is $\dfrac{x^2}{4} + \dfrac{y^2}{16} + \dfrac{z^2}{c^2} = 1$, since the horizontal trace in the

plane $z = 0$ must be the original ellipse. The traces of the ellipsoid in the yz-plane must be circles since the surface is obtained

by rotation about the x-axis. Therefore, $c^2 = 16$ and the equation of the ellipsoid is $\dfrac{x^2}{4} + \dfrac{y^2}{16} + \dfrac{z^2}{16} = 1 \quad \Leftrightarrow$

$4x^2 + y^2 + z^2 = 16$.

PROBLEMS PLUS

1. Since three-dimensional situations are often difficult to visualize and work with, let us first try to find an analogous problem in two dimensions. The analogue of a cube is a square and the analogue of a sphere is a circle. Thus a similar problem in two dimensions is the following: if five circles with the same radius r are contained in a square of side 1 m so that the circles touch each other and four of the circles touch two sides of the square, find r.

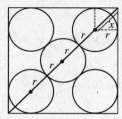

The diagonal of the square is $\sqrt{2}$. The diagonal is also $4r + 2x$. But x is the diagonal of a smaller square of side r. Therefore

$$x = \sqrt{2}\,r \quad \Rightarrow \quad \sqrt{2} = 4r + 2x = 4r + 2\sqrt{2}\,r = (4 + 2\sqrt{2})r \quad \Rightarrow \quad r = \frac{\sqrt{2}}{4 + 2\sqrt{2}}.$$

Let's use these ideas to solve the original three-dimensional problem. The diagonal of the cube is $\sqrt{1^2 + 1^2 + 1^2} = \sqrt{3}$. The diagonal of the cube is also $4r + 2x$ where x is the diagonal of a smaller cube with edge r. Therefore

$$x = \sqrt{r^2 + r^2 + r^2} = \sqrt{3}\,r \quad \Rightarrow \quad \sqrt{3} = 4r + 2x = 4r + 2\sqrt{3}\,r = (4 + 2\sqrt{3})r. \text{ Thus } r = \frac{\sqrt{3}}{4 + 2\sqrt{3}} = \frac{2\sqrt{3} - 3}{2}.$$

The radius of each ball is $\left(\sqrt{3} - \frac{3}{2}\right)$ m.

3. (a) We find the line of intersection L as in Example 12.5.7(b). Observe that the point $(-1, c, c)$ lies on both planes. Now since L lies in both planes, it is perpendicular to both of the normal vectors $\mathbf{n}_1$ and $\mathbf{n}_2$, and thus parallel to their cross product

$$\mathbf{n}_1 \times \mathbf{n}_2 = \begin{vmatrix} \mathbf{i} & \mathbf{j} & \mathbf{k} \\ c & 1 & 1 \\ 1 & -c & c \end{vmatrix} = \langle 2c, -c^2 + 1, -c^2 - 1 \rangle$$

So symmetric equations of L can be written as $\dfrac{x + 1}{-2c} = \dfrac{y - c}{c^2 - 1} = \dfrac{z - c}{c^2 + 1}$, provided that $c \neq 0, \pm 1$.

If $c = 0$, then the two planes are given by $y + z = 0$ and $x = -1$, so symmetric equations of L are $x = -1$, $y = -z$. If $c = -1$, then the two planes are given by $-x + y + z = -1$ and $x + y + z = -1$, and they intersect in the line $x = 0$, $y = -z - 1$. If $c = 1$, then the two planes are given by $x + y + z = 1$ and $x - y + z = 1$, and they intersect in the line $y = 0$, $x = 1 - z$.

(b) If we set $z = t$ in the symmetric equations and solve for x and y separately, we get $x + 1 = \dfrac{(t - c)(-2c)}{c^2 + 1}$,

$y - c = \dfrac{(t - c)(c^2 - 1)}{c^2 + 1} \quad \Rightarrow \quad x = \dfrac{-2ct + (c^2 - 1)}{c^2 + 1}$, $y = \dfrac{(c^2 - 1)t + 2c}{c^2 + 1}$. Eliminating c from these equations, we have $x^2 + y^2 = t^2 + 1$. So the curve traced out by L in the plane $z = t$ is a circle with center at $(0, 0, t)$ and radius $\sqrt{t^2 + 1}$.

(c) The area of a horizontal cross-section of the solid is $A(z) = \pi(z^2 + 1)$, so $V = \int_0^1 A(z)\,dz = \pi\left[\frac{1}{3}z^3 + z\right]_0^1 = \frac{4\pi}{3}$.

5. $\mathbf{v}_3 = \text{proj}_{\mathbf{v}_1} \mathbf{v}_2 = \dfrac{\mathbf{v}_1 \cdot \mathbf{v}_2}{|\mathbf{v}_1|^2}\,\mathbf{v}_1 = \dfrac{5}{2^2}\,\mathbf{v}_1$ so $|\mathbf{v}_3| = \dfrac{5}{2^2}\,|\mathbf{v}_1| = \dfrac{5}{2}$,

$\mathbf{v}_4 = \text{proj}_{\mathbf{v}_2} \mathbf{v}_3 = \dfrac{\mathbf{v}_2 \cdot \mathbf{v}_3}{|\mathbf{v}_2|^2}\,\mathbf{v}_2 = \dfrac{\mathbf{v}_2 \cdot \frac{5}{2^2}\mathbf{v}_1}{|\mathbf{v}_2|^2}\,\mathbf{v}_2 = \dfrac{5}{2^2 \cdot 3^2}(\mathbf{v}_1 \cdot \mathbf{v}_2)\,\mathbf{v}_2 = \dfrac{5^2}{2^2 \cdot 3^2}\,\mathbf{v}_2 \;\Rightarrow\; |\mathbf{v}_4| = \dfrac{5^2}{2^2 \cdot 3^2}\,|\mathbf{v}_2| = \dfrac{5^2}{2^2 \cdot 3}$,

$\mathbf{v}_5 = \text{proj}_{\mathbf{v}_3} \mathbf{v}_4 = \dfrac{\mathbf{v}_3 \cdot \mathbf{v}_4}{|\mathbf{v}_3|^2}\,\mathbf{v}_3 = \dfrac{\frac{5}{2^2}\mathbf{v}_1 \cdot \frac{5^2}{2^2 3^2}\mathbf{v}_2}{\left(\frac{5}{2}\right)^2}\left(\dfrac{5}{2^2}\mathbf{v}_1\right) = \dfrac{5^2}{2^4 \cdot 3^2}(\mathbf{v}_1 \cdot \mathbf{v}_2)\,\mathbf{v}_1 = \dfrac{5^3}{2^4 \cdot 3^2}\,\mathbf{v}_1 \;\Rightarrow$

$|\mathbf{v}_5| = \dfrac{5^3}{2^4 \cdot 3^2}\,|\mathbf{v}_1| = \dfrac{5^3}{2^3 \cdot 3^2}$. Similarly, $|\mathbf{v}_6| = \dfrac{5^4}{2^4 \cdot 3^3}$, $|\mathbf{v}_7| = \dfrac{5^5}{2^5 \cdot 3^4}$, and in general, $|\mathbf{v}_n| = \dfrac{5^{n-2}}{2^{n-2} \cdot 3^{n-3}} = 3\left(\frac{5}{6}\right)^{n-2}$.

Thus

$$\sum_{n=1}^{\infty} |\mathbf{v}_n| = |\mathbf{v}_1| + |\mathbf{v}_2| + \sum_{n=3}^{\infty} 3\left(\tfrac{5}{6}\right)^{n-2} = 2 + 3 + \sum_{n=1}^{\infty} 3\left(\tfrac{5}{6}\right)^n$$

$$= 5 + \sum_{n=1}^{\infty} \tfrac{5}{2}\left(\tfrac{5}{6}\right)^{n-1} = 5 + \dfrac{\frac{5}{2}}{1 - \frac{5}{6}} \quad \text{[sum of a geometric series]} \quad = 5 + 15 = 20$$

7. (a) When $\theta = \theta_s$, the block is not moving, so the sum of the forces on the block

must be $\mathbf{0}$, thus $\mathbf{N} + \mathbf{F} + \mathbf{W} = \mathbf{0}$. This relationship is illustrated
geometrically in the figure. Since the vectors form a right triangle, we have

$$\tan(\theta_s) = \dfrac{|\mathbf{F}|}{|\mathbf{N}|} = \dfrac{\mu_s n}{n} = \mu_s.$$

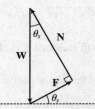

(b) We place the block at the origin and sketch the force vectors acting on the block, including the additional horizontal force $\mathbf{H}$, with initial points at the origin. We then rotate this system so that $\mathbf{F}$ lies along the positive x-axis and the inclined plane is parallel to the x-axis. (See the following figure.)

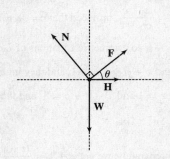
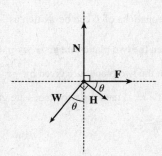

$|\mathbf{F}|$ is maximal, so $|\mathbf{F}| = \mu_s n$ for $\theta > \theta_s$. Then the vectors, in terms of components parallel and perpendicular to the inclined plane, are

$$\mathbf{N} = n\,\mathbf{j} \qquad \mathbf{F} = (\mu_s n)\,\mathbf{i}$$

$$\mathbf{W} = (-mg\sin\theta)\,\mathbf{i} + (-mg\cos\theta)\,\mathbf{j} \qquad \mathbf{H} = (h_{\min}\cos\theta)\,\mathbf{i} + (-h_{\min}\sin\theta)\,\mathbf{j}$$

Equating components, we have

$$\mu_s n - mg\sin\theta + h_{\min}\cos\theta = 0 \quad \Rightarrow \quad h_{\min}\cos\theta + \mu_s n = mg\sin\theta \tag{1}$$

$$n - mg\cos\theta - h_{\min}\sin\theta = 0 \quad \Rightarrow \quad h_{\min}\sin\theta + mg\cos\theta = n \tag{2}$$

(c) Since **(2)** is solved for n, we substitute into **(1)**:

$$h_{\min}\cos\theta + \mu_s(h_{\min}\sin\theta + mg\cos\theta) = mg\sin\theta \quad \Rightarrow$$

$$h_{\min}\cos\theta + h_{\min}\mu_s\sin\theta = mg\sin\theta - mg\mu_s\cos\theta \quad \Rightarrow$$

$$h_{\min} = mg\left(\frac{\sin\theta - \mu_s\cos\theta}{\cos\theta + \mu_s\sin\theta}\right) = mg\left(\frac{\tan\theta - \mu_s}{1 + \mu_s\tan\theta}\right)$$

From part (a) we know $\mu_s = \tan\theta_s$, so this becomes $h_{\min} = mg\left(\dfrac{\tan\theta - \tan\theta_s}{1 + \tan\theta_s\tan\theta}\right)$ and using a trigonometric identity,

this is $mg\tan(\theta - \theta_s)$ as desired.

Note for $\theta = \theta_s$, $h_{\min} = mg\tan 0 = 0$, which makes sense since the block is at rest for θ_s, thus no additional force **H** is necessary to prevent it from moving. As θ increases, the factor $\tan(\theta - \theta_s)$, and hence the value of $h_{\min}$, increases slowly for small values of $\theta - \theta_s$ but much more rapidly as $\theta - \theta_s$ becomes significant. This seems reasonable, as the steeper the inclined plane, the less the horizontal components of the various forces affect the movement of the block, so we would need a much larger magnitude of horizontal force to keep the block motionless. If we allow $\theta \to 90°$, corresponding to the inclined plane being placed vertically, the value of $h_{\min}$ is quite large; this is to be expected, as it takes a great amount of horizontal force to keep an object from moving vertically. In fact, without friction (so $\theta_s = 0$), we would have $\theta \to 90° \quad \Rightarrow \quad h_{\min} \to \infty$, and it would be impossible to keep the block from slipping.

(d) Since $h_{\max}$ is the largest value of h that keeps the block from slipping, the force of friction is keeping the block from moving *up* the inclined plane; thus, **F** is directed *down* the plane. Our system of forces is similar to that in part (b), then, except that we have $\mathbf{F} = -(\mu_s n)\,\mathbf{i}$. (Note that $|\mathbf{F}|$ is again maximal.) Following our procedure in parts (b) and (c), we equate components:

$$-\mu_s n - mg\sin\theta + h_{\max}\cos\theta = 0 \quad \Rightarrow \quad h_{\max}\cos\theta - \mu_s n = mg\sin\theta$$

$$n - mg\cos\theta - h_{\max}\sin\theta = 0 \quad \Rightarrow \quad h_{\max}\sin\theta + mg\cos\theta = n$$

Then substituting,

$$h_{\max}\cos\theta - \mu_s(h_{\max}\sin\theta + mg\cos\theta) = mg\sin\theta \quad \Rightarrow$$

$$h_{\max}\cos\theta - h_{\max}\mu_s\sin\theta = mg\sin\theta + mg\mu_s\cos\theta \quad \Rightarrow$$

$$h_{\max} = mg\left(\frac{\sin\theta + \mu_s\cos\theta}{\cos\theta - \mu_s\sin\theta}\right) = mg\left(\frac{\tan\theta + \mu_s}{1 - \mu_s\tan\theta}\right)$$

$$= mg\left(\frac{\tan\theta + \tan\theta_s}{1 - \tan\theta_s\tan\theta}\right) = mg\tan(\theta + \theta_s)$$

We would expect $h_{\max}$ to increase as θ increases, with similar behavior as we established for $h_{\min}$, but with $h_{\max}$ values always larger than $h_{\min}$. We can see that this is the case if we graph $h_{\max}$ as a function of θ, as the curve is the graph of

$h_{\min}$ translated $2\theta_s$ to the left, so the equation does seem reasonable. Notice that the equation predicts $h_{\max} \to \infty$ as $\theta \to (90° - \theta_s)$. In fact, as $h_{\max}$ increases, the normal force increases as well. When $(90° - \theta_s) \le \theta \le 90°$, the horizontal force is completely counteracted by the sum of the normal and frictional forces, so no part of the horizontal force contributes to moving the block up the plane no matter how large its magnitude.

13 □ VECTOR FUNCTIONS

13.1 Vector Functions and Space Curves

1. The component functions $\ln(t+1)$, $\dfrac{t}{\sqrt{9-t^2}}$, and 2^t are all defined when $t+1>0 \;\Rightarrow\; t>-1$ and $9-t^2>0 \;\Rightarrow$

$-3 < t < 3$, so the domain of $\mathbf{r}$ is $(-1,3)$.

3. $\displaystyle\lim_{t\to 0} e^{-3t} = e^0 = 1$, $\displaystyle\lim_{t\to 0}\frac{t^2}{\sin^2 t} = \lim_{t\to 0}\frac{1}{\dfrac{\sin^2 t}{t^2}} = \frac{1}{\displaystyle\lim_{t\to 0}\frac{\sin^2 t}{t^2}} = \frac{1}{\left(\displaystyle\lim_{t\to 0}\frac{\sin t}{t}\right)^2} = \frac{1}{1^2} = 1$,

and $\displaystyle\lim_{t\to 0}\cos 2t = \cos 0 = 1$. Thus

$$\lim_{t\to 0}\left(e^{-3t}\,\mathbf{i} + \frac{t^2}{\sin^2 t}\,\mathbf{j} + \cos 2t\,\mathbf{k}\right) = \left[\lim_{t\to 0} e^{-3t}\right]\mathbf{i} + \left[\lim_{t\to 0}\frac{t^2}{\sin^2 t}\right]\mathbf{j} + \left[\lim_{t\to 0}\cos 2t\right]\mathbf{k} = \mathbf{i} + \mathbf{j} + \mathbf{k}.$$

5. $\displaystyle\lim_{t\to\infty}\frac{1+t^2}{1-t^2} = \lim_{t\to\infty}\frac{(1/t^2)+1}{(1/t^2)-1} = \frac{0+1}{0-1} = -1$, $\displaystyle\lim_{t\to\infty}\tan^{-1} t = \frac{\pi}{2}$, $\displaystyle\lim_{t\to\infty}\frac{1-e^{-2t}}{t} = \lim_{t\to\infty}\frac{1}{t} - \frac{1}{te^{2t}} = 0 - 0 = 0$. Thus

$$\lim_{t\to\infty}\left\langle \frac{1+t^2}{1-t^2}, \tan^{-1} t, \frac{1-e^{-2t}}{t}\right\rangle = \left\langle -1, \frac{\pi}{2}, 0\right\rangle.$$

7. The corresponding parametric equations for this curve are $x = \sin t$, $y = t$.

We can make a table of values, or we can eliminate the parameter: $t = y \;\Rightarrow$

$x = \sin y$, with $y \in \mathbb{R}$. By comparing different values of t, we find the direction in

which t increases as indicated in the graph.

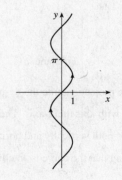

9. The corresponding parametric equations are $x = t$, $y = 2-t$, $z = 2t$, which are

parametric equations of a line through the point $(0,2,0)$ and with direction vector

$\langle 1, -1, 2\rangle$.

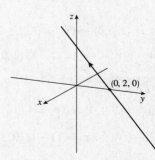

11. The corresponding parametric equations are $x = 3$, $y = t$, $z = 2 - t^2$.

Eliminating the parameter in y and z gives $z = 2 - y^2$. Because $x = 3$, the

curve is a parabola in the vertical plane $x = 3$ with vertex $(3, 0, 2)$.

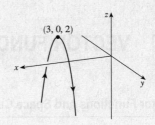

13. The parametric equations are $x = t^2$, $y = t^4$, $z = t^6$. These are positive

for $t \neq 0$ and 0 when $t = 0$. So the curve lies entirely in the first octant.

The projection of the graph onto the xy-plane is $y = x^2$, $y > 0$, a half parabola.

The projection onto the xz-plane is $z = x^3$, $z > 0$, a half cubic, and the projection

onto the yz-plane is $y^3 = z^2$.

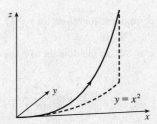

15. The projection of the curve onto the xy-plane is given by $\mathbf{r}(t) = \langle t, \sin t, 0 \rangle$ [we use 0 for the z-component] whose graph

is the curve $y = \sin x$, $z = 0$. Similarly, the projection onto the xz-plane is $\mathbf{r}(t) = \langle t, 0, 2\cos t \rangle$, whose graph is the cosine

wave $z = 2 \cos x$, $y = 0$, and the projection onto the yz-plane is $\mathbf{r}(t) = \langle 0, \sin t, 2 \cos t \rangle$ whose graph is the ellipse

$y^2 + \frac{1}{4}z^2 = 1$, $x = 0$.

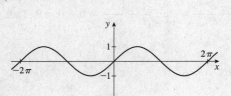 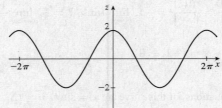 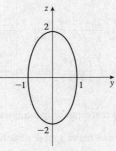

xy-plane $\qquad\qquad\qquad$ xz-plane $\qquad\qquad\qquad$ yz-plane

From the projection onto the yz-plane we see that the curve lies on an elliptical

cylinder with axis the x-axis. The other two projections show that the curve

oscillates both vertically and horizontally as we move in the x-direction,

suggesting that the curve is an elliptical helix that spirals along the cylinder.

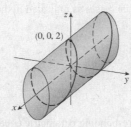

17. Taking $\mathbf{r}_0 = \langle 2, 0, 0 \rangle$ and $\mathbf{r}_1 = \langle 6, 2, -2 \rangle$, we have from Equation 12.5.4

$\mathbf{r}(t) = (1 - t)\,\mathbf{r}_0 + t\,\mathbf{r}_1 = (1 - t)\,\langle 2, 0, 0 \rangle + t\,\langle 6, 2, -2 \rangle,\ 0 \le t \le 1$ or $\mathbf{r}(t) = \langle 2 + 4t, 2t, -2t \rangle,\ 0 \le t \le 1$.

Parametric equations are $x = 2 + 4t$, $y = 2t$, $z = -2t$, $0 \le t \le 1$.

19. Taking $\mathbf{r}_0 = \langle 0, -1, 1 \rangle$ and $\mathbf{r}_1 = \langle \frac{1}{2}, \frac{1}{3}, \frac{1}{4} \rangle$, we have

$\mathbf{r}(t) = (1 - t)\,\mathbf{r}_0 + t\,\mathbf{r}_1 = (1 - t)\,\langle 0, -1, 1 \rangle + t\,\langle \frac{1}{2}, \frac{1}{3}, \frac{1}{4} \rangle,\ 0 \le t \le 1$ or $\mathbf{r}(t) = \langle \frac{1}{2}t, -1 + \frac{4}{3}t, 1 - \frac{3}{4}t \rangle,\ 0 \le t \le 1$.

Parametric equations are $x = \frac{1}{2}t$, $y = -1 + \frac{4}{3}t$, $z = 1 - \frac{3}{4}t$, $0 \le t \le 1$.

21. $x = t \cos t,\ y = t,\ z = t \sin t,\ t \geq 0.$ At any point (x, y, z) on the curve, $x^2 + z^2 = t^2 \cos^2 t + t^2 \sin^2 t = t^2 = y^2$ so the curve lies on the circular cone $x^2 + z^2 = y^2$ with axis the y-axis. Also notice that $y \geq 0$; the graph is II.

23. $x = t,\ y = 1/(1 + t^2),\ z = t^2.$ At any point on the curve we have $z = x^2$, so the curve lies on a parabolic cylinder parallel to the y-axis. Notice that $0 < y \leq 1$ and $z \geq 0$. Also the curve passes through $(0, 1, 0)$ when $t = 0$ and $y \to 0,\ z \to \infty$ as $t \to \pm\infty$, so the graph must be V.

25. $x = \cos 8t,\ y = \sin 8t,\ z = e^{0.8t},\ t \geq 0.$ $x^2 + y^2 = \cos^2 8t + \sin^2 8t = 1$, so the curve lies on a circular cylinder with axis the z-axis. A point (x, y, z) on the curve lies directly above the point $(x, y, 0)$, which moves counterclockwise around the unit circle in the xy-plane as t increases. The curve starts at $(1, 0, 1)$, when $t = 0$, and $z \to \infty$ (at an increasing rate) as $t \to \infty$, so the graph is IV.

27. If $x = t \cos t,\ y = t \sin t,\ z = t$, then $x^2 + y^2 = t^2 \cos^2 t + t^2 \sin^2 t = t^2 = z^2$, so the curve lies on the cone $z^2 = x^2 + y^2$. Since $z = t$, the curve is a spiral on this cone.

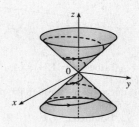

29. Here $x = 2t,\ y = e^t,\ z = e^{2t}$. Then $t = x/2 \ \Rightarrow \ y = e^t = e^{x/2}$, so the curve lies on the cylinder $y = e^{x/2}$. Also $z = e^{2t} = e^x$, so the curve lies on the cylinder $z = e^x$. Since $z = e^{2t} = \left(e^t\right)^2 = y^2$, the curve also lies on the parabolic cylinder $z = y^2$.

31. Parametric equations for the curve are $x = t,\ y = 0,\ z = 2t - t^2$. Substituting into the equation of the paraboloid gives $2t - t^2 = t^2 \ \Rightarrow \ 2t = 2t^2 \ \Rightarrow \ t = 0, 1$. Since $\mathbf{r}(0) = \mathbf{0}$ and $\mathbf{r}(1) = \mathbf{i} + \mathbf{k}$, the points of intersection are $(0, 0, 0)$ and $(1, 0, 1)$.

33. $\mathbf{r}(t) = \langle \cos t \sin 2t,\ \sin t \sin 2t,\ \cos 2t \rangle$.

We include both a regular plot and a plot showing a tube of radius 0.08 around the curve.

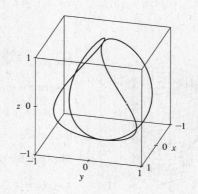 

35. $\mathbf{r}(t) = \langle \sin 3t \cos t, \frac{1}{4}t, \sin 3t \sin t \rangle$

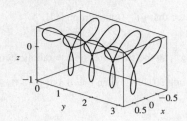

37. $\mathbf{r}(t) = \langle \cos 2t, \cos 3t, \cos 4t \rangle$

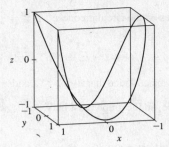

39.

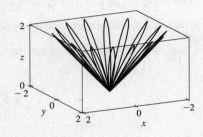

$x = (1 + \cos 16t) \cos t$, $y = (1 + \cos 16t) \sin t$, $z = 1 + \cos 16t$. At any point on the graph,

$$x^2 + y^2 = (1 + \cos 16t)^2 \cos^2 t + (1 + \cos 16t)^2 \sin^2 t$$
$$= (1 + \cos 16t)^2 = z^2, \text{ so the graph lies on the cone } x^2 + y^2 = z^2.$$

From the graph at left, we see that this curve looks like the projection of a leaved two-dimensional curve onto a cone.

41. If $t = -1$, then $x = 1$, $y = 4$, $z = 0$, so the curve passes through the point $(1, 4, 0)$. If $t = 3$, then $x = 9$, $y = -8$, $z = 28$, so the curve passes through the point $(9, -8, 28)$. For the point $(4, 7, -6)$ to be on the curve, we require $y = 1 - 3t = 7 \Rightarrow$ $t = -2$. But then $z = 1 + (-2)^3 = -7 \neq -6$, so $(4, 7, -6)$ is not on the curve.

43. Both equations are solved for z, so we can substitute to eliminate z: $\sqrt{x^2 + y^2} = 1 + y \Rightarrow x^2 + y^2 = 1 + 2y + y^2 \Rightarrow$ $x^2 = 1 + 2y \Rightarrow y = \frac{1}{2}(x^2 - 1)$. We can form parametric equations for the curve C of intersection by choosing a parameter $x = t$, then $y = \frac{1}{2}(t^2 - 1)$ and $z = 1 + y = 1 + \frac{1}{2}(t^2 - 1) = \frac{1}{2}(t^2 + 1)$. Thus a vector function representing C is $\mathbf{r}(t) = t\,\mathbf{i} + \frac{1}{2}(t^2 - 1)\,\mathbf{j} + \frac{1}{2}(t^2 + 1)\,\mathbf{k}$.

45. The projection of the curve C of intersection onto the xy-plane is the circle $x^2 + y^2 = 1$, $z = 0$, so we can write $x = \cos t$, $y = \sin t$, $0 \le t \le 2\pi$. Since C also lies on the surface $z = x^2 - y^2$, we have $z = x^2 - y^2 = \cos^2 t - \sin^2 t$ or $\cos 2t$. Thus parametric equations for C are $x = \cos t$, $y = \sin t$, $z = \cos 2t$, $0 \le t \le 2\pi$, and the corresponding vector function is $\mathbf{r}(t) = \cos t\,\mathbf{i} + \sin t\,\mathbf{j} + \cos 2t\,\mathbf{k}$, $0 \le t \le 2\pi$.

47.

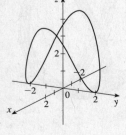 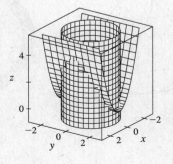

The projection of the curve C of intersection onto the xy-plane is the circle $x^2 + y^2 = 4$, $z = 0$. Then we can write $x = 2\cos t$, $y = 2\sin t$, $0 \le t \le 2\pi$. Since C also lies on the surface $z = x^2$, we have $z = x^2 = (2\cos t)^2 = 4\cos^2 t$. Then parametric equations for C are $x = 2\cos t$, $y = 2\sin t$, $z = 4\cos^2 t$, $0 \le t \le 2\pi$.

49. For the particles to collide, we require $\mathbf{r}_1(t) = \mathbf{r}_2(t)$ ⇔ $\langle t^2, 7t - 12, t^2 \rangle = \langle 4t - 3, t^2, 5t - 6 \rangle$. Equating components

gives $t^2 = 4t - 3$, $7t - 12 = t^2$, and $t^2 = 5t - 6$. From the first equation, $t^2 - 4t + 3 = 0$ ⇔ $(t - 3)(t - 1) = 0$ so $t = 1$

or $t = 3$. $t = 1$ does not satisfy the other two equations, but $t = 3$ does. The particles collide when $t = 3$, at the

point $(9, 9, 9)$.

51. (a) We plot the parametric equations for $0 \le t \le 2\pi$ in the first figure. We get a better idea of the shape of the curve if we plot

it simultaneously with the hyperboloid of one sheet from part (b), as shown in the second figure.

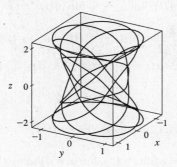 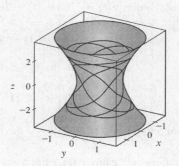

(b) Here $x = \frac{27}{26} \sin 8t - \frac{8}{39} \sin 18t$, $y = -\frac{27}{26} \cos 8t + \frac{8}{39} \cos 18t$, $z = \frac{144}{65} \sin 5t$.

For any point on the curve,

$$x^2 + y^2 = \left(\frac{27}{26} \sin 8t - \frac{8}{39} \sin 18t\right)^2 + \left(-\frac{27}{26} \cos 8t + \frac{8}{39} \cos 18t\right)^2$$

$$= \frac{27^2}{26^2} \sin^2 8t - 2 \cdot \frac{27 \cdot 8}{26 \cdot 39} \sin 8t \sin 18t + \frac{64}{39^2} \sin^2 18t$$

$$+ \frac{27^2}{26^2} \cos^2 8t - 2 \cdot \frac{27 \cdot 8}{26 \cdot 39} \cos 8t \cos 18t + \frac{64}{39^2} \cos^2 18t$$

$$= \frac{27^2}{26^2} \left(\sin^2 8t + \cos^2 8t\right) + \frac{64}{39^2} \left(\sin^2 18t + \cos^2 18t\right) - \frac{72}{169} \left(\sin 8t \sin 18t + \cos 8t \cos 18t\right)$$

$$= \frac{27^2}{26^2} + \frac{64}{39^2} - \frac{72}{169} \cos(18t - 8t) = \frac{27^2}{26^2} + \frac{64}{39^2} - \frac{72}{169} \cos 10t$$

using the trigonometric identities $\sin^2 \theta + \cos^2 \theta = 1$ and $\cos(x - y) = \cos x \cos y + \sin x \sin y$. Also

$z^2 = \frac{144^2}{65^2} \sin^2 5t$, and the identity $\sin^2 x = \dfrac{1 - \cos 2x}{2}$ gives $z^2 = \frac{144^2}{65^2} \cdot \frac{1}{2} [1 - \cos(2 \cdot 5t)] = \frac{144^2}{2 \cdot 65^2} - \frac{144^2}{2 \cdot 65^2} \cos 10t$.

Then

$$144(x^2 + y^2) - 25z^2 = 144 \left(\frac{27^2}{26^2} + \frac{64}{39^2} - \frac{72}{169} \cos 10t\right) - 25 \left(\frac{144^2}{2 \cdot 65^2} - \frac{144^2}{2 \cdot 65^2} \cos 10t\right)$$

$$= 144 \left(\frac{27^2}{26^2} + \frac{64}{39^2} - \frac{25 \cdot 144}{2 \cdot 65^2} - \frac{72}{169} \cos 10t + \frac{25 \cdot 144}{2 \cdot 65^2} \cos 10t\right)$$

$$= 144 \left(\frac{27^2}{26^2} + \frac{64}{39^2} - \frac{72}{169} - \frac{72}{169} \cos 10t + \frac{72}{169} \cos 10t\right) = 144 \left(\frac{25}{36}\right) = 100$$

Thus the curve lies on the surface $144(x^2 + y^2) - 25z^2 = 100$ or $144x^2 + 144y^2 - 25z^2 = 100$, a hyperboloid of one

sheet with axis the z-axis.

53. Let $\mathbf{u}(t) = \langle u_1(t), u_2(t), u_3(t) \rangle$ and $\mathbf{v}(t) = \langle v_1(t), v_2(t), v_3(t) \rangle$. In each part of this problem the basic procedure is to use Equation 1 and then analyze the individual component functions using the limit properties we have already developed for real-valued functions.

(a) $\lim\limits_{t \to a} \mathbf{u}(t) + \lim\limits_{t \to a} \mathbf{v}(t) = \left\langle \lim\limits_{t \to a} u_1(t), \lim\limits_{t \to a} u_2(t), \lim\limits_{t \to a} u_3(t) \right\rangle + \left\langle \lim\limits_{t \to a} v_1(t), \lim\limits_{t \to a} v_2(t), \lim\limits_{t \to a} v_3(t) \right\rangle$ and the limits of these component functions must each exist since the vector functions both possess limits as $t \to a$. Then adding the two vectors and using the addition property of limits for real-valued functions, we have that

$$\lim\limits_{t \to a} \mathbf{u}(t) + \lim\limits_{t \to a} \mathbf{v}(t) = \left\langle \lim\limits_{t \to a} u_1(t) + \lim\limits_{t \to a} v_1(t), \lim\limits_{t \to a} u_2(t) + \lim\limits_{t \to a} v_2(t), \lim\limits_{t \to a} u_3(t) + \lim\limits_{t \to a} v_3(t) \right\rangle .$$

$$= \left\langle \lim\limits_{t \to a} [u_1(t) + v_1(t)], \lim\limits_{t \to a} [u_2(t) + v_2(t)], \lim\limits_{t \to a} [u_3(t) + v_3(t)] \right\rangle$$

$$= \lim\limits_{t \to a} \langle u_1(t) + v_1(t), u_2(t) + v_2(t), u_3(t) + v_3(t) \rangle \qquad \text{[using (1) backward]}$$

$$= \lim\limits_{t \to a} [\mathbf{u}(t) + \mathbf{v}(t)]$$

(b) $\lim\limits_{t \to a} c\mathbf{u}(t) = \lim\limits_{t \to a} \langle cu_1(t), cu_2(t), cu_3(t) \rangle = \left\langle \lim\limits_{t \to a} cu_1(t), \lim\limits_{t \to a} cu_2(t), \lim\limits_{t \to a} cu_3(t) \right\rangle$

$$= \left\langle c \lim\limits_{t \to a} u_1(t), c \lim\limits_{t \to a} u_2(t), c \lim\limits_{t \to a} u_3(t) \right\rangle = c \left\langle \lim\limits_{t \to a} u_1(t), \lim\limits_{t \to a} u_2(t), \lim\limits_{t \to a} u_3(t) \right\rangle$$

$$= c \lim\limits_{t \to a} \langle u_1(t), u_2(t), u_3(t) \rangle = c \lim\limits_{t \to a} \mathbf{u}(t)$$

(c) $\lim\limits_{t \to a} \mathbf{u}(t) \cdot \lim\limits_{t \to a} \mathbf{v}(t) = \left\langle \lim\limits_{t \to a} u_1(t), \lim\limits_{t \to a} u_2(t), \lim\limits_{t \to a} u_3(t) \right\rangle \cdot \left\langle \lim\limits_{t \to a} v_1(t), \lim\limits_{t \to a} v_2(t), \lim\limits_{t \to a} v_3(t) \right\rangle$

$$= \left[\lim\limits_{t \to a} u_1(t) \right] \left[\lim\limits_{t \to a} v_1(t) \right] + \left[\lim\limits_{t \to a} u_2(t) \right] \left[\lim\limits_{t \to a} v_2(t) \right] + \left[\lim\limits_{t \to a} u_3(t) \right] \left[\lim\limits_{t \to a} v_3(t) \right]$$

$$= \lim\limits_{t \to a} u_1(t)v_1(t) + \lim\limits_{t \to a} u_2(t)v_2(t) + \lim\limits_{t \to a} u_3(t)v_3(t)$$

$$= \lim\limits_{t \to a} [u_1(t)v_1(t) + u_2(t)v_2(t) + u_3(t)v_3(t)] = \lim\limits_{t \to a} [\mathbf{u}(t) \cdot \mathbf{v}(t)]$$

(d) $\lim\limits_{t \to a} \mathbf{u}(t) \times \lim\limits_{t \to a} \mathbf{v}(t) = \left\langle \lim\limits_{t \to a} u_1(t), \lim\limits_{t \to a} u_2(t), \lim\limits_{t \to a} u_3(t) \right\rangle \times \left\langle \lim\limits_{t \to a} v_1(t), \lim\limits_{t \to a} v_2(t), \lim\limits_{t \to a} v_3(t) \right\rangle$

$$= \left\langle \left[\lim\limits_{t \to a} u_2(t) \right] \left[\lim\limits_{t \to a} v_3(t) \right] - \left[\lim\limits_{t \to a} u_3(t) \right] \left[\lim\limits_{t \to a} v_2(t) \right], \right.$$

$$\left[\lim\limits_{t \to a} u_3(t) \right] \left[\lim\limits_{t \to a} v_1(t) \right] - \left[\lim\limits_{t \to a} u_1(t) \right] \left[\lim\limits_{t \to a} v_3(t) \right],$$

$$\left. \left[\lim\limits_{t \to a} u_1(t) \right] \left[\lim\limits_{t \to a} v_2(t) \right] - \left[\lim\limits_{t \to a} u_2(t) \right] \left[\lim\limits_{t \to a} v_1(t) \right] \right\rangle$$

$$= \left\langle \lim\limits_{t \to a} [u_2(t)v_3(t) - u_3(t)v_2(t)], \lim\limits_{t \to a} [u_3(t)v_1(t) - u_1(t)v_3(t)], \right.$$

$$\left. \lim\limits_{t \to a} [u_1(t)v_2(t) - u_2(t)v_1(t)] \right\rangle$$

$$= \lim\limits_{t \to a} \langle u_2(t)v_3(t) - u_3(t)v_2(t), u_3(t)v_1(t) - u_1(t)v_3(t), u_1(t)v_2(t) - u_2(t)v_1(t) \rangle$$

$$= \lim\limits_{t \to a} [\mathbf{u}(t) \times \mathbf{v}(t)]$$

13.2 Derivatives and Integrals of Vector Functions

1. (a)

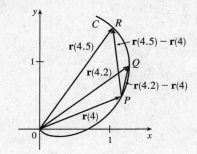

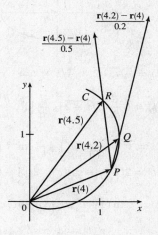

(b) $\dfrac{\mathbf{r}(4.5) - \mathbf{r}(4)}{0.5} = 2[\mathbf{r}(4.5) - \mathbf{r}(4)]$, so we draw a vector in the same

direction but with twice the length of the vector $\mathbf{r}(4.5) - \mathbf{r}(4)$.

$\dfrac{\mathbf{r}(4.2) - \mathbf{r}(4)}{0.2} = 5[\mathbf{r}(4.2) - \mathbf{r}(4)]$, so we draw a vector in the same

direction but with 5 times the length of the vector $\mathbf{r}(4.2) - \mathbf{r}(4)$.

(c) By Definition 1, $\mathbf{r}'(4) = \lim\limits_{h \to 0} \dfrac{\mathbf{r}(4+h) - \mathbf{r}(4)}{h}$. $\quad \mathbf{T}(4) = \dfrac{\mathbf{r}'(4)}{|\mathbf{r}'(4)|}$.

(d) $\mathbf{T}(4)$ is a unit vector in the same direction as $\mathbf{r}'(4)$, that is, parallel to the

tangent line to the curve at $\mathbf{r}(4)$ with length 1.

3. $\mathbf{r}(t) = \langle t - 2, t^2 + 1 \rangle$,

$\mathbf{r}(-1) = \langle -3, 2 \rangle$.

Since $(x + 2)^2 = t^2 = y - 1 \;\Rightarrow$

$y = (x + 2)^2 + 1$, the curve is a

parabola.

(a), (c)

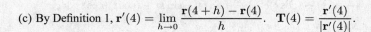

(b) $\mathbf{r}'(t) = \langle 1, 2t \rangle$,

$\mathbf{r}'(-1) = \langle 1, -2 \rangle$

5. $\mathbf{r}(t) = e^{2t}\,\mathbf{i} + e^t\,\mathbf{j}$, $\mathbf{r}(0) = \mathbf{i} + \mathbf{j}$.

Since $x = e^{2t} = (e^t)^2 = y^2$, the

curve is part of a parabola. Note

that here $x > 0$, $y > 0$.

(a), (c)

(b) $\mathbf{r}'(t) = 2e^{2t}\,\mathbf{i} + e^t\,\mathbf{j}$,

$\mathbf{r}'(0) = 2\,\mathbf{i} + \mathbf{j}$

7. $\mathbf{r}(t) = 4\sin t\,\mathbf{i} - 2\cos t\,\mathbf{j}$, $\mathbf{r}(3\pi/4) = 4(\sqrt{2}/2)\,\mathbf{i} - 2(-\sqrt{2}/2)\,\mathbf{j} = 2\sqrt{2}\,\mathbf{i} + \sqrt{2}\,\mathbf{j}$.

Here $(x/4)^2 + (y/2)^2 = \sin^2 t + \cos^2 t = 1$, so the curve is the ellipse $\dfrac{x^2}{16} + \dfrac{y^2}{4} = 1$.

(a), (c)

(b) $\mathbf{r}'(t) = 4\cos t\,\mathbf{i} + 2\sin t\,\mathbf{j}$,

$\mathbf{r}'(3\pi/4) = -2\sqrt{2}\,\mathbf{i} + \sqrt{2}\,\mathbf{j}$.

9. $\mathbf{r}(t) = \left\langle \sqrt{t-2}, 3, 1/t^2 \right\rangle \quad\Rightarrow$

$\mathbf{r}'(t) = \left\langle \dfrac{d}{dt}\left[\sqrt{t-2}\right], \dfrac{d}{dt}\left[3\right], \dfrac{d}{dt}\left[1/t^2\right] \right\rangle = \left\langle \tfrac{1}{2}(t-2)^{-1/2}, 0, -2t^{-3} \right\rangle = \left\langle \dfrac{1}{2\sqrt{t-2}}, 0, -\dfrac{2}{t^3} \right\rangle$

11. $\mathbf{r}(t) = t^2\,\mathbf{i} + \cos(t^2)\,\mathbf{j} + \sin^2 t\,\mathbf{k} \quad\Rightarrow$

$\mathbf{r}'(t) = 2t\,\mathbf{i} + \left[-\sin(t^2)\cdot 2t\right]\mathbf{j} + (2\sin t\cdot\cos t)\,\mathbf{k} = 2t\,\mathbf{i} - 2t\sin(t^2)\,\mathbf{j} + 2\sin t\cos t\,\mathbf{k}$

13. $\mathbf{r}(t) = t\sin t\,\mathbf{i} + e^t\cos t\,\mathbf{j} + \sin t\cos t\,\mathbf{k} \quad\Rightarrow$

$$\mathbf{r}'(t) = \left[t\cdot\cos t + (\sin t)\cdot 1\right]\mathbf{i} + \left[e^t(-\sin t) + (\cos t)e^t\right]\mathbf{j} + \left[(\sin t)(-\sin t) + (\cos t)(\cos t)\right]\mathbf{k}$$
$$= (t\cos t + \sin t)\,\mathbf{i} + e^t(\cos t - \sin t)\,\mathbf{j} + (\cos^2 t - \sin^2 t)\,\mathbf{k}$$

15. $\mathbf{r}'(t) = \mathbf{0} + \mathbf{b} + 2t\,\mathbf{c} = \mathbf{b} + 2t\,\mathbf{c}$ by Formulas 1 and 3 of Theorem 3.

17. $\mathbf{r}(t) = \left\langle t^2 - 2t, 1 + 3t, \tfrac{1}{3}t^3 + \tfrac{1}{2}t^2 \right\rangle \quad\Rightarrow\quad \mathbf{r}'(t) = \left\langle 2t - 2, 3, t^2 + t \right\rangle \quad\Rightarrow\quad \mathbf{r}'(2) = \langle 2, 3, 6 \rangle$.

So $|\mathbf{r}'(2)| = \sqrt{2^2 + 3^2 + 6^2} = \sqrt{49} = 7$ and $\mathbf{T}(2) = \dfrac{\mathbf{r}'(2)}{|\mathbf{r}'(2)|} = \tfrac{1}{7}\langle 2, 3, 6 \rangle = \left\langle \tfrac{2}{7}, \tfrac{3}{7}, \tfrac{6}{7} \right\rangle$.

19. $\mathbf{r}'(t) = -\sin t\,\mathbf{i} + 3\,\mathbf{j} + 4\cos 2t\,\mathbf{k} \quad\Rightarrow\quad \mathbf{r}'(0) = 3\,\mathbf{j} + 4\,\mathbf{k}$. Thus

$$\mathbf{T}(0) = \dfrac{\mathbf{r}'(0)}{|\mathbf{r}'(0)|} = \dfrac{1}{\sqrt{0^2 + 3^2 + 4^2}}\,(3\,\mathbf{j} + 4\,\mathbf{k}) = \tfrac{1}{5}(3\,\mathbf{j} + 4\,\mathbf{k}) = \tfrac{3}{5}\,\mathbf{j} + \tfrac{4}{5}\,\mathbf{k}.$$

21. $\mathbf{r}(t) = \left\langle t, t^2, t^3 \right\rangle \quad\Rightarrow\quad \mathbf{r}'(t) = \left\langle 1, 2t, 3t^2 \right\rangle$. Then $\mathbf{r}'(1) = \langle 1, 2, 3 \rangle$ and $|\mathbf{r}'(1)| = \sqrt{1^2 + 2^2 + 3^2} = \sqrt{14}$, so

$$\mathbf{T}(1) = \dfrac{\mathbf{r}'(1)}{|\mathbf{r}'(1)|} = \tfrac{1}{\sqrt{14}}\,\langle 1, 2, 3 \rangle = \left\langle \tfrac{1}{\sqrt{14}}, \tfrac{2}{\sqrt{14}}, \tfrac{3}{\sqrt{14}} \right\rangle. \quad \mathbf{r}''(t) = \langle 0, 2, 6t \rangle, \text{ so}$$

$$\mathbf{r}'(t) \times \mathbf{r}''(t) = \begin{vmatrix} \mathbf{i} & \mathbf{j} & \mathbf{k} \\ 1 & 2t & 3t^2 \\ 0 & 2 & 6t \end{vmatrix} = \begin{vmatrix} 2t & 3t^2 \\ 2 & 6t \end{vmatrix}\mathbf{i} - \begin{vmatrix} 1 & 3t^2 \\ 0 & 6t \end{vmatrix}\mathbf{j} + \begin{vmatrix} 1 & 2t \\ 0 & 2 \end{vmatrix}\mathbf{k}$$

$$= (12t^2 - 6t^2)\,\mathbf{i} - (6t - 0)\,\mathbf{j} + (2 - 0)\,\mathbf{k} = \left\langle 6t^2, -6t, 2 \right\rangle$$

23. The vector equation for the curve is $\mathbf{r}(t) = \left\langle t^2 + 1, 4\sqrt{t}, e^{t^2-t} \right\rangle$, so $\mathbf{r}'(t) = \left\langle 2t, 2/\sqrt{t}, (2t-1)e^{t^2-t} \right\rangle$. The point $(2, 4, 1)$

corresponds to $t = 1$, so the tangent vector there is $\mathbf{r}'(1) = \langle 2, 2, 1 \rangle$. Thus, the tangent line goes through the point $(2, 4, 1)$

and is parallel to the vector $\langle 2, 2, 1 \rangle$. Parametric equations are $x = 2 + 2t$, $y = 4 + 2t$, $z = 1 + t$.

25. The vector equation for the curve is $\mathbf{r}(t) = \langle e^{-t}\cos t, e^{-t}\sin t, e^{-t}\rangle$, so

$$\mathbf{r}'(t) = \langle e^{-t}(-\sin t) + (\cos t)(-e^{-t}), e^{-t}\cos t + (\sin t)(-e^{-t}), (-e^{-t})\rangle$$
$$= \langle -e^{-t}(\cos t + \sin t), e^{-t}(\cos t - \sin t), -e^{-t}\rangle$$

The point $(1, 0, 1)$ corresponds to $t = 0$, so the tangent vector there is

$\mathbf{r}'(0) = \langle -e^0(\cos 0 + \sin 0), e^0(\cos 0 - \sin 0), -e^0\rangle = \langle -1, 1, -1\rangle$. Thus, the tangent line is parallel to the vector

$\langle -1, 1, -1\rangle$ and parametric equations are $x = 1 + (-1)t = 1 - t$, $y = 0 + 1 \cdot t = t$, $z = 1 + (-1)t = 1 - t$.

27. First we parametrize the curve C of intersection. The projection of C onto the xy-plane is contained in the circle

$x^2 + y^2 = 25$, $z = 0$, so we can write $x = 5\cos t$, $y = 5\sin t$. C also lies on the cylinder $y^2 + z^2 = 20$, and $z \geq 0$

near the point $(3, 4, 2)$, so we can write $z = \sqrt{20 - y^2} = \sqrt{20 - 25\sin^2 t}$. A vector equation then for C is

$\mathbf{r}(t) = \left\langle 5\cos t, 5\sin t, \sqrt{20 - 25\sin^2 t}\right\rangle \ \Rightarrow \ \mathbf{r}'(t) = \left\langle -5\sin t, 5\cos t, \frac{1}{2}(20 - 25\sin^2 t)^{-1/2}(-50\sin t\cos t)\right\rangle$.

The point $(3, 4, 2)$ corresponds to $t = \cos^{-1}\left(\frac{3}{5}\right)$, so the tangent vector there is

$\mathbf{r}'\left(\cos^{-1}\left(\frac{3}{5}\right)\right) = \left\langle -5\left(\frac{4}{5}\right), 5\left(\frac{3}{5}\right), \frac{1}{2}\left(20 - 25\left(\frac{4}{5}\right)^2\right)^{-1/2}\left(-50\left(\frac{4}{5}\right)\left(\frac{3}{5}\right)\right)\right\rangle = \langle -4, 3, -6\rangle$.

The tangent line is parallel to this vector and passes through $(3, 4, 2)$, so a vector equation for the line

is $\mathbf{r}(t) = (3 - 4t)\mathbf{i} + (4 + 3t)\mathbf{j} + (2 - 6t)\mathbf{k}$.

29. $\mathbf{r}(t) = \langle t, e^{-t}, 2t - t^2\rangle \ \Rightarrow \ \mathbf{r}'(t) = \langle 1, -e^{-t}, 2 - 2t\rangle$. At $(0, 1, 0)$,

$t = 0$ and $\mathbf{r}'(0) = \langle 1, -1, 2\rangle$. Thus, parametric equations of the tangent

line are $x = t$, $y = 1 - t$, $z = 2t$.

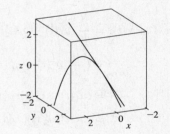

31. $\mathbf{r}(t) = \langle t\cos t, t, t\sin t\rangle \ \Rightarrow \ \mathbf{r}'(t) = \langle \cos t - t\sin t, 1, t\cos t + \sin t\rangle$.

At $(-\pi, \pi, 0)$, $t = \pi$ and $\mathbf{r}'(\pi) = \langle -1, 1, -\pi\rangle$. Thus, parametric equations

of the tangent line are $x = -\pi - t$, $y = \pi + t$, $z = -\pi t$.

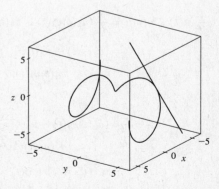

33. The angle of intersection of the two curves is the angle between the two tangent vectors to the curves at the point of

intersection. Since $\mathbf{r}_1'(t) = \langle 1, 2t, 3t^2\rangle$ and $t = 0$ at $(0, 0, 0)$, $\mathbf{r}_1'(0) = \langle 1, 0, 0\rangle$ is a tangent vector to $\mathbf{r}_1$ at $(0, 0, 0)$. Similarly,

$\mathbf{r}_2'(t) = \langle \cos t, 2\cos 2t, 1\rangle$ and since $\mathbf{r}_2(0) = \langle 0, 0, 0\rangle$, $\mathbf{r}_2'(0) = \langle 1, 2, 1\rangle$ is a tangent vector to $\mathbf{r}_2$ at $(0, 0, 0)$. If θ is the angle

between these two tangent vectors, then $\cos\theta = \frac{1}{\sqrt{1}\sqrt{6}}\langle 1, 0, 0\rangle \cdot \langle 1, 2, 1\rangle = \frac{1}{\sqrt{6}}$ and $\theta = \cos^{-1}\left(\frac{1}{\sqrt{6}}\right) \approx 66°$.

35. $\int_0^2 (t\,\mathbf{i} - t^3\,\mathbf{j} + 3t^5\,\mathbf{k})\,dt = \left(\int_0^2 t\,dt\right)\mathbf{i} - \left(\int_0^2 t^3\,dt\right)\mathbf{j} + \left(\int_0^2 3t^5\,dt\right)\mathbf{k}$

$$= \left[\tfrac{1}{2}t^2\right]_0^2 \mathbf{i} - \left[\tfrac{1}{4}t^4\right]_0^2 \mathbf{j} + \left[\tfrac{1}{2}t^6\right]_0^2 \mathbf{k}$$

$$= \tfrac{1}{2}(4 - 0)\,\mathbf{i} - \tfrac{1}{4}(16 - 0)\,\mathbf{j} + \tfrac{1}{2}(64 - 0)\,\mathbf{k} = 2\,\mathbf{i} - 4\,\mathbf{j} + 32\,\mathbf{k}$$

37. $\int_0^1 \left(\dfrac{1}{t+1}\,\mathbf{i} + \dfrac{1}{t^2+1}\,\mathbf{j} + \dfrac{t}{t^2+1}\,\mathbf{k}\right)dt = \left(\int_0^1 \dfrac{1}{t+1}\,dt\right)\mathbf{i} + \left(\int_0^1 \dfrac{1}{t^2+1}\,dt\right)\mathbf{j} + \left(\int_0^1 \dfrac{t}{t^2+1}\,dt\right)\mathbf{k}$

$$= \left[\ln|t+1|\right]_0^1 \mathbf{i} + \left[\tan^{-1} t\right]_0^1 \mathbf{j} + \left[\tfrac{1}{2}\ln(t^2+1)\right]_0^1 \mathbf{k}$$

$$= (\ln 2 - \ln 1)\,\mathbf{i} + (\tfrac{\pi}{4} - 0)\,\mathbf{j} + \tfrac{1}{2}(\ln 2 - \ln 1)\,\mathbf{k} = \ln 2\,\mathbf{i} + \tfrac{\pi}{4}\,\mathbf{j} + \tfrac{1}{2}\ln 2\,\mathbf{k}$$

39. $\int (\sec^2 t\,\mathbf{i} + t(t^2+1)^3\,\mathbf{j} + t^2 \ln t\,\mathbf{k})\,dt = \left(\int \sec^2 t\,dt\right)\mathbf{i} + \left(\int t(t^2+1)^3\,dt\right)\mathbf{j} + \left(\int t^2 \ln t\,dt\right)\mathbf{k}$

$$= \tan t\,\mathbf{i} + \tfrac{1}{8}(t^2+1)^4\,\mathbf{j} + \left(\tfrac{1}{3}t^3 \ln t - \tfrac{1}{9}t^3\right)\mathbf{k} + \mathbf{C},$$

where $\mathbf{C}$ is a vector constant of integration. [For the z-component, integrate by parts with $u = \ln t$, $dv = t^2\,dt$.]

41. $\mathbf{r}'(t) = 2t\,\mathbf{i} + 3t^2\,\mathbf{j} + \sqrt{t}\,\mathbf{k} \;\Rightarrow\; \mathbf{r}(t) = t^2\,\mathbf{i} + t^3\,\mathbf{j} + \tfrac{2}{3}t^{3/2}\,\mathbf{k} + \mathbf{C}$, where $\mathbf{C}$ is a constant vector.

But $\mathbf{i} + \mathbf{j} = \mathbf{r}(1) = \mathbf{i} + \mathbf{j} + \tfrac{2}{3}\mathbf{k} + \mathbf{C}$. Thus $\mathbf{C} = -\tfrac{2}{3}\mathbf{k}$ and $\mathbf{r}(t) = t^2\,\mathbf{i} + t^3\,\mathbf{j} + \left(\tfrac{2}{3}t^{3/2} - \tfrac{2}{3}\right)\mathbf{k}$.

For Exercises 43–46, let $\mathbf{u}(t) = \langle u_1(t), u_2(t), u_3(t) \rangle$ and $\mathbf{v}(t) = \langle v_1(t), v_2(t), v_3(t) \rangle$. In each of these exercises, the procedure is to apply Theorem 2 so that the corresponding properties of derivatives of real-valued functions can be used.

43. $\dfrac{d}{dt}\left[\mathbf{u}(t) + \mathbf{v}(t)\right] = \dfrac{d}{dt}\langle u_1(t) + v_1(t), u_2(t) + v_2(t), u_3(t) + v_3(t) \rangle$

$$= \left\langle \dfrac{d}{dt}[u_1(t) + v_1(t)], \dfrac{d}{dt}[u_2(t) + v_2(t)], \dfrac{d}{dt}[u_3(t) + v_3(t)] \right\rangle$$

$$= \langle u_1'(t) + v_1'(t), u_2'(t) + v_2'(t), u_3'(t) + v_3'(t) \rangle$$

$$= \langle u_1'(t), u_2'(t), u_3'(t) \rangle + \langle v_1'(t), v_2'(t), v_3'(t) \rangle = \mathbf{u}'(t) + \mathbf{v}'(t)$$

45. $\dfrac{d}{dt}\left[\mathbf{u}(t) \times \mathbf{v}(t)\right] = \dfrac{d}{dt}\langle u_2(t)v_3(t) - u_3(t)v_2(t), u_3(t)v_1(t) - u_1(t)v_3(t), u_1(t)v_2(t) - u_2(t)v_1(t) \rangle$

$$= \langle u_2'v_3(t) + u_2(t)v_3'(t) - u_3'(t)v_2(t) - u_3(t)v_2'(t),$$

$$u_3'(t)v_1(t) + u_3(t)v_1'(t) - u_1'(t)v_3(t) - u_1(t)v_3'(t),$$

$$u_1'(t)v_2(t) + u_1(t)v_2'(t) - u_2'(t)v_1(t) - u_2(t)v_1'(t) \rangle$$

$$= \langle u_2'(t)v_3(t) - u_3'(t)v_2(t), u_3'(t)v_1(t) - u_1'(t)v_3(t), u_1'(t)v_2(t) - u_2'(t)v_1(t) \rangle$$

$$+ \langle u_2(t)v_3'(t) - u_3(t)v_2'(t), u_3(t)v_1'(t) - u_1(t)v_3'(t), u_1(t)v_2'(t) - u_2(t)v_1'(t) \rangle$$

$$= \mathbf{u}'(t) \times \mathbf{v}(t) + \mathbf{u}(t) \times \mathbf{v}'(t)$$

Alternate solution: Let $\mathbf{r}(t) = \mathbf{u}(t) \times \mathbf{v}(t)$. Then

$$\mathbf{r}(t+h) - \mathbf{r}(t) = [\mathbf{u}(t+h) \times \mathbf{v}(t+h)] - [\mathbf{u}(t) \times \mathbf{v}(t)]$$

$$= [\mathbf{u}(t+h) \times \mathbf{v}(t+h)] - [\mathbf{u}(t) \times \mathbf{v}(t)] + [\mathbf{u}(t+h) \times \mathbf{v}(t)] - [\mathbf{u}(t+h) \times \mathbf{v}(t)]$$

$$= \mathbf{u}(t+h) \times [\mathbf{v}(t+h) - \mathbf{v}(t)] + [\mathbf{u}(t+h) - \mathbf{u}(t)] \times \mathbf{v}(t)$$

(Be careful of the order of the cross product.) Dividing through by h and taking the limit as $h \to 0$ we have

$$\mathbf{r}'(t) = \lim_{h \to 0} \frac{\mathbf{u}(t+h) \times [\mathbf{v}(t+h) - \mathbf{v}(t)]}{h} + \lim_{h \to 0} \frac{[\mathbf{u}(t+h) - \mathbf{u}(t)] \times \mathbf{v}(t)}{h} = \mathbf{u}(t) \times \mathbf{v}'(t) + \mathbf{u}'(t) \times \mathbf{v}(t)$$

by Exercise 13.1.53(a) and Definition 1.

47. $\dfrac{d}{dt}[\mathbf{u}(t) \cdot \mathbf{v}(t)] = \mathbf{u}'(t) \cdot \mathbf{v}(t) + \mathbf{u}(t) \cdot \mathbf{v}'(t)$ [by Formula 4 of Theorem 3]

$$= \langle \cos t, -\sin t, 1 \rangle \cdot \langle t, \cos t, \sin t \rangle + \langle \sin t, \cos t, t \rangle \cdot \langle 1, -\sin t, \cos t \rangle$$

$$= t \cos t - \cos t \sin t + \sin t + \sin t - \cos t \sin t + t \cos t$$

$$= 2t \cos t + 2 \sin t - 2 \cos t \sin t$$

49. By Formula 4 of Theorem 3, $f'(t) = \mathbf{u}'(t) \cdot \mathbf{v}(t) + \mathbf{u}(t) \cdot \mathbf{v}'(t)$, and $\mathbf{v}'(t) = \langle 1, 2t, 3t^2 \rangle$, so

$$f'(2) = \mathbf{u}'(2) \cdot \mathbf{v}(2) + \mathbf{u}(2) \cdot \mathbf{v}'(2) = \langle 3, 0, 4 \rangle \cdot \langle 2, 4, 8 \rangle + \langle 1, 2, -1 \rangle \cdot \langle 1, 4, 12 \rangle = 6 + 0 + 32 + 1 + 8 - 12 = 35.$$

51. $\mathbf{r}(t) = \mathbf{a} \cos \omega t + \mathbf{b} \sin \omega t \quad \Rightarrow \quad \mathbf{r}'(t) = -\mathbf{a}\omega \sin \omega t + \mathbf{b}\omega \cos \omega t$ by Formulas 1 and 3 of Theorem 3. Then

$$\mathbf{r}(t) \times \mathbf{r}'(t) = (\mathbf{a} \cos \omega t + \mathbf{b} \sin \omega t) \times (-\mathbf{a}\omega \sin \omega t + \mathbf{b}\omega \cos \omega t)$$

$$= (\mathbf{a} \cos \omega t + \mathbf{b} \sin \omega t) \times (-\mathbf{a}\omega \sin \omega t) + (\mathbf{a} \cos \omega t + \mathbf{b} \sin \omega t) \times (\mathbf{b}\omega \cos \omega t)$$

$$\text{[by Property 3 of Theorem 12.4.11]}$$

$$= \mathbf{a} \cos \omega t \times (-\mathbf{a}\omega \sin \omega t) + \mathbf{b} \sin \omega t \times (-\mathbf{a}\omega \sin \omega t) + \mathbf{a} \cos \omega t \times \mathbf{b}\omega \cos \omega t + \mathbf{b} \sin \omega t \times \mathbf{b}\omega \cos \omega t$$

$$\text{[by Property 4]}$$

$$= (\cos \omega t)(-\omega \sin \omega t)(\mathbf{a} \times \mathbf{a}) + (\sin \omega t)(-\omega \sin \omega t)(\mathbf{b} \times \mathbf{a}) + (\cos \omega t)(\omega \cos \omega t)(\mathbf{a} \times \mathbf{b})$$

$$+ (\sin \omega t)(\omega \cos \omega t)(\mathbf{b} \times \mathbf{b}) \qquad \text{[by Property 2]}$$

$$= \mathbf{0} + (\omega \sin^2 \omega t)(\mathbf{a} \times \mathbf{b}) + (\omega \cos^2 \omega t)(\mathbf{a} \times \mathbf{b}) + \mathbf{0} \qquad \text{[by Property 1 and Example 12.4.2]}$$

$$= \omega (\sin^2 \omega t + \cos^2 \omega t)(\mathbf{a} \times \mathbf{b}) = \omega (\mathbf{a} \times \mathbf{b}) = \omega \mathbf{a} \times \mathbf{b} \qquad \text{[by Property 2]}$$

53. $\dfrac{d}{dt}[\mathbf{r}(t) \times \mathbf{r}'(t)] = \mathbf{r}'(t) \times \mathbf{r}'(t) + \mathbf{r}(t) \times \mathbf{r}''(t)$ by Formula 5 of Theorem 3. But $\mathbf{r}'(t) \times \mathbf{r}'(t) = \mathbf{0}$ (by Example 12.4.2).

Thus, $\dfrac{d}{dt}[\mathbf{r}(t) \times \mathbf{r}'(t)] = \mathbf{r}(t) \times \mathbf{r}''(t)$.

55. $\dfrac{d}{dt}|\mathbf{r}(t)| = \dfrac{d}{dt}[\mathbf{r}(t) \cdot \mathbf{r}(t)]^{1/2} = \frac{1}{2}[\mathbf{r}(t) \cdot \mathbf{r}(t)]^{-1/2}[2\mathbf{r}(t) \cdot \mathbf{r}'(t)] = \dfrac{1}{|\mathbf{r}(t)|}\mathbf{r}(t) \cdot \mathbf{r}'(t)$

57. Since $\mathbf{u}(t) = \mathbf{r}(t) \cdot [\mathbf{r}'(t) \times \mathbf{r}''(t)]$,

$$\mathbf{u}'(t) = \mathbf{r}'(t) \cdot [\mathbf{r}'(t) \times \mathbf{r}''(t)] + \mathbf{r}(t) \cdot \dfrac{d}{dt}[\mathbf{r}'(t) \times \mathbf{r}''(t)]$$

$$= 0 + \mathbf{r}(t) \cdot [\mathbf{r}''(t) \times \mathbf{r}''(t) + \mathbf{r}'(t) \times \mathbf{r}'''(t)] \qquad \text{[since } \mathbf{r}'(t) \perp \mathbf{r}'(t) \times \mathbf{r}''(t)]$$

$$= \mathbf{r}(t) \cdot [\mathbf{r}'(t) \times \mathbf{r}'''(t)] \qquad \text{[since } \mathbf{r}''(t) \times \mathbf{r}''(t) = \mathbf{0}]$$

13.3 Arc Length and Curvature

1. $\mathbf{r}(t) = \langle t, 3\cos t, 3\sin t \rangle \Rightarrow \mathbf{r}'(t) = \langle 1, -3\sin t, 3\cos t \rangle \Rightarrow$

$|\mathbf{r}'(t)| = \sqrt{1^2 + (-3\sin t)^2 + (3\cos t)^2} = \sqrt{1 + 9(\sin^2 t + \cos^2 t)} = \sqrt{10}.$

Then using Formula 3, we have $L = \int_{-5}^{5} |\mathbf{r}'(t)|\, dt = \int_{-5}^{5} \sqrt{10}\, dt = \sqrt{10}\, t\big]_{-5}^{5} = 10\sqrt{10}.$

3. $\mathbf{r}(t) = \sqrt{2}\, t\,\mathbf{i} + e^t\mathbf{j} + e^{-t}\mathbf{k} \Rightarrow \mathbf{r}'(t) = \sqrt{2}\,\mathbf{i} + e^t\mathbf{j} - e^{-t}\mathbf{k} \Rightarrow$

$|\mathbf{r}'(t)| = \sqrt{\left(\sqrt{2}\right)^2 + (e^t)^2 + (-e^{-t})^2} = \sqrt{2 + e^{2t} + e^{-2t}} = \sqrt{(e^t + e^{-t})^2} = e^t + e^{-t}$ [since $e^t + e^{-t} > 0$].

Then $L = \int_0^1 |\mathbf{r}'(t)|\, dt = \int_0^1 (e^t + e^{-t})\, dt = \left[e^t - e^{-t}\right]_0^1 = e - e^{-1}.$

5. $\mathbf{r}(t) = \mathbf{i} + t^2\mathbf{j} + t^3\mathbf{k} \Rightarrow \mathbf{r}'(t) = 2t\,\mathbf{j} + 3t^2\,\mathbf{k} \Rightarrow |\mathbf{r}'(t)| = \sqrt{4t^2 + 9t^4} = t\sqrt{4 + 9t^2}$ [since $t \geq 0$].

Then $L = \int_0^1 |\mathbf{r}'(t)|\, dt = \int_0^1 t\sqrt{4 + 9t^2}\, dt = \frac{1}{18} \cdot \frac{2}{3}(4 + 9t^2)^{3/2}\Big]_0^1 = \frac{1}{27}(13^{3/2} - 4^{3/2}) = \frac{1}{27}(13^{3/2} - 8).$

7. $\mathbf{r}(t) = \langle t^2, t^3, t^4 \rangle \Rightarrow \mathbf{r}'(t) = \langle 2t, 3t^2, 4t^3 \rangle \Rightarrow |\mathbf{r}'(t)| = \sqrt{(2t)^2 + (3t^2)^2 + (4t^3)^2} = \sqrt{4t^2 + 9t^4 + 16t^6}$, so

$L = \int_0^2 |\mathbf{r}'(t)|\, dt = \int_0^2 \sqrt{4t^2 + 9t^4 + 16t^6}\, dt \approx 18.6833.$

9. $\mathbf{r}(t) = \langle \cos \pi t, 2t, \sin 2\pi t \rangle \Rightarrow \mathbf{r}'(t) = \langle -\pi \sin \pi t, 2, 2\pi \cos 2\pi t \rangle \Rightarrow |\mathbf{r}'(t)| = \sqrt{\pi^2 \sin^2 \pi t + 4 + 4\pi^2 \cos^2 2\pi t}.$

The point $(1, 0, 0)$ corresponds to $t = 0$ and $(1, 4, 0)$ corresponds to $t = 2$, so the length is

$L = \int_0^2 |\mathbf{r}'(t)|\, dt = \int_0^2 \sqrt{\pi^2 \sin^2 \pi t + 4 + 4\pi^2 \cos^2 2\pi t}\, dt \approx 10.3311.$

11. The projection of the curve C onto the xy-plane is the curve $x^2 = 2y$ or $y = \frac{1}{2}x^2$, $z = 0$. Then we can choose the parameter

$x = t \Rightarrow y = \frac{1}{2}t^2$. Since C also lies on the surface $3z = xy$, we have $z = \frac{1}{3}xy = \frac{1}{3}(t)(\frac{1}{2}t^2) = \frac{1}{6}t^3$. Then parametric

equations for C are $x = t$, $y = \frac{1}{2}t^2$, $z = \frac{1}{6}t^3$ and the corresponding vector equation is $\mathbf{r}(t) = \langle t, \frac{1}{2}t^2, \frac{1}{6}t^3 \rangle$. The origin

corresponds to $t = 0$ and the point $(6, 18, 36)$ corresponds to $t = 6$, so

$$L = \int_0^6 |\mathbf{r}'(t)|\, dt = \int_0^6 \left|\langle 1, t, \tfrac{1}{2}t^2 \rangle\right|\, dt = \int_0^6 \sqrt{1^2 + t^2 + \left(\tfrac{1}{2}t^2\right)^2}\, dt = \int_0^6 \sqrt{1 + t^2 + \tfrac{1}{4}t^4}\, dt$$

$$= \int_0^6 \sqrt{(1 + \tfrac{1}{2}t^2)^2}\, dt = \int_0^6 (1 + \tfrac{1}{2}t^2)\, dt = \left[t + \tfrac{1}{6}t^3\right]_0^6 = 6 + 36 = 42$$

13. (a) $\mathbf{r}(t) = (5 - t)\,\mathbf{i} + (4t - 3)\,\mathbf{j} + 3t\,\mathbf{k} \Rightarrow \mathbf{r}'(t) = -\mathbf{i} + 4\mathbf{j} + 3\mathbf{k}$ and $\frac{ds}{dt} = |\mathbf{r}'(t)| = \sqrt{1 + 16 + 9} = \sqrt{26}$. The point

$P(4, 1, 3)$ corresponds to $t = 1$, so the arc length function from P is

$s(t) = \int_1^t |\mathbf{r}'(u)|\, du = \int_1^t \sqrt{26}\, du = \sqrt{26}\, u\big|_1^t = \sqrt{26}\,(t - 1).$ Since $s = \sqrt{26}\,(t - 1)$, we have $t = \dfrac{s}{\sqrt{26}} + 1.$

Substituting for t in the original equation, the reparametrization of the curve with respect to arc length is

$$\mathbf{r}(t(s)) = \left[5 - \left(\frac{s}{\sqrt{26}} + 1\right)\right]\mathbf{i} + \left[4\left(\frac{s}{\sqrt{26}} + 1\right) - 3\right]\mathbf{j} + 3\left(\frac{s}{\sqrt{26}} + 1\right)\mathbf{k}$$

$$= \left(4 - \frac{s}{\sqrt{26}}\right)\mathbf{i} + \left(\frac{4s}{\sqrt{26}} + 1\right)\mathbf{j} + \left(\frac{3s}{\sqrt{26}} + 3\right)\mathbf{k}$$

(b) The point 4 units along the curve from P has position vector

$$\mathbf{r}(t(4)) = \left(4 - \frac{4}{\sqrt{26}}\right)\mathbf{i} + \left(\frac{4(4)}{\sqrt{26}} + 1\right)\mathbf{j} + \left(\frac{3(4)}{\sqrt{26}} + 3\right)\mathbf{k}, \text{ so the point is } \left(4 - \frac{4}{\sqrt{26}}, \frac{16}{\sqrt{26}} + 1, \frac{12}{\sqrt{26}} + 3\right).$$

15. Here $\mathbf{r}(t) = \langle 3\sin t, 4t, 3\cos t\rangle$, so $\mathbf{r}'(t) = \langle 3\cos t, 4, -3\sin t\rangle$ and $|\mathbf{r}'(t)| = \sqrt{9\cos^2 t + 16 + 9\sin^2 t} = \sqrt{25} = 5$.

The point $(0, 0, 3)$ corresponds to $t = 0$, so the arc length function beginning at $(0, 0, 3)$ and measuring in the positive

direction is given by $s(t) = \int_0^t |\mathbf{r}'(u)|\, du = \int_0^t 5\, du = 5t$. $s(t) = 5 \ \Rightarrow \ 5t = 5 \ \Rightarrow \ t = 1$, thus your location after

moving 5 units along the curve is $(3\sin 1, 4, 3\cos 1)$.

17. (a) $\mathbf{r}(t) = \langle t, 3\cos t, 3\sin t\rangle \ \Rightarrow \ \mathbf{r}'(t) = \langle 1, -3\sin t, 3\cos t\rangle \ \Rightarrow \ |\mathbf{r}'(t)| = \sqrt{1 + 9\sin^2 t + 9\cos^2 t} = \sqrt{10}$.

Then $\mathbf{T}(t) = \dfrac{\mathbf{r}'(t)}{|\mathbf{r}'(t)|} = \dfrac{1}{\sqrt{10}} \langle 1, -3\sin t, 3\cos t\rangle$ or $\left\langle \frac{1}{\sqrt{10}}, -\frac{3}{\sqrt{10}}\sin t, \frac{3}{\sqrt{10}}\cos t\right\rangle$.

$\mathbf{T}'(t) = \dfrac{1}{\sqrt{10}} \langle 0, -3\cos t, -3\sin t\rangle \ \Rightarrow \ |\mathbf{T}'(t)| = \dfrac{1}{\sqrt{10}}\sqrt{0 + 9\cos^2 t + 9\sin^2 t} = \dfrac{3}{\sqrt{10}}$. Thus

$\mathbf{N}(t) = \dfrac{\mathbf{T}'(t)}{|\mathbf{T}'(t)|} = \dfrac{1/\sqrt{10}}{3/\sqrt{10}} \langle 0, -3\cos t, -3\sin t\rangle = \langle 0, -\cos t, -\sin t\rangle$.

(b) $\kappa(t) = \dfrac{|\mathbf{T}'(t)|}{|\mathbf{r}'(t)|} = \dfrac{3/\sqrt{10}}{\sqrt{10}} = \dfrac{3}{10}$

19. (a) $\mathbf{r}(t) = \langle \sqrt{2}\,t, e^t, e^{-t}\rangle \ \Rightarrow \ \mathbf{r}'(t) = \langle \sqrt{2}, e^t, -e^{-t}\rangle \ \Rightarrow \ |\mathbf{r}'(t)| = \sqrt{2 + e^{2t} + e^{-2t}} = \sqrt{(e^t + e^{-t})^2} = e^t + e^{-t}$.

Then

$\mathbf{T}(t) = \dfrac{\mathbf{r}'(t)}{|\mathbf{r}'(t)|} = \dfrac{1}{e^t + e^{-t}} \langle \sqrt{2}, e^t, -e^{-t}\rangle = \dfrac{1}{e^{2t} + 1} \langle \sqrt{2}\,e^t, e^{2t}, -1\rangle \quad \left[\text{after multiplying by } \dfrac{e^t}{e^t}\right] \quad$ and

$\mathbf{T}'(t) = \dfrac{1}{e^{2t} + 1} \langle \sqrt{2}\,e^t, 2e^{2t}, 0\rangle - \dfrac{2e^{2t}}{(e^{2t} + 1)^2} \langle \sqrt{2}\,e^t, e^{2t}, -1\rangle$

$\qquad = \dfrac{1}{(e^{2t} + 1)^2} \left[(e^{2t} + 1)\langle \sqrt{2}\,e^t, 2e^{2t}, 0\rangle - 2e^{2t}\langle \sqrt{2}\,e^t, e^{2t}, -1\rangle\right] = \dfrac{1}{(e^{2t} + 1)^2} \langle \sqrt{2}\,e^t\,(1 - e^{2t}), 2e^{2t}, 2e^{2t}\rangle$

Then

$$|\mathbf{T}'(t)| = \dfrac{1}{(e^{2t} + 1)^2}\sqrt{2e^{2t}(1 - 2e^{2t} + e^{4t}) + 4e^{4t} + 4e^{4t}} = \dfrac{1}{(e^{2t} + 1)^2}\sqrt{2e^{2t}(1 + 2e^{2t} + e^{4t})}$$

$$= \dfrac{1}{(e^{2t} + 1)^2}\sqrt{2e^{2t}(1 + e^{2t})^2} = \dfrac{\sqrt{2}\,e^t(1 + e^{2t})}{(e^{2t} + 1)^2} = \dfrac{\sqrt{2}\,e^t}{e^{2t} + 1}$$

Therefore

$$\mathbf{N}(t) = \dfrac{\mathbf{T}'(t)}{|\mathbf{T}'(t)|} = \dfrac{e^{2t} + 1}{\sqrt{2}\,e^t}\dfrac{1}{(e^{2t} + 1)^2}\langle \sqrt{2}\,e^t(1 - e^{2t}), 2e^{2t}, 2e^{2t}\rangle$$

$$= \dfrac{1}{\sqrt{2}\,e^t(e^{2t} + 1)}\langle \sqrt{2}\,e^t(1 - e^{2t}), 2e^{2t}, 2e^{2t}\rangle = \dfrac{1}{e^{2t} + 1}\langle 1 - e^{2t}, \sqrt{2}\,e^t, \sqrt{2}\,e^t\rangle$$

(b) $\kappa(t) = \dfrac{|\mathbf{T}'(t)|}{|\mathbf{r}'(t)|} = \dfrac{\sqrt{2}\,e^t}{e^{2t} + 1}\cdot\dfrac{1}{e^t + e^{-t}} = \dfrac{\sqrt{2}\,e^t}{e^{3t} + 2e^t + e^{-t}} = \dfrac{\sqrt{2}\,e^{2t}}{e^{4t} + 2e^{2t} + 1} = \dfrac{\sqrt{2}\,e^{2t}}{(e^{2t} + 1)^2}$

21. $\mathbf{r}(t) = t^3\,\mathbf{j} + t^2\,\mathbf{k}$ $\Rightarrow$ $\mathbf{r}'(t) = 3t^2\,\mathbf{j} + 2t\,\mathbf{k}$, $\mathbf{r}''(t) = 6t\,\mathbf{j} + 2\,\mathbf{k}$, $|\mathbf{r}'(t)| = \sqrt{0^2 + (3t^2)^2 + (2t)^2} = \sqrt{9t^4 + 4t^2}$,

$\mathbf{r}'(t) \times \mathbf{r}''(t) = -6t^2\,\mathbf{i}$, $|\mathbf{r}'(t) \times \mathbf{r}''(t)| = 6t^2$. Then $\kappa(t) = \dfrac{|\mathbf{r}'(t) \times \mathbf{r}''(t)|}{|\mathbf{r}'(t)|^3} = \dfrac{6t^2}{\left(\sqrt{9t^4 + 4t^2}\,\right)^3} = \dfrac{6t^2}{(9t^4 + 4t^2)^{3/2}}$.

23. $\mathbf{r}(t) = \sqrt{6}\,t^2\,\mathbf{i} + 2t\,\mathbf{j} + 2t^3\,\mathbf{k}$ $\Rightarrow$ $\mathbf{r}'(t) = 2\sqrt{6}\,t\,\mathbf{i} + 2\,\mathbf{j} + 6t^2\,\mathbf{k}$, $\mathbf{r}''(t) = 2\sqrt{6}\,\mathbf{i} + 12t\,\mathbf{k}$,

$|\mathbf{r}'(t)| = \sqrt{24t^2 + 4 + 36t^4} = \sqrt{4(9t^4 + 6t^2 + 1)} = \sqrt{4(3t^2 + 1)^2} = 2(3t^2 + 1)$,

$\mathbf{r}'(t) \times \mathbf{r}''(t) = 24t\,\mathbf{i} - 12\sqrt{6}\,t^2\,\mathbf{j} - 4\sqrt{6}\,\mathbf{k}$,

$|\mathbf{r}'(t) \times \mathbf{r}''(t)| = \sqrt{576t^2 + 864t^4 + 96} = \sqrt{96(9t^4 + 6t^2 + 1)} = \sqrt{96(3t^2 + 1)^2} = 4\sqrt{6}\,(3t^2 + 1)$.

Then $\kappa(t) = \dfrac{|\mathbf{r}'(t) \times \mathbf{r}''(t)|}{|\mathbf{r}'(t)|^3} = \dfrac{4\sqrt{6}\,(3t^2 + 1)}{8(3t^2 + 1)^3} = \dfrac{\sqrt{6}}{2(3t^2 + 1)^2}$.

25. $\mathbf{r}(t) = \langle t, t^2, t^3 \rangle$ $\Rightarrow$ $\mathbf{r}'(t) = \langle 1, 2t, 3t^2 \rangle$. The point $(1, 1, 1)$ corresponds to $t = 1$, and $\mathbf{r}'(1) = \langle 1, 2, 3 \rangle$ $\Rightarrow$

$|\mathbf{r}'(1)| = \sqrt{1 + 4 + 9} = \sqrt{14}$. $\mathbf{r}''(t) = \langle 0, 2, 6t \rangle$ $\Rightarrow$ $\mathbf{r}''(1) = \langle 0, 2, 6 \rangle$. $\mathbf{r}'(1) \times \mathbf{r}''(1) = \langle 6, -6, 2 \rangle$, so

$|\mathbf{r}'(1) \times \mathbf{r}''(1)| = \sqrt{36 + 36 + 4} = \sqrt{76}$. Then $\kappa(1) = \dfrac{|\mathbf{r}'(1) \times \mathbf{r}''(1)|}{|\mathbf{r}'(1)|^3} = \dfrac{\sqrt{76}}{\sqrt{14}^3} = \dfrac{1}{7}\sqrt{\dfrac{19}{14}}$.

27. $f(x) = x^4$, $f'(x) = 4x^3$, $f''(x) = 12x^2$, $\kappa(x) = \dfrac{|f''(x)|}{[1 + (f'(x))^2]^{3/2}} = \dfrac{|12x^2|}{[1 + (4x^3)^2]^{3/2}} = \dfrac{12x^2}{(1 + 16x^6)^{3/2}}$

29. $f(x) = xe^x$, $f'(x) = xe^x + e^x$, $f''(x) = xe^x + 2e^x$,

$\kappa(x) = \dfrac{|f''(x)|}{[1 + (f'(x))^2]^{3/2}} = \dfrac{|xe^x + 2e^x|}{[1 + (xe^x + e^x)^2]^{3/2}} = \dfrac{|x + 2|\,e^x}{[1 + (xe^x + e^x)^2]^{3/2}}$

31. Since $y' = y'' = e^x$, the curvature is $\kappa(x) = \dfrac{|y''(x)|}{[1 + (y'(x))^2]^{3/2}} = \dfrac{e^x}{(1 + e^{2x})^{3/2}} = e^x(1 + e^{2x})^{-3/2}$.

To find the maximum curvature, we first find the critical numbers of $\kappa(x)$:

$\kappa'(x) = e^x(1 + e^{2x})^{-3/2} + e^x\left(-\tfrac{3}{2}\right)(1 + e^{2x})^{-5/2}(2e^{2x}) = e^x\dfrac{1 + e^{2x} - 3e^{2x}}{(1 + e^{2x})^{5/2}} = e^x\dfrac{1 - 2e^{2x}}{(1 + e^{2x})^{5/2}}$.

$\kappa'(x) = 0$ when $1 - 2e^{2x} = 0$, so $e^{2x} = \tfrac{1}{2}$ or $x = -\tfrac{1}{2}\ln 2$. And since $1 - 2e^{2x} > 0$ for $x < -\tfrac{1}{2}\ln 2$ and $1 - 2e^{2x} < 0$

for $x > -\tfrac{1}{2}\ln 2$, the maximum curvature is attained at the point $\left(-\tfrac{1}{2}\ln 2, e^{(-\ln 2)/2}\right) = \left(-\tfrac{1}{2}\ln 2, \tfrac{1}{\sqrt{2}}\right)$.

Since $\lim\limits_{x \to \infty} e^x(1 + e^{2x})^{-3/2} = 0$, $\kappa(x)$ approaches 0 as $x \to \infty$.

33. (a) C appears to be changing direction more quickly at P than Q, so we would expect the curvature to be greater at P.

(b) First we sketch approximate osculating circles at P and Q. Using the

axes scale as a guide, we measure the radius of the osculating circle

at P to be approximately 0.8 units, thus $\rho = \dfrac{1}{\kappa}$ $\Rightarrow$

$\kappa = \dfrac{1}{\rho} \approx \dfrac{1}{0.8} \approx 1.3$. Similarly, we estimate the radius of the

osculating circle at Q to be 1.4 units, so $\kappa = \dfrac{1}{\rho} \approx \dfrac{1}{1.4} \approx 0.7$.

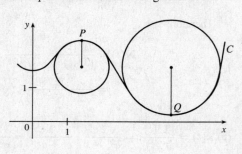

35. $y = x^{-2} \implies y' = -2x^{-3}, \quad y'' = 6x^{-4},$ and

$$\kappa(x) = \frac{|y''|}{[1 + (y')^2]^{3/2}} = \frac{|6x^{-4}|}{[1 + (-2x^{-3})^2]^{3/2}} = \frac{6}{x^4(1 + 4x^{-6})^{3/2}}.$$

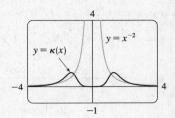

The appearance of the two humps in this graph is perhaps a little surprising, but it is

explained by the fact that $y = x^{-2}$ increases asymptotically at the origin from both

directions, and so its graph has very little bend there. [Note that $\kappa(0)$ is undefined.]

37. $\mathbf{r}(t) = \langle te^t, e^{-t}, \sqrt{2}\,t \rangle \implies \mathbf{r}'(t) = \langle (t+1)e^t, -e^{-t}, \sqrt{2} \rangle, \quad \mathbf{r}''(t) = \langle (t+2)e^t, e^{-t}, 0 \rangle.$ Then

$\mathbf{r}'(t) \times \mathbf{r}''(t) = \langle -\sqrt{2}\,e^{-t}, \sqrt{2}(t+2)e^t, 2t+3 \rangle, \quad |\mathbf{r}'(t) \times \mathbf{r}''(t)| = \sqrt{2e^{-2t} + 2(t+2)^2 e^{2t} + (2t+3)^2},$

$|\mathbf{r}'(t)| = \sqrt{(t+1)^2 e^{2t} + e^{-2t} + 2},$ and $\kappa(t) = \dfrac{|\mathbf{r}'(t) \times \mathbf{r}''(t)|}{|\mathbf{r}'(t)|^3} = \dfrac{\sqrt{2e^{-2t} + 2(t+2)^2 e^{2t} + (2t+3)^2}}{[(t+1)^2 e^{2t} + e^{-2t} + 2]^{3/2}}.$

We plot the space curve and its curvature function for $-5 \leq t \leq 5$ below.

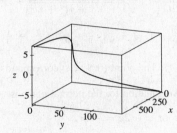

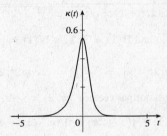

From the graph of $\kappa(t)$ we see that curvature is maximized for $t = 0$, so the curve bends most sharply at the point $(0, 1, 0)$.

The curve bends more gradually as we move away from this point, becoming almost linear. This is reflected in the curvature

graph, where $\kappa(t)$ becomes nearly 0 as $|t|$ increases.

39. Notice that the curve b has two inflection points at which the graph appears almost straight. We would expect the curvature to

be 0 or nearly 0 at these values, but the curve a isn't near 0 there. Thus, a must be the graph of $y = f(x)$ rather than the graph

of curvature, and b is the graph of $y = \kappa(x)$.

41. Using a CAS, we find (after simplifying)

$\kappa(t) = \dfrac{6\sqrt{4\cos^2 t - 12\cos t + 13}}{(17 - 12\cos t)^{3/2}}.$ (To compute cross

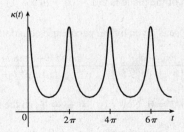

products in Maple, use the `VectorCalculus` or

`LinearAlgebra` package and the `CrossProduct(a,b)`

command; in Mathematica, use `Cross[a,b]`.) Curvature is

largest at integer multiples of 2π.

43. $x = t^2 \implies \dot{x} = 2t \implies \ddot{x} = 2, \quad y = t^3 \implies \dot{y} = 3t^2 \implies \ddot{y} = 6t.$

Then $\kappa(t) = \dfrac{|\dot{x}\ddot{y} - \dot{y}\ddot{x}|}{[\dot{x}^2 + \dot{y}^2]^{3/2}} = \dfrac{|(2t)(6t) - (3t^2)(2)|}{[(2t)^2 + (3t^2)^2]^{3/2}} = \dfrac{|12t^2 - 6t^2|}{(4t^2 + 9t^4)^{3/2}} = \dfrac{6t^2}{(4t^2 + 9t^4)^{3/2}}.$

45. $x = e^t \cos t \;\Rightarrow\; \dot{x} = e^t(\cos t - \sin t) \;\Rightarrow\; \ddot{x} = e^t(-\sin t - \cos t) + e^t(\cos t - \sin t) = -2e^t \sin t,$

$y = e^t \sin t \;\Rightarrow\; \dot{y} = e^t(\cos t + \sin t) \;\Rightarrow\; \ddot{y} = e^t(-\sin t + \cos t) + e^t(\cos t + \sin t) = 2e^t \cos t.$ Then

$$\kappa(t) = \frac{|\dot{x}\ddot{y} - \dot{y}\ddot{x}|}{[\dot{x}^2 + \dot{y}^2]^{3/2}} = \frac{\left|e^t(\cos t - \sin t)(2e^t \cos t) - e^t(\cos t + \sin t)(-2e^t \sin t)\right|}{\left([e^t(\cos t - \sin t)]^2 + [e^t(\cos t + \sin t)]^2\right)^{3/2}}$$

$$= \frac{\left|2e^{2t}(\cos^2 t - \sin t \cos t + \sin t \cos t + \sin^2 t)\right|}{\left[e^{2t}(\cos^2 t - 2\cos t \sin t + \sin^2 t + \cos^2 t + 2\cos t \sin t + \sin^2 t)\right]^{3/2}} = \frac{\left|2e^{2t}(1)\right|}{[e^{2t}(1+1)]^{3/2}} = \frac{2e^{2t}}{e^{3t}(2)^{3/2}} = \frac{1}{\sqrt{2}\,e^t}$$

47. $\left(1, \frac{2}{3}, 1\right)$ corresponds to $t = 1$. $\mathbf{T}(t) = \dfrac{\mathbf{r}'(t)}{|\mathbf{r}'(t)|} = \dfrac{\langle 2t, 2t^2, 1\rangle}{\sqrt{4t^2 + 4t^4 + 1}} = \dfrac{\langle 2t, 2t^2, 1\rangle}{2t^2 + 1}$, so $\mathbf{T}(1) = \left\langle \frac{2}{3}, \frac{2}{3}, \frac{1}{3}\right\rangle$.

$\mathbf{T}'(t) = -4t(2t^2 + 1)^{-2}\,\langle 2t, 2t^2, 1\rangle + (2t^2 + 1)^{-1}\,\langle 2, 4t, 0\rangle$ \quad [by Formula 3 of Theorem 13.2.3]

$\qquad = (2t^2 + 1)^{-2}\,\langle -8t^2 + 4t^2 + 2, -8t^3 + 8t^3 + 4t, -4t\rangle = 2(2t^2 + 1)^{-2}\,\langle 1 - 2t^2, 2t, -2t\rangle$

$\mathbf{N}(t) = \dfrac{\mathbf{T}'(t)}{|\mathbf{T}'(t)|} = \dfrac{2(2t^2 + 1)^{-2}\,\langle 1 - 2t^2, 2t, -2t\rangle}{2(2t^2 + 1)^{-2}\sqrt{(1 - 2t^2)^2 + (2t)^2 + (-2t)^2}} = \dfrac{\langle 1 - 2t^2, 2t, -2t\rangle}{\sqrt{1 - 4t^2 + 4t^4 + 8t^2}} = \dfrac{\langle 1 - 2t^2, 2t, -2t\rangle}{1 + 2t^2}$

$\mathbf{N}(1) = \left\langle -\frac{1}{3}, \frac{2}{3}, -\frac{2}{3}\right\rangle$ and $\mathbf{B}(1) = \mathbf{T}(1) \times \mathbf{N}(1) = \left\langle -\frac{4}{9} - \frac{2}{9}, -\left(-\frac{4}{9} + \frac{1}{9}\right), \frac{4}{9} + \frac{2}{9}\right\rangle = \left\langle -\frac{2}{3}, \frac{1}{3}, \frac{2}{3}\right\rangle$.

49. $\mathbf{r}(t) = \langle \sin 2t, -\cos 2t, 4t\rangle \;\Rightarrow\; \mathbf{r}'(t) = \langle 2\cos 2t, 2\sin 2t, 4\rangle$. The point $(0, 1, 2\pi)$ corresponds to $t = \pi/2$, and the

normal plane there has normal vector $\mathbf{r}'(\pi/2) = \langle -2, 0, 4\rangle$. An equation for the normal plane is

$-2(x - 0) + 0(y - 1) + 4(z - 2\pi) = 0$ or $-2x + 4z = 8\pi$ or $x - 2z = -4\pi$.

$\mathbf{T}(t) = \dfrac{\mathbf{r}'(t)}{|\mathbf{r}'(t)|} = \dfrac{\langle 2\cos 2t, 2\sin 2t, 4\rangle}{\sqrt{4\cos^2 2t + 4\sin^2 2t + 16}} = \dfrac{1}{2\sqrt{5}}\,\langle 2\cos 2t, 2\sin 2t, 4\rangle = \dfrac{1}{\sqrt{5}}\,\langle \cos 2t, \sin 2t, 2\rangle \;\Rightarrow$

$\mathbf{T}'(t) = \frac{1}{\sqrt{5}}\,\langle -2\sin 2t, 2\cos 2t, 0\rangle \;\Rightarrow\; |\mathbf{T}'(t)| = \frac{1}{\sqrt{5}}\sqrt{4\sin^2 2t + 4\cos^2 2t} = \frac{2}{\sqrt{5}},$ and

$\mathbf{N}(t) = \dfrac{\mathbf{T}'(t)}{|\mathbf{T}'(t)|} = \langle -\sin 2t, \cos 2t, 0\rangle.$ Then $\mathbf{T}(\pi/2) = \frac{1}{\sqrt{5}}\,\langle -1, 0, 2\rangle,\ \mathbf{N}(\pi/2) = \langle 0, -1, 0\rangle,$ and

$\mathbf{B}(\pi/2) = \mathbf{T}(\pi/2) \times \mathbf{N}(\pi/2) = \frac{1}{\sqrt{5}}\,\langle 2, 0, 1\rangle.$ Since $\mathbf{B}(\pi/2)$ is normal to the osculating plane, so is $\langle 2, 0, 1\rangle$, and an

equation of the plane is $2(x - 0) + 0(y - 1) + 1(z - 2\pi) = 0$ or $2x + z = 2\pi.$

51. The ellipse is given by the parametric equations $x = 2\cos t,\ y = 3\sin t$, so using the result from Exercise 42,

$$\kappa(t) = \frac{|\dot{x}\ddot{y} - \ddot{x}\dot{y}|}{[\dot{x}^2 + \dot{y}^2]^{3/2}} = \frac{|(-2\sin t)(-3\sin t) - (3\cos t)(-2\cos t)|}{(4\sin^2 t + 9\cos^2 t)^{3/2}} = \frac{6}{(4\sin^2 t + 9\cos^2 t)^{3/2}}.$$

At $(2, 0)$, $t = 0$. Now $\kappa(0) = \frac{6}{27} = \frac{2}{9}$, so the radius of the osculating circle is

$1/\kappa(0) = \frac{9}{2}$ and its center is $\left(-\frac{5}{2}, 0\right)$. Its equation is therefore $\left(x + \frac{5}{2}\right)^2 + y^2 = \frac{81}{4}$.

At $(0, 3)$, $t = \frac{\pi}{2}$, and $\kappa\left(\frac{\pi}{2}\right) = \frac{6}{8} = \frac{3}{4}$. So the radius of the osculating circle is $\frac{4}{3}$ and

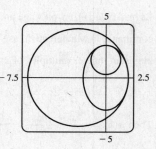

its center is $\left(0, \frac{5}{3}\right)$. Hence its equation is $x^2 + \left(y - \frac{5}{3}\right)^2 = \frac{16}{9}$.

53. Here $\mathbf{r}(t) = \langle t^3, 3t, t^4 \rangle$, and $\mathbf{r}'(t) = \langle 3t^2, 3, 4t^3 \rangle$ is normal to the normal plane for any t. The given plane has normal vector

$\langle 6, 6, -8 \rangle$, and the planes are parallel when their normal vectors are parallel. Thus we need to find a value for t where

$\langle 3t^2, 3, 4t^3 \rangle = k \langle 6, 6, -8 \rangle$ for some $k \neq 0$. From the y-component we see that $k = \frac{1}{2}$, and

$\langle 3t^2, 3, 4t^3 \rangle = \frac{1}{2} \langle 6, 6, -8 \rangle = \langle 3, 3, -4 \rangle$ for $t = -1$. Thus the planes are parallel at the point $(-1, -3, 1)$.

55. First we parametrize the curve of intersection. We can choose $y = t$; then $x = y^2 = t^2$ and $z = x^2 = t^4$, and the curve is

given by $\mathbf{r}(t) = \langle t^2, t, t^4 \rangle$. $\mathbf{r}'(t) = \langle 2t, 1, 4t^3 \rangle$ and the point $(1, 1, 1)$ corresponds to $t = 1$, so $\mathbf{r}'(1) = \langle 2, 1, 4 \rangle$ is a normal

vector for the normal plane. Thus an equation of the normal plane is

$2(x - 1) + 1(y - 1) + 4(z - 1) = 0$ or $2x + y + 4z = 7$. $\mathbf{T}(t) = \dfrac{\mathbf{r}'(t)}{|\mathbf{r}'(t)|} = \dfrac{1}{\sqrt{4t^2 + 1 + 16t^6}} \langle 2t, 1, 4t^3 \rangle$ and

$\mathbf{T}'(t) = -\frac{1}{2}(4t^2 + 1 + 16t^6)^{-3/2}(8t + 96t^5) \langle 2t, 1, 4t^3 \rangle + (4t^2 + 1 + 16t^6)^{-1/2} \langle 2, 0, 12t^2 \rangle$. A normal vector for

the osculating plane is $\mathbf{B}(1) = \mathbf{T}(1) \times \mathbf{N}(1)$, but $\mathbf{r}'(1) = \langle 2, 1, 4 \rangle$ is parallel to $\mathbf{T}(1)$ and

$\mathbf{T}'(1) = -\frac{1}{2}(21)^{-3/2}(104)\langle 2, 1, 4 \rangle + (21)^{-1/2}\langle 2, 0, 12 \rangle = \frac{2}{21\sqrt{21}} \langle -31, -26, 22 \rangle$ is parallel to $\mathbf{N}(1)$ as is $\langle -31, -26, 22 \rangle$,

so $\langle 2, 1, 4 \rangle \times \langle -31, -26, 22 \rangle = \langle 126, -168, -21 \rangle$ is normal to the osculating plane. Thus an equation for the osculating

plane is $126(x - 1) - 168(y - 1) - 21(z - 1) = 0$ or $6x - 8y - z = -3$.

57. $\mathbf{r}(t) = \langle e^t \cos t, e^t \sin t, e^t \rangle$ $\Rightarrow$ $\mathbf{r}'(t) = \langle e^t(\cos t - \sin t), e^t(\cos t + \sin t), e^t \rangle$ so

$$|\mathbf{r}'(t)| = \sqrt{e^{2t}(\cos t - \sin t)^2 + e^{2t}(\cos t + \sin t)^2 + e^{2t}}$$

$$= \sqrt{e^{2t} \left[2(\cos^2 t + \sin^2 t) - 2\cos t \sin t + 2\cos t \sin t + 1 \right]} = \sqrt{3e^{2t}} = \sqrt{3}\, e^t$$

and $\mathbf{T}(t) = \dfrac{\mathbf{r}'(t)}{|\mathbf{r}'(t)|} = \dfrac{1}{\sqrt{3}\, e^t} \langle e^t(\cos t - \sin t), e^t(\cos t + \sin t), e^t \rangle = \dfrac{1}{\sqrt{3}} \langle \cos t - \sin t, \cos t + \sin t, 1 \rangle$. The vector

$\mathbf{k} = \langle 0, 0, 1 \rangle$ is parallel to the z-axis, so for any t, the angle α between $\mathbf{T}(t)$ and the z-axis is given by

$$\cos \alpha = \frac{\mathbf{T}(t) \cdot \mathbf{k}}{|\mathbf{T}(t)|\, |\mathbf{k}|} = \frac{\frac{1}{\sqrt{3}} \langle \cos t - \sin t, \cos t + \sin t, 1 \rangle \cdot \langle 0, 0, 1 \rangle}{\frac{1}{\sqrt{3}} \sqrt{(\cos t - \sin t)^2 + (\cos t + \sin t)^2 + 1}\, \sqrt{1}} = \frac{1}{\sqrt{2(\cos^2 t + \sin^2 t) + 1}} = \frac{1}{\sqrt{3}}. \text{ Thus the angle}$$

is constant; specifically, $\alpha = \cos^{-1}(1/\sqrt{3}) \approx 54.7\,°$.

$\mathbf{N}(t) = \dfrac{\mathbf{T}'(t)}{|\mathbf{T}'(t)|} = \dfrac{(1/\sqrt{3})\,\langle -\sin t - \cos t, -\sin t + \cos t, 0 \rangle}{(1/\sqrt{3})\sqrt{2\,(\sin^2 t + \cos^2 t)}} = \dfrac{1}{\sqrt{2}} \langle -\sin t - \cos t, -\sin t + \cos t, 0 \rangle$, and the angle β

made with the z-axis is given by $\cos \beta = \dfrac{\mathbf{N}(t) \cdot \mathbf{k}}{|\mathbf{N}(t)|\, |\mathbf{k}|} = 0$, so $\beta = 90\,°$.

$\mathbf{B}(t) = \mathbf{T}(t) \times \mathbf{N}(t) = \frac{1}{\sqrt{6}} \langle \sin t - \cos t, -\sin t - \cos t, 2 \rangle$ and the angle γ made with the z-axis is given by

$$\cos \gamma = \frac{\mathbf{B}(t) \cdot \mathbf{k}}{|\mathbf{B}(t)|\, |\mathbf{k}|} = \frac{\frac{1}{\sqrt{6}} \langle \sin t - \cos t, -\sin t - \cos t, 2 \rangle \cdot \langle 0, 0, 1 \rangle}{\frac{1}{\sqrt{6}} \sqrt{(\sin t - \cos t)^2 + (-\sin t - \cos t)^2 + 4}\, \sqrt{1}} = \frac{2}{\sqrt{6}} \text{ or equivalently } \frac{\sqrt{6}}{3}. \text{ Again the angle is}$$

constant; specifically, $\gamma = \cos^{-1}(2/\sqrt{6}) \approx 35.3\,°$.

59. $\kappa = \left|\dfrac{d\mathbf{T}}{ds}\right| = \left|\dfrac{d\mathbf{T}/dt}{ds/dt}\right| = \dfrac{|d\mathbf{T}/dt|}{ds/dt}$ and $\mathbf{N} = \dfrac{d\mathbf{T}/dt}{|d\mathbf{T}/dt|}$, so $\kappa\mathbf{N} = \dfrac{\left|\dfrac{d\mathbf{T}}{dt}\right|}{\left|\dfrac{d\mathbf{T}}{dt}\right|}\dfrac{\dfrac{d\mathbf{T}}{dt}}{\dfrac{ds}{dt}} = \dfrac{d\mathbf{T}/dt}{ds/dt} = \dfrac{d\mathbf{T}}{ds}$ by the Chain Rule.

61. (a) $|\mathbf{B}| = 1 \;\Rightarrow\; \mathbf{B}\cdot\mathbf{B} = 1 \;\Rightarrow\; \dfrac{d}{ds}(\mathbf{B}\cdot\mathbf{B}) = 0 \;\Rightarrow\; 2\dfrac{d\mathbf{B}}{ds}\cdot\mathbf{B} = 0 \;\Rightarrow\; \dfrac{d\mathbf{B}}{ds} \perp \mathbf{B}$

(b) $\mathbf{B} = \mathbf{T}\times\mathbf{N} \;\Rightarrow\;$

$$\dfrac{d\mathbf{B}}{ds} = \dfrac{d}{ds}(\mathbf{T}\times\mathbf{N}) = \dfrac{d}{dt}(\mathbf{T}\times\mathbf{N})\dfrac{1}{ds/dt} = \dfrac{d}{dt}(\mathbf{T}\times\mathbf{N})\dfrac{1}{|\mathbf{r}'(t)|} = [(\mathbf{T}'\times\mathbf{N}) + (\mathbf{T}\times\mathbf{N}')]\dfrac{1}{|\mathbf{r}'(t)|}$$

$$= \left[\left(\mathbf{T}'\times\dfrac{\mathbf{T}'}{|\mathbf{T}'|}\right) + (\mathbf{T}\times\mathbf{N}')\right]\dfrac{1}{|\mathbf{r}'(t)|} = \dfrac{\mathbf{T}\times\mathbf{N}'}{|\mathbf{r}'(t)|} \;\Rightarrow\; \dfrac{d\mathbf{B}}{ds} \perp \mathbf{T}$$

(c) $\mathbf{B} = \mathbf{T}\times\mathbf{N} \;\Rightarrow\; \mathbf{T}\perp\mathbf{N}, \mathbf{B}\perp\mathbf{T}$ and $\mathbf{B}\perp\mathbf{N}$. So $\mathbf{B}, \mathbf{T}$ and $\mathbf{N}$ form an orthogonal set of vectors in the three-dimensional space $\mathbb{R}^3$. From parts (a) and (b), $d\mathbf{B}/ds$ is perpendicular to both $\mathbf{B}$ and $\mathbf{T}$, so $d\mathbf{B}/ds$ is parallel to $\mathbf{N}$. Therefore, $d\mathbf{B}/ds = -\tau(s)\mathbf{N}$, where $\tau(s)$ is a scalar.

(d) Since $\mathbf{B} = \mathbf{T}\times\mathbf{N}, \mathbf{T}\perp\mathbf{N}$ and both $\mathbf{T}$ and $\mathbf{N}$ are unit vectors, $\mathbf{B}$ is a unit vector mutually perpendicular to both $\mathbf{T}$ and $\mathbf{N}$. For a plane curve, $\mathbf{T}$ and $\mathbf{N}$ always lie in the plane of the curve, so that $\mathbf{B}$ is a constant unit vector always perpendicular to the plane. Thus $d\mathbf{B}/ds = \mathbf{0}$, but $d\mathbf{B}/ds = -\tau(s)\mathbf{N}$ and $\mathbf{N}\ne\mathbf{0}$, so $\tau(s) = 0$.

63. (a) $\mathbf{r}' = s'\mathbf{T} \;\Rightarrow\; \mathbf{r}'' = s''\mathbf{T} + s'\mathbf{T}' = s''\mathbf{T} + s'\dfrac{d\mathbf{T}}{ds}s' = s''\mathbf{T} + \kappa(s')^2\mathbf{N}$ by the first Serret-Frenet formula.

(b) Using part (a), we have

$$\mathbf{r}'\times\mathbf{r}'' = (s'\mathbf{T})\times[s''\mathbf{T} + \kappa(s')^2\mathbf{N}]$$

$$= [(s'\mathbf{T})\times(s''\mathbf{T})] + [(s'\mathbf{T})\times(\kappa(s')^2\mathbf{N})] \qquad \text{[by Property 3 of Theorem 12.4.11]}$$

$$= (s's'')(\mathbf{T}\times\mathbf{T}) + \kappa(s')^3(\mathbf{T}\times\mathbf{N}) = \mathbf{0} + \kappa(s')^3\mathbf{B} = \kappa(s')^3\mathbf{B}$$

(c) Using part (a), we have

$$\mathbf{r}''' = [s''\mathbf{T} + \kappa(s')^2\mathbf{N}]' = s'''\mathbf{T} + s''\mathbf{T}' + \kappa'(s')^2\mathbf{N} + 2\kappa s's''\mathbf{N} + \kappa(s')^2\mathbf{N}'$$

$$= s'''\mathbf{T} + s''\dfrac{d\mathbf{T}}{ds}s' + \kappa'(s')^2\mathbf{N} + 2\kappa s's''\mathbf{N} + \kappa(s')^2\dfrac{d\mathbf{N}}{ds}s'$$

$$= s'''\mathbf{T} + s''s'\kappa\mathbf{N} + \kappa'(s')^2\mathbf{N} + 2\kappa s's''\mathbf{N} + \kappa(s')^3(-\kappa\mathbf{T} + \tau\mathbf{B}) \qquad \text{[by the second formula]}$$

$$= [s''' - \kappa^2(s')^3]\mathbf{T} + [3\kappa s's'' + \kappa'(s')^2]\mathbf{N} + \kappa\tau(s')^3\mathbf{B}$$

(d) Using parts (b) and (c) and the facts that $\mathbf{B}\cdot\mathbf{T} = 0, \mathbf{B}\cdot\mathbf{N} = 0$, and $\mathbf{B}\cdot\mathbf{B} = 1$, we get

$$\dfrac{(\mathbf{r}'\times\mathbf{r}'')\cdot\mathbf{r}'''}{|\mathbf{r}'\times\mathbf{r}''|^2} = \dfrac{\kappa(s')^3\mathbf{B}\cdot\{[s''' - \kappa^2(s')^3]\mathbf{T} + [3\kappa s's'' + \kappa'(s')^2]\mathbf{N} + \kappa\tau(s')^3\mathbf{B}\}}{|\kappa(s')^3\mathbf{B}|^2} = \dfrac{\kappa(s')^3\kappa\tau(s')^3}{[\kappa(s')^3]^2} = \tau.$$

65. $\mathbf{r} = \langle t, \tfrac{1}{2}t^2, \tfrac{1}{3}t^3\rangle \;\Rightarrow\; \mathbf{r}' = \langle 1, t, t^2\rangle, \mathbf{r}'' = \langle 0, 1, 2t\rangle, \mathbf{r}''' = \langle 0, 0, 2\rangle \;\Rightarrow\; \mathbf{r}'\times\mathbf{r}'' = \langle t^2, -2t, 1\rangle \;\Rightarrow\;$

$$\tau = \dfrac{(\mathbf{r}'\times\mathbf{r}'')\cdot\mathbf{r}'''}{|\mathbf{r}'\times\mathbf{r}''|^2} = \dfrac{\langle t^2, -2t, 1\rangle\cdot\langle 0, 0, 2\rangle}{t^4 + 4t^2 + 1} = \dfrac{2}{t^4 + 4t^2 + 1}$$

67. For one helix, the vector equation is $\mathbf{r}(t) = \langle 10\cos t, 10\sin t, 34t/(2\pi) \rangle$ (measuring in angstroms), because the radius of each helix is 10 angstroms, and z increases by 34 angstroms for each increase of 2π in t. Using the arc length formula, letting t go from 0 to $2.9 \times 10^8 \times 2\pi$, we find the approximate length of each helix to be

$$L = \int_0^{2.9\times10^8\times2\pi} |\mathbf{r}'(t)|\,dt = \int_0^{2.9\times10^8\times2\pi} \sqrt{(-10\sin t)^2 + (10\cos t)^2 + \left(\tfrac{34}{2\pi}\right)^2}\,dt = \sqrt{100 + \left(\tfrac{34}{2\pi}\right)^2}\,t\,\bigg]_0^{2.9\times10^8\times2\pi}$$

$$= 2.9 \times 10^8 \times 2\pi \sqrt{100 + \left(\tfrac{34}{2\pi}\right)^2} \approx 2.07 \times 10^{10}\ \text{Å} \text{ — more than two meters!}$$

13.4 Motion in Space: Velocity and Acceleration

1. (a) If $\mathbf{r}(t) = x(t)\,\mathbf{i} + y(t)\,\mathbf{j} + z(t)\,\mathbf{k}$ is the position vector of the particle at time t, then the average velocity over the time interval $[0, 1]$ is

$$\mathbf{v}_{\text{ave}} = \frac{\mathbf{r}(1) - \mathbf{r}(0)}{1 - 0} = \frac{(4.5\,\mathbf{i} + 6.0\,\mathbf{j} + 3.0\,\mathbf{k}) - (2.7\,\mathbf{i} + 9.8\,\mathbf{j} + 3.7\,\mathbf{k})}{1} = 1.8\,\mathbf{i} - 3.8\,\mathbf{j} - 0.7\,\mathbf{k}$$

Similarly, over the other intervals we have

$$[0.5, 1]: \quad \mathbf{v}_{\text{ave}} = \frac{\mathbf{r}(1) - \mathbf{r}(0.5)}{1 - 0.5} = \frac{(4.5\,\mathbf{i} + 6.0\,\mathbf{j} + 3.0\,\mathbf{k}) - (3.5\,\mathbf{i} + 7.2\,\mathbf{j} + 3.3\,\mathbf{k})}{0.5} = 2.0\,\mathbf{i} - 2.4\,\mathbf{j} - 0.6\,\mathbf{k}$$

$$[1, 2]: \quad \mathbf{v}_{\text{ave}} = \frac{\mathbf{r}(2) - \mathbf{r}(1)}{2 - 1} = \frac{(7.3\,\mathbf{i} + 7.8\,\mathbf{j} + 2.7\,\mathbf{k}) - (4.5\,\mathbf{i} + 6.0\,\mathbf{j} + 3.0\,\mathbf{k})}{1} = 2.8\,\mathbf{i} + 1.8\,\mathbf{j} - 0.3\,\mathbf{k}$$

$$[1, 1.5]: \quad \mathbf{v}_{\text{ave}} = \frac{\mathbf{r}(1.5) - \mathbf{r}(1)}{1.5 - 1} = \frac{(5.9\,\mathbf{i} + 6.4\,\mathbf{j} + 2.8\,\mathbf{k}) - (4.5\,\mathbf{i} + 6.0\,\mathbf{j} + 3.0\,\mathbf{k})}{0.5} = 2.8\,\mathbf{i} + 0.8\,\mathbf{j} - 0.4\,\mathbf{k}$$

(b) We can estimate the velocity at $t = 1$ by averaging the average velocities over the time intervals $[0.5, 1]$ and $[1, 1.5]$:

$\mathbf{v}(1) \approx \tfrac{1}{2}[(2\,\mathbf{i} - 2.4\,\mathbf{j} - 0.6\,\mathbf{k}) + (2.8\,\mathbf{i} + 0.8\,\mathbf{j} - 0.4\,\mathbf{k})] = 2.4\,\mathbf{i} - 0.8\,\mathbf{j} - 0.5\,\mathbf{k}$. Then the speed is

$|\mathbf{v}(1)| \approx \sqrt{(2.4)^2 + (-0.8)^2 + (-0.5)^2} \approx 2.58$.

3. $\mathbf{r}(t) = \langle -\tfrac{1}{2}t^2, t \rangle \quad \Rightarrow$ At $t = 2$:

$\mathbf{v}(t) = \mathbf{r}'(t) = \langle -t, 1 \rangle$ $\mathbf{v}(2) = \langle -2, 1 \rangle$

$\mathbf{a}(t) = \mathbf{r}''(t) = \langle -1, 0 \rangle$ $\mathbf{a}(2) = \langle -1, 0 \rangle$

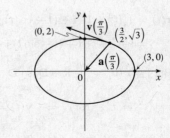

$|\mathbf{v}(t)| = \sqrt{t^2 + 1}$

Notice that $x = -\tfrac{1}{2}y^2$, so the path is a parabola.

5. $\mathbf{r}(t) = 3\cos t\,\mathbf{i} + 2\sin t\,\mathbf{j} \quad \Rightarrow$ At $t = \pi/3$:

$\mathbf{v}(t) = -3\sin t\,\mathbf{i} + 2\cos t\,\mathbf{j}$ $\mathbf{v}\left(\tfrac{\pi}{3}\right) = -\tfrac{3\sqrt{3}}{2}\,\mathbf{i} + \mathbf{j}$

$\mathbf{a}(t) = -3\cos t\,\mathbf{i} - 2\sin t\,\mathbf{j}$ $\mathbf{a}\left(\tfrac{\pi}{3}\right) = -\tfrac{3}{2}\,\mathbf{i} - \sqrt{3}\,\mathbf{j}$

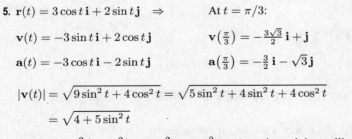

$|\mathbf{v}(t)| = \sqrt{9\sin^2 t + 4\cos^2 t} = \sqrt{5\sin^2 t + 4\sin^2 t + 4\cos^2 t}$

$\qquad\;\; = \sqrt{4 + 5\sin^2 t}$

Notice that $x^2/9 + y^2/4 = \sin^2 t + \cos^2 t = 1$, so the path is an ellipse.

7. $\mathbf{r}(t) = t\,\mathbf{i} + t^2\,\mathbf{j} + 2\,\mathbf{k}$ ⇒   At $t = 1$:

$\mathbf{v}(t) = \mathbf{i} + 2t\,\mathbf{j}$ $\mathbf{v}(1) = \mathbf{i} + 2\,\mathbf{j}$

$\mathbf{a}(t) = 2\,\mathbf{j}$ $\mathbf{a}(1) = 2\,\mathbf{j}$

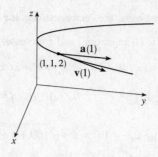

$|\mathbf{v}(t)| = \sqrt{1 + 4t^2}$

Here $x = t$, $y = t^2$ ⇒ $y = x^2$ and $z = 2$, so the path of the particle is a

parabola in the plane $z = 2$.

9. $\mathbf{r}(t) = \langle t^2 + t, t^2 - t, t^3 \rangle$ ⇒ $\mathbf{v}(t) = \mathbf{r}'(t) = \langle 2t + 1, 2t - 1, 3t^2 \rangle$, $\quad \mathbf{a}(t) = \mathbf{v}'(t) = \langle 2, 2, 6t \rangle$,

$|\mathbf{v}(t)| = \sqrt{(2t + 1)^2 + (2t - 1)^2 + (3t^2)^2} = \sqrt{9t^4 + 8t^2 + 2}$.

11. $\mathbf{r}(t) = \sqrt{2}\,t\,\mathbf{i} + e^t\,\mathbf{j} + e^{-t}\,\mathbf{k}$ ⇒ $\mathbf{v}(t) = \mathbf{r}'(t) = \sqrt{2}\,\mathbf{i} + e^t\,\mathbf{j} - e^{-t}\,\mathbf{k}$, $\quad \mathbf{a}(t) = \mathbf{v}'(t) = e^t\,\mathbf{j} + e^{-t}\,\mathbf{k}$,

$|\mathbf{v}(t)| = \sqrt{2 + e^{2t} + e^{-2t}} = \sqrt{(e^t + e^{-t})^2} = e^t + e^{-t}$.

13. $\mathbf{r}(t) = e^t \langle \cos t, \sin t, t \rangle$ ⇒

$\mathbf{v}(t) = \mathbf{r}'(t) = e^t \langle \cos t, \sin t, t \rangle + e^t \langle -\sin t, \cos t, 1 \rangle = e^t \langle \cos t - \sin t, \sin t + \cos t, t + 1 \rangle$

$\mathbf{a}(t) = \mathbf{v}'(t) = e^t \langle \cos t - \sin t - \sin t - \cos t, \sin t + \cos t + \cos t - \sin t, t + 1 + 1 \rangle$

$\quad = e^t \langle -2 \sin t, 2 \cos t, t + 2 \rangle$

$|\mathbf{v}(t)| = e^t \sqrt{\cos^2 t + \sin^2 t - 2 \cos t \sin t + \sin^2 t + \cos^2 t + 2 \sin t \cos t + t^2 + 2t + 1}$

$\quad = e^t \sqrt{t^2 + 2t + 3}$

15. $\mathbf{a}(t) = 2\,\mathbf{i} + 2t\,\mathbf{k}$ ⇒ $\mathbf{v}(t) = \int \mathbf{a}(t)\,dt = \int (2\,\mathbf{i} + 2t\,\mathbf{k})\,dt = 2t\,\mathbf{i} + t^2\,\mathbf{k} + \mathbf{C}$. Then $\mathbf{v}(0) = \mathbf{C}$ but we were given that

$\mathbf{v}(0) = 3\,\mathbf{i} - \mathbf{j}$, so $\mathbf{C} = 3\,\mathbf{i} - \mathbf{j}$ and $\mathbf{v}(t) = 2t\,\mathbf{i} + t^2\,\mathbf{k} + 3\,\mathbf{i} - \mathbf{j} = (2t + 3)\,\mathbf{i} - \mathbf{j} + t^2\,\mathbf{k}$.

$\mathbf{r}(t) = \int \mathbf{v}(t)\,dt = \int \left[(2t + 3)\,\mathbf{i} - \mathbf{j} + t^2\,\mathbf{k} \right] dt = (t^2 + 3t)\,\mathbf{i} - t\,\mathbf{j} + \frac{1}{3}t^3\,\mathbf{k} + \mathbf{D}$. Here $\mathbf{r}(0) = \mathbf{D}$ and we were given that

$\mathbf{r}(0) = \mathbf{j} + \mathbf{k}$, so $\mathbf{D} = \mathbf{j} + \mathbf{k}$ and $\mathbf{r}(t) = (t^2 + 3t)\,\mathbf{i} + (1 - t)\,\mathbf{j} + \left(\frac{1}{3}t^3 + 1 \right)\mathbf{k}$.

17. (a) $\mathbf{a}(t) = 2t\,\mathbf{i} + \sin t\,\mathbf{j} + \cos 2t\,\mathbf{k}$ ⇒   (b)

$\mathbf{v}(t) = \int (2t\,\mathbf{i} + \sin t\,\mathbf{j} + \cos 2t\,\mathbf{k})\,dt = t^2\,\mathbf{i} - \cos t\,\mathbf{j} + \frac{1}{2}\sin 2t\,\mathbf{k} + \mathbf{C}$

and $\mathbf{i} = \mathbf{v}(0) = -\mathbf{j} + \mathbf{C}$, so $\mathbf{C} = \mathbf{i} + \mathbf{j}$

and $\mathbf{v}(t) = (t^2 + 1)\,\mathbf{i} + (1 - \cos t)\,\mathbf{j} + \frac{1}{2}\sin 2t\,\mathbf{k}$.

$\mathbf{r}(t) = \int [(t^2 + 1)\,\mathbf{i} + (1 - \cos t)\,\mathbf{j} + \frac{1}{2}\sin 2t\,\mathbf{k}]\,dt$

$\quad = \left(\frac{1}{3}t^3 + t \right)\mathbf{i} + (t - \sin t)\,\mathbf{j} - \frac{1}{4}\cos 2t\,\mathbf{k} + \mathbf{D}$

But $\mathbf{j} = \mathbf{r}(0) = -\frac{1}{4}\mathbf{k} + \mathbf{D}$, so $\mathbf{D} = \mathbf{j} + \frac{1}{4}\mathbf{k}$ and $\mathbf{r}(t) = \left(\frac{1}{3}t^3 + t \right)\mathbf{i} + (t - \sin t + 1)\,\mathbf{j} + \left(\frac{1}{4} - \frac{1}{4}\cos 2t \right)\mathbf{k}$.

19. $\mathbf{r}(t) = \langle t^2, 5t, t^2 - 16t \rangle$ ⇒ $\mathbf{v}(t) = \langle 2t, 5, 2t - 16 \rangle$, $|\mathbf{v}(t)| = \sqrt{4t^2 + 25 + 4t^2 - 64t + 256} = \sqrt{8t^2 - 64t + 281}$

and $\dfrac{d}{dt}\,|\mathbf{v}(t)| = \frac{1}{2}(8t^2 - 64t + 281)^{-1/2}(16t - 64)$. This is zero if and only if the numerator is zero, that is,

$16t - 64 = 0$ or $t = 4$. Since $\frac{d}{dt}|\mathbf{v}(t)| < 0$ for $t < 4$ and $\frac{d}{dt}|\mathbf{v}(t)| > 0$ for $t > 4$, the minimum speed of $\sqrt{153}$ is attained at $t = 4$ units of time.

21. $|\mathbf{F}(t)| = 20$ N in the direction of the positive z-axis, so $\mathbf{F}(t) = 20\,\mathbf{k}$. Also $m = 4$ kg, $\mathbf{r}(0) = \mathbf{0}$ and $\mathbf{v}(0) = \mathbf{i} - \mathbf{j}$.

Since $20\mathbf{k} = \mathbf{F}(t) = 4\,\mathbf{a}(t)$, $\mathbf{a}(t) = 5\,\mathbf{k}$. Then $\mathbf{v}(t) = 5t\,\mathbf{k} + \mathbf{c}_1$ where $\mathbf{c}_1 = \mathbf{i} - \mathbf{j}$ so $\mathbf{v}(t) = \mathbf{i} - \mathbf{j} + 5t\,\mathbf{k}$ and the

speed is $|\mathbf{v}(t)| = \sqrt{1 + 1 + 25t^2} = \sqrt{25t^2 + 2}$. Also $\mathbf{r}(t) = t\,\mathbf{i} - t\,\mathbf{j} + \frac{5}{2}t^2\,\mathbf{k} + \mathbf{c}_2$ and $\mathbf{0} = \mathbf{r}(0)$, so $\mathbf{c}_2 = \mathbf{0}$

and $\mathbf{r}(t) = t\,\mathbf{i} - t\,\mathbf{j} + \frac{5}{2}t^2\,\mathbf{k}$.

23. $|\mathbf{v}(0)| = 200$ m/s and, since the angle of elevation is $60°$, a unit vector in the direction of the velocity is

$(\cos 60°)\mathbf{i} + (\sin 60°)\mathbf{j} = \frac{1}{2}\mathbf{i} + \frac{\sqrt{3}}{2}\mathbf{j}$. Thus $\mathbf{v}(0) = 200\left(\frac{1}{2}\mathbf{i} + \frac{\sqrt{3}}{2}\mathbf{j}\right) = 100\,\mathbf{i} + 100\sqrt{3}\,\mathbf{j}$ and if we set up the axes so that the

projectile starts at the origin, then $\mathbf{r}(0) = \mathbf{0}$. Ignoring air resistance, the only force is that due to gravity, so

$\mathbf{F}(t) = m\mathbf{a}(t) = -mg\,\mathbf{j}$ where $g \approx 9.8$ m/s^2. Thus $\mathbf{a}(t) = -9.8\,\mathbf{j}$ and, integrating, we have $\mathbf{v}(t) = -9.8t\,\mathbf{j} + \mathbf{C}$. But

$100\,\mathbf{i} + 100\sqrt{3}\,\mathbf{j} = \mathbf{v}(0) = \mathbf{C}$, so $\mathbf{v}(t) = 100\,\mathbf{i} + \left(100\sqrt{3} - 9.8t\right)\mathbf{j}$ and then (integrating again)

$\mathbf{r}(t) = 100t\,\mathbf{i} + \left(100\sqrt{3}\,t - 4.9t^2\right)\mathbf{j} + \mathbf{D}$ where $\mathbf{0} = \mathbf{r}(0) = \mathbf{D}$. Thus the position function of the projectile is

$\mathbf{r}(t) = 100t\,\mathbf{i} + \left(100\sqrt{3}\,t - 4.9t^2\right)\mathbf{j}$.

(a) Parametric equations for the projectile are $x(t) = 100t$, $y(t) = 100\sqrt{3}\,t - 4.9t^2$. The projectile reaches the ground when

$y(t) = 0$ (and $t > 0$) $\Rightarrow$ $100\sqrt{3}\,t - 4.9t^2 = t\left(100\sqrt{3} - 4.9t\right) = 0$ $\Rightarrow$ $t = \frac{100\sqrt{3}}{4.9} \approx 35.3$ s. So the range is

$x\left(\frac{100\sqrt{3}}{4.9}\right) = 100\left(\frac{100\sqrt{3}}{4.9}\right) \approx 3535$ m.

(b) The maximum height is reached when $y(t)$ has a critical number (or equivalently, when the vertical component

of velocity is 0): $y'(t) = 0$ $\Rightarrow$ $100\sqrt{3} - 9.8t = 0$ $\Rightarrow$ $t = \frac{100\sqrt{3}}{9.8} \approx 17.7$ s. Thus the maximum height is

$y\left(\frac{100\sqrt{3}}{9.8}\right) = 100\sqrt{3}\left(\frac{100\sqrt{3}}{9.8}\right) - 4.9\left(\frac{100\sqrt{3}}{9.8}\right)^2 \approx 1531$ m.

(c) From part (a), impact occurs at $t = \frac{100\sqrt{3}}{4.9}$ s. Thus, the velocity at impact is

$\mathbf{v}\left(\frac{100\sqrt{3}}{4.9}\right) = 100\,\mathbf{i} + \left[100\sqrt{3} - 9.8\left(\frac{100\sqrt{3}}{4.9}\right)\right]\mathbf{j} = 100\,\mathbf{i} - 100\sqrt{3}\,\mathbf{j}$ and the speed is

$\left|\mathbf{v}\left(\frac{100\sqrt{3}}{4.9}\right)\right| = \sqrt{10,000 + 30,000} = 200$ m/s.

25. As in Example 5, $\mathbf{r}(t) = (v_0 \cos 45°)t\,\mathbf{i} + \left[(v_0 \sin 45°)t - \frac{1}{2}gt^2\right]\mathbf{j} = \frac{1}{2}\left[v_0\sqrt{2}\,t\,\mathbf{i} + \left(v_0\sqrt{2}\,t - gt^2\right)\mathbf{j}\right]$. The ball lands when

$y = 0$ (and $t > 0$) $\Rightarrow$ $t = \frac{v_0\sqrt{2}}{g}$ s. Now since it lands 90 m away, $90 = x = \frac{1}{2}v_0\sqrt{2}\,\frac{v_0\sqrt{2}}{g}$ or $v_0^2 = 90g$ and the initial

velocity is $v_0 = \sqrt{90g} \approx 30$ m/s.

27. As in Example 5, $\mathbf{r}(t) = (v_0 \cos 36°)t\,\mathbf{i} + \left[(v_0 \sin 36°)t - \frac{1}{2}gt^2\right]\mathbf{j}$ and then

$\mathbf{v}(t) = \mathbf{r}'(t) = (v_0 \cos 36°)\,\mathbf{i} + \left[(v_0 \sin 36°) - gt\right]\mathbf{j}$. The shell reaches its maximum height when the vertical component of velocity is zero, so $(v_0 \sin 36°) - gt = 0 \quad \Rightarrow \quad t = \dfrac{v_0 \sin 36°}{g}$. The vertical height of the shell at that time is 1600 ft, so

$$(v_0 \sin 36°)\left(\frac{v_0 \sin 36°}{g}\right) - \tfrac{1}{2}g\left(\frac{v_0 \sin 36°}{g}\right)^2 = 1600 \quad \Rightarrow \quad \left(\frac{v_0^2 \sin^2 36°}{g}\right) - \frac{1}{2}\left(\frac{v_0^2 \sin^2 36°}{g}\right) = 1600 \quad \Rightarrow$$

$$\frac{v_0^2 \sin^2 36°}{2g} = 1600 \quad \Rightarrow \quad v_0^2 = \frac{1600(2g)}{\sin^2 36°} \quad \Rightarrow \quad v_0 = \sqrt{\frac{3200g}{\sin^2 36°}} \approx \frac{\sqrt{3200(32)}}{\sin 36°} \approx 544 \text{ ft/s}.$$

29. Place the catapult at the origin and assume the catapult is 100 meters from the city, so the city lies between $(100, 0)$ and $(600, 0)$. The initial speed is $v_0 = 80$ m/s and let θ be the angle the catapult is set at. As in Example 5, the trajectory of the catapulted rock is given by $\mathbf{r}(t) = (80 \cos \theta)t\,\mathbf{i} + \left[(80 \sin \theta)t - 4.9t^2\right]\mathbf{j}$. The top of the near city wall is at $(100, 15)$, which the rock will hit when $(80 \cos \theta)\,t = 100 \quad \Rightarrow \quad t = \dfrac{5}{4 \cos \theta}$ and $(80 \sin \theta)t - 4.9t^2 = 15 \quad \Rightarrow$

$$80 \sin \theta \cdot \frac{5}{4 \cos \theta} - 4.9\left(\frac{5}{4 \cos \theta}\right)^2 = 15 \quad \Rightarrow \quad 100 \tan \theta - 7.65625 \sec^2 \theta = 15. \text{ Replacing } \sec^2 \theta \text{ with } \tan^2 \theta + 1 \text{ gives}$$

$7.65625 \tan^2 \theta - 100 \tan \theta + 22.65625 = 0$. Using the quadratic formula, we have $\tan \theta \approx 0.230635, 12.8306 \quad \Rightarrow$ $\theta \approx 13.0°, 85.5°$. So for $13.0° < \theta < 85.5°$, the rock will land beyond the near city wall. The base of the far wall is located at $(600, 0)$ which the rock hits if $(80 \cos \theta)t = 600 \quad \Rightarrow \quad t = \dfrac{15}{2 \cos \theta}$ and $(80 \sin \theta)t - 4.9t^2 = 0 \quad \Rightarrow$

$$80 \sin \theta \cdot \frac{15}{2 \cos \theta} - 4.9\left(\frac{15}{2 \cos \theta}\right)^2 = 0 \quad \Rightarrow \quad 600 \tan \theta - 275.625 \sec^2 \theta = 0 \quad \Rightarrow$$

$275.625 \tan^2 \theta - 600 \tan \theta + 275.625 = 0$. Solutions are $\tan \theta \approx 0.658678, 1.51819 \quad \Rightarrow \quad \theta \approx 33.4°, 56.6°$. Thus the rock lands beyond the enclosed city ground for $33.4° < \theta < 56.6°$, and the angles that allow the rock to land on city ground are $13.0° < \theta < 33.4°, 56.6° < \theta < 85.5°$. If you consider that the rock can hit the far wall and bounce back into the city, we calculate the angles that cause the rock to hit the top of the wall at $(600, 15)$: $(80 \cos \theta)t = 600 \quad \Rightarrow \quad t = \dfrac{15}{2 \cos \theta}$ and

$(80 \sin \theta)t - 4.9t^2 = 15 \quad \Rightarrow \quad 600 \tan \theta - 275.625 \sec^2 \theta = 15 \quad \Rightarrow \quad 275.625 \tan^2 \theta - 600 \tan \theta + 290.625 = 0$. Solutions are $\tan \theta \approx 0.727506, 1.44936 \quad \Rightarrow \quad \theta \approx 36.0°, 55.4°$, so the catapult should be set with angle θ where $13.0° < \theta < 36.0°, 55.4° < \theta < 85.5°$.

31. Here $\mathbf{a}(t) = -4\mathbf{j} - 32\mathbf{k}$ so $\mathbf{v}(t) = -4t\,\mathbf{j} - 32t\,\mathbf{k} + \mathbf{v}_0 = -4t\,\mathbf{j} - 32t\,\mathbf{k} + 50\mathbf{i} + 80\mathbf{k} = 50\mathbf{i} - 4t\,\mathbf{j} + (80 - 32t)\mathbf{k}$ and $\mathbf{r}(t) = 50t\,\mathbf{i} - 2t^2\,\mathbf{j} + (80t - 16t^2)\mathbf{k}$ (note that $\mathbf{r}_0 = \mathbf{0}$). The ball lands when the z-component of $\mathbf{r}(t)$ is zero and $t > 0$: $80t - 16t^2 = 16t(5 - t) = 0 \quad \Rightarrow \quad t = 5$. The position of the ball then is $\mathbf{r}(5) = 50(5)\,\mathbf{i} - 2(5)^2\,\mathbf{j} + [80(5) - 16(5)^2]\mathbf{k} = 250\mathbf{i} - 50\mathbf{j}$ or equivalently the point $(250, -50, 0)$. This is a distance of $\sqrt{250^2 + (-50)^2 + 0^2} = \sqrt{65,000} \approx 255$ ft from the origin at an angle of $\tan^{-1}\left(\frac{50}{250}\right) \approx 11.3°$ from the eastern direction toward the south. The speed of the ball is $|\mathbf{v}(5)| = |50\mathbf{i} - 20\mathbf{j} - 80\mathbf{k}| = \sqrt{50^2 + (-20)^2 + (-80)^2} = \sqrt{9300} \approx 96.4$ ft/s.

33. (a) After t seconds, the boat will be $5t$ meters west of point A. The velocity

of the water at that location is $\frac{3}{400}(5t)(40-5t)\,\mathbf{j}$. The velocity of the

boat in still water is $5\,\mathbf{i}$, so the resultant velocity of the boat is

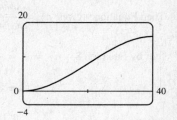

$\mathbf{v}(t) = 5\,\mathbf{i} + \frac{3}{400}(5t)(40-5t)\,\mathbf{j} = 5\,\mathbf{i} + \left(\frac{3}{2}t - \frac{3}{16}t^2\right)\mathbf{j}$. Integrating, we obtain

$\mathbf{r}(t) = 5t\,\mathbf{i} + \left(\frac{3}{4}t^2 - \frac{1}{16}t^3\right)\mathbf{j} + \mathbf{C}$. If we place the origin at A (and consider $\mathbf{j}$

to coincide with the northern direction) then $\mathbf{r}(0) = \mathbf{0} \;\Rightarrow\; \mathbf{C} = \mathbf{0}$ and we have $\mathbf{r}(t) = 5t\,\mathbf{i} + \left(\frac{3}{4}t^2 - \frac{1}{16}t^3\right)\mathbf{j}$. The boat

reaches the east bank after 8 s, and it is located at $\mathbf{r}(8) = 5(8)\mathbf{i} + \left(\frac{3}{4}(8)^2 - \frac{1}{16}(8)^3\right)\mathbf{j} = 40\,\mathbf{i} + 16\,\mathbf{j}$. Thus the boat is 16 m

downstream.

(b) Let α be the angle north of east that the boat heads. Then the velocity of the boat in still water is given by

$5(\cos\alpha)\,\mathbf{i} + 5(\sin\alpha)\,\mathbf{j}$. At t seconds, the boat is $5(\cos\alpha)t$ meters from the west bank, at which point the velocity

of the water is $\frac{3}{400}[5(\cos\alpha)t][40 - 5(\cos\alpha)t]\,\mathbf{j}$. The resultant velocity of the boat is given by

$\mathbf{v}(t) = 5(\cos\alpha)\,\mathbf{i} + \left[5\sin\alpha + \frac{3}{400}(5t\cos\alpha)(40 - 5t\cos\alpha)\right]\mathbf{j} = (5\cos\alpha)\,\mathbf{i} + \left(5\sin\alpha + \frac{3}{2}t\cos\alpha - \frac{3}{16}t^2\cos^2\alpha\right)\mathbf{j}$.

Integrating, $\mathbf{r}(t) = (5t\cos\alpha)\,\mathbf{i} + \left(5t\sin\alpha + \frac{3}{4}t^2\cos\alpha - \frac{1}{16}t^3\cos^2\alpha\right)\mathbf{j}$ (where we have again placed

the origin at A). The boat will reach the east bank when $5t\cos\alpha = 40 \;\Rightarrow\; t = \dfrac{40}{5\cos\alpha} = \dfrac{8}{\cos\alpha}$.

In order to land at point $B(40, 0)$ we need $5t\sin\alpha + \frac{3}{4}t^2\cos\alpha - \frac{1}{16}t^3\cos^2\alpha = 0 \;\Rightarrow\;$

$5\left(\dfrac{8}{\cos\alpha}\right)\sin\alpha + \frac{3}{4}\left(\dfrac{8}{\cos\alpha}\right)^2\cos\alpha - \frac{1}{16}\left(\dfrac{8}{\cos\alpha}\right)^3\cos^2\alpha = 0 \;\Rightarrow\; \dfrac{1}{\cos\alpha}(40\sin\alpha + 48 - 32) = 0 \;\Rightarrow\;$

$40\sin\alpha + 16 = 0 \;\Rightarrow\; \sin\alpha = -\frac{2}{5}$. Thus $\alpha = \sin^{-1}\left(-\frac{2}{5}\right) \approx -23.6°$, so the boat should head 23.6° south of

east (upstream). The path does seem realistic. The boat initially heads

upstream to counteract the effect of the current. Near the center of the river,

the current is stronger and the boat is pushed downstream. When the boat

nears the eastern bank, the current is slower and the boat is able to progress

upstream to arrive at point B.

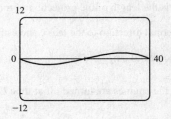

35. If $\mathbf{r}'(t) = \mathbf{c} \times \mathbf{r}(t)$ then $\mathbf{r}'(t)$ is perpendicular to both $\mathbf{c}$ and $\mathbf{r}(t)$. Remember that $\mathbf{r}'(t)$ points in the direction of motion, so if

$\mathbf{r}'(t)$ is always perpendicular to $\mathbf{c}$, the path of the particle must lie in a plane perpendicular to $\mathbf{c}$. But $\mathbf{r}'(t)$ is also perpendicular

to the position vector $\mathbf{r}(t)$ which confines the path to a sphere centered at the origin. Considering both restrictions, the path

must be contained in a circle that lies in a plane perpendicular to $\mathbf{c}$, and the circle is centered on a line through the origin in the

direction of $\mathbf{c}$.

37. $\mathbf{r}(t) = (t^2 + 1)\,\mathbf{i} + t^3\,\mathbf{j} \;\Rightarrow\; \mathbf{r}'(t) = 2t\,\mathbf{i} + 3t^2\,\mathbf{j}$,

$|\mathbf{r}'(t)| = \sqrt{(2t)^2 + (3t^2)^2} = \sqrt{4t^2 + 9t^4} = t\sqrt{4 + 9t^2}$ [since $t \geq 0$], $\quad \mathbf{r}''(t) = 2\,\mathbf{i} + 6t\,\mathbf{j}, \quad \mathbf{r}'(t) \times \mathbf{r}''(t) = 6t^2\,\mathbf{k}$.

[continued]

Then Equation 9 gives $a_T = \dfrac{\mathbf{r}'(t) \cdot \mathbf{r}''(t)}{|\mathbf{r}'(t)|} = \dfrac{(2t)(2) + (3t^2)(6t)}{t\sqrt{4 + 9t^2}} = \dfrac{4t + 18t^3}{t\sqrt{4 + 9t^2}} = \dfrac{4 + 18t^2}{\sqrt{4 + 9t^2}}$

$$\left[\text{or by Equation 8,} \quad a_T = v' = \frac{d}{dt}\left[t\sqrt{4 + 9t^2} \right] = t \cdot \tfrac{1}{2}\left(4 + 9t^2\right)^{-1/2}(18t) + \left(4 + 9t^2\right)^{1/2} \cdot 1 \right.$$
$$\left. = \left(4 + 9t^2\right)^{-1/2}\left(9t^2 + 4 + 9t^2\right) = (4 + 18t^2)/\sqrt{4 + 9t^2} \quad\quad\quad\quad \right]$$

and Equation 10 gives $a_N = \dfrac{|\mathbf{r}'(t) \times \mathbf{r}''(t)|}{|\mathbf{r}'(t)|} = \dfrac{6t^2}{t\sqrt{4 + 9t^2}} = \dfrac{6t}{\sqrt{4 + 9t^2}}.$

39. $\mathbf{r}(t) = \cos t\,\mathbf{i} + \sin t\,\mathbf{j} + t\,\mathbf{k} \;\Rightarrow\; \mathbf{r}'(t) = -\sin t\,\mathbf{i} + \cos t\,\mathbf{j} + \mathbf{k}, \quad |\mathbf{r}'(t)| = \sqrt{\sin^2 t + \cos^2 t + 1} = \sqrt{2},$

$\mathbf{r}''(t) = -\cos t\,\mathbf{i} - \sin t\,\mathbf{j}, \quad \mathbf{r}'(t) \times \mathbf{r}''(t) = \sin t\,\mathbf{i} - \cos t\,\mathbf{j} + \mathbf{k}.$

Then $a_T = \dfrac{\mathbf{r}'(t) \cdot \mathbf{r}''(t)}{|\mathbf{r}'(t)|} = \dfrac{\sin t\,\cos t - \sin t\,\cos t}{\sqrt{2}} = 0$ and $a_N = \dfrac{|\mathbf{r}'(t) \times \mathbf{r}''(t)|}{|\mathbf{r}'(t)|} = \dfrac{\sqrt{\sin^2 t + \cos^2 t + 1}}{\sqrt{2}} = \dfrac{\sqrt{2}}{\sqrt{2}} = 1.$

41. $\mathbf{r}(t) = \ln t\,\mathbf{i} + (t^2 + 3t)\,\mathbf{j} + 4\sqrt{t}\,\mathbf{k} \;\Rightarrow\; \mathbf{r}'(t) = (1/t)\,\mathbf{i} + (2t + 3)\,\mathbf{j} + (2/\sqrt{t})\,\mathbf{k} \;\Rightarrow$

$\mathbf{r}''(t) = (-1/t^2)\,\mathbf{i} + 2\,\mathbf{j} - (1/t^{3/2})\,\mathbf{k}.$ The point $(0, 4, 4)$ corresponds to $t = 1$, where

$\mathbf{r}'(1) = \mathbf{i} + 5\,\mathbf{j} + 2\,\mathbf{k}, \quad \mathbf{r}''(1) = -\mathbf{i} + 2\,\mathbf{j} - \mathbf{k}, \quad$ and $\quad \mathbf{r}'(1) \times \mathbf{r}''(1) = -9\,\mathbf{i} - \mathbf{j} + 7\,\mathbf{k}.$ Thus at the point $(0, 4, 4)$,

$a_T = \dfrac{\mathbf{r}'(1) \cdot \mathbf{r}''(1)}{|\mathbf{r}'(1)|} = \dfrac{-1 + 10 - 2}{\sqrt{1 + 25 + 4}} = \dfrac{7}{\sqrt{30}}$ and $a_N = \dfrac{|\mathbf{r}'(1) \times \mathbf{r}''(1)|}{|\mathbf{r}'(1)|} = \dfrac{\sqrt{81 + 1 + 49}}{\sqrt{30}} = \sqrt{\dfrac{131}{30}}.$

43. The tangential component of $\mathbf{a}$ is the length of the projection of $\mathbf{a}$ onto $\mathbf{T}$, so we sketch

the scalar projection of $\mathbf{a}$ in the tangential direction to the curve and estimate its length to

be 4.5 (using the fact that $\mathbf{a}$ has length 10 as a guide). Similarly, the normal component of

$\mathbf{a}$ is the length of the projection of $\mathbf{a}$ onto $\mathbf{N}$, so we sketch the scalar projection of $\mathbf{a}$ in the

normal direction to the curve and estimate its length to be 9.0. Thus $a_T \approx 4.5 \text{ cm/s}^2$ and

$a_N \approx 9.0 \text{ cm/s}^2.$

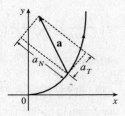

45. If the engines are turned off at time t, then the spacecraft will continue to travel in the direction of $\mathbf{v}(t)$, so we need a t such

that for some scalar $s > 0$, $\mathbf{r}(t) + s\,\mathbf{v}(t) = \langle 6, 4, 9 \rangle.$ $\quad \mathbf{v}(t) = \mathbf{r}'(t) = \mathbf{i} + \dfrac{1}{t}\,\mathbf{j} + \dfrac{8t}{(t^2 + 1)^2}\,\mathbf{k} \;\Rightarrow$

$\mathbf{r}(t) + s\,\mathbf{v}(t) = \left\langle 3 + t + s,\, 2 + \ln t + \dfrac{s}{t},\, 7 - \dfrac{4}{t^2 + 1} + \dfrac{8st}{(t^2 + 1)^2} \right\rangle \;\Rightarrow\; 3 + t + s = 6 \;\Rightarrow\; s = 3 - t,$

so $7 - \dfrac{4}{t^2 + 1} + \dfrac{8(3 - t)t}{(t^2 + 1)^2} = 9 \;\Leftrightarrow\; \dfrac{24t - 12t^2 - 4}{(t^2 + 1)^2} = 2 \;\Leftrightarrow\; t^4 + 8t^2 - 12t + 3 = 0.$

It is easily seen that $t = 1$ is a root of this polynomial. Also $2 + \ln 1 + \dfrac{3 - 1}{1} = 4$, so $t = 1$ is the desired solution.

13 Review

TRUE-FALSE QUIZ

1. **True.** If we reparametrize the curve by replacing $u = t^3$, we have $\mathbf{r}(u) = u\,\mathbf{i} + 2u\,\mathbf{j} + 3u\,\mathbf{k}$, which is a line through the origin with direction vector $\mathbf{i} + 2\,\mathbf{j} + 3\,\mathbf{k}$.

3. **False.** The vector function represents a line, but the line does not pass through the origin; the x-component is 0 only for $t = 0$ which corresponds to the point $(0, 3, 0)$ not $(0, 0, 0)$.

5. **False.** By Formula 5 of Theorem 13.2.3, $\dfrac{d}{dt}\left[\mathbf{u}(t) \times \mathbf{v}(t)\right] = \mathbf{u}'(t) \times \mathbf{v}(t) + \mathbf{u}(t) \times \mathbf{v}'(t)$.

7. **False.** κ is the magnitude of the rate of change of the unit tangent vector $\mathbf{T}$ with respect to arc length s, not with respect to t.

9. **True.** At an inflection point where f is twice continuously differentiable we must have $f''(x) = 0$, and by Equation 13.3.11, the curvature is 0 there.

11. **False.** If $\mathbf{r}(t)$ is the position of a moving particle at time t and $|\mathbf{r}(t)| = 1$ then the particle lies on the unit circle or the unit sphere, but this does not mean that the speed $|\mathbf{r}'(t)|$ must be constant. As a counterexample, let $\mathbf{r}(t) = \langle t, \sqrt{1 - t^2}\rangle$, then $\mathbf{r}'(t) = \langle 1, -t/\sqrt{1 - t^2}\rangle$ and $|\mathbf{r}(t)| = \sqrt{t^2 + 1 - t^2} = 1$ but $|\mathbf{r}'(t)| = \sqrt{1 + t^2/(1 - t^2)} = 1/\sqrt{1 - t^2}$ which is not constant.

13. **True.** See the discussion preceding Example 7 in Section 13.3.

EXERCISES

1. (a) The corresponding parametric equations for the curve are $x = t$, $y = \cos \pi t$, $z = \sin \pi t$. Since $y^2 + z^2 = 1$, the curve is contained in a circular cylinder with axis the x-axis. Since $x = t$, the curve is a helix.

   (b) $\mathbf{r}(t) = t\,\mathbf{i} + \cos \pi t\,\mathbf{j} + \sin \pi t\,\mathbf{k} \ \Rightarrow$

   $\mathbf{r}'(t) = \mathbf{i} - \pi \sin \pi t\,\mathbf{j} + \pi \cos \pi t\,\mathbf{k} \ \Rightarrow$

   $\mathbf{r}''(t) = -\pi^2 \cos \pi t\,\mathbf{j} - \pi^2 \sin \pi t\,\mathbf{k}$

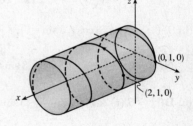

3. The projection of the curve C of intersection onto the xy-plane is the circle $x^2 + y^2 = 16, z = 0$. So we can write $x = 4\cos t, \ y = 4\sin t, \ 0 \le t \le 2\pi$. From the equation of the plane, we have $z = 5 - x = 5 - 4\cos t$, so parametric equations for C are $x = 4\cos t, \ y = 4\sin t, \ z = 5 - 4\cos t, \ 0 \le t \le 2\pi$, and the corresponding vector function is $\mathbf{r}(t) = 4\cos t\,\mathbf{i} + 4\sin t\,\mathbf{j} + (5 - 4\cos t)\,\mathbf{k}, \ 0 \le t \le 2\pi$.

5. $\int_0^1 (t^2\,\mathbf{i} + t\cos \pi t\,\mathbf{j} + \sin \pi t\,\mathbf{k})\,dt = \left(\int_0^1 t^2\,dt\right)\mathbf{i} + \left(\int_0^1 t\cos \pi t\,dt\right)\mathbf{j} + \left(\int_0^1 \sin \pi t\,dt\right)\mathbf{k}$

$\qquad = \left[\tfrac{1}{3}t^3\right]_0^1 \mathbf{i} + \left(\left[\tfrac{t}{\pi}\sin \pi t\right]_0^1 - \int_0^1 \tfrac{1}{\pi}\sin \pi t\,dt\right)\mathbf{j} + \left[-\tfrac{1}{\pi}\cos \pi t\right]_0^1 \mathbf{k}$

$\qquad = \tfrac{1}{3}\mathbf{i} + \left[\tfrac{1}{\pi^2}\cos \pi t\right]_0^1 \mathbf{j} + \tfrac{2}{\pi}\mathbf{k} = \tfrac{1}{3}\mathbf{i} - \tfrac{2}{\pi^2}\mathbf{j} + \tfrac{2}{\pi}\mathbf{k}$

where we integrated by parts in the y-component.

7. $\mathbf{r}(t) = \langle t^2, t^3, t^4 \rangle \;\Rightarrow\; \mathbf{r}'(t) = \langle 2t, 3t^2, 4t^3 \rangle \;\Rightarrow\; |\mathbf{r}'(t)| = \sqrt{4t^2 + 9t^4 + 16t^6}$ and

$L = \int_0^3 |\mathbf{r}'(t)|\, dt = \int_0^3 \sqrt{4t^2 + 9t^4 + 16t^6}\, dt$. Using Simpson's Rule with $f(t) = \sqrt{4t^2 + 9t^4 + 16t^6}$ and $n = 6$ we

have $\Delta t = \frac{3-0}{6} = \frac{1}{2}$ and

$$L \approx \tfrac{\Delta t}{3}\left[f(0) + 4f\!\left(\tfrac{1}{2}\right) + 2f(1) + 4f\!\left(\tfrac{3}{2}\right) + 2f(2) + 4f\!\left(\tfrac{5}{2}\right) + f(3) \right]$$

$$= \tfrac{1}{6}\left[\sqrt{0+0+0} + 4\cdot\sqrt{4\left(\tfrac{1}{2}\right)^2 + 9\left(\tfrac{1}{2}\right)^4 + 16\left(\tfrac{1}{2}\right)^6} + 2\cdot\sqrt{4(1)^2 + 9(1)^4 + 16(1)^6}\right.$$

$$+ 4\cdot\sqrt{4\left(\tfrac{3}{2}\right)^2 + 9\left(\tfrac{3}{2}\right)^4 + 16\left(\tfrac{3}{2}\right)^6} + 2\cdot\sqrt{4(2)^2 + 9(2)^4 + 16(2)^6}$$

$$\left. + 4\cdot\sqrt{4\left(\tfrac{5}{2}\right)^2 + 9\left(\tfrac{5}{2}\right)^4 + 16\left(\tfrac{5}{2}\right)^6} + \sqrt{4(3)^2 + 9(3)^4 + 16(3)^6} \right]$$

$$\approx 86.631$$

9. The angle of intersection of the two curves, θ, is the angle between their respective tangents at the point of intersection.

For both curves the point $(1, 0, 0)$ occurs when $t = 0$.

$\mathbf{r}_1'(t) = -\sin t\, \mathbf{i} + \cos t\, \mathbf{j} + \mathbf{k} \;\Rightarrow\; \mathbf{r}_1'(0) = \mathbf{j} + \mathbf{k}$ and $\mathbf{r}_2'(t) = \mathbf{i} + 2t\, \mathbf{j} + 3t^2\, \mathbf{k} \;\Rightarrow\; \mathbf{r}_2'(0) = \mathbf{i}$.

$\mathbf{r}_1'(0) \cdot \mathbf{r}_2'(0) = (\mathbf{j} + \mathbf{k}) \cdot \mathbf{i} = 0$. Therefore, the curves intersect in a right angle, that is, $\theta = 90°$.

11. (a) $\mathbf{r}(t) = \langle \sin^3 t, \cos^3 t, \sin^2 t \rangle \;\Rightarrow\; \mathbf{r}'(t) = \langle 3\sin^2 t \cos t, -3\cos^2 t \sin t, 2\sin t \cos t \rangle$,

$$|\mathbf{r}'(t)| = \sqrt{9\sin^4 t \cos^2 t + 9\cos^4 t \sin^2 t + 4\sin^2 t \cos^2 t}$$

$$= \sqrt{\sin^2 t \cos^2 t\, (9\sin^2 t + 9\cos^2 t + 4)} = \sqrt{13}\,\sin t \cos t \quad [\text{since } 0 \le t \le \pi/2 \;\Rightarrow\; \sin t, \cos t \ge 0]$$

Then $\mathbf{T}(t) = \dfrac{\mathbf{r}'(t)}{|\mathbf{r}'(t)|} = \dfrac{1}{\sqrt{13}\,\sin t \cos t} \langle 3\sin^2 t \cos t, -3\cos^2 t \sin t, 2\sin t \cos t \rangle = \tfrac{1}{\sqrt{13}} \langle 3\sin t, -3\cos t, 2 \rangle$.

(b) $\mathbf{T}'(t) = \tfrac{1}{\sqrt{13}} \langle 3\cos t, 3\sin t, 0 \rangle$, $|\mathbf{T}'(t)| = \tfrac{1}{\sqrt{13}}\sqrt{9\cos^2 t + 9\sin^2 t + 0} = \tfrac{3}{\sqrt{13}}$, and

$\mathbf{N}(t) = \dfrac{\mathbf{T}'(t)}{|\mathbf{T}'(t)|} = \tfrac{1}{3} \langle 3\cos t, 3\sin t, 0 \rangle = \langle \cos t, \sin t, 0 \rangle$.

(c) $\mathbf{B}(t) = \mathbf{T}(t) \times \mathbf{N}(t) = \tfrac{1}{\sqrt{13}} \langle 3\sin t, -3\cos t, 2 \rangle \times \langle \cos t, \sin t, 0 \rangle = \tfrac{1}{\sqrt{13}} \langle -2\sin t, 2\cos t, 3 \rangle$

(d) $\kappa(t) = \dfrac{|\mathbf{T}'(t)|}{|\mathbf{r}'(t)|} = \dfrac{3/\sqrt{13}}{\sqrt{13}\,\sin t \cos t} = \dfrac{3}{13\sin t \cos t}$ or $\tfrac{3}{13}\sec t \csc t$

13. $y' = 4x^3$, $y'' = 12x^2$ and $\kappa(x) = \dfrac{|y''|}{[1+(y')^2]^{3/2}} = \dfrac{|12x^2|}{(1+16x^6)^{3/2}}$, so $\kappa(1) = \dfrac{12}{17^{3/2}}$.

15. $\mathbf{r}(t) = \langle \sin 2t, t, \cos 2t \rangle \;\Rightarrow\; \mathbf{r}'(t) = \langle 2\cos 2t, 1, -2\sin 2t \rangle \;\Rightarrow\; \mathbf{T}(t) = \tfrac{1}{\sqrt{5}} \langle 2\cos 2t, 1, -2\sin 2t \rangle \;\Rightarrow\;$

$\mathbf{T}'(t) = \tfrac{1}{\sqrt{5}} \langle -4\sin 2t, 0, -4\cos 2t \rangle \;\Rightarrow\; \mathbf{N}(t) = \langle -\sin 2t, 0, -\cos 2t \rangle$. So $\mathbf{N} = \mathbf{N}(\pi) = \langle 0, 0, -1 \rangle$ and

$\mathbf{B} = \mathbf{T} \times \mathbf{N} = \tfrac{1}{\sqrt{5}} \langle -1, 2, 0 \rangle$. So a normal to the osculating plane is $\langle -1, 2, 0 \rangle$ and an equation is

$-1(x - 0) + 2(y - \pi) + 0(z - 1) = 0$ or $x - 2y + 2\pi = 0$.

17. $\mathbf{r}(t) = t\ln t\,\mathbf{i} + t\,\mathbf{j} + e^{-t}\,\mathbf{k}, \quad \mathbf{v}(t) = \mathbf{r}'(t) = (1+\ln t)\,\mathbf{i} + \mathbf{j} - e^{-t}\,\mathbf{k},$

$|\mathbf{v}(t)| = \sqrt{(1+\ln t)^2 + 1^2 + (-e^{-t})^2} = \sqrt{2 + 2\ln t + (\ln t)^2 + e^{-2t}}, \quad \mathbf{a}(t) = \mathbf{v}'(t) = \tfrac{1}{t}\,\mathbf{i} + e^{-t}\,\mathbf{k}$

19. $\mathbf{v}(t) = \int \mathbf{a}(t)\,dt = \int (6t\,\mathbf{i} + 12t^2\,\mathbf{j} - 6t\,\mathbf{k})\,dt = 3t^2\,\mathbf{i} + 4t^3\,\mathbf{j} - 3t^2\,\mathbf{k} + \mathbf{C}$, but $\mathbf{i} - \mathbf{j} + 3\,\mathbf{k} = \mathbf{v}(0) = \mathbf{0} + \mathbf{C}$,

so $\mathbf{C} = \mathbf{i} - \mathbf{j} + 3\,\mathbf{k}$ and $\mathbf{v}(t) = (3t^2 + 1)\,\mathbf{i} + (4t^3 - 1)\,\mathbf{j} + (3 - 3t^2)\,\mathbf{k}$.

$\mathbf{r}(t) = \int \mathbf{v}(t)\,dt = (t^3 + t)\,\mathbf{i} + (t^4 - t)\,\mathbf{j} + (3t - t^3)\,\mathbf{k} + \mathbf{D}$.

But $\mathbf{r}(0) = \mathbf{0}$, so $\mathbf{D} = \mathbf{0}$ and $\mathbf{r}(t) = (t^3 + t)\,\mathbf{i} + (t^4 - t)\,\mathbf{j} + (3t - t^3)\,\mathbf{k}$.

21. Example 13.4.5 showed that the maximum horizontal range is achieved with an angle of elevation of $45°$. In this case, however, the projectile would hit the top of the tunnel using that angle. The horizontal range will be maximized with the largest angle of elevation that keeps the projectile within a height of 30 m. From Example 13.4.5 we know that the position function of the projectile is $\mathbf{r}(t) = (v_0\cos\alpha)t\,\mathbf{i} + \left[(v_0\sin\alpha)t - \tfrac{1}{2}gt^2\right]\mathbf{j}$ and the velocity is

$\mathbf{v}(t) = \mathbf{r}'(t) = (v_0\cos\alpha)\,\mathbf{i} + [(v_0\sin\alpha) - gt]\,\mathbf{j}$. The projectile achieves its maximum height when the vertical component of velocity is zero, so $(v_0\sin\alpha) - gt = 0 \implies t = \dfrac{v_0\sin\alpha}{g}$. We want the vertical height of the projectile at that time to be

30 m: $(v_0\sin\alpha)\left(\dfrac{v_0\sin\alpha}{g}\right) - \tfrac{1}{2}g\left(\dfrac{v_0\sin\alpha}{g}\right)^2 = 30 \implies$

$\left(\dfrac{v_0^2\sin^2\alpha}{g}\right) - \dfrac{1}{2}\left(\dfrac{v_0^2\sin^2\alpha}{g}\right) = 30 \implies \dfrac{v_0^2\sin^2\alpha}{2g} = 30 \implies \sin^2\alpha = \dfrac{30(2g)}{v_0^2} = \dfrac{60(9.8)}{40^2} = 0.3675 \implies$

$\sin\alpha = \sqrt{0.3675}$. Thus the desired angle of elevation is $\alpha = \sin^{-1}\sqrt{0.3675} \approx 37.3°$.

From the same example, the horizontal distance traveled is $d = \dfrac{v_0^2\sin 2\alpha}{g} \approx \dfrac{40^2\sin(74.6°)}{9.8} \approx 157.4$ m.

23. (a) Instead of proceeding directly, we use Formula 3 of Theorem 13.2.3: $\mathbf{r}(t) = t\,\mathbf{R}(t) \implies$

$\mathbf{v} = \mathbf{r}'(t) = \mathbf{R}(t) + t\,\mathbf{R}'(t) = \cos\omega t\,\mathbf{i} + \sin\omega t\,\mathbf{j} + t\,\mathbf{v}_d$.

(b) Using the same method as in part (a) and starting with $\mathbf{v} = \mathbf{R}(t) + t\,\mathbf{R}'(t)$, we have

$\mathbf{a} = \mathbf{v}' = \mathbf{R}'(t) + \mathbf{R}'(t) + t\,\mathbf{R}''(t) = 2\,\mathbf{R}'(t) + t\,\mathbf{R}''(t) = 2\,\mathbf{v}_d + t\,\mathbf{a}_d$.

(c) Here we have $\mathbf{r}(t) = e^{-t}\cos\omega t\,\mathbf{i} + e^{-t}\sin\omega t\,\mathbf{j} = e^{-t}\,\mathbf{R}(t)$. So, as in parts (a) and (b),

$\mathbf{v} = \mathbf{r}'(t) = e^{-t}\,\mathbf{R}'(t) - e^{-t}\,\mathbf{R}(t) = e^{-t}[\mathbf{R}'(t) - \mathbf{R}(t)] \implies$

$\mathbf{a} = \mathbf{v}' = e^{-t}[\mathbf{R}''(t) - \mathbf{R}'(t)] - e^{-t}[\mathbf{R}'(t) - \mathbf{R}(t)] = e^{-t}[\mathbf{R}''(t) - 2\,\mathbf{R}'(t) + \mathbf{R}(t)]$

$= e^{-t}\,\mathbf{a}_d - 2e^{-t}\,\mathbf{v}_d + e^{-t}\,\mathbf{R}$

Thus, the Coriolis acceleration (the sum of the "extra" terms not involving $\mathbf{a}_d$) is $-2e^{-t}\,\mathbf{v}_d + e^{-t}\,\mathbf{R}$.

PROBLEMS PLUS

1. (a) $\mathbf{r}(t) = R\cos\omega t\,\mathbf{i} + R\sin\omega t\,\mathbf{j} \Rightarrow \mathbf{v} = \mathbf{r}'(t) = -\omega R\sin\omega t\,\mathbf{i} + \omega R\cos\omega t\,\mathbf{j}$, so $\mathbf{r} = R(\cos\omega t\,\mathbf{i} + \sin\omega t\,\mathbf{j})$ and

$\mathbf{v} = \omega R(-\sin\omega t\,\mathbf{i} + \cos\omega t\,\mathbf{j})$. $\mathbf{v}\cdot\mathbf{r} = \omega R^2(-\cos\omega t\sin\omega t + \sin\omega t\cos\omega t) = 0$, so $\mathbf{v}\perp\mathbf{r}$. Since $\mathbf{r}$ points along a

radius of the circle, and $\mathbf{v}\perp\mathbf{r}$, $\mathbf{v}$ is tangent to the circle. Because it is a velocity vector, $\mathbf{v}$ points in the direction of motion.

(b) In (a), we wrote $\mathbf{v}$ in the form $\omega R\,\mathbf{u}$, where $\mathbf{u}$ is the unit vector $-\sin\omega t\,\mathbf{i} + \cos\omega t\,\mathbf{j}$. Clearly $|\mathbf{v}| = \omega R\,|\mathbf{u}| = \omega R$. At

speed ωR, the particle completes one revolution, a distance $2\pi R$, in time $T = \dfrac{2\pi R}{\omega R} = \dfrac{2\pi}{\omega}$.

(c) $\mathbf{a} = \dfrac{d\mathbf{v}}{dt} = -\omega^2 R\cos\omega t\,\mathbf{i} - \omega^2 R\sin\omega t\,\mathbf{j} = -\omega^2 R(\cos\omega t\,\mathbf{i} + \sin\omega t\,\mathbf{j})$, so $\mathbf{a} = -\omega^2\mathbf{r}$. This shows that $\mathbf{a}$ is proportional

to $\mathbf{r}$ and points in the opposite direction (toward the origin). Also, $|\mathbf{a}| = \omega^2\,|\mathbf{r}| = \omega^2 R$.

(d) By Newton's Second Law (see Section 13.4), $\mathbf{F} = m\mathbf{a}$, so $|\mathbf{F}| = m\,|\mathbf{a}| = mR\omega^2 = \dfrac{m\,(\omega R)^2}{R} = \dfrac{m\,|\mathbf{v}|^2}{R}$.

3. (a) The projectile reaches maximum height when $0 = \dfrac{dy}{dt} = \dfrac{d}{dt}\left[(v_0\sin\alpha)t - \tfrac{1}{2}gt^2\right] = v_0\sin\alpha - gt$; that is, when

$t = \dfrac{v_0\sin\alpha}{g}$ and $y = (v_0\sin\alpha)\left(\dfrac{v_0\sin\alpha}{g}\right) - \dfrac{1}{2}g\left(\dfrac{v_0\sin\alpha}{g}\right)^2 = \dfrac{v_0^2\sin^2\alpha}{2g}$. This is the maximum height attained when

the projectile is fired with an angle of elevation α. This maximum height is largest when $\alpha = 90°$. In that case, $\sin\alpha = 1$

and the maximum height is $\dfrac{v_0^2}{2g}$.

(b) Let $R = v_0^2/g$. We are asked to consider the parabola $x^2 + 2Ry - R^2 = 0$ which can be rewritten as $y = -\dfrac{1}{2R}x^2 + \dfrac{R}{2}$.

The points on or inside this parabola are those for which $-R \le x \le R$ and $0 \le y \le \dfrac{-1}{2R}x^2 + \dfrac{R}{2}$. When the projectile is

fired at angle of elevation α, the points (x, y) along its path satisfy the relations $x = (v_0\cos\alpha)\,t$ and

$y = (v_0\sin\alpha)t - \tfrac{1}{2}gt^2$, where $0 \le t \le (2v_0\sin\alpha)/g$ (as in Example 13.4.5). Thus

$|x| \le \left|v_0\cos\alpha\left(\dfrac{2v_0\sin\alpha}{g}\right)\right| = \left|\dfrac{v_0^2}{g}\sin 2\alpha\right| \le \left|\dfrac{v_0^2}{g}\right| = |R|$. This shows that $-R \le x \le R$.

For t in the specified range, we also have $y = t(v_0\sin\alpha - \tfrac{1}{2}gt) = \tfrac{1}{2}gt\left(\dfrac{2v_0\sin\alpha}{g} - t\right) \ge 0$ and

$y = (v_0\sin\alpha)\dfrac{x}{v_0\cos\alpha} - \dfrac{g}{2}\left(\dfrac{x}{v_0\cos\alpha}\right)^2 = (\tan\alpha)\,x - \dfrac{g}{2v_0^2\cos^2\alpha}x^2 = -\dfrac{1}{2R\cos^2\alpha}x^2 + (\tan\alpha)\,x$. Thus

$$y - \left(\dfrac{-1}{2R}x^2 + \dfrac{R}{2}\right) = \dfrac{-1}{2R\cos^2\alpha}x^2 + \dfrac{1}{2R}x^2 + (\tan\alpha)\,x - \dfrac{R}{2}$$

$$= \dfrac{x^2}{2R}\left(1 - \dfrac{1}{\cos^2\alpha}\right) + (\tan\alpha)\,x - \dfrac{R}{2} = \dfrac{x^2(1 - \sec^2\alpha) + 2R(\tan\alpha)\,x - R^2}{2R}$$

$$= \dfrac{-(\tan^2\alpha)\,x^2 + 2R(\tan\alpha)\,x - R^2}{2R} = \dfrac{-[(\tan\alpha)\,x - R]^2}{2R} \le 0$$

We have shown that every target that can be hit by the projectile lies on or inside the parabola $y = -\dfrac{1}{2R} x^2 + \dfrac{R}{2}$.

Now let (a, b) be any point on or inside the parabola $y = -\dfrac{1}{2R} x^2 + \dfrac{R}{2}$. Then $-R \leq a \leq R$ and $0 \leq b \leq -\dfrac{1}{2R} a^2 + \dfrac{R}{2}$.

We seek an angle α such that (a, b) lies in the path of the projectile; that is, we wish to find an angle α such that

$b = -\dfrac{1}{2R \cos^2 \alpha} a^2 + (\tan \alpha)\, a$ or equivalently $b = \dfrac{-1}{2R} (\tan^2 \alpha + 1)a^2 + (\tan \alpha)\, a$. Rearranging this equation we get

$\dfrac{a^2}{2R} \tan^2 \alpha - a \tan \alpha + \left(\dfrac{a^2}{2R} + b \right) = 0$ or $a^2(\tan \alpha)^2 - 2aR(\tan \alpha) + (a^2 + 2bR) = 0 \ (\star)$. This quadratic equation

for $\tan \alpha$ has real solutions exactly when the discriminant is nonnegative. Now $B^2 - 4AC \geq 0 \ \Leftrightarrow$

$(-2aR)^2 - 4a^2(a^2 + 2bR) \geq 0 \ \Leftrightarrow \ 4a^2(R^2 - a^2 - 2bR) \geq 0 \ \Leftrightarrow \ -a^2 - 2bR + R^2 \geq 0 \ \Leftrightarrow$

$b \leq \dfrac{1}{2R} (R^2 - a^2) \ \Leftrightarrow \ b \leq \dfrac{-1}{2R} a^2 + \dfrac{R}{2}$. This condition is satisfied since (a, b) is on or inside the parabola

$y = -\dfrac{1}{2R} x^2 + \dfrac{R}{2}$. It follows that (a, b) lies in the path of the projectile when $\tan \alpha$ satisfies $(\star)$, that is, when

$\tan \alpha = \dfrac{2aR \pm \sqrt{4a^2(R^2 - a^2 - 2bR)}}{2a^2} = \dfrac{R \pm \sqrt{R^2 - 2bR - a^2}}{a}$.

(c)

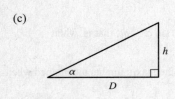

If the gun is pointed at a target with height h at a distance D downrange, then

$\tan \alpha = h/D$. When the projectile reaches a distance D downrange (remember

we are assuming that it doesn't hit the ground first), we have $D = x = (v_0 \cos \alpha)t$,

so $t = \dfrac{D}{v_0 \cos \alpha}$ and $y = (v_0 \sin \alpha)t - \frac{1}{2}gt^2 = D \tan \alpha - \dfrac{gD^2}{2v_0^2 \cos^2 \alpha}$.

Meanwhile, the target, whose x-coordinate is also D, has fallen from height h to height

$h - \frac{1}{2}gt^2 = D \tan \alpha - \dfrac{gD^2}{2v_0^2 \cos^2 \alpha}$. Thus the projectile hits the target.

5. (a) $\mathbf{a} = -g\mathbf{j} \ \Rightarrow \ \mathbf{v} = \mathbf{v}_0 - gt\mathbf{j} = 2\mathbf{i} - gt\mathbf{j} \ \Rightarrow \ \mathbf{s} = \mathbf{s}_0 + 2t\mathbf{i} - \frac{1}{2}gt^2\mathbf{j} = 3.5\mathbf{j} + 2t\mathbf{i} - \frac{1}{2}gt^2\mathbf{j} \ \Rightarrow$

$\mathbf{s} = 2t\mathbf{i} + \left(3.5 - \frac{1}{2}gt^2 \right)\mathbf{j}$. Therefore $y = 0$ when $t = \sqrt{7/g}$ seconds. At that instant, the ball is $2\sqrt{7/g} \approx 0.94$ ft to the

right of the table top. Its coordinates (relative to an origin on the floor directly under the table's edge) are $(0.94, 0)$. At

impact, the velocity is $\mathbf{v} = 2\mathbf{i} - \sqrt{7g}\,\mathbf{j}$, so the speed is $|\mathbf{v}| = \sqrt{4 + 7g} \approx 15$ ft/s.

(b) The slope of the curve when $t = \sqrt{\dfrac{7}{g}}$ is $\dfrac{dy}{dx} = \dfrac{dy/dt}{dx/dt} = \dfrac{-gt}{2} = \dfrac{-g\sqrt{7/g}}{2} = \dfrac{-\sqrt{7g}}{2}$. Thus $\cot \theta = \dfrac{\sqrt{7g}}{2}$

and $\theta \approx 7.6°$.

(c) From (a), $|\mathbf{v}| = \sqrt{4 + 7g}$. So the ball rebounds with speed $0.8\sqrt{4 + 7g} \approx 12.08$ ft/s at angle of inclination

$90° - \theta \approx 82.3886°$. By Example 13.4.5, the horizontal distance traveled between bounces is $d = \dfrac{v_0^2 \sin 2\alpha}{g}$, where

$v_0 \approx 12.08$ ft/s and $\alpha \approx 82.3886°$. Therefore, $d \approx 1.197$ ft. So the ball strikes the floor at about

$2\sqrt{7/g} + 1.197 \approx 2.13$ ft to the right of the table's edge.

7. The trajectory of the projectile is given by $\mathbf{r}(t) = (v\cos\alpha)t\,\mathbf{i} + \left[(v\sin\alpha)t - \frac{1}{2}gt^2\right]\mathbf{j}$, so

$\mathbf{v}(t) = \mathbf{r}'(t) = v\cos\alpha\,\mathbf{i} + (v\sin\alpha - gt)\,\mathbf{j}$ and

$$|\mathbf{v}(t)| = \sqrt{(v\cos\alpha)^2 + (v\sin\alpha - gt)^2} = \sqrt{v^2 - (2vg\sin\alpha)\,t + g^2t^2} = \sqrt{g^2\left(t^2 - \frac{2v}{g}(\sin\alpha)\,t + \frac{v^2}{g^2}\right)}$$

$$= g\sqrt{\left(t - \frac{v}{g}\sin\alpha\right)^2 + \frac{v^2}{g^2} - \frac{v^2}{g^2}\sin^2\alpha} = g\sqrt{\left(t - \frac{v}{g}\sin\alpha\right)^2 + \frac{v^2}{g^2}\cos^2\alpha}$$

The projectile hits the ground when $(v\sin\alpha)t - \frac{1}{2}gt^2 = 0 \;\Rightarrow\; t = \frac{2v}{g}\sin\alpha$, so the distance traveled by the projectile is

$$L(\alpha) = \int_0^{(2v/g)\sin\alpha} |\mathbf{v}(t)|\,dt = \int_0^{(2v/g)\sin\alpha} g\sqrt{\left(t - \frac{v}{g}\sin\alpha\right)^2 + \frac{v^2}{g^2}\cos^2\alpha}\,dt$$

$$= g\left[\frac{t - (v/g)\sin\alpha}{2}\sqrt{\left(t - \frac{v}{g}\sin\alpha\right)^2 + \left(\frac{v}{g}\cos\alpha\right)^2}\right.$$

$$\left. + \frac{[(v/g)\cos\alpha]^2}{2}\ln\left(t - \frac{v}{g}\sin\alpha + \sqrt{\left(t - \frac{v}{g}\sin\alpha\right)^2 + \left(\frac{v}{g}\cos\alpha\right)^2}\right)\right]_0^{(2v/g)\sin\alpha}$$

[using Formula 21 in the Table of Integrals]

$$= \frac{g}{2}\left[\frac{v}{g}\sin\alpha\sqrt{\left(\frac{v}{g}\sin\alpha\right)^2 + \left(\frac{v}{g}\cos\alpha\right)^2} + \left(\frac{v}{g}\cos\alpha\right)^2\ln\left(\frac{v}{g}\sin\alpha + \sqrt{\left(\frac{v}{g}\sin\alpha\right)^2 + \left(\frac{v}{g}\cos\alpha\right)^2}\right)\right.$$

$$\left. + \frac{v}{g}\sin\alpha\sqrt{\left(\frac{v}{g}\sin\alpha\right)^2 + \left(\frac{v}{g}\cos\alpha\right)^2} - \left(\frac{v}{g}\cos\alpha\right)^2\ln\left(-\frac{v}{g}\sin\alpha + \sqrt{\left(\frac{v}{g}\sin\alpha\right)^2 + \left(\frac{v}{g}\cos\alpha\right)^2}\right)\right]$$

$$= \frac{g}{2}\left[\frac{v}{g}\sin\alpha\cdot\frac{v}{g} + \frac{v^2}{g^2}\cos^2\alpha\ln\left(\frac{v}{g}\sin\alpha + \frac{v}{g}\right) + \frac{v}{g}\sin\alpha\cdot\frac{v}{g} - \frac{v^2}{g^2}\cos^2\alpha\ln\left(-\frac{v}{g}\sin\alpha + \frac{v}{g}\right)\right]$$

$$= \frac{v^2}{g}\sin\alpha + \frac{v^2}{2g}\cos^2\alpha\ln\left(\frac{(v/g)\sin\alpha + v/g}{-(v/g)\sin\alpha + v/g}\right) = \frac{v^2}{g}\sin\alpha + \frac{v^2}{2g}\cos^2\alpha\ln\left(\frac{1 + \sin\alpha}{1 - \sin\alpha}\right)$$

We want to maximize $L(\alpha)$ for $0 \le \alpha \le \pi/2$.

$$L'(\alpha) = \frac{v^2}{g}\cos\alpha + \frac{v^2}{2g}\left[\cos^2\alpha\cdot\frac{1 - \sin\alpha}{1 + \sin\alpha}\cdot\frac{2\cos\alpha}{(1 - \sin\alpha)^2} - 2\cos\alpha\sin\alpha\ln\left(\frac{1 + \sin\alpha}{1 - \sin\alpha}\right)\right]$$

$$= \frac{v^2}{g}\cos\alpha + \frac{v^2}{2g}\left[\cos^2\alpha\cdot\frac{2}{\cos\alpha} - 2\cos\alpha\sin\alpha\ln\left(\frac{1 + \sin\alpha}{1 - \sin\alpha}\right)\right]$$

$$= \frac{v^2}{g}\cos\alpha + \frac{v^2}{g}\cos\alpha\left[1 - \sin\alpha\ln\left(\frac{1 + \sin\alpha}{1 - \sin\alpha}\right)\right] = \frac{v^2}{g}\cos\alpha\left[2 - \sin\alpha\ln\left(\frac{1 + \sin\alpha}{1 - \sin\alpha}\right)\right]$$

$L(\alpha)$ has critical points for $0 < \alpha < \pi/2$ when $L'(\alpha) = 0 \;\Rightarrow\; 2 - \sin\alpha\ln\left(\frac{1 + \sin\alpha}{1 - \sin\alpha}\right) = 0$ [since $\cos\alpha \ne 0$].

Solving by graphing (or using a CAS) gives $\alpha \approx 0.9855$. Compare values at the critical point and the endpoints:

$L(0) = 0$, $L(\pi/2) = v^2/g$, and $L(0.9855) \approx 1.20v^2/g$. Thus the distance traveled by the projectile is maximized for $\alpha \approx 0.9855$ or $\approx 56°$.

9. We can write the vector equation as $\mathbf{r}(t) = \mathbf{a}t^2 + \mathbf{b}t + \mathbf{c}$ where $\mathbf{a} = \langle a_1, a_2, a_3 \rangle$, $\mathbf{b} = \langle b_1, b_2, b_3 \rangle$, and $\mathbf{c} = \langle c_1, c_2, c_3 \rangle$.

Then $\mathbf{r}'(t) = 2t\,\mathbf{a} + \mathbf{b}$ which says that each tangent vector is the sum of a scalar multiple of $\mathbf{a}$ and the vector $\mathbf{b}$. Thus the tangent vectors are all parallel to the plane determined by $\mathbf{a}$ and $\mathbf{b}$ so the curve must be parallel to this plane. [Here we assume that $\mathbf{a}$ and $\mathbf{b}$ are nonparallel. Otherwise the tangent vectors are all parallel and the curve lies along a single line.] A normal vector for the plane is $\mathbf{a} \times \mathbf{b} = \langle a_2 b_3 - a_3 b_2, a_3 b_1 - a_1 b_3, a_1 b_2 - a_2 b_1 \rangle$. The point (c_1, c_2, c_3) lies on the plane (when $t = 0$), so an equation of the plane is

$$(a_2 b_3 - a_3 b_2)(x - c_1) + (a_3 b_1 - a_1 b_3)(y - c_2) + (a_1 b_2 - a_2 b_1)(z - c_3) = 0$$

or

$$(a_2 b_3 - a_3 b_2)x + (a_3 b_1 - a_1 b_3)y + (a_1 b_2 - a_2 b_1)z = a_2 b_3 c_1 - a_3 b_2 c_1 + a_3 b_1 c_2 - a_1 b_3 c_2 + a_1 b_2 c_3 - a_2 b_1 c_3$$

14 ☐ PARTIAL DERIVATIVES

14.1 Functions of Several Variables

1. (a) From Table 1, $f(-15, 40) = -27$, which means that if the temperature is $-15°$C and the wind speed is 40 km/h, then the air would feel equivalent to approximately $-27°$C without wind.

   (b) The question is asking: when the temperature is $-20°$C, what wind speed gives a wind-chill index of $-30°$C? From Table 1, the speed is 20 km/h.

   (c) The question is asking: when the wind speed is 20 km/h, what temperature gives a wind-chill index of $-49°$C? From Table 1, the temperature is $-35°$C.

   (d) The function $W = f(-5, v)$ means that we fix T at -5 and allow v to vary, resulting in a function of one variable. In other words, the function gives wind-chill index values for different wind speeds when the temperature is $-5°$C. From Table 1 (look at the row corresponding to $T = -5$), the function decreases and appears to approach a constant value as v increases.

   (e) The function $W = f(T, 50)$ means that we fix v at 50 and allow T to vary, again giving a function of one variable. In other words, the function gives wind-chill index values for different temperatures when the wind speed is 50 km/h . From Table 1 (look at the column corresponding to $v = 50$), the function increases almost linearly as T increases.

3. $P(120, 20) = 1.47(120)^{0.65}(20)^{0.35} \approx 94.2$, so when the manufacturer invests \$20 million in capital and 120,000 hours of labor are completed yearly, the monetary value of the production is about \$94.2 million.

5. (a) $f(160, 70) = 0.1091(160)^{0.425}(70)^{0.725} \approx 20.5$, which means that the surface area of a person 70 inches (5 feet 10 inches) tall who weighs 160 pounds is approximately 20.5 square feet.

   (b) Answers will vary depending on the height and weight of the reader.

7. (a) According to Table 4, $f(40, 15) = 25$, which means that if a 40-knot wind has been blowing in the open sea for 15 hours, it will create waves with estimated heights of 25 feet.

   (b) $h = f(30, t)$ means we fix v at 30 and allow t to vary, resulting in a function of one variable. Thus here, $h = f(30, t)$ gives the wave heights produced by 30-knot winds blowing for t hours. From the table (look at the row corresponding to $v = 30$), the function increases but at a declining rate as t increases. In fact, the function values appear to be approaching a limiting value of approximately 19, which suggests that 30-knot winds cannot produce waves higher than about 19 feet.

   (c) $h = f(v, 30)$ means we fix t at 30, again giving a function of one variable. So, $h = f(v, 30)$ gives the wave heights produced by winds of speed v blowing for 30 hours. From the table (look at the column corresponding to $t = 30$), the function appears to increase at an increasing rate, with no apparent limiting value. This suggests that faster winds (lasting 30 hours) always create higher waves.

9. (a) $g(2, -1) = \cos(2 + 2(-1)) = \cos(0) = 1$

(b) $x + 2y$ is defined for all choices of values for x and y and the cosine function is defined for all input values, so the domain of g is $\mathbb{R}^2$.

(c) The range of the cosine function is $[-1, 1]$ and $x + 2y$ generates all possible input values for the cosine function, so the range of $\cos(x + 2y)$ is $[-1, 1]$.

11. (a) $f(1, 1, 1) = \sqrt{1} + \sqrt{1} + \sqrt{1} + \ln(4 - 1^2 - 1^2 - 1^2) = 3 + \ln 1 = 3$

(b) $\sqrt{x}$, $\sqrt{y}$, $\sqrt{z}$ are defined only when $x \geq 0$, $y \geq 0$, $z \geq 0$, and $\ln(4 - x^2 - y^2 - z^2)$ is defined when
$4 - x^2 - y^2 - z^2 > 0 \;\Leftrightarrow\; x^2 + y^2 + z^2 < 4$, thus the domain is
$\{(x, y, z) \mid x^2 + y^2 + z^2 < 4, \; x \geq 0, \; y \geq 0, \; z \geq 0\}$, the portion of the interior of a sphere of radius 2, centered at the origin, that is in the first octant.

13. $\sqrt{x - 2}$ is defined only when $x - 2 \geq 0$, or $x \geq 2$, and $\sqrt{y - 1}$ is defined only when $y - 1 \geq 0$, or $y \geq 1$. So the domain of f is $\{(x, y) \mid x \geq 2, \; y \geq 1\}$.

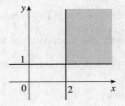

15. $\ln(9 - x^2 - 9y^2)$ is defined only when
$9 - x^2 - 9y^2 > 0$, or $\frac{1}{9}x^2 + y^2 < 1$. So the domain of f
is $\{(x, y) \mid \frac{1}{9}x^2 + y^2 < 1\}$, the interior of an ellipse.

17. g is not defined if $x + y = 0 \;\Leftrightarrow\; y = -x$ (and is defined otherwise). Thus the domain of g is
$\{(x, y) \mid y \neq -x\}$, the set of all points in $\mathbb{R}^2$ that are not on the line $y = -x$.

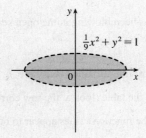

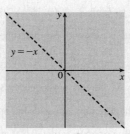

19. $\sqrt{y - x^2}$ is defined only when $y - x^2 \geq 0$, or $y \geq x^2$.

In addition, f is not defined if $1 - x^2 = 0$ ⇔

$x = \pm 1$. Thus the domain of f is

$\{(x, y) \mid y \geq x^2,\ x \neq \pm 1\}$.

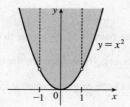

21. f is defined only when $4 - x^2 \geq 0$ ⇔ $-2 \leq x \leq 2$

and $9 - y^2 \geq 0$ ⇔ $-3 \leq y \leq 3$ and $1 - z^2 \geq 0$

⇔ $-1 \leq z \leq 1$. Thus the domain of f is

$\{(x, y, z) \mid -2 \leq x \leq 2,\ -3 \leq y \leq 3,\ -1 \leq z \leq 1\}$,

a solid rectangular box with vertices $(\pm 2, \pm 3, \pm 1)$

(all combinations).

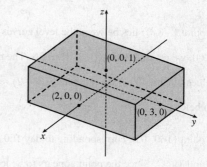

23. The graph of f has equation $z = y$, a plane which

intersects the yz-plane in the line $z = y$, $x = 0$. The

portion of this plane in the first octant is shown.

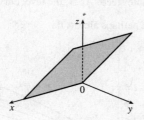

25. $z = 10 - 4x - 5y$ or $4x + 5y + z = 10$, a plane with

intercepts 2.5, 2, and 10.

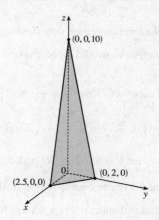

27. $z = \sin x$, a cylinder.

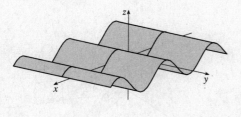

29. $z = x^2 + 4y^2 + 1$, an elliptic paraboloid opening upward

with vertex at $(0, 0, 1)$.

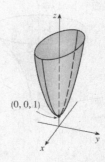

31. $z = \sqrt{4 - 4x^2 - y^2}$ so $4x^2 + y^2 + z^2 = 4$ or $x^2 + \dfrac{y^2}{4} + \dfrac{z^2}{4} = 1$

and $z \geq 0$, the top half of an ellipsoid.

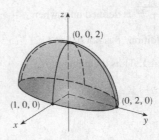

33. The point $(-3, 3)$ lies between the level curves with z-values 50 and 60. Since the point is a little closer to the level curve with $z = 60$, we estimate that $f(-3, 3) \approx 56$. The point $(3, -2)$ appears to be just about halfway between the level curves with z-values 30 and 40, so we estimate $f(3, -2) \approx 35$. The graph rises as we approach the origin, gradually from above, steeply from below.

35. The point $(160, 10)$, corresponding to day 160 and a depth of 10 m, lies between the isothermals with temperature values of 8 and $12°$C. Since the point appears to be located about three-fourths the distance from the $8°$C isothermal to the $12°$C isothermal, we estimate the temperature at that point to be approximately $11°$C. The point $(180, 5)$ lies between the 16 and $20°$C isothermals, very close to the $20°$C level curve, so we estimate the temperature there to be about $19.5°$C.

37. Near A, the level curves are very close together, indicating that the terrain is quite steep. At B, the level curves are much farther apart, so we would expect the terrain to be much less steep than near A, perhaps almost flat.

39. The level curves of $B(m, h) = \dfrac{m}{h^2}$ are $\dfrac{m}{h^2} = k \iff m = kh^2$ or

equivalently $h = \sqrt{m/k} = \dfrac{1}{\sqrt{k}} \sqrt{m}$ since $m > 0$, $h > 0$. We draw the

level curves for $k = 18.5, 25, 30,$ and 40.

The shaded region corresponds to BMI values between 18.5 and 25, those considered optimal. For a mass of 62 kg and a height of 152 cm

(1.52 m), the BMI is $B(62, 1.52) = \dfrac{62}{1.52^2} \approx 26.8$, which is outside the optimal range.

41.

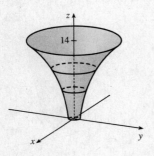

43.

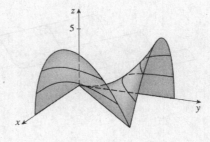

45. The level curves are $x^2 - y^2 = k$. When $k = 0$ the level curve is the pair of lines $y = \pm x$, and when $k \neq 0$ the level curves are a family of hyperbolas (oriented differently for $k > 0$ than for $k < 0$).

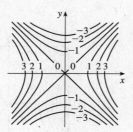

47. The level curves are $\sqrt{x} + y = k$ or $y = -\sqrt{x} + k$, a family of vertical translations of the graph of the root function $y = -\sqrt{x}$.

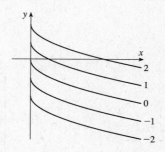

49. The level curves are $ye^x = k$ or $y = ke^{-x}$, a family of exponential curves.

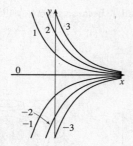

51. The level curves are $\sqrt[3]{x^2 + y^2} = k$ or $x^2 + y^2 = k^3$ ($k \geq 0$), a family of circles centered at the origin with radius $k^{3/2}$.

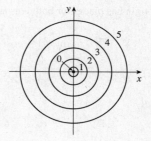

53. The contour map consists of the level curves $k = x^2 + 9y^2$, a family of ellipses with major axis the x-axis. (Or, if $k = 0$, the origin.)

The graph of $f(x, y)$ is the surface $z = x^2 + 9y^2$, an elliptic paraboloid.

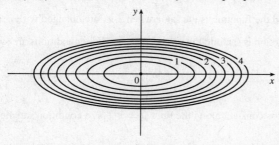

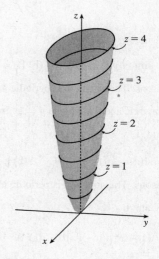

If we visualize lifting each ellipse $k = x^2 + 9y^2$ of the contour map to the plane $z = k$, we have horizontal traces that indicate the shape of the graph of f.

55. The isothermals are given by $k = 100/(1 + x^2 + 2y^2)$ or

$x^2 + 2y^2 = (100 - k)/k$ $[0 < k \leq 100]$, a family of ellipses.

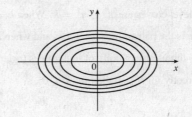

57. $f(x, y) = xy^2 - x^3$

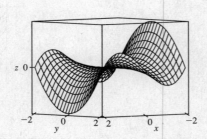

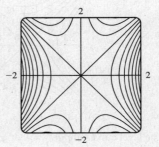

The traces parallel to the yz-plane (such as the left-front trace in the graph above) are parabolas; those parallel to the xz-plane (such as the right-front trace) are cubic curves. The surface is called a monkey saddle because a monkey sitting on the surface near the origin has places for both legs and tail to rest.

59. $f(x, y) = e^{-(x^2+y^2)/3} \left(\sin(x^2) + \cos(y^2) \right)$

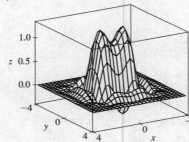

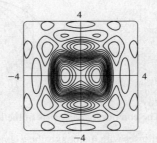

61. $z = \sin(xy)$ (a) C   (b) II

Reasons: This function is periodic in both x and y, and the function is the same when x is interchanged with y, so its graph is symmetric about the plane $y = x$. In addition, the function is 0 along the x- and y-axes. These conditions are satisfied only by C and II.

63. $z = \sin(x - y)$ (a) F   (b) I

Reasons: This function is periodic in both x and y but is constant along the lines $y = x + k$, a condition satisfied only by F and I.

65. $z = (1 - x^2)(1 - y^2)$ (a) B   (b) VI

Reasons: This function is 0 along the lines $x = \pm 1$ and $y = \pm 1$. The only contour map in which this could occur is VI. Also note that the trace in the xz-plane is the parabola $z = 1 - x^2$ and the trace in the yz-plane is the parabola $z = 1 - y^2$, so the graph is B.

67. $k = x + 3y + 5z$ is a family of parallel planes with normal vector $\langle 1, 3, 5 \rangle$.

69. Equations for the level surfaces are $k = y^2 + z^2$. For $k > 0$, we have a family of circular cylinders with axis the x-axis and radius $\sqrt{k}$. When $k = 0$ the level surface is the x-axis. (There are no level surfaces for $k < 0$.)

71. (a) The graph of g is the graph of f shifted upward 2 units.

(b) The graph of g is the graph of f stretched vertically by a factor of 2.

(c) The graph of g is the graph of f reflected about the xy-plane.

(d) The graph of $g(x, y) = -f(x, y) + 2$ is the graph of f reflected about the xy-plane and then shifted upward 2 units.

73. $f(x, y) = 3x - x^4 - 4y^2 - 10xy$

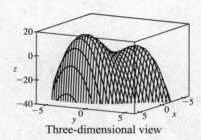
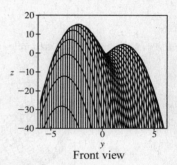

Three-dimensional view   Front view

It does appear that the function has a maximum value, at the higher of the two "hilltops." From the front view graph, the maximum value appears to be approximately 15. Both hilltops could be considered local maximum points, as the values of f there are larger than at the neighboring points. There does not appear to be any local minimum point; although the valley shape between the two peaks looks like a minimum of some kind, some neighboring points have lower function values.

75.

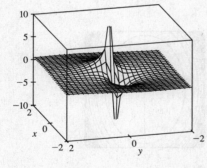

$f(x, y) = \dfrac{x + y}{x^2 + y^2}$. As both x and y become large, the function values appear to approach 0, regardless of which direction is considered. As (x, y) approaches the origin, the graph exhibits asymptotic behavior. From some directions, $f(x, y) \to \infty$, while in others $f(x, y) \to -\infty$. (These are the vertical spikes visible in the graph.) If the graph is examined carefully, however, one can see that $f(x, y)$ approaches 0 along the line $y = -x$.

77. $f(x, y) = e^{cx^2 + y^2}$. First, if $c = 0$, the graph is the cylindrical surface $z = e^{y^2}$ (whose level curves are parallel lines). When $c > 0$, the vertical trace above the y-axis remains fixed while the sides of the surface in the x-direction "curl" upward, giving the graph a shape resembling an elliptic paraboloid. The level curves of the surface are ellipses centered at the origin.

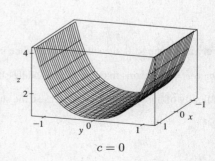

$c = 0$

[continued]

For $0 < c < 1$, the ellipses have major axis the x-axis and the eccentricity increases as $c \to 0$.

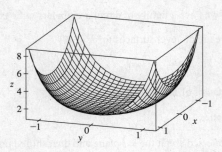 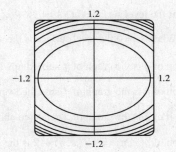

$c = 0.5$ (level curves in increments of 1)

For $c = 1$ the level curves are circles centered at the origin.

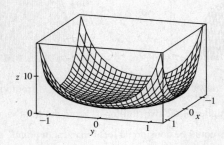 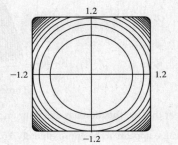

$c = 1$ (level curves in increments of 1)

When $c > 1$, the level curves are ellipses with major axis the y-axis, and the eccentricity increases as c increases.

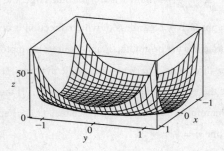 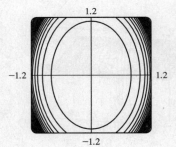

$c = 2$ (level curves in increments of 4)

For values of $c < 0$, the sides of the surface in the x-direction curl downward and approach the xy-plane (while the vertical trace $x = 0$ remains fixed), giving a saddle-shaped appearance to the graph near the point $(0, 0, 1)$. The level curves consist of a family of hyperbolas. As c decreases, the surface becomes flatter in the x-direction and the surface's approach to the curve in the trace $x = 0$ becomes steeper, as the graphs demonstrate.

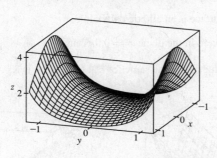 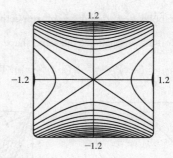

$c = -0.5$ (level curves in increments of 0.25)

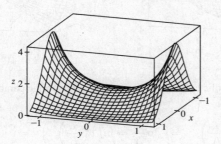 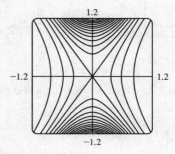

$c = -2$ (level curves in increments of 0.25)

79. $z = x^2 + y^2 + cxy$. When $c < -2$, the surface intersects the plane $z = k \neq 0$ in a hyperbola. (See the following graph.)

It intersects the plane $x = y$ in the parabola $z = (2 + c)x^2$, and the plane $x = -y$ in the parabola $z = (2 - c)x^2$. These

parabolas open in opposite directions, so the surface is a hyperbolic paraboloid.

When $c = -2$ the surface is $z = x^2 + y^2 - 2xy = (x - y)^2$. So the surface is constant along each line $x - y = k$. That

is, the surface is a cylinder with axis $x - y = 0$, $z = 0$. The shape of the cylinder is determined by its intersection with the

plane $x + y = 0$, where $z = 4x^2$, and hence the cylinder is parabolic with minima of 0 on the line $y = x$.

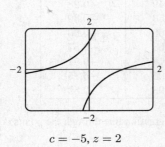 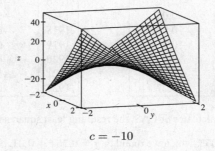 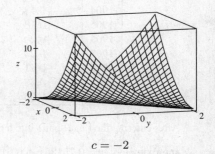

$c = -5$, $z = 2$ $\qquad\qquad$ $c = -10$ $\qquad\qquad$ $c = -2$

When $-2 < c \leq 0$, $z \geq 0$ for all x and y. If x and y have the same sign, then

$x^2 + y^2 + cxy \geq x^2 + y^2 - 2xy = (x - y)^2 \geq 0$. If they have opposite signs, then $cxy \geq 0$. The intersection with the surface and the plane $z = k > 0$ is an ellipse (see graph below). The intersection with the surface and the planes $x = 0$ and

$y = 0$ are parabolas $z = y^2$ and $z = x^2$ respectively, so the surface is an elliptic paraboloid.

When $c > 0$ the graphs have the same shape, but are reflected in the plane $x = 0$, because

$x^2 + y^2 + cxy = (-x)^2 + y^2 + (-c)(-x)y$. That is, the value of z is the same for c at (x, y) as it is for $-c$ at $(-x, y)$.

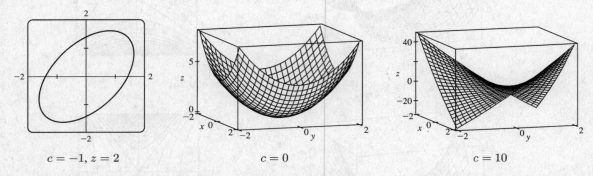

| $c = -1, z = 2$ | $c = 0$ | $c = 10$ |

So the surface is an elliptic paraboloid for $0 < c < 2$, a parabolic cylinder for $c = 2$, and a hyperbolic paraboloid for $c > 2$.

81. (a) $P = bL^\alpha K^{1-\alpha}$ $\Rightarrow$ $\dfrac{P}{K} = bL^\alpha K^{-\alpha}$ $\Rightarrow$ $\dfrac{P}{K} = b\left(\dfrac{L}{K}\right)^\alpha$ $\Rightarrow$ $\ln\dfrac{P}{K} = \ln\left(b\left(\dfrac{L}{K}\right)^\alpha\right)$ $\Rightarrow$

$\ln\dfrac{P}{K} = \ln b + \alpha \ln\left(\dfrac{L}{K}\right)$

(b) We list the values for $\ln(L/K)$ and $\ln(P/K)$ for the years 1899–1922. (Historically, these values were rounded to

2 decimal places.)

Year	$x = \ln(L/K)$	$y = \ln(P/K)$	Year	$x = \ln(L/K)$	$y = \ln(P/K)$
1899	0	0	1911	−0.38	−0.34
1900	−0.02	−0.06	1912	−0.38	−0.24
1901	−0.04	−0.02	1913	−0.41	−0.25
1902	−0.04	0	1914	−0.47	−0.37
1903	−0.07	−0.05	1915	−0.53	−0.34
1904	−0.13	−0.12	1916	−0.49	−0.28
1905	−0.18	−0.04	1917	−0.53	−0.39
1906	−0.20	−0.07	1918	−0.60	−0.50
1907	−0.23	−0.15	1919	−0.68	−0.57
1908	−0.41	−0.38	1920	−0.74	−0.57
1909	−0.33	−0.24	1921	−1.05	−0.85
1910	−0.35	−0.27	1922	−0.98	−0.59

After entering the (x, y) pairs into a calculator or CAS, the resulting least squares regression line through the points is

approximately $y = 0.75136x + 0.01053$, which we round to $y = 0.75x + 0.01$.

(c) Comparing the regression line from part (b) to the equation $y = \ln b + \alpha x$ with $x = \ln(L/K)$ and $y = \ln(P/K)$, we have

$\alpha = 0.75$ and $\ln b = 0.01$ $\Rightarrow$ $b = e^{0.01} \approx 1.01$. Thus, the Cobb-Douglas production function is

$P = bL^\alpha K^{1-\alpha} = 1.01L^{0.75}K^{0.25}$.

14.2   Limits and Continuity

1. In general, we can't say anything about $f(3,1)$! $\lim\limits_{(x,y)\to(3,1)} f(x,y) = 6$ means that the values of $f(x,y)$ approach 6 as

(x,y) approaches, but is not equal to, $(3,1)$. If f is continuous, we know that $\lim\limits_{(x,y)\to(a,b)} f(x,y) = f(a,b)$, so

$\lim\limits_{(x,y)\to(3,1)} f(x,y) = f(3,1) = 6$.

3. We make a table of values of

$f(x,y) = \dfrac{x^2y^3 + x^3y^2 - 5}{2 - xy}$ for a set

of (x,y) points near the origin.

y \ x	−0.2	−0.1	−0.05	0	0.05	0.1	0.2
−0.2	−2.551	−2.525	−2.513	−2.500	−2.488	−2.475	−2.451
−0.1	−2.525	−2.513	−2.506	−2.500	−2.494	−2.488	−2.475
−0.05	−2.513	−2.506	−2.503	−2.500	−2.497	−2.494	−2.488
0	−2.500	−2.500	−2.500		−2.500	−2.500	−2.500
0.05	−2.488	−2.494	−2.497	−2.500	−2.503	−2.506	−2.513
0.1	−2.475	−2.488	−2.494	−2.500	−2.506	−2.513	−2.525
0.2	−2.451	−2.475	−2.488	−2.500	−2.513	−2.525	−2.551

As the table shows, the values of $f(x,y)$ seem to approach -2.5 as (x,y) approaches the origin from a variety of different

directions. This suggests that $\lim\limits_{(x,y)\to(0,0)} f(x,y) = -2.5$. Since f is a rational function, it is continuous on its domain. f is

defined at $(0,0)$, so we can use direct substitution to establish that $\lim\limits_{(x,y)\to(0,0)} f(x,y) = \dfrac{0^2 0^3 + 0^3 0^2 - 5}{2 - 0 \cdot 0} = -\dfrac{5}{2}$, verifying

our guess.

5. $f(x,y) = x^2 y^3 - 4y^2$ is a polynomial, and hence continuous, so we can find the limit by direct substitution:

$\lim\limits_{(x,y)\to(3,2)} f(x,y) = f(3,2) = (3)^2 (2)^3 - 4(2)^2 = 56$.

7. $x - y$ is a polynomial and therefore continuous. Since $\sin t$ is a continuous function, the composition $\sin(x - y)$ is also

continuous. The function y is a polynomial, and hence continuous, and the product of continuous functions is continuous, so

$f(x,y) = y\sin(x - y)$ is a continuous function. Then $\lim\limits_{(x,y)\to(\pi,\pi/2)} f(x,y) = f\left(\pi, \tfrac{\pi}{2}\right) = \tfrac{\pi}{2} \sin\left(\pi - \tfrac{\pi}{2}\right) = \tfrac{\pi}{2} \sin\tfrac{\pi}{2} = \tfrac{\pi}{2}$.

9. $f(x,y) = (x^4 - 4y^2)/(x^2 + 2y^2)$.   First approach $(0,0)$ along the x-axis. Then $f(x,0) = x^4/x^2 = x^2$ for $x \neq 0$, so

$f(x,y) \to 0$. Now approach $(0,0)$ along the y-axis. For $y \neq 0$, $f(0,y) = -4y^2/2y^2 = -2$, so $f(x,y) \to -2$. Since f has

two different limits along two different lines, the limit does not exist.

11. $f(x,y) = (y^2 \sin^2 x)/(x^4 + y^4)$.   On the x-axis, $f(x,0) = 0$ for $x \neq 0$, so $f(x,y) \to 0$ as $(x,y) \to (0,0)$ along the

x-axis. Approaching $(0,0)$ along the line $y = x$, $f(x,x) = \dfrac{x^2 \sin^2 x}{x^4 + x^4} = \dfrac{\sin^2 x}{2x^2} = \dfrac{1}{2}\left(\dfrac{\sin x}{x}\right)^2$ for $x \neq 0$ and

$\lim\limits_{x\to 0} \dfrac{\sin x}{x} = 1$, so $f(x,y) \to \tfrac{1}{2}$. Since f has two different limits along two different lines, the limit does not exist.

13. $f(x,y) = \dfrac{xy}{\sqrt{x^2+y^2}}$. We can see that the limit along any line through $(0,0)$ is 0, as well as along other paths through

$(0,0)$ such as $x = y^2$ and $y = x^2$. So we suspect that the limit exists and equals 0; we use the Squeeze Theorem to prove our

assertion. Since $|y| \leq \sqrt{x^2+y^2}$, we have $\dfrac{|y|}{\sqrt{x^2+y^2}} \leq 1$ and so $0 \leq \left| \dfrac{xy}{\sqrt{x^2+y^2}} \right| \leq |x|$. Now $|x| \to 0$ as $(x,y) \to (0,0)$,

so $\left| \dfrac{xy}{\sqrt{x^2+y^2}} \right| \to 0$ and hence $\displaystyle \lim_{(x,y)\to(0,0)} f(x,y) = 0$.

15. Let $f(x,y) = \dfrac{xy^2 \cos y}{x^2+y^4}$. Then $f(x,0) = 0$ for $x \neq 0$, so $f(x,y) \to 0$ as $(x,y) \to (0,0)$ along the x-axis. Approaching

$(0,0)$ along the y-axis or the line $y = x$ also gives a limit of 0. But $f(y^2,y) = \dfrac{y^2 y^2 \cos y}{(y^2)^2 + y^4} = \dfrac{y^4 \cos y}{2y^4} = \dfrac{\cos y}{2}$ for $y \neq 0$,

so $f(x,y) \to \frac{1}{2} \cos 0 = \frac{1}{2}$ as $(x,y) \to (0,0)$ along the parabola $x = y^2$. Thus the limit doesn't exist.

17. $\displaystyle \lim_{(x,y)\to(0,0)} \frac{x^2+y^2}{\sqrt{x^2+y^2+1}-1} = \lim_{(x,y)\to(0,0)} \frac{x^2+y^2}{\sqrt{x^2+y^2+1}-1} \cdot \frac{\sqrt{x^2+y^2+1}+1}{\sqrt{x^2+y^2+1}+1}$

$= \displaystyle \lim_{(x,y)\to(0,0)} \frac{(x^2+y^2)\left(\sqrt{x^2+y^2+1}+1\right)}{x^2+y^2} = \lim_{(x,y)\to(0,0)} \left(\sqrt{x^2+y^2+1}+1\right) = 2$

19. e^{y^2} is a composition of continuous functions and hence continuous. xz is a continuous function and $\tan t$ is continuous for

$t \neq \frac{\pi}{2} + n\pi$ (n an integer), so the composition $\tan(xz)$ is continuous for $xz \neq \frac{\pi}{2} + n\pi$. Thus the product

$f(x,y,z) = e^{y^2} \tan(xz)$ is a continuous function for $xz \neq \frac{\pi}{2} + n\pi$. If $x = \pi$ and $z = \frac{1}{3}$ then $xz \neq \frac{\pi}{2} + n\pi$, so

$\displaystyle \lim_{(x,y,z)\to(\pi,0,1/3)} f(x,y,z) = f\left(\pi,0,1/3\right) = e^{0^2} \tan(\pi \cdot 1/3) = 1 \cdot \tan(\pi/3) = \sqrt{3}$.

21. $f(x,y,z) = \dfrac{xy + yz^2 + xz^2}{x^2 + y^2 + z^4}$. Then $f(x,0,0) = 0/x^2 = 0$ for $x \neq 0$, so as $(x,y,z) \to (0,0,0)$ along the x-axis,

$f(x,y,z) \to 0$. But $f(x,x,0) = x^2/(2x^2) = \frac{1}{2}$ for $x \neq 0$, so as $(x,y,z) \to (0,0,0)$ along the line $y = x$, $z = 0$,

$f(x,y,z) \to \frac{1}{2}$. Thus the limit doesn't exist.

23.

From the ridges on the graph, we see that as $(x,y) \to (0,0)$ along the lines under the two ridges, $f(x,y)$ approaches different values. So the limit does not exist.

25. $h(x,y) = g(f(x,y)) = (2x + 3y - 6)^2 + \sqrt{2x + 3y - 6}$. Since f is a polynomial, it is continuous on $\mathbb{R}^2$ and g is

continuous on its domain $\{t \mid t \geq 0\}$. Thus h is continuous on its domain

$\{(x,y) \mid 2x + 3y - 6 \geq 0\} = \{(x,y) \mid y \geq -\frac{2}{3}x + 2\}$, which consists of all points on or above the line $y = -\frac{2}{3}x + 2$.

27.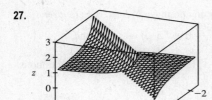

From the graph, it appears that f is discontinuous along the line $y = x$.

If we consider $f(x, y) = e^{1/(x-y)}$ as a composition of functions,

$g(x, y) = 1/(x - y)$ is a rational function and therefore continuous except

where $x - y = 0 \iff y = x$. Since the function $h(t) = e^t$ is continuous

everywhere, the composition $h(g(x, y)) = e^{1/(x-y)} = f(x, y)$ is

continuous except along the line $y = x$, as we suspected.

29. The functions xy and $1 + e^{x-y}$ are continuous everywhere, and $1 + e^{x-y}$ is never zero, so $F(x, y) = \dfrac{xy}{1 + e^{x-y}}$ is continuous

on its domain $\mathbb{R}^2$.

31. $F(x, y) = \dfrac{1 + x^2 + y^2}{1 - x^2 - y^2}$ is a rational function and thus is continuous on its domain

$\{(x, y) \mid 1 - x^2 - y^2 \neq 0\} = \{(x, y) \mid x^2 + y^2 \neq 1\}$.

33. $\sqrt{x}$ is continuous on its domain $\{(x, y) \mid x \geq 0\}$ and $\sqrt{1 - x^2 - y^2}$ is continuous on its domain

$\{(x, y) \mid 1 - x^2 - y^2 \geq 0\} = \{(x, y) \mid x^2 + y^2 \leq 1\}$, so the sum $G(x, y) = \sqrt{x} + \sqrt{1 - x^2 - y^2}$ is continuous for $x \geq 0$

and $x^2 + y^2 \leq 1$, that is, $\{(x, y) \mid x^2 + y^2 \leq 1, \, x \geq 0\}$. This is the right half of the unit disk.

35. $f(x, y, z) = h(g(x, y, z))$ where $g(x, y, z) = x^2 + y^2 + z^2$, a polynomial that is continuous

everywhere, and $h(t) = \arcsin t$, continuous on $[-1, 1]$. Thus f is continuous on its domain

$\{(x, y, z) \mid -1 \leq x^2 + y^2 + z^2 \leq 1\} = \{(x, y, z) \mid x^2 + y^2 + z^2 \leq 1\}$, so f is continuous on the unit ball.

37. $f(x, y) = \begin{cases} \dfrac{x^2 y^3}{2x^2 + y^2} & \text{if } (x, y) \neq (0, 0) \\ 1 & \text{if } (x, y) = (0, 0) \end{cases}$ The first piece of f is a rational function defined everywhere except at the

origin, so f is continuous on $\mathbb{R}^2$ except possibly at the origin. Since $x^2 \leq 2x^2 + y^2$, we have $\left| x^2 y^3 / (2x^2 + y^2) \right| \leq \left| y^3 \right|$.

We know that $\left| y^3 \right| \to 0$ as $(x, y) \to (0, 0)$. So, by the Squeeze Theorem, $\displaystyle\lim_{(x,y) \to (0,0)} f(x, y) = \lim_{(x,y) \to (0,0)} \dfrac{x^2 y^3}{2x^2 + y^2} = 0$.

But $f(0, 0) = 1$, so f is discontinuous at $(0, 0)$. Therefore, f is continuous on the set $\{(x, y) \mid (x, y) \neq (0, 0)\}$.

39. $\displaystyle\lim_{(x,y) \to (0,0)} \dfrac{x^3 + y^3}{x^2 + y^2} = \lim_{r \to 0^+} \dfrac{(r \cos \theta)^3 + (r \sin \theta)^3}{r^2} = \lim_{r \to 0^+} (r \cos^3 \theta + r \sin^3 \theta) = 0$

41. $\displaystyle\lim_{(x,y) \to (0,0)} \dfrac{e^{-x^2 - y^2} - 1}{x^2 + y^2} = \lim_{r \to 0^+} \dfrac{e^{-r^2} - 1}{r^2} = \lim_{r \to 0^+} \dfrac{e^{-r^2}(-2r)}{2r}$ [using l'Hospital's Rule]

$$= \lim_{r \to 0^+} -e^{-r^2} = -e^0 = -1$$

43. $f(x, y) = \begin{cases} \dfrac{\sin(xy)}{xy} & \text{if } (x, y) \neq (0, 0) \\ 1 & \text{if } (x, y) = (0, 0) \end{cases}$

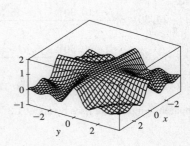

From the graph, it appears that f is continuous everywhere. We know

xy is continuous on $\mathbb{R}^2$ and $\sin t$ is continuous everywhere, so

$\sin(xy)$ is continuous on $\mathbb{R}^2$ and $\dfrac{\sin(xy)}{xy}$ is continuous on $\mathbb{R}^2$

except possibly where $xy = 0$. To show that f is continuous at those points, consider any point (a, b) in $\mathbb{R}^2$ where $ab = 0$. Because xy is continuous, $xy \to ab = 0$ as $(x, y) \to (a, b)$. If we let $t = xy$, then $t \to 0$ as $(x, y) \to (a, b)$ and

$$\lim_{(x,y)\to(a,b)} \frac{\sin(xy)}{xy} = \lim_{t\to 0} \frac{\sin(t)}{t} = 1 \text{ by Equation 2.4.2 [ET 3.3.2]. Thus } \lim_{(x,y)\to(a,b)} f(x, y) = f(a, b) \text{ and } f \text{ is continuous}$$

on $\mathbb{R}^2$.

45. Since $|\mathbf{x} - \mathbf{a}|^2 = |\mathbf{x}|^2 + |\mathbf{a}|^2 - 2 |\mathbf{x}|\, |\mathbf{a}| \cos\theta \geq |\mathbf{x}|^2 + |\mathbf{a}|^2 - 2 |\mathbf{x}|\, |\mathbf{a}| = (|\mathbf{x}| - |\mathbf{a}|)^2$, we have $\bigl|\, |\mathbf{x}| - |\mathbf{a}|\, \bigr| \leq |\mathbf{x} - \mathbf{a}|$. Let $\epsilon > 0$ be given and set $\delta = \epsilon$. Then if $0 < |\mathbf{x} - \mathbf{a}| < \delta$, $\bigl|\, |\mathbf{x}| - |\mathbf{a}|\, \bigr| \leq |\mathbf{x} - \mathbf{a}| < \delta = \epsilon$. Hence $\lim_{\mathbf{x}\to\mathbf{a}} |\mathbf{x}| = |\mathbf{a}|$ and $f(\mathbf{x}) = |\mathbf{x}|$ is continuous on $\mathbb{R}^n$.

14.3 Partial Derivatives

1. (a) $\partial T/\partial x$ represents the rate of change of T when we fix y and t and consider T as a function of the single variable x, which describes how quickly the temperature changes when longitude changes but latitude and time are constant. $\partial T/\partial y$ represents the rate of change of T when we fix x and t and consider T as a function of y, which describes how quickly the temperature changes when latitude changes but longitude and time are constant. $\partial T/\partial t$ represents the rate of change of T when we fix x and y and consider T as a function of t, which describes how quickly the temperature changes over time for a constant longitude and latitude.

(b) $f_x(158, 21, 9)$ represents the rate of change of temperature at longitude $158°$W, latitude $21°$N at 9:00 AM when only longitude varies. Since the air is warmer to the west than to the east, increasing longitude results in an increased air temperature, so we would expect $f_x(158, 21, 9)$ to be positive. $f_y(158, 21, 9)$ represents the rate of change of temperature at the same time and location when only latitude varies. Since the air is warmer to the south and cooler to the north, increasing latitude results in a decreased air temperature, so we would expect $f_y(158, 21, 9)$ to be negative. $f_t(158, 21, 9)$ represents the rate of change of temperature at the same time and location when only time varies. Since typically air temperature increases from the morning to the afternoon as the sun warms it, we would expect $f_t(158, 21, 9)$ to be positive.

3. (a) By Definition 4, $f_T(-15, 30) = \lim\limits_{h\to 0} \dfrac{f(-15 + h, 30) - f(-15, 30)}{h}$, which we can approximate by considering $h = 5$ and $h = -5$ and using the values given in the table:

$$f_T(-15, 30) \approx \frac{f(-10, 30) - f(-15, 30)}{5} = \frac{-20 - (-26)}{5} = \frac{6}{5} = 1.2,$$

$$f_T(-15, 30) \approx \frac{f(-20, 30) - f(-15, 30)}{-5} = \frac{-33 - (-26)}{-5} = \frac{-7}{-5} = 1.4. \text{ Averaging these values, we estimate}$$

$f_T(-15, 30)$ to be approximately 1.3. Thus, when the actual temperature is $-15°$C and the wind speed is 30 km/h, the apparent temperature rises by about $1.3°$C for every degree that the actual temperature rises.

Similarly, $f_v(-15, 30) = \lim\limits_{h \to 0} \dfrac{f(-15, 30 + h) - f(-15, 30)}{h}$ which we can approximate by considering $h = 10$

and $h = -10$: $f_v(-15, 30) \approx \dfrac{f(-15, 40) - f(-15, 30)}{10} = \dfrac{-27 - (-26)}{10} = \dfrac{-1}{10} = -0.1$,

$f_v(-15, 30) \approx \dfrac{f(-15, 20) - f(-15, 30)}{-10} = \dfrac{-24 - (-26)}{-10} = \dfrac{2}{-10} = -0.2$. Averaging these values, we estimate

$f_v(-15, 30)$ to be approximately -0.15. Thus, when the actual temperature is $-15°$C and the wind speed is 30 km/h, the

apparent temperature decreases by about $0.15°$C for every km/h that the wind speed increases.

(b) For a fixed wind speed v, the values of the wind-chill index W increase as temperature T increases (look at a column of

the table), so $\dfrac{\partial W}{\partial T}$ is positive. For a fixed temperature T, the values of W decrease (or remain constant) as v increases

(look at a row of the table), so $\dfrac{\partial W}{\partial v}$ is negative (or perhaps 0).

(c) For fixed values of T, the function values $f(T, v)$ appear to become constant (or nearly constant) as v increases, so the

corresponding rate of change is 0 or near 0 as v increases. This suggests that $\lim\limits_{v \to \infty} (\partial W / \partial v) = 0$.

5. (a) If we start at $(1, 2)$ and move in the positive x-direction, the graph of f increases. Thus $f_x(1, 2)$ is positive.

(b) If we start at $(1, 2)$ and move in the positive y-direction, the graph of f decreases. Thus $f_y(1, 2)$ is negative.

7. (a) $f_{xx} = \frac{\partial}{\partial x}(f_x)$, so f_{xx} is the rate of change of f_x in the x-direction. f_x is negative at $(-1, 2)$ and if we move in the

positive x-direction, the surface becomes less steep. Thus the values of f_x are increasing and $f_{xx}(-1, 2)$ is positive.

(b) f_{yy} is the rate of change of f_y in the y-direction. f_y is negative at $(-1, 2)$ and if we move in the positive y-direction, the

surface becomes steeper. Thus the values of f_y are decreasing, and $f_{yy}(-1, 2)$ is negative.

9. First of all, if we start at the point $(3, -3)$ and move in the positive y-direction, we see that both b and c decrease, while a

increases. Both b and c have a low point at about $(3, -1.5)$, while a is 0 at this point. So a is definitely the graph of f_y, and

one of b and c is the graph of f. To see which is which, we start at the point $(-3, -1.5)$ and move in the positive x-direction.

b traces out a line with negative slope, while c traces out a parabola opening downward. This tells us that b is the x-derivative

of c. So c is the graph of f, b is the graph of f_x, and a is the graph of f_y.

11. $f(x, y) = 16 - 4x^2 - y^2 \;\Rightarrow\; f_x(x, y) = -8x$ and $f_y(x, y) = -2y \;\Rightarrow\; f_x(1, 2) = -8$ and $f_y(1, 2) = -4$. The graph

of f is the paraboloid $z = 16 - 4x^2 - y^2$ and the vertical plane $y = 2$ intersects it in the parabola $z = 12 - 4x^2$, $y = 2$

(the curve C_1 in the first figure). The slope of the tangent line
to this parabola at $(1, 2, 8)$ is $f_x(1, 2) = -8$. Similarly the
plane $x = 1$ intersects the paraboloid in the parabola
$z = 12 - y^2$, $x = 1$ (the curve C_2 in the second figure) and
the slope of the tangent line at $(1, 2, 8)$ is $f_y(1, 2) = -4$.

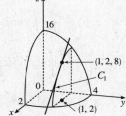
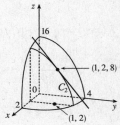

13. $f(x, y) = x^2 y^3 \quad \Rightarrow \quad f_x = 2xy^3, \quad f_y = 3x^2 y^2$

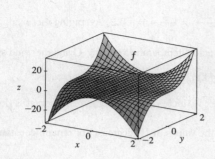

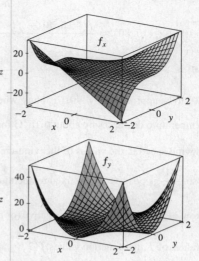

Note that traces of f in planes parallel to the xz-plane are parabolas which open downward for $y < 0$ and upward for $y > 0$, and the traces of f_x in these planes are straight lines, which have negative slopes for $y < 0$ and positive slopes for $y > 0$. The traces of f in planes parallel to the yz-plane are cubic curves, and the traces of f_y in these planes are parabolas.

15. $f(x, y) = x^4 + 5xy^3 \quad \Rightarrow \quad f_x(x, y) = 4x^3 + 5y^3, \quad f_y(x, y) = 0 + 5x \cdot 3y^2 = 15xy^2$

17. $f(x, t) = t^2 e^{-x} \quad \Rightarrow \quad f_x(x, t) = t^2 \cdot e^{-x}(-1) = -t^2 e^{-x}, \quad f_t(x, t) = 2te^{-x}$

19. $z = \ln(x + t^2) \quad \Rightarrow \quad \dfrac{\partial z}{\partial x} = \dfrac{1}{x + t^2}, \quad \dfrac{\partial z}{\partial t} = \dfrac{2t}{x + t^2}$

21. $f(x, y) = x/y = xy^{-1} \quad \Rightarrow \quad f_x(x, y) = y^{-1} = 1/y, \quad f_y(x, y) = -xy^{-2} = -x/y^2$

23. $f(x, y) = \dfrac{ax + by}{cx + dy} \quad \Rightarrow \quad f_x(x, y) = \dfrac{(cx + dy)(a) - (ax + by)(c)}{(cx + dy)^2} = \dfrac{(ad - bc)y}{(cx + dy)^2}$,

$f_y(x, y) = \dfrac{(cx + dy)(b) - (ax + by)(d)}{(cx + dy)^2} = \dfrac{(bc - ad)x}{(cx + dy)^2}$

25. $g(u, v) = (u^2 v - v^3)^5 \quad \Rightarrow \quad g_u(u, v) = 5(u^2 v - v^3)^4 \cdot 2uv = 10uv(u^2 v - v^3)^4$,

$g_v(u, v) = 5(u^2 v - v^3)^4 (u^2 - 3v^2) = 5(u^2 - 3v^2)(u^2 v - v^3)^4$

27. $R(p, q) = \tan^{-1}(pq^2) \quad \Rightarrow \quad R_p(p, q) = \dfrac{1}{1 + (pq^2)^2} \cdot q^2 = \dfrac{q^2}{1 + p^2 q^4}, \quad R_q(p, q) = \dfrac{1}{1 + (pq^2)^2} \cdot 2pq = \dfrac{2pq}{1 + p^2 q^4}$

29. $F(x, y) = \displaystyle\int_y^x \cos(e^t)\, dt \quad \Rightarrow \quad F_x(x, y) = \dfrac{\partial}{\partial x} \displaystyle\int_y^x \cos(e^t)\, dt = \cos(e^x)$ by the Fundamental Theorem of Calculus, Part 1;

$F_y(x, y) = \dfrac{\partial}{\partial y} \displaystyle\int_y^x \cos(e^t)\, dt = \dfrac{\partial}{\partial y}\left[-\displaystyle\int_x^y \cos(e^t)\, dt\right] = -\dfrac{\partial}{\partial y}\displaystyle\int_x^y \cos(e^t)\, dt = -\cos(e^y)$.

31. $f(x, y, z) = x^3 yz^2 + 2yz \quad \Rightarrow \quad f_x(x, y, z) = 3x^2 yz^2, \quad f_y(x, y, z) = x^3 z^2 + 2z, \quad f_z(x, y, z) = 2x^3 yz + 2y$

33. $w = \ln(x + 2y + 3z)$ $\Rightarrow$ $\dfrac{\partial w}{\partial x} = \dfrac{1}{x + 2y + 3z}$, $\dfrac{\partial w}{\partial y} = \dfrac{2}{x + 2y + 3z}$, $\dfrac{\partial w}{\partial z} = \dfrac{3}{x + 2y + 3z}$

35. $p = \sqrt{t^4 + u^2 \cos v}$ $\Rightarrow$ $\dfrac{\partial p}{\partial t} = \frac{1}{2}(t^4 + u^2 \cos v)^{-1/2}(4t^3) = \dfrac{2t^3}{\sqrt{t^4 + u^2 \cos v}}$,

$\dfrac{\partial p}{\partial u} = \frac{1}{2}(t^4 + u^2 \cos v)^{-1/2}(2u \cos v) = \dfrac{u \cos v}{\sqrt{t^4 + u^2 \cos v}}$, $\dfrac{\partial p}{\partial v} = \frac{1}{2}(t^4 + u^2 \cos v)^{-1/2}[u^2(-\sin v)] = -\dfrac{u^2 \sin v}{2\sqrt{t^4 + u^2 \cos v}}$

37. $h(x, y, z, t) = x^2 y \cos(z/t)$ $\Rightarrow$ $h_x(x, y, z, t) = 2xy \cos(z/t)$, $h_y(x, y, z, t) = x^2 \cos(z/t)$,

$h_z(x, y, z, t) = -x^2 y \sin(z/t)(1/t) = (-x^2 y/t) \sin(z/t)$, $h_t(x, y, z, t) = -x^2 y \sin(z/t)(-zt^{-2}) = (x^2 y z/t^2) \sin(z/t)$

39. $u = \sqrt{x_1^2 + x_2^2 + \cdots + x_n^2}$. For each $i = 1, \ldots, n$, $u_{x_i} = \frac{1}{2}\left(x_1^2 + x_2^2 + \cdots + x_n^2\right)^{-1/2}(2x_i) = \dfrac{x_i}{\sqrt{x_1^2 + x_2^2 + \cdots + x_n^2}}$.

41. $R(s, t) = te^{s/t}$ $\Rightarrow$ $R_t(s, t) = t \cdot e^{s/t}(-s/t^2) + e^{s/t} \cdot 1 = \left(1 - \dfrac{s}{t}\right)e^{s/t}$, so $R_t(0, 1) = \left(1 - \dfrac{0}{1}\right)e^{0/1} = 1$.

43. $f(x, y, z) = \ln \dfrac{1 - \sqrt{x^2 + y^2 + z^2}}{1 + \sqrt{x^2 + y^2 + z^2}}$ $\Rightarrow$

$f_y(x, y, z) = \dfrac{1}{\dfrac{1 - \sqrt{x^2 + y^2 + z^2}}{1 + \sqrt{x^2 + y^2 + z^2}}} \cdot$

$\dfrac{\left(1 + \sqrt{x^2 + y^2 + z^2}\right)\left(-\frac{1}{2}(x^2 + y^2 + z^2)^{-1/2} \cdot 2y\right) - \left(1 - \sqrt{x^2 + y^2 + z^2}\right)\left(\frac{1}{2}(x^2 + y^2 + z^2)^{-1/2} \cdot 2y\right)}{\left(1 + \sqrt{x^2 + y^2 + z^2}\right)^2}$

$= \dfrac{1 + \sqrt{x^2 + y^2 + z^2}}{1 - \sqrt{x^2 + y^2 + z^2}} \cdot \dfrac{-y(x^2 + y^2 + z^2)^{-1/2}\left(1 + \sqrt{x^2 + y^2 + z^2} + 1 - \sqrt{x^2 + y^2 + z^2}\right)}{\left(1 + \sqrt{x^2 + y^2 + z^2}\right)^2}$

$= \dfrac{-y(x^2 + y^2 + z^2)^{-1/2}(2)}{\left(1 - \sqrt{x^2 + y^2 + z^2}\right)\left(1 + \sqrt{x^2 + y^2 + z^2}\right)} = \dfrac{-2y}{\sqrt{x^2 + y^2 + z^2}\,[1 - (x^2 + y^2 + z^2)]}$

so $f_y(1, 2, 2) = \dfrac{-2(2)}{\sqrt{1^2 + 2^2 + 2^2}\,[1 - (1^2 + 2^2 + 2^2)]} = \dfrac{-4}{\sqrt{9}\,(1 - 9)} = \dfrac{1}{6}$.

45. $f(x, y) = xy^2 - x^3 y$ $\Rightarrow$

$f_x(x, y) = \displaystyle\lim_{h \to 0} \dfrac{f(x + h, y) - f(x, y)}{h} = \lim_{h \to 0} \dfrac{(x + h)y^2 - (x + h)^3 y - (xy^2 - x^3 y)}{h}$

$= \displaystyle\lim_{h \to 0} \dfrac{h(y^2 - 3x^2 y - 3xyh - yh^2)}{h} = \lim_{h \to 0}(y^2 - 3x^2 y - 3xyh - yh^2) = y^2 - 3x^2 y$

$f_y(x, y) = \displaystyle\lim_{h \to 0} \dfrac{f(x, y + h) - f(x, y)}{h} = \lim_{h \to 0} \dfrac{x(y + h)^2 - x^3(y + h) - (xy^2 - x^3 y)}{h} = \lim_{h \to 0} \dfrac{h(2xy + xh - x^3)}{h}$

$= \displaystyle\lim_{h \to 0}(2xy + xh - x^3) = 2xy - x^3$

47. $x^2 + 2y^2 + 3z^2 = 1 \implies \dfrac{\partial}{\partial x}(x^2 + 2y^2 + 3z^2) = \dfrac{\partial}{\partial x}(1) \implies 2x + 0 + 6z\dfrac{\partial z}{\partial x} = 0 \implies 6z\dfrac{\partial z}{\partial x} = -2x \implies$

$\dfrac{\partial z}{\partial x} = \dfrac{-2x}{6z} = -\dfrac{x}{3z}$, and $\dfrac{\partial}{\partial y}(x^2 + 2y^2 + 3z^2) = \dfrac{\partial}{\partial y}(1) \implies 0 + 4y + 6z\dfrac{\partial z}{\partial y} = 0 \implies 6z\dfrac{\partial z}{\partial y} = -4y \implies$

$\dfrac{\partial z}{\partial y} = \dfrac{-4y}{6z} = -\dfrac{2y}{3z}.$

49. $e^z = xyz \implies \dfrac{\partial}{\partial x}(e^z) = \dfrac{\partial}{\partial x}(xyz) \implies e^z\dfrac{\partial z}{\partial x} = y\left(x\dfrac{\partial z}{\partial x} + z \cdot 1\right) \implies e^z\dfrac{\partial z}{\partial x} - xy\dfrac{\partial z}{\partial x} = yz \implies$

$(e^z - xy)\dfrac{\partial z}{\partial x} = yz$, so $\dfrac{\partial z}{\partial x} = \dfrac{yz}{e^z - xy}.$

$\dfrac{\partial}{\partial y}(e^z) = \dfrac{\partial}{\partial y}(xyz) \implies e^z\dfrac{\partial z}{\partial y} = x\left(y\dfrac{\partial z}{\partial x} + z \cdot 1\right) \implies e^z\dfrac{\partial z}{\partial y} - xy\dfrac{\partial z}{\partial y} = xz \implies (e^z - xy)\dfrac{\partial z}{\partial y} = xz$, so

$\dfrac{\partial z}{\partial y} = \dfrac{xz}{e^z - xy}.$

51. (a) $z = f(x) + g(y) \implies \dfrac{\partial z}{\partial x} = f'(x), \quad \dfrac{\partial z}{\partial y} = g'(y)$

(b) $z = f(x + y)$. Let $u = x + y$. Then $\dfrac{\partial z}{\partial x} = \dfrac{df}{du}\dfrac{\partial u}{\partial x} = \dfrac{df}{du}(1) = f'(u) = f'(x + y)$,

$\dfrac{\partial z}{\partial y} = \dfrac{df}{du}\dfrac{\partial u}{\partial y} = \dfrac{df}{du}(1) = f'(u) = f'(x + y).$

53. $f(x, y) = x^4y - 2x^3y^2 \implies f_x(x, y) = 4x^3y - 6x^2y^2, \ f_y(x, y) = x^4 - 4x^3y$. Then $f_{xx}(x, y) = 12x^2y - 12xy^2$,

$f_{xy}(x, y) = 4x^3 - 12x^2y, \ f_{yx}(x, y) = 4x^3 - 12x^2y, \ \text{and} \ f_{yy}(x, y) = -4x^3.$

55. $z = \dfrac{y}{2x + 3y} = y(2x + 3y)^{-1} \implies z_x = y(-1)(2x + 3y)^{-2}(2) = -\dfrac{2y}{(2x + 3y)^2}$,

$z_y = \dfrac{(2x + 3y) \cdot 1 - y \cdot 3}{(2x + 3y)^2} = \dfrac{2x}{(2x + 3y)^2}$. Then $z_{xx} = -2y(-2)(2x + 3y)^{-3}(2) = \dfrac{8y}{(2x + 3y)^3}$,

$z_{xy} = -\dfrac{(2x + 3y)^2 \cdot 2 - 2y \cdot 2(2x + 3y)(3)}{[(2x + 3y)^2]^2} = -\dfrac{(2x + 3y)(4x + 6y - 12y)}{(2x + 3y)^4} = \dfrac{6y - 4x}{(2x + 3y)^3}$,

$z_{yx} = \dfrac{(2x + 3y)^2 \cdot 2 - 2x \cdot 2(2x + 3y)(2)}{[(2x + 3y)^2]^2} = \dfrac{6y - 4x}{(2x + 3y)^3}, \ z_{yy} = 2x(-2)(2x + 3y)^{-3}(3) = -\dfrac{12x}{(2x + 3y)^3}.$

57. $v = \sin(s^2 - t^2) \implies v_s = \cos(s^2 - t^2) \cdot 2s = 2s\cos(s^2 - t^2), \ v_t = \cos(s^2 - t^2) \cdot (-2t) = -2t\cos(s^2 - t^2)$. Then

$v_{ss} = 2s\left[-\sin(s^2 - t^2) \cdot 2s\right] + \cos(s^2 - t^2) \cdot 2 = 2\cos(s^2 - t^2) - 4s^2\sin(s^2 - t^2)$,

$v_{st} = 2s\left[-\sin(s^2 - t^2) \cdot (-2t)\right] = 4st\sin(s^2 - t^2), \ v_{ts} = -2t\left[-\sin(s^2 - t^2) \cdot 2s\right] = 4st\sin(s^2 - t^2)$,

$v_{tt} = -2t \cdot \left[-\sin(s^2 - t^2) \cdot (-2t)\right] + \cos(s^2 - t^2) \cdot (-2) = -2\cos(s^2 - t^2) - 4t^2\sin(s^2 - t^2).$

59. $u = x^4y^3 - y^4 \implies u_x = 4x^3y^3, \ u_{xy} = 12x^3y^2$ and $u_y = 3x^4y^2 - 4y^3, \ u_{yx} = 12x^3y^2.$

Thus $u_{xy} = u_{yx}.$

61. $u = \cos(x^2 y) \quad \Rightarrow \quad u_x = -\sin(x^2 y) \cdot 2xy = -2xy\sin(x^2 y),$

$u_{xy} = -2xy \cdot \cos(x^2 y) \cdot x^2 + \sin(x^2 y) \cdot (-2x) = -2x^3 y\cos(x^2 y) - 2x\sin(x^2 y)$ and

$u_y = -\sin(x^2 y) \cdot x^2 = -x^2 \sin(x^2 y), \quad u_{yx} = -x^2 \cdot \cos(x^2 y) \cdot 2xy + \sin(x^2 y) \cdot (-2x) = -2x^3 y\cos(x^2 y) - 2x\sin(x^2 y).$

Thus $u_{xy} = u_{yx}$.

63. $f(x, y) = x^4 y^2 - x^3 y \quad \Rightarrow \quad f_x = 4x^3 y^2 - 3x^2 y, \; f_{xx} = 12x^2 y^2 - 6xy, \; f_{xxx} = 24xy^2 - 6y$ and

$f_{xy} = 8x^3 y - 3x^2, \; f_{xyx} = 24x^2 y - 6x.$

65. $f(x, y, z) = e^{xyz^2} \quad \Rightarrow \quad f_x = e^{xyz^2} \cdot yz^2 = yz^2 e^{xyz^2}, \; f_{xy} = yz^2 \cdot e^{xyz^2}(xz^2) + e^{xyz^2} \cdot z^2 = (xyz^4 + z^2)e^{xyz^2},$

$f_{xyz} = (xyz^4 + z^2) \cdot e^{xyz^2}(2xyz) + e^{xyz^2} \cdot (4xyz^3 + 2z) = (2x^2 y^2 z^5 + 6xyz^3 + 2z)e^{xyz^2}.$

67. $W = \sqrt{u + v^2} \quad \Rightarrow \quad \dfrac{\partial W}{\partial v} = \frac{1}{2}(u + v^2)^{-1/2}(2v) = v(u + v^2)^{-1/2},$

$\dfrac{\partial^2 W}{\partial u \, \partial v} = v\left(-\frac{1}{2}\right)(u + v^2)^{-3/2}(1) = -\frac{1}{2}v(u + v^2)^{-3/2}, \quad \dfrac{\partial^3 W}{\partial u^2 \, \partial v} = -\frac{1}{2}v\left(-\frac{3}{2}\right)(u + v^2)^{-5/2}(1) = \frac{3}{4}v(u + v^2)^{-5/2}.$

69. $w = \dfrac{x}{y + 2z} = x(y + 2z)^{-1} \quad \Rightarrow \quad \dfrac{\partial w}{\partial x} = (y + 2z)^{-1}, \quad \dfrac{\partial^2 w}{\partial y \, \partial x} = -(y + 2z)^{-2}(1) = -(y + 2z)^{-2},$

$\dfrac{\partial^3 w}{\partial z \, \partial y \, \partial x} = -(-2)(y + 2z)^{-3}(2) = 4(y + 2z)^{-3} = \dfrac{4}{(y + 2z)^3} \quad$ and $\quad \dfrac{\partial w}{\partial y} = x(-1)(y + 2z)^{-2}(1) = -x(y + 2z)^{-2},$

$\dfrac{\partial^2 w}{\partial x \, \partial y} = -(y + 2z)^{-2}, \quad \dfrac{\partial^3 w}{\partial x^2 \, \partial y} = 0.$

71. Assuming that the third partial derivatives of f are continuous (easily verified), we can write $f_{xzy} = f_{yxz}$. Then

$f(x, y, z) = xy^2 z^3 + \arcsin\left(x\sqrt{z}\right) \quad \Rightarrow \quad f_y = 2xyz^3 + 0, \; f_{yx} = 2yz^3,$ and $f_{yxz} = 6yz^2 = f_{xzy}.$

73. By Definition 4, $f_x(3, 2) = \displaystyle\lim_{h \to 0} \dfrac{f(3 + h, 2) - f(3, 2)}{h}$ which we can approximate by considering $h = 0.5$ and $h = -0.5$:

$f_x(3, 2) \approx \dfrac{f(3.5, 2) - f(3, 2)}{0.5} = \dfrac{22.4 - 17.5}{0.5} = 9.8, \; f_x(3, 2) \approx \dfrac{f(2.5, 2) - f(3, 2)}{-0.5} = \dfrac{10.2 - 17.5}{-0.5} = 14.6.$ Averaging

these values, we estimate $f_x(3, 2)$ to be approximately 12.2. Similarly, $f_x(3, 2.2) = \displaystyle\lim_{h \to 0} \dfrac{f(3 + h, 2.2) - f(3, 2.2)}{h}$ which

we can approximate by considering $h = 0.5$ and $h = -0.5$: $f_x(3, 2.2) \approx \dfrac{f(3.5, 2.2) - f(3, 2.2)}{0.5} = \dfrac{26.1 - 15.9}{0.5} = 20.4,$

$f_x(3, 2.2) \approx \dfrac{f(2.5, 2.2) - f(3, 2.2)}{-0.5} = \dfrac{9.3 - 15.9}{-0.5} = 13.2.$ Averaging these values, we have $f_x(3, 2.2) \approx 16.8.$

To estimate $f_{xy}(3, 2)$, we first need an estimate for $f_x(3, 1.8)$:

$f_x(3, 1.8) \approx \dfrac{f(3.5, 1.8) - f(3, 1.8)}{0.5} = \dfrac{20.0 - 18.1}{0.5} = 3.8, \; f_x(3, 1.8) \approx \dfrac{f(2.5, 1.8) - f(3, 1.8)}{-0.5} = \dfrac{12.5 - 18.1}{-0.5} = 11.2.$

Averaging these values, we get $f_x(3, 1.8) \approx 7.5.$ Now $f_{xy}(x, y) = \dfrac{\partial}{\partial y}[f_x(x, y)]$ and $f_x(x, y)$ is itself a function of two

variables, so Definition 4 says that $f_{xy}(x, y) = \dfrac{\partial}{\partial y}[f_x(x, y)] = \displaystyle\lim_{h \to 0} \dfrac{f_x(x, y + h) - f_x(x, y)}{h} \quad \Rightarrow$

$f_{xy}(3, 2) = \lim_{h \to 0} \dfrac{f_x(3, 2 + h) - f_x(3, 2)}{h}$. We can estimate this value using our previous work with $h = 0.2$ and $h = -0.2$:

$f_{xy}(3, 2) \approx \dfrac{f_x(3, 2.2) - f_x(3, 2)}{0.2} = \dfrac{16.8 - 12.2}{0.2} = 23,\ f_{xy}(3, 2) \approx \dfrac{f_x(3, 1.8) - f_x(3, 2)}{-0.2} = \dfrac{7.5 - 12.2}{-0.2} = 23.5.$

Averaging these values, we estimate $f_{xy}(3, 2)$ to be approximately 23.25.

75. $u = e^{-\alpha^2 k^2 t} \sin kx \ \Rightarrow \ u_x = ke^{-\alpha^2 k^2 t} \cos kx,\ u_{xx} = -k^2 e^{-\alpha^2 k^2 t} \sin kx,\ $ and $\ u_t = -\alpha^2 k^2 e^{-\alpha^2 k^2 t} \sin kx.$ Thus

$\alpha^2 u_{xx} = u_t.$

77. $u = \dfrac{1}{\sqrt{x^2 + y^2 + z^2}} \ \Rightarrow \ u_x = \left(-\tfrac{1}{2}\right)(x^2 + y^2 + z^2)^{-3/2}(2x) = -x(x^2 + y^2 + z^2)^{-3/2}$ and

$u_{xx} = -(x^2 + y^2 + z^2)^{-3/2} - x\left(-\tfrac{3}{2}\right)(x^2 + y^2 + z^2)^{-5/2}(2x) = \dfrac{2x^2 - y^2 - z^2}{(x^2 + y^2 + z^2)^{5/2}}.$

By symmetry, $u_{yy} = \dfrac{2y^2 - x^2 - z^2}{(x^2 + y^2 + z^2)^{5/2}}$ and $u_{zz} = \dfrac{2z^2 - x^2 - y^2}{(x^2 + y^2 + z^2)^{5/2}}.$

Thus $u_{xx} + u_{yy} + u_{zz} = \dfrac{2x^2 - y^2 - z^2 + 2y^2 - x^2 - z^2 + 2z^2 - x^2 - y^2}{(x^2 + y^2 + z^2)^{5/2}} = 0.$

79. Let $v = x + at,\ w = x - at.$ Then $u_t = \dfrac{\partial[f(v) + g(w)]}{\partial t} = \dfrac{df(v)}{dv}\dfrac{\partial v}{\partial t} + \dfrac{dg(w)}{dw}\dfrac{\partial w}{\partial t} = af'(v) - ag'(w)$ and

$u_{tt} = \dfrac{\partial[af'(v) - ag'(w)]}{\partial t} = a[af''(v) + ag''(w)] = a^2[f''(v) + g''(w)].$ Similarly, by using the Chain Rule we have

$u_x = f'(v) + g'(w)$ and $u_{xx} = f''(v) + g''(w).$ Thus $u_{tt} = a^2 u_{xx}.$

81. $c(x, t) = \dfrac{1}{\sqrt{4\pi Dt}} e^{-x^2/(4Dt)} \ \Rightarrow$

$\dfrac{\partial c}{\partial t} = \dfrac{1}{\sqrt{4\pi Dt}} \cdot e^{-x^2/(4Dt)}\left[-x^2(-1)(4Dt)^{-2}(4D)\right] + e^{-x^2/(4Dt)} \cdot \left(-\tfrac{1}{2}\right)(4\pi Dt)^{-3/2}(4\pi D)$

$= (4\pi Dt)^{-3/2}\left(4\pi Dt \cdot \dfrac{x^2}{4Dt^2} - 2\pi D\right)e^{-x^2/(4Dt)} = \dfrac{2\pi D}{(4\pi Dt)^{3/2}}\left(\dfrac{x^2}{2Dt} - 1\right)e^{-x^2/(4Dt)},$

$\dfrac{\partial c}{\partial x} = \dfrac{1}{\sqrt{4\pi Dt}} e^{-x^2/(4Dt)} \cdot \dfrac{-2x}{4Dt} = \dfrac{-2\pi x}{(4\pi Dt)^{3/2}} e^{-x^2/(4Dt)},$ and

$\dfrac{\partial^2 c}{\partial x^2} = \dfrac{-2\pi}{(4\pi Dt)^{3/2}}\left[x \cdot e^{-x^2/(4Dt)} \cdot \dfrac{-2x}{4Dt} + e^{-x^2/(4Dt)} \cdot 1\right]$

$= \dfrac{-2\pi}{(4\pi Dt)^{3/2}}\left(-\dfrac{x^2}{2Dt} + 1\right)e^{-x^2/(4Dt)} = \dfrac{2\pi}{(4\pi Dt)^{3/2}}\left(\dfrac{x^2}{2Dt} - 1\right)e^{-x^2/(4Dt)}.$

Thus $\dfrac{\partial c}{\partial t} = \dfrac{2\pi D}{(4\pi Dt)^{3/2}}\left(\dfrac{x^2}{2Dt} - 1\right)e^{-x^2/(4Dt)} = D\left[\dfrac{2\pi}{(4\pi Dt)^{3/2}}\left(\dfrac{x^2}{2Dt} - 1\right)e^{-x^2/(4Dt)}\right] = D\dfrac{\partial^2 c}{\partial x^2}.$

83. By the Chain Rule, taking the partial derivative of both sides with respect to R_1 gives

$\dfrac{\partial R^{-1}}{\partial R}\dfrac{\partial R}{\partial R_1} = \dfrac{\partial\left[(1/R_1) + (1/R_2) + (1/R_3)\right]}{\partial R_1}$ or $-R^{-2}\dfrac{\partial R}{\partial R_1} = -R_1^{-2}.$ Thus $\dfrac{\partial R}{\partial R_1} = \dfrac{R^2}{R_1^2}.$

85. If we fix $K = K_0$, $P(L, K_0)$ is a function of a single variable L, and $\dfrac{dP}{dL} = \alpha\dfrac{P}{L}$ is a separable differential equation. Then

$$\frac{dP}{P} = \alpha\frac{dL}{L} \quad\Rightarrow\quad \int\frac{dP}{P} = \int\alpha\frac{dL}{L} \quad\Rightarrow\quad \ln|P| = \alpha\ln|L| + C\,(K_0), \text{ where } C(K_0) \text{ can depend on } K_0. \text{ Then}$$

$$|P| = e^{\alpha\ln|L| + C(K_0)}, \text{ and since } P > 0 \text{ and } L > 0, \text{ we have } P = e^{\alpha\ln L}e^{C(K_0)} = e^{C(K_0)}e^{\ln L^\alpha} = C_1(K_0)L^\alpha \text{ where}$$

$$C_1(K_0) = e^{C(K_0)}.$$

87. $\left(P + \dfrac{n^2a}{V^2}\right)(V - nb) = nRT \quad\Rightarrow\quad T = \dfrac{1}{nR}\left(P + \dfrac{n^2a}{V^2}\right)(V - nb), \text{ so } \dfrac{\partial T}{\partial P} = \dfrac{1}{nR}(1)(V - nb) = \dfrac{V - nb}{nR}.$

We can also write $P + \dfrac{n^2a}{V^2} = \dfrac{nRT}{V - nb} \quad\Rightarrow\quad P = \dfrac{nRT}{V - nb} - \dfrac{n^2a}{V^2} = nRT(V - nb)^{-1} - n^2aV^{-2},$ so

$$\frac{\partial P}{\partial V} = -nRT(V - nb)^{-2}(1) + 2n^2aV^{-3} = \frac{2n^2a}{V^3} - \frac{nRT}{(V - nb)^2}.$$

89. By Exercise 88, $PV = mRT \quad\Rightarrow\quad P = \dfrac{mRT}{V}, \text{ so } \dfrac{\partial P}{\partial T} = \dfrac{mR}{V}.$ Also, $PV = mRT \quad\Rightarrow\quad V = \dfrac{mRT}{P} \text{ and } \dfrac{\partial V}{\partial T} = \dfrac{mR}{P}.$

Since $T = \dfrac{PV}{mR}$, we have $T\dfrac{\partial P}{\partial T}\dfrac{\partial V}{\partial T} = \dfrac{PV}{mR}\cdot\dfrac{mR}{V}\cdot\dfrac{mR}{P} = mR.$

91. (a) $S = f(w, h) = 0.1091w^{0.425}h^{0.725} \quad\Rightarrow\quad \dfrac{\partial S}{\partial w} = 0.1091(0.425)w^{0.425 - 1}h^{0.725} = 0.0463675w^{-0.575}h^{0.725},$ so

$\dfrac{\partial S}{\partial w}(160, 70) = 0.0463675(160)^{-0.575}(70)^{0.725} \approx 0.0545.$ This means that for a person 70 inches tall who weighs 160

pounds, an increase in weight (while height remains constant) causes the surface area to increase at a rate of about 0.0545

square feet (about 7.85 square inches) per pound.

(b) $\dfrac{\partial S}{\partial h} = 0.1091(0.725)w^{0.425}h^{0.725 - 1} = 0.0790975w^{0.425}h^{-0.275},$ so

$\dfrac{\partial S}{\partial h}(160, 70) = 0.0790975(160)^{0.425}(70)^{-0.275} \approx 0.213.$ This means that for a person 70 inches tall who weighs 160

pounds, an increase in height (while weight remains unchanged at 160 pounds) causes the surface area to increase at a rate

of about 0.213 square feet (about 30.7 square inches) per inch of height.

93. $P(v, x, m) = Av^3 + \dfrac{B(mg/x)^2}{v} = Av^3 + Bm^2g^2x^{-2}v^{-1}.$

$\partial P/\partial v = 3Av^2 - \dfrac{B(mg/x)^2}{v^2}$ is the rate of change of the power needed during flapping mode with respect to the bird's

velocity when the mass and fraction of flapping time remain constant. $\partial P/\partial x = -2Bm^2g^2x^{-3}v^{-1} = -\dfrac{2Bm^2g^2}{x^3v}$ is the

rate at which the power changes with respect to the fraction of time spent in flapping mode when the mass and velocity are

held constant. $\partial P/\partial m = 2Bmg^2x^{-2}v^{-1} = \dfrac{2Bmg^2}{x^2v}$ is the rate of change of the power with respect to mass when the

velocity and fraction of flapping time remain constant.

95. $\frac{\partial K}{\partial m} = \frac{1}{2}v^2$, $\frac{\partial K}{\partial v} = mv$, $\frac{\partial^2 K}{\partial v^2} = m$. Thus $\frac{\partial K}{\partial m} \cdot \frac{\partial^2 K}{\partial v^2} = \frac{1}{2}v^2 m = K$.

97. $f_x(x, y) = x + 4y \;\Rightarrow\; f_{xy}(x, y) = 4$ and $f_y(x, y) = 3x - y \;\Rightarrow\; f_{yx}(x, y) = 3$. Since f_{xy} and f_{yx} are continuous

everywhere but $f_{xy}(x, y) \neq f_{yx}(x, y)$, Clairaut's Theorem implies that such a function $f(x, y)$ does not exist.

99. By the geometry of partial derivatives, the slope of the tangent line is $f_x(1, 2)$. By implicit differentiation of

$4x^2 + 2y^2 + z^2 = 16$, we get $8x + 2z\,(\partial z/\partial x) = 0 \;\Rightarrow\; \partial z/\partial x = -4x/z$, so when $x = 1$ and $z = 2$ we have

$\partial z/\partial x = -2$. So the slope is $f_x(1, 2) = -2$. Thus the tangent line is given by $z - 2 = -2(x - 1)$, $y = 2$. Taking the

parameter to be $t = x - 1$, we can write parametric equations for this line: $x = 1 + t$, $y = 2$, $z = 2 - 2t$.

101. By Clairaut's Theorem, $f_{xyy} = (f_{xy})_y = (f_{yx})_y = f_{yxy} = (f_y)_{xy} = (f_y)_{yx} = f_{yyx}$.

103. Let $g(x) = f(x, 0) = x(x^2)^{-3/2}e^0 = x\,|x|^{-3}$. But we are using the point $(1, 0)$, so near $(1, 0)$, $g(x) = x^{-2}$. Then

$g'(x) = -2x^{-3}$ and $g'(1) = -2$, so using (1) we have $f_x(1, 0) = g'(1) = -2$.

105. (a)

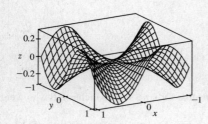

(b) For $(x, y) \neq (0, 0)$,

$$f_x(x, y) = \frac{(3x^2y - y^3)(x^2 + y^2) - (x^3y - xy^3)(2x)}{(x^2 + y^2)^2}$$

$$= \frac{x^4y + 4x^2y^3 - y^5}{(x^2 + y^2)^2}$$

and by symmetry $f_y(x, y) = \dfrac{x^5 - 4x^3y^2 - xy^4}{(x^2 + y^2)^2}$.

(c) $f_x(0, 0) = \lim\limits_{h \to 0} \dfrac{f(h, 0) - f(0, 0)}{h} = \lim\limits_{h \to 0} \dfrac{(0/h^2) - 0}{h} = 0$ and $f_y(0, 0) = \lim\limits_{h \to 0} \dfrac{f(0, h) - f(0, 0)}{h} = 0$.

(d) By (3), $f_{xy}(0, 0) = \dfrac{\partial f_x}{\partial y} = \lim\limits_{h \to 0} \dfrac{f_x(0, h) - f_x(0, 0)}{h} = \lim\limits_{h \to 0} \dfrac{(-h^5 - 0)/h^4}{h} = -1$ while by (2),

$$f_{yx}(0, 0) = \frac{\partial f_y}{\partial x} = \lim_{h \to 0} \frac{f_y(h, 0) - f_y(0, 0)}{h} = \lim_{h \to 0} \frac{h^5/h^4}{h} = 1.$$

(e) For $(x, y) \neq (0, 0)$, we use a CAS to compute

$$f_{xy}(x, y) = \frac{x^6 + 9x^4y^2 - 9x^2y^4 - y^6}{(x^2 + y^2)^3}$$

Now as $(x, y) \to (0, 0)$ along the x-axis, $f_{xy}(x, y) \to 1$ while as

$(x, y) \to (0, 0)$ along the y-axis, $f_{xy}(x, y) \to -1$. Thus f_{xy} isn't

continuous at $(0, 0)$ and Clairaut's Theorem doesn't apply, so there is

no contradiction. The graphs of f_{xy} and f_{yx} are identical except at the

origin, where we observe the discontinuity.

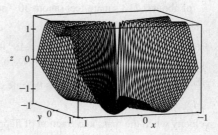

14.4 Tangent Planes and Linear Approximations

1. $z = f(x, y) = 2x^2 + y^2 - 5y \implies f_x(x, y) = 4x, \ f_y(x, y) = 2y - 5$, so $f_x(1, 2) = 4, \ f_y(1, 2) = -1$.

By Equation 2, an equation of the tangent plane is $z - (-4) = f_x(1, 2)(x - 1) + f_y(1, 2)(y - 2) \implies$

$z + 4 = 4(x - 1) + (-1)(y - 2)$ or $z = 4x - y - 6$.

3. $z = f(x, y) = e^{x-y} \implies f_x(x, y) = e^{x-y}(1) = e^{x-y}, \ f_y(x, y) = e^{x-y}(-1) = -e^{x-y},$ so $f_x(2, 2) = 1$ and

$f_y(2, 2) = -1$. Thus an equation of the tangent plane is $z - 1 = f_x(2, 2)(x - 2) + f_y(2, 2)(y - 2) \implies$

$z - 1 = 1(x - 2) + (-1)(y - 2)$ or $z = x - y + 1$.

5. $z = f(x, y) = x \sin(x + y) \implies f_x(x, y) = x \cdot \cos(x + y) + \sin(x + y) \cdot 1 = x \cos(x + y) + \sin(x + y),$

$f_y(x, y) = x \cos(x + y)$, so $f_x(-1, 1) = (-1)\cos 0 + \sin 0 = -1, \ f_y(-1, 1) = (-1)\cos 0 = -1$ and an equation of the

tangent plane is $z - 0 = (-1)(x + 1) + (-1)(y - 1)$ or $x + y + z = 0$.

7. $z = f(x, y) = x^2 + xy + 3y^2$, so $f_x(x, y) = 2x + y \implies f_x(1, 1) = 3, \ f_y(x, y) = x + 6y \implies f_y(1, 1) = 7$ and an

equation of the tangent plane is $z - 5 = 3(x - 1) + 7(y - 1)$ or $z = 3x + 7y - 5$. After zooming in, the surface and the

tangent plane become almost indistinguishable. (Here, the tangent plane is below the surface.) If we zoom in farther, the

surface and the tangent plane will appear to coincide.

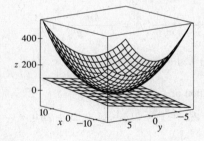 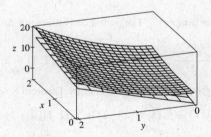

9. $f(x, y) = \dfrac{1 + \cos^2(x - y)}{1 + \cos^2(x + y)}.$ A CAS gives

$f_x(x, y) = -\dfrac{2 \cos(x - y) \sin(x - y)}{1 + \cos^2(x + y)} + \dfrac{2 \left[1 + \cos^2(x - y)\right] \cos(x + y) \sin(x + y)}{\left[1 + \cos^2(x + y)\right]^2}$ and

$f_y(x, y) = \dfrac{2 \cos(x - y) \sin(x - y)}{1 + \cos^2(x + y)} + \dfrac{2 \left[1 + \cos^2(x - y)\right] \cos(x + y) \sin(x + y)}{\left[1 + \cos^2(x + y)\right]^2}$. We use the CAS to evaluate these at

$(\pi/3, \pi/6)$, giving $f_x(\pi/3, \pi/6) = -\sqrt{3}/2$ and $f_y(\pi/3, \pi/6) = \sqrt{3}/2$. Substituting into Equation 2, an equation of the

tangent plane is $z = -\frac{\sqrt{3}}{2}\left(x - \frac{\pi}{3}\right) + \frac{\sqrt{3}}{2}\left(y - \frac{\pi}{6}\right) + \frac{7}{4}$. The surface and tangent plane are shown in the first graph below.

After zooming in, the surface and the tangent plane become almost indistinguishable, as shown in the second graph. (Here, the tangent plane is above the surface.) If we zoom in farther, the surface and the tangent plane will appear to coincide.

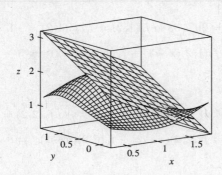 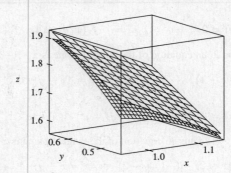

11. $f(x, y) = 1 + x \ln(xy - 5)$. The partial derivatives are $f_x(x, y) = x \cdot \dfrac{1}{xy - 5}(y) + \ln(xy - 5) \cdot 1 = \dfrac{xy}{xy - 5} + \ln(xy - 5)$

and $f_y(x, y) = x \cdot \dfrac{1}{xy - 5}(x) = \dfrac{x^2}{xy - 5}$, so $f_x(2, 3) = 6$ and $f_y(2, 3) = 4$. Both f_x and f_y are continuous functions for

$xy > 5$, so by Theorem 8, f is differentiable at $(2, 3)$. By Equation 3, the linearization of f at $(2, 3)$ is given by

$L(x, y) = f(2, 3) + f_x(2, 3)(x - 2) + f_y(2, 3)(y - 3) = 1 + 6(x - 2) + 4(y - 3) = 6x + 4y - 23$.

13. $f(x, y) = x^2 e^y$. The partial derivatives are $f_x(x, y) = 2xe^y$ and $f_y(x, y) = x^2 e^y$, so $f_x(1, 0) = 2$ and $f_y(1, 0) = 1$. Both

f_x and f_y are continuous functions, so by Theorem 8, f is differentiable at $(1, 0)$. By Equation 3, the linearization of f at

$(1, 0)$ is given by $L(x, y) = f(1, 0) + f_x(1, 0)(x - 1) + f_y(1, 0)(y - 0) = 1 + 2(x - 1) + 1(y - 0) = 2x + y - 1$.

15. $f(x, y) = 4 \arctan(xy)$. The partial derivatives are $f_x(x, y) = 4 \cdot \dfrac{1}{1 + (xy)^2}(y) = \dfrac{4y}{1 + x^2 y^2}$, and

$f_y(x, y) = \dfrac{4x}{1 + x^2 y^2}$, so $f_x(1, 1) = 2$ and $f_y(1, 1) = 2$. Both f_x and f_y are continuous

functions, so f is differentiable at $(1, 1)$ by Theorem 8. The linearization of f at $(1, 1)$ is

$L(x, y) = f(1, 1) + f_x(1, 1)(x - 1) + f_y(1, 1)(y - 1) = 4(\pi/4) + 2(x - 1) + 2(y - 1) = 2x + 2y + \pi - 4$.

17. Let $f(x, y) = e^x \cos(xy)$. Then $f_x(x, y) = e^x[-\sin(xy)](y) + e^x \cos(xy) = e^x[\cos(xy) - y \sin(xy)]$ and

$f_y(x, y) = e^x[-\sin(xy)](x) = -xe^x \sin(xy)$. Both f_x and f_y are continuous functions, so by Theorem 8, f is differentiable

at $(0, 0)$. We have $f_x(0, 0) = e^0(\cos 0 - 0) = 1$, $f_y(0, 0) = 0$ and the linear approximation of f at $(0, 0)$ is

$f(x, y) \approx f(0, 0) + f_x(0, 0)(x - 0) + f_y(0, 0)(y - 0) = 1 + 1x + 0y = x + 1$.

19. We can estimate $f(2.2, 4.9)$ using a linear approximation of f at $(2, 5)$, given by

$f(x, y) \approx f(2, 5) + f_x(2, 5)(x - 2) + f_y(2, 5)(y - 5) = 6 + 1(x - 2) + (-1)(y - 5) = x - y + 9$. Thus

$f(2.2, 4.9) \approx 2.2 - 4.9 + 9 = 6.3$.

21. $f(x, y, z) = \sqrt{x^2 + y^2 + z^2} \Rightarrow f_x(x, y, z) = \dfrac{x}{\sqrt{x^2 + y^2 + z^2}}$, $f_y(x, y, z) = \dfrac{y}{\sqrt{x^2 + y^2 + z^2}}$, and

$f_z(x, y, z) = \dfrac{z}{\sqrt{x^2 + y^2 + z^2}}$, so $f_x(3, 2, 6) = \frac{3}{7}$, $f_y(3, 2, 6) = \frac{2}{7}$, $f_z(3, 2, 6) = \frac{6}{7}$. Then the linear approximation of f

at $(3, 2, 6)$ is given by

$$f(x, y, z) \approx f(3, 2, 6) + f_x(3, 2, 6)(x - 3) + f_y(3, 2, 6)(y - 2) + f_z(3, 2, 6)(z - 6)$$

$$= 7 + \tfrac{3}{7}(x - 3) + \tfrac{2}{7}(y - 2) + \tfrac{6}{7}(z - 6) = \tfrac{3}{7}x + \tfrac{2}{7}y + \tfrac{6}{7}z$$

Thus $\sqrt{(3.02)^2 + (1.97)^2 + (5.99)^2} = f(3.02, 1.97, 5.99) \approx \tfrac{3}{7}(3.02) + \tfrac{2}{7}(1.97) + \tfrac{6}{7}(5.99) \approx 6.9914.$

23. From the table, $f(94, 80) = 127$. To estimate $f_T(94, 80)$ and $f_H(94, 80)$ we follow the procedure used in Section 14.3. Since

$f_T(94, 80) = \lim\limits_{h \to 0} \dfrac{f(94 + h, 80) - f(94, 80)}{h}$, we approximate this quantity with $h = \pm 2$ and use the values given in the

table:

$$f_T(94, 80) \approx \frac{f(96, 80) - f(94, 80)}{2} = \frac{135 - 127}{2} = 4, \quad f_T(94, 80) \approx \frac{f(92, 80) - f(94, 80)}{-2} = \frac{119 - 127}{-2} = 4$$

Averaging these values gives $f_T(94, 80) \approx 4$. Similarly, $f_H(94, 80) = \lim\limits_{h \to 0} \dfrac{f(94, 80 + h) - f(94, 80)}{h}$, so we use $h = \pm 5$:

$$f_H(94, 80) \approx \frac{f(94, 85) - f(94, 80)}{5} = \frac{132 - 127}{5} = 1, \quad f_H(94, 80) \approx \frac{f(94, 75) - f(94, 80)}{-5} = \frac{122 - 127}{-5} = 1$$

Averaging these values gives $f_H(94, 80) \approx 1$. The linear approximation, then, is

$$f(T, H) \approx f(94, 80) + f_T(94, 80)(T - 94) + f_H(94, 80)(H - 80)$$

$$\approx 127 + 4(T - 94) + 1(H - 80) \qquad [\text{or } 4T + H - 329]$$

Thus when $T = 95$ and $H = 78$, $f(95, 78) \approx 127 + 4(95 - 94) + 1(78 - 80) = 129$, so we estimate the heat index to be

approximately $129°\text{F}$.

25. $z = e^{-2x} \cos 2\pi t \quad \Rightarrow$

$$dz = \frac{\partial z}{\partial x}\, dx + \frac{\partial z}{\partial t}\, dt = e^{-2x}(-2) \cos 2\pi t \, dx + e^{-2x}(-\sin 2\pi t)(2\pi) \, dt = -2e^{-2x} \cos 2\pi t \, dx - 2\pi e^{-2x} \sin 2\pi t \, dt$$

27. $m = p^5 q^3 \quad \Rightarrow \quad dm = \dfrac{\partial m}{\partial p}\, dp + \dfrac{\partial m}{\partial q}\, dq = 5p^4 q^3 \, dp + 3p^5 q^2 \, dq$

29. $R = \alpha \beta^2 \cos \gamma \quad \Rightarrow \quad dR = \dfrac{\partial R}{\partial \alpha}\, d\alpha + \dfrac{\partial R}{\partial \beta}\, d\beta + \dfrac{\partial R}{\partial \gamma}\, d\gamma = \beta^2 \cos \gamma \, d\alpha + 2\alpha \beta \cos \gamma \, d\beta - \alpha \beta^2 \sin \gamma \, d\gamma$

31. $dx = \Delta x = 0.05$, $dy = \Delta y = 0.1$, $z = 5x^2 + y^2$, $z_x = 10x$, $z_y = 2y$. Thus when $x = 1$ and $y = 2$,

$dz = z_x(1, 2)\, dx + z_y(1, 2)\, dy = (10)(0.05) + (4)(0.1) = 0.9$ while

$\Delta z = f(1.05, 2.1) - f(1, 2) = 5(1.05)^2 + (2.1)^2 - 5 - 4 = 0.9225.$

33. $dA = \dfrac{\partial A}{\partial x}\, dx + \dfrac{\partial A}{\partial y}\, dy = y\, dx + x\, dy$ and $|\Delta x| \leq 0.1$, $|\Delta y| \leq 0.1$. We use $dx = 0.1$, $dy = 0.1$ with $x = 30$, $y = 24$; then

the maximum error in the area is about $dA = 24(0.1) + 30(0.1) = 5.4 \text{ cm}^2.$

35. The volume of a can is $V = \pi r^2 h$ and $\Delta V \approx dV$ is an estimate of the amount of tin. Here $dV = 2\pi r h\, dr + \pi r^2\, dh$, so put

$dr = 0.04$, $dh = 0.08$ (0.04 on top, 0.04 on bottom) and then $\Delta V \approx dV = 2\pi(48)(0.04) + \pi(16)(0.08) \approx 16.08$ cm^3.

Thus the amount of tin is about 16 cm^3.

37. $T = \dfrac{mgR}{2r^2 + R^2}$, so the differential of T is

$$dT = \frac{\partial T}{\partial R}\, dR + \frac{\partial T}{\partial r}\, dr = \frac{(2r^2 + R^2)(mg) - mgR(2R)}{(2r^2 + R^2)^2}\, dR + \frac{(2r^2 + R^2)(0) - mgR(4r)}{(2r^2 + R^2)^2}\, dr$$

$$= \frac{mg(2r^2 - R^2)}{(2r^2 + R^2)^2}\, dR - \frac{4mgRr}{(2r^2 + R^2)^2}\, dr$$

Here we have $\Delta R = 0.1$ and $\Delta r = 0.1$, so we take $dR = 0.1$, $dr = 0.1$ with $R = 3$, $r = 0.7$. Then the change in the

tension T is approximately

$$dT = \frac{mg[2(0.7)^2 - (3)^2]}{[2(0.7)^2 + (3)^2]^2}\, (0.1) - \frac{4mg(3)(0.7)}{[2(0.7)^2 + (3)^2]^2}\, (0.1)$$

$$= -\frac{0.802mg}{(9.98)^2} - \frac{0.84mg}{(9.98)^2} = -\frac{1.642}{99.6004}\, mg \approx -0.0165mg$$

Because the change is negative, tension decreases.

39. First we find $\dfrac{\partial R}{\partial R_1}$ implicitly by taking partial derivatives of both sides with respect to R_1:

$$\frac{\partial}{\partial R_1}\left(\frac{1}{R}\right) = \frac{\partial\,[(1/R_1) + (1/R_2) + (1/R_3)]}{\partial R_1} \quad\Rightarrow\quad -R^{-2}\frac{\partial R}{\partial R_1} = -R_1^{-2} \quad\Rightarrow\quad \frac{\partial R}{\partial R_1} = \frac{R^2}{R_1^2}. \text{ Then by symmetry,}$$

$\dfrac{\partial R}{\partial R_2} = \dfrac{R^2}{R_2^2}$, $\dfrac{\partial R}{\partial R_3} = \dfrac{R^2}{R_3^2}$. When $R_1 = 25$, $R_2 = 40$ and $R_3 = 50$, $\dfrac{1}{R} = \dfrac{17}{200} \Leftrightarrow R = \dfrac{200}{17}\,\Omega$. Since the possible error

for each R_i is 0.5%, the maximum error of R is attained by setting $\Delta R_i = 0.005 R_i$. So

$$\Delta R \approx dR = \frac{\partial R}{\partial R_1}\Delta R_1 + \frac{\partial R}{\partial R_2}\Delta R_2 + \frac{\partial R}{\partial R_3}\Delta R_3 = (0.005)R^2\left(\frac{1}{R_1} + \frac{1}{R_2} + \frac{1}{R_3}\right) = (0.005)R = \tfrac{1}{17} \approx 0.059\,\Omega.$$

41. (a) $B(m, h) = m/h^2 \quad\Rightarrow\quad B_m(m, h) = 1/h^2$ and $B_h(m, h) = -2m/h^3$. Since

$h > 0$, both B_m and B_h are continuous functions, so B is differentiable at $(23, 1.10)$. We

have $B(23, 1.10) = 23/(1.10)^2 \approx 19.01$, $B_m(23, 1.10) = 1/(1.10)^2 \approx 0.8264$, and

$B_h(23, 1.10) = -2(23)/(1.10)^3 \approx -34.56$, so the linear approximation of B at $(23, 1.10)$ is

$B(m, h) \approx B(23, 1.10) + B_m(23, 1.10)(m-23) + B_h(23, 1.10)(h-1.10) \approx 19.01 + 0.8264(m-23) - 34.56(h-1.10)$

or $B(m, h) \approx 0.8264m - 34.56h + 38.02$.

(b) From part (a), for values near $m = 23$ and $h = 1.10$, $B(m, h) \approx 0.8264m - 34.56h + 38.02$. If m

increases by 1 kg to 24 kg and h increases by 0.03 m to 1.13 m, we estimate the BMI to be

$B(24, 1.13) \approx 0.8264(24) - 34.56(1.13) + 38.02 \approx 18.801$. This is very close to the actual computed BMI:

$B(24, 1.13) = 24/(1.13)^2 \approx 18.796$.

43. $\Delta z = f(a + \Delta x, b + \Delta y) - f(a, b) = (a + \Delta x)^2 + (b + \Delta y)^2 - (a^2 + b^2)$

$\quad = a^2 + 2a\,\Delta x + (\Delta x)^2 + b^2 + 2b\,\Delta y + (\Delta y)^2 - a^2 - b^2 = 2a\,\Delta x + (\Delta x)^2 + 2b\,\Delta y + (\Delta y)^2$

But $f_x(a, b) = 2a$ and $f_y(a, b) = 2b$ and so $\Delta z = f_x(a, b)\,\Delta x + f_y(a, b)\,\Delta y + \Delta x\,\Delta x + \Delta y\,\Delta y$, which is Definition 7

with $\varepsilon_1 = \Delta x$ and $\varepsilon_2 = \Delta y$. Hence f is differentiable.

45. To show that f is continuous at (a, b) we need to show that $\displaystyle\lim_{(x,y) \to (a,b)} f(x, y) = f(a, b)$ or

equivalently $\displaystyle\lim_{(\Delta x, \Delta y) \to (0,0)} f(a + \Delta x, b + \Delta y) = f(a, b)$. Since f is differentiable at (a, b),

$f(a + \Delta x, b + \Delta y) - f(a, b) = \Delta z = f_x(a, b)\,\Delta x + f_y(a, b)\,\Delta y + \varepsilon_1\,\Delta x + \varepsilon_2\,\Delta y$, where ε_1 and $\varepsilon_2 \to 0$ as

$(\Delta x, \Delta y) \to (0, 0)$. Thus $f(a + \Delta x, b + \Delta y) = f(a, b) + f_x(a, b)\,\Delta x + f_y(a, b)\,\Delta y + \varepsilon_1\,\Delta x + \varepsilon_2\,\Delta y$. Taking the limit of

both sides as $(\Delta x, \Delta y) \to (0, 0)$ gives $\displaystyle\lim_{(\Delta x, \Delta y) \to (0,0)} f(a + \Delta x, b + \Delta y) = f(a, b)$. Thus f is continuous at (a, b).

14.5   The Chain Rule

1. $z = xy^3 - x^2 y$, $x = t^2 + 1$, $y = t^2 - 1$ $\Rightarrow$

$\dfrac{dz}{dt} = \dfrac{\partial z}{\partial x}\dfrac{dx}{dt} + \dfrac{\partial z}{\partial y}\dfrac{dy}{dt} = (y^3 - 2xy)(2t) + (3xy^2 - x^2)(2t) = 2t(y^3 - 2xy + 3xy^2 - x^2)$

3. $z = \sin x \cos y$, $x = \sqrt{t}$, $y = 1/t$ $\Rightarrow$

$\dfrac{dz}{dt} = \dfrac{\partial z}{\partial x}\dfrac{dx}{dt} + \dfrac{\partial z}{\partial y}\dfrac{dy}{dt} = (\cos x \cos y)\left(\tfrac{1}{2}t^{-1/2}\right) + (-\sin x \sin y)\left(-t^{-2}\right) = \dfrac{1}{2\sqrt{t}}\cos x \cos y + \dfrac{1}{t^2}\sin x \sin y$

5. $w = xe^{y/z}$, $x = t^2$, $y = 1 - t$, $z = 1 + 2t$ $\Rightarrow$

$\dfrac{dw}{dt} = \dfrac{\partial w}{\partial x}\dfrac{dx}{dt} + \dfrac{\partial w}{\partial y}\dfrac{dy}{dt} + \dfrac{\partial w}{\partial z}\dfrac{dz}{dt} = e^{y/z} \cdot 2t + xe^{y/z}\left(\dfrac{1}{z}\right) \cdot (-1) + xe^{y/z}\left(-\dfrac{y}{z^2}\right) \cdot 2 = e^{y/z}\left(2t - \dfrac{x}{z} - \dfrac{2xy}{z^2}\right)$

7. $z = (x - y)^5$, $x = s^2 t$, $y = st^2$ $\Rightarrow$

$\dfrac{\partial z}{\partial s} = \dfrac{\partial z}{\partial x}\dfrac{\partial x}{\partial s} + \dfrac{\partial z}{\partial y}\dfrac{\partial y}{\partial s} = 5(x - y)^4(1) \cdot 2st + 5(x - y)^4(-1) \cdot t^2 = 5(x - y)^4\left(2st - t^2\right)$

$\dfrac{\partial z}{\partial t} = \dfrac{\partial z}{\partial x}\dfrac{\partial x}{\partial t} + \dfrac{\partial z}{\partial y}\dfrac{\partial y}{\partial t} = 5(x - y)^4(1) \cdot s^2 + 5(x - y)^4(-1) \cdot 2st = 5(x - y)^4\left(s^2 - 2st\right)$

9. $z = \ln(3x + 2y)$, $x = s \sin t$, $y = t \cos s$ $\Rightarrow$

$\dfrac{\partial z}{\partial s} = \dfrac{\partial z}{\partial x}\dfrac{\partial x}{\partial s} + \dfrac{\partial z}{\partial y}\dfrac{\partial y}{\partial s} = \dfrac{3}{3x + 2y}(\sin t) + \dfrac{2}{3x + 2y}(-t \sin s) = \dfrac{3 \sin t - 2t \sin s}{3x + 2y}$

$\dfrac{\partial z}{\partial t} = \dfrac{\partial z}{\partial x}\dfrac{\partial x}{\partial t} + \dfrac{\partial z}{\partial y}\dfrac{\partial y}{\partial t} = \dfrac{3}{3x + 2y}(s \cos t) + \dfrac{2}{3x + 2y}(\cos s) = \dfrac{3s \cos t + 2 \cos s}{3x + 2y}$

11. $z = e^r \cos\theta$, $r = st$, $\theta = \sqrt{s^2 + t^2}$ $\Rightarrow$

$$\frac{\partial z}{\partial s} = \frac{\partial z}{\partial r}\frac{\partial r}{\partial s} + \frac{\partial z}{\partial \theta}\frac{\partial \theta}{\partial s} = e^r \cos\theta \cdot t + e^r(-\sin\theta) \cdot \tfrac{1}{2}(s^2 + t^2)^{-1/2}(2s) = te^r \cos\theta - e^r \sin\theta \cdot \frac{s}{\sqrt{s^2 + t^2}}$$

$$= e^r\left(t\cos\theta - \frac{s}{\sqrt{s^2 + t^2}}\sin\theta\right)$$

$$\frac{\partial z}{\partial t} = \frac{\partial z}{\partial r}\frac{\partial r}{\partial t} + \frac{\partial z}{\partial \theta}\frac{\partial \theta}{\partial t} = e^r \cos\theta \cdot s + e^r(-\sin\theta) \cdot \tfrac{1}{2}(s^2 + t^2)^{-1/2}(2t) = se^r \cos\theta - e^r \sin\theta \cdot \frac{t}{\sqrt{s^2 + t^2}}$$

$$= e^r\left(s\cos\theta - \frac{t}{\sqrt{s^2 + t^2}}\sin\theta\right)$$

13. Let $x = g(t)$ and $y = h(t)$. Then $p(t) = f(x, y)$ and the Chain Rule (2) gives $\dfrac{dp}{dt} = \dfrac{\partial f}{\partial x}\dfrac{dx}{dt} + \dfrac{\partial f}{\partial y}\dfrac{dy}{dt}$. When $t = 2$,

$x = g(2) = 4$ and $y = h(2) = 5$, so $p'(2) = f_x(4, 5)\,g'(2) + f_y(4, 5)\,h'(2) = (2)(-3) + (8)(6) = 42$.

15. $g(u, v) = f(x(u, v), y(u, v))$ where $x = e^u + \sin v$, $y = e^u + \cos v$ $\Rightarrow$

$\dfrac{\partial x}{\partial u} = e^u$, $\dfrac{\partial x}{\partial v} = \cos v$, $\dfrac{\partial y}{\partial u} = e^u$, $\dfrac{\partial y}{\partial v} = -\sin v$. By the Chain Rule (3), $\dfrac{\partial g}{\partial u} = \dfrac{\partial f}{\partial x}\dfrac{\partial x}{\partial u} + \dfrac{\partial f}{\partial y}\dfrac{\partial y}{\partial u}$. Then

$g_u(0, 0) = f_x(x(0, 0), y(0, 0))\,x_u(0, 0) + f_y(x(0, 0), y(0, 0))\,y_u(0, 0) = f_x(1, 2)(e^0) + f_y(1, 2)(e^0) = 2(1) + 5(1) = 7$.

Similarly, $\dfrac{\partial g}{\partial v} = \dfrac{\partial f}{\partial x}\dfrac{\partial x}{\partial v} + \dfrac{\partial f}{\partial y}\dfrac{\partial y}{\partial v}$. Then

$g_v(0, 0) = f_x(x(0, 0), y(0, 0))\,x_v(0, 0) + f_y(x(0, 0), y(0, 0))\,y_v(0, 0) = f_x(1, 2)(\cos 0) + f_y(1, 2)(-\sin 0)$

$\qquad = 2(1) + 5(0) = 2$

17.

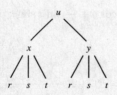

$u = f(x, y)$, $x = x(r, s, t)$, $y = y(r, s, t)$ $\Rightarrow$

$$\frac{\partial u}{\partial r} = \frac{\partial u}{\partial x}\frac{\partial x}{\partial r} + \frac{\partial u}{\partial y}\frac{\partial y}{\partial r}, \quad \frac{\partial u}{\partial s} = \frac{\partial u}{\partial x}\frac{\partial x}{\partial s} + \frac{\partial u}{\partial y}\frac{\partial y}{\partial s},$$

$$\frac{\partial u}{\partial t} = \frac{\partial u}{\partial x}\frac{\partial x}{\partial t} + \frac{\partial u}{\partial y}\frac{\partial y}{\partial t}$$

19.

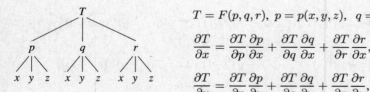

$T = F(p, q, r)$, $p = p(x, y, z)$, $q = q(x, y, z)$, $r = r(x, y, z)$ $\Rightarrow$

$$\frac{\partial T}{\partial x} = \frac{\partial T}{\partial p}\frac{\partial p}{\partial x} + \frac{\partial T}{\partial q}\frac{\partial q}{\partial x} + \frac{\partial T}{\partial r}\frac{\partial r}{\partial x},$$

$$\frac{\partial T}{\partial y} = \frac{\partial T}{\partial p}\frac{\partial p}{\partial y} + \frac{\partial T}{\partial q}\frac{\partial q}{\partial y} + \frac{\partial T}{\partial r}\frac{\partial r}{\partial y},$$

$$\frac{\partial T}{\partial z} = \frac{\partial T}{\partial p}\frac{\partial p}{\partial z} + \frac{\partial T}{\partial q}\frac{\partial q}{\partial z} + \frac{\partial T}{\partial r}\frac{\partial r}{\partial z}$$

21. $z = x^4 + x^2 y,\ x = s + 2t - u,\ y = stu^2\ \Rightarrow$

$$\frac{\partial z}{\partial s} = \frac{\partial z}{\partial x}\frac{\partial x}{\partial s} + \frac{\partial z}{\partial y}\frac{\partial y}{\partial s} = (4x^3 + 2xy)(1) + (x^2)(tu^2),$$

$$\frac{\partial z}{\partial t} = \frac{\partial z}{\partial x}\frac{\partial x}{\partial t} + \frac{\partial z}{\partial y}\frac{\partial y}{\partial t} = (4x^3 + 2xy)(2) + (x^2)(su^2),$$

$$\frac{\partial z}{\partial u} = \frac{\partial z}{\partial x}\frac{\partial x}{\partial u} + \frac{\partial z}{\partial y}\frac{\partial y}{\partial u} = (4x^3 + 2xy)(-1) + (x^2)(2stu).$$

When $s = 4, t = 2$, and $u = 1$ we have $x = 7$ and $y = 8$,

so $\dfrac{\partial z}{\partial s} = (1484)(1) + (49)(2) = 1582,\ \dfrac{\partial z}{\partial t} = (1484)(2) + (49)(4) = 3164,\ \dfrac{\partial z}{\partial u} = (1484)(-1) + (49)(16) = -700.$

23. $w = xy + yz + zx,\ x = r\cos\theta,\ y = r\sin\theta,\ z = r\theta\ \Rightarrow$

$$\frac{\partial w}{\partial r} = \frac{\partial w}{\partial x}\frac{\partial x}{\partial r} + \frac{\partial w}{\partial y}\frac{\partial y}{\partial r} + \frac{\partial w}{\partial z}\frac{\partial z}{\partial r} = (y + z)(\cos\theta) + (x + z)(\sin\theta) + (y + x)(\theta),$$

$$\frac{\partial w}{\partial \theta} = \frac{\partial w}{\partial x}\frac{\partial x}{\partial \theta} + \frac{\partial w}{\partial y}\frac{\partial y}{\partial \theta} + \frac{\partial w}{\partial z}\frac{\partial z}{\partial \theta} = (y + z)(-r\sin\theta) + (x + z)(r\cos\theta) + (y + x)(r).$$

When $r = 2$ and $\theta = \pi/2$ we have $x = 0,\ y = 2$, and $z = \pi$, so $\dfrac{\partial w}{\partial r} = (2 + \pi)(0) + (0 + \pi)(1) + (2 + 0)(\pi/2) = 2\pi$ and

$$\frac{\partial w}{\partial \theta} = (2 + \pi)(-2) + (0 + \pi)(0) + (2 + 0)(2) = -2\pi.$$

25. $N = \dfrac{p + q}{p + r},\ p = u + vw,\ q = v + uw,\ r = w + uv\ \Rightarrow$

$$\frac{\partial N}{\partial u} = \frac{\partial N}{\partial p}\frac{\partial p}{\partial u} + \frac{\partial N}{\partial q}\frac{\partial q}{\partial u} + \frac{\partial N}{\partial r}\frac{\partial r}{\partial u}$$

$$= \frac{(p + r)(1) - (p + q)(1)}{(p + r)^2}\,(1) + \frac{(p + r)(1) - (p + q)(0)}{(p + r)^2}\,(w) + \frac{(p + r)(0) - (p + q)(1)}{(p + r)^2}\,(v)$$

$$= \frac{(r - q) + (p + r)w - (p + q)v}{(p + r)^2},$$

$$\frac{\partial N}{\partial v} = \frac{\partial N}{\partial p}\frac{\partial p}{\partial v} + \frac{\partial N}{\partial q}\frac{\partial q}{\partial v} + \frac{\partial N}{\partial r}\frac{\partial r}{\partial v} = \frac{r - q}{(p + r)^2}\,(w) + \frac{p + r}{(p + r)^2}\,(1) + \frac{-(p + q)}{(p + r)^2}\,(u) = \frac{(r - q)w + (p + r) - (p + q)u}{(p + r)^2},$$

$$\frac{\partial N}{\partial w} = \frac{\partial N}{\partial p}\frac{\partial p}{\partial w} + \frac{\partial N}{\partial q}\frac{\partial q}{\partial w} + \frac{\partial N}{\partial r}\frac{\partial r}{\partial w} = \frac{r - q}{(p + r)^2}\,(v) + \frac{p + r}{(p + r)^2}\,(u) + \frac{-(p + q)}{(p + r)^2}\,(1) = \frac{(r - q)v + (p + r)u - (p + q)}{(p + r)^2}.$$

When $u = 2, v = 3$, and $w = 4$ we have $p = 14, q = 11$, and $r = 10$, so $\dfrac{\partial N}{\partial u} = \dfrac{-1 + (24)(4) - (25)(3)}{(24)^2} = \dfrac{20}{576} = \dfrac{5}{144}$,

$$\frac{\partial N}{\partial v} = \frac{(-1)(4) + 24 - (25)(2)}{(24)^2} = \frac{-30}{576} = -\frac{5}{96}, \text{ and } \frac{\partial N}{\partial w} = \frac{(-1)(3) + (24)(2) - 25}{(24)^2} = \frac{20}{576} = \frac{5}{144}.$$

27. $y\cos x = x^2 + y^2$, so let $F(x, y) = y\cos x - x^2 - y^2 = 0$. Then by Equation 6

$$\frac{dy}{dx} = -\frac{F_x}{F_y} = -\frac{-y\sin x - 2x}{\cos x - 2y} = \frac{2x + y\sin x}{\cos x - 2y}.$$

29. $\tan^{-1}(x^2 y) = x + xy^2$, so let $F(x, y) = \tan^{-1}(x^2 y) - x - xy^2 = 0$. Then

$$F_x(x, y) = \frac{1}{1 + (x^2 y)^2}(2xy) - 1 - y^2 = \frac{2xy}{1 + x^4 y^2} - 1 - y^2 = \frac{2xy - (1 + y^2)(1 + x^4 y^2)}{1 + x^4 y^2},$$

$$F_y(x, y) = \frac{1}{1 + (x^2 y)^2}(x^2) - 2xy = \frac{x^2}{1 + x^4 y^2} - 2xy = \frac{x^2 - 2xy(1 + x^4 y^2)}{1 + x^4 y^2}$$

and $\quad \dfrac{dy}{dx} = -\dfrac{F_x}{F_y} = -\dfrac{[2xy - (1 + y^2)(1 + x^4 y^2)]/(1 + x^4 y^2)}{[x^2 - 2xy(1 + x^4 y^2)]/(1 + x^4 y^2)} = \dfrac{(1 + y^2)(1 + x^4 y^2) - 2xy}{x^2 - 2xy(1 + x^4 y^2)}$

$$= \frac{1 + x^4 y^2 + y^2 + x^4 y^4 - 2xy}{x^2 - 2xy - 2x^5 y^3}$$

31. $x^2 + 2y^2 + 3z^2 = 1$, so let $F(x, y, z) = x^2 + 2y^2 + 3z^2 - 1 = 0$. Then by Equations 7

$$\frac{\partial z}{\partial x} = -\frac{F_x}{F_z} = -\frac{2x}{6z} = -\frac{x}{3z} \quad \text{and} \quad \frac{\partial z}{\partial y} = -\frac{F_y}{F_z} = -\frac{4y}{6z} = -\frac{2y}{3z}.$$

33. $e^z = xyz$, so let $F(x, y, z) = e^z - xyz = 0$. Then $\dfrac{\partial z}{\partial x} = -\dfrac{F_x}{F_z} = -\dfrac{-yz}{e^z - xy} = \dfrac{yz}{e^z - xy}$ and

$$\frac{\partial z}{\partial y} = -\frac{F_y}{F_z} = -\frac{-xz}{e^z - xy} = \frac{xz}{e^z - xy}.$$

35. Since x and y are each functions of t, $T(x, y)$ is a function of t, so by the Chain Rule, $\dfrac{dT}{dt} = \dfrac{\partial T}{\partial x}\dfrac{dx}{dt} + \dfrac{\partial T}{\partial y}\dfrac{dy}{dt}$. After

3 seconds, $x = \sqrt{1 + t} = \sqrt{1 + 3} = 2$, $y = 2 + \frac{1}{3}t = 2 + \frac{1}{3}(3) = 3$, $\dfrac{dx}{dt} = \dfrac{1}{2\sqrt{1 + t}} = \dfrac{1}{2\sqrt{1 + 3}} = \dfrac{1}{4}$, and $\dfrac{dy}{dt} = \dfrac{1}{3}$.

Then $\dfrac{dT}{dt} = T_x(2, 3)\dfrac{dx}{dt} + T_y(2, 3)\dfrac{dy}{dt} = 4\left(\frac{1}{4}\right) + 3\left(\frac{1}{3}\right) = 2$. Thus the temperature is rising at a rate of $2°\text{C/s}$.

37. $C = 1449.2 + 4.6T - 0.055T^2 + 0.00029T^3 + 0.016D$, so $\dfrac{\partial C}{\partial T} = 4.6 - 0.11T + 0.00087T^2$ and $\dfrac{\partial C}{\partial D} = 0.016$.

According to the graph, the diver is experiencing a temperature of approximately $12.5°\text{C}$ at $t = 20$ minutes, so

$\dfrac{\partial C}{\partial T} = 4.6 - 0.11(12.5) + 0.00087(12.5)^2 \approx 3.36$. By sketching tangent lines at $t = 20$ to the graphs given, we estimate

$\dfrac{dD}{dt} \approx \dfrac{1}{2}$ and $\dfrac{dT}{dt} \approx -\dfrac{1}{10}$. Then, by the Chain Rule, $\dfrac{dC}{dt} = \dfrac{\partial C}{\partial T}\dfrac{dT}{dt} + \dfrac{\partial C}{\partial D}\dfrac{dD}{dt} \approx (3.36)\left(-\frac{1}{10}\right) + (0.016)\left(\frac{1}{2}\right) \approx -0.33$.

Thus the speed of sound experienced by the diver is decreasing at a rate of approximately 0.33 m/s per minute.

39. (a) $V = \ell wh$, so by the Chain Rule,

$$\frac{dV}{dt} = \frac{\partial V}{\partial \ell}\frac{d\ell}{dt} + \frac{\partial V}{\partial w}\frac{dw}{dt} + \frac{\partial V}{\partial h}\frac{dh}{dt} = wh\frac{d\ell}{dt} + \ell h\frac{dw}{dt} + \ell w\frac{dh}{dt} = 2 \cdot 2 \cdot 2 + 1 \cdot 2 \cdot 2 + 1 \cdot 2 \cdot (-3) = 6 \text{ m}^3/\text{s}.$$

(b) $S = 2(\ell w + \ell h + wh)$, so by the Chain Rule,

$$\frac{dS}{dt} = \frac{\partial S}{\partial \ell}\frac{d\ell}{dt} + \frac{\partial S}{\partial w}\frac{dw}{dt} + \frac{\partial S}{\partial h}\frac{dh}{dt} = 2(w + h)\frac{d\ell}{dt} + 2(\ell + h)\frac{dw}{dt} + 2(\ell + w)\frac{dh}{dt}$$

$$= 2(2 + 2)2 + 2(1 + 2)2 + 2(1 + 2)(-3) = 10 \text{ m}^2/\text{s}$$

(c) $L^2 = \ell^2 + w^2 + h^2 \;\Rightarrow\; 2L\dfrac{dL}{dt} = 2\ell\dfrac{d\ell}{dt} + 2w\dfrac{dw}{dt} + 2h\dfrac{dh}{dt} = 2(1)(2) + 2(2)(2) + 2(2)(-3) = 0 \;\Rightarrow$

$dL/dt = 0$ m/s.

41. $\dfrac{dP}{dt} = 0.05,\ \dfrac{dT}{dt} = 0.15,\ V = 8.31\dfrac{T}{P}$ and $\dfrac{dV}{dt} = \dfrac{8.31}{P}\dfrac{dT}{dt} - 8.31\dfrac{T}{P^2}\dfrac{dP}{dt}$. Thus when $P = 20$ and $T = 320$,

$\dfrac{dV}{dt} = 8.31\left[\dfrac{0.15}{20} - \dfrac{(0.05)(320)}{400}\right] \approx -0.27$ L/s.

43. Let x be the length of the first side of the triangle and y the length of the second side. The area A of the triangle is given by

$A = \frac{1}{2}xy\sin\theta$ where θ is the angle between the two sides. Thus A is a function of x, y, and θ, and x, y, and θ are each in turn

functions of time t. We are given that $\dfrac{dx}{dt} = 3$, $\dfrac{dy}{dt} = -2$, and because A is constant, $\dfrac{dA}{dt} = 0$. By the Chain Rule,

$\dfrac{dA}{dt} = \dfrac{\partial A}{\partial x}\dfrac{dx}{dt} + \dfrac{\partial A}{\partial y}\dfrac{dy}{dt} + \dfrac{\partial A}{\partial \theta}\dfrac{d\theta}{dt} \;\Rightarrow\; \dfrac{dA}{dt} = \frac{1}{2}y\sin\theta\cdot\dfrac{dx}{dt} + \frac{1}{2}x\sin\theta\cdot\dfrac{dy}{dt} + \frac{1}{2}xy\cos\theta\cdot\dfrac{d\theta}{dt}$. When $x = 20$, $y = 30$,

and $\theta = \pi/6$ we have

$$0 = \tfrac{1}{2}(30)\left(\sin\tfrac{\pi}{6}\right)(3) + \tfrac{1}{2}(20)\left(\sin\tfrac{\pi}{6}\right)(-2) + \tfrac{1}{2}(20)(30)\left(\cos\tfrac{\pi}{6}\right)\dfrac{d\theta}{dt}$$

$$= 45\cdot\tfrac{1}{2} - 20\cdot\tfrac{1}{2} + 300\cdot\dfrac{\sqrt{3}}{2}\cdot\dfrac{d\theta}{dt} = \tfrac{25}{2} + 150\sqrt{3}\,\dfrac{d\theta}{dt}$$

Solving for $\dfrac{d\theta}{dt}$ gives $\dfrac{d\theta}{dt} = \dfrac{-25/2}{150\sqrt{3}} = -\dfrac{1}{12\sqrt{3}}$, so the angle between the sides is decreasing at a rate of

$1/(12\sqrt{3}) \approx 0.048$ rad/s.

45. (a) By the Chain Rule, $\dfrac{\partial z}{\partial r} = \dfrac{\partial z}{\partial x}\cos\theta + \dfrac{\partial z}{\partial y}\sin\theta$, $\dfrac{\partial z}{\partial \theta} = \dfrac{\partial z}{\partial x}(-r\sin\theta) + \dfrac{\partial z}{\partial y}r\cos\theta$.

(b) $\left(\dfrac{\partial z}{\partial r}\right)^2 = \left(\dfrac{\partial z}{\partial x}\right)^2\cos^2\theta + 2\dfrac{\partial z}{\partial x}\dfrac{\partial z}{\partial y}\cos\theta\sin\theta + \left(\dfrac{\partial z}{\partial y}\right)^2\sin^2\theta$,

$\left(\dfrac{\partial z}{\partial \theta}\right)^2 = \left(\dfrac{\partial z}{\partial x}\right)^2 r^2\sin^2\theta - 2\dfrac{\partial z}{\partial x}\dfrac{\partial z}{\partial y}r^2\cos\theta\sin\theta + \left(\dfrac{\partial z}{\partial y}\right)^2 r^2\cos^2\theta$. Thus

$\left(\dfrac{\partial z}{\partial r}\right)^2 + \dfrac{1}{r^2}\left(\dfrac{\partial z}{\partial \theta}\right)^2 = \left[\left(\dfrac{\partial z}{\partial x}\right)^2 + \left(\dfrac{\partial z}{\partial y}\right)^2\right](\cos^2\theta + \sin^2\theta) = \left(\dfrac{\partial z}{\partial x}\right)^2 + \left(\dfrac{\partial z}{\partial y}\right)^2$.

47. Let $u = x - y$ and $v = x + y$. Then $z = \dfrac{1}{x}[f(u) + g(v)]$ and

$$\dfrac{\partial z}{\partial x} = \dfrac{1}{x}\left[\dfrac{df}{du}\dfrac{\partial u}{\partial x} + \dfrac{dg}{dv}\dfrac{\partial v}{\partial x}\right] + [f(u) + g(v)]\left(-\dfrac{1}{x^2}\right)$$

$$= \dfrac{1}{x}[f'(u)(1) + g'(v)(1)] - \dfrac{1}{x^2}[f(u) + g(v)] = \dfrac{1}{x}[f'(u) + g'(v)] - \dfrac{1}{x^2}[f(u) + g(v)]$$

$$\dfrac{\partial z}{\partial y} = \dfrac{1}{x}\left[\dfrac{df}{du}\dfrac{\partial u}{\partial y} + \dfrac{dg}{dv}\dfrac{\partial v}{\partial y}\right] = \dfrac{1}{x}[f'(u)(-1) + g'(v)(1)] = \dfrac{1}{x}[-f'(u) + g'(v)]$$

[continued]

$$\frac{\partial^2 z}{\partial y^2} = \frac{1}{x}\left[\frac{d}{du}\left[-f'(u)\right]\frac{\partial u}{\partial y} + \frac{d}{dv}\left[g'(v)\right]\frac{\partial v}{\partial y}\right] = \frac{1}{x}\left[-f''(u)(-1) + g''(v)(1)\right] = \frac{1}{x}\left[f''(u) + g''(v)\right]$$

Thus

$$\frac{\partial}{\partial x}\left(x^2\frac{\partial z}{\partial x}\right) = \frac{\partial}{\partial x}\left(x\left[f'(u) + g'(v)\right] - \left[f(u) + g(v)\right]\right)$$

$$= x\left[f''(u)(1) + g''(v)(1)\right] + \left[f'(u) + g'(v)\right](1) - \left[f'(u)(1) + g'(v)(1)\right]$$

$$= x\left[f''(u) + g''(v)\right] + f'(u) + g'(v) - f'(u) - g'(v) = x\left[f''(u) + g''(v)\right]$$

$$= x^2 \cdot \frac{1}{x}\left[f''(u) + g''(v)\right] = x^2\frac{\partial^2 z}{\partial y^2}$$

49. Let $u = x + at$, $v = x - at$. Then $z = f(u) + g(v)$, so $\partial z/\partial u = f'(u)$ and $\partial z/\partial v = g'(v)$.

Thus $\dfrac{\partial z}{\partial t} = \dfrac{\partial z}{\partial u}\dfrac{\partial u}{\partial t} + \dfrac{\partial z}{\partial v}\dfrac{\partial v}{\partial t} = af'(u) - ag'(v)$ and

$$\frac{\partial^2 z}{\partial t^2} = a\frac{\partial}{\partial t}\left[f'(u) - g'(v)\right] = a\left(\frac{df'(u)}{du}\frac{\partial u}{\partial t} - \frac{dg'(v)}{dv}\frac{\partial v}{\partial t}\right) = a^2 f''(u) + a^2 g''(v).$$

Similarly, $\dfrac{\partial z}{\partial x} = f'(u) + g'(v)$ and $\dfrac{\partial^2 z}{\partial x^2} = f''(u) + g''(v)$. Thus $\dfrac{\partial^2 z}{\partial t^2} = a^2\dfrac{\partial^2 z}{\partial x^2}$.

51. $\dfrac{\partial z}{\partial s} = \dfrac{\partial z}{\partial x}2s + \dfrac{\partial z}{\partial y}2r$. Then

$$\frac{\partial^2 z}{\partial r\,\partial s} = \frac{\partial}{\partial r}\left(\frac{\partial z}{\partial x}2s\right) + \frac{\partial}{\partial r}\left(\frac{\partial z}{\partial y}2r\right)$$

$$= \frac{\partial^2 z}{\partial x^2}\frac{\partial x}{\partial r}2s + \frac{\partial}{\partial y}\left(\frac{\partial z}{\partial x}\right)\frac{\partial y}{\partial r}2s + \frac{\partial z}{\partial x}\frac{\partial}{\partial r}2s + \frac{\partial^2 z}{\partial y^2}\frac{\partial y}{\partial r}2r + \frac{\partial}{\partial x}\left(\frac{\partial z}{\partial y}\right)\frac{\partial x}{\partial r}2r + \frac{\partial z}{\partial y}2$$

$$= 4rs\frac{\partial^2 z}{\partial x^2} + \frac{\partial^2 z}{\partial y\,\partial x}4s^2 + 0 + 4rs\frac{\partial^2 z}{\partial y^2} + \frac{\partial^2 z}{\partial x\,\partial y}4r^2 + 2\frac{\partial z}{\partial y}$$

By the continuity of the partials, $\dfrac{\partial^2 z}{\partial r\,\partial s} = 4rs\dfrac{\partial^2 z}{\partial x^2} + 4rs\dfrac{\partial^2 z}{\partial y^2} + (4r^2 + 4s^2)\dfrac{\partial^2 z}{\partial x\,\partial y} + 2\dfrac{\partial z}{\partial y}$.

53. $\dfrac{\partial z}{\partial r} = \dfrac{\partial z}{\partial x}\cos\theta + \dfrac{\partial z}{\partial y}\sin\theta$ and $\dfrac{\partial z}{\partial\theta} = -\dfrac{\partial z}{\partial x}r\sin\theta + \dfrac{\partial z}{\partial y}r\cos\theta$. Then

$$\frac{\partial^2 z}{\partial r^2} = \cos\theta\left(\frac{\partial^2 z}{\partial x^2}\cos\theta + \frac{\partial^2 z}{\partial y\,\partial x}\sin\theta\right) + \sin\theta\left(\frac{\partial^2 z}{\partial y^2}\sin\theta + \frac{\partial^2 z}{\partial x\,\partial y}\cos\theta\right)$$

$$= \cos^2\theta\frac{\partial^2 z}{\partial x^2} + 2\cos\theta\sin\theta\frac{\partial^2 z}{\partial x\,\partial y} + \sin^2\theta\frac{\partial^2 z}{\partial y^2}$$

and

$$\frac{\partial^2 z}{\partial\theta^2} = -r\cos\theta\frac{\partial z}{\partial x} + (-r\sin\theta)\left(\frac{\partial^2 z}{\partial x^2}(-r\sin\theta) + \frac{\partial^2 z}{\partial y\,\partial x}r\cos\theta\right)$$

$$-r\sin\theta\frac{\partial z}{\partial y} + r\cos\theta\left(\frac{\partial^2 z}{\partial y^2}r\cos\theta + \frac{\partial^2 z}{\partial x\,\partial y}(-r\sin\theta)\right)$$

$$= -r\cos\theta\frac{\partial z}{\partial x} - r\sin\theta\frac{\partial z}{\partial y} + r^2\sin^2\theta\frac{\partial^2 z}{\partial x^2} - 2r^2\cos\theta\sin\theta\frac{\partial^2 z}{\partial x\,\partial y} + r^2\cos^2\theta\frac{\partial^2 z}{\partial y^2}$$

Thus

$$\frac{\partial^2 z}{\partial r^2} + \frac{1}{r^2}\frac{\partial^2 z}{\partial \theta^2} + \frac{1}{r}\frac{\partial z}{\partial r} = \left(\cos^2\theta + \sin^2\theta\right)\frac{\partial^2 z}{\partial x^2} + \left(\sin^2\theta + \cos^2\theta\right)\frac{\partial^2 z}{\partial y^2}$$

$$-\frac{1}{r}\cos\theta\,\frac{\partial z}{\partial x} - \frac{1}{r}\sin\theta\,\frac{\partial z}{\partial y} + \frac{1}{r}\left(\cos\theta\,\frac{\partial z}{\partial x} + \sin\theta\,\frac{\partial z}{\partial y}\right)$$

$$= \frac{\partial^2 z}{\partial x^2} + \frac{\partial^2 z}{\partial y^2} \text{ as desired.}$$

55. (a) Since f is a polynomial, it has continuous second-order partial derivatives, and

$$f(tx, ty) = (tx)^2(ty) + 2(tx)(ty)^2 + 5(ty)^3 = t^3x^2y + 2t^3xy^2 + 5t^3y^3 = t^3(x^2y + 2xy^2 + 5y^3) = t^3 f(x, y).$$

Thus, f is homogeneous of degree 3.

(b) Differentiating both sides of $f(tx, ty) = t^n f(x, y)$ with respect to t using the Chain Rule, we get

$$\frac{\partial}{\partial t}f(tx, ty) = \frac{\partial}{\partial t}\left[t^n f(x, y)\right] \quad \Leftrightarrow$$

$$\frac{\partial}{\partial(tx)}f(tx, ty)\cdot\frac{\partial(tx)}{\partial t} + \frac{\partial}{\partial(ty)}f(tx, ty)\cdot\frac{\partial(ty)}{\partial t} = x\frac{\partial}{\partial(tx)}f(tx, ty) + y\frac{\partial}{\partial(ty)}f(tx, ty) = nt^{n-1}f(x, y).$$

Setting $t = 1$: $x\frac{\partial}{\partial x}f(x, y) + y\frac{\partial}{\partial y}f(x, y) = nf(x, y).$

57. Differentiating both sides of $f(tx, ty) = t^n f(x, y)$ with respect to x using the Chain Rule, we get

$$\frac{\partial}{\partial x}f(tx, ty) = \frac{\partial}{\partial x}\left[t^n f(x, y)\right] \quad \Leftrightarrow$$

$$\frac{\partial}{\partial(tx)}f(tx, ty)\cdot\frac{\partial(tx)}{\partial x} + \frac{\partial}{\partial(ty)}f(tx, ty)\cdot\frac{\partial(ty)}{\partial x} = t^n\frac{\partial}{\partial x}f(x, y) \quad \Leftrightarrow \quad tf_x(tx, ty) = t^n f_x(x, y).$$

Thus $f_x(tx, ty) = t^{n-1}f_x(x, y)$.

59. Given a function defined implicitly by $F(x, y) = 0$, where F is differentiable and $F_y \neq 0$, we know that $\dfrac{dy}{dx} = -\dfrac{F_x}{F_y}$. Let

$$G(x, y) = -\frac{F_x}{F_y} \text{ so } \frac{dy}{dx} = G(x, y). \text{ Differentiating both sides with respect to } x \text{ and using the Chain Rule gives}$$

$$\frac{d^2 y}{dx^2} = \frac{\partial G}{\partial x}\frac{dx}{dx} + \frac{\partial G}{\partial y}\frac{dy}{dx} \text{ where } \frac{\partial G}{\partial x} = \frac{\partial}{\partial x}\left(-\frac{F_x}{F_y}\right) = -\frac{F_y F_{xx} - F_x F_{yx}}{F_y^2}, \frac{\partial G}{\partial y} = \frac{\partial}{\partial y}\left(-\frac{F_x}{F_y}\right) = -\frac{F_y F_{xy} - F_x F_{yy}}{F_y^2}.$$

Thus

$$\frac{d^2 y}{dx^2} = \left(-\frac{F_y F_{xx} - F_x F_{yx}}{F_y^2}\right)(1) + \left(-\frac{F_y F_{xy} - F_x F_{yy}}{F_y^2}\right)\left(-\frac{F_x}{F_y}\right)$$

$$= -\frac{F_{xx}F_y^2 - F_{yx}F_x F_y - F_{xy}F_y F_x + F_{yy}F_x^2}{F_y^3}$$

But F has continuous second derivatives, so by Clauraut's Theorem, $F_{yx} = F_{xy}$ and we have

$$\frac{d^2 y}{dx^2} = -\frac{F_{xx}F_y^2 - 2F_{xy}F_x F_y + F_{yy}F_x^2}{F_y^3} \text{ as desired.}$$

14.6 Directional Derivatives and the Gradient Vector

1. We can approximate the directional derivative of the pressure function at K in the direction of S by the average rate of change

of pressure between the points where the red line intersects the contour lines closest to K (extend the red line slightly at the

left). In the direction of S, the pressure changes from 1000 millibars to 996 millibars and we estimate the distance between

these two points to be approximately 50 km (using the fact that the distance from K to S is 300 km). Then the rate of change

of pressure in the direction given is approximately $\frac{996 - 1000}{50} = -0.08$ millibar/km.

3. $D_{\mathbf{u}}\, f(-20, 30) = \nabla f(-20, 30) \cdot \mathbf{u} = f_T(-20, 30)\left(\frac{1}{\sqrt{2}}\right) + f_v(-20, 30)\left(\frac{1}{\sqrt{2}}\right).$

$f_T(-20, 30) = \lim\limits_{h \to 0} \dfrac{f(-20 + h, 30) - f(-20, 30)}{h}$, so we can approximate $f_T(-20, 30)$ by considering $h = \pm 5$ and

using the values given in the table: $f_T(-20, 30) \approx \dfrac{f(-15, 30) - f(-20, 30)}{5} = \dfrac{-26 - (-33)}{5} = 1.4$,

$f_T(-20, 30) \approx \dfrac{f(-25, 30) - f(-20, 30)}{-5} = \dfrac{-39 - (-33)}{-5} = 1.2$. Averaging these values gives $f_T(-20, 30) \approx 1.3$.

Similarly, $f_v(-20, 30) = \lim\limits_{h \to 0} \dfrac{f(-20, 30 + h) - f(-20, 30)}{h}$, so we can approximate $f_v(-20, 30)$ with $h = \pm 10$:

$f_v(-20, 30) \approx \dfrac{f(-20, 40) - f(-20, 30)}{10} = \dfrac{-34 - (-33)}{10} = -0.1$,

$f_v(-20, 30) \approx \dfrac{f(-20, 20) - f(-20, 30)}{-10} = \dfrac{-30 - (-33)}{-10} = -0.3$. Averaging these values gives $f_v(-20, 30) \approx -0.2$.

Then $D_{\mathbf{u}} f(-20, 30) \approx 1.3\left(\frac{1}{\sqrt{2}}\right) + (-0.2)\left(\frac{1}{\sqrt{2}}\right) \approx 0.778$.

5. $f(x, y) = y \cos(xy) \quad \Rightarrow \quad f_x(x, y) = y[-\sin(xy)](y) = -y^2 \sin(xy)$ and

$f_y(x, y) = y[-\sin(xy)](x) + [\cos(xy)](1) = \cos(xy) - xy \sin(xy)$. If $\mathbf{u}$ is a unit vector in the direction of $\theta = \pi/4$, then

from Equation 6, $D_{\mathbf{u}} f(0, 1) = f_x(0, 1) \cos\left(\frac{\pi}{4}\right) + f_y(0, 1) \sin\left(\frac{\pi}{4}\right) = 0 \cdot \frac{\sqrt{2}}{2} + 1 \cdot \frac{\sqrt{2}}{2} = \frac{\sqrt{2}}{2}$.

7. $f(x, y) = x/y = xy^{-1}$

(a) $\nabla f(x, y) = \dfrac{\partial f}{\partial x}\, \mathbf{i} + \dfrac{\partial f}{\partial y}\, \mathbf{j} = y^{-1}\, \mathbf{i} + (-xy^{-2})\, \mathbf{j} = \dfrac{1}{y}\, \mathbf{i} - \dfrac{x}{y^2}\, \mathbf{j}$

(b) $\nabla f(2, 1) = \dfrac{1}{1}\, \mathbf{i} - \dfrac{2}{1^2}\, \mathbf{j} = \mathbf{i} - 2\, \mathbf{j}$

(c) By Equation 9, $D_{\mathbf{u}}\, f(2, 1) = \nabla f(2, 1) \cdot \mathbf{u} = (\mathbf{i} - 2\, \mathbf{j}) \cdot \left(\frac{3}{5}\, \mathbf{i} + \frac{4}{5}\, \mathbf{j}\right) = \frac{3}{5} - \frac{8}{5} = -1$.

9. $f(x, y, z) = x^2 yz - xyz^3$

(a) $\nabla f(x, y, z) = \langle f_x(x, y, z), f_y(x, y, z), f_z(x, y, z)\rangle = \langle 2xyz - yz^3,\, x^2 z - xz^3,\, x^2 y - 3xyz^2\rangle$

(b) $\nabla f(2, -1, 1) = \langle -4 + 1, 4 - 2, -4 + 6\rangle = \langle -3, 2, 2\rangle$

(c) By Equation 14, $D_{\mathbf{u}} f(2, -1, 1) = \nabla f(2, -1, 1) \cdot \mathbf{u} = \langle -3, 2, 2\rangle \cdot \langle 0, \frac{4}{5}, -\frac{3}{5}\rangle = 0 + \frac{8}{5} - \frac{6}{5} = \frac{2}{5}$.

11. $f(x, y) = e^x \sin y \implies \nabla f(x, y) = \langle e^x \sin y, e^x \cos y \rangle$, $\nabla f(0, \pi/3) = \left\langle \frac{\sqrt{3}}{2}, \frac{1}{2} \right\rangle$, and a

unit vector in the direction of **v** is $\mathbf{u} = \frac{1}{\sqrt{(-6)^2 + 8^2}} \langle -6, 8 \rangle = \frac{1}{10} \langle -6, 8 \rangle = \langle -\frac{3}{5}, \frac{4}{5} \rangle$, so

$$D_{\mathbf{u}} f(0, \pi/3) = \nabla f(0, \pi/3) \cdot \mathbf{u} = \left\langle \frac{\sqrt{3}}{2}, \frac{1}{2} \right\rangle \cdot \langle -\frac{3}{5}, \frac{4}{5} \rangle = -\frac{3\sqrt{3}}{10} + \frac{4}{10} = \frac{4 - 3\sqrt{3}}{10}.$$

13. $g(s, t) = s\sqrt{t} \implies \nabla g(s, t) = (\sqrt{t})\,\mathbf{i} + (s/(2\sqrt{t}))\,\mathbf{j}$, $\nabla g(2, 4) = 2\,\mathbf{i} + \frac{1}{2}\,\mathbf{j}$, and a unit vector in the direction of **v** is

$\mathbf{u} = \frac{1}{\sqrt{2^2 + (-1)^2}} (2\,\mathbf{i} - \mathbf{j}) = \frac{1}{\sqrt{5}}(2\,\mathbf{i} - \mathbf{j})$, so $D_{\mathbf{u}} g(2, 4) = \nabla g(2, 4) \cdot \mathbf{u} = (2\,\mathbf{i} + \frac{1}{2}\,\mathbf{j}) \cdot \frac{1}{\sqrt{5}}(2\,\mathbf{i} - \mathbf{j}) = \frac{1}{\sqrt{5}} \left(4 - \frac{1}{2} \right) = \frac{7}{2\sqrt{5}}$ or

$\frac{7\sqrt{5}}{10}$.

15. $f(x, y, z) = x^2 y + y^2 z \implies \nabla f(x, y, z) = \langle 2xy, x^2 + 2yz, y^2 \rangle$, $\nabla f(1, 2, 3) = \langle 4, 13, 4 \rangle$, and a unit

vector in the direction of **v** is $\mathbf{u} = \frac{1}{\sqrt{4 + 1 + 4}} \langle 2, -1, 2 \rangle = \frac{1}{3} \langle 2, -1, 2 \rangle$, so

$$D_{\mathbf{u}} f(1, 2, 3) = \nabla f(1, 2, 3) \cdot \mathbf{u} = \langle 4, 13, 4 \rangle \cdot \frac{1}{3} \langle 2, -1, 2 \rangle = \frac{1}{3}(8 - 13 + 8) = \frac{3}{3} = 1.$$

17. $h(r, s, t) = \ln(3r + 6s + 9t) \implies \nabla h(r, s, t) = \langle 3/(3r + 6s + 9t), 6/(3r + 6s + 9t), 9/(3r + 6s + 9t) \rangle$,

$\nabla h(1, 1, 1) = \left\langle \frac{1}{6}, \frac{1}{3}, \frac{1}{2} \right\rangle$, and a unit vector in the direction of $\mathbf{v} = 4\,\mathbf{i} + 12\,\mathbf{j} + 6\,\mathbf{k}$.

is $\mathbf{u} = \frac{1}{\sqrt{16 + 144 + 36}} (4\,\mathbf{i} + 12\,\mathbf{j} + 6\,\mathbf{k}) = \frac{2}{7}\,\mathbf{i} + \frac{6}{7}\,\mathbf{j} + \frac{3}{7}\,\mathbf{k}$, so

$$D_{\mathbf{u}} h(1, 1, 1) = \nabla h(1, 1, 1) \cdot \mathbf{u} = \left\langle \frac{1}{6}, \frac{1}{3}, \frac{1}{2} \right\rangle \cdot \left\langle \frac{2}{7}, \frac{6}{7}, \frac{3}{7} \right\rangle = \frac{1}{21} + \frac{2}{7} + \frac{3}{14} = \frac{23}{42}.$$

19. $f(x, y) = \sqrt{xy} \implies \nabla f(x, y) = \left\langle \frac{1}{2}(xy)^{-1/2}(y), \frac{1}{2}(xy)^{-1/2}(x) \right\rangle = \left\langle \frac{y}{2\sqrt{xy}}, \frac{x}{2\sqrt{xy}} \right\rangle$, so $\nabla f(2, 8) = \langle 1, \frac{1}{4} \rangle$.

The unit vector in the direction of $\overrightarrow{PQ} = \langle 5 - 2, 4 - 8 \rangle = \langle 3, -4 \rangle$ is $\mathbf{u} = \langle \frac{3}{5}, -\frac{4}{5} \rangle$, so

$$D_{\mathbf{u}} f(2, 8) = \nabla f(2, 8) \cdot \mathbf{u} = \langle 1, \frac{1}{4} \rangle \cdot \langle \frac{3}{5}, -\frac{4}{5} \rangle = \frac{2}{5}.$$

21. $f(x, y) = 4y\sqrt{x} \implies \nabla f(x, y) = \left\langle 4y \cdot \frac{1}{2} x^{-1/2}, 4\sqrt{x} \right\rangle = \langle 2y/\sqrt{x}, 4\sqrt{x} \rangle$.

$\nabla f(4, 1) = \langle 1, 8 \rangle$ is the direction of maximum rate of change, and the maximum rate is $|\nabla f(4, 1)| = \sqrt{1 + 64} = \sqrt{65}$.

23. $f(x, y) = \sin(xy) \implies \nabla f(x, y) = \langle y \cos(xy), x \cos(xy) \rangle$, $\nabla f(1, 0) = \langle 0, 1 \rangle$. Thus the maximum rate of change is

$|\nabla f(1, 0)| = 1$ in the direction $\langle 0, 1 \rangle$.

25. $f(x, y, z) = x/(y + z) = x(y + z)^{-1} \implies$

$$\nabla f(x, y, z) = \langle 1/(y + z), -x(y + z)^{-2}(1), -x(y + z)^{-2}(1) \rangle = \left\langle \frac{1}{y + z}, -\frac{x}{(y + z)^2}, -\frac{x}{(y + z)^2} \right\rangle,$$

$\nabla f(8, 1, 3) = \langle \frac{1}{4}, -\frac{8}{4^2}, -\frac{8}{4^2} \rangle = \langle \frac{1}{4}, -\frac{1}{2}, -\frac{1}{2} \rangle$. Thus the maximum rate of change is

$|\nabla f(8, 1, 3)| = \sqrt{\frac{1}{16} + \frac{1}{4} + \frac{1}{4}} = \sqrt{\frac{9}{16}} = \frac{3}{4}$ in the direction $\langle \frac{1}{4}, -\frac{1}{2}, -\frac{1}{2} \rangle$ or equivalently $\langle 1, -2, -2 \rangle$.

27. (a) As in the proof of Theorem 15, $D_{\mathbf{u}} f = |\nabla f| \cos\theta$. Since the minimum value of $\cos\theta$ is -1 occurring when $\theta = \pi$, the

minimum value of $D_{\mathbf{u}} f$ is $-|\nabla f|$ occurring when $\theta = \pi$, that is when $\mathbf{u}$ is in the opposite direction of ∇f

(assuming $\nabla f \neq \mathbf{0}$).

(b) $f(x,y) = x^4 y - x^2 y^3 \ \Rightarrow \ \nabla f(x,y) = \langle 4x^3 y - 2xy^3, x^4 - 3x^2 y^2 \rangle$, so f decreases fastest at the point $(2,-3)$ in the

direction $-\nabla f(2,-3) = -\langle 12, -92 \rangle = \langle -12, 92 \rangle$.

29. The direction of fastest change is $\nabla f(x,y) = (2x-2)\,\mathbf{i} + (2y-4)\,\mathbf{j}$, so we need to find all points (x,y) where $\nabla f(x,y)$ is

parallel to $\mathbf{i}+\mathbf{j}$ $\Leftrightarrow$ $(2x-2)\,\mathbf{i} + (2y-4)\,\mathbf{j} = k\,(\mathbf{i}+\mathbf{j})$ $\Leftrightarrow$ $k = 2x-2$ and $k = 2y-4$. Then $2x-2 = 2y-4$ $\Rightarrow$

$y = x+1$, so the direction of fastest change is $\mathbf{i}+\mathbf{j}$ at all points on the line $y = x+1$.

31. $T = \dfrac{k}{\sqrt{x^2+y^2+z^2}}$ and $120 = T(1,2,2) = \dfrac{k}{3}$ so $k = 360$.

(a) $\mathbf{u} = \dfrac{\langle 1,-1,1\rangle}{\sqrt{3}}$,

$D_{\mathbf{u}} T(1,2,2) = \nabla T(1,2,2) \cdot \mathbf{u} = \left[-360\left(x^2+y^2+z^2\right)^{-3/2}\langle x,y,z\rangle\right]_{(1,2,2)} \cdot \mathbf{u} = -\frac{40}{3}\langle 1,2,2\rangle \cdot \frac{1}{\sqrt{3}}\langle 1,-1,1\rangle = -\frac{40}{3\sqrt{3}}$

(b) From (a), $\nabla T = -360\left(x^2+y^2+z^2\right)^{-3/2}\langle x,y,z\rangle$, and since $\langle x,y,z\rangle$ is the position vector of the point (x,y,z), the

vector $-\langle x,y,z\rangle$, and thus ∇T, always points toward the origin.

33. $\nabla V(x,y,z) = \langle 10x - 3y + yz, xz - 3x, xy\rangle$, $\nabla V(3,4,5) = \langle 38, 6, 12\rangle$

(a) $D_{\mathbf{u}} V(3,4,5) = \langle 38, 6, 12\rangle \cdot \frac{1}{\sqrt{3}}\langle 1,1,-1\rangle = \frac{32}{\sqrt{3}}$

(b) $\nabla V(3,4,5) = \langle 38, 6, 12\rangle$, or equivalently, $\langle 19, 3, 6\rangle$.

(c) $|\nabla V(3,4,5)| = \sqrt{38^2 + 6^2 + 12^2} = \sqrt{1624} = 2\sqrt{406}$

35. A unit vector in the direction of $\overrightarrow{AB}$ is $\mathbf{i}$ and a unit vector in the direction of $\overrightarrow{AC}$ is $\mathbf{j}$. Thus $D_{\overrightarrow{AB}} f(1,3) = f_x(1,3) = 3$ and

$D_{\overrightarrow{AC}} f(1,3) = f_y(1,3) = 26$. Therefore $\nabla f(1,3) = \langle f_x(1,3), f_y(1,3)\rangle = \langle 3, 26\rangle$, and by definition,

$D_{\overrightarrow{AD}} f(1,3) = \nabla f \cdot \mathbf{u}$ where $\mathbf{u}$ is a unit vector in the direction of $\overrightarrow{AD}$, which is $\langle \frac{5}{13}, \frac{12}{13}\rangle$. Therefore,

$D_{\overrightarrow{AD}} f(1,3) = \langle 3, 26\rangle \cdot \langle \frac{5}{13}, \frac{12}{13}\rangle = 3 \cdot \frac{5}{13} + 26 \cdot \frac{12}{13} = \frac{327}{13}$.

37. (a) $\nabla(au + bv) = \left\langle \dfrac{\partial(au+bv)}{\partial x}, \dfrac{\partial(au+bv)}{\partial y}\right\rangle = \left\langle a\dfrac{\partial u}{\partial x} + b\dfrac{\partial v}{\partial x}, a\dfrac{\partial u}{\partial y} + b\dfrac{\partial v}{\partial y}\right\rangle = a\left\langle \dfrac{\partial u}{\partial x}, \dfrac{\partial u}{\partial y}\right\rangle + b\left\langle \dfrac{\partial v}{\partial x}, \dfrac{\partial v}{\partial y}\right\rangle$

$= a\,\nabla u + b\,\nabla v$

(b) $\nabla(uv) = \left\langle v\dfrac{\partial u}{\partial x} + u\dfrac{\partial v}{\partial x}, v\dfrac{\partial u}{\partial y} + u\dfrac{\partial v}{\partial y}\right\rangle = v\left\langle \dfrac{\partial u}{\partial x}, \dfrac{\partial u}{\partial y}\right\rangle + u\left\langle \dfrac{\partial v}{\partial x}, \dfrac{\partial v}{\partial y}\right\rangle = v\,\nabla u + u\,\nabla v$

(c) $\nabla\left(\dfrac{u}{v}\right) = \left\langle \dfrac{v\,\dfrac{\partial u}{\partial x} - u\,\dfrac{\partial v}{\partial x}}{v^2}, \dfrac{v\,\dfrac{\partial u}{\partial y} - u\,\dfrac{\partial v}{\partial y}}{v^2} \right\rangle = \dfrac{v\left\langle \dfrac{\partial u}{\partial x}, \dfrac{\partial u}{\partial y}\right\rangle - u\left\langle \dfrac{\partial v}{\partial x}, \dfrac{\partial v}{\partial y}\right\rangle}{v^2} = \dfrac{v\,\nabla u - u\,\nabla v}{v^2}$

(d) $\nabla u^n = \left\langle \dfrac{\partial(u^n)}{\partial x}, \dfrac{\partial(u^n)}{\partial y}\right\rangle = \left\langle nu^{n-1}\dfrac{\partial u}{\partial x}, nu^{n-1}\dfrac{\partial u}{\partial y}\right\rangle = nu^{n-1}\nabla u$

39. $f(x,y) = x^3 + 5x^2 y + y^3 \quad \Rightarrow$

$D_{\mathbf{u}}f(x,y) = \nabla f(x,y)\cdot\mathbf{u} = \left\langle 3x^2 + 10xy, 5x^2 + 3y^2\right\rangle\cdot\left\langle\frac{3}{5},\frac{4}{5}\right\rangle = \frac{9}{5}x^2 + 6xy + 4x^2 + \frac{12}{5}y^2 = \frac{29}{5}x^2 + 6xy + \frac{12}{5}y^2$. Then

$D_{\mathbf{u}}^2 f(x,y) = D_{\mathbf{u}}\left[D_{\mathbf{u}}f(x,y)\right] = \nabla\left[D_{\mathbf{u}}f(x,y)\right]\cdot\mathbf{u} = \left\langle\frac{58}{5}x + 6y, 6x + \frac{24}{5}y\right\rangle\cdot\left\langle\frac{3}{5},\frac{4}{5}\right\rangle$

$\qquad = \frac{174}{25}x + \frac{18}{5}y + \frac{24}{5}x + \frac{96}{25}y = \frac{294}{25}x + \frac{186}{25}y$

and $D_{\mathbf{u}}^2 f(2,1) = \frac{294}{25}(2) + \frac{186}{25}(1) = \frac{774}{25}$.

41. Let $F(x,y,z) = 2(x-2)^2 + (y-1)^2 + (z-3)^2$. Then $2(x-2)^2 + (y-1)^2 + (z-3)^2 = 10$ is a level surface of F.

$F_x(x,y,z) = 4(x-2) \quad\Rightarrow\quad F_x(3,3,5) = 4$, $F_y(x,y,z) = 2(y-1) \quad\Rightarrow\quad F_y(3,3,5) = 4$, and

$F_z(x,y,z) = 2(z-3) \quad\Rightarrow\quad F_z(3,3,5) = 4$.

(a) Equation 19 gives an equation of the tangent plane at $(3,3,5)$ as $4(x-3) + 4(y-3) + 4(z-5) = 0 \quad\Leftrightarrow$

$\qquad 4x + 4y + 4z = 44$ or equivalently $x + y + z = 11$.

(b) By Equation 20, the normal line has symmetric equations $\dfrac{x-3}{4} = \dfrac{y-3}{4} = \dfrac{z-5}{4}$ or equivalently

$\qquad x - 3 = y - 3 = z - 5$. Corresponding parametric equations are $x = 3 + t$, $y = 3 + t$, $z = 5 + t$.

43. Let $F(x,y,z) = xy^2z^3$. Then $xy^2z^3 = 8$ is a level surface of F and $\nabla F(x,y,z) = \left\langle y^2z^3, 2xyz^3, 3xy^2z^2\right\rangle$.

(a) $\nabla F(2,2,1) = \left\langle 4, 8, 24\right\rangle$ is a normal vector for the tangent plane at $(2,2,1)$, so an equation of the tangent plane is

$\qquad 4(x-2) + 8(y-2) + 24(z-1) = 0$ or $4x + 8y + 24z = 48$ or equivalently $x + 2y + 6z = 12$.

(b) The normal line has direction $\nabla F(2,2,1) = \left\langle 4, 8, 24\right\rangle$ or equivalently $\left\langle 1, 2, 6\right\rangle$, so parametric equations are $x = 2 + t$,

$\qquad y = 2 + 2t$, $z = 1 + 6t$, and symmetric equations are $x - 2 = \dfrac{y-2}{2} = \dfrac{z-1}{6}$.

45. Let $F(x,y,z) = x + y + z - e^{xyz}$. Then $x + y + z = e^{xyz}$ is the level surface $F(x,y,z) = 0$,

and $\nabla F(x,y,z) = \left\langle 1 - yze^{xyz}, 1 - xze^{xyz}, 1 - xye^{xyz}\right\rangle$.

(a) $\nabla F(0,0,1) = \left\langle 1, 1, 1\right\rangle$ is a normal vector for the tangent plane at $(0,0,1)$, so an equation of the tangent plane

$\qquad$ is $1(x-0) + 1(y-0) + 1(z-1) = 0$ or $x + y + z = 1$.

(b) The normal line has direction $\left\langle 1, 1, 1\right\rangle$, so parametric equations are $x = t$, $y = t$, $z = 1 + t$, and symmetric equations are

$\qquad x = y = z - 1$.

47. $F(x, y, z) = xy + yz + zx$, $\nabla F(x, y, z) = \langle y + z, x + z, y + x \rangle$,

$\nabla F(1, 1, 1) = \langle 2, 2, 2 \rangle$, so an equation of the tangent plane is $2x + 2y + 2z = 6$

or $x + y + z = 3$, and the normal line is given by $x - 1 = y - 1 = z - 1$ or

$x = y = z$. To graph the surface we solve for z: $z = \dfrac{3 - xy}{x + y}$.

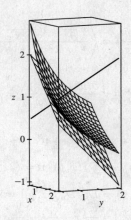

49. $f(x, y) = xy \Rightarrow \nabla f(x, y) = \langle y, x \rangle$, $\nabla f(3, 2) = \langle 2, 3 \rangle$. $\nabla f(3, 2)$

is perpendicular to the tangent line, so the tangent line has equation

$\nabla f(3, 2) \cdot \langle x - 3, y - 2 \rangle = 0 \Rightarrow \langle 2, 3 \rangle \cdot \langle x - 3, x - 2 \rangle = 0 \Rightarrow$

$2(x - 3) + 3(y - 2) = 0$ or $2x + 3y = 12$.

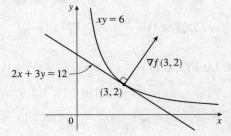

51. $\nabla F(x_0, y_0, z_0) = \left\langle \dfrac{2x_0}{a^2}, \dfrac{2y_0}{b^2}, \dfrac{2z_0}{c^2} \right\rangle$. Thus an equation of the tangent plane at (x_0, y_0, z_0) is

$\dfrac{2x_0}{a^2} x + \dfrac{2y_0}{b^2} y + \dfrac{2z_0}{c^2} z = 2\left(\dfrac{x_0^2}{a^2} + \dfrac{y_0^2}{b^2} + \dfrac{z_0^2}{c^2} \right) = 2(1) = 2$ since (x_0, y_0, z_0) is a point on the ellipsoid. Hence

$\dfrac{x_0}{a^2} x + \dfrac{y_0}{b^2} y + \dfrac{z_0}{c^2} z = 1$ is an equation of the tangent plane.

53. $\nabla F(x_0, y_0, z_0) = \left\langle \dfrac{2x_0}{a^2}, \dfrac{2y_0}{b^2}, \dfrac{-1}{c} \right\rangle$, so an equation of the tangent plane is $\dfrac{2x_0}{a^2} x + \dfrac{2y_0}{b^2} y - \dfrac{1}{c} z = \dfrac{2x_0^2}{a^2} + \dfrac{2y_0^2}{b^2} - \dfrac{z_0}{c}$

or $\dfrac{2x_0}{a^2} x + \dfrac{2y_0}{b^2} y = \dfrac{z}{c} + 2\left(\dfrac{x_0^2}{a^2} + \dfrac{y_0^2}{b^2} \right) - \dfrac{z_0}{c}$. But $\dfrac{z_0}{c} = \dfrac{x_0^2}{a^2} + \dfrac{y_0^2}{b^2}$, so the equation can be written as

$\dfrac{2x_0}{a^2} x + \dfrac{2y_0}{b^2} y = \dfrac{z + z_0}{c}$.

55. The hyperboloid $x^2 - y^2 - z^2 = 1$ is a level surface of $F(x, y, z) = x^2 - y^2 - z^2$ and $\nabla F(x, y, z) = \langle 2x, -2y, -2z \rangle$ is a

normal vector to the surface and hence a normal vector for the tangent plane at (x, y, z). The tangent plane is parallel to the

plane $z = x + y$ or $x + y - z = 0$ if and only if the corresponding normal vectors are parallel, so we need a point (x_0, y_0, z_0)

on the hyperboloid where $\langle 2x_0, -2y_0, -2z_0 \rangle = c \langle 1, 1, -1 \rangle$ or equivalently $\langle x_0, -y_0, -z_0 \rangle = k \langle 1, 1, -1 \rangle$ for some $k \neq 0$.

Then we must have $x_0 = k$, $y_0 = -k$, $z_0 = k$ and substituting into the equation of the hyperboloid gives

$k^2 - (-k)^2 - k^2 = 1 \Leftrightarrow -k^2 = 1$, an impossibility. Thus there is no such point on the hyperboloid.

57. Let (x_0, y_0, z_0) be a point on the cone [other than $(0, 0, 0)$]. The cone is a level surface of $F(x, y, z) = x^2 + y^2 - z^2$ and

$\nabla F(x, y, z) = \langle 2x, 2y, -2z \rangle$, so $\nabla F(x_0, y_0, z_0) = \langle 2x_0, 2y_0, -2z_0 \rangle$ is a normal vector to the cone at this point and an equation of the tangent plane there is $2x_0\,(x - x_0) + 2y_0\,(y - y_0) - 2z_0\,(z - z_0) = 0$ or

$x_0 x + y_0 y - z_0 z = x_0^2 + y_0^2 - z_0^2$. But $x_0^2 + y_0^2 = z_0^2$ so the tangent plane is given by $x_0 x + y_0 y - z_0 z = 0$, a plane which always contains the origin.

59. Let $F(x, y, z) = x^2 + y^2 - z$. Then the paraboloid is the level surface $F(x, y, z) = 0$ and $\nabla F(x, y, z) = \langle 2x, 2y, -1 \rangle$, so

$\nabla F(1, 1, 2) = \langle 2, 2, -1 \rangle$ is a normal vector to the surface. Thus the normal line at $(1, 1, 2)$ is given by $x = 1 + 2t$,

$y = 1 + 2t$, $z = 2 - t$. Substitution into the equation of the paraboloid $z = x^2 + y^2$ gives $2 - t = (1 + 2t)^2 + (1 + 2t)^2$ $\Leftrightarrow$

$2 - t = 2 + 8t + 8t^2$ $\Leftrightarrow$ $8t^2 + 9t = 0$ $\Leftrightarrow$ $t(8t + 9) = 0$. Thus the line intersects the paraboloid when $t = 0$,

corresponding to the given point $(1, 1, 2)$, or when $t = -\frac{9}{8}$, corresponding to the point $\left(-\frac{5}{4}, -\frac{5}{4}, \frac{25}{8}\right)$.

61. Let (x_0, y_0, z_0) be a point on the surface. Then an equation of the tangent plane at the point is

$$\frac{x}{2\sqrt{x_0}} + \frac{y}{2\sqrt{y_0}} + \frac{z}{2\sqrt{z_0}} = \frac{\sqrt{x_0} + \sqrt{y_0} + \sqrt{z_0}}{2}. \text{ But } \sqrt{x_0} + \sqrt{y_0} + \sqrt{z_0} = \sqrt{c}, \text{ so the equation is}$$

$$\frac{x}{\sqrt{x_0}} + \frac{y}{\sqrt{y_0}} + \frac{z}{\sqrt{z_0}} = \sqrt{c}. \text{ The } x\text{-}, y\text{-}, \text{ and } z\text{-intercepts are } \sqrt{cx_0}, \sqrt{cy_0} \text{ and } \sqrt{cz_0} \text{ respectively. (The } x\text{-intercept is found by}$$

setting $y = z = 0$ and solving the resulting equation for x, and the y- and z-intercepts are found similarly.) So the sum of the

intercepts is $\sqrt{c}\left(\sqrt{x_0} + \sqrt{y_0} + \sqrt{z_0}\right) = c$, a constant.

63. If $f(x, y, z) = z - x^2 - y^2$ and $g(x, y, z) = 4x^2 + y^2 + z^2$, then the tangent line is perpendicular to both ∇f and ∇g

at $(-1, 1, 2)$. The vector $\mathbf{v} = \nabla f \times \nabla g$ will therefore be parallel to the tangent line.

We have $\nabla f(x, y, z) = \langle -2x, -2y, 1 \rangle$ $\Rightarrow$ $\nabla f(-1, 1, 2) = \langle 2, -2, 1 \rangle$, and $\nabla g(x, y, z) = \langle 8x, 2y, 2z \rangle$ $\Rightarrow$

$\nabla g(-1, 1, 2) = \langle -8, 2, 4 \rangle$. Hence $\mathbf{v} = \nabla f \times \nabla g = \begin{vmatrix} \mathbf{i} & \mathbf{j} & \mathbf{k} \\ 2 & -2 & 1 \\ -8 & 2 & 4 \end{vmatrix} = -10\,\mathbf{i} - 16\,\mathbf{j} - 12\,\mathbf{k}.$

Parametric equations are: $x = -1 - 10t$, $y = 1 - 16t$, $z = 2 - 12t$.

65. Parametric equations for the helix are $x = \cos \pi t$, $y = \sin \pi t$, $z = t$, and substituting into the equation of the paraboloid

gives $t = \cos^2 \pi t + \sin^2 \pi t$ $\Rightarrow$ $t = 1$. Thus the helix intersects the surface at the point $(\cos \pi, \sin \pi, 1) = (-1, 0, 1)$. Here

$\mathbf{r}'(t) = \langle -\pi \sin \pi t, \pi \cos \pi t, 1 \rangle$, so the tangent vector to the helix at that point is $\mathbf{r}'(1) = \langle -\pi \sin \pi, \pi \cos \pi, 1 \rangle = \langle 0, -\pi, 1 \rangle$.

The paraboloid $z = x^2 + y^2$ $\Leftrightarrow$ $x^2 + y^2 - z = 0$ is a level surface of $F(x, y, z) = x^2 + y^2 - z$ and

$\nabla F(x, y, z) = \langle 2x, 2y, -1 \rangle$, so a normal vector to the tangent plane at $(-1, 0, 1)$ is $\nabla F(-1, 0, 1) = \langle -2, 0, -1 \rangle$. The angle

θ between $\mathbf{r}'(1)$ and $\nabla F(-1, 0, 1)$ is given by

$$\cos \theta = \frac{\langle 0, -\pi, 1 \rangle \cdot \langle -2, 0, -1 \rangle}{|\langle 0, -\pi, 1 \rangle|\,|\langle -2, 0, -1 \rangle|} = \frac{0 + 0 - 1}{\sqrt{0 + \pi^2 + 1}\,\sqrt{4 + 0 + 1}} = \frac{-1}{\sqrt{5(\pi^2 + 1)}} \quad \Rightarrow$$

$\theta = \cos^{-1} \dfrac{-1}{\sqrt{5(\pi^2 + 1)}} \approx 97.8°$. Because $\nabla F(-1, 0, 1)$ is perpendicular to the tangent plane, the angle of intersection

between the helix and the paraboloid is approximately $97.8° - 90° = 7.8°$.

67. (a) The direction of the normal line of F is given by ∇F, and that of G by ∇G. Assuming that

$\nabla F \neq 0 \neq \nabla G$, the two normal lines are perpendicular at P if $\nabla F \cdot \nabla G = 0$ at P $\Leftrightarrow$

$\langle \partial F/\partial x, \partial F/\partial y, \partial F/\partial z \rangle \cdot \langle \partial G/\partial x, \partial G/\partial y, \partial G/\partial z \rangle = 0$ at P $\Leftrightarrow$ $F_x G_x + F_y G_y + F_z G_z = 0$ at P.

(b) Here $F = x^2 + y^2 - z^2$ and $G = x^2 + y^2 + z^2 - r^2$, so

$\nabla F \cdot \nabla G = \langle 2x, 2y, -2z \rangle \cdot \langle 2x, 2y, 2z \rangle = 4x^2 + 4y^2 - 4z^2 = 4F = 0$, since the point (x, y, z) lies on the graph of

$F = 0$. To see that this is true without using calculus, note that $G = 0$ is the equation of a sphere centered at the origin and

$F = 0$ is the equation of a right circular cone with vertex at the origin (which is generated by lines through the origin). At

any point of intersection, the sphere's normal line (which passes through the origin) lies on the cone, and thus is

perpendicular to the cone's normal line. So the surfaces with equations $F = 0$ and $G = 0$ are everywhere orthogonal.

69. Let $\mathbf{u} = \langle a, b \rangle$ and $\mathbf{v} = \langle c, d \rangle$. Then we know that at the given point, $D_{\mathbf{u}} f = \nabla f \cdot \mathbf{u} = a f_x + b f_y$ and

$D_{\mathbf{v}} f = \nabla f \cdot \mathbf{v} = c f_x + d f_y$. But these are just two linear equations in the two unknowns f_x and f_y, and since $\mathbf{u}$ and $\mathbf{v}$ are

not parallel, we can solve the equations to find $\nabla f = \langle f_x, f_y \rangle$ at the given point. In fact,

$\nabla f = \left\langle \dfrac{d\, D_{\mathbf{u}} f - b\, D_{\mathbf{v}} f}{ad - bc}, \dfrac{a\, D_{\mathbf{v}} f - c\, D_{\mathbf{u}} f}{ad - bc} \right\rangle$.

14.7 Maximum and Minimum Values

1. (a) First we compute $D(1,1) = f_{xx}(1,1)\, f_{yy}(1,1) - [f_{xy}(1,1)]^2 = (4)(2) - (1)^2 = 7$. Since $D(1,1) > 0$ and

$f_{xx}(1,1) > 0$, f has a local minimum at $(1,1)$ by the Second Derivatives Test.

(b) $D(1,1) = f_{xx}(1,1)\, f_{yy}(1,1) - [f_{xy}(1,1)]^2 = (4)(2) - (3)^2 = -1$. Since $D(1,1) < 0$, f has a saddle point at $(1,1)$ by

the Second Derivatives Test.

3. In the figure, a point at approximately $(1,1)$ is enclosed by level curves which are oval in shape and indicate that as we move

away from the point in any direction the values of f are increasing. Hence we would expect a local minimum at or near $(1,1)$.

The level curves near $(0,0)$ resemble hyperbolas, and as we move away from the origin, the values of f increase in some

directions and decrease in others, so we would expect to find a saddle point there.

To verify our predictions, we have $f(x,y) = 4 + x^3 + y^3 - 3xy$ $\Rightarrow$ $f_x(x,y) = 3x^2 - 3y$, $f_y(x,y) = 3y^2 - 3x$. We

have critical points where these partial derivatives are equal to 0: $3x^2 - 3y = 0$, $3y^2 - 3x = 0$. Substituting $y = x^2$ from the

first equation into the second equation gives $3(x^2)^2 - 3x = 0$ $\Rightarrow$ $3x(x^3 - 1) = 0$ $\Rightarrow$ $x = 0$ or $x = 1$. Then we have

two critical points, $(0,0)$ and $(1,1)$. The second partial derivatives are $f_{xx}(x,y) = 6x$, $f_{xy}(x,y) = -3$, and $f_{yy}(x,y) = 6y$,

so $D(x,y) = f_{xx}(x,y)\, f_{yy}(x,y) - [f_{xy}(x,y)]^2 = (6x)(6y) - (-3)^2 = 36xy - 9$. Then $D(0,0) = 36(0)(0) - 9 = -9$,

and $D(1,1) = 36(1)(1) - 9 = 27$. Since $D(0,0) < 0$, f has a saddle point at $(0,0)$ by the Second Derivatives Test. Since

$D(1,1) > 0$ and $f_{xx}(1,1) > 0$, f has a local minimum at $(1,1)$.

5. $f(x,y) = x^2 + xy + y^2 + y$ $\Rightarrow$ $f_x = 2x + y$, $f_y = x + 2y + 1$, $f_{xx} = 2$, $f_{xy} = 1$, $f_{yy} = 2$. Then $f_x = 0$ implies

$y = -2x$, and substitution into $f_y = x + 2y + 1 = 0$ gives $x + 2(-2x) + 1 = 0$ $\Rightarrow$ $-3x = -1$ $\Rightarrow$ $x = \frac{1}{3}$.

Then $y = -\frac{2}{3}$ and the only critical point is $\left(\frac{1}{3}, -\frac{2}{3}\right)$.

$D(x, y) = f_{xx}f_{yy} - (f_{xy})^2 = (2)(2) - (1)^2 = 3$, and since

$D\left(\frac{1}{3}, -\frac{2}{3}\right) = 3 > 0$ and $f_{xx}\left(\frac{1}{3}, -\frac{2}{3}\right) = 2 > 0$, $f\left(\frac{1}{3}, -\frac{2}{3}\right) = -\frac{1}{3}$ is a local

minimum by the Second Derivatives Test.

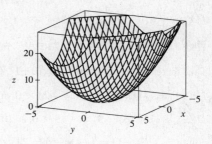

7. $f(x, y) = (x - y)(1 - xy) = x - y - x^2y + xy^2 \;\Rightarrow\; f_x = 1 - 2xy + y^2, \; f_y = -1 - x^2 + 2xy, \; f_{xx} = -2y,$

$f_{xy} = -2x + 2y, \; f_{yy} = 2x$. Then $f_x = 0$ implies $1 - 2xy + y^2 = 0$ and $f_y = 0$ implies $-1 - x^2 + 2xy = 0$. Adding the

two equations gives $1 + y^2 - 1 - x^2 = 0 \;\Rightarrow\; y^2 = x^2 \;\Rightarrow\; y = \pm x$, but if $y = -x$ then $f_x = 0$ implies

$1 + 2x^2 + x^2 = 0 \;\Rightarrow\; 3x^2 = -1$ which has no real solution. If $y = x$

then substitution into $f_x = 0$ gives $1 - 2x^2 + x^2 = 0 \;\Rightarrow\; x^2 = 1 \;\Rightarrow$

$x = \pm 1$, so the critical points are $(1, 1)$ and $(-1, -1)$. Now

$D(1, 1) = (-2)(2) - 0^2 = -4 < 0$ and

$D(-1, -1) = (2)(-2) - 0^2 = -4 < 0$, so $(1, 1)$ and $(-1, -1)$ are

saddle points.

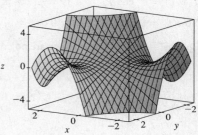

9. $f(x, y) = x^2 + y^4 + 2xy \;\Rightarrow\; f_x = 2x + 2y, \; f_y = 4y^3 + 2x, \; f_{xx} = 2, \; f_{xy} = 2, \; f_{yy} = 12y^2$. Then $f_x = 0$ implies

$y = -x$, and substitution into $f_y = 4y^3 + 2x = 0$ gives $-4x^3 + 2x = 0 \;\Rightarrow\; 2x\left(1 - 2x^2\right) = 0 \;\Rightarrow\; x = 0$ or

$x = \pm\frac{1}{\sqrt{2}}$. Thus the critical points are $(0, 0)$, $\left(\frac{1}{\sqrt{2}}, -\frac{1}{\sqrt{2}}\right)$, and $\left(-\frac{1}{\sqrt{2}}, \frac{1}{\sqrt{2}}\right)$. Now

$D(x, y) = f_{xx}f_{yy} - (f_{xy})^2 = (2)(12y^2) - (2)^2 = 24y^2 - 4,$

so $D(0, 0) = -4 < 0$ and $(0, 0)$ is a saddle point.

$D\left(\frac{1}{\sqrt{2}}, -\frac{1}{\sqrt{2}}\right) = D\left(-\frac{1}{\sqrt{2}}, \frac{1}{\sqrt{2}}\right) = 24\left(\frac{1}{2}\right) - 4 = 8 > 0$ and

$f_{xx}\left(\frac{1}{\sqrt{2}}, -\frac{1}{\sqrt{2}}\right) = f_{xx}\left(-\frac{1}{\sqrt{2}}, \frac{1}{\sqrt{2}}\right) = 2 > 0$, so $f\left(\frac{1}{\sqrt{2}}, -\frac{1}{\sqrt{2}}\right) = -\frac{1}{4}$

and $f\left(-\frac{1}{\sqrt{2}}, \frac{1}{\sqrt{2}}\right) = -\frac{1}{4}$ are local minima.

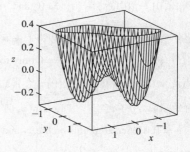

11. $f(x, y) = x^3 - 3x + 3xy^2 \;\Rightarrow\; f_x = 3x^2 - 3 + 3y^2, \; f_y = 6xy, \; f_{xx} = 6x, \; f_{xy} = 6y, \; f_{yy} = 6x$. Then $f_y = 0$ implies

$x = 0$ or $y = 0$. If $x = 0$, substitution into $f_x = 0$ gives $3y^2 = 3 \;\Rightarrow\; y = \pm 1$, and if $y = 0$, substitution into $f_x = 0$

gives $x = \pm 1$. Thus the critical points are $(0, \pm 1)$ and $(\pm 1, 0)$.

$D(0, \pm 1) = 0 - 36 < 0$, so $(0, \pm 1)$ are saddle points.

$D(\pm 1, 0) = 36 - 0 > 0$, $f_{xx}(1, 0) = 6 > 0$, and $f_{xx}(-1, 0) = -6 < 0$,

so $f(1, 0) = -2$ is a local minimum and $f(-1, 0) = 2$ is a local maximum.

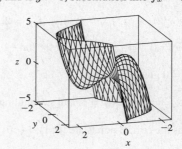

13. $f(x, y) = x^4 - 2x^2 + y^3 - 3y \implies f_x = 4x^3 - 4x, \ f_y = 3y^2 - 3, \ f_{xx} = 12x^2 - 4, \ f_{xy} = 0, \ f_{yy} = 6y$.

Then $f_x = 0$ implies $4x(x^2 - 1) = 0 \implies x = 0$ or $x = \pm 1$, and $f_y = 0$ implies $3(y^2 - 1) = 0 \implies y = \pm 1$.

Thus there are six critical points: $(0, \pm 1)$, $(\pm 1, 1)$, and $(\pm 1, -1)$.

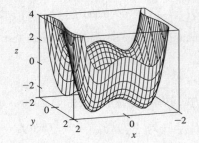

$D(0, 1) = (-4)(6) - (0)^2 = -24 < 0$ and

$D(\pm 1, -1) = (8)(-6) = -48 < 0$, so $(0, 1)$ and $(\pm 1, -1)$ are saddle

points. $\quad D(0, -1) = (-4)(-6) = 24 > 0$ and $f_{xx}(0, -1) = -4 < 0$, so

$f(0, -1) = 2$ is a local maximum. $\quad D(\pm 1, 1) = (8)(6) = 48 > 0$ and

$f_{xx}(\pm 1, 1) = 8 > 0$, so $f(\pm 1, 1) = -3$ are local minima.

15. $f(x, y) = e^x \cos y \implies f_x = e^x \cos y, \ f_y = -e^x \sin y$.

Now $f_x = 0$ implies $\cos y = 0$ or $y = \frac{\pi}{2} + n\pi$ for n an integer.

But $\sin\left(\frac{\pi}{2} + n\pi\right) \neq 0$, so there are no critical points.

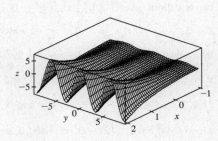

17. $f(x, y) = xy + e^{-xy} \implies f_x = y - ye^{-xy}, \ f_y = x - xe^{-xy}, \ f_{xx} = y^2 e^{-xy},$

$f_{xy} = 1 - \left[y(-xe^{-xy}) + e^{-xy}(1)\right] = 1 + (xy - 1)e^{-xy}, \ f_{yy} = x^2 e^{-xy}$. Then $f_x = 0$ implies $y(1 - e^{-xy}) = 0 \implies$

$y = 0$ or $e^{-xy} = 1 \implies x = 0$ or $y = 0$. If $x = 0$ then $f_y = 0$ for any y-value, so all points of the form $(0, y_0)$ are

critical points. If $y = 0$, then $f_y = x - xe^0 = 0$ for any x-value, so all points of the form $(x_0, 0)$ are critical points. We have

$D(x_0, 0) = (0)(x_0^2) - (0)^2 = 0$ and $D(0, y_0) = (y_0^2)(0) - (0)^2 = 0$, so the Second Derivatives Test gives no information.

Notice that if we let $t = xy$, then $f(x, y) = g(t) = t + e^{-t} \implies$

$g'(t) = 1 - e^{-t}$. Now $g'(t) = 0$ only for $t = 0$, and $g'(t) < 0$ for $t < 0$,

$g'(t) > 0$ for $t > 0$. Thus $g(0) = 1$ is a local and absolute minimum, so

$f(x, y) = xy + e^{-xy} \geq 1$ for all (x, y) with equality if and only if $x = 0$

or $y = 0$. Hence all points on the x- and y-axes are local (and absolute)

minima, where $f(x, y) = 1$.

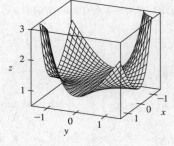

19. $f(x, y) = y^2 - 2y \cos x \implies f_x = 2y \sin x, \ f_y = 2y - 2\cos x,$

$f_{xx} = 2y \cos x, \ f_{xy} = 2 \sin x, \ f_{yy} = 2$. Then $f_x = 0$ implies $y = 0$ or

$\sin x = 0 \implies x = 0, \pi,$ or 2π for $-1 \leq x \leq 7$. Substituting $y = 0$ into

$f_y = 0$ gives $\cos x = 0 \implies x = \frac{\pi}{2}$ or $\frac{3\pi}{2}$, substituting $x = 0$ or $x = 2\pi$

into $f_y = 0$ gives $y = 1$, and substituting $x = \pi$ into $f_y = 0$ gives $y = -1$.

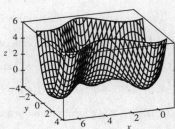

Thus the critical points are $(0, 1)$, $\left(\frac{\pi}{2}, 0\right)$, $(\pi, -1)$, $\left(\frac{3\pi}{2}, 0\right)$, and $(2\pi, 1)$.

$D\left(\frac{\pi}{2},0\right) = D\left(\frac{3\pi}{2},0\right) = -4 < 0$ so $\left(\frac{\pi}{2},0\right)$ and $\left(\frac{3\pi}{2},0\right)$ are saddle points. $D(0,1) = D(\pi,-1) = D(2\pi,1) = 4 > 0$ and

$f_{xx}(0,1) = f_{xx}(\pi,-1) = f_{xx}(2\pi,1) = 2 > 0$, so $f(0,1) = f(\pi,-1) = f(2\pi,1) = -1$ are local minima.

21. $f(x,y) = x^2 + 4y^2 - 4xy + 2 \;\Rightarrow\; f_x = 2x - 4y,\; f_y = 8y - 4x,\; f_{xx} = 2,\; f_{xy} = -4,\; f_{yy} = 8$. Then $f_x = 0$

and $f_y = 0$ each implies $y = \frac{1}{2}x$, so all points of the form $\left(x_0, \frac{1}{2}x_0\right)$ are critical points and for each of these we have

$D\left(x_0, \frac{1}{2}x_0\right) = (2)(8) - (-4)^2 = 0$. The Second Derivatives Test gives no information, but

$f(x,y) = x^2 + 4y^2 - 4xy + 2 = (x - 2y)^2 + 2 \geq 2$ with equality if and only if $y = \frac{1}{2}x$. Thus $f\left(x_0, \frac{1}{2}x_0\right) = 2$ are all local

(and absolute) minima.

23. $f(x,y) = x^2 + y^2 + x^{-2}y^{-2}$

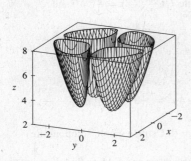
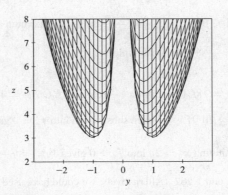
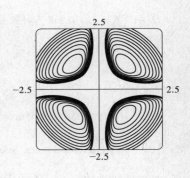

From the graphs, there appear to be local minima of about $f(1,\pm1) = f(-1,\pm1) \approx 3$ (and no local maxima or saddle

points). $f_x = 2x - 2x^{-3}y^{-2},\; f_y = 2y - 2x^{-2}y^{-3},\; f_{xx} = 2 + 6x^{-4}y^{-2},\; f_{xy} = 4x^{-3}y^{-3},\; f_{yy} = 2 + 6x^{-2}y^{-4}$. Then

$f_x = 0$ implies $2x^4y^2 - 2 = 0$ or $x^4y^2 = 1$ or $y^2 = x^{-4}$. Note that neither x nor y can be zero. Now $f_y = 0$ implies

$2x^2y^4 - 2 = 0$, and with $y^2 = x^{-4}$ this implies $2x^{-6} - 2 = 0$ or $x^6 = 1$. Thus $x = \pm1$ and if $x = 1$, $y = \pm1$; if $x = -1$,

$y = \pm1$. So the critical points are $(1,1),\; (1,-1),(-1,1)$ and $(-1,-1)$. Now $D(1,\pm1) = D(-1,\pm1) = 64 - 16 > 0$ and

$f_{xx} > 0$ always, so $f(1,\pm1) = f(-1,\pm1) = 3$ are local minima.

25. $f(x,y) = \sin x + \sin y + \sin(x+y),\; 0 \leq x \leq 2\pi,\; 0 \leq y \leq 2\pi$

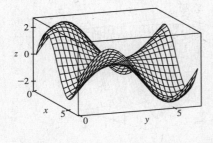
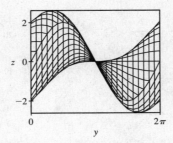
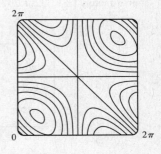

From the graphs it appears that f has a local maximum at about $(1,1)$ with value approximately 2.6, a local minimum

at about $(5,5)$ with value approximately -2.6, and a saddle point at about $(3,3)$.

$f_x = \cos x + \cos(x+y),\; f_y = \cos y + \cos(x+y),\; f_{xx} = -\sin x - \sin(x+y),\; f_{yy} = -\sin y - \sin(x+y),$

$f_{xy} = -\sin(x+y)$. Setting $f_x = 0$ and $f_y = 0$ and subtracting gives $\cos x - \cos y = 0$ or $\cos x = \cos y$. Thus $x = y$

or $x = 2\pi - y$. If $x = y$, $f_x = 0$ becomes $\cos x + \cos 2x = 0$ or $2\cos^2 x + \cos x - 1 = 0$, a quadratic in $\cos x$. Thus

$\cos x = -1$ or $\frac{1}{2}$ and $x = \pi$, $\frac{\pi}{3}$, or $\frac{5\pi}{3}$, giving the critical points (π, π), $\left(\frac{\pi}{3}, \frac{\pi}{3}\right)$ and $\left(\frac{5\pi}{3}, \frac{5\pi}{3}\right)$. Similarly if

$x = 2\pi - y$, $f_x = 0$ becomes $(\cos x) + 1 = 0$ and the resulting critical point is (π, π). Now

$D(x, y) = \sin x \sin y + \sin x \sin(x+y) + \sin y \sin(x+y)$. So $D(\pi, \pi) = 0$ and the Second Derivatives Test doesn't apply.

However, along the line $y = x$ we have $f(x, x) = 2\sin x + \sin 2x = 2\sin x + 2\sin x \cos x = 2\sin x(1 + \cos x)$, and

$f(x, x) > 0$ for $0 < x < \pi$ while $f(x, x) < 0$ for $\pi < x < 2\pi$. Thus every disk with center (π, π) contains points where f is

positive as well as points where f is negative, so the graph crosses its tangent plane ($z = 0$) there and (π, π) is a saddle point.

$D\left(\frac{\pi}{3}, \frac{\pi}{3}\right) = \frac{9}{4} > 0$ and $f_{xx}\left(\frac{\pi}{3}, \frac{\pi}{3}\right) < 0$ so $f\left(\frac{\pi}{3}, \frac{\pi}{3}\right) = \frac{3\sqrt{3}}{2}$ is a local maximum while $D\left(\frac{5\pi}{3}, \frac{5\pi}{3}\right) = \frac{9}{4} > 0$ and

$f_{xx}\left(\frac{5\pi}{3}, \frac{5\pi}{3}\right) > 0$, so $f\left(\frac{5\pi}{3}, \frac{5\pi}{3}\right) = -\frac{3\sqrt{3}}{2}$ is a local minimum.

27. $f(x, y) = x^4 + y^4 - 4x^2y + 2y \Rightarrow f_x(x, y) = 4x^3 - 8xy$ and $f_y(x, y) = 4y^3 - 4x^2 + 2$. $f_x = 0 \Rightarrow$

$4x(x^2 - 2y) = 0$, so $x = 0$ or $x^2 = 2y$. If $x = 0$ then substitution into $f_y = 0$ gives $4y^3 = -2 \Rightarrow y = -\frac{1}{\sqrt[3]{2}}$, so

$\left(0, -\frac{1}{\sqrt[3]{2}}\right)$ is a critical point. Substituting $x^2 = 2y$ into $f_y = 0$ gives $4y^3 - 8y + 2 = 0$. Using a graph, solutions are

approximately $y = -1.526$, 0.259, and 1.267. (Alternatively, we could have used a calculator or a CAS to find these roots.)

We have $x^2 = 2y \Rightarrow x = \pm\sqrt{2y}$, so $y = -1.526$ gives no real-valued solution for x, but

$y = 0.259 \Rightarrow x \approx \pm0.720$ and $y = 1.267 \Rightarrow x \approx \pm1.592$. Thus to three decimal places, the critical points are

$\left(0, -\frac{1}{\sqrt[3]{2}}\right) \approx (0, -0.794)$, $(\pm0.720, 0.259)$, and $(\pm1.592, 1.267)$. Now since $f_{xx} = 12x^2 - 8y$, $f_{xy} = -8x$, $f_{yy} = 12y^2$,

and $D = (12x^2 - 8y)(12y^2) - 64x^2$, we have $D(0, -0.794) > 0$, $f_{xx}(0, -0.794) > 0$, $D(\pm0.720, 0.259) < 0$,

$D(\pm1.592, 1.267) > 0$, and $f_{xx}(\pm1.592, 1.267) > 0$. Therefore $f(0, -0.794) \approx -1.191$ and $f(\pm1.592, 1.267) \approx -1.310$

are local minima, and $(\pm0.720, 0.259)$ are saddle points. There is no highest point on the graph, but the lowest points are

approximately $(\pm1.592, 1.267, -1.310)$.

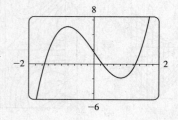

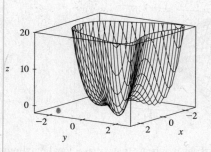

29. $f(x, y) = x^4 + y^3 - 3x^2 + y^2 + x - 2y + 1$ $\Rightarrow$ $f_x(x, y) = 4x^3 - 6x + 1$ and $f_y(x, y) = 3y^2 + 2y - 2$. From the

graphs, we see that to three decimal places, $f_x = 0$ when $x \approx -1.301, 0.170,$ or 1.131, and $f_y = 0$ when $y \approx -1.215$ or

0.549. (Alternatively, we could have used a calculator or a CAS to find these roots. We could also use the quadratic formula to

find the solutions of $f_y = 0$.) So, to three decimal places, f has critical points at $(-1.301, -1.215)$, $(-1.301, 0.549)$,

$(0.170, -1.215)$, $(0.170, 0.549)$, $(1.131, -1.215)$, and $(1.131, 0.549)$. Now since $f_{xx} = 12x^2 - 6$, $f_{xy} = 0$, $f_{yy} = 6y + 2$,

and $D = (12x^2 - 6)(6y + 2)$, we have $D(-1.301, -1.215) < 0$, $D(-1.301, 0.549) > 0$, $f_{xx}(-1.301, 0.549) > 0$,

$D(0.170, -1.215) > 0$, $f_{xx}(0.170, -1.215) < 0$, $D(0.170, 0.549) < 0$, $D(1.131, -1.215) < 0$, $D(1.131, 0.549) > 0$, and

$f_{xx}(1.131, 0.549) > 0$. Therefore, to three decimal places, $f(-1.301, 0.549) \approx -3.145$ and $f(1.131, 0.549) \approx -0.701$ are

local minima, $f(0.170, -1.215) \approx 3.197$ is a local maximum, and $(-1.301, -1.215)$, $(0.170, 0.549)$, and $(1.131, -1.215)$

are saddle points. There is no highest or lowest point on the graph.

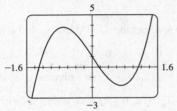
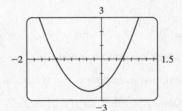

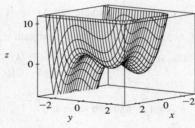

31. Since f is a polynomial it is continuous on D, so an absolute maximum and minimum exist. Here $f_x = 2x - 2$, $f_y = 2y$, and

setting $f_x = f_y = 0$ gives $(1, 0)$ as the only critical point (which is inside D), where $f(1, 0) = -1$. Along L_1: $x = 0$ and

$f(0, y) = y^2$ for $-2 \le y \le 2$, a quadratic function which attains its minimum at $y = 0$, where $f(0, 0) = 0$, and its maximum

at $y = \pm 2$, where $f(0, \pm 2) = 4$. Along L_2: $y = x - 2$ for $0 \le x \le 2$, and $f(x, x - 2) = 2x^2 - 6x + 4 = 2\left(x - \frac{3}{2}\right)^2 - \frac{1}{2}$,

a quadratic which attains its minimum at $x = \frac{3}{2}$, where $f\left(\frac{3}{2}, -\frac{1}{2}\right) = -\frac{1}{2}$, and its maximum at $x = 0$, where $f(0, -2) = 4$.

Along L_3: $y = 2 - x$ for $0 \le x \le 2$, and

$f(x, 2 - x) = 2x^2 - 6x + 4 = 2\left(x - \frac{3}{2}\right)^2 - \frac{1}{2}$, a quadratic which attains

its minimum at $x = \frac{3}{2}$, where $f\left(\frac{3}{2}, \frac{1}{2}\right) = -\frac{1}{2}$, and its maximum at $x = 0$,

where $f(0, 2) = 4$. Thus the absolute maximum of f on D is $f(0, \pm 2) = 4$

and the absolute minimum is $f(1, 0) = -1$.

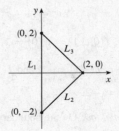

33. $f_x(x, y) = 2x + 2xy$, $f_y(x, y) = 2y + x^2$, and setting $f_x = f_y = 0$

gives $(0, 0)$ as the only critical point in D, with $f(0, 0) = 4$.

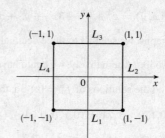

On L_1: $y = -1$, $f(x, -1) = 5$, a constant.

On L_2: $x = 1$, $f(1, y) = y^2 + y + 5$, a quadratic in y which attains its

maximum at $(1, 1)$, $f(1, 1) = 7$ and its minimum at $\left(1, -\frac{1}{2}\right)$, $f\left(1, -\frac{1}{2}\right) = \frac{19}{4}$.

On L_3: $f(x, 1) = 2x^2 + 5$ which attains its maximum at $(-1, 1)$ and $(1, 1)$

with $f(\pm 1, 1) = 7$ and its minimum at $(0, 1)$, $f(0, 1) = 5$.

On L_4: $f(-1, y) = y^2 + y + 5$ with maximum at $(-1, 1)$, $f(-1, 1) = 7$ and minimum at $\left(-1, -\frac{1}{2}\right)$, $f\left(-1, -\frac{1}{2}\right) = \frac{19}{4}$.

Thus the absolute maximum is attained at both $(\pm 1, 1)$ with $f(\pm 1, 1) = 7$ and the absolute minimum on D is attained at

$(0, 0)$ with $f(0, 0) = 4$.

35. $f(x, y) = x^2 + 2y^2 - 2x - 4y + 1$ $\Rightarrow$ $f_x = 2x - 2$, $f_y = 4y - 4$. Setting $f_x = 0$ and $f_y = 0$ gives $(1, 1)$ as the only

critical point (which is inside D), where $f(1, 1) = -2$. Along L_1: $y = 0$, so $f(x, 0) = x^2 - 2x + 1 = (x - 1)^2$, $0 \le x \le 2$,

which has a maximum value both at $x = 0$ and $x = 2$ where $f(0, 0) = f(2, 0) = 1$ and a minimum value at $x = 1$, where

$f(1, 0) = 0$. Along L_2: $x = 2$, so $f(2, y) = 2y^2 - 4y + 1 = 2(y - 1)^2 - 1$, $0 \le y \le 3$, which has a maximum value at

$y = 3$ where $f(2, 3) = 7$ and a minimum value at $y = 1$ where $f(2, 1) = -1$. Along L_3: $y = 3$, so

$f(x, 3) = x^2 - 2x + 7 = (x - 1)^2 + 6$, $0 \le x \le 2$, which has a maximum value both at $x = 0$ and $x = 2$ where

$f(0, 3) = f(2, 3) = 7$ and a minimum value at $x = 1$, where $f(1, 3) = 6$. Along L_4: $x = 0$, so

$f(0, y) = 2y^2 - 4y + 1 = 2(y - 1)^2 - 1$, $0 \le y \le 3$, which has a

maximum value at $y = 3$ where $f(0, 3) = 7$ and a minimum value at $y = 1$

where $f(0, 1) = -1$. Thus the absolute maximum is attained at both $(0, 3)$

and $(2, 3)$, where $f(0, 3) = f(2, 3) = 7$, and the absolute minimum is

$f(1, 1) = -2$.

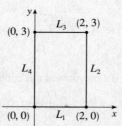

37. $f_x(x, y) = 6x^2$ and $f_y(x, y) = 4y^3$. And so $f_x = 0$ and $f_y = 0$ only occur when $x = y = 0$. Hence, the only critical point

inside the disk is at $x = y = 0$ where $f(0, 0) = 0$. Now on the circle $x^2 + y^2 = 1$, $y^2 = 1 - x^2$ so let

$g(x) = f(x, y) = 2x^3 + (1 - x^2)^2 = x^4 + 2x^3 - 2x^2 + 1$, $-1 \le x \le 1$. Then $g'(x) = 4x^3 + 6x^2 - 4x = 0$ $\Rightarrow$ $x = 0$,

-2, or $\frac{1}{2}$. $f(0, \pm 1) = g(0) = 1$, $f\left(\frac{1}{2}, \pm \frac{\sqrt{3}}{2}\right) = g\left(\frac{1}{2}\right) = \frac{13}{16}$, and $(-2, -3)$ is not in D. Checking the endpoints, we get

$f(-1, 0) = g(-1) = -2$ and $f(1, 0) = g(1) = 2$. Thus the absolute maximum and minimum of f on D are $f(1, 0) = 2$ and

$f(-1, 0) = -2$.

Another method: On the boundary $x^2 + y^2 = 1$ we can write $x = \cos\theta$, $y = \sin\theta$, so $f(\cos\theta, \sin\theta) = 2\cos^3\theta + \sin^4\theta$,

$0 \le \theta \le 2\pi$.

39. $f(x, y) = -(x^2 - 1)^2 - (x^2 y - x - 1)^2$ $\Rightarrow$ $f_x(x, y) = -2(x^2 - 1)(2x) - 2(x^2 y - x - 1)(2xy - 1)$ and

$f_y(x, y) = -2(x^2 y - x - 1)x^2$. Setting $f_y(x, y) = 0$ gives either $x = 0$ or $x^2 y - x - 1 = 0$.

There are no critical points for $x = 0$, since $f_x(0, y) = -2$, so we set $x^2 y - x - 1 = 0$ $\Leftrightarrow$ $y = \dfrac{x + 1}{x^2}$ $[x \neq 0]$,

so $f_x\left(x, \dfrac{x+1}{x^2}\right) = -2(x^2 - 1)(2x) - 2\left(x^2\,\dfrac{x+1}{x^2} - x - 1\right)\left(2x\,\dfrac{x+1}{x^2} - 1\right) = -4x(x^2 - 1)$. Therefore

$f_x(x, y) = f_y(x, y) = 0$ at the points $(1, 2)$ and $(-1, 0)$. To classify these critical points, we calculate

$f_{xx}(x, y) = -12x^2 - 12x^2 y^2 + 12xy + 4y + 2$, $f_{yy}(x, y) = -2x^4$,

and $f_{xy}(x, y) = -8x^3 y + 6x^2 + 4x$. In order to use the Second Derivatives

Test we calculate

$D(-1, 0) = f_{xx}(-1, 0)\,f_{yy}(-1, 0) - [f_{xy}(-1, 0)]^2 = 16 > 0$,

$f_{xx}(-1, 0) = -10 < 0$, $D(1, 2) = 16 > 0$, and $f_{xx}(1, 2) = -26 < 0$, so

both $(-1, 0)$ and $(1, 2)$ give local maxima.

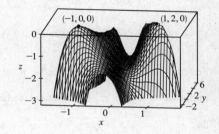

41. Let d be the distance from $(2, 0, -3)$ to any point (x, y, z) on the plane $x + y + z = 1$, so $d = \sqrt{(x-2)^2 + y^2 + (z+3)^2}$

where $z = 1 - x - y$, and we minimize $d^2 = f(x, y) = (x - 2)^2 + y^2 + (4 - x - y)^2$. Then

$f_x(x, y) = 2(x - 2) + 2(4 - x - y)(-1) = 4x + 2y - 12$, $f_y(x, y) = 2y + 2(4 - x - y)(-1) = 2x + 4y - 8$. Solving

$4x + 2y - 12 = 0$ and $2x + 4y - 8 = 0$ simultaneously gives $x = \frac{8}{3}$, $y = \frac{2}{3}$, so the only critical point is $\left(\frac{8}{3}, \frac{2}{3}\right)$. An absolute

minimum exists (since there is a minimum distance from the point to the plane) and it must occur at a critical point, so the

shortest distance occurs for $x = \frac{8}{3}$, $y = \frac{2}{3}$ for which $d = \sqrt{\left(\frac{8}{3} - 2\right)^2 + \left(\frac{2}{3}\right)^2 + \left(4 - \frac{8}{3} - \frac{2}{3}\right)^2} = \sqrt{\frac{4}{3}} = \frac{2}{\sqrt{3}}$.

43. Let d be the distance from the point $(4, 2, 0)$ to any point (x, y, z) on the cone, so $d = \sqrt{(x-4)^2 + (y-2)^2 + z^2}$ where

$z^2 = x^2 + y^2$, and we minimize $d^2 = (x - 4)^2 + (y - 2)^2 + x^2 + y^2 = f(x, y)$. Then

$f_x(x, y) = 2(x - 4) + 2x = 4x - 8$, $f_y(x, y) = 2(y - 2) + 2y = 4y - 4$, and the critical points occur when

$f_x = 0$ $\Rightarrow$ $x = 2$, $f_y = 0$ $\Rightarrow$ $y = 1$. Thus the only critical point is $(2, 1)$. An absolute minimum exists (since there is a

minimum distance from the cone to the point) which must occur at a critical point, so the points on the cone closest

to $(4, 2, 0)$ are $\left(2, 1, \pm\sqrt{5}\right)$.

45. Let x, y, z be the positive numbers. Then $x + y + z = 100$ $\Rightarrow$ $z = 100 - x - y$, and we want to maximize

$xyz = xy(100 - x - y) = 100xy - x^2 y - xy^2 = f(x, y)$ for $0 < x, y, z < 100$. $f_x = 100y - 2xy - y^2$,

$f_y = 100x - x^2 - 2xy$, $f_{xx} = -2y$, $f_{yy} = -2x$, $f_{xy} = 100 - 2x - 2y$. Then $f_x = 0$ implies $y(100 - 2x - y) = 0$ $\Rightarrow$

$y = 100 - 2x$ (since $y > 0$). Substituting into $f_y = 0$ gives $x[100 - x - 2(100 - 2x)] = 0$ $\Rightarrow$ $3x - 100 = 0$ (since

$x > 0$) $\Rightarrow$ $x = \frac{100}{3}$. Then $y = 100 - 2\left(\frac{100}{3}\right) = \frac{100}{3}$, and the only critical point is $\left(\frac{100}{3}, \frac{100}{3}\right)$.

$D\left(\frac{100}{3}, \frac{100}{3}\right) = \left(-\frac{200}{3}\right)\left(-\frac{200}{3}\right) - \left(-\frac{100}{3}\right)^2 = \frac{10{,}000}{3} > 0$ and $f_{xx}\left(\frac{100}{3}, \frac{100}{3}\right) = -\frac{200}{3} < 0$. Thus $f\left(\frac{100}{3}, \frac{100}{3}\right)$

is a local maximum. It is also the absolute maximum (compare to the values of f as x, y, or $z \to 0$ or 100), so the numbers are $x = y = z = \frac{100}{3}$.

47. Center the sphere at the origin so that its equation is $x^2 + y^2 + z^2 = r^2$, and orient the inscribed rectangular box so that its edges are parallel to the coordinate axes. Any vertex of the box satisfies $x^2 + y^2 + z^2 = r^2$, so take (x, y, z) to be the vertex in the first octant. Then the box has length $2x$, width $2y$, and height $2z = 2\sqrt{r^2 - x^2 - y^2}$ with volume given by

$$V(x, y) = (2x)(2y)\left(2\sqrt{r^2 - x^2 - y^2}\right) = 8xy\sqrt{r^2 - x^2 - y^2} \text{ for } 0 < x < r, 0 < y < r. \text{ Then}$$

$$V_x = (8xy) \cdot \tfrac{1}{2}(r^2 - x^2 - y^2)^{-1/2}(-2x) + \sqrt{r^2 - x^2 - y^2} \cdot 8y = \frac{8y(r^2 - 2x^2 - y^2)}{\sqrt{r^2 - x^2 - y^2}} \text{ and } V_y = \frac{8x(r^2 - x^2 - 2y^2)}{\sqrt{r^2 - x^2 - y^2}}.$$

Setting $V_x = 0$ gives $y = 0$ or $2x^2 + y^2 = r^2$, but $y > 0$ so only the latter solution applies. Similarly, $V_y = 0$ with $x > 0$ implies $x^2 + 2y^2 = r^2$. Substituting, we have $2x^2 + y^2 = x^2 + 2y^2 \Rightarrow x^2 = y^2 \Rightarrow y = x$. Then $x^2 + 2y^2 = r^2 \Rightarrow 3x^2 = r^2 \Rightarrow x = \sqrt{r^2/3} = r/\sqrt{3} = y$. Thus the only critical point is $(r/\sqrt{3}, r/\sqrt{3})$. There must be a maximum volume and here it must occur at a critical point, so the maximum volume occurs when $x = y = r/\sqrt{3}$ and the maximum volume is $V\left(\frac{r}{\sqrt{3}}, \frac{r}{\sqrt{3}}\right) = 8\left(\frac{r}{\sqrt{3}}\right)\left(\frac{r}{\sqrt{3}}\right)\sqrt{r^2 - \left(\frac{r}{\sqrt{3}}\right)^2 - \left(\frac{r}{\sqrt{3}}\right)^2} = \frac{8}{3\sqrt{3}}r^3$.

49. Maximize $f(x, y) = \frac{xy}{3}(6 - x - 2y)$, then the maximum volume is $V = xyz$.

$f_x = \tfrac{1}{3}(6y - 2xy - y^2) = \tfrac{1}{3}y(6 - 2x - 2y)$ and $f_y = \tfrac{1}{3}x(6 - x - 4y)$. Setting $f_x = 0$ and $f_y = 0$ gives the critical point $(2, 1)$ which geometrically must give a maximum. Thus the volume of the largest such box is $V = (2)(1)\left(\frac{2}{3}\right) = \frac{4}{3}$.

51. Let the dimensions be x, y, and z; then $4x + 4y + 4z = c$ and the volume is

$V = xyz = xy(\tfrac{1}{4}c - x - y) = \tfrac{1}{4}cxy - x^2y - xy^2$, $x > 0$, $y > 0$. Then $V_x = \tfrac{1}{4}cy - 2xy - y^2$ and $V_y = \tfrac{1}{4}cx - x^2 - 2xy$, so $V_x = 0 = V_y$ when $2x + y = \tfrac{1}{4}c$ and $x + 2y = \tfrac{1}{4}c$. Solving, we get $x = \tfrac{1}{12}c$, $y = \tfrac{1}{12}c$ and $z = \tfrac{1}{4}c - x - y = \tfrac{1}{12}c$. From the geometrical nature of the problem, this critical point must give an absolute maximum. Thus the box is a cube with edge length $\tfrac{1}{12}c$.

53. Let the dimensions be x, y and z, then minimize $xy + 2(xz + yz)$ if $xyz = 32{,}000$ cm^3. Then

$f(x, y) = xy + [64{,}000(x + y)/xy] = xy + 64{,}000(x^{-1} + y^{-1})$, $f_x = y - 64{,}000x^{-2}$, $f_y = x - 64{,}000y^{-2}$. And $f_x = 0$ implies $y = 64{,}000/x^2$; substituting into $f_y = 0$ implies $x^3 = 64{,}000$ or $x = 40$ and then $y = 40$. Now $D(x, y) = [(2)(64{,}000)]^2 x^{-3}y^{-3} - 1 > 0$ for $(40, 40)$ and $f_{xx}(40, 40) > 0$ so this is indeed a minimum. Thus the dimensions of the box are $x = y = 40$ cm, $z = 20$ cm.

55. Let x, y, z be the dimensions of the rectangular box. Then the volume of the box is xyz and

$$L = \sqrt{x^2 + y^2 + z^2} \Rightarrow L^2 = x^2 + y^2 + z^2 \Rightarrow z = \sqrt{L^2 - x^2 - y^2}.$$

Substituting, we have volume $V(x, y) = xy\sqrt{L^2 - x^2 - y^2}$ $(x, y > 0)$.

$$V_x = xy \cdot \tfrac{1}{2}(L^2 - x^2 - y^2)^{-1/2}(-2x) + y\sqrt{L^2 - x^2 - y^2} = y\sqrt{L^2 - x^2 - y^2} - \frac{x^2 y}{\sqrt{L^2 - x^2 - y^2}},$$

$V_y = x\sqrt{L^2 - x^2 - y^2} - \dfrac{xy^2}{\sqrt{L^2 - x^2 - y^2}}$. $V_x = 0$ implies $y(L^2 - x^2 - y^2) = x^2 y$ $\Rightarrow$ $y(L^2 - 2x^2 - y^2) = 0$ $\Rightarrow$

$2x^2 + y^2 = L^2$ (since $y > 0$), and $V_y = 0$ implies $x(L^2 - x^2 - y^2) = xy^2$ $\Rightarrow$ $x(L^2 - x^2 - 2y^2) = 0$ $\Rightarrow$

$x^2 + 2y^2 = L^2$ (since $x > 0$). Substituting $y^2 = L^2 - 2x^2$ into $x^2 + 2y^2 = L^2$ gives $x^2 + 2L^2 - 4x^2 = L^2$ $\Rightarrow$

$3x^2 = L^2$ $\Rightarrow$ $x = L/\sqrt{3}$ (since $x > 0$) and then $y = \sqrt{L^2 - 2(L/\sqrt{3})^2} = L/\sqrt{3}$.

So the only critical point is $(L/\sqrt{3}, L/\sqrt{3})$ which, from the geometrical nature of the problem, must give an absolute

maximum. Thus the maximum volume is $V(L/\sqrt{3}, L/\sqrt{3}) = (L/\sqrt{3})^2\sqrt{L^2 - (L/\sqrt{3})^2 - (L/\sqrt{3})^2} = L^3/(3\sqrt{3})$

cubic units.

57. (a) We are given that $p_1 + p_2 + p_3 = 1$ $\Rightarrow$ $p_3 = 1 - p_1 - p_2$, so

$$H = -p_1 \ln p_1 - p_2 \ln p_2 - p_3 \ln p_3 = -p_1 \ln p_1 - p_2 \ln p_2 - (1 - p_1 - p_2)\ln(1 - p_1 - p_2).$$

(b) Because p_i is a proportion we have $0 \le p_i \le 1$, but H is undefined unless

$p_1 > 0$, $p_2 > 0$, and $1 - p_1 - p_2 > 0$ $\Leftrightarrow$ $p_1 + p_2 < 1$. This last

restriction forces $p_1 < 1$ and $p_2 < 1$, so the domain of H is

$\{(p_1, p_2) \mid 0 < p_1 < 1,\ p_2 < 1 - p_1\}$. It is the interior of the triangle

drawn in the figure.

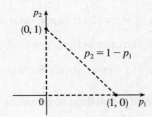

(c) $\quad H_{p_1} = -[p_1 \cdot (1/p_1) + (\ln p_1) \cdot 1] - [(1 - p_1 - p_2) \cdot (-1)/(1 - p_1 - p_2) + \ln(1 - p_1 - p_2) \cdot (-1)]$

$\qquad\qquad = -1 - \ln p_1 + 1 + \ln(1 - p_1 - p_2) = \ln(1 - p_1 - p_2) - \ln p_1$

Similarly $H_{p_2} = \ln(1 - p_1 - p_2) - \ln p_2$. Then $H_{p_1} = 0$ implies

$\ln(1 - p_1 - p_2) = \ln p_1$ $\Rightarrow$ $1 - p_1 - p_2 = p_1$ $\Rightarrow$ $p_2 = 1 - 2p_1$, and $H_{p_2} = 0$ implies

$\ln(1 - p_1 - p_2) = \ln p_2$ $\Rightarrow$ $p_1 = 1 - 2p_2$. Substituting, we have $p_1 = 1 - 2(1 - 2p_1)$ $\Rightarrow$

$3p_1 = 1$ $\Rightarrow$ $p_1 = \tfrac{1}{3}$, and then $p_2 = 1 - 2\left(\tfrac{1}{3}\right) = \tfrac{1}{3}$. Thus the only critical point is $\left(\tfrac{1}{3}, \tfrac{1}{3}\right)$.

$$D(p_1, p_2) = H_{p_1 p_1} H_{p_2 p_2} - (H_{p_1 p_2})^2 = \left(\frac{-1}{1 - p_1 - p_2} - \frac{1}{p_1}\right)\left(\frac{-1}{1 - p_1 - p_2} - \frac{1}{p_2}\right) - \left(\frac{-1}{1 - p_1 - p_2}\right)^2,\ \text{so}$$

$D\left(\tfrac{1}{3}, \tfrac{1}{3}\right) = (-6)(-6) - (-3)^2 = 27 > 0$ and $H_{p_1 p_1}\left(\tfrac{1}{3}, \tfrac{1}{3}\right) = -6 < 0$. Thus

$H\left(\tfrac{1}{3}, \tfrac{1}{3}\right) = -\tfrac{1}{3}\ln\tfrac{1}{3} - \tfrac{1}{3}\ln\tfrac{1}{3} - \tfrac{1}{3}\ln\tfrac{1}{3} = -\ln\tfrac{1}{3} = \ln 3$ is a local maximum. Here it is also the absolute maximum, so the

maximum value of H is $\ln 3$, which occurs for $p_1 = p_2 = p_3 = \tfrac{1}{3}$ (all three species have equal proportion in the

ecosystem).

59. Note that here the variables are m and b, and $f(m, b) = \sum\limits_{i=1}^{n} [y_i - (mx_i + b)]^2$. Then $f_m = \sum\limits_{i=1}^{n} -2x_i[y_i - (mx_i + b)] = 0$

implies $\sum\limits_{i=1}^{n} (x_iy_i - mx_i^2 - bx_i) = 0$ or $\sum\limits_{i=1}^{n} x_iy_i = m \sum\limits_{i=1}^{n} x_i^2 + b \sum\limits_{i=1}^{n} x_i$ and $f_b = \sum\limits_{i=1}^{n} -2[y_i - (mx_i + b)] = 0$ implies

$\sum\limits_{i=1}^{n} y_i = m \sum\limits_{i=1}^{n} x_i + \sum\limits_{i=1}^{n} b = m \left(\sum\limits_{i=1}^{n} x_i \right) + nb$. Thus we have the two desired equations.

Now $f_{mm} = \sum\limits_{i=1}^{n} 2x_i^2$, $f_{bb} = \sum\limits_{i=1}^{n} 2 = 2n$ and $f_{mb} = \sum\limits_{i=1}^{n} 2x_i$. And $f_{mm}(m, b) > 0$ always and

$D(m, b) = 4n \left(\sum\limits_{i=1}^{n} x_i^2 \right) - 4 \left(\sum\limits_{i=1}^{n} x_i \right)^2 = 4 \left[n \left(\sum\limits_{i=1}^{n} x_i^2 \right) - \left(\sum\limits_{i=1}^{n} x_i \right)^2 \right] > 0$ always so the solutions of these two

equations do indeed minimize $\sum\limits_{i=1}^{n} d_i^2$.

14.8 Lagrange Multipliers

1. At the extreme values of f, the level curves of f just touch the curve $g(x, y) = 8$ with a common tangent line. (See Figure 1 and the accompanying discussion.) We can observe several such occurrences on the contour map, but the level curve $f(x, y) = c$ with the largest value of c which still intersects the curve $g(x, y) = 8$ is approximately $c = 59$, and the smallest value of c corresponding to a level curve which intersects $g(x, y) = 8$ appears to be $c = 30$. Thus we estimate the maximum value of f subject to the constraint $g(x, y) = 8$ to be about 59 and the minimum to be 30.

3. We want to find the extreme values of $f(x, y) = x^2 - y^2$ subject to the constraint $g(x, y) = x^2 + y^2 = 1$. Then $\nabla f = \lambda \nabla g \ \Rightarrow \ \langle 2x, -2y \rangle = \lambda \langle 2x, 2y \rangle$, so we solve the equations $2x = 2\lambda x$, $-2y = 2\lambda y$, and $x^2 + y^2 = 1$. From the first equation we have $2x(\lambda - 1) = 0 \ \Rightarrow \ x = 0$ or $\lambda = 1$. If $x = 0$ then substitution into the constraint gives $y^2 = 1 \ \Rightarrow \ y = \pm 1$. If $\lambda = 1$ then substitution into the second equation gives $-2y = 2y \ \Rightarrow \ y = 0$, and from the constraint we must have $x = \pm 1$. Thus the possible points for the extreme values of f are $(0, \pm 1)$ and $(\pm 1, 0)$. Evaluating f at these points, we see that the maximum value of f is $f(\pm 1, 0) = 1$ and the minimum is $f(0, \pm 1) = -1$.

5. $f(x, y) = xy$, $g(x, y) = 4x^2 + y^2 = 8$, and $\nabla f = \lambda \nabla g \ \Rightarrow \ \langle y, x \rangle = \langle 8\lambda x, 2\lambda y \rangle$, so $y = 8\lambda x$, $x = 2\lambda y$, and $4x^2 + y^2 = 8$. First note that if $x = 0$ then $y = 0$ by the first equation, and if $y = 0$ then $x = 0$ by the second equation. But this contradicts the third equation, so $x \neq 0$ and $y \neq 0$. Then from the first two equations we have $\frac{y}{8x} = \lambda = \frac{x}{2y} \ \Rightarrow \ 2y^2 = 8x^2 \ \Rightarrow \ y^2 = 4x^2$, and substitution into the third equation gives $4x^2 + 4x^2 = 8 \ \Rightarrow \ x = \pm 1$. If $x = \pm 1$ then $y^2 = 4 \ \Rightarrow \ y = \pm 2$, so f has possible extreme values at $(1, \pm 2)$ and $(-1, \pm 2)$. Evaluating f at these points, we see that the maximum value is $f(1, 2) = f(-1, -2) = 2$ and the minimum is $f(1, -2) = f(-1, 2) = -2$.

7. $f(x, y, z) = 2x + 2y + z$, $g(x, y, z) = x^2 + y^2 + z^2 = 9$, and $\nabla f = \lambda \nabla g$ $\Rightarrow$ $\langle 2, 2, 1 \rangle = \langle 2\lambda x, 2\lambda y, 2\lambda z \rangle$, so $2\lambda x = 2$,

$2\lambda y = 2$, $2\lambda z = 1$, and $x^2 + y^2 + z^2 = 9$. The first three equations imply $x = \dfrac{1}{\lambda}$, $y = \dfrac{1}{\lambda}$, and $z = \dfrac{1}{2\lambda}$. But substitution into

the fourth equation gives $\left(\dfrac{1}{\lambda}\right)^2 + \left(\dfrac{1}{\lambda}\right)^2 + \left(\dfrac{1}{2\lambda}\right)^2 = 9$ $\Rightarrow$ $\dfrac{9}{4\lambda^2} = 9$ $\Rightarrow$ $\lambda = \pm\frac{1}{2}$, so f has possible extreme values at

the points $(2, 2, 1)$ and $(-2, -2, -1)$. The maximum value of f on $x^2 + y^2 + z^2 = 9$ is $f(2, 2, 1) = 9$, and the minimum is

$f(-2, -2, -1) = -9$.

9. $f(x, y, z) = xy^2 z$, $g(x, y, z) = x^2 + y^2 + z^2 = 4$, and $\nabla f = \lambda \nabla g$ $\Rightarrow$ $\langle y^2 z, 2xyz, xy^2 \rangle = \lambda \langle 2x, 2y, 2z \rangle$. Then

$y^2 z = 2\lambda x$, $2xyz = 2\lambda y$, $xy^2 = 2\lambda z$, and $x^2 + y^2 + z^2 = 4$.

Case 1: If $\lambda = 0$, then the first equation implies that $y = 0$ or $z = 0$. If $y = 0$, then any values of x and z satisfy the first three

equations, so from the fourth equation all points $(x, 0, z)$ such that $x^2 + z^2 = 4$ are possible points. If $z = 0$ then from the

third equation $x = 0$ or $y = 0$, and from the fourth equation, the possible points are $(0, \pm 2, 0)$, $(\pm 2, 0, 0)$. The f-value in all

these cases is 0.

Case 2: If $\lambda \neq 0$ but any one of x, y, z is zero, the first three equations imply that all three coordinates must be zero,

contradicting the fourth equation. Thus if $\lambda \neq 0$, none of x, y, z is zero and from the first three equations we have

$\dfrac{y^2 z}{2x} = \lambda = xz = \dfrac{xy^2}{2z}$. This gives $y^2 z = 2x^2 z$ $\Rightarrow$ $y^2 = 2x^2$ and $2y^2 z^2 = 2x^2 y^2$ $\Rightarrow$ $z^2 = x^2$. Substituting into the

fourth equation, we have $x^2 + 2x^2 + x^2 = 4$ $\Rightarrow$ $x^2 = 1$ $\Rightarrow$ $x = \pm 1$, so $y = \pm\sqrt{2}$ and $z = \pm 1$, giving possible points

$(\pm 1, \pm\sqrt{2}, \pm 1)$ (all combinations). The value of f is 2 when x and z are the same sign and -2 when they are opposite.

Thus the maximum of f subject to the constraint is $f(1, \pm\sqrt{2}, 1) = f(-1, \pm\sqrt{2}, -1) = 2$ and the minimum is

$f(1, \pm\sqrt{2}, -1) = f(-1, \pm\sqrt{2}, 1) = -2$.

11. $f(x, y, z) = x^2 + y^2 + z^2$, $g(x, y, z) = x^4 + y^4 + z^4 = 1$ $\Rightarrow$ $\nabla f = \langle 2x, 2y, 2z \rangle$, $\lambda \nabla g = \langle 4\lambda x^3, 4\lambda y^3, 4\lambda z^3 \rangle$.

Case 1: If $x \neq 0$, $y \neq 0$, and $z \neq 0$, then $\nabla f = \lambda \nabla g$ implies $\lambda = 1/(2x^2) = 1/(2y^2) = 1/(2z^2)$ or $x^2 = y^2 = z^2$ and

$3x^4 = 1$ or $x = \pm\frac{1}{\sqrt[4]{3}}$ giving the points $\left(\pm\frac{1}{\sqrt[4]{3}}, \frac{1}{\sqrt[4]{3}}, \frac{1}{\sqrt[4]{3}}\right)$, $\left(\pm\frac{1}{\sqrt[4]{3}}, -\frac{1}{\sqrt[4]{3}}, \frac{1}{\sqrt[4]{3}}\right)$, $\left(\pm\frac{1}{\sqrt[4]{3}}, \frac{1}{\sqrt[4]{3}}, -\frac{1}{\sqrt[4]{3}}\right)$, $\left(\pm\frac{1}{\sqrt[4]{3}}, -\frac{1}{\sqrt[4]{3}}, -\frac{1}{\sqrt[4]{3}}\right)$

all with an f-value of $\sqrt{3}$.

Case 2: If one of the variables equals zero and the other two are not zero, then the squares of the two nonzero coordinates are

equal with common value $\frac{1}{\sqrt{2}}$ and corresponding f-value of $\sqrt{2}$.

Case 3: If exactly two of the variables are zero, then the third variable has value ± 1 with the corresponding f-value of 1.

Thus on $x^4 + y^4 + z^4 = 1$, the maximum value of f is $\sqrt{3}$ and the minimum value is 1.

13. $f(x, y, z, t) = x + y + z + t$, $g(x, y, z, t) = x^2 + y^2 + z^2 + t^2 = 1$ $\Rightarrow$ $\langle 1, 1, 1, 1 \rangle = \langle 2\lambda x, 2\lambda y, 2\lambda z, 2\lambda t \rangle$, so

$\lambda = 1/(2x) = 1/(2y) = 1/(2z) = 1/(2t)$ and $x = y = z = t$. But $x^2 + y^2 + z^2 + t^2 = 1$, so the possible points are

$\left(\pm\frac{1}{2}, \pm\frac{1}{2}, \pm\frac{1}{2}, \pm\frac{1}{2}\right)$. Thus the maximum value of f is $f\left(\frac{1}{2}, \frac{1}{2}, \frac{1}{2}, \frac{1}{2}\right) = 2$ and the minimum value is

$f\left(-\frac{1}{2}, -\frac{1}{2}, -\frac{1}{2}, -\frac{1}{2}\right) = -2$.

15. $f(x, y) = x^2 + y^2$, $g(x, y) = xy = 1$, and $\nabla f = \lambda \nabla g \Rightarrow \langle 2x, 2y \rangle = \langle \lambda y, \lambda x \rangle$, so $2x = \lambda y$, $2y = \lambda x$, and $xy = 1$.

From the last equation, $x \neq 0$ and $y \neq 0$, so $2x = \lambda y \Rightarrow \lambda = 2x/y$. Substituting, we have $2y = (2x/y)\,x \Rightarrow$

$y^2 = x^2 \Rightarrow y = \pm x$. But $xy = 1$, so $x = y = \pm 1$ and the possible points for the extreme values of f are $(1, 1)$ and

$(-1, -1)$. Here there is no maximum value, since the constraint $xy = 1 \Leftrightarrow y = 1/x$ allows x or y to become arbitrarily

large, and hence $f(x, y) = x^2 + y^2$ can be made arbitrarily large. The minimum value is $f(1, 1) = f(-1, -1) = 2$.

17. $f(x, y, z) = x + y + z$, $g(x, y, z) = x^2 + z^2 = 2$, $h(x, y, z) = x + y = 1$, and $\nabla f = \lambda \nabla g + \mu \nabla h \Rightarrow$

$\langle 1, 1, 1 \rangle = \langle 2\lambda x, 0, 2\lambda z \rangle + \langle \mu, \mu, 0 \rangle$. Then $1 = 2\lambda x + \mu$, $1 = \mu$, $1 = 2\lambda z$, $x^2 + z^2 = 2$, and $x + y = 1$. Substituting

$\mu = 1$ into the first equation gives $\lambda = 0$ or $x = 0$. But $\lambda = 0$ contradicts $1 = 2\lambda z$, so $x = 0$. Then $x + y = 1 \Rightarrow y = 1$

and $x^2 + z^2 = 2 \Rightarrow z = \pm\sqrt{2}$, so the possible points are $\left(0, 1, \pm\sqrt{2}\right)$. The maximum value of f subject to the

constraints is $f(0, 1, \sqrt{2}) = 1 + \sqrt{2} \approx 2.41$ and the minimum is $f(0, 1, -\sqrt{2}) = 1 - \sqrt{2} \approx -0.41$.

Note: Since $x + y = 1$ is one of the constraints, we could have solved the problem by solving $f(x, z) = 1 + z$ subject to

$x^2 + z^2 = 2$.

19. $f(x, y, z) = yz + xy$, $g(x, y, z) = xy = 1$, $h(x, y, z) = y^2 + z^2 = 1 \Rightarrow \nabla f = \langle y, x + z, y \rangle$, $\lambda \nabla g = \langle \lambda y, \lambda x, 0 \rangle$,

$\mu \nabla h = \langle 0, 2\mu y, 2\mu z \rangle$. Then $y = \lambda y$ implies $\lambda = 1$ [$y \neq 0$ since $g(x, y, z) = 1$], $x + z = \lambda x + 2\mu y$ and $y = 2\mu z$. Thus

$\mu = z/(2y) = y/(2y)$ or $y^2 = z^2$, and so $y^2 + z^2 = 1$ implies $y = \pm\frac{1}{\sqrt{2}}$, $z = \pm\frac{1}{\sqrt{2}}$. Then $xy = 1$ implies $x = \pm\sqrt{2}$ and

the possible points are $\left(\pm\sqrt{2}, \pm\frac{1}{\sqrt{2}}, \frac{1}{\sqrt{2}}\right)$, $\left(\pm\sqrt{2}, \pm\frac{1}{\sqrt{2}}, -\frac{1}{\sqrt{2}}\right)$. Hence the maximum of f subject to the constraints is

$f\left(\pm\sqrt{2}, \pm\frac{1}{\sqrt{2}}, \pm\frac{1}{\sqrt{2}}\right) = \frac{3}{2}$ and the minimum is $f\left(\pm\sqrt{2}, \pm\frac{1}{\sqrt{2}}, \mp\frac{1}{\sqrt{2}}\right) = \frac{1}{2}$.

Note: Since $xy = 1$ is one of the constraints we could have solved the problem by solving $f(y, z) = yz + 1$ subject to

$y^2 + z^2 = 1$.

21. $f(x, y) = x^2 + y^2 + 4x - 4y$. For the interior of the region, we find the critical points: $f_x = 2x + 4$, $f_y = 2y - 4$, so the

only critical point is $(-2, 2)$ (which is inside the region) and $f(-2, 2) = -8$. For the boundary, we use Lagrange multipliers.

$g(x, y) = x^2 + y^2 = 9$, so $\nabla f = \lambda \nabla g \Rightarrow \langle 2x + 4, 2y - 4 \rangle = \langle 2\lambda x, 2\lambda y \rangle$. Thus $2x + 4 = 2\lambda x$ and $2y - 4 = 2\lambda y$.

Adding the two equations gives $2x + 2y = 2\lambda x + 2\lambda y \Rightarrow x + y = \lambda(x + y) \Rightarrow (x + y)(\lambda - 1) = 0$, so

$x + y = 0 \Rightarrow y = -x$ or $\lambda - 1 = 0 \Rightarrow \lambda = 1$. But $\lambda = 1$ leads to a contradiction in $2x + 4 = 2\lambda x$, so $y = -x$ and

$x^2 + y^2 = 9$ implies $2y^2 = 9 \Rightarrow y = \pm\frac{3}{\sqrt{2}}$. We have $f\left(\frac{3}{\sqrt{2}}, -\frac{3}{\sqrt{2}}\right) = 9 + 12\sqrt{2} \approx 25.97$ and

$f\left(-\frac{3}{\sqrt{2}}, \frac{3}{\sqrt{2}}\right) = 9 - 12\sqrt{2} \approx -7.97$, so the maximum value of f on the disk $x^2 + y^2 \leq 9$ is $f\left(\frac{3}{\sqrt{2}}, -\frac{3}{\sqrt{2}}\right) = 9 + 12\sqrt{2}$ and

the minimum is $f(-2, 2) = -8$.

23. $f(x, y) = e^{-xy}$.   For the interior of the region, we find the critical points: $f_x = -ye^{-xy}$, $f_y = -xe^{-xy}$, so the only

critical point is $(0, 0)$, and $f(0, 0) = 1$. For the boundary, we use Lagrange multipliers. $g(x, y) = x^2 + 4y^2 = 1$ $\Rightarrow$

$\lambda \nabla g = \langle 2\lambda x, 8\lambda y \rangle$, so setting $\nabla f = \lambda \nabla g$ we get $-ye^{-xy} = 2\lambda x$ and $-xe^{-xy} = 8\lambda y$. The first of these gives

$e^{-xy} = -2\lambda x / y$, and then the second gives $-x(-2\lambda x / y) = 8\lambda y$ $\Rightarrow$ $x^2 = 4y^2$. Solving this last equation with the

constraint $x^2 + 4y^2 = 1$ gives $x = \pm \frac{1}{\sqrt{2}}$ and $y = \pm \frac{1}{2\sqrt{2}}$. Now $f\left(\pm \frac{1}{\sqrt{2}}, \mp \frac{1}{2\sqrt{2}}\right) = e^{1/4} \approx 1.284$ and

$f\left(\pm \frac{1}{\sqrt{2}}, \pm \frac{1}{2\sqrt{2}}\right) = e^{-1/4} \approx 0.779$. The former are the maxima on the region and the latter are the minima.

25. (a) $f(x, y) = x$, $g(x, y) = y^2 + x^4 - x^3 = 0$ $\Rightarrow$ $\nabla f = \langle 1, 0 \rangle = \lambda \nabla g = \lambda \langle 4x^3 - 3x^2, 2y \rangle$. Then

$1 = \lambda(4x^3 - 3x^2)$ **(1)** and $0 = 2\lambda y$ **(2)**. We have $\lambda \neq 0$ from **(1)**, so **(2)** gives $y = 0$. Then, from the constraint equation,

$x^4 - x^3 = 0$ $\Rightarrow$ $x^3(x - 1) = 0$ $\Rightarrow$ $x = 0$ or $x = 1$. But $x = 0$ contradicts **(1)**, so the only possible extreme value

subject to the constraint is $f(1, 0) = 1$. (The question remains whether this is indeed the minimum of f.)

(b) The constraint is $y^2 + x^4 - x^3 = 0$ $\Leftrightarrow$ $y^2 = x^3 - x^4$. The left side is non-negative, so we must have $x^3 - x^4 \geq 0$

which is true only for $0 \leq x \leq 1$. Therefore the minimum possible value for $f(x, y) = x$ is 0 which occurs for $x = y = 0$.

However, $\lambda \nabla g(0, 0) = \lambda \langle 0 - 0, 0 \rangle = \langle 0, 0 \rangle$ and $\nabla f(0, 0) = \langle 1, 0 \rangle$, so $\nabla f(0, 0) \neq \lambda \nabla g(0, 0)$ for all values of λ.

(c) Here $\nabla g(0, 0) = \mathbf{0}$ but the method of Lagrange multipliers requires that $\nabla g \neq \mathbf{0}$ everywhere on the constraint curve.

27. $P(L, K) = bL^\alpha K^{1-\alpha}$, $g(L, K) = mL + nK = p$ $\Rightarrow$ $\nabla P = \langle \alpha bL^{\alpha-1}K^{1-\alpha}, (1-\alpha)bL^\alpha K^{-\alpha} \rangle$, $\lambda \nabla g = \langle \lambda m, \lambda n \rangle$.

Then $\alpha b(K/L)^{1-\alpha} = \lambda m$ and $(1-\alpha)b(L/K)^\alpha = \lambda n$ and $mL + nK = p$, so $\alpha b(K/L)^{1-\alpha}/m = (1-\alpha)b(L/K)^\alpha/n$ or

$n\alpha/[m(1-\alpha)] = (L/K)^\alpha(L/K)^{1-\alpha}$ or $L = Kn\alpha/[m(1-\alpha)]$. Substituting into $mL + nK = p$ gives $K = (1-\alpha)p/n$

and $L = \alpha p/m$ for the maximum production.

29. Let the sides of the rectangle be x and y. Then $f(x, y) = xy$, $g(x, y) = 2x + 2y = p$ $\Rightarrow$ $\nabla f(x, y) = \langle y, x \rangle$,

$\lambda \nabla g = \langle 2\lambda, 2\lambda \rangle$. Then $\lambda = \frac{1}{2}y = \frac{1}{2}x$ implies $x = y$ and the rectangle with maximum area is a square with side length $\frac{1}{4}p$.

31. The distance from $(2, 0, -3)$ to a point (x, y, z) on the plane is $d = \sqrt{(x-2)^2 + y^2 + (z+3)^2}$, so we seek to minimize

$d^2 = f(x, y, z) = (x-2)^2 + y^2 + (z+3)^2$ subject to the constraint that (x, y, z) lies on the plane $x + y + z = 1$, that is,

that $g(x, y, z) = x + y + z = 1$.   Then $\nabla f = \lambda \nabla g$ $\Rightarrow$ $\langle 2(x-2), 2y, 2(z+3) \rangle = \langle \lambda, \lambda, \lambda \rangle$, so $x = (\lambda + 4)/2$,

$y = \lambda/2$, $z = (\lambda - 6)/2$. Substituting into the constraint equation gives $\frac{\lambda+4}{2} + \frac{\lambda}{2} + \frac{\lambda-6}{2} = 1$ $\Rightarrow$ $3\lambda - 2 = 2$ $\Rightarrow$

$\lambda = \frac{4}{3}$, so $x = \frac{8}{3}$, $y = \frac{2}{3}$, and $z = -\frac{7}{3}$. This must correspond to a minimum, so the shortest distance is

$d = \sqrt{\left(\frac{8}{3} - 2\right)^2 + \left(\frac{2}{3}\right)^2 + \left(-\frac{7}{3} + 3\right)^2} = \sqrt{\frac{4}{3}} = \frac{2}{\sqrt{3}}$.

33. Let $f(x, y, z) = d^2 = (x-4)^2 + (y-2)^2 + z^2$. Then we want to minimize f subject to the constraint

$g(x, y, z) = x^2 + y^2 - z^2 = 0$. $\nabla f = \lambda \nabla g$ ⟹ $\langle 2(x-4), 2(y-2), 2z \rangle = \langle 2\lambda x, 2\lambda y, -2\lambda z \rangle$, so $x - 4 = \lambda x$,

$y - 2 = \lambda y$, and $z = -\lambda z$. From the last equation we have $z + \lambda z = 0$ ⟹ $z(1 + \lambda) = 0$, so either $z = 0$ or $\lambda = -1$. But

from the constraint equation we have $z = 0$ ⟹ $x^2 + y^2 = 0$ ⟹ $x = y = 0$ which is not possible from the first two

equations. So $\lambda = -1$ and $x - 4 = \lambda x$ ⟹ $x = 2$, $y - 2 = \lambda y$ ⟹ $y = 1$, and $x^2 + y^2 - z^2 = 0$ ⟹

$4 + 1 - z^2 = 0$ ⟹ $z = \pm\sqrt{5}$. This must correspond to a minimum, so the points on the cone closest to $(4, 2, 0)$

are $\left(2, 1, \pm\sqrt{5}\right)$.

35. $f(x, y, z) = xyz$, $g(x, y, z) = x + y + z = 100$ ⟹ $\nabla f = \langle yz, xz, xy \rangle = \lambda \nabla g = \langle \lambda, \lambda, \lambda \rangle$. Then $\lambda = yz = xz = xy$

implies $x = y = z = \frac{100}{3}$.

37. If the dimensions are $2x$, $2y$, and $2z$, then maximize $f(x, y, z) = (2x)(2y)(2z) = 8xyz$ subject to

$g(x, y, z) = x^2 + y^2 + z^2 = r^2$ $(x > 0, y > 0, z > 0)$. Then $\nabla f = \lambda \nabla g$ ⟹ $\langle 8yz, 8xz, 8xy \rangle = \lambda \langle 2x, 2y, 2z \rangle$ ⟹

$8yz = 2\lambda x$, $8xz = 2\lambda y$, and $8xy = 2\lambda z$, so $\lambda = \dfrac{4yz}{x} = \dfrac{4xz}{y} = \dfrac{4xy}{z}$. This gives $x^2 z = y^2 z$ ⟹ $x^2 = y^2$ (since $z \neq 0$)

and $xy^2 = xz^2$ ⟹ $z^2 = y^2$, so $x^2 = y^2 = z^2$ ⟹ $x = y = z$, and substituting into the constraint

equation gives $3x^2 = r^2$ ⟹ $x = r/\sqrt{3} = y = z$. Thus the largest volume of such a box is

$f\left(\dfrac{r}{\sqrt{3}}, \dfrac{r}{\sqrt{3}}, \dfrac{r}{\sqrt{3}}\right) = 8\left(\dfrac{r}{\sqrt{3}}\right)\left(\dfrac{r}{\sqrt{3}}\right)\left(\dfrac{r}{\sqrt{3}}\right) = \dfrac{8}{3\sqrt{3}} r^3$.

39. $f(x, y, z) = xyz$, $g(x, y, z) = x + 2y + 3z = 6$ ⟹ $\nabla f = \langle yz, xz, xy \rangle = \lambda \nabla g = \langle \lambda, 2\lambda, 3\lambda \rangle$.

Then $\lambda = yz = \frac{1}{2}xz = \frac{1}{3}xy$ implies $x = 2y$, $z = \frac{2}{3}y$. But $2y + 2y + 2y = 6$ so $y = 1$, $x = 2$, $z = \frac{2}{3}$ and the volume

is $V = \frac{4}{3}$.

41. $f(x, y, z) = xyz$, $g(x, y, z) = 4(x + y + z) = c$ ⟹ $\nabla f = \langle yz, xz, xy \rangle$, $\lambda \nabla g = \langle 4\lambda, 4\lambda, 4\lambda \rangle$. Thus

$4\lambda = yz = xz = xy$ or $x = y = z = \frac{1}{12}c$ are the dimensions giving the maximum volume.

43. If the dimensions of the box are given by x, y, and z, then we need to find the maximum value of $f(x, y, z) = xyz$

$[x, y, z > 0]$ subject to the constraint $L = \sqrt{x^2 + y^2 + z^2}$ or $g(x, y, z) = x^2 + y^2 + z^2 = L^2$. $\nabla f = \lambda \nabla g$ ⟹

$\langle yz, xz, xy \rangle = \lambda \langle 2x, 2y, 2z \rangle$, so $yz = 2\lambda x$ ⟹ $\lambda = \dfrac{yz}{2x}$, $xz = 2\lambda y$ ⟹ $\lambda = \dfrac{xz}{2y}$, and $xy = 2\lambda z$ ⟹ $\lambda = \dfrac{xy}{2z}$.

Thus $\lambda = \dfrac{yz}{2x} = \dfrac{xz}{2y}$ ⟹ $x^2 = y^2$ [since $z \neq 0$] ⟹ $x = y$ and $\lambda = \dfrac{yz}{2x} = \dfrac{xy}{2z}$ ⟹ $x = z$ [since $y \neq 0$].

Substituting into the constraint equation gives $x^2 + x^2 + x^2 = L^2$ ⟹ $x^2 = L^2/3$ ⟹ $x = L/\sqrt{3} = y = z$ and the

maximum volume is $\left(L/\sqrt{3}\right)^3 = L^3/(3\sqrt{3})$.

45. We need to find the extreme values of $f(x, y, z) = x^2 + y^2 + z^2$ subject to the two constraints $g(x, y, z) = x + y + 2z = 2$

and $h(x, y, z) = x^2 + y^2 - z = 0$. $\nabla f = \langle 2x, 2y, 2z \rangle$, $\lambda \nabla g = \langle \lambda, \lambda, 2\lambda \rangle$ and $\mu \nabla h = \langle 2\mu x, 2\mu y, -\mu \rangle$. Thus we need

$2x = \lambda + 2\mu x$ **(1)**, $2y = \lambda + 2\mu y$ **(2)**, $2z = 2\lambda - \mu$ **(3)**, $x + y + 2z = 2$ **(4)**, and $x^2 + y^2 - z = 0$ **(5)**.

From **(1)** and **(2)**, $2(x - y) = 2\mu(x - y)$, so if $x \neq y$, $\mu = 1$. Putting this in **(3)** gives $2z = 2\lambda - 1$ or $\lambda = z + \frac{1}{2}$, but putting

$\mu = 1$ into **(1)** says $\lambda = 0$. Hence $z + \frac{1}{2} = 0$ or $z = -\frac{1}{2}$. Then **(4)** and **(5)** become $x + y - 3 = 0$ and $x^2 + y^2 + \frac{1}{2} = 0$. The

last equation cannot be true, so this case gives no solution. So we must have $x = y$. Then **(4)** and **(5)** become $2x + 2z = 2$ and

$2x^2 - z = 0$ which imply $z = 1 - x$ and $z = 2x^2$. Thus $2x^2 = 1 - x$ or $2x^2 + x - 1 = (2x - 1)(x + 1) = 0$ so $x = \frac{1}{2}$ or

$x = -1$. The two points to check are $\left(\frac{1}{2}, \frac{1}{2}, \frac{1}{2} \right)$ and $(-1, -1, 2)$: $f\left(\frac{1}{2}, \frac{1}{2}, \frac{1}{2} \right) = \frac{3}{4}$ and $f(-1, -1, 2) = 6$. Thus $\left(\frac{1}{2}, \frac{1}{2}, \frac{1}{2} \right)$ is

the point on the ellipse nearest the origin and $(-1, -1, 2)$ is the one farthest from the origin.

47. $f(x, y, z) = y e^{x-z}$, $g(x, y, z) = 9x^2 + 4y^2 + 36z^2 = 36$, $h(x, y, z) = xy + yz = 1$. $\nabla f = \lambda \nabla g + \mu \nabla h$ $\Rightarrow$

$\langle y e^{x-z}, e^{x-z}, -y e^{x-z} \rangle = \lambda \langle 18x, 8y, 72z \rangle + \mu \langle y, x + z, y \rangle$, so $y e^{x-z} = 18\lambda x + \mu y$, $e^{x-z} = 8\lambda y + \mu(x + z)$,

$-y e^{x-z} = 72\lambda z + \mu y$, $9x^2 + 4y^2 + 36z^2 = 36$, $xy + yz = 1$. Using a CAS to solve these 5 equations simultaneously for x,

y, z, λ, and μ (in Maple, use the `allvalues` command), we get 4 real-valued solutions:

$$x \approx 0.222444, \qquad y \approx -2.157012, \qquad z \approx -0.686049, \qquad \lambda \approx -0.200401, \qquad \mu \approx 2.108584$$

$$x \approx -1.951921, \qquad y \approx -0.545867, \qquad z \approx 0.119973, \qquad \lambda \approx 0.003141, \qquad \mu \approx -0.076238$$

$$x \approx 0.155142, \qquad y \approx 0.904622, \qquad z \approx 0.950293, \qquad \lambda \approx -0.012447, \qquad \mu \approx 0.489938$$

$$x \approx 1.138731, \qquad y \approx 1.768057, \qquad z \approx -0.573138, \qquad \lambda \approx 0.317141, \qquad \mu \approx 1.862675$$

Substituting these values into f gives $f(0.222444, -2.157012, -0.686049) \approx -5.3506$,

$f(-1.951921, -0.545867, 0.119973) \approx -0.0688$, $f(0.155142, 0.904622, 0.950293) \approx 0.4084$,

$f(1.138731, 1.768057, -0.573138) \approx 9.7938$. Thus the maximum is approximately 9.7938, and the minimum is

approximately -5.3506.

49. (a) We wish to maximize $f(x_1, x_2, \ldots, x_n) = \sqrt[n]{x_1 x_2 \cdots x_n}$ subject to

$g(x_1, x_2, \ldots, x_n) = x_1 + x_2 + \cdots + x_n = c$ and $x_i > 0$.

$\nabla f = \left\langle \frac{1}{n}(x_1 x_2 \cdots x_n)^{\frac{1}{n} - 1}(x_2 \cdots x_n), \frac{1}{n}(x_1 x_2 \cdots x_n)^{\frac{1}{n} - 1}(x_1 x_3 \cdots x_n), \ldots, \frac{1}{n}(x_1 x_2 \cdots x_n)^{\frac{1}{n} - 1}(x_1 \cdots x_{n-1}) \right\rangle$

and $\lambda \nabla g = \langle \lambda, \lambda, \ldots, \lambda \rangle$, so we need to solve the system of equations

$$\frac{1}{n}(x_1 x_2 \cdots x_n)^{\frac{1}{n} - 1}(x_2 \cdots x_n) = \lambda \quad \Rightarrow \quad x_1^{1/n} x_2^{1/n} \cdots x_n^{1/n} = n\lambda x_1$$

$$\frac{1}{n}(x_1 x_2 \cdots x_n)^{\frac{1}{n} - 1}(x_1 x_3 \cdots x_n) = \lambda \quad \Rightarrow \quad x_1^{1/n} x_2^{1/n} \cdots x_n^{1/n} = n\lambda x_2$$

$$\vdots$$

$$\frac{1}{n}(x_1 x_2 \cdots x_n)^{\frac{1}{n} - 1}(x_1 \cdots x_{n-1}) = \lambda \quad \Rightarrow \quad x_1^{1/n} x_2^{1/n} \cdots x_n^{1/n} = n\lambda x_n$$

This implies $n\lambda x_1 = n\lambda x_2 = \cdots = n\lambda x_n$. Note $\lambda \neq 0$, otherwise we can't have all $x_i > 0$. Thus $x_1 = x_2 = \cdots = x_n$.

[continued]

But $x_1 + x_2 + \cdots + x_n = c \quad \Rightarrow \quad nx_1 = c \quad \Rightarrow \quad x_1 = \dfrac{c}{n} = x_2 = x_3 = \cdots = x_n$. Then the only point where f can

have an extreme value is $\left(\dfrac{c}{n}, \dfrac{c}{n}, \ldots, \dfrac{c}{n}\right)$. Since we can choose values for $(x_1, x_2, \ldots, x_n)$ that make f as close to

zero (but not equal) as we like, f has no minimum value. Thus the maximum value is

$$f\left(\frac{c}{n}, \frac{c}{n}, \ldots, \frac{c}{n}\right) = \sqrt[n]{\frac{c}{n} \cdot \frac{c}{n} \cdot \cdots \cdot \frac{c}{n}} = \frac{c}{n}.$$

(b) From part (a), $\dfrac{c}{n}$ is the maximum value of f. Thus $f(x_1, x_2, \ldots, x_n) = \sqrt[n]{x_1 x_2 \cdots x_n} \leq \dfrac{c}{n}$. But

$x_1 + x_2 + \cdots + x_n = c$, so $\sqrt[n]{x_1 x_2 \cdots x_n} \leq \dfrac{x_1 + x_2 + \cdots + x_n}{n}$. These two means are equal when f attains its

maximum value $\dfrac{c}{n}$, but this can occur only at the point $\left(\dfrac{c}{n}, \dfrac{c}{n}, \ldots, \dfrac{c}{n}\right)$ we found in part (a). So the means are equal only

when $x_1 = x_2 = x_3 = \cdots = x_n = \dfrac{c}{n}$.

14 Review

TRUE-FALSE QUIZ

1. True. $f_y(a, b) = \lim\limits_{h \to 0} \dfrac{f(a, b+h) - f(a, b)}{h}$ from Equation 14.3.3. Let $h = y - b$. As $h \to 0$, $y \to b$. Then by substituting,

we get $f_y(a, b) = \lim\limits_{y \to b} \dfrac{f(a, y) - f(a, b)}{y - b}$.

3. False. $f_{xy} = \dfrac{\partial^2 f}{\partial y\, \partial x}$.

5. False. See Example 14.2.3.

7. True. If f has a local minimum and f is differentiable at (a, b) then by Theorem 14.7.2, $f_x(a, b) = 0$ and $f_y(a, b) = 0$, so

$\nabla f(a, b) = \langle f_x(a, b), f_y(a, b)\rangle = \langle 0, 0\rangle = \mathbf{0}$.

9. False. $\nabla f(x, y) = \langle 0, 1/y\rangle$.

11. True. $\nabla f = \langle \cos x, \cos y\rangle$, so $|\nabla f| = \sqrt{\cos^2 x + \cos^2 y}$. But $|\cos \theta| \leq 1$, so $|\nabla f| \leq \sqrt{2}$. Now

$D_{\mathbf{u}} f(x, y) = \nabla f \cdot \mathbf{u} = |\nabla f|\, |\mathbf{u}| \cos \theta$, but $\mathbf{u}$ is a unit vector, so $|D_{\mathbf{u}} f(x, y)| \leq \sqrt{2} \cdot 1 \cdot 1 = \sqrt{2}$.

EXERCISES

1. $\ln(x + y + 1)$ is defined only when $x + y + 1 > 0 \quad \Leftrightarrow \quad y > -x - 1$,

so the domain of f is $\{(x, y) \mid y > -x - 1\}$, all those points above the

line $y = -x - 1$.

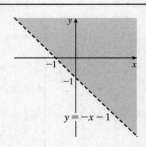

3. $z = f(x, y) = 1 - y^2$, a parabolic cylinder

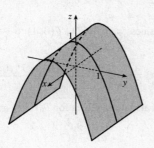

5. The level curves are $\sqrt{4x^2 + y^2} = k$ or $4x^2 + y^2 = k^2$, $k \geq 0$, a family of ellipses.

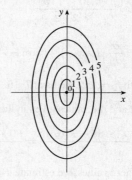

7.

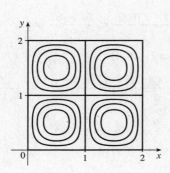

9. f is a rational function, so it is continuous on its domain. Since f is defined at $(1, 1)$, we use direct substitution to evaluate the limit:

$$\lim_{(x,y) \to (1,1)} \frac{2xy}{x^2 + 2y^2} = \frac{2(1)(1)}{1^2 + 2(1)^2} = \frac{2}{3}.$$

11. (a) $T_x(6, 4) = \lim\limits_{h \to 0} \dfrac{T(6 + h, 4) - T(6, 4)}{h}$, so we can approximate $T_x(6, 4)$ by considering $h = \pm 2$ and using the values given in the table: $T_x(6, 4) \approx \dfrac{T(8, 4) - T(6, 4)}{2} = \dfrac{86 - 80}{2} = 3$,

$T_x(6, 4) \approx \dfrac{T(4, 4) - T(6, 4)}{-2} = \dfrac{72 - 80}{-2} = 4$. Averaging these values, we estimate $T_x(6, 4)$ to be approximately

$3.5°\text{C/m}$. Similarly, $T_y(6, 4) = \lim\limits_{h \to 0} \dfrac{T(6, 4 + h) - T(6, 4)}{h}$, which we can approximate with $h = \pm 2$:

$T_y(6, 4) \approx \dfrac{T(6, 6) - T(6, 4)}{2} = \dfrac{75 - 80}{2} = -2.5$, $T_y(6, 4) \approx \dfrac{T(6, 2) - T(6, 4)}{-2} = \dfrac{87 - 80}{-2} = -3.5$. Averaging these

values, we estimate $T_y(6, 4)$ to be approximately $-3.0°\text{C/m}$.

(b) Here $\mathbf{u} = \left\langle \frac{1}{\sqrt{2}}, \frac{1}{\sqrt{2}} \right\rangle$, so by Equation 14.6.9, $D_{\mathbf{u}} T(6, 4) = \nabla T(6, 4) \cdot \mathbf{u} = T_x(6, 4) \frac{1}{\sqrt{2}} + T_y(6, 4) \frac{1}{\sqrt{2}}$. Using our

estimates from part (a), we have $D_{\mathbf{u}} T(6, 4) \approx (3.5) \frac{1}{\sqrt{2}} + (-3.0) \frac{1}{\sqrt{2}} = \frac{1}{2\sqrt{2}} \approx 0.35$. This means that as we move

through the point $(6, 4)$ in the direction of $\mathbf{u}$, the temperature increases at a rate of approximately $0.35°\text{C/m}$.

Alternatively, we can use Definition 14.6.2: $D_{\mathbf{u}} T(6, 4) = \lim\limits_{h \to 0} \dfrac{T\left(6 + h \frac{1}{\sqrt{2}}, 4 + h \frac{1}{\sqrt{2}}\right) - T(6, 4)}{h}$, which we can estimate with $h = \pm 2\sqrt{2}$. Then $D_{\mathbf{u}} T(6,4) \approx \dfrac{T(8,6) - T(6,4)}{2\sqrt{2}} = \dfrac{80 - 80}{2\sqrt{2}} = 0$,

$D_{\mathbf{u}} T(6,4) \approx \dfrac{T(4,2) - T(6,4)}{-2\sqrt{2}} = \dfrac{74 - 80}{-2\sqrt{2}} = \dfrac{3}{\sqrt{2}}$. Averaging these values, we have $D_{\mathbf{u}} T(6,4) \approx \dfrac{3}{2\sqrt{2}} \approx 1.1^\circ\text{C/m}$.

(c) $T_{xy}(x,y) = \dfrac{\partial}{\partial y}\left[T_x(x,y)\right] = \lim\limits_{h \to 0} \dfrac{T_x(x, y+h) - T_x(x,y)}{h}$, so $T_{xy}(6,4) = \lim\limits_{h \to 0} \dfrac{T_x(6, 4+h) - T_x(6,4)}{h}$ which we can

estimate with $h = \pm 2$. We have $T_x(6,4) \approx 3.5$ from part (a), but we will also need values for $T_x(6,6)$ and $T_x(6,2)$. If we

use $h = \pm 2$ and the values given in the table, we have

$T_x(6,6) \approx \dfrac{T(8,6) - T(6,6)}{2} = \dfrac{80 - 75}{2} = 2.5$, $T_x(6,6) \approx \dfrac{T(4,6) - T(6,6)}{-2} = \dfrac{68 - 75}{-2} = 3.5$.

Averaging these values, we estimate $T_x(6,6) \approx 3.0$. Similarly,

$T_x(6,2) \approx \dfrac{T(8,2) - T_x(6,2)}{2} = \dfrac{90 - 87}{2} = 1.5$, $T_x(6,2) \approx \dfrac{T(4,2) - T(6,2)}{-2} = \dfrac{74 - 87}{-2} = 6.5$.

Averaging these values, we estimate $T_x(6,2) \approx 4.0$. Finally, we estimate $T_{xy}(6,4)$:

$T_{xy}(6,4) \approx \dfrac{T_x(6,6) - T_x(6,4)}{2} = \dfrac{3.0 - 3.5}{2} = -0.25$, $T_{xy}(6,4) \approx \dfrac{T_x(6,2) - T_x(6,4)}{-2} = \dfrac{4.0 - 3.5}{-2} = -0.25$.

Averaging these values, we have $T_{xy}(6,4) \approx -0.25$.

13. $f(x,y) = (5y^3 + 2x^2 y)^8 \quad \Rightarrow \quad f_x = 8(5y^3 + 2x^2 y)^7 (4xy) = 32xy(5y^3 + 2x^2 y)^7$,

$f_y = 8(5y^3 + 2x^2 y)^7 (15y^2 + 2x^2) = (16x^2 + 120y^2)(5y^3 + 2x^2 y)^7$

15. $F(\alpha, \beta) = \alpha^2 \ln(\alpha^2 + \beta^2) \quad \Rightarrow \quad F_\alpha = \alpha^2 \cdot \dfrac{1}{\alpha^2 + \beta^2}(2\alpha) + \ln(\alpha^2 + \beta^2) \cdot 2\alpha = \dfrac{2\alpha^3}{\alpha^2 + \beta^2} + 2\alpha \ln(\alpha^2 + \beta^2)$,

$F_\beta = \alpha^2 \cdot \dfrac{1}{\alpha^2 + \beta^2}(2\beta) = \dfrac{2\alpha^2 \beta}{\alpha^2 + \beta^2}$

17. $S(u, v, w) = u \arctan(v\sqrt{w}) \quad \Rightarrow \quad S_u = \arctan(v\sqrt{w})$, $S_v = u \cdot \dfrac{1}{1 + (v\sqrt{w})^2}(\sqrt{w}) = \dfrac{u\sqrt{w}}{1 + v^2 w}$,

$S_w = u \cdot \dfrac{1}{1 + (v\sqrt{w})^2}\left(v \cdot \tfrac{1}{2} w^{-1/2}\right) = \dfrac{uv}{2\sqrt{w}\,(1 + v^2 w)}$

19. $f(x,y) = 4x^3 - xy^2 \quad \Rightarrow \quad f_x = 12x^2 - y^2$, $f_y = -2xy$, $f_{xx} = 24x$, $f_{yy} = -2x$, $f_{xy} = f_{yx} = -2y$

21. $f(x,y,z) = x^k y^l z^m \quad \Rightarrow \quad f_x = kx^{k-1} y^l z^m$, $f_y = lx^k y^{l-1} z^m$, $f_z = mx^k y^l z^{m-1}$, $f_{xx} = k(k-1)x^{k-2} y^l z^m$,

$f_{yy} = l(l-1)x^k y^{l-2} z^m$, $f_{zz} = m(m-1)x^k y^l z^{m-2}$, $f_{xy} = f_{yx} = klx^{k-1} y^{l-1} z^m$, $f_{xz} = f_{zx} = kmx^{k-1} y^l z^{m-1}$,

$f_{yz} = f_{zy} = lmx^k y^{l-1} z^{m-1}$

23. $z = xy + xe^{y/x} \quad \Rightarrow \quad \dfrac{\partial z}{\partial x} = y - \dfrac{y}{x} e^{y/x} + e^{y/x}$, $\dfrac{\partial z}{\partial y} = x + e^{y/x}$ and

$x \dfrac{\partial z}{\partial x} + y \dfrac{\partial z}{\partial y} = x\left(y - \dfrac{y}{x} e^{y/x} + e^{y/x}\right) + y\left(x + e^{y/x}\right) = xy - ye^{y/x} + xe^{y/x} + xy + ye^{y/x} = xy + xy + xe^{y/x} = xy + z$.

25. (a) $z_x = 6x + 2 \Rightarrow z_x(1, -2) = 8$ and $z_y = -2y \Rightarrow z_y(1, -2) = 4$, so an equation of the tangent plane is

$z - 1 = 8(x - 1) + 4(y + 2)$ or $z = 8x + 4y + 1$.

(b) A normal vector to the tangent plane (and the surface) at $(1, -2, 1)$ is $\langle 8, 4, -1 \rangle$. Then parametric equations for the normal

line there are $x = 1 + 8t$, $y = -2 + 4t$, $z = 1 - t$, and symmetric equations are $\dfrac{x - 1}{8} = \dfrac{y + 2}{4} = \dfrac{z - 1}{-1}$.

27. (a) Let $F(x, y, z) = x^2 + 2y^2 - 3z^2$. Then $F_x = 2x$, $F_y = 4y$, $F_z = -6z$, so $F_x(2, -1, 1) = 4$, $F_y(2, -1, 1) = -4$,

$F_z(2, -1, 1) = -6$. From Equation 14.6.19, an equation of the tangent plane is $4(x - 2) - 4(y + 1) - 6(z - 1) = 0$

or, equivalently, $2x - 2y - 3z = 3$.

(b) From Equations 14.6.20, symmetric equations for the normal line are $\dfrac{x - 2}{4} = \dfrac{y + 1}{-4} = \dfrac{z - 1}{-6}$.

29. (a) Let $F(x, y, z) = x + 2y + 3z - \sin(xyz)$. Then $F_x = 1 - yz\cos(xyz)$, $F_y = 2 - xz\cos(xyz)$, $F_z = 3 - xy\cos(xyz)$,

so $F_x(2, -1, 0) = 1$, $F_y(2, -1, 0) = 2$, $F_z(2, -1, 0) = 5$. From Equation 14.6.19, an equation of the tangent plane is

$1(x - 2) + 2(y + 1) + 5(z - 0) = 0$ or $x + 2y + 5z = 0$.

(b) From Equations 14.6.20, symmetric equations for the normal line are $\dfrac{x - 2}{1} = \dfrac{y + 1}{2} = \dfrac{z}{5}$ or $x - 2 = \dfrac{y + 1}{2} = \dfrac{z}{5}$.

Parametric equations are $x = 2 + t$, $y = -1 + 2t$, $z = 5t$.

31. The hyperboloid is a level surface of the function $F(x, y, z) = x^2 + 4y^2 - z^2$, so a normal vector to the surface at (x_0, y_0, z_0)

is $\nabla F(x_0, y_0, z_0) = \langle 2x_0, 8y_0, -2z_0 \rangle$. A normal vector for the plane $2x + 2y + z = 5$ is $\langle 2, 2, 1 \rangle$. For the planes to be

parallel, we need the normal vectors to be parallel, so $\langle 2x_0, 8y_0, -2z_0 \rangle = k\langle 2, 2, 1 \rangle$, or $x_0 = k$, $y_0 = \frac{1}{4}k$, and $z_0 = -\frac{1}{2}k$.

But $x_0^2 + 4y_0^2 - z_0^2 = 4 \Rightarrow k^2 + \frac{1}{4}k^2 - \frac{1}{4}k^2 = 4 \Rightarrow k^2 = 4 \Rightarrow k = \pm 2$. So there are two such points:

$(2, \frac{1}{2}, -1)$ and $(-2, -\frac{1}{2}, 1)$.

33. $f(x, y, z) = x^3\sqrt{y^2 + z^2} \Rightarrow f_x(x, y, z) = 3x^2\sqrt{y^2 + z^2}$, $f_y(x, y, z) = \dfrac{yx^3}{\sqrt{y^2 + z^2}}$, $f_z(x, y, z) = \dfrac{zx^3}{\sqrt{y^2 + z^2}}$,

so $f(2, 3, 4) = 8(5) = 40$, $f_x(2, 3, 4) = 3(4)\sqrt{25} = 60$, $f_y(2, 3, 4) = \dfrac{3(8)}{\sqrt{25}} = \dfrac{24}{5}$, and $f_z(2, 3, 4) = \dfrac{4(8)}{\sqrt{25}} = \dfrac{32}{5}$. Then the

linear approximation of f at $(2, 3, 4)$ is

$$f(x, y, z) \approx f(2, 3, 4) + f_x(2, 3, 4)(x - 2) + f_y(2, 3, 4)(y - 3) + f_z(2, 3, 4)(z - 4)$$

$$= 40 + 60(x - 2) + \tfrac{24}{5}(y - 3) + \tfrac{32}{5}(z - 4) = 60x + \tfrac{24}{5}y + \tfrac{32}{5}z - 120$$

Then $(1.98)^3\sqrt{(3.01)^2 + (3.97)^2} = f(1.98, 3.01, 3.97) \approx 60(1.98) + \tfrac{24}{5}(3.01) + \tfrac{32}{5}(3.97) - 120 = 38.656$.

35. $\dfrac{du}{dp} = \dfrac{\partial u}{\partial x}\dfrac{dx}{dp} + \dfrac{\partial u}{\partial y}\dfrac{dy}{dp} + \dfrac{\partial u}{\partial z}\dfrac{dz}{dp} = 2xy^3(1 + 6p) + 3x^2y^2(pe^p + e^p) + 4z^3(p\cos p + \sin p)$

37. By the Chain Rule, $\dfrac{\partial z}{\partial s} = \dfrac{\partial z}{\partial x}\dfrac{\partial x}{\partial s} + \dfrac{\partial z}{\partial y}\dfrac{\partial y}{\partial s}$. When $s = 1$ and $t = 2$, $x = g(1, 2) = 3$ and $y = h(1, 2) = 6$, so

$\dfrac{\partial z}{\partial s} = f_x(3, 6)g_s(1, 2) + f_y(3, 6)h_s(1, 2) = (7)(-1) + (8)(-5) = -47$. Similarly, $\dfrac{\partial z}{\partial t} = \dfrac{\partial z}{\partial x}\dfrac{\partial x}{\partial t} + \dfrac{\partial z}{\partial y}\dfrac{\partial y}{\partial t}$, so

$\dfrac{\partial z}{\partial t} = f_x(3, 6)g_t(1, 2) + f_y(3, 6)h_t(1, 2) = (7)(4) + (8)(10) = 108$.

39. $\dfrac{\partial z}{\partial x} = 2xf'(x^2 - y^2)$, $\quad \dfrac{\partial z}{\partial y} = 1 - 2yf'(x^2 - y^2)$ $\quad \left[\text{where } f' = \dfrac{df}{d(x^2 - y^2)}\right]$. Then

$y\dfrac{\partial z}{\partial x} + x\dfrac{\partial z}{\partial y} = 2xyf'(x^2 - y^2) + x - 2xyf'(x^2 - y^2) = x$.

41. $\dfrac{\partial z}{\partial x} = \dfrac{\partial z}{\partial u}y + \dfrac{\partial z}{\partial v}\dfrac{-y}{x^2}$ and

$$\dfrac{\partial^2 z}{\partial x^2} = y\dfrac{\partial}{\partial x}\left(\dfrac{\partial z}{\partial u}\right) + \dfrac{2y}{x^3}\dfrac{\partial z}{\partial v} + \dfrac{-y}{x^2}\dfrac{\partial}{\partial x}\left(\dfrac{\partial z}{\partial v}\right) = \dfrac{2y}{x^3}\dfrac{\partial z}{\partial v} + y\left(\dfrac{\partial^2 z}{\partial u^2}y + \dfrac{\partial^2 z}{\partial v\,\partial u}\dfrac{-y}{x^2}\right) + \dfrac{-y}{x^2}\left(\dfrac{\partial^2 z}{\partial v^2}\dfrac{-y}{x^2} + \dfrac{\partial^2 z}{\partial u\,\partial v}y\right)$$

$$= \dfrac{2y}{x^3}\dfrac{\partial z}{\partial v} + y^2\dfrac{\partial^2 z}{\partial u^2} - \dfrac{2y^2}{x^2}\dfrac{\partial^2 z}{\partial u\,\partial v} + \dfrac{y^2}{x^4}\dfrac{\partial^2 z}{\partial v^2}$$

Also $\dfrac{\partial z}{\partial y} = x\dfrac{\partial z}{\partial u} + \dfrac{1}{x}\dfrac{\partial z}{\partial v}$ and

$$\dfrac{\partial^2 z}{\partial y^2} = x\dfrac{\partial}{\partial y}\left(\dfrac{\partial z}{\partial u}\right) + \dfrac{1}{x}\dfrac{\partial}{\partial y}\left(\dfrac{\partial z}{\partial v}\right) = x\left(\dfrac{\partial^2 z}{\partial u^2}x + \dfrac{\partial^2 z}{\partial v\,\partial u}\dfrac{1}{x}\right) + \dfrac{1}{x}\left(\dfrac{\partial^2 z}{\partial v^2}\dfrac{1}{x} + \dfrac{\partial^2 z}{\partial u\,\partial v}x\right) = x^2\dfrac{\partial^2 z}{\partial u^2} + 2\dfrac{\partial^2 z}{\partial u\,\partial v} + \dfrac{1}{x^2}\dfrac{\partial^2 z}{\partial v^2}$$

Thus

$$x^2\dfrac{\partial^2 z}{\partial x^2} - y^2\dfrac{\partial^2 z}{\partial y^2} = \dfrac{2y}{x}\dfrac{\partial z}{\partial v} + x^2y^2\dfrac{\partial^2 z}{\partial u^2} - 2y^2\dfrac{\partial^2 z}{\partial u\,\partial v} + \dfrac{y^2}{x^2}\dfrac{\partial^2 z}{\partial v^2} - x^2y^2\dfrac{\partial^2 z}{\partial u^2} - 2y^2\dfrac{\partial^2 z}{\partial u\,\partial v} - \dfrac{y^2}{x^2}\dfrac{\partial^2 z}{\partial v^2}$$

$$= \dfrac{2y}{x}\dfrac{\partial z}{\partial v} - 4y^2\dfrac{\partial^2 z}{\partial u\,\partial v} = 2v\dfrac{\partial z}{\partial v} - 4uv\dfrac{\partial^2 z}{\partial u\,\partial v}$$

since $y = xv = \dfrac{uv}{y}$ or $y^2 = uv$.

43. $f(x, y, z) = x^2 e^{yz^2} \quad \Rightarrow \quad \nabla f = \langle f_x, f_y, f_z\rangle = \left\langle 2xe^{yz^2}, x^2 e^{yz^2}\cdot z^2, x^2 e^{yz^2}\cdot 2yz\right\rangle = \left\langle 2xe^{yz^2}, x^2 z^2 e^{yz^2}, 2x^2 yz e^{yz^2}\right\rangle$

45. $f(x, y) = x^2 e^{-y} \quad \Rightarrow \quad \nabla f = \langle 2xe^{-y}, -x^2 e^{-y}\rangle$, $\nabla f(-2, 0) = \langle -4, -4\rangle$. The direction is given by $\langle 4, -3\rangle$, so

$\mathbf{u} = \dfrac{1}{\sqrt{4^2 + (-3)^2}}\langle 4, -3\rangle = \tfrac{1}{5}\langle 4, -3\rangle$ and $D_{\mathbf{u}}f(-2, 0) = \nabla f(-2, 0)\cdot\mathbf{u} = \langle -4, -4\rangle\cdot\tfrac{1}{5}\langle 4, -3\rangle = \tfrac{1}{5}(-16 + 12) = -\tfrac{4}{5}$.

47. $\nabla f = \left\langle 2xy, x^2 + 1/(2\sqrt{y})\right\rangle$, $|\nabla f(2, 1)| = \left|\left\langle 4, \tfrac{9}{2}\right\rangle\right|$. Thus the maximum rate of change of f at $(2, 1)$ is $\dfrac{\sqrt{145}}{2}$ in the

direction $\left\langle 4, \tfrac{9}{2}\right\rangle$.

49. First we draw a line passing through Homestead and the eye of the hurricane. We can approximate the directional derivative at

Homestead in the direction of the eye of the hurricane by the average rate of change of wind speed between the points where this line intersects the contour lines closest to Homestead. In the direction of the eye of the hurricane, the wind speed changes

from 45 to 50 knots. We estimate the distance between these two points to be approximately 8 miles, so the rate of change of

wind speed in the direction given is approximately $\frac{50-45}{8} = \frac{5}{8} = 0.625$ knot/mi.

51. $f(x,y) = x^2 - xy + y^2 + 9x - 6y + 10 \Rightarrow f_x = 2x - y + 9,$

$f_y = -x + 2y - 6,\ f_{xx} = 2 = f_{yy},\ f_{xy} = -1.$ Then $f_x = 0$ and $f_y = 0$ imply

$y = 1,\ x = -4.$ Thus the only critical point is $(-4, 1)$ and $f_{xx}(-4, 1) > 0,$

$D(-4, 1) = 3 > 0,$ so $f(-4, 1) = -11$ is a local minimum.

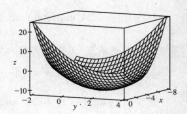

53. $f(x,y) = 3xy - x^2y - xy^2 \Rightarrow f_x = 3y - 2xy - y^2,\ f_y = 3x - x^2 - 2xy,$

$f_{xx} = -2y,\ f_{yy} = -2x,\ f_{xy} = 3 - 2x - 2y.$ Then $f_x = 0$ implies

$y(3 - 2x - y) = 0$ so $y = 0$ or $y = 3 - 2x.$ Substituting into $f_y = 0$ implies

$x(3 - x) = 0$ or $3x(-1 + x) = 0.$ Hence the critical points are $(0, 0),\ (3, 0),$

$(0, 3)$ and $(1, 1).$ $D(0, 0) = D(3, 0) = D(0, 3) = -9 < 0$ so $(0, 0),\ (3, 0),$ and

$(0, 3)$ are saddle points. $D(1, 1) = 3 > 0$ and $f_{xx}(1, 1) = -2 < 0,$ so

$f(1, 1) = 1$ is a local maximum.

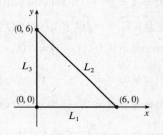

55. First solve inside $D.$ Here $f_x = 4y^2 - 2xy^2 - y^3,\ f_y = 8xy - 2x^2y - 3xy^2.$

Then $f_x = 0$ implies $y = 0$ or $y = 4 - 2x,$ but $y = 0$ isn't inside $D.$ Substituting

$y = 4 - 2x$ into $f_y = 0$ implies $x = 0,\ x = 2$ or $x = 1,$ but $x = 0$ isn't inside $D,$

and when $x = 2,\ y = 0$ but $(2, 0)$ isn't inside $D.$ Thus the only critical point inside

D is $(1, 2)$ and $f(1, 2) = 4.$ Secondly we consider the boundary of $D.$

On L_1: $f(x, 0) = 0$ and so $f = 0$ on $L_1.$ On L_2: $x = -y + 6$ and

$f(-y + 6, y) = y^2(6 - y)(-2) = -2(6y^2 - y^3)$ which has critical points

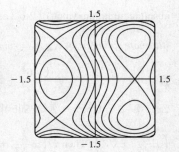

at $y = 0$ and $y = 4.$ Then $f(6, 0) = 0$ while $f(2, 4) = -64.$ On L_3: $f(0, y) = 0,$ so $f = 0$ on $L_3.$ Thus on D the absolute

maximum of f is $f(1, 2) = 4$ while the absolute minimum is $f(2, 4) = -64.$

57. $f(x,y) = x^3 - 3x + y^4 - 2y^2$

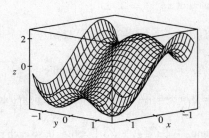

From the graphs, it appears that f has a local maximum $f(-1, 0) \approx 2,$ local minima $f(1, \pm1) \approx -3,$ and saddle points at

$(-1, \pm1)$ and $(1, 0).$

To find the exact quantities, we calculate $f_x = 3x^2 - 3 = 0 \iff x = \pm 1$ and $f_y = 4y^3 - 4y = 0 \iff$
$y = 0, \pm 1$, giving the critical points estimated above. Also $f_{xx} = 6x$, $f_{xy} = 0$, $f_{yy} = 12y^2 - 4$, so using the Second
Derivatives Test, $D(-1, 0) = 24 > 0$ and $f_{xx}(-1, 0) = -6 < 0$ indicating a local maximum $f(-1, 0) = 2$;
$D(1, \pm 1) = 48 > 0$ and $f_{xx}(1, \pm 1) = 6 > 0$ indicating local minima $f(1, \pm 1) = -3$; and $D(-1, \pm 1) = -48$ and
$D(1, 0) = -24$, indicating saddle points.

59. $f(x, y) = x^2 y$, $g(x, y) = x^2 + y^2 = 1 \Rightarrow \nabla f = \langle 2xy, x^2 \rangle = \lambda \nabla g = \langle 2\lambda x, 2\lambda y \rangle$. Then $2xy = 2\lambda x$ implies $x = 0$ or
$y = \lambda$. If $x = 0$ then $x^2 + y^2 = 1$ gives $y = \pm 1$ and we have possible points $(0, \pm 1)$ where $f(0, \pm 1) = 0$. If $y = \lambda$ then
$x^2 = 2\lambda y$ implies $x^2 = 2y^2$ and substitution into $x^2 + y^2 = 1$ gives $3y^2 = 1 \Rightarrow y = \pm\frac{1}{\sqrt{3}}$ and $x = \pm\sqrt{\frac{2}{3}}$. The
corresponding possible points are $\left(\pm\sqrt{\frac{2}{3}}, \pm\frac{1}{\sqrt{3}}\right)$. The absolute maximum is $f\left(\pm\sqrt{\frac{2}{3}}, \frac{1}{\sqrt{3}}\right) = \frac{2}{3\sqrt{3}}$ while the absolute
minimum is $f\left(\pm\sqrt{\frac{2}{3}}, -\frac{1}{\sqrt{3}}\right) = -\frac{2}{3\sqrt{3}}$.

61. $f(x, y, z) = xyz$, $g(x, y, z) = x^2 + y^2 + z^2 = 3$. $\nabla f = \lambda \nabla g \Rightarrow \langle yz, xz, xy \rangle = \lambda \langle 2x, 2y, 2z \rangle$. If any of x, y, or z is
zero, then $x = y = z = 0$ which contradicts $x^2 + y^2 + z^2 = 3$. Then $\lambda = \frac{yz}{2x} = \frac{xz}{2y} = \frac{xy}{2z} \Rightarrow 2y^2 z = 2x^2 z \Rightarrow$
$y^2 = x^2$, and similarly $2yz^2 = 2x^2 y \Rightarrow z^2 = x^2$. Substituting into the constraint equation gives $x^2 + x^2 + x^2 = 3 \Rightarrow$
$x^2 = 1 = y^2 = z^2$. Thus the possible points are $(1, 1, \pm 1)$, $(1, -1, \pm 1)$, $(-1, 1, \pm 1)$, $(-1, -1, \pm 1)$. The absolute maximum
is $f(1, 1, 1) = f(1, -1, -1) = f(-1, 1, -1) = f(-1, -1, 1) = 1$ and the absolute
minimum is $f(1, 1, -1) = f(1, -1, 1) = f(-1, 1, 1) = f(-1, -1, -1) = -1$.

63. $f(x, y, z) = x^2 + y^2 + z^2$, $g(x, y, z) = xy^2 z^3 = 2 \Rightarrow \nabla f = \langle 2x, 2y, 2z \rangle = \lambda \nabla g = \langle \lambda y^2 z^3, 2\lambda xyz^3, 3\lambda xy^2 z^2 \rangle$.
Since $xy^2 z^3 = 2$, $x \neq 0$, $y \neq 0$ and $z \neq 0$, so $2x = \lambda y^2 z^3$ **(1)**, $1 = \lambda xz^3$ **(2)**, $2 = 3\lambda xy^2 z$ **(3)**. Then **(2)** and **(3)** imply
$\frac{1}{xz^3} = \frac{2}{3xy^2 z}$ or $y^2 = \frac{2}{3}z^2$ so $y = \pm z\sqrt{\frac{2}{3}}$. Similarly **(1)** and **(3)** imply $\frac{2x}{y^2 z^3} = \frac{2}{3xy^2 z}$ or $3x^2 = z^2$ so $x = \pm\frac{1}{\sqrt{3}}z$. But
$xy^2 z^3 = 2$ so x and z must have the same sign, that is, $x = \frac{1}{\sqrt{3}}z$. Thus $g(x, y, z) = 2$ implies $\frac{1}{\sqrt{3}}z\left(\frac{2}{3}z^2\right)z^3 = 2$ or
$z = \pm 3^{1/4}$ and the possible points are $(\pm 3^{-1/4}, 3^{-1/4}\sqrt{2}, \pm 3^{1/4})$, $(\pm 3^{-1/4}, -3^{-1/4}\sqrt{2}, \pm 3^{1/4})$. However at each of these
points f takes on the same value, $2\sqrt{3}$. But $(2, 1, 1)$ also satisfies $g(x, y, z) = 2$ and $f(2, 1, 1) = 6 > 2\sqrt{3}$. Thus f has an
absolute minimum value of $2\sqrt{3}$ and no absolute maximum subject to the constraint $xy^2 z^3 = 2$.

Alternate solution: $g(x, y, z) = xy^2 z^3 = 2$ implies $y^2 = \frac{2}{xz^3}$, so minimize $f(x, z) = x^2 + \frac{2}{xz^3} + z^2$. Then
$f_x = 2x - \frac{2}{x^2 z^3}$, $f_z = -\frac{6}{xz^4} + 2z$, $f_{xx} = 2 + \frac{4}{x^3 z^3}$, $f_{zz} = \frac{24}{xz^5} + 2$ and $f_{xz} = \frac{6}{x^2 z^4}$. Now $f_x = 0$ implies
$2x^3 z^3 - 2 = 0$ or $z = 1/x$. Substituting into $f_y = 0$ implies $-6x^3 + 2x^{-1} = 0$ or $x = \frac{1}{\sqrt[4]{3}}$, so the two critical points are
$\left(\pm\frac{1}{\sqrt[4]{3}}, \pm\sqrt[4]{3}\right)$. Then $D\left(\pm\frac{1}{\sqrt[4]{3}}, \pm\sqrt[4]{3}\right) = (2 + 4)\left(2 + \frac{24}{3}\right) - \left(\frac{6}{\sqrt{3}}\right)^2 > 0$ and $f_{xx}\left(\pm\frac{1}{\sqrt[4]{3}}, \pm\sqrt[4]{3}\right) = 6 > 0$, so each point
is a minimum. Finally, $y^2 = \frac{2}{xz^3}$, so the four points closest to the origin are $\left(\pm\frac{1}{\sqrt[4]{3}}, \frac{\sqrt{2}}{\sqrt[4]{3}}, \pm\sqrt[4]{3}\right)$, $\left(\pm\frac{1}{\sqrt[4]{3}}, -\frac{\sqrt{2}}{\sqrt[4]{3}}, \pm\sqrt[4]{3}\right)$.

65.

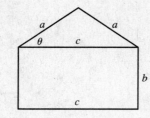

The area of the triangle is $\frac{1}{2}ca\sin\theta$ and the area of the rectangle is bc. Thus, the area of the whole object is $f(a, b, c) = \frac{1}{2}ca\sin\theta + bc$. The perimeter of the object is $g(a, b, c) = 2a + 2b + c = P$. To simplify $\sin\theta$ in terms of a, b, and c notice that $a^2\sin^2\theta + \left(\frac{1}{2}c\right)^2 = a^2 \Rightarrow \sin\theta = \frac{1}{2a}\sqrt{4a^2 - c^2}$.

Thus $f(a, b, c) = \frac{c}{4}\sqrt{4a^2 - c^2} + bc$. (Instead of using θ, we could just have used the Pythagorean Theorem.) As a result, by Lagrange's method, we must find a, b, c, and λ by solving $\nabla f = \lambda\nabla g$ which gives the following equations: $ca(4a^2 - c^2)^{-1/2} = 2\lambda$ **(1)**, $c = 2\lambda$ **(2)**, $\frac{1}{4}(4a^2 - c^2)^{1/2} - \frac{1}{4}c^2(4a^2 - c^2)^{-1/2} + b = \lambda$ **(3)**, and $2a + 2b + c = P$ **(4)**. From **(2)**, $\lambda = \frac{1}{2}c$ and so **(1)** produces $ca(4a^2 - c^2)^{-1/2} = c \Rightarrow (4a^2 - c^2)^{1/2} = a \Rightarrow$

$4a^2 - c^2 = a^2 \Rightarrow c = \sqrt{3}\,a$ **(5)**. Similarly, since $(4a^2 - c^2)^{1/2} = a$ and $\lambda = \frac{1}{2}c$, **(3)** gives $\frac{a}{4} - \frac{c^2}{4a} + b = \frac{c}{2}$, so from

(5), $\frac{a}{4} - \frac{3a}{4} + b = \frac{\sqrt{3}\,a}{2} \Rightarrow -\frac{a}{2} - \frac{\sqrt{3}\,a}{2} = -b \Rightarrow b = \frac{a}{2}\left(1 + \sqrt{3}\right)$ **(6)**. Substituting **(5)** and **(6)** into **(4)** we get:

$2a + a\left(1 + \sqrt{3}\right) + \sqrt{3}\,a = P \Rightarrow 3a + 2\sqrt{3}\,a = P \Rightarrow a = \frac{P}{3 + 2\sqrt{3}} = \frac{2\sqrt{3} - 3}{3}P$ and thus

$b = \frac{\left(2\sqrt{3} - 3\right)\left(1 + \sqrt{3}\right)}{6}P = \frac{3 - \sqrt{3}}{6}P$ and $c = \left(2 - \sqrt{3}\right)P$.

PROBLEMS PLUS

1. The areas of the smaller rectangles are $A_1 = xy$, $A_2 = (L-x)y$,

$A_3 = (L-x)(W-y)$, $A_4 = x(W-y)$. For $0 \leq x \leq L, 0 \leq y \leq W$, let

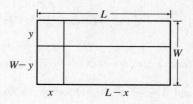

$$
\begin{aligned}
f(x,y) &= A_1^2 + A_2^2 + A_3^2 + A_4^2 \\
&= x^2y^2 + (L-x)^2y^2 + (L-x)^2(W-y)^2 + x^2(W-y)^2 \\
&= [x^2 + (L-x)^2][y^2 + (W-y)^2]
\end{aligned}
$$

Then we need to find the maximum and minimum values of $f(x,y)$. Here

$f_x(x,y) = [2x - 2(L-x)][y^2 + (W-y)^2] = 0 \quad \Rightarrow \quad 4x - 2L = 0$ or $x = \frac{1}{2}L$, and

$f_y(x,y) = [x^2 + (L-x)^2][2y - 2(W-y)] = 0 \quad \Rightarrow \quad 4y - 2W = 0$ or $y = W/2$. Also

$f_{xx} = 4[y^2 + (W-y)^2]$, $f_{yy} = 4[x^2 + (L-x)^2]$, and $f_{xy} = (4x - 2L)(4y - 2W)$. Then

$D = 16[y^2 + (W-y)^2][x^2 + (L-x)^2] - (4x - 2L)^2(4y - 2W)^2$. Thus when $x = \frac{1}{2}L$ and $y = \frac{1}{2}W$, $D > 0$ and

$f_{xx} = 2W^2 > 0$. Thus a minimum of f occurs at $\left(\frac{1}{2}L, \frac{1}{2}W\right)$ and this minimum value is $f\left(\frac{1}{2}L, \frac{1}{2}W\right) = \frac{1}{4}L^2W^2$.

There are no other critical points, so the maximum must occur on the boundary. Now along the width of the rectangle let

$g(y) = f(0,y) = f(L,y) = L^2[y^2 + (W-y)^2]$, $0 \leq y \leq W$. Then $g'(y) = L^2[2y - 2(W-y)] = 0 \quad \Leftrightarrow \quad y = \frac{1}{2}W$.

And $g\left(\frac{1}{2}\right) = \frac{1}{2}L^2W^2$. Checking the endpoints, we get $g(0) = g(W) = L^2W^2$. Along the length of the rectangle let

$h(x) = f(x,0) = f(x,W) = W^2[x^2 + (L-x)^2]$, $0 \leq x \leq L$. By symmetry $h'(x) = 0 \quad \Leftrightarrow \quad x = \frac{1}{2}L$ and

$h\left(\frac{1}{2}L\right) = \frac{1}{2}L^2W^2$. At the endpoints we have $h(0) = h(L) = L^2W^2$. Therefore L^2W^2 is the maximum value of f.

This maximum value of f occurs when the "cutting" lines correspond to sides of the rectangle.

3. (a) The area of a trapezoid is $\frac{1}{2}h(b_1 + b_2)$, where h is the height (the distance between the two parallel sides) and b_1, b_2 are

the lengths of the bases (the parallel sides). From the figure in the text, we see that $h = x \sin \theta$, $b_1 = w - 2x$, and

$b_2 = w - 2x + 2x \cos \theta$. Therefore the cross-sectional area of the rain gutter is

$$
\begin{aligned}
A(x,\theta) &= \frac{1}{2}x \sin \theta \, [(w - 2x) + (w - 2x + 2x \cos \theta)] = (x \sin \theta)(w - 2x + x \cos \theta) \\
&= wx \sin \theta - 2x^2 \sin \theta + x^2 \sin \theta \cos \theta, \quad 0 < x \leq \tfrac{1}{2}w, 0 < \theta \leq \tfrac{\pi}{2}
\end{aligned}
$$

We look for the critical points of A: $\partial A / \partial x = w \sin \theta - 4x \sin \theta + 2x \sin \theta \cos \theta$ and

$\partial A / \partial \theta = wx \cos \theta - 2x^2 \cos \theta + x^2(\cos^2 \theta - \sin^2 \theta)$, so $\partial A / \partial x = 0 \quad \Leftrightarrow \quad \sin \theta \, (w - 4x + 2x \cos \theta) = 0 \quad \Leftrightarrow$

$\cos \theta = \dfrac{4x - w}{2x} = 2 - \dfrac{w}{2x} \quad (0 < \theta \leq \tfrac{\pi}{2} \quad \Rightarrow \quad \sin \theta > 0)$. If, in addition, $\partial A / \partial \theta = 0$, then

$$0 = wx \cos\theta - 2x^2 \cos\theta + x^2(2\cos^2\theta - 1)$$

$$= wx\left(2 - \frac{w}{2x}\right) - 2x^2\left(2 - \frac{w}{2x}\right) + x^2\left[2\left(2 - \frac{w}{2x}\right)^2 - 1\right]$$

$$= 2wx - \tfrac{1}{2}w^2 - 4x^2 + wx + x^2\left[8 - \frac{4w}{x} + \frac{w^2}{2x^2} - 1\right] = -wx + 3x^2 = x(3x - w)$$

Since $x > 0$, we must have $x = \frac{1}{3}w$, in which case $\cos\theta = \frac{1}{2}$, so $\theta = \frac{\pi}{3}$, $\sin\theta = \frac{\sqrt{3}}{2}$, $k = \frac{\sqrt{3}}{6}w$, $b_1 = \frac{1}{3}w$, $b_2 = \frac{2}{3}w$,

and $A = \frac{\sqrt{3}}{12}w^2$. As in Example 14.7.6, we can argue from the physical nature of this problem that we have found a local

maximum of A. Now checking the boundary of A, let

$g(\theta) = A(w/2, \theta) = \frac{1}{2}w^2\sin\theta - \frac{1}{2}w^2\sin\theta + \frac{1}{4}w^2\sin\theta\cos\theta = \frac{1}{8}w^2\sin 2\theta$, $0 < \theta \le \frac{\pi}{2}$. Clearly g is maximized when

$\sin 2\theta = 1$ in which case $A = \frac{1}{8}w^2$. Also along the line $\theta = \frac{\pi}{2}$, let $h(x) = A\left(x, \frac{\pi}{2}\right) = wx - 2x^2$, $0 < x < \frac{1}{2}w$ $\Rightarrow$

$h'(x) = w - 4x = 0$ $\Leftrightarrow$ $x = \frac{1}{4}w$, and $h\left(\frac{1}{4}w\right) = w\left(\frac{1}{4}w\right) - 2\left(\frac{1}{4}w\right)^2 = \frac{1}{8}w^2$. Since $\frac{1}{8}w^2 < \frac{\sqrt{3}}{12}w^2$, we conclude that

the local maximum found earlier was an absolute maximum.

(b) If the metal were bent into a semi-circular gutter of radius r, we would have $w = \pi r$ and $A = \frac{1}{2}\pi r^2 = \frac{1}{2}\pi\left(\frac{w}{\pi}\right)^2 = \frac{w^2}{2\pi}$.

Since $\frac{w^2}{2\pi} > \frac{\sqrt{3}\,w^2}{12}$, it *would* be better to bend the metal into a gutter with a semicircular cross-section.

5. Let $g(x, y) = xf\left(\frac{y}{x}\right)$. Then $g_x(x, y) = f\left(\frac{y}{x}\right) + xf'\left(\frac{y}{x}\right)\left(-\frac{y}{x^2}\right) = f\left(\frac{y}{x}\right) - \frac{y}{x}f'\left(\frac{y}{x}\right)$ and

$g_y(x, y) = xf'\left(\frac{y}{x}\right)\left(\frac{1}{x}\right) = f'\left(\frac{y}{x}\right)$. Thus the tangent plane at (x_0, y_0, z_0) on the surface has equation

$$z - x_0 f\left(\frac{y_0}{x_0}\right) = \left[f\left(\frac{y_0}{x_0}\right) - y_0 x_0^{-1} f'\left(\frac{y_0}{x_0}\right)\right](x - x_0) + f'\left(\frac{y_0}{x_0}\right)(y - y_0) \quad \Rightarrow$$

$\left[f\left(\frac{y_0}{x_0}\right) - y_0 x_0^{-1} f'\left(\frac{y_0}{x_0}\right)\right]x + \left[f'\left(\frac{y_0}{x_0}\right)\right]y - z = 0$. But any plane whose equation is of the form $ax + by + cz = 0$

passes through the origin. Thus the origin is the common point of intersection.

7. Since we are minimizing the area of the ellipse, and the circle lies above the x-axis,

the ellipse will intersect the circle for only one value of y. This y-value must

satisfy both the equation of the circle and the equation of the ellipse. Now

$\frac{x^2}{a^2} + \frac{y^2}{b^2} = 1$ $\Rightarrow$ $x^2 = \frac{a^2}{b^2}(b^2 - y^2)$. Substituting into the equation of the

circle gives $\frac{a^2}{b^2}(b^2 - y^2) + y^2 - 2y = 0$ $\Rightarrow$ $\left(\frac{b^2 - a^2}{b^2}\right)y^2 - 2y + a^2 = 0$.

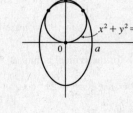

In order for there to be only one solution to this quadratic equation, the discriminant must be 0, so $4 - 4a^2\dfrac{b^2 - a^2}{b^2} = 0$ $\Rightarrow$

$b^2 - a^2b^2 + a^4 = 0$. The area of the ellipse is $A(a, b) = \pi ab$, and we minimize this function subject to the constraint

$g(a, b) = b^2 - a^2b^2 + a^4 = 0$.

Now $\nabla A = \lambda \nabla g \iff \pi b = \lambda(4a^3 - 2ab^2), \pi a = \lambda(2b - 2ba^2) \implies \lambda = \dfrac{\pi b}{2a(2a^2 - b^2)}$ **(1)**,

$\lambda = \dfrac{\pi a}{2b(1 - a^2)}$ **(2)**, $b^2 - a^2b^2 + a^4 = 0$ **(3)**. Comparing **(1)** and **(2)** gives $\dfrac{\pi b}{2a(2a^2 - b^2)} = \dfrac{\pi a}{2b(1 - a^2)} \implies$

$2\pi b^2 = 4\pi a^4 \iff a^2 = \dfrac{1}{\sqrt{2}}\, b$. Substitute this into **(3)** to get $b = \dfrac{3}{\sqrt{2}} \implies a = \sqrt{\dfrac{3}{2}}$.

15 ◻ MULTIPLE INTEGRALS

15.1 Double Integrals over Rectangles

1. (a) The subrectangles are shown in the figure.

The surface is the graph of $f(x, y) = xy$ and $\Delta A = 4$, so we estimate

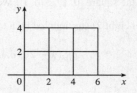

$$V \approx \sum_{i=1}^{3} \sum_{j=1}^{2} f(x_i, y_j)\, \Delta A$$

$$= f(2,2)\,\Delta A + f(2,4)\,\Delta A + f(4,2)\,\Delta A + f(4,4)\,\Delta A + f(6,2)\,\Delta A + f(6,4)\,\Delta A$$

$$= 4(4) + 8(4) + 8(4) + 16(4) + 12(4) + 24(4) = 288$$

(b) $V \approx \sum_{i=1}^{3} \sum_{j=1}^{2} f(\overline{x}_i, \overline{y}_j)\, \Delta A = f(1,1)\,\Delta A + f(1,3)\,\Delta A + f(3,1)\,\Delta A + f(3,3)\,\Delta A + f(5,1)\,\Delta A + f(5,3)\,\Delta A$

$$= 1(4) + 3(4) + 3(4) + 9(4) + 5(4) + 15(4) = 144$$

3. (a) The subrectangles are shown in the figure. Since $\Delta A = 1 \cdot \frac{1}{2} = \frac{1}{2}$, we estimate

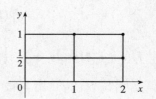

$$\iint_R xe^{-xy}\, dA \approx \sum_{i=1}^{2} \sum_{j=1}^{2} f(x_{ij}^*, y_{ij}^*)\, \Delta A$$

$$= f\!\left(1, \tfrac{1}{2}\right)\Delta A + f(1,1)\,\Delta A + f\!\left(2, \tfrac{1}{2}\right)\Delta A + f(2,1)\,\Delta A$$

$$= e^{-1/2}\!\left(\tfrac{1}{2}\right) + e^{-1}\!\left(\tfrac{1}{2}\right) + 2e^{-1}\!\left(\tfrac{1}{2}\right) + 2e^{-2}\!\left(\tfrac{1}{2}\right) \approx 0.990$$

(b) $\iint_R xe^{-xy}\, dA \approx \sum_{i=1}^{2} \sum_{j=1}^{2} f(\overline{x}_i, \overline{y}_j)\, \Delta A$

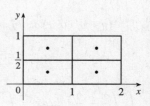

$$= f\!\left(\tfrac{1}{2}, \tfrac{1}{4}\right)\Delta A + f\!\left(\tfrac{1}{2}, \tfrac{3}{4}\right)\Delta A + f\!\left(\tfrac{3}{2}, \tfrac{1}{4}\right)\Delta A + f\!\left(\tfrac{3}{2}, \tfrac{3}{4}\right)\Delta A$$

$$= \tfrac{1}{2}e^{-1/8}\!\left(\tfrac{1}{2}\right) + \tfrac{1}{2}e^{-3/8}\!\left(\tfrac{1}{2}\right) + \tfrac{3}{2}e^{-3/8}\!\left(\tfrac{1}{2}\right) + \tfrac{3}{2}e^{-9/8}\!\left(\tfrac{1}{2}\right) \approx 1.151$$

5. The values of $f(x, y) = \sqrt{52 - x^2 - y^2}$ get smaller as we move farther from the origin, so on any of the subrectangles in the problem, the function will have its largest value at the lower left corner of the subrectangle and its smallest value at the upper right corner, and any other value will lie between these two. So using these subrectangles we have $U < V < L$. (Note that this is true no matter how R is divided into subrectangles.)

7. (a) With $m = n = 2$, we have $\Delta A = 4$. Using the contour map to estimate the value of f at the center of each subrectangle, we have

$$\iint_R f(x, y)\, dA \approx \sum_{i=1}^{2} \sum_{j=1}^{2} f(\overline{x}_i, \overline{y}_j)\, \Delta A = \Delta A[f(1,1) + f(1,3) + f(3,1) + f(3,3)] \approx 4(27 + 4 + 14 + 17) = 248$$

(b) $f_{\text{ave}} = \frac{1}{A(R)} \iint_R f(x, y)\, dA \approx \frac{1}{16}(248) = 15.5$

9. $z = \sqrt{2} > 0$, so we can interpret the double integral as the volume of the solid S that lies below the plane $z = \sqrt{2}$ and above

the rectangle $[2, 6] \times [-1, 5]$. S is a rectangular solid, so $\iint_R \sqrt{2}\, dA = 4 \cdot 6 \cdot \sqrt{2} = 24\sqrt{2}$.

11. $z = 4 - 2y \geq 0$ for $0 \leq y \leq 1$, so we can interpret the integral as the

volume of the solid S that lies below the plane $z = 4 - 2y$ and above

the square $[0, 1] \times [0, 1]$. We can picture S as a rectangular solid (with

height 2) surmounted by a triangular cylinder; thus

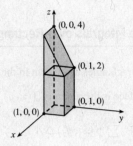

$$\iint_R (4 - 2y)\, dA = (1)(1)(2) + \tfrac{1}{2}(1)(1)(2) = 3$$

13. $\int_0^2 (x + 3x^2 y^2)\, dx = \left[\dfrac{x^2}{2} + 3\dfrac{x^3}{3} y^2\right]_{x=0}^{x=2} = \left[\tfrac{1}{2}x^2 + x^3 y^2\right]_{x=0}^{x=2} = \left[\tfrac{1}{2}(2)^2 + (2)^3 y^2\right] - \left[\tfrac{1}{2}(0)^2 + (0)^3 y^2\right] = 2 + 8y^2$,

$\int_0^3 (x + 3x^2 y^2)\, dy = \left[xy + 3x^2 \dfrac{y^3}{3}\right]_{y=0}^{y=3} = \left[xy + x^2 y^3\right]_{y=0}^{y=3} = \left[x(3) + x^2(3)^3\right] - \left[x(0) + x^2(0)^3\right] = 3x + 27x^2$

15. $\int_1^4 \int_0^2 (6x^2 y - 2x)\, dy\, dx = \int_1^4 \left[3x^2 y^2 - 2xy\right]_{y=0}^{y=2} dx = \int_1^4 \left[(12x^2 - 4x) - (0 - 0)\right] dx$

$\qquad\qquad = \int_1^4 (12x^2 - 4x)\, dx = \left[4x^3 - 2x^2\right]_1^4 = (256 - 32) - (4 - 2) = 222$

17. $\int_0^1 \int_1^2 (x + e^{-y})\, dx\, dy = \int_0^1 \left[\tfrac{1}{2}x^2 + xe^{-y}\right]_{x=1}^{x=2} dy = \int_0^1 \left[(2 + 2e^{-y}) - (\tfrac{1}{2} + e^{-y})\right] dy$

$\qquad\qquad = \int_0^1 (\tfrac{3}{2} + e^{-y})\, dy = \left[\tfrac{3}{2}y - e^{-y}\right]_0^1 = (\tfrac{3}{2} - e^{-1}) - (0 - 1) = \tfrac{5}{2} - e^{-1}$

19. $\int_{-3}^3 \int_0^{\pi/2} (y + y^2 \cos x)\, dx\, dy = \int_{-3}^3 \left[xy + y^2 \sin x\right]_{x=0}^{x=\pi/2} dy = \int_{-3}^3 \left(\tfrac{\pi}{2}y + y^2\right) dy$

$\qquad\qquad = \left[\tfrac{\pi}{4}y^2 + \tfrac{1}{3}y^3\right]_{-3}^3 = \left[\left(\tfrac{9\pi}{4} + 9\right) - \left(\tfrac{9\pi}{4} - 9\right)\right] = 18$

21. $\int_1^4 \int_1^2 \left(\dfrac{x}{y} + \dfrac{y}{x}\right) dy\, dx = \int_1^4 \left[x \ln|y| + \dfrac{1}{x} \cdot \dfrac{1}{2} y^2\right]_{y=1}^{y=2} dx = \int_1^4 \left(x \ln 2 + \dfrac{3}{2x}\right) dx = \left[\tfrac{1}{2}x^2 \ln 2 + \tfrac{3}{2}\ln|x|\right]_1^4$

$\qquad\qquad = (8 \ln 2 + \tfrac{3}{2}\ln 4) - (\tfrac{1}{2}\ln 2 + 0) = \tfrac{15}{2}\ln 2 + \tfrac{3}{2}\ln 4$ or $\tfrac{15}{2}\ln 2 + 3\ln(4^{1/2}) = \tfrac{21}{2}\ln 2$

23. $\int_0^3 \int_0^{\pi/2} t^2 \sin^3 \phi\, d\phi\, dt = \int_0^{\pi/2} \sin^3 \phi\, d\phi \int_0^3 t^2\, dt$ [by Equation 11] $= \int_0^{\pi/2} (1 - \cos^2 \phi) \sin \phi\, d\phi \int_0^3 t^2\, dt$

$\qquad\qquad = \left[\tfrac{1}{3}\cos^3 \phi - \cos \phi\right]_0^{\pi/2} \left[\tfrac{1}{3}t^3\right]_0^3 = \left[(0 - 0) - (\tfrac{1}{3} - 1)\right] \cdot \tfrac{1}{3}(27 - 0) = \tfrac{2}{3}(9) = 6$

25. $\int_0^1 \int_0^1 v(u + v^2)^4\, du\, dv = \int_0^1 \left[\tfrac{1}{5}v(u + v^2)^5\right]_{u=0}^{u=1} dv = \tfrac{1}{5}\int_0^1 v \left[(1 + v^2)^5 - (0 + v^2)^5\right] dv$

$\qquad\qquad = \tfrac{1}{5}\int_0^1 \left[v(1 + v^2)^5 - v^{11}\right] dv = \tfrac{1}{5}\left[\tfrac{1}{2} \cdot \tfrac{1}{6}(1 + v^2)^6 - \tfrac{1}{12}v^{12}\right]_0^1$

$\qquad\qquad$ [substitute $t = 1 + v^2 \Rightarrow dt = 2v\, dv$ in the first term]

$\qquad\qquad = \tfrac{1}{60}\left[(2^6 - 1) - (1 - 0)\right] = \tfrac{1}{60}(63 - 1) = \tfrac{31}{30}$

27. $\iint_R x \sec^2 y \, dA = \int_0^2 \int_0^{\pi/4} x \sec^2 y \, dy \, dx = \int_0^2 x \, dx \int_0^{\pi/4} \sec^2 y \, dy = \left[\frac{1}{2}x^2\right]_0^2 \left[\tan y\right]_0^{\pi/4}$

$= (2 - 0)(\tan\frac{\pi}{4} - \tan 0) = 2(1 - 0) = 2$

29. $\iint_R \frac{xy^2}{x^2 + 1} \, dA = \int_0^1 \int_{-3}^3 \frac{xy^2}{x^2+1} \, dy \, dx = \int_0^1 \frac{x}{x^2+1} \, dx \int_{-3}^3 y^2 \, dy = \left[\frac{1}{2}\ln(x^2+1)\right]_0^1 \left[\frac{1}{3}y^3\right]_{-3}^3$

$= \frac{1}{2}(\ln 2 - \ln 1) \cdot \frac{1}{3}(27 + 27) = 9\ln 2$

31. $\int_0^{\pi/6} \int_0^{\pi/3} x \sin(x + y) \, dy \, dx$

$= \int_0^{\pi/6} \left[-x\cos(x+y)\right]_{y=0}^{y=\pi/3} \, dx = \int_0^{\pi/6} \left[x\cos x - x\cos\left(x + \frac{\pi}{3}\right)\right] dx$

$= x\left[\sin x - \sin\left(x + \frac{\pi}{3}\right)\right]_0^{\pi/6} - \int_0^{\pi/6} \left[\sin x - \sin\left(x + \frac{\pi}{3}\right)\right] dx$ [by integrating by parts separately for each term]

$= \frac{\pi}{6}\left[\frac{1}{2} - 1\right] - \left[-\cos x + \cos\left(x + \frac{\pi}{3}\right)\right]_0^{\pi/6} = -\frac{\pi}{12} - \left[-\frac{\sqrt{3}}{2} + 0 - \left(-1 + \frac{1}{2}\right)\right] = \frac{\sqrt{3}-1}{2} - \frac{\pi}{12}$

33. $\iint_R ye^{-xy} \, dA = \int_0^3 \int_0^2 ye^{-xy} \, dx \, dy = \int_0^3 \left[-e^{-xy}\right]_{x=0}^{x=2} \, dy = \int_0^3 (-e^{-2y} + 1) \, dy = \left[\frac{1}{2}e^{-2y} + y\right]_0^3$

$= \frac{1}{2}e^{-6} + 3 - \left(\frac{1}{2} + 0\right) = \frac{1}{2}e^{-6} + \frac{5}{2}$

35. $z = f(x, y) = 4 - x - 2y \geq 0$ for $0 \leq x \leq 1$ and $0 \leq y \leq 1$. So the solid

is the region in the first octant which lies below the plane $z = 4 - x - 2y$

and above $[0, 1] \times [0, 1]$.

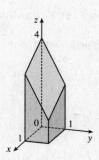

37. The solid lies under the plane $4x + 6y - 2z + 15 = 0$ or $z = 2x + 3y + \frac{15}{2}$ so

$V = \iint_R \left(2x + 3y + \frac{15}{2}\right) dA = \int_{-1}^1 \int_{-1}^2 \left(2x + 3y + \frac{15}{2}\right) dx \, dy = \int_{-1}^1 \left[x^2 + 3xy + \frac{15}{2}x\right]_{x=-1}^{x=2} dy$

$= \int_{-1}^1 \left[(19 + 6y) - \left(-\frac{13}{2} - 3y\right)\right] dy = \int_{-1}^1 \left(\frac{51}{2} + 9y\right) dy = \left[\frac{51}{2}y + \frac{9}{2}y^2\right]_{-1}^1 = 30 - (-21) = 51$

39. $V = \int_{-2}^2 \int_{-1}^1 \left(1 - \frac{1}{4}x^2 - \frac{1}{9}y^2\right) dx \, dy = 4\int_0^2 \int_0^1 \left(1 - \frac{1}{4}x^2 - \frac{1}{9}y^2\right) dx \, dy$

$= 4\int_0^2 \left[x - \frac{1}{12}x^3 - \frac{1}{9}y^2 x\right]_{x=0}^{x=1} dy = 4\int_0^2 \left(\frac{11}{12} - \frac{1}{9}y^2\right) dy = 4\left[\frac{11}{12}y - \frac{1}{27}y^3\right]_0^2 = 4 \cdot \frac{83}{54} = \frac{166}{27}$

41. The solid lies under the surface $z = 1 + x^2 y e^y$ and above the rectangle $R = [-1, 1] \times [0, 1]$, so its volume is

$V = \iint_R (1 + x^2 y e^y) \, dA = \int_0^1 \int_{-1}^1 (1 + x^2 y e^y) \, dx \, dy = \int_0^1 \left[x + \frac{1}{3}x^3 y e^y\right]_{x=-1}^{x=1} dy$

$= \int_0^1 \left(2 + \frac{2}{3}y e^y\right) dy = \left[2y + \frac{2}{3}(y - 1)e^y\right]_0^1$ [by integrating by parts in the second term]

$= (2 + 0) - \left(0 - \frac{2}{3}e^0\right) = 2 + \frac{2}{3} = \frac{8}{3}$

43. The solid lies below the surface $z = 2 + x^2 + (y-2)^2$ and above the plane $z = 1$ for $-1 \le x \le 1, 0 \le y \le 4$. The volume

of the solid is the difference in volumes between the solid that lies under $z = 2 + x^2 + (y-2)^2$ over the rectangle

$R = [-1, 1] \times [0, 4]$ and the solid that lies under $z = 1$ over R.

$$V = \int_0^4 \int_{-1}^1 [2 + x^2 + (y-2)^2]\, dx\, dy - \int_0^4 \int_{-1}^1 (1)\, dx\, dy$$

$$= \int_0^4 \left[2x + \tfrac{1}{3}x^3 + x(y-2)^2 \right]_{x=-1}^{x=1}\, dy - \int_{-1}^1 dx \int_0^4 dy$$

$$= \int_0^4 \left[(2 + \tfrac{1}{3} + (y-2)^2) - (-2 - \tfrac{1}{3} - (y-2)^2) \right] dy - [x]_{-1}^1\, [y]_0^4$$

$$= \int_0^4 \left[\tfrac{14}{3} + 2(y-2)^2 \right] dy - [1-(-1)][4-0] = \left[\tfrac{14}{3}y + \tfrac{2}{3}(y-2)^3 \right]_0^4 - (2)(4)$$

$$= \left[\left(\tfrac{56}{3} + \tfrac{16}{3} \right) - \left(0 - \tfrac{16}{3} \right) \right] - 8 = \tfrac{88}{3} - 8 = \tfrac{64}{3}$$

45. In Maple, we can calculate the integral by defining the integrand as `f`

and then using the command `int(int(f,x=0..1),y=0..1);`.

In Mathematica, we can use the command

      `Integrate[f,{x,0,1},{y,0,1}]`

We find that $\iint_R x^5 y^3 e^{xy}\, dA = 21e - 57 \approx 0.0839$. We can use `plot3d`

(in Maple) or `Plot3D` (in Mathematica) to graph the function.

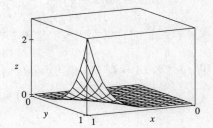

47. R is the rectangle $[-1, 1] \times [0, 5]$. Thus, $A(R) = 2 \cdot 5 = 10$ and

$$f_{\text{ave}} = \frac{1}{A(R)} \iint_R f(x, y)\, dA = \tfrac{1}{10} \int_0^5 \int_{-1}^1 x^2 y\, dx\, dy = \tfrac{1}{10} \int_0^5 \left[\tfrac{1}{3}x^3 y \right]_{x=-1}^{x=1}\, dy = \tfrac{1}{10} \int_0^5 \tfrac{2}{3}y\, dy = \tfrac{1}{10} \left[\tfrac{1}{3}y^2 \right]_0^5 = \tfrac{5}{6}.$$

49. $\displaystyle \iint_R \frac{xy}{1+x^4}\, dA = \int_{-1}^1 \int_0^1 \frac{xy}{1+x^4}\, dy\, dx = \int_{-1}^1 \frac{x}{1+x^4}\, dx \int_0^1 y\, dy$ [by Equation 11]   but $f(x) = \dfrac{x}{1+x^4}$ is an odd

function so $\displaystyle \int_{-1}^1 f(x)\, dx = 0$ (by Theorem 4.5.6 [ET 5.5.7]). Thus $\displaystyle \iint_R \frac{xy}{1+x^4}\, dA = 0 \cdot \int_0^1 y\, dy = 0.$

51. Let $f(x, y) = \dfrac{x-y}{(x+y)^3}$. Then a CAS gives $\int_0^1 \int_0^1 f(x, y)\, dy\, dx = \tfrac{1}{2}$ and $\int_0^1 \int_0^1 f(x, y)\, dx\, dy = -\tfrac{1}{2}$.

To explain the seeming violation of Fubini's Theorem, note that f has an infinite discontinuity at $(0, 0)$ and thus does not

satisfy the conditions of Fubini's Theorem. In fact, both iterated integrals involve improper integrals which diverge at their

lower limits of integration.

15.2   Double Integrals over General Regions

1. $\int_1^5 \int_0^x (8x - 2y)\, dy\, dx = \int_1^5 \left[8xy - y^2 \right]_{y=0}^{y=x}\, dx = \int_1^5 [8x(x) - (x)^2 - 8x(0) + (0)^2]\, dx$

$$= \int_1^5 7x^2\, dx = \tfrac{7}{3}x^3 \Big]_1^5 = \tfrac{7}{3}(125 - 1) = \tfrac{868}{3}$$

3. $\int_0^1 \int_0^y xe^{y^3}\, dx\, dy = \int_0^1 \left[\tfrac{1}{2}x^2 e^{y^3} \right]_{x=0}^{x=y}\, dy = \int_0^1 \tfrac{1}{2}e^{y^3} \left[(y)^2 - (0)^2 \right] dy$

$$= \tfrac{1}{2} \int_0^1 y^2 e^{y^3}\, dy = \tfrac{1}{2} \left[\tfrac{1}{3}e^{y^3} \right]_0^1 = \tfrac{1}{2} \cdot \tfrac{1}{3} \left(e^1 - e^0 \right) = \tfrac{1}{6}(e - 1)$$

5. $\int_0^1 \int_0^{s^2} \cos(s^3)\, dt\, ds = \int_0^1 \left[t \cos(s^3) \right]_{t=0}^{t=s^2} ds = \int_0^1 s^2 \cos(s^3)\, ds = \frac{1}{3} \sin(s^3) \big]_0^1 = \frac{1}{3}(\sin 1 - \sin 0) = \frac{1}{3}\sin 1$

7. $\iint_D \frac{y}{x^2+1}\, dA = \int_0^4 \int_0^{\sqrt{x}} \frac{y}{x^2+1}\, dy\, dx = \int_0^4 \left[\frac{1}{x^2+1} \cdot \frac{y^2}{2} \right]_{y=0}^{y=\sqrt{x}} dx = \frac{1}{2} \int_0^4 \frac{x}{x^2+1}\, dx$

$$= \frac{1}{2} \left[\frac{1}{2} \ln |x^2+1| \right]_0^4 = \frac{1}{4} \left[\ln(x^2+1) \right]_0^4 = \frac{1}{4}(\ln 17 - \ln 1) = \frac{1}{4}\ln 17$$

9. $\iint_D e^{-y^2}\, dA = \int_0^3 \int_0^y e^{-y^2}\, dx\, dy = \int_0^3 \left[x e^{-y^2} \right]_{x=0}^{x=y} dy = \int_0^3 \left(y e^{-y^2} - 0 \right) dy = \int_0^3 y e^{-y^2}\, dy$

$$= -\frac{1}{2} e^{-y^2} \Big]_0^3 = -\frac{1}{2}\left(e^{-9} - e^0 \right) = \frac{1}{2}\left(1 - e^{-9} \right)$$

11. (a) At the right we sketch an example of a region D that can be described as lying

between the graphs of two continuous functions of x (a type I region) but not as

lying between graphs of two continuous functions of y (a type II region). The

regions shown in Figures 6 and 8 in the text are additional examples.

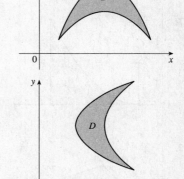

(b) Now we sketch an example of a region D that can be described as lying between

the graphs of two continuous functions of y but not as lying between graphs of two

continuous functions of x. The first region shown in Figure 7 is another example.

13.

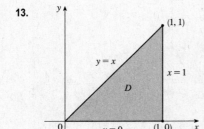

As a type I region, D lies between the lower boundary $y = 0$ and the upper

boundary $y = x$ for $0 \le x \le 1$, so $D = \{(x,y) \mid 0 \le x \le 1, 0 \le y \le x\}$. If we

describe D as a type II region, D lies between the left boundary $x = y$ and the

right boundary $x = 1$ for $0 \le y \le 1$, so $D = \{(x,y) \mid 0 \le y \le 1, y \le x \le 1\}$.

Thus $\iint_D x\, dA = \int_0^1 \int_0^x x\, dy\, dx = \int_0^1 \left[xy \right]_{y=0}^{y=x} dx = \int_0^1 x^2\, dx = \frac{1}{3} x^3 \big]_0^1 = \frac{1}{3}(1-0) = \frac{1}{3}$ or

$\iint_D x\, dA = \int_0^1 \int_y^1 x\, dx\, dy = \int_0^1 \left[\frac{1}{2} x^2 \right]_{x=y}^{x=1} dy = \frac{1}{2} \int_0^1 (1-y^2)\, dy = \frac{1}{2} \left[y - \frac{1}{3} y^3 \right]_0^1 = \frac{1}{2} \left[(1 - \frac{1}{3}) - 0 \right] = \frac{1}{3}$.

15.

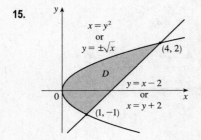

The curves $y = x - 2$ or $x = y + 2$ and $x = y^2$ intersect when $y + 2 = y^2$ $\Leftrightarrow$

$y^2 - y - 2 = 0$ $\Leftrightarrow$ $(y-2)(y+1) = 0$ $\Leftrightarrow$ $y = -1, y = 2$, so the points of

intersection are $(1, -1)$ and $(4, 2)$. If we describe D as a type I region, the upper

boundary curve is $y = \sqrt{x}$ but the lower boundary curve consists of two parts,

$y = -\sqrt{x}$ for $0 \le x \le 1$ and $y = x - 2$ for $1 \le x \le 4$.

[continued]

Thus $D = \{(x, y) \mid 0 \le x \le 1,\ -\sqrt{x} \le y \le \sqrt{x}\} \cup \{(x, y) \mid 1 \le x \le 4,\ x - 2 \le y \le \sqrt{x}\}$ and

$\iint_D y\, dA = \int_0^1 \int_{-\sqrt{x}}^{\sqrt{x}} y\, dy\, dx + \int_1^4 \int_{x-2}^{\sqrt{x}} y\, dy\, dx$. If we describe D as a type II region, D is enclosed by the left boundary

$x = y^2$ and the right boundary $x = y + 2$ for $-1 \le y \le 2$, so $D = \{(x, y) \mid -1 \le y \le 2,\ y^2 \le x \le y + 2\}$ and

$\iint_D y\, dA = \int_{-1}^2 \int_{y^2}^{y+2} y\, dx\, dy$. In either case, the resulting iterated integrals are not difficult to evaluate but the region D is

more simply described as a type II region, giving one iterated integral rather than a sum of two, so we evaluate the latter

integral:

$$\iint_D y\, dA = \int_{-1}^2 \int_{y^2}^{y+2} y\, dx\, dy = \int_{-1}^2 \left[xy\right]_{x=y^2}^{x=y+2} dy = \int_{-1}^2 (y + 2 - y^2) y\, dy = \int_{-1}^2 (y^2 + 2y - y^3)\, dy$$

$$= \left[\tfrac{1}{3}y^3 + y^2 - \tfrac{1}{4}y^4\right]_{-1}^2 = \left(\tfrac{8}{3} + 4 - 4\right) - \left(-\tfrac{1}{3} + 1 - \tfrac{1}{4}\right) = \tfrac{9}{4}$$

17.

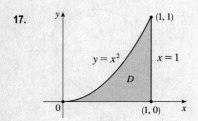

$\int_0^1 \int_0^{x^2} x \cos y\, dy\, dx = \int_0^1 \left[x \sin y\right]_{y=0}^{y=x^2} dx = \int_0^1 x \sin x^2\, dx$

$\qquad\qquad\qquad = -\tfrac{1}{2}\cos x^2 \Big]_0^1 = -\tfrac{1}{2}(\cos 1 - \cos 0) = \tfrac{1}{2}(1 - \cos 1)$

19.

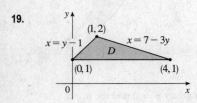

$\iint_D y^2\, dA = \int_1^2 \int_{y-1}^{7-3y} y^2\, dx\, dy = \int_1^2 \left[xy^2\right]_{x=y-1}^{x=7-3y} dy$

$\qquad\qquad = \int_1^2 \left[(7 - 3y) - (y - 1)\right] y^2\, dy = \int_1^2 (8y^2 - 4y^3)\, dy$

$\qquad\qquad = \left[\tfrac{8}{3}y^3 - y^4\right]_1^2 = \tfrac{64}{3} - 16 - \tfrac{8}{3} + 1 = \tfrac{11}{3}$

21.

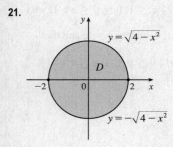

$\displaystyle\int_{-2}^2 \int_{-\sqrt{4-x^2}}^{\sqrt{4-x^2}} (2x - y)\, dy\, dx$

$\qquad = \int_{-2}^2 \left[2xy - \tfrac{1}{2}y^2\right]_{y=-\sqrt{4-x^2}}^{y=\sqrt{4-x^2}} dx$

$\qquad = \int_{-2}^2 \left[2x\sqrt{4-x^2} - \tfrac{1}{2}(4-x^2) + 2x\sqrt{4-x^2} + \tfrac{1}{2}(4-x^2)\right] dx$

$\qquad = \int_{-2}^2 4x\sqrt{4-x^2}\, dx = -\tfrac{4}{3}(4 - x^2)^{3/2}\Big]_{-2}^2 = 0$

$\qquad$ [Or, note that $4x\sqrt{4-x^2}$ is an odd function, so $\int_{-2}^2 4x\sqrt{4-x^2}\, dx = 0$.]

23.

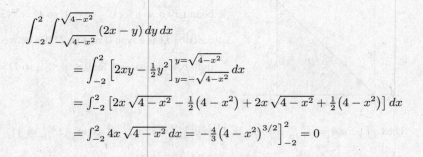

$V = \int_0^1 \int_{x^2}^{\sqrt{x}} (3x + 2y)\, dy\, dx = \int_0^1 \left[3xy + y^2\right]_{y=x^2}^{y=\sqrt{x}} dx$

$\qquad = \int_0^1 \left[(3x\sqrt{x} + x) - (3x^3 + x^4)\right] dx = \int_0^1 (3x^{3/2} + x - 3x^3 - x^4)\, dx$

$\qquad = \left[3 \cdot \tfrac{2}{5}x^{5/2} + \tfrac{1}{2}x^2 - \tfrac{3}{4}x^4 - \tfrac{1}{5}x^5\right]_0^1 = \tfrac{6}{5} + \tfrac{1}{2} - \tfrac{3}{4} - \tfrac{1}{5} - 0 = \tfrac{3}{4}$

25.

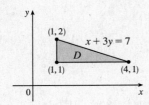

$V = \int_1^2 \int_1^{7-3y} xy\, dx\, dy = \int_1^2 \left[\frac{1}{2}x^2 y\right]_{x=1}^{x=7-3y} dy$

$= \frac{1}{2} \int_1^2 y\left[(7-3y)^2 - 1\right] dy = \frac{1}{2}\int_1^2 (48y - 42y^2 + 9y^3)\, dy$

$= \frac{1}{2}\left[24y^2 - 14y^3 + \frac{9}{4}y^4\right]_1^2 = \frac{31}{8}$

27.

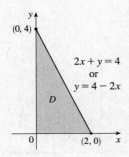

$V = \int_0^2 \int_0^{4-2x} (4 - 2x - y)\, dy\, dx = \int_0^2 \left[4y - 2xy - \frac{1}{2}y^2\right]_{y=0}^{y=4-2x} dx$

$= \int_0^2 \left[4(4-2x) - 2x(4-2x) - \frac{1}{2}(4-2x)^2 - 0\right] dx$

$= \int_0^2 (2x^2 - 8x + 8)\, dx = \left[\frac{2}{3}x^3 - 4x^2 + 8x\right]_0^2 = \frac{16}{3} - 16 + 16 - 0 = \frac{16}{3}$

29.

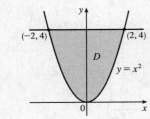

$V = \int_{-2}^2 \int_{x^2}^4 x^2\, dy\, dx$

$= \int_{-2}^2 \left[x^2 y\right]_{y=x^2}^{y=4} dx = \int_{-2}^2 (4x^2 - x^4)\, dx$

$= \left[\frac{4}{3}x^3 - \frac{1}{5}x^5\right]_{-2}^2 = \frac{32}{3} - \frac{32}{5} + \frac{32}{3} - \frac{32}{5} = \frac{128}{15}$

31.

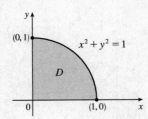

$V = \int_0^1 \int_0^{\sqrt{1-x^2}} y\, dy\, dx = \int_0^1 \left[\frac{y^2}{2}\right]_{y=0}^{y=\sqrt{1-x^2}} dx$

$= \int_0^1 \frac{1-x^2}{2}\, dx = \frac{1}{2}\left[x - \frac{1}{3}x^3\right]_0^1 = \frac{1}{3}$

33.

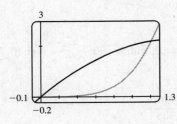

From the graph, it appears that the two curves intersect at $x = 0$ and at $x \approx 1.213$. Thus the desired integral is

$\iint_D x\, dA \approx \int_0^{1.213} \int_{x^4}^{3x - x^2} x\, dy\, dx = \int_0^{1.213} \left[xy\right]_{y=x^4}^{y=3x-x^2} dx$

$= \int_0^{1.213} (3x^2 - x^3 - x^5)\, dx = \left[x^3 - \frac{1}{4}x^4 - \frac{1}{6}x^6\right]_0^{1.213}$

≈ 0.713

35. The region of integration is bounded by the curves $y = 1 - x^2$ and $y = x^2 - 1$ which intersect at $(\pm 1, 0)$ with $1 - x^2 \geq x^2 - 1$ on $[-1, 1]$. Within this region, the plane $z = 2x + 2y + 10$ is above the plane $z = 2 - x - y$, so

$$V = \int_{-1}^{1} \int_{x^2-1}^{1-x^2} (2x + 2y + 10) \, dy \, dx - \int_{-1}^{1} \int_{x^2-1}^{1-x^2} (2 - x - y) \, dy \, dx$$

$$= \int_{-1}^{1} \int_{x^2-1}^{1-x^2} (2x + 2y + 10 - (2 - x - y)) \, dy \, dx$$

$$= \int_{-1}^{1} \int_{x^2-1}^{1-x^2} (3x + 3y + 8) \, dy \, dx = \int_{-1}^{1} \left[3xy + \tfrac{3}{2}y^2 + 8y \right]_{y=x^2-1}^{y=1-x^2} dx$$

$$= \int_{-1}^{1} \left[3x(1 - x^2) + \tfrac{3}{2}(1 - x^2)^2 + 8(1 - x^2) - 3x(x^2 - 1) - \tfrac{3}{2}(x^2 - 1)^2 - 8(x^2 - 1) \right] dx$$

$$= \int_{-1}^{1} (-6x^3 - 16x^2 + 6x + 16) \, dx = \left[-\tfrac{3}{2}x^4 - \tfrac{16}{3}x^3 + 3x^2 + 16x \right]_{-1}^{1}$$

$$= -\tfrac{3}{2} - \tfrac{16}{3} + 3 + 16 + \tfrac{3}{2} - \tfrac{16}{3} - 3 + 16 = \tfrac{64}{3}$$

37. The region of integration is bounded by the curves $y = x^2$ and

$y = 1 - x^2$ which intersect at $\left(\pm\frac{1}{\sqrt{2}}, \frac{1}{2} \right)$.

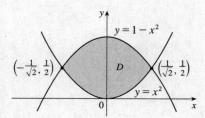

The solid lies under the graph of $z = 3$ and above the graph of $z = y$,

so its volume is

$$V = \int_{-1/\sqrt{2}}^{1/\sqrt{2}} \int_{x^2}^{1-x^2} 3 \, dy \, dx - \int_{-1/\sqrt{2}}^{1/\sqrt{2}} \int_{x^2}^{1-x^2} y \, dy \, dx = \int_{-1/\sqrt{2}}^{1/\sqrt{2}} \int_{x^2}^{1-x^2} (3 - y) \, dy \, dx$$

$$= \int_{-1/\sqrt{2}}^{1/\sqrt{2}} \left[3y - \tfrac{1}{2}y^2 \right]_{y=x^2}^{y=1-x^2} dx = \int_{-1/\sqrt{2}}^{1/\sqrt{2}} \left[\left(3(1 - x^2) - \tfrac{1}{2}(1 - x^2)^2 \right) - \left(3x^2 - \tfrac{1}{2}(x^2)^2 \right) \right] dx$$

$$= \int_{-1/\sqrt{2}}^{1/\sqrt{2}} \left(\tfrac{5}{2} - 5x^2 \right) dx = \left[\tfrac{5}{2}x - \tfrac{5}{3}x^3 \right]_{-1/\sqrt{2}}^{1/\sqrt{2}} = \left(\tfrac{5}{2\sqrt{2}} - \tfrac{5}{6\sqrt{2}} \right) - \left(-\tfrac{5}{2\sqrt{2}} + \tfrac{5}{6\sqrt{2}} \right)$$

$$= \tfrac{10}{3\sqrt{2}} \text{ or } \tfrac{5\sqrt{2}}{3}$$

39. The solid lies below the plane $z = 1 - x - y$

or $x + y + z = 1$ and above the region

$D = \{(x, y) \mid 0 \le x \le 1, 0 \le y \le 1 - x\}$

in the xy-plane. The solid is a tetrahedron.

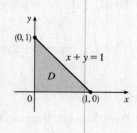

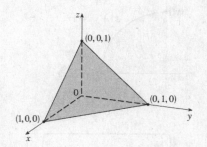

41. The two bounding curves $y = x^3 - x$ and $y = x^2 + x$ intersect at the origin and at $x = 2$, with $x^2 + x > x^3 - x$ on $(0, 2)$.

Using a CAS, we find that the volume of the solid is

$$V = \int_0^2 \int_{x^3 - x}^{x^2 + x} (x^3 y^4 + xy^2) \, dy \, dx = \frac{13{,}984{,}735{,}616}{14{,}549{,}535}$$

43. The two surfaces intersect in the circle $x^2 + y^2 = 1$, $z = 0$ and the region of integration is the disk D: $x^2 + y^2 \le 1$.

Using a CAS, the volume is $\displaystyle\iint_D (1 - x^2 - y^2) \, dA = \int_{-1}^{1} \int_{-\sqrt{1-x^2}}^{\sqrt{1-x^2}} (1 - x^2 - y^2) \, dy \, dx = \frac{\pi}{2}$.

45.

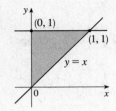

Because the region of integration is

$$D = \{(x,y) \mid 0 \le x \le y, 0 \le y \le 1\} = \{(x,y) \mid x \le y \le 1, 0 \le x \le 1\}$$

we have $\int_0^1 \int_0^y f(x,y)\, dx\, dy = \iint_D f(x,y)\, dA = \int_0^1 \int_x^1 f(x,y)\, dy\, dx$.

47.

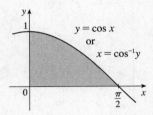

Because the region of integration is

$$D = \{(x,y) \mid 0 \le y \le \cos x, 0 \le x \le \pi/2\}$$
$$= \{(x,y) \mid 0 \le x \le \cos^{-1} y, 0 \le y \le 1\}$$

we have

$$\int_0^{\pi/2} \int_0^{\cos x} f(x,y)\, dy\, dx = \iint_D f(x,y)\, dA = \int_0^1 \int_0^{\cos^{-1} y} f(x,y)\, dx\, dy.$$

49.

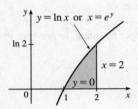

Because the region of integration is

$$D = \{(x,y) \mid 0 \le y \le \ln x, 1 \le x \le 2\} = \{(x,y) \mid e^y \le x \le 2, 0 \le y \le \ln 2\}$$

we have

$$\int_1^2 \int_0^{\ln x} f(x,y)\, dy\, dx = \iint_D f(x,y)\, dA = \int_0^{\ln 2} \int_{e^y}^2 f(x,y)\, dx\, dy$$

51.

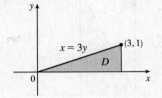

$$\int_0^1 \int_{3y}^3 e^{x^2}\, dx\, dy = \int_0^3 \int_0^{x/3} e^{x^2}\, dy\, dx = \int_0^3 \left[e^{x^2} y\right]_{y=0}^{y=x/3} dx$$

$$= \int_0^3 \left(\frac{x}{3}\right) e^{x^2}\, dx = \tfrac{1}{6} e^{x^2}\Big]_0^3 = \frac{e^9 - 1}{6}$$

53.

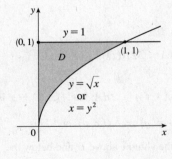

$$\int_0^1 \int_{\sqrt{x}}^1 \sqrt{y^3 + 1}\, dy\, dx = \int_0^1 \int_0^{y^2} \sqrt{y^3 + 1}\, dx\, dy = \int_0^1 \sqrt{y^3 + 1}\, [x]_{x=0}^{x=y^2}\, dy$$

$$= \int_0^1 y^2 \sqrt{y^3 + 1}\, dy = \tfrac{2}{9} \left(y^3 + 1\right)^{3/2}\Big]_0^1$$

$$= \tfrac{2}{9} \left(2^{3/2} - 1^{3/2}\right) = \tfrac{2}{9} \left(2\sqrt{2} - 1\right)$$

55.

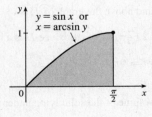

$$\int_0^1 \int_{\arcsin y}^{\pi/2} \cos x \sqrt{1 + \cos^2 x}\, dx\, dy$$

$$= \int_0^{\pi/2} \int_0^{\sin x} \cos x \sqrt{1 + \cos^2 x}\, dy\, dx$$

$$= \int_0^{\pi/2} \cos x \sqrt{1 + \cos^2 x}\, [y]_{y=0}^{y=\sin x}\, dx$$

$$= \int_0^{\pi/2} \cos x \sqrt{1 + \cos^2 x}\, \sin x\, dx \qquad \begin{bmatrix} \text{Let } u = \cos x,\, du = -\sin x\, dx, \\ dx = du/(-\sin x) \end{bmatrix}$$

$$= \int_1^0 -u \sqrt{1 + u^2}\, du = -\tfrac{1}{3} \left(1 + u^2\right)^{3/2}\Big]_1^0$$

$$= \tfrac{1}{3} \left(\sqrt{8} - 1\right) = \tfrac{1}{3} \left(2\sqrt{2} - 1\right)$$

57. $D = \{(x,y) \mid 0 \leq x \leq 1, \; -x + 1 \leq y \leq 1\} \cup \{(x,y) \mid -1 \leq x \leq 0, \; x + 1 \leq y \leq 1\}$

$\cup \{(x,y) \mid 0 \leq x \leq 1, \; -1 \leq y \leq x - 1\} \cup \{(x,y) \mid -1 \leq x \leq 0, \; -1 \leq y \leq -x - 1\}$, all type I.

$$\iint_D x^2 \, dA = \int_0^1 \int_{1-x}^1 x^2 \, dy \, dx + \int_{-1}^0 \int_{x+1}^1 x^2 \, dy \, dx + \int_0^1 \int_{-1}^{x-1} x^2 \, dy \, dx + \int_{-1}^0 \int_{-1}^{-x-1} x^2 \, dy \, dx$$

$$= 4 \int_0^1 \int_{1-x}^1 x^2 \, dy \, dx \qquad \text{[by symmetry of the regions and because } f(x,y) = x^2 \geq 0]$$

$$= 4 \int_0^1 x^3 \, dx = 4 \left[\tfrac{1}{4} x^4 \right]_0^1 = 1$$

59. Since $x^2 + y^2 \leq 1$ on S, we must have $0 \leq x^2 \leq 1$ and $0 \leq y^2 \leq 1$, so $0 \leq x^2 y^2 \leq 1$ $\Rightarrow$ $3 \leq 4 - x^2 y^2 \leq 4$ $\Rightarrow$

$\sqrt{3} \leq \sqrt{4 - x^2 y^2} \leq 2$. Here we have $A(S) = \tfrac{1}{2}\pi(1)^2 = \tfrac{\pi}{2}$, so by Property 11,

$\sqrt{3} A(S) \leq \iint_S \sqrt{4 - x^2 y^2} \, dA \leq 2 A(S)$ $\Rightarrow$ $\tfrac{\sqrt{3}}{2}\pi \leq \iint_S \sqrt{4 - x^2 y^2} \, dA \leq \pi$ or we can say

$2.720 < \iint_S \sqrt{4 - x^2 y^2} \, dA < 3.142$. (We have rounded the lower bound down and the upper bound up to preserve the

inequalities.)

61. The average value of a function f of two variables defined on a rectangle R was

defined in Section 15.1 as $f_{ave} = \frac{1}{A(R)} \iint_R f(x,y) \, dA$. Extending this definition

to general regions D, we have $f_{ave} = \frac{1}{A(D)} \iint_D f(x,y) \, dA$.

Here $D = \{(x,y) \mid 0 \leq x \leq 1, 0 \leq y \leq 3x\}$, so $A(D) = \tfrac{1}{2}(1)(3) = \tfrac{3}{2}$ and

$$f_{ave} = \frac{1}{A(D)} \iint_D f(x,y) \, dA = \frac{1}{3/2} \int_0^1 \int_0^{3x} xy \, dy \, dx$$

$$= \tfrac{2}{3} \int_0^1 \left[\tfrac{1}{2} xy^2 \right]_{y=0}^{y=3x} dx = \tfrac{1}{3} \int_0^1 9x^3 \, dx = \tfrac{3}{4} x^4 \Big]_0^1 = \tfrac{3}{4}$$

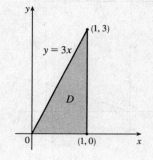

63. Since $m \leq f(x,y) \leq M$, $\iint_D m \, dA \leq \iint_D f(x,y) \, dA \leq \iint_D M \, dA$ by (8) $\Rightarrow$

$m \iint_D 1 \, dA \leq \iint_D f(x,y) \, dA \leq M \iint_D 1 \, dA$ by (7) $\Rightarrow$ $mA(D) \leq \iint_D f(x,y) \, dA \leq MA(D)$ by (10).

65.

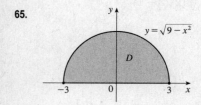

First we can write $\iint_D (x + 2) \, dA = \iint_D x \, dA + \iint_D 2 \, dA$. But $f(x,y) = x$ is

an odd function with respect to x [that is, $f(-x,y) = -f(x,y)$] and D is

symmetric with respect to x. Consequently, the volume above D and below the

graph of f is the same as the volume below D and above the graph of f, so

$\iint_D x \, dA = 0$. Also, $\iint_D 2 \, dA = 2 \cdot A(D) = 2 \cdot \tfrac{1}{2}\pi(3)^2 = 9\pi$ since D is a half

disk of radius 3. Thus $\iint_D (x + 2) \, dA = 0 + 9\pi = 9\pi$.

67. We can write $\iint_D (2x + 3y) \, dA = \iint_D 2x \, dA + \iint_D 3y \, dA$. $\iint_D 2x \, dA$ represents the volume of the solid lying under the

plane $z = 2x$ and above the rectangle D. This solid region is a triangular cylinder with length b and whose cross-section is a triangle with width a and height $2a$. (See the first figure.)

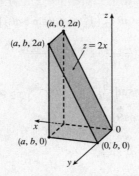

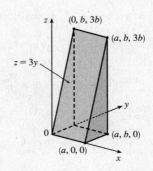

Thus its volume is $\frac{1}{2} \cdot a \cdot 2a \cdot b = a^2 b$. Similarly, $\iint_D 3y \, dA$ represents the volume of a triangular cylinder with length a, triangular cross-section with width b and height $3b$, and volume $\frac{1}{2} \cdot b \cdot 3b \cdot a = \frac{3}{2}ab^2$. (See the second figure.) Thus

$$\iint_D (2x + 3y) \, dA = a^2 b + \tfrac{3}{2}ab^2$$

69. $\iint_D \left(ax^3 + by^3 + \sqrt{a^2 - x^2}\right) dA = \iint_D ax^3 \, dA + \iint_D by^3 \, dA + \iint_D \sqrt{a^2 - x^2} \, dA$. Now ax^3 is odd with respect to x and by^3 is odd with respect to y, and the region of integration is symmetric with respect to both x and y, so $\iint_D ax^3 \, dA = \iint_D by^3 \, dA = 0$.

$\iint_D \sqrt{a^2 - x^2} \, dA$ represents the volume of the solid region under the graph of $z = \sqrt{a^2 - x^2}$ and above the rectangle D, namely a half circular cylinder with radius a and length $2b$ (see the figure) whose volume is

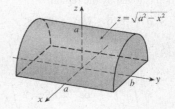

$\frac{1}{2} \cdot \pi r^2 h = \frac{1}{2}\pi a^2 (2b) = \pi a^2 b$. Thus

$\iint_D \left(ax^3 + by^3 + \sqrt{a^2 - x^2}\right) dA = 0 + 0 + \pi a^2 b = \pi a^2 b$.

15.3 Double Integrals in Polar Coordinates

1. The region R is more easily described by polar coordinates: $R = \{(r, \theta) \mid 2 \le r \le 5, 0 \le \theta \le 2\pi\}$.
Thus $\iint_R f(x, y) \, dA = \int_0^{2\pi} \int_2^5 f(r \cos \theta, r \sin \theta) \, r \, dr \, d\theta$.

3. The region R is more easily described by polar coordinates: $R = \{(r, \theta) \mid 0 \le r \le 1, \pi \le \theta \le 2\pi\}$.
Thus $\iint_R f(x, y) \, dA = \int_\pi^{2\pi} \int_0^1 f(r \cos \theta, r \sin \theta) \, r \, dr \, d\theta$.

5. The integral $\int_{\pi/4}^{3\pi/4} \int_1^2 r \, dr \, d\theta$ represents the area of the region
$R = \{(r, \theta) \mid 1 \le r \le 2, \pi/4 \le \theta \le 3\pi/4\}$, the top quarter portion of a ring (annulus).

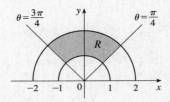

$\int_{\pi/4}^{3\pi/4} \int_1^2 r \, dr \, d\theta = \left(\int_{\pi/4}^{3\pi/4} d\theta\right)\left(\int_1^2 r \, dr\right)$

$= \left[\theta\right]_{\pi/4}^{3\pi/4} \left[\frac{1}{2}r^2\right]_1^2 = \left(\frac{3\pi}{4} - \frac{\pi}{4}\right) \cdot \frac{1}{2}(4 - 1) = \frac{\pi}{2} \cdot \frac{3}{2} = \frac{3\pi}{4}$

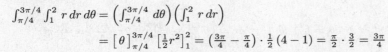

7. The half disk D can be described in polar coordinates as $D = \{(r, \theta) \mid 0 \le r \le 5, 0 \le \theta \le \pi\}$. Then

$$\iint_D x^2 y \, dA = \int_0^\pi \int_0^5 (r\cos\theta)^2 (r\sin\theta) \, r \, dr \, d\theta = \left(\int_0^\pi \cos^2\theta \sin\theta \, d\theta\right)\left(\int_0^5 r^4 \, dr\right)$$

$$= \left[-\tfrac{1}{3}\cos^3\theta\right]_0^\pi \left[\tfrac{1}{5}r^5\right]_0^5 = -\tfrac{1}{3}(-1-1)\cdot 625 = \tfrac{1250}{3}$$

9. $\iint_R \sin(x^2 + y^2) \, dA = \int_0^{\pi/2} \int_1^3 \sin(r^2) \, r \, dr \, d\theta = \int_0^{\pi/2} d\theta \int_1^3 r \sin(r^2) \, dr = \left[\theta\right]_0^{\pi/2} \left[-\tfrac{1}{2}\cos(r^2)\right]_1^3$

$$= \left(\tfrac{\pi}{2}\right)\left[-\tfrac{1}{2}(\cos 9 - \cos 1)\right] = \tfrac{\pi}{4}(\cos 1 - \cos 9)$$

11. $\iint_D e^{-x^2 - y^2} \, dA = \int_{-\pi/2}^{\pi/2} \int_0^2 e^{-r^2} r \, dr \, d\theta = \int_{-\pi/2}^{\pi/2} d\theta \int_0^2 r e^{-r^2} \, dr$

$$= \left[\theta\right]_{-\pi/2}^{\pi/2} \left[-\tfrac{1}{2}e^{-r^2}\right]_0^2 = \pi\left(-\tfrac{1}{2}\right)(e^{-4} - e^0) = \tfrac{\pi}{2}(1 - e^{-4})$$

13. R is the region shown in the figure, and can be described

by $R = \{(r, \theta) \mid 0 \le \theta \le \pi/4, 1 \le r \le 2\}$. Thus

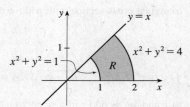

$\iint_R \arctan(y/x) \, dA = \int_0^{\pi/4} \int_1^2 \arctan(\tan\theta) \, r \, dr \, d\theta$ since $y/x = \tan\theta$.

Also, $\arctan(\tan\theta) = \theta$ for $0 \le \theta \le \pi/4$, so the integral becomes

$\int_0^{\pi/4} \int_1^2 \theta \, r \, dr \, d\theta = \int_0^{\pi/4} \theta \, d\theta \int_1^2 r \, dr = \left[\tfrac{1}{2}\theta^2\right]_0^{\pi/4} \left[\tfrac{1}{2}r^2\right]_1^2 = \tfrac{\pi^2}{32} \cdot \tfrac{3}{2} = \tfrac{3}{64}\pi^2$.

15. One loop is given by the region

$D = \{(r, \theta) \mid -\pi/6 \le \theta \le \pi/6, 0 \le r \le \cos 3\theta\}$, so the area is

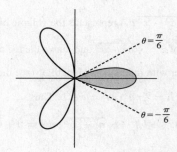

$$\iint_D dA = \int_{-\pi/6}^{\pi/6} \int_0^{\cos 3\theta} r \, dr \, d\theta = \int_{-\pi/6}^{\pi/6} \left[\tfrac{1}{2}r^2\right]_{r=0}^{r=\cos 3\theta} d\theta$$

$$= \int_{-\pi/6}^{\pi/6} \tfrac{1}{2}\cos^2 3\theta \, d\theta = 2\int_0^{\pi/6} \tfrac{1}{2}\left(\frac{1 + \cos 6\theta}{2}\right) d\theta$$

$$= \tfrac{1}{2}\left[\theta + \tfrac{1}{6}\sin 6\theta\right]_0^{\pi/6} = \frac{\pi}{12}$$

17. In polar coordinates the circle $(x-1)^2 + y^2 = 1 \Leftrightarrow x^2 + y^2 = 2x$ is $r^2 = 2r\cos\theta \Rightarrow r = 2\cos\theta$,

and the circle $x^2 + y^2 = 1$ is $r = 1$. The curves intersect in the first quadrant when

$2\cos\theta = 1 \Rightarrow \cos\theta = \tfrac{1}{2} \Rightarrow \theta = \pi/3$, so the portion of the region in the first quadrant is given by

$D = \{(r, \theta) \mid 1 \le r \le 2\cos\theta, 0 \le \theta \le \pi/3\}$. By symmetry, the total area

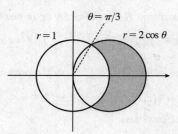

is twice the area of D:

$2A(D) = 2\iint_D dA = 2\int_0^{\pi/3} \int_1^{2\cos\theta} r \, dr \, d\theta = 2\int_0^{\pi/3} \left[\tfrac{1}{2}r^2\right]_{r=1}^{r=2\cos\theta} d\theta$

$$= \int_0^{\pi/3} (4\cos^2\theta - 1) \, d\theta = \int_0^{\pi/3} \left[4 \cdot \tfrac{1}{2}(1 + \cos 2\theta) - 1\right] d\theta$$

$$= \int_0^{\pi/3} (1 + 2\cos 2\theta) \, d\theta = \left[\theta + \sin 2\theta\right]_0^{\pi/3} = \frac{\pi}{3} + \frac{\sqrt{3}}{2}$$

19. $V = \iint_{x^2 + y^2 \le 25} (x^2 + y^2) \, dA = \int_0^{2\pi} \int_0^5 r^2 \cdot r \, dr \, d\theta = \int_0^{2\pi} d\theta \int_0^5 r^3 \, dr = \left[\theta\right]_0^{2\pi} \left[\tfrac{1}{4}r^4\right]_0^5 = 2\pi\left(\tfrac{625}{4}\right) = \tfrac{625}{2}\pi$

21. $2x + y + z = 4 \iff z = 4 - 2x - y$, so the volume of the solid is

$$V = \iint_{x^2 + y^2 \le 1} (4 - 2x - y)\, dA = \int_0^{2\pi} \int_0^1 (4 - 2r\cos\theta - r\sin\theta)\, r\, dr\, d\theta$$

$$= \int_0^{2\pi} \int_0^1 \left[4r - r^2 (2\cos\theta + \sin\theta)\right] dr\, d\theta = \int_0^{2\pi} \left[2r^2 - \tfrac{1}{3} r^3 (2\cos\theta + \sin\theta)\right]_{r=0}^{r=1} d\theta$$

$$= \int_0^{2\pi} \left[2 - \tfrac{1}{3}(2\cos\theta + \sin\theta)\right] d\theta = \left[2\theta - \tfrac{1}{3}(2\sin\theta - \cos\theta)\right]_0^{2\pi} = 4\pi + \tfrac{1}{3} - 0 - \tfrac{1}{3} = 4\pi$$

23. By symmetry,

$$V = 2 \iint_{x^2 + y^2 \le a^2} \sqrt{a^2 - x^2 - y^2}\, dA = 2 \int_0^{2\pi} \int_0^a \sqrt{a^2 - r^2}\, r\, dr\, d\theta = 2 \int_0^{2\pi} d\theta \int_0^a r\sqrt{a^2 - r^2}\, dr$$

$$= 2 \big[\theta\big]_0^{2\pi} \left[-\tfrac{1}{3}(a^2 - r^2)^{3/2}\right]_0^a = 2(2\pi)\left(0 + \tfrac{1}{3}a^3\right) = \tfrac{4}{3}\pi a^3$$

25. The cone $z = \sqrt{x^2 + y^2}$ intersects the sphere $x^2 + y^2 + z^2 = 1$ when $x^2 + y^2 + \left(\sqrt{x^2 + y^2}\right)^2 = 1$ or $x^2 + y^2 = \tfrac{1}{2}$. So

$$V = \iint_{x^2 + y^2 \le 1/2} \left(\sqrt{1 - x^2 - y^2} - \sqrt{x^2 + y^2}\right) dA = \int_0^{2\pi} \int_0^{1/\sqrt{2}} \left(\sqrt{1 - r^2} - r\right) r\, dr\, d\theta$$

$$= \int_0^{2\pi} d\theta \int_0^{1/\sqrt{2}} \left(r\sqrt{1 - r^2} - r^2\right) dr = \big[\theta\big]_0^{2\pi} \left[-\tfrac{1}{3}(1 - r^2)^{3/2} - \tfrac{1}{3}r^3\right]_0^{1/\sqrt{2}} = 2\pi\left(-\tfrac{1}{3}\right)\left(\tfrac{1}{\sqrt{2}} - 1\right) = \tfrac{\pi}{3}\left(2 - \sqrt{2}\right)$$

27. The given solid is the region inside the cylinder $x^2 + y^2 = 4$ between the surfaces $z = \sqrt{64 - 4x^2 - 4y^2}$

and $z = -\sqrt{64 - 4x^2 - 4y^2}$. So

$$V = \iint_{x^2 + y^2 \le 4} \left[\sqrt{64 - 4x^2 - 4y^2} - \left(-\sqrt{64 - 4x^2 - 4y^2}\right)\right] dA = \iint_{x^2 + y^2 \le 4} 2 \cdot 2\sqrt{16 - x^2 - y^2}\, dA$$

$$= 4\int_0^{2\pi} \int_0^2 \sqrt{16 - r^2}\, r\, dr\, d\theta = 4\int_0^{2\pi} d\theta \int_0^2 r\sqrt{16 - r^2}\, dr = 4\big[\theta\big]_0^{2\pi} \left[-\tfrac{1}{3}(16 - r^2)^{3/2}\right]_0^2$$

$$= 8\pi\left(-\tfrac{1}{3}\right)\left(12^{3/2} - 16^{2/3}\right) = \tfrac{8\pi}{3}\left(64 - 24\sqrt{3}\right)$$

29.

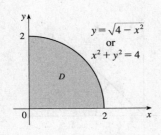

$y = \sqrt{4 - x^2}$
or
$x^2 + y^2 = 4$

D

$$\int_0^2 \int_0^{\sqrt{4 - x^2}} e^{-x^2 - y^2}\, dy\, dx = \int_0^{\pi/2} \int_0^2 e^{-r^2} r\, dr\, d\theta$$

$$= \int_0^{\pi/2} d\theta \int_0^2 r e^{-r^2}\, dr = \big[\theta\big]_0^{\pi/2} \left[-\tfrac{1}{2}e^{-r^2}\right]_0^2$$

$$= \tfrac{\pi}{2}\left[-\tfrac{1}{2}\left(e^{-4} - 1\right)\right] = \tfrac{\pi}{4}\left(1 - e^{-4}\right)$$

31. The region D of integration is shown in the figure. In polar coordinates the line $x = \sqrt{3}\, y$ is $\theta = \pi/6$, so

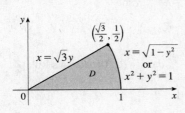

$\left(\tfrac{\sqrt{3}}{2}, \tfrac{1}{2}\right)$

$x = \sqrt{3}\,y$

$x = \sqrt{1 - y^2}$
or
$x^2 + y^2 = 1$

D

$$\int_0^{1/2} \int_{\sqrt{3}\,y}^{\sqrt{1 - y^2}} xy^2\, dx\, dy = \int_0^{\pi/6} \int_0^1 (r\cos\theta)(r\sin\theta)^2\, r\, dr\, d\theta$$

$$= \int_0^{\pi/6} \sin^2\theta \cos\theta\, d\theta \int_0^1 r^4\, dr$$

$$= \left[\tfrac{1}{3}\sin^3\theta\right]_0^{\pi/6} \left[\tfrac{1}{5}r^5\right]_0^1$$

$$= \left[\tfrac{1}{3}\left(\tfrac{1}{2}\right)^3 - 0\right]\left[\tfrac{1}{5} - 0\right] = \tfrac{1}{120}$$

33. $D = \{(r, \theta) \mid 0 \le r \le 1, 0 \le \theta \le 2\pi\}$, so

$\iint_D e^{(x^2 + y^2)^2} \, dA = \int_0^{2\pi} \int_0^1 e^{(r^2)^2} \, r \, dr \, d\theta = \int_0^{2\pi} d\theta \int_0^1 re^{r^4} \, dr = 2\pi \int_0^1 re^{r^4} \, dr$. Using a calculator, we estimate

$2\pi \int_0^1 re^{r^4} \, dr \approx 4.5951$.

35. The surface of the water in the pool is a circular disk D with radius 20 ft. If we place D on coordinate axes with the origin at

the center of D and define $f(x, y)$ to be the depth of the water at (x, y), then the volume of water in the pool is the volume of

the solid that lies above $D = \{(x, y) \mid x^2 + y^2 \le 400\}$ and below the graph of $f(x, y)$. We can associate north with the

positive y-direction, so we are given that the depth is constant in the x-direction and the depth increases linearly in the

y-direction from $f(0, -20) = 2$ to $f(0, 20) = 7$. The trace in the yz-plane is a line segment from $(0, -20, 2)$ to $(0, 20, 7)$.

The slope of this line is $\frac{7-2}{20-(-20)} = \frac{1}{8}$, so an equation of the line is $z - 7 = \frac{1}{8}(y - 20)$ $\Rightarrow$ $z = \frac{1}{8}y + \frac{9}{2}$. Since $f(x, y)$ is

independent of x, $f(x, y) = \frac{1}{8}y + \frac{9}{2}$. Thus the volume is given by $\iint_D f(x, y) \, dA$, which is most conveniently evaluated

using polar coordinates. Then $D = \{(r, \theta) \mid 0 \le r \le 20, 0 \le \theta \le 2\pi\}$ and substituting $x = r\cos\theta$, $y = r\sin\theta$ the integral

becomes

$$\int_0^{2\pi} \int_0^{20} \left(\frac{1}{8} r \sin\theta + \frac{9}{2} \right) r \, dr \, d\theta = \int_0^{2\pi} \left[\frac{1}{24} r^3 \sin\theta + \frac{9}{4} r^2 \right]_{r=0}^{r=20} d\theta = \int_0^{2\pi} \left(\frac{1000}{3} \sin\theta + 900 \right) d\theta$$

$$= \left[-\frac{1000}{3} \cos\theta + 900\theta \right]_0^{2\pi} = 1800\pi$$

Thus the pool contains $1800\pi \approx 5655$ ft^3 of water.

37. As in Exercise 15.2.61, $f_{\text{ave}} = \frac{1}{A(D)} \iint_D f(x, y) dA$. Here $D = \{(r, \theta) \mid a \le r \le b, 0 \le \theta \le 2\pi\}$,

so $A(D) = \pi b^2 - \pi a^2 = \pi(b^2 - a^2)$ and

$$f_{\text{ave}} = \frac{1}{A(D)} \iint_D \frac{1}{\sqrt{x^2 + y^2}} \, dA = \frac{1}{\pi(b^2 - a^2)} \int_0^{2\pi} \int_a^b \frac{1}{\sqrt{r^2}} \, r \, dr \, d\theta = \frac{1}{\pi(b^2 - a^2)} \int_0^{2\pi} d\theta \int_a^b dr$$

$$= \frac{1}{\pi(b^2 - a^2)} \left[\theta \right]_0^{2\pi} \left[r \right]_a^b = \frac{1}{\pi(b^2 - a^2)} (2\pi)(b - a) = \frac{2(b - a)}{(b + a)(b - a)} = \frac{2}{a + b}$$

39.

$$\int_{1/\sqrt{2}}^1 \int_{\sqrt{1-x^2}}^x xy \, dy \, dx + \int_1^{\sqrt{2}} \int_0^x xy \, dy \, dx + \int_{\sqrt{2}}^2 \int_0^{\sqrt{4-x^2}} xy \, dy \, dx$$

$$= \int_0^{\pi/4} \int_1^2 r^3 \cos\theta \sin\theta \, dr \, d\theta = \int_0^{\pi/4} \left[\frac{r^4}{4} \cos\theta \sin\theta \right]_{r=1}^{r=2} d\theta$$

$$= \frac{15}{4} \int_0^{\pi/4} \sin\theta \cos\theta \, d\theta = \frac{15}{4} \left[\frac{\sin^2\theta}{2} \right]_0^{\pi/4} = \frac{15}{16}$$

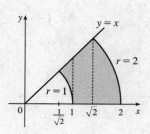

41. (a) We integrate by parts with $u = x$ and $dv = xe^{-x^2} \, dx$. Then $du = dx$ and $v = -\frac{1}{2}e^{-x^2}$, so

$$\int_0^\infty x^2 e^{-x^2} \, dx = \lim_{t \to \infty} \int_0^t x^2 e^{-x^2} \, dx = \lim_{t \to \infty} \left(-\frac{1}{2} x e^{-x^2} \Big|_0^t + \int_0^t \frac{1}{2} e^{-x^2} \, dx \right)$$

$$= \lim_{t \to \infty} \left(-\frac{1}{2} t e^{-t^2} \right) + \frac{1}{2} \int_0^\infty e^{-x^2} \, dx = 0 + \frac{1}{2} \int_0^\infty e^{-x^2} \, dx \qquad \text{[by l'Hospital's Rule]}$$

$$= \frac{1}{4} \int_{-\infty}^\infty e^{-x^2} \, dx \qquad \text{[since } e^{-x^2} \text{ is an even function]}$$

$$= \frac{1}{4} \sqrt{\pi} \qquad \text{[by Exercise 40(c)]}$$

(b) Let $u = \sqrt{x}$. Then $u^2 = x \;\Rightarrow\; dx = 2u\,du \;\Rightarrow$

$\int_0^\infty \sqrt{x}\,e^{-x}\,dx = \lim\limits_{t\to\infty} \int_0^t \sqrt{x}\,e^{-x}\,dx = \lim\limits_{t\to\infty} \int_0^{\sqrt{t}} u e^{-u^2} 2u\,du = 2\int_0^\infty u^2 e^{-u^2}\,du = 2\left(\tfrac{1}{4}\sqrt{\pi}\right)$ [by part(a)] $= \tfrac{1}{2}\sqrt{\pi}$.

15.4   Applications of Double Integrals

1. $Q = \iint_D \sigma(x,y)\,dA = \int_0^5 \int_2^5 (2x + 4y)\,dy\,dx = \int_0^5 \left[2xy + 2y^2\right]_{y=2}^{y=5} dx$

$= \int_0^5 (10x + 50 - 4x - 8)\,dx = \int_0^5 (6x + 42)\,dx = \left[3x^2 + 42x\right]_0^5 = 75 + 210 = 285\;\text{C}$

3. $m = \iint_D \rho(x,y)\,dA = \int_1^3 \int_1^4 ky^2\,dy\,dx = k\int_1^3 dx \int_1^4 y^2\,dy = k\,[x]_1^3 \left[\tfrac{1}{3}y^3\right]_1^4 = k(2)(21) = 42k,$

$\overline{x} = \tfrac{1}{m}\iint_D x\rho(x,y)\,dA = \tfrac{1}{42k}\int_1^3 \int_1^4 kxy^2\,dy\,dx = \tfrac{1}{42}\int_1^3 x\,dx \int_1^4 y^2\,dy = \tfrac{1}{42}\left[\tfrac{1}{2}x^2\right]_1^3 \left[\tfrac{1}{3}y^3\right]_1^4 = \tfrac{1}{42}(4)(21) = 2,$

$\overline{y} = \tfrac{1}{m}\iint_D y\rho(x,y)\,dA = \tfrac{1}{42k}\int_1^3 \int_1^4 ky^3\,dy\,dx = \tfrac{1}{42}\int_1^3 dx \int_1^4 y^3\,dy = \tfrac{1}{42}[x]_1^3 \left[\tfrac{1}{4}y^4\right]_1^4 = \tfrac{1}{42}(2)\left(\tfrac{255}{4}\right) = \tfrac{85}{28}$

Hence $m = 42k,\; (\overline{x}, \overline{y}) = \left(2, \tfrac{85}{28}\right).$

5. $m = \int_0^2 \int_{x/2}^{3-x} (x + y)\,dy\,dx = \int_0^2 \left[xy + \tfrac{1}{2}y^2\right]_{y=x/2}^{y=3-x} dx = \int_0^2 \left[x(3 - x) + \tfrac{1}{2}(3 - x)^2 - \tfrac{1}{2}x^2 - \tfrac{1}{8}x^2\right] dx$

$= \int_0^2 \left(-\tfrac{9}{8}x^2 + \tfrac{9}{2}\right) dx = \left[-\tfrac{9}{8}\left(\tfrac{1}{3}x^3\right) + \tfrac{9}{2}x\right]_0^2 = 6,$

$M_y = \int_0^2 \int_{x/2}^{3-x} (x^2 + xy)\,dy\,dx = \int_0^2 \left[x^2 y + \tfrac{1}{2}xy^2\right]_{y=x/2}^{y=3-x} dx = \int_0^2 \left(\tfrac{9}{2}x - \tfrac{9}{8}x^3\right) dx = \tfrac{9}{2},$

$M_x = \int_0^2 \int_{x/2}^{3-y} (xy + y^2)\,dy\,dx = \int_0^2 \left[\tfrac{1}{2}xy^2 + \tfrac{1}{3}y^3\right]_{y=x/2}^{y=3-x} dx = \int_0^2 \left(9 - \tfrac{9}{2}x\right) dx = 9.$

Hence $m = 6,\; (\overline{x}, \overline{y}) = \left(\dfrac{M_y}{m}, \dfrac{M_x}{m}\right) = \left(\dfrac{3}{4}, \dfrac{3}{2}\right).$

7. $m = \int_{-1}^1 \int_0^{1-x^2} ky\,dy\,dx = k\int_{-1}^1 \left[\tfrac{1}{2}y^2\right]_{y=0}^{y=1-x^2} dx = \tfrac{1}{2}k\int_{-1}^1 (1 - x^2)^2\,dx = \tfrac{1}{2}k\int_{-1}^1 (1 - 2x^2 + x^4)\,dx$

$= \tfrac{1}{2}k\left[x - \tfrac{2}{3}x^3 + \tfrac{1}{5}x^5\right]_{-1}^1 = \tfrac{1}{2}k\left(1 - \tfrac{2}{3} + \tfrac{1}{5} + 1 - \tfrac{2}{3} + \tfrac{1}{5}\right) = \tfrac{8}{15}k,$

$M_y = \int_{-1}^1 \int_0^{1-x^2} kxy\,dy\,dx = k\int_{-1}^1 \left[\tfrac{1}{2}xy^2\right]_{y=0}^{y=1-x^2} dx = \tfrac{1}{2}k\int_{-1}^1 x(1 - x^2)^2\,dx = \tfrac{1}{2}k\int_{-1}^1 (x - 2x^3 + x^5)\,dx$

$= \tfrac{1}{2}k\left[\tfrac{1}{2}x^2 - \tfrac{1}{2}x^4 + \tfrac{1}{6}x^6\right]_{-1}^1 = \tfrac{1}{2}k\left(\tfrac{1}{2} - \tfrac{1}{2} + \tfrac{1}{6} - \tfrac{1}{2} + \tfrac{1}{2} - \tfrac{1}{6}\right) = 0,$

$M_x = \int_{-1}^1 \int_0^{1-x^2} ky^2\,dy\,dx = k\int_{-1}^1 \left[\tfrac{1}{3}y^3\right]_{y=0}^{y=1-x^2} dx = \tfrac{1}{3}k\int_{-1}^1 (1 - x^2)^3\,dx = \tfrac{1}{3}k\int_{-1}^1 (1 - 3x^2 + 3x^4 - x^6)\,dx$

$= \tfrac{1}{3}k\left[x - x^3 + \tfrac{3}{5}x^5 - \tfrac{1}{7}x^7\right]_{-1}^1 = \tfrac{1}{3}k\left(1 - 1 + \tfrac{3}{5} - \tfrac{1}{7} + 1 - 1 + \tfrac{3}{5} - \tfrac{1}{7}\right) = \tfrac{32}{105}k.$

Hence $m = \tfrac{8}{15}k,\; (\overline{x}, \overline{y}) = \left(0, \dfrac{32k/105}{8k/15}\right) = \left(0, \tfrac{4}{7}\right).$

9. $m = \int_0^1 \int_0^{e^{-x}} xy\,dy\,dx = \int_0^1 x\left[\tfrac{1}{2}y^2\right]_{y=0}^{y=e^{-x}} dx = \tfrac{1}{2}\int_0^1 x\left(e^{-x}\right)^2 dx = \tfrac{1}{2}\int_0^1 xe^{-2x}\,dx$ $\begin{bmatrix}\text{integrate by parts with}\\ u = x,\, dv = e^{-2x}\,dx\end{bmatrix}$

$= \tfrac{1}{2}\left[-\tfrac{1}{4}(2x + 1)e^{-2x}\right]_0^1 = -\tfrac{1}{8}\left(3e^{-2} - 1\right) = \tfrac{1}{8} - \tfrac{3}{8}e^{-2},$

[continued]

$M_y = \int_0^1 \int_0^{e^{-x}} x^2 y \, dy \, dx = \int_0^1 x^2 \left[\frac{1}{2}y^2\right]_{y=0}^{y=e^{-x}} dx = \frac{1}{2}\int_0^1 x^2 e^{-2x}\, dx$ [integrate by parts twice]

$\quad = \frac{1}{2}\left[-\frac{1}{4}\left(2x^2 + 2x + 1\right)e^{-2x}\right]_0^1 = -\frac{1}{8}\left(5e^{-2} - 1\right) = \frac{1}{8} - \frac{5}{8}e^{-2}$,

$M_x = \int_0^1 \int_0^{e^{-x}} xy^2 \, dy \, dx = \int_0^1 x \left[\frac{1}{3}y^3\right]_{y=0}^{y=e^{-x}} dx = \frac{1}{3}\int_0^1 xe^{-3x}\, dx$

$\quad = \frac{1}{3}\left[-\frac{1}{9}(3x+1)e^{-3x}\right]_0^1 = -\frac{1}{27}\left(4e^{-3} - 1\right) = \frac{1}{27} - \frac{4}{27}e^{-3}$.

Hence $m = \frac{1}{8}\left(1 - 3e^{-2}\right)$, $(\overline{x}, \overline{y}) = \left(\dfrac{\frac{1}{8}\left(1 - 5e^{-2}\right)}{\frac{1}{8}\left(1 - 3e^{-2}\right)}, \dfrac{\frac{1}{27}\left(1 - 4e^{-3}\right)}{\frac{1}{8}\left(1 - 3e^{-2}\right)}\right) = \left(\dfrac{e^2 - 5}{e^2 - 3}, \dfrac{8\left(e^3 - 4\right)}{27\left(e^3 - 3e\right)}\right)$.

11. $\rho(x, y) = ky$, $m = \iint_D ky \, dA = \int_0^{\pi/2}\int_0^1 k(r\sin\theta)\, r\, dr\, d\theta = k\int_0^{\pi/2}\sin\theta\, d\theta \int_0^1 r^2\, dr$

$\quad = k\left[-\cos\theta\right]_0^{\pi/2}\left[\frac{1}{3}r^3\right]_0^1 = k(1)\left(\frac{1}{3}\right) = \frac{1}{3}k$,

$M_y = \iint_D x \cdot ky \, dA = \int_0^{\pi/2}\int_0^1 k(r\cos\theta)(r\sin\theta)\, r\, dr\, d\theta = k\int_0^{\pi/2}\sin\theta\cos\theta\, d\theta \int_0^1 r^3\, dr$

$\quad = k\left[\frac{1}{2}\sin^2\theta\right]_0^{\pi/2}\left[\frac{1}{4}r^4\right]_0^1 = k\left(\frac{1}{2}\right)\left(\frac{1}{4}\right) = \frac{1}{8}k$,

$M_x = \iint_D y \cdot ky \, dA = \int_0^{\pi/2}\int_0^1 k(r\sin\theta)^2\, r\, dr\, d\theta = k\int_0^{\pi/2}\sin^2\theta\, d\theta \int_0^1 r^3\, dr$

$\quad = k\left[\frac{1}{2}\theta - \frac{1}{4}\sin 2\theta\right]_0^{\pi/2}\left[\frac{1}{4}r^4\right]_0^1 = k\left(\frac{\pi}{4}\right)\left(\frac{1}{4}\right) = \frac{\pi}{16}k$.

Hence $(\overline{x}, \overline{y}) = \left(\dfrac{k/8}{k/3}, \dfrac{k\pi/16}{k/3}\right) = \left(\dfrac{3}{8}, \dfrac{3\pi}{16}\right)$.

13.

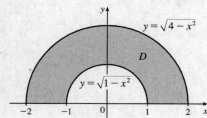

$\rho(x, y) = k\sqrt{x^2 + y^2} = kr$,

$m = \iint_D \rho(x, y)\, dA = \int_0^{\pi}\int_1^2 kr \cdot r\, dr\, d\theta$

$\quad = k\int_0^{\pi} d\theta \int_1^2 r^2\, dr = k(\pi)\left[\frac{1}{3}r^3\right]_1^2 = \frac{7}{3}\pi k$,

$M_y = \iint_D x\rho(x, y)\, dA = \int_0^{\pi}\int_1^2 (r\cos\theta)(kr)\, r\, dr\, d\theta = k\int_0^{\pi}\cos\theta\, d\theta \int_1^2 r^3\, dr$

$\quad = k\left[\sin\theta\right]_0^{\pi}\left[\frac{1}{4}r^4\right]_1^2 = k(0)\left(\frac{15}{4}\right) = 0$ [this is to be expected as the region and density function are symmetric about the y-axis]

$M_x = \iint_D y\rho(x, y)\, dA = \int_0^{\pi}\int_1^2 (r\sin\theta)(kr)\, r\, dr\, d\theta = k\int_0^{\pi}\sin\theta\, d\theta \int_1^2 r^3\, dr$

$\quad = k\left[-\cos\theta\right]_0^{\pi}\left[\frac{1}{4}r^4\right]_1^2 = k(1+1)\left(\frac{15}{4}\right) = \frac{15}{2}k$.

Hence $(\overline{x}, \overline{y}) = \left(0, \dfrac{15k/2}{7\pi k/3}\right) = \left(0, \dfrac{45}{14\pi}\right)$.

15. Placing the vertex opposite the hypotenuse at $(0, 0)$, $\rho(x, y) = k(x^2 + y^2)$. Then

$m = \int_0^a \int_0^{a-x} k(x^2 + y^2)\, dy\, dx = k\int_0^a \left[ax^2 - x^3 + \frac{1}{3}(a-x)^3\right] dx = k\left[\frac{1}{3}ax^3 - \frac{1}{4}x^4 - \frac{1}{12}(a-x)^4\right]_0^a = \frac{1}{6}ka^4$.

By symmetry, $M_y = M_x = \int_0^a \int_0^{a-x} ky(x^2 + y^2)\, dy\, dx = k\int_0^a \left[\frac{1}{2}(a-x)^2 x^2 + \frac{1}{4}(a-x)^4\right] dx$

$\quad\quad = k\left[\frac{1}{6}a^2 x^3 - \frac{1}{4}ax^4 + \frac{1}{10}x^5 - \frac{1}{20}(a-x)^5\right]_0^a = \frac{1}{15}ka^5$

Hence $(\overline{x}, \overline{y}) = \left(\frac{2}{5}a, \frac{2}{5}a\right)$.

17. $I_x = \iint_D y^2 \rho(x,y) dA = \int_1^3 \int_1^4 y^2 \cdot ky^2 \, dy \, dx = k \int_1^3 dx \int_1^4 y^4 \, dy = k \, [x]_1^3 \left[\frac{1}{5}y^5\right]_1^4 = k(2)\left(\frac{1023}{5}\right) = 409.2k,$

$I_y = \iint_D x^2 \rho(x,y) \, dA = \int_1^3 \int_1^4 x^2 \cdot ky^2 \, dy \, dx = k \int_1^3 x^2 \, dx \int_1^4 y^2 \, dy = k \left[\frac{1}{3}x^3\right]_1^3 \left[\frac{1}{3}y^3\right]_1^4 = k\left(\frac{26}{3}\right)(21) = 182k,$

and $I_0 = I_x + I_y = 409.2k + 182k = 591.2k.$

19. As in Exercise 15, we place the vertex opposite the hypotenuse at $(0,0)$ and the equal sides along the positive axes.

$I_x = \int_0^a \int_0^{a-x} y^2 k(x^2 + y^2) \, dy \, dx = k \int_0^a \int_0^{a-x}(x^2y^2 + y^4) \, dy \, dx = k \int_0^a \left[\frac{1}{3}x^2y^3 + \frac{1}{5}y^5\right]_{y=0}^{y=a-x} \, dx$

$= k \int_0^a \left[\frac{1}{3}x^2(a-x)^3 + \frac{1}{5}(a-x)^5\right] dx = k \left[\frac{1}{3}\left(\frac{1}{3}a^3x^3 - \frac{3}{4}a^2x^4 + \frac{3}{5}ax^5 - \frac{1}{6}x^6\right) - \frac{1}{30}(a-x)^6\right]_0^a = \frac{7}{180}ka^6,$

$I_y = \int_0^a \int_0^{a-x} x^2 k(x^2 + y^2) \, dy \, dx = k \int_0^a \int_0^{a-x}(x^4 + x^2y^2) \, dy \, dx = k \int_0^a \left[x^4y + \frac{1}{3}x^2y^3\right]_{y=0}^{y=a-x} \, dx$

$= k \int_0^a \left[x^4(a-x) + \frac{1}{3}x^2(a-x)^3\right] dx = k \left[\frac{1}{5}ax^5 - \frac{1}{6}x^6 + \frac{1}{3}\left(\frac{1}{3}a^3x^3 - \frac{3}{4}a^2x^4 + \frac{3}{5}ax^5 - \frac{1}{6}x^6\right)\right]_0^a = \frac{7}{180}ka^6,$

and $I_0 = I_x + I_y = \frac{7}{90}ka^6.$

21. $I_x = \iint_D y^2 \rho(x,y) dA = \int_0^h \int_0^b \rho y^2 \, dx \, dy = \rho \int_0^b dx \int_0^h y^2 \, dy = \rho \, [x]_0^b \left[\frac{1}{3}y^3\right]_0^h = \rho b \left(\frac{1}{3}h^3\right) = \frac{1}{3}\rho bh^3,$

$I_y = \iint_D x^2 \rho(x,y) dA = \int_0^h \int_0^b \rho x^2 \, dx \, dy = \rho \int_0^b x^2 \, dx \int_0^h dy = \rho \left[\frac{1}{3}x^3\right]_0^b [y]_0^h = \frac{1}{3}\rho b^3 h,$

and $m = \rho$ (area of rectangle) $= \rho bh$ since the lamina is homogeneous. Hence $\overline{\overline{x}}^2 = \dfrac{I_y}{m} = \dfrac{\frac{1}{3}\rho b^3 h}{\rho bh} = \dfrac{b^2}{3} \;\Rightarrow\; \overline{\overline{x}} = \dfrac{b}{\sqrt{3}}$

and $\overline{\overline{y}}^2 = \dfrac{I_x}{m} = \dfrac{\frac{1}{3}\rho bh^3}{\rho bh} = \dfrac{h^2}{3} \;\Rightarrow\; \overline{\overline{y}} = \dfrac{h}{\sqrt{3}}.$

23. In polar coordinates, the region is $D = \left\{(r,\theta) \mid 0 \le r \le a, 0 \le \theta \le \frac{\pi}{2}\right\}$, so

$I_x = \iint_D y^2 \rho \, dA = \int_0^{\pi/2} \int_0^a \rho(r\sin\theta)^2 \, r \, dr \, d\theta = \rho \int_0^{\pi/2} \sin^2\theta \, d\theta \int_0^a r^3 \, dr$

$= \rho\left[\frac{1}{2}\theta - \frac{1}{4}\sin 2\theta\right]_0^{\pi/2} \left[\frac{1}{4}r^4\right]_0^a = \rho\left(\frac{\pi}{4}\right)\left(\frac{1}{4}a^4\right) = \frac{1}{16}\rho a^4 \pi,$

$I_y = \iint_D x^2 \rho \, dA = \int_0^{\pi/2} \int_0^a \rho(r\cos\theta)^2 \, r \, dr \, d\theta = \rho \int_0^{\pi/2} \cos^2\theta \, d\theta \int_0^a r^3 \, dr$

$= \rho\left[\frac{1}{2}\theta + \frac{1}{4}\sin 2\theta\right]_0^{\pi/2} \left[\frac{1}{4}r^4\right]_0^a = \rho\left(\frac{\pi}{4}\right)\left(\frac{1}{4}a^4\right) = \frac{1}{16}\rho a^4 \pi,$

and $m = \rho \cdot A(D) = \rho \cdot \frac{1}{4}\pi a^2$ since the lamina is homogeneous. Hence $\overline{\overline{x}}^2 = \overline{\overline{y}}^2 = \dfrac{\frac{1}{16}\rho a^4 \pi}{\frac{1}{4}\rho a^2 \pi} = \dfrac{a^2}{4} \;\Rightarrow\; \overline{\overline{x}} = \overline{\overline{y}} = \dfrac{a}{2}.$

25. The right loop of the curve is given by $D = \{(r,\theta) \mid 0 \le r \le \cos 2\theta, \; -\pi/4 \le \theta \le \pi/4\}$. Using a CAS, we

find $m = \iint_D \rho(x,y) \, dA = \iint_D (x^2 + y^2) \, dA = \int_{-\pi/4}^{\pi/4} \int_0^{\cos 2\theta} r^2 \, r \, dr \, d\theta = \dfrac{3\pi}{64}.$ Then

$\overline{x} = \dfrac{1}{m} \iint_D x\rho(x,y) \, dA = \dfrac{64}{3\pi} \int_{-\pi/4}^{\pi/4} \int_0^{\cos 2\theta} (r\cos\theta) \, r^2 \, r \, dr \, d\theta = \dfrac{64}{3\pi} \int_{-\pi/4}^{\pi/4} \int_0^{\cos 2\theta} r^4 \cos\theta \, dr \, d\theta = \dfrac{16384\sqrt{2}}{10395\pi}$ and

$\overline{y} = \dfrac{1}{m} \iint_D y\rho(x,y) \, dA = \dfrac{64}{3\pi} \int_{-\pi/4}^{\pi/4} \int_0^{\cos 2\theta} (r\sin\theta) \, r^2 \, r \, dr \, d\theta = \dfrac{64}{3\pi} \int_{-\pi/4}^{\pi/4} \int_0^{\cos 2\theta} r^4 \sin\theta \, dr \, d\theta = 0,$ so

$(\overline{x}, \overline{y}) = \left(\dfrac{16384\sqrt{2}}{10395\pi}, 0\right).$

[continued]

The moments of inertia are

$$I_x = \iint_D y^2 \rho(x,y)\, dA = \int_{-\pi/4}^{\pi/4} \int_0^{\cos 2\theta} (r\sin\theta)^2\, r^2\, r\, dr\, d\theta = \int_{-\pi/4}^{\pi/4} \int_0^{\cos 2\theta} r^5 \sin^2\theta\, dr\, d\theta = \frac{5\pi}{384} - \frac{4}{105},$$

$$I_y = \iint_D x^2 \rho(x,y)\, dA = \int_{-\pi/4}^{\pi/4} \int_0^{\cos 2\theta} (r\cos\theta)^2\, r^2\, r\, dr\, d\theta = \int_{-\pi/4}^{\pi/4} \int_0^{\cos 2\theta} r^5 \cos^2\theta\, dr\, d\theta = \frac{5\pi}{384} + \frac{4}{105}, \text{ and}$$

$$I_0 = I_x + I_y = \frac{5\pi}{192}.$$

27. (a) $f(x,y)$ is a joint density function, so we know $\iint_{\mathbb{R}^2} f(x,y)\, dA = 1$. Since $f(x,y) = 0$ outside the

rectangle $[0,1] \times [0,2]$, we can say

$$\iint_{\mathbb{R}^2} f(x,y)\, dA = \int_{-\infty}^{\infty} \int_{-\infty}^{\infty} f(x,y)\, dy\, dx = \int_0^1 \int_0^2 Cx(1+y)\, dy\, dx$$

$$= C \int_0^1 x \big[y + \tfrac{1}{2}y^2\big]_{y=0}^{y=2}\, dx = C \int_0^1 4x\, dx = C\big[2x^2\big]_0^1 = 2C$$

Then $2C = 1 \ \Rightarrow \ C = \frac{1}{2}$.

(b) $P(X \le 1, Y \le 1) = \int_{-\infty}^1 \int_{-\infty}^1 f(x,y)\, dy\, dx = \int_0^1 \int_0^1 \tfrac{1}{2}x(1+y)\, dy\, dx$

$$= \int_0^1 \tfrac{1}{2}x\big[y + \tfrac{1}{2}y^2\big]_{y=0}^{y=1}\, dx = \int_0^1 \tfrac{1}{2}x\big(\tfrac{3}{2}\big)\, dx = \tfrac{3}{4}\big[\tfrac{1}{2}x^2\big]_0^1 = \tfrac{3}{8} \text{ or } 0.375$$

(c) $P(X+Y \le 1) = P((X,Y) \in D)$ where D is the triangular region shown in

the figure. Thus

$$P(X+Y \le 1) = \iint_D f(x,y)\, dA = \int_0^1 \int_0^{1-x} \tfrac{1}{2}x(1+y)\, dy\, dx$$

$$= \int_0^1 \tfrac{1}{2}x\big[y + \tfrac{1}{2}y^2\big]_{y=0}^{y=1-x}\, dx = \int_0^1 \tfrac{1}{2}x\big(\tfrac{1}{2}x^2 - 2x + \tfrac{3}{2}\big)\, dx$$

$$= \tfrac{1}{4}\int_0^1 (x^3 - 4x^2 + 3x)\, dx = \tfrac{1}{4}\Big[\frac{x^4}{4} - 4\frac{x^3}{3} + 3\frac{x^2}{2}\Big]_0^1$$

$$= \tfrac{5}{48} \approx 0.1042$$

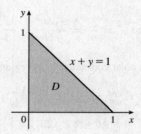

29. (a) $f(x,y) \ge 0$, so f is a joint density function if $\iint_{\mathbb{R}^2} f(x,y)\, dA = 1$. Here, $f(x,y) = 0$ outside the first quadrant, so

$$\iint_{\mathbb{R}^2} f(x,y)\, dA = \int_0^\infty \int_0^\infty 0.1 e^{-(0.5x+0.2y)}\, dy\, dx = 0.1 \int_0^\infty \int_0^\infty e^{-0.5x} e^{-0.2y}\, dy\, dx = 0.1 \int_0^\infty e^{-0.5x}\, dx \int_0^\infty e^{-0.2y}\, dy$$

$$= 0.1 \lim_{t\to\infty} \int_0^t e^{-0.5x}\, dx \lim_{t\to\infty} \int_0^t e^{-0.2y}\, dy = 0.1 \lim_{t\to\infty} \big[-2e^{-0.5x}\big]_0^t \lim_{t\to\infty} \big[-5e^{-0.2y}\big]_0^t$$

$$= 0.1 \lim_{t\to\infty} \big[-2(e^{-0.5t} - 1)\big] \lim_{t\to\infty} \big[-5(e^{-0.2t} - 1)\big] = (0.1)\cdot(-2)(0-1)\cdot(-5)(0-1) = 1$$

Thus $f(x,y)$ is a joint density function.

(b) (i) No restriction is placed on X, so

$$P(Y \ge 1) = \int_{-\infty}^\infty \int_1^\infty f(x,y)\, dy\, dx = \int_0^\infty \int_1^\infty 0.1 e^{-(0.5x+0.2y)}\, dy\, dx$$

$$= 0.1 \int_0^\infty e^{-0.5x}\, dx \int_1^\infty e^{-0.2y}\, dy = 0.1 \lim_{t\to\infty} \int_0^t e^{-0.5x}\, dx \lim_{t\to\infty} \int_1^t e^{-0.2y}\, dy$$

$$= 0.1 \lim_{t\to\infty} \big[-2e^{-0.5x}\big]_0^t \lim_{t\to\infty} \big[-5e^{-0.2y}\big]_1^t = 0.1 \lim_{t\to\infty} \big[-2(e^{-0.5t} - 1)\big] \lim_{t\to\infty} \big[-5(e^{-0.2t} - e^{-0.2})\big]$$

$$(0.1)\cdot(-2)(0-1)\cdot(-5)(0 - e^{-0.2}) = e^{-0.2} \approx 0.8187$$

(ii) $P(X \le 2, Y \le 4) = \int_{-\infty}^{2} \int_{-\infty}^{4} f(x, y)\, dy\, dx = \int_{0}^{2} \int_{0}^{4} 0.1 e^{-(0.5x + 0.2y)}\, dy\, dx$

$$= 0.1 \int_{0}^{2} e^{-0.5x}\, dx \int_{0}^{4} e^{-0.2y}\, dy = 0.1 \left[-2e^{-0.5x} \right]_{0}^{2} \left[-5e^{-0.2y} \right]_{0}^{4}$$

$$= (0.1) \cdot (-2)(e^{-1} - 1) \cdot (-5)(e^{-0.8} - 1)$$

$$= (e^{-1} - 1)(e^{-0.8} - 1) = 1 + e^{-1.8} - e^{-0.8} - e^{-1} \approx 0.3481$$

(c) The expected value of X is given by

$$\mu_1 = \iint_{\mathbb{R}^2} x\, f(x, y)\, dA = \int_{0}^{\infty} \int_{0}^{\infty} x \left[0.1 e^{-(0.5x + 0.2y)} \right] dy\, dx$$

$$= 0.1 \int_{0}^{\infty} x e^{-0.5x}\, dx \int_{0}^{\infty} e^{-0.2y}\, dy = 0.1 \lim_{t \to \infty} \int_{0}^{t} x e^{-0.5x}\, dx \lim_{t \to \infty} \int_{0}^{t} e^{-0.2y}\, dy$$

To evaluate the first integral, we integrate by parts with $u = x$ and $dv = e^{-0.5x}\, dx$ (or we can use Formula 96

in the Table of Integrals): $\int x e^{-0.5x}\, dx = -2x e^{-0.5x} - \int -2e^{-0.5x}\, dx = -2x e^{-0.5x} - 4e^{-0.5x} = -2(x + 2)e^{-0.5x}$.

Thus

$$\mu_1 = 0.1 \lim_{t \to \infty} \left[-2(x + 2)e^{-0.5x} \right]_{0}^{t} \lim_{t \to \infty} \left[-5e^{-0.2y} \right]_{0}^{t}$$

$$= 0.1 \lim_{t \to \infty} (-2) \left[(t + 2)e^{-0.5t} - 2 \right] \lim_{t \to \infty} (-5) \left[e^{-0.2t} - 1 \right]$$

$$= 0.1(-2) \left(\lim_{t \to \infty} \frac{t + 2}{e^{0.5t}} - 2 \right)(-5)(-1) = 2 \qquad \text{[by l'Hospital's Rule]}$$

The expected value of Y is given by

$$\mu_2 = \iint_{\mathbb{R}^2} y\, f(x, y)\, dA = \int_{0}^{\infty} \int_{0}^{\infty} y \left[0.1 e^{-(0.5 + 0.2y)} \right] dy\, dx$$

$$= 0.1 \int_{0}^{\infty} e^{-0.5x}\, dx \int_{0}^{\infty} y e^{-0.2y}\, dy = 0.1 \lim_{t \to \infty} \int_{0}^{t} e^{-0.5x}\, dx \lim_{t \to \infty} \int_{0}^{t} y e^{-0.2y}\, dy$$

To evaluate the second integral, we integrate by parts with $u = y$ and $dv = e^{-0.2y}\, dy$ (or again we can use Formula 96 in

the Table of Integrals) which gives $\int y e^{-0.2y}\, dy = -5y e^{-0.2y} + \int 5 e^{-0.2y}\, dy = -5(y + 5)e^{-0.2y}$. Then

$$\mu_2 = 0.1 \lim_{t \to \infty} \left[-2e^{-0.5x} \right]_{0}^{t} \lim_{t \to \infty} \left[-5(y + 5)e^{-0.2y} \right]_{0}^{t}$$

$$= 0.1 \lim_{t \to \infty} \left[-2(e^{-0.5t} - 1) \right] \lim_{t \to \infty} \left(-5 \left[(t + 5)e^{-0.2t} - 5 \right] \right)$$

$$= 0.1(-2)(-1) \cdot (-5) \left(\lim_{t \to \infty} \frac{t + 5}{e^{0.2t}} - 5 \right) = 5 \qquad \text{[by l'Hospital's Rule]}$$

31. (a) The random variables X and Y are normally distributed with $\mu_1 = 45$, $\mu_2 = 20$, $\sigma_1 = 0.5$, and $\sigma_2 = 0.1$.

The individual density functions for X and Y, then, are $f_1(x) = \dfrac{1}{0.5 \sqrt{2\pi}} e^{-(x - 45)^2 / 0.5}$ and

$f_2(y) = \dfrac{1}{0.1 \sqrt{2\pi}} e^{-(y - 20)^2 / 0.02}$. Since X and Y are independent, the joint density function is the product

$$f(x, y) = f_1(x) f_2(y) = \frac{1}{0.5 \sqrt{2\pi}} e^{-(x - 45)^2 / 0.5} \frac{1}{0.1 \sqrt{2\pi}} e^{-(y - 20)^2 / 0.02} = \frac{10}{\pi} e^{-2(x - 45)^2 - 50(y - 20)^2}.$$

Then $P(40 \le X \le 50, 20 \le Y \le 25) = \int_{40}^{50} \int_{20}^{25} f(x, y)\, dy\, dx = \dfrac{10}{\pi} \int_{40}^{50} \int_{20}^{25} e^{-2(x - 45)^2 - 50(y - 20)^2}\, dy\, dx$.

Using a CAS or calculator to evaluate the integral, we get $P(40 \le X \le 50, 20 \le Y \le 25) \approx 0.500$.

(b) $P(4(X-45)^2 + 100(Y-20)^2 \leq 2) = \iint_D \frac{10}{\pi} e^{-2(x-45)^2 - 50(y-20)^2}\, dA$, where D is the region enclosed by the ellipse

$4(x-45)^2 + 100(y-20)^2 = 2$. Solving for y gives $y = 20 \pm \frac{1}{10}\sqrt{2 - 4(x-45)^2}$, the upper and lower halves of the

ellipse, and these two halves meet where $y = 20$ [since the ellipse is centered at $(45, 20)$] $\Rightarrow$ $4(x-45)^2 = 2$ $\Rightarrow$

$x = 45 \pm \frac{1}{\sqrt{2}}$. Thus

$$\iint_D \frac{10}{\pi} e^{-2(x-45)^2 - 50(y-20)^2}\, dA = \frac{10}{\pi} \int_{45-1/\sqrt{2}}^{45+1/\sqrt{2}} \int_{20-\frac{1}{10}\sqrt{2-4(x-45)^2}}^{20+\frac{1}{10}\sqrt{2-4(x-45)^2}} e^{-2(x-45)^2 - 50(y-20)^2}\, dy\, dx.$$

Using a CAS or calculator to evaluate the integral, we get $P(4(X-45)^2 + 100(Y-20)^2 \leq 2) \approx 0.632$.

33. (a) If $f(P, A)$ is the probability that an individual at A will be infected by an individual at P, and $k\, dA$ is the number of

infected individuals in an element of area dA, then $f(P, A)k\, dA$ is the number of infections that should result from

exposure of the individual at A to infected people in the element of area dA. Integration over D gives the number of

infections of the person at A due to all the infected people in D. In rectangular coordinates (with the origin at the city's

center), the exposure of a person at A is

$$E = \iint_D k f(P, A)\, dA = k \iint_D \frac{1}{20}\left[20 - d(P, A)\right] dA = k \iint_D \left[1 - \frac{1}{20}\sqrt{(x-x_0)^2 + (y-y_0)^2}\right] dA$$

(b) If $A = (0, 0)$, then

$$E = k \iint_D \left[1 - \frac{1}{20}\sqrt{x^2 + y^2}\right] dA$$

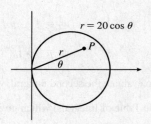

$r = 20\cos\theta$

$$= k \int_0^{2\pi} \int_0^{10} \left(1 - \frac{1}{20}r\right) r\, dr\, d\theta = 2\pi k \left[\frac{1}{2}r^2 - \frac{1}{60}r^3\right]_0^{10}$$

$$= 2\pi k\left(50 - \frac{50}{3}\right) = \frac{200}{3}\pi k \approx 209k$$

For A at the edge of the city, it is convenient to use a polar coordinate system centered at A. Then the polar equation for

the circular boundary of the city becomes $r = 20\cos\theta$ instead of $r = 10$, and the distance from A to a point P in the city

is again r (see the figure). So

$$E = k \int_{-\pi/2}^{\pi/2} \int_0^{20\cos\theta} \left(1 - \frac{1}{20}r\right) r\, dr\, d\theta = k \int_{-\pi/2}^{\pi/2} \left[\frac{1}{2}r^2 - \frac{1}{60}r^3\right]_{r=0}^{r=20\cos\theta} d\theta$$

$$= k \int_{-\pi/2}^{\pi/2}\left(200\cos^2\theta - \frac{400}{3}\cos^3\theta\right) d\theta = 200k \int_{-\pi/2}^{\pi/2}\left[\frac{1}{2} + \frac{1}{2}\cos 2\theta - \frac{2}{3}(1 - \sin^2\theta)\cos\theta\right] d\theta$$

$$= 200k\left[\frac{1}{2}\theta + \frac{1}{4}\sin 2\theta - \frac{2}{3}\sin\theta + \frac{2}{3}\cdot\frac{1}{3}\sin^3\theta\right]_{-\pi/2}^{\pi/2} = 200k\left[\frac{\pi}{4} + 0 - \frac{2}{3} + \frac{2}{9} + \frac{\pi}{4} + 0 - \frac{2}{3} + \frac{2}{9}\right]$$

$$= 200k\left(\frac{\pi}{2} - \frac{8}{9}\right) \approx 136k$$

Therefore the risk of infection is much lower at the edge of the city than in the middle, so it is better to live at the edge.

15.5 Surface Area

1. Here $z = f(x, y) = 5x + 3y + 6$ and D is the rectangle $[1, 4] \times [2, 6]$, so by Formula 2 the area of the surface is

$$A(S) = \iint_D \sqrt{[f_x(x, y)]^2 + [f_y(x, y)]^2 + 1}\, dA = \iint_D \sqrt{5^2 + 3^2 + 1}\, dA = \sqrt{35} \iint_D dA$$
$$= \sqrt{35}\, A(D) = \sqrt{35}\, (3)(4) = 12\sqrt{35}$$

3. The surface S is given by $z = f(x, y) = 6 - 3x - 2y$ which intersects the xy-plane in the line $3x + 2y = 6$, so D is the

triangular region given by $\left\{(x, y) \mid 0 \le x \le 2, 0 \le y \le 3 - \frac{3}{2}x\right\}$. By Formula 2, the surface area of S is

$$A(S) = \iint_D \sqrt{(-3)^2 + (-2)^2 + 1}\, dA = \sqrt{14} \iint_D dA = \sqrt{14}\, A(D) = \sqrt{14}\left(\frac{1}{2} \cdot 2 \cdot 3\right) = 3\sqrt{14}$$

5. The paraboloid intersects the plane $z = -2$ when $1 - x^2 - y^2 = -2 \Leftrightarrow x^2 + y^2 = 3$, so $D = \left\{(x, y) \mid x^2 + y^2 \le 3\right\}$.

Here $z = f(x, y) = 1 - x^2 - y^2 \Rightarrow f_x = -2x, f_y = -2y$ and

$$A(S) = \iint_D \sqrt{(-2x)^2 + (-2y)^2 + 1}\, dA = \iint_D \sqrt{4(x^2 + y^2) + 1}\, dA = \int_0^{2\pi} \int_0^{\sqrt{3}} \sqrt{4r^2 + 1}\, r\, dr\, d\theta$$

$$= \int_0^{2\pi} d\theta \int_0^{\sqrt{3}} r\sqrt{4r^2 + 1}\, dr = \left[\theta\right]_0^{2\pi} \left[\frac{1}{12}(4r^2 + 1)^{3/2}\right]_0^{\sqrt{3}} = 2\pi \cdot \frac{1}{12}\left(13^{3/2} - 1\right) = \frac{\pi}{6}\left(13\sqrt{13} - 1\right)$$

7. $z = f(x, y) = y^2 - x^2$ with $1 \le x^2 + y^2 \le 4$. Then

$$A(S) = \iint_D \sqrt{4x^2 + 4y^2 + 1}\, dA = \int_0^{2\pi} \int_1^2 \sqrt{4r^2 + 1}\, r\, dr\, d\theta = \int_0^{2\pi} d\theta \int_1^2 r\sqrt{4r^2 + 1}\, dr$$

$$= \left[\theta\right]_0^{2\pi} \left[\frac{1}{12}(4r^2 + 1)^{3/2}\right]_1^2 = \frac{\pi}{6}\left(17\sqrt{17} - 5\sqrt{5}\right)$$

9. $z = f(x, y) = xy$ with $x^2 + y^2 \le 1$, so $f_x = y, f_y = x \Rightarrow$

$$A(S) = \iint_D \sqrt{y^2 + x^2 + 1}\, dA = \int_0^{2\pi} \int_0^1 \sqrt{r^2 + 1}\, r\, dr\, d\theta = \int_0^{2\pi} \left[\frac{1}{3}(r^2 + 1)^{3/2}\right]_{r=0}^{r=1} d\theta$$

$$= \int_0^{2\pi} \frac{1}{3}\left(2\sqrt{2} - 1\right) d\theta = \frac{2\pi}{3}\left(2\sqrt{2} - 1\right)$$

11. $z = \sqrt{a^2 - x^2 - y^2}$, $z_x = -x(a^2 - x^2 - y^2)^{-1/2}$, $z_y = -y(a^2 - x^2 - y^2)^{-1/2}$,

$$A(S) = \iint_D \sqrt{\frac{x^2 + y^2}{a^2 - x^2 - y^2} + 1}\, dA$$

$$= \int_{-\pi/2}^{\pi/2} \int_0^{a\cos\theta} \sqrt{\frac{r^2}{a^2 - r^2} + 1}\, r\, dr\, d\theta$$

$$= \int_{-\pi/2}^{\pi/2} \int_0^{a\cos\theta} \frac{ar}{\sqrt{a^2 - r^2}}\, dr\, d\theta$$

$$= \int_{-\pi/2}^{\pi/2} \left[-a\sqrt{a^2 - r^2}\right]_{r=0}^{r=a\cos\theta} d\theta$$

$$= \int_{-\pi/2}^{\pi/2} -a\left(\sqrt{a^2 - a^2\cos^2\theta} - a\right) d\theta = 2a^2 \int_0^{\pi/2} \left(1 - \sqrt{1 - \cos^2\theta}\right) d\theta$$

$$= 2a^2 \int_0^{\pi/2} d\theta - 2a^2 \int_0^{\pi/2} \sqrt{\sin^2\theta}\, d\theta = a^2\pi - 2a^2 \int_0^{\pi/2} \sin\theta\, d\theta = a^2(\pi - 2)$$

$r = a\cos\theta$

13. $z = f(x, y) = (1 + x^2 + y^2)^{-1}$, $f_x = -2x(1 + x^2 + y^2)^{-2}$, $f_y = -2y(1 + x^2 + y^2)^{-2}$. Then

$$A(S) = \iint_{x^2+y^2 \leq 1} \sqrt{[-2x(1 + x^2 + y^2)^{-2}]^2 + [-2y(1 + x^2 + y^2)^{-2}]^2 + 1}\ dA$$

$$= \iint_{x^2+y^2 \leq 1} \sqrt{4(x^2 + y^2)(1 + x^2 + y^2)^{-4} + 1}\ dA$$

Converting to polar coordinates we have

$$A(S) = \int_0^{2\pi} \int_0^1 \sqrt{4r^2(1 + r^2)^{-4} + 1}\ r\ dr\ d\theta = \int_0^{2\pi} d\theta \int_0^1 r \sqrt{4r^2(1 + r^2)^{-4} + 1}\ dr$$

$$= 2\pi \int_0^1 r \sqrt{4r^2(1 + r^2)^{-4} + 1}\ dr \approx 3.6258 \text{ using a calculator.}$$

15. (a) The midpoints of the four squares are $\left(\frac{1}{4}, \frac{1}{4}\right)$, $\left(\frac{1}{4}, \frac{3}{4}\right)$, $\left(\frac{3}{4}, \frac{1}{4}\right)$, and $\left(\frac{3}{4}, \frac{3}{4}\right)$. Here $f(x, y) = x^2 + y^2$, so the Midpoint Rule

gives

$$A(S) = \iint_D \sqrt{[f_x(x, y)]^2 + [f_y(x, y)]^2 + 1}\ dA = \iint_D \sqrt{(2x)^2 + (2y)^2 + 1}\ dA$$

$$\approx \frac{1}{4}\left(\sqrt{\left[2\left(\frac{1}{4}\right)\right]^2 + \left[2\left(\frac{1}{4}\right)\right]^2 + 1} + \sqrt{\left[2\left(\frac{1}{4}\right)\right]^2 + \left[2\left(\frac{3}{4}\right)\right]^2 + 1} \right.$$

$$\left. + \sqrt{\left[2\left(\frac{3}{4}\right)\right]^2 + \left[2\left(\frac{1}{4}\right)\right]^2 + 1} + \sqrt{\left[2\left(\frac{3}{4}\right)\right]^2 + \left[2\left(\frac{3}{4}\right)\right]^2 + 1} \right)$$

$$= \frac{1}{4}\left(\sqrt{\frac{3}{2}} + 2\sqrt{\frac{7}{2}} + \sqrt{\frac{11}{2}} \right) \approx 1.8279$$

(b) A CAS estimates the integral to be $A(S) = \iint_D \sqrt{1 + (2x)^2 + (2y)^2}\ dA = \int_0^1 \int_0^1 \sqrt{1 + 4x^2 + 4y^2}\ dy\ dx \approx 1.8616$.

This agrees with the Midpoint estimate only in the first decimal place.

17. $z = 1 + 2x + 3y + 4y^2$, so

$$A(S) = \iint_D \sqrt{1 + \left(\frac{\partial z}{\partial x}\right)^2 + \left(\frac{\partial z}{\partial y}\right)^2}\ dA = \int_1^4 \int_0^1 \sqrt{1 + 4 + (3 + 8y)^2}\ dy\ dx = \int_1^4 \int_0^1 \sqrt{14 + 48y + 64y^2}\ dy\ dx.$$

Using a CAS, we have $\int_1^4 \int_0^1 \sqrt{14 + 48y + 64y^2}\ dy\ dx = \frac{45}{8}\sqrt{14} + \frac{15}{16}\ln\left(11\sqrt{5} + 3\sqrt{14}\sqrt{5}\right) - \frac{15}{16}\ln\left(3\sqrt{5} + \sqrt{14}\sqrt{5}\right)$

or $\frac{45}{8}\sqrt{14} + \frac{15}{16}\ln\dfrac{11\sqrt{5} + 3\sqrt{70}}{3\sqrt{5} + \sqrt{70}}$.

19. $f(x, y) = 1 + x^2 y^2 \implies f_x = 2xy^2$, $f_y = 2x^2 y$. We use a CAS (with precision reduced to five significant digits, to speed

up the calculation) to estimate the integral

$$A(S) = \int_{-1}^1 \int_{-\sqrt{1-x^2}}^{\sqrt{1-x^2}} \sqrt{f_x^2 + f_y^2 + 1}\ dy\ dx = \int_{-1}^1 \int_{-\sqrt{1-x^2}}^{\sqrt{1-x^2}} \sqrt{4x^2 y^4 + 4x^4 y^2 + 1}\ dy\ dx, \text{ and find that } A(S) \approx 3.3213.$$

21. Here $z = f(x, y) = ax + by + c$, $f_x(x, y) = a$, $f_y(x, y) = b$, so

$$A(S) = \iint_D \sqrt{a^2 + b^2 + 1}\ dA = \sqrt{a^2 + b^2 + 1} \iint_D dA = \sqrt{a^2 + b^2 + 1}\ A(D).$$

23. If we project the surface onto the xz-plane, then the surface lies "above" the disk $x^2 + z^2 \leq 25$ in the xz-plane.

We have $y = f(x, z) = x^2 + z^2$ and, adapting Formula 2, the area of the surface is

$$A(S) = \iint\limits_{x^2+z^2\leq 25} \sqrt{[f_x(x,z)]^2 + [f_z(x,z)]^2 + 1}\, dA = \iint\limits_{x^2+z^2\leq 25} \sqrt{4x^2 + 4z^2 + 1}\, dA$$

Converting to polar coordinates $x = r\cos\theta$, $z = r\sin\theta$ we have

$$A(S) = \int_0^{2\pi}\int_0^5 \sqrt{4r^2+1}\, r\, dr\, d\theta = \int_0^{2\pi} d\theta \int_0^5 r(4r^2+1)^{1/2}\, dr = \left[\theta\right]_0^{2\pi}\left[\tfrac{1}{12}(4r^2+1)^{3/2}\right]_0^5 = \tfrac{\pi}{6}\left(101\sqrt{101}-1\right)$$

15.6 Triple Integrals

1. $\iiint_B xyz^2\, dV = \int_0^1\int_0^3\int_{-1}^2 xyz^2\, dy\, dz\, dx = \int_0^1\int_0^3 \left[\tfrac{1}{2}xy^2z^2\right]_{y=-1}^{y=2} dz\, dx = \int_0^1\int_0^3 \tfrac{3}{2}xz^2\, dz\, dx$

$\qquad = \int_0^1 \left[\tfrac{1}{2}xz^3\right]_{z=0}^{z=3} dx = \int_0^1 \tfrac{27}{2}x\, dx = \tfrac{27}{4}x^2\big]_0^1 = \tfrac{27}{4}$

3. $\int_0^2\int_0^{z^2}\int_0^{y-z} (2x-y)\, dx\, dy\, dz = \int_0^2\int_0^{z^2}\left[x^2-xy\right]_{x=0}^{x=y-z} dy\, dz = \int_0^2\int_0^{z^2}\left[(y-z)^2-(y-z)y\right] dy\, dz$

$\qquad = \int_0^2\int_0^{z^2} (z^2-yz)\, dy\, dz = \int_0^2\left[yz^2-\tfrac{1}{2}y^2z\right]_{y=0}^{y=z^2} dz = \int_0^2\left(z^4-\tfrac{1}{2}z^5\right) dz$

$\qquad = \left[\tfrac{1}{5}z^5-\tfrac{1}{12}z^6\right]_0^2 = \tfrac{32}{5}-\tfrac{64}{12} = \tfrac{16}{15}$

5. $\int_1^2\int_0^{2z}\int_0^{\ln x} xe^{-y}\, dy\, dx\, dz = \int_1^2\int_0^{2z}\left[-xe^{-y}\right]_{y=0}^{y=\ln x} dx\, dz = \int_1^2\int_0^{2z}\left(-xe^{-\ln x}+xe^0\right) dx\, dz$

$\qquad = \int_1^2\int_0^{2z}(-1+x)\, dx\, dz = \int_1^2\left[-x+\tfrac{1}{2}x^2\right]_{x=0}^{x=2z} dz$

$\qquad = \int_1^2(-2z+2z^2)\, dz = \left[-z^2+\tfrac{2}{3}z^3\right]_1^2 = -4+\tfrac{16}{3}+1-\tfrac{2}{3} = \tfrac{5}{3}$

7. $\int_0^\pi\int_0^1\int_0^{\sqrt{1-z^2}} z\sin x\, dy\, dz\, dx = \int_0^\pi\int_0^1\left[yz\sin x\right]_{y=0}^{y=\sqrt{1-z^2}} dz\, dx = \int_0^\pi\int_0^1 z\sqrt{1-z^2}\,\sin x\, dz\, dx$

$\qquad = \int_0^\pi \sin x\left[-\tfrac{1}{3}(1-z^2)^{3/2}\right]_{z=0}^{z=1} dx = \int_0^\pi \tfrac{1}{3}\sin x\, dx = -\tfrac{1}{3}\cos x\big]_0^\pi = -\tfrac{1}{3}(-1-1) = \tfrac{2}{3}$

9. $\iiint_E y\, dV = \int_0^3\int_0^x\int_{x-y}^{x+y} y\, dz\, dy\, dx = \int_0^3\int_0^x \left[yz\right]_{z=x-y}^{z=x+y} dy\, dx = \int_0^3\int_0^x 2y^2\, dy\, dx$

$\qquad = \int_0^3\left[\tfrac{2}{3}y^3\right]_{y=0}^{y=x} dx = \int_0^3 \tfrac{2}{3}x^3\, dx = \tfrac{1}{6}x^4\big]_0^3 = \tfrac{81}{6} = \tfrac{27}{2}$

11. $\iiint_E \dfrac{z}{x^2+z^2}\, dV = \int_1^4\int_y^4\int_0^z \dfrac{z}{x^2+z^2}\, dx\, dz\, dy = \int_1^4\int_y^4\left[z\cdot\tfrac{1}{z}\tan^{-1}\tfrac{x}{z}\right]_{x=0}^{x=z} dz\, dy$

$\qquad = \int_1^4\int_y^4\left[\tan^{-1}(1)-\tan^{-1}(0)\right] dz\, dy = \int_1^4\int_y^4\left(\tfrac{\pi}{4}-0\right) dz\, dy = \tfrac{\pi}{4}\int_1^4 \left[z\right]_{z=y}^{z=4} dy$

$\qquad = \tfrac{\pi}{4}\int_1^4(4-y)\, dy = \tfrac{\pi}{4}\left[4y-\tfrac{1}{2}y^2\right]_1^4 = \tfrac{\pi}{4}\left(16-8-4+\tfrac{1}{2}\right) = \tfrac{9\pi}{8}$

13. Here $E = \{(x,y,z) \mid 0\leq x\leq 1, 0\leq y\leq\sqrt{x}, 0\leq z\leq 1+x+y\}$, so

$$\iiint_E 6xy\, dV = \int_0^1\int_0^{\sqrt{x}}\int_0^{1+x+y} 6xy\, dz\, dy\, dx = \int_0^1\int_0^{\sqrt{x}}\left[6xyz\right]_{z=0}^{z=1+x+y} dy\, dx$$

$$= \int_0^1\int_0^{\sqrt{x}} 6xy(1+x+y)\, dy\, dx = \int_0^1\left[3xy^2+3x^2y^2+2xy^3\right]_{y=0}^{y=\sqrt{x}} dx$$

$$= \int_0^1\left(3x^2+3x^3+2x^{5/2}\right) dx = \left[x^3+\tfrac{3}{4}x^4+\tfrac{4}{7}x^{7/2}\right]_0^1 = \tfrac{65}{28}$$

15.

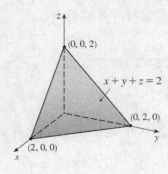

Here $T = \{(x, y, z) \mid 0 \le x \le 2,\ 0 \le y \le 2 - x,\ 0 \le z \le 2 - x - y\}$.

Thus,

$$\iiint_T y^2\, dV = \int_0^2 \int_0^{2-x} \int_0^{2-x-y} y^2\, dz\, dy\, dx$$

$$= \int_0^2 \int_0^{2-x} y^2(2 - x - y)\, dy\, dx$$

$$= \int_0^2 \int_0^{2-x} \left[(2 - x)y^2 - y^3\right] dy\, dx$$

$$= \int_0^2 \left[(2 - x)\left(\tfrac{1}{3}y^3\right) - \tfrac{1}{4}y^4\right]_{y=0}^{y=2-x} dx$$

$$= \int_0^2 \left[\tfrac{1}{3}(2 - x)^4 - \tfrac{1}{4}(2 - x)^4\right] dx = \int_0^2 \tfrac{1}{12}(2 - x)^4\, dx$$

$$= \left[\tfrac{1}{12}\left(-\tfrac{1}{5}\right)(2 - x)^5\right]_0^2 = -\tfrac{1}{60}(0 - 32) = \tfrac{8}{15}$$

17.

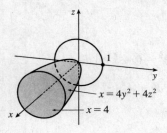

The projection of E onto the yz-plane is the disk $y^2 + z^2 \le 1$. Using polar coordinates $y = r\cos\theta$ and $z = r\sin\theta$, we get

$$\iiint_E x\, dV = \iint_D \left[\int_{4y^2 + 4z^2}^4 x\, dx\right] dA = \tfrac{1}{2} \iint_D \left[4^2 - (4y^2 + 4z^2)^2\right] dA$$

$$= 8\int_0^{2\pi}\int_0^1 (1 - r^4)\, r\, dr\, d\theta = 8\int_0^{2\pi} d\theta \int_0^1 (r - r^5)\, dr$$

$$= 8(2\pi)\left[\tfrac{1}{2}r^2 - \tfrac{1}{6}r^6\right]_0^1 = \tfrac{16\pi}{3}$$

19. The plane $2x + y + z = 4$ intersects the xy-plane when

$2x + y + 0 = 4 \ \Rightarrow\ y = 4 - 2x$, so

$E = \{(x, y, z) \mid 0 \le x \le 2,\ 0 \le y \le 4 - 2x,\ 0 \le z \le 4 - 2x - y\}$ and

$$V = \int_0^2 \int_0^{4-2x} \int_0^{4-2x-y} dz\, dy\, dx = \int_0^2 \int_0^{4-2x} (4 - 2x - y)\, dy\, dx$$

$$= \int_0^2 \left[4y - 2xy - \tfrac{1}{2}y^2\right]_{y=0}^{y=4-2x} dx$$

$$= \int_0^2 \left[4(4 - 2x) - 2x(4 - 2x) - \tfrac{1}{2}(4 - 2x)^2\right] dx$$

$$= \int_0^2 (2x^2 - 8x + 8)\, dx = \left[\tfrac{2}{3}x^3 - 4x^2 + 8x\right]_0^2 = \tfrac{16}{3}$$

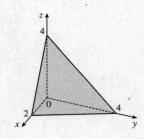

21. The plane $y + z = 1$ intersects the xy-plane in the line $y = 1$, so

$E = \{(x, y, z) \mid -1 \le x \le 1,\ x^2 \le y \le 1,\ 0 \le z \le 1 - y\}$ and

$$V = \iiint_E dV = \int_{-1}^1 \int_{x^2}^1 \int_0^{1-y} dz\, dy\, dx = \int_{-1}^1 \int_{x^2}^1 (1 - y)\, dy\, dx$$

$$= \int_{-1}^1 \left[y - \tfrac{1}{2}y^2\right]_{y=x^2}^{y=1} dx = \int_{-1}^1 \left(\tfrac{1}{2} - x^2 + \tfrac{1}{2}x^4\right) dx$$

$$= \left[\tfrac{1}{2}x - \tfrac{1}{3}x^3 + \tfrac{1}{10}x^5\right]_{-1}^1 = \tfrac{1}{2} - \tfrac{1}{3} + \tfrac{1}{10} + \tfrac{1}{2} - \tfrac{1}{3} + \tfrac{1}{10} = \tfrac{8}{15}$$

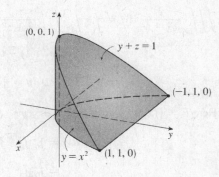

23. (a) The wedge can be described as the region

$$D = \{(x, y, z) \mid y^2 + z^2 \leq 1, 0 \leq x \leq 1, 0 \leq y \leq x\}$$
$$= \{(x, y, z) \mid 0 \leq x \leq 1, 0 \leq y \leq x, 0 \leq z \leq \sqrt{1 - y^2}\}$$

So the integral expressing the volume of the wedge is

$$\iiint_D dV = \int_0^1 \int_0^x \int_0^{\sqrt{1-y^2}} dz\, dy\, dx.$$

(b) A CAS gives $\int_0^1 \int_0^x \int_0^{\sqrt{1-y^2}} dz\, dy\, dx = \frac{\pi}{4} - \frac{1}{3}$.

(Or use Formulas 30 and 87 from the Table of Integrals.)

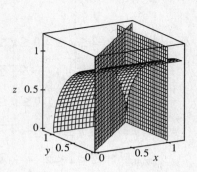

25. Here $f(x, y, z) = \cos(xyz)$ and $\Delta V = \frac{1}{2} \cdot \frac{1}{2} \cdot \frac{1}{2} = \frac{1}{8}$, so the Midpoint Rule gives

$$\iiint_B f(x, y, z)\, dV \approx \sum_{i=1}^l \sum_{j=1}^m \sum_{k=1}^n f(\overline{x}_i, \overline{y}_j, \overline{z}_k)\, \Delta V$$

$$= \frac{1}{8} \left[f\!\left(\tfrac{1}{4}, \tfrac{1}{4}, \tfrac{1}{4}\right) + f\!\left(\tfrac{1}{4}, \tfrac{1}{4}, \tfrac{3}{4}\right) + f\!\left(\tfrac{1}{4}, \tfrac{3}{4}, \tfrac{1}{4}\right) + f\!\left(\tfrac{1}{4}, \tfrac{3}{4}, \tfrac{3}{4}\right) \right.$$
$$\left. + f\!\left(\tfrac{3}{4}, \tfrac{1}{4}, \tfrac{1}{4}\right) + f\!\left(\tfrac{3}{4}, \tfrac{1}{4}, \tfrac{3}{4}\right) + f\!\left(\tfrac{3}{4}, \tfrac{3}{4}, \tfrac{1}{4}\right) + f\!\left(\tfrac{3}{4}, \tfrac{3}{4}, \tfrac{3}{4}\right) \right]$$

$$= \frac{1}{8} \left[\cos \tfrac{1}{64} + \cos \tfrac{3}{64} + \cos \tfrac{3}{64} + \cos \tfrac{9}{64} + \cos \tfrac{3}{64} + \cos \tfrac{9}{64} + \cos \tfrac{9}{64} + \cos \tfrac{27}{64} \right] \approx 0.985$$

27. $E = \{(x, y, z) \mid 0 \leq x \leq 1, 0 \leq z \leq 1 - x, 0 \leq y \leq 2 - 2z\}$,

the solid bounded by the three coordinate planes and the planes

$z = 1 - x$, $y = 2 - 2z$.

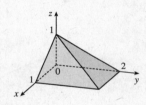

29.

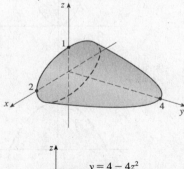

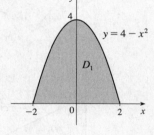

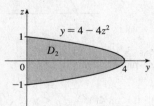

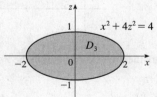

If D_1, D_2, D_3 are the projections of E on the xy-, yz-, and xz-planes, then

$$D_1 = \{(x, y) \mid -2 \leq x \leq 2, 0 \leq y \leq 4 - x^2\} = \{(x, y) \mid 0 \leq y \leq 4, -\sqrt{4 - y} \leq x \leq \sqrt{4 - y}\}$$

$$D_2 = \{(y, z) \mid 0 \leq y \leq 4, -\tfrac{1}{2}\sqrt{4 - y} \leq z \leq \tfrac{1}{2}\sqrt{4 - y}\} = \{(y, z) \mid -1 \leq z \leq 1, 0 \leq y \leq 4 - 4z^2\}$$

$$D_3 = \{(x, z) \mid x^2 + 4z^2 \leq 4\}$$

[continued]

Therefore

$$E = \left\{ (x,y,z) \mid -2 \le x \le 2, 0 \le y \le 4 - x^2, \ -\tfrac{1}{2}\sqrt{4 - x^2 - y} \le z \le \tfrac{1}{2}\sqrt{4 - x^2 - y} \right\}$$

$$= \left\{ (x,y,z) \mid 0 \le y \le 4, \ -\sqrt{4 - y} \le x \le \sqrt{4 - y}, \ -\tfrac{1}{2}\sqrt{4 - x^2 - y} \le z \le \tfrac{1}{2}\sqrt{4 - x^2 - y} \right\}$$

$$= \left\{ (x,y,z) \mid -1 \le z \le 1, 0 \le y \le 4 - 4z^2, \ -\sqrt{4 - y - 4z^2} \le x \le \sqrt{4 - y - 4z^2} \right\}$$

$$= \left\{ (x,y,z) \mid 0 \le y \le 4, \ -\tfrac{1}{2}\sqrt{4 - y} \le z \le \tfrac{1}{2}\sqrt{4 - y}, \ -\sqrt{4 - y - 4z^2} \le x \le \sqrt{4 - y - 4z^2} \right\}$$

$$= \left\{ (x,y,z) \mid -2 \le x \le 2, \ -\tfrac{1}{2}\sqrt{4 - x^2} \le z \le \tfrac{1}{2}\sqrt{4 - x^2}, 0 \le y \le 4 - x^2 - 4z^2 \right\}$$

$$= \left\{ (x,y,z) \mid -1 \le z \le 1, \ -\sqrt{4 - 4z^2} \le x \le \sqrt{4 - 4z^2}, 0 \le y \le 4 - x^2 - 4z^2 \right\}$$

Then

$$\iiint_E f(x,y,z)\,dV = \int_{-2}^{2} \int_{0}^{4-x^2} \int_{-\sqrt{4-x^2-y}/2}^{\sqrt{4-x^2-y}/2} f(x,y,z)\,dz\,dy\,dx = \int_{0}^{4} \int_{-\sqrt{4-y}}^{\sqrt{4-y}} \int_{-\sqrt{4-x^2-y}/2}^{\sqrt{4-x^2-y}/2} f(x,y,z)\,dz\,dx\,dy$$

$$= \int_{-1}^{1} \int_{0}^{4-4z^2} \int_{-\sqrt{4-y-4z^2}}^{\sqrt{4-y-4z^2}} f(x,y,z)\,dx\,dy\,dz = \int_{0}^{4} \int_{-\sqrt{4-y}/2}^{\sqrt{4-y}/2} \int_{-\sqrt{4-y-4z^2}}^{\sqrt{4-y-4z^2}} f(x,y,z)\,dx\,dz\,dy$$

$$= \int_{-2}^{2} \int_{-\sqrt{4-x^2}/2}^{\sqrt{4-x^2}/2} \int_{0}^{4-x^2-4z^2} f(x,y,z)\,dy\,dz\,dx = \int_{-1}^{1} \int_{-\sqrt{4-4z^2}}^{\sqrt{4-4z^2}} \int_{0}^{4-x^2-4z^2} f(x,y,z)\,dy\,dx\,dz$$

31.

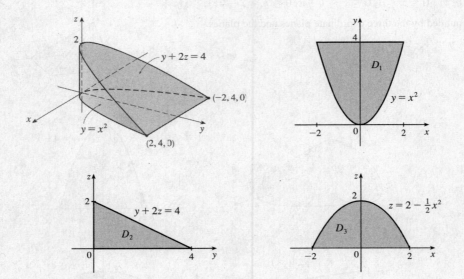

If D_1, D_2, and D_3 are the projections of E on the xy-, yz-, and xz-planes, then

$$D_1 = \left\{ (x,y) \mid -2 \le x \le 2, x^2 \le y \le 4 \right\} = \left\{ (x,y) \mid 0 \le y \le 4, -\sqrt{y} \le x \le \sqrt{y} \right\},$$

$$D_2 = \left\{ (y,z) \mid 0 \le y \le 4, 0 \le z \le 2 - \tfrac{1}{2}y \right\} = \left\{ (y,z) \mid 0 \le z \le 2, 0 \le y \le 4 - 2z \right\}, \text{ and}$$

$$D_3 = \left\{ (x,z) \mid -2 \le x \le 2, 0 \le z \le 2 - \tfrac{1}{2}x^2 \right\} = \left\{ (x,z) \mid 0 \le z \le 2, -\sqrt{4 - 2z} \le x \le \sqrt{4 - 2z} \right\}$$

Therefore
$$E = \left\{(x,y,z) \mid -2 \le x \le 2,\, x^2 \le y \le 4,\, 0 \le z \le 2 - \tfrac{1}{2}y\right\}$$

$$= \left\{(x,y,z) \mid 0 \le y \le 4,\, -\sqrt{y} \le x \le \sqrt{y},\, 0 \le z \le 2 - \tfrac{1}{2}y\right\}$$

$$= \left\{(x,y,z) \mid 0 \le y \le 4,\, 0 \le z \le 2 - \tfrac{1}{2}y,\, -\sqrt{y} \le x \le \sqrt{y}\right\}$$

$$= \left\{(x,y,z) \mid 0 \le z \le 2,\, 0 \le y \le 4 - 2z,\, -\sqrt{y} \le x \le \sqrt{y}\right\}$$

$$= \left\{(x,y,z) \mid -2 \le x \le 2,\, 0 \le z \le 2 - \tfrac{1}{2}x^2,\, x^2 \le y \le 4 - 2z\right\}$$

$$= \left\{(x,y,z) \mid 0 \le z \le 2,\, -\sqrt{4-2z} \le x \le \sqrt{4-2z},\, x^2 \le y \le 4 - 2z\right\}$$

Then
$$\iiint_E f(x,y,z)\,dV = \int_{-2}^{2}\int_{x^2}^{4}\int_{0}^{2-y/2} f(x,y,z)\,dz\,dy\,dx = \int_{0}^{4}\int_{-\sqrt{y}}^{\sqrt{y}}\int_{0}^{2-y/2} f(x,y,z)\,dz\,dx\,dy$$

$$= \int_{0}^{4}\int_{0}^{2-y/2}\int_{-\sqrt{y}}^{\sqrt{y}} f(x,y,z)\,dx\,dz\,dy = \int_{0}^{2}\int_{0}^{4-2z}\int_{-\sqrt{y}}^{\sqrt{y}} f(x,y,z)\,dx\,dy\,dz$$

$$= \int_{-2}^{2}\int_{0}^{2-x^2/2}\int_{x^2}^{4-2z} f(x,y,z)\,dy\,dz\,dx = \int_{0}^{2}\int_{-\sqrt{4-2z}}^{\sqrt{4-2z}}\int_{x^2}^{4-2z} f(x,y,z)\,dy\,dx\,dz$$

33.

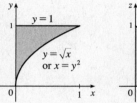

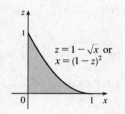

The diagrams show the projections of E onto the xy-, yz-, and xz-planes. Therefore

$$\int_{0}^{1}\int_{\sqrt{x}}^{1}\int_{0}^{1-y} f(x,y,z)\,dz\,dy\,dx = \int_{0}^{1}\int_{0}^{y^2}\int_{0}^{1-y} f(x,y,z)\,dz\,dx\,dy = \int_{0}^{1}\int_{0}^{1-z}\int_{0}^{y^2} f(x,y,z)\,dx\,dy\,dz$$

$$= \int_{0}^{1}\int_{0}^{1-y}\int_{0}^{y^2} f(x,y,z)\,dx\,dz\,dy = \int_{0}^{1}\int_{0}^{1-\sqrt{x}}\int_{\sqrt{x}}^{1-z} f(x,y,z)\,dy\,dz\,dx$$

$$= \int_{0}^{1}\int_{0}^{(1-z)^2}\int_{\sqrt{x}}^{1-z} f(x,y,z)\,dy\,dx\,dz$$

35.

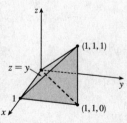

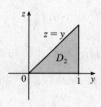

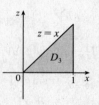

$$\int_{0}^{1}\int_{y}^{1}\int_{0}^{y} f(x,y,z)\,dz\,dx\,dy = \iiint_E f(x,y,z)\,dV \text{ where } E = \{(x,y,z) \mid 0 \le z \le y,\, y \le x \le 1,\, 0 \le y \le 1\}.$$

If D_1, D_2, and D_3 are the projections of E onto the xy-, yz- and xz-planes then

$$D_1 = \{(x,y) \mid 0 \le y \le 1,\, y \le x \le 1\} = \{(x,y) \mid 0 \le x \le 1,\, 0 \le y \le x\},$$

$$D_2 = \{(y,z) \mid 0 \le y \le 1,\, 0 \le z \le y\} = \{(y,z) \mid 0 \le z \le 1,\, z \le y \le 1\}, \text{ and}$$

$$D_3 = \{(x,z) \mid 0 \le x \le 1,\, 0 \le z \le x\} = \{(x,z) \mid 0 \le z \le 1,\, z \le x \le 1\}.$$

[continued]

Thus we also have

$$E = \{(x, y, z) \mid 0 \le x \le 1, 0 \le y \le x, 0 \le z \le y\} = \{(x, y, z) \mid 0 \le y \le 1, 0 \le z \le y, y \le x \le 1\}$$

$$= \{(x, y, z) \mid 0 \le z \le 1, z \le y \le 1, y \le x \le 1\} = \{(x, y, z) \mid 0 \le x \le 1, 0 \le z \le x, z \le y \le x\}$$

$$= \{(x, y, z) \mid 0 \le z \le 1, z \le x \le 1, z \le y \le x\}.$$

Then

$$\int_0^1 \int_y^1 \int_0^y f(x, y, z)\, dz\, dx\, dy = \int_0^1 \int_0^x \int_0^y f(x, y, z)\, dz\, dy\, dx = \int_0^1 \int_0^y \int_y^1 f(x, y, z)\, dx\, dz\, dy$$

$$= \int_0^1 \int_z^1 \int_y^1 f(x, y, z)\, dx\, dy\, dz = \int_0^1 \int_0^x \int_z^x f(x, y, z)\, dy\, dz\, dx$$

$$= \int_0^1 \int_z^1 \int_z^x f(x, y, z)\, dy\, dx\, dz.$$

37. The region C is the solid bounded by a circular cylinder of radius 2 with axis the z-axis for $-2 \le z \le 2$. We can write

$\iiint_C (4 + 5x^2 yz^2)\, dV = \iiint_C 4\, dV + \iiint_C 5x^2 yz^2\, dV$, but $f(x, y, z) = 5x^2 yz^2$ is an odd function with

respect to y. Since C is symmetrical about the xz-plane, we have $\iiint_C 5x^2 yz^2\, dV = 0$. Thus

$\iiint_C (4 + 5x^2 yz^2)\, dV = \iiint_C 4\, dV = 4 \cdot V(E) = 4 \cdot \pi(2)^2(4) = 64\pi.$

39. The projection of E onto the xy-plane is the disk $D = \{(x, y) \mid x^2 + y^2 \le 1\}$.

$$m = \iiint_E \rho(x, y, z)\, dV = \iint_D \left[\int_0^{1-x^2-y^2} 3\, dz \right] dA = \iint_D 3(1 - x^2 - y^2)\, dA$$

$$= 3 \int_0^1 \int_0^{2\pi} (1 - r^2)\, r\, dr\, d\theta = 3 \int_0^{2\pi} d\theta \int_0^1 (r - r^3)\, dr$$

$$= 3 \left[\theta \right]_0^{2\pi} \left[\tfrac{1}{2}r^2 - \tfrac{1}{4}r^4 \right]_0^1 = 3(2\pi)\left(\tfrac{1}{2} - \tfrac{1}{4} \right) = \tfrac{3}{2}\pi$$

$$M_{yz} = \iiint_E x\rho(x, y, z)\, dV = \iint_D \left[\int_0^{1-x^2-y^2} 3x\, dz \right] dA = \iint_D 3x(1 - x^2 - y^2)\, dA$$

$$= 3 \int_0^1 \int_0^{2\pi} (r\cos\theta)(1 - r^2)\, r\, dr\, d\theta = 3 \int_0^{2\pi} \cos\theta\, d\theta \int_0^1 (r^2 - r^4)\, dr$$

$$= 3 \left[\sin\theta \right]_0^{2\pi} \left[\tfrac{1}{3}r^3 - \tfrac{1}{5}r^5 \right]_0^1 = 3(0)\left(\tfrac{1}{3} - \tfrac{1}{5} \right) = 0$$

$$M_{xz} = \iiint_E y\rho(x, y, z)\, dV = \iint_D \left[\int_0^{1-x^2-y^2} 3y\, dz \right] dA = \iint_D 3y(1 - x^2 - y^2)\, dA$$

$$= 3 \int_0^1 \int_0^{2\pi} (r\sin\theta)(1 - r^2)\, r\, dr\, d\theta = 3 \int_0^{2\pi} \sin\theta\, d\theta \int_0^1 (r^2 - r^4)\, dr$$

$$= 3 \left[-\cos\theta \right]_0^{2\pi} \left[\tfrac{1}{3}r^3 - \tfrac{1}{5}r^5 \right]_0^1 = 3(0)\left(\tfrac{1}{3} - \tfrac{1}{5} \right) = 0$$

$$M_{xy} = \iiint_E z\rho(x, y, z)\, dV = \iint_D \left[\int_0^{1-x^2-y^2} 3z\, dz \right] dA = \iint_D \left[\tfrac{3}{2}z^2 \right]_{z=0}^{z=1-x^2-y^2} dA$$

$$= \tfrac{3}{2} \iint_D (1 - x^2 - y^2)^2\, dA = \tfrac{3}{2} \int_0^1 \int_0^{2\pi} (1 - r^2)^2\, r\, dr\, d\theta$$

$$= \tfrac{3}{2} \int_0^{2\pi} d\theta \int_0^1 (r - 2r^3 + r^5)\, dr = \tfrac{3}{2} \left[\theta \right]_0^{2\pi} \left[\tfrac{1}{2}r^2 - \tfrac{1}{2}r^4 + \tfrac{1}{6}r^6 \right]_0^1$$

$$= \tfrac{3}{2}(2\pi)\left(\tfrac{1}{2} - \tfrac{1}{2} + \tfrac{1}{6} \right) = \tfrac{1}{2}\pi$$

Thus the mass is $\tfrac{3}{2}\pi$ and the center of mass is $(\overline{x}, \overline{y}, \overline{z}) = \left(\dfrac{M_{yz}}{m}, \dfrac{M_{xz}}{m}, \dfrac{M_{xy}}{m} \right) = \left(0, 0, \dfrac{1}{3} \right)$.

41. $m = \int_0^a \int_0^a \int_0^a (x^2 + y^2 + z^2)\, dx\, dy\, dz = \int_0^a \int_0^a \left[\frac{1}{3}x^3 + xy^2 + xz^2\right]_{x=0}^{x=a} dy\, dz = \int_0^a \int_0^a \left(\frac{1}{3}a^3 + ay^2 + az^2\right) dy\, dz$

$= \int_0^a \left[\frac{1}{3}a^3 y + \frac{1}{3}ay^3 + ayz^2\right]_{y=0}^{y=a} dz = \int_0^a \left(\frac{2}{3}a^4 + a^2 z^2\right) dz = \left[\frac{2}{3}a^4 z + \frac{1}{3}a^2 z^3\right]_0^a = \frac{2}{3}a^5 + \frac{1}{3}a^5 = a^5$

$M_{yz} = \int_0^a \int_0^a \int_0^a \left[x^3 + x(y^2 + z^2)\right] dx\, dy\, dz = \int_0^a \int_0^a \left[\frac{1}{4}a^4 + \frac{1}{2}a^2(y^2 + z^2)\right] dy\, dz$

$= \int_0^a \left(\frac{1}{4}a^5 + \frac{1}{6}a^5 + \frac{1}{2}a^3 z^2\right) dz = \frac{1}{4}a^6 + \frac{1}{3}a^6 = \frac{7}{12}a^6 = M_{xz} = M_{xy}$ by symmetry of E and $\rho(x, y, z)$.

Hence $(\overline{x}, \overline{y}, \overline{z}) = \left(\frac{7}{12}a, \frac{7}{12}a, \frac{7}{12}a\right)$.

43. $I_x = \int_0^L \int_0^L \int_0^L k(y^2 + z^2)\, dz\, dy\, dx = k \int_0^L \int_0^L \left(Ly^2 + \frac{1}{3}L^3\right) dy\, dx = k \int_0^L \frac{2}{3}L^4\, dx = \frac{2}{3}kL^5$

By symmetry, $I_x = I_y = I_z = \frac{2}{3}kL^5$.

45. $I_z = \iiint_E (x^2 + y^2)\, \rho(x, y, z)\, dV = \iint_{x^2+y^2 \le a^2} \left[\int_0^h k(x^2 + y^2)\, dz\right] dA = \iint_{x^2+y^2 \le a^2} k(x^2 + y^2)h\, dA$

$= kh \int_0^{2\pi} \int_0^a (r^2)\, r\, dr\, d\theta = kh \int_0^{2\pi} d\theta \int_0^a r^3\, dr = kh(2\pi)\left[\frac{1}{4}r^4\right]_0^a = 2\pi kh \cdot \frac{1}{4}a^4 = \frac{1}{2}\pi kha^4$

47. (a) $m = \int_{-1}^1 \int_{x^2}^1 \int_0^{1-y} \sqrt{x^2 + y^2}\, dz\, dy\, dx$

(b) $(\overline{x}, \overline{y}, \overline{z})$ where $\overline{x} = \frac{1}{m} \int_{-1}^1 \int_{x^2}^1 \int_0^{1-y} x\, \sqrt{x^2 + y^2}\, dz\, dy\, dx$, $\overline{y} = \frac{1}{m} \int_{-1}^1 \int_{x^2}^1 \int_0^{1-y} y\, \sqrt{x^2 + y^2}\, dz\, dy\, dx$, and

$\overline{z} = \frac{1}{m} \int_{-1}^1 \int_{x^2}^1 \int_0^{1-y} z\, \sqrt{x^2 + y^2}\, dz\, dy\, dx$.

(c) $I_z = \int_{-1}^1 \int_{x^2}^1 \int_0^{1-y} (x^2 + y^2)\sqrt{x^2 + y^2}\, dz\, dy\, dx = \int_{-1}^1 \int_{x^2}^1 \int_0^{1-y} (x^2 + y^2)^{3/2}\, dz\, dy\, dx$

49. (a) $m = \int_0^1 \int_0^{\sqrt{1-x^2}} \int_0^y (1 + x + y + z)\, dz\, dy\, dx = \frac{3\pi}{32} + \frac{11}{24}$

(b) $(\overline{x}, \overline{y}, \overline{z}) = \Bigg(m^{-1} \int_0^1 \int_0^{\sqrt{1-x^2}} \int_0^y x(1 + x + y + z)\, dz\, dy\, dx,$

$m^{-1} \int_0^1 \int_0^{\sqrt{1-x^2}} \int_0^y y(1 + x + y + z)\, dz\, dy\, dx,$

$m^{-1} \int_0^1 \int_0^{\sqrt{1-x^2}} \int_0^y z(1 + x + y + z)\, dz\, dy\, dx \Bigg)$

$= \left(\frac{28}{9\pi + 44}, \frac{30\pi + 128}{45\pi + 220}, \frac{45\pi + 208}{135\pi + 660}\right)$

(c) $I_z = \int_0^1 \int_0^{\sqrt{1-x^2}} \int_0^y (x^2 + y^2)(1 + x + y + z)\, dz\, dy\, dx = \frac{68 + 15\pi}{240}$

51. (a) $f(x, y, z)$ is a joint density function, so we know $\iiint_{\mathbb{R}^3} f(x, y, z)\, dV = 1$. Here we have

$\iiint_{\mathbb{R}^3} f(x, y, z)\, dV = \int_{-\infty}^{\infty} \int_{-\infty}^{\infty} \int_{-\infty}^{\infty} f(x, y, z)\, dz\, dy\, dx = \int_0^2 \int_0^2 \int_0^2 Cxyz\, dz\, dy\, dx$

$= C \int_0^2 x\, dx \int_0^2 y\, dy \int_0^2 z\, dz = C\left[\frac{1}{2}x^2\right]_0^2 \left[\frac{1}{2}y^2\right]_0^2 \left[\frac{1}{2}z^2\right]_0^2 = 8C$

Then we must have $8C = 1 \;\Rightarrow\; C = \frac{1}{8}$.

(b) $P(X \le 1, Y \le 1, Z \le 1) = \int_{-\infty}^1 \int_{-\infty}^1 \int_{-\infty}^1 f(x, y, z)\, dz\, dy\, dx = \int_0^1 \int_0^1 \int_0^1 \frac{1}{8}xyz\, dz\, dy\, dx$

$= \frac{1}{8} \int_0^1 x\, dx \int_0^1 y\, dy \int_0^1 z\, dz = \frac{1}{8}\left[\frac{1}{2}x^2\right]_0^1 \left[\frac{1}{2}y^2\right]_0^1 \left[\frac{1}{2}z^2\right]_0^1 = \frac{1}{8}\left(\frac{1}{2}\right)^3 = \frac{1}{64}$

(c) $P(X + Y + Z \le 1) = P((X, Y, Z) \in E)$ where E is the solid region in the first octant bounded by the coordinate planes

and the plane $x + y + z = 1$. The plane $x + y + z = 1$ meets the xy-plane in the line $x + y = 1$, so we have

$$P(X + Y + Z \le 1) = \iiint_E f(x, y, z)\, dV = \int_0^1 \int_0^{1-x} \int_0^{1-x-y} \tfrac{1}{8} xyz \, dz \, dy \, dx$$

$$= \tfrac{1}{8} \int_0^1 \int_0^{1-x} xy \left[\tfrac{1}{2} z^2\right]_{z=0}^{z=1-x-y} dy \, dx = \tfrac{1}{16} \int_0^1 \int_0^{1-x} xy(1-x-y)^2 \, dy \, dx$$

$$= \tfrac{1}{16} \int_0^1 \int_0^{1-x} [(x^3 - 2x^2 + x)y + (2x^2 - 2x)y^2 + xy^3] \, dy \, dx$$

$$= \tfrac{1}{16} \int_0^1 \left[(x^3 - 2x^2 + x)\tfrac{1}{2} y^2 + (2x^2 - 2x)\tfrac{1}{3} y^3 + x\left(\tfrac{1}{4} y^4\right)\right]_{y=0}^{y=1-x} dx$$

$$= \tfrac{1}{192} \int_0^1 (x - 4x^2 + 6x^3 - 4x^4 + x^5) \, dx = \tfrac{1}{192}\left(\tfrac{1}{30}\right) = \tfrac{1}{5760}$$

53. $V(E) = L^3 \;\Rightarrow\; f_{\text{ave}} = \dfrac{1}{L^3} \displaystyle\int_0^L \int_0^L \int_0^L xyz \, dx \, dy \, dz = \dfrac{1}{L^3} \int_0^L x \, dx \int_0^L y \, dy \int_0^L z \, dz$

$$= \dfrac{1}{L^3} \left[\dfrac{x^2}{2}\right]_0^L \left[\dfrac{y^2}{2}\right]_0^L \left[\dfrac{z^2}{2}\right]_0^L = \dfrac{1}{L^3} \dfrac{L^2}{2} \dfrac{L^2}{2} \dfrac{L^2}{2} = \dfrac{L^3}{8}$$

55. (a) The triple integral will attain its maximum when the integrand $1 - x^2 - 2y^2 - 3z^2$ is positive in the region E and negative

everywhere else. For if E contains some region F where the integrand is negative, the integral could be increased by

excluding F from E, and if E fails to contain some part G of the region where the integrand is positive, the integral could

be increased by including G in E. So we require that $x^2 + 2y^2 + 3z^2 \le 1$. This describes the region bounded by the

ellipsoid $x^2 + 2y^2 + 3z^2 = 1$.

(b) The maximum value of $\iiint_E (1 - x^2 - 2y^2 - 3z^2) \, dV$ occurs when E is the solid region bounded by the ellipsoid

$x^2 + 2y^2 + 3z^2 = 1$. The projection of E on the xy-plane is the planar region bounded by the ellipse $x^2 + 2y^2 = 1$, so

$$E = \left\{ (x, y, z) \mid -1 \le x \le 1,\, -\sqrt{\tfrac{1}{2}(1 - x^2)} \le y \le \sqrt{\tfrac{1}{2}(1 - x^2)},\, -\sqrt{\tfrac{1}{3}(1 - x^2 - 2y^2)} \le z \le \sqrt{\tfrac{1}{3}(1 - x^2 - 2y^2)} \right\}$$

and

$$\iiint_E (1 - x^2 - 2y^2 - 3z^2) \, dV = \int_{-1}^1 \int_{-\sqrt{\frac{1}{2}(1-x^2)}}^{\sqrt{\frac{1}{2}(1-x^2)}} \int_{-\sqrt{\frac{1}{3}(1-x^2-2y^2)}}^{\sqrt{\frac{1}{3}(1-x^2-2y^2)}} (1 - x^2 - 2y^2 - 3z^2) \, dz \, dy \, dx = \dfrac{4\sqrt{6}}{45} \pi$$

using a CAS.

15.7 Triple Integrals in Cylindrical Coordinates

1. (a)

From Equations 1, $x = r\cos\theta = 4\cos\dfrac{\pi}{3} = 4 \cdot \dfrac{1}{2} = 2$,

$y = r\sin\theta = 4\sin\dfrac{\pi}{3} = 4 \cdot \dfrac{\sqrt{3}}{2} = 2\sqrt{3}$, $z = -2$, so the point is

$(2, 2\sqrt{3}, -2)$ in rectangular coordinates.

(b)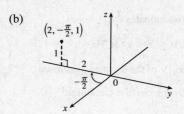

$x = 2\cos\left(-\frac{\pi}{2}\right) = 0, \; y = 2\sin\left(-\frac{\pi}{2}\right) = -2,$

and $z = 1$, so the point is $(0, -2, 1)$ in rectangular coordinates.

3. (a) From Equations 2 we have $r^2 = (-1)^2 + 1^2 = 2$ so $r = \sqrt{2}$; $\tan\theta = \frac{1}{-1} = -1$ and the point $(-1, 1)$ is in the second

quadrant of the xy-plane, so $\theta = \frac{3\pi}{4} + 2n\pi$; $z = 1$. Thus, one set of cylindrical coordinates is $\left(\sqrt{2}, \frac{3\pi}{4}, 1\right)$.

(b) $r^2 = (-2)^2 + \left(2\sqrt{3}\right)^2 = 16$ so $r = 4$; $\tan\theta = \frac{2\sqrt{3}}{-2} = -\sqrt{3}$ and the point $\left(-2, 2\sqrt{3}\right)$ is in the second quadrant of the

xy-plane, so $\theta = \frac{2\pi}{3} + 2n\pi$; $z = 3$. Thus, one set of cylindrical coordinates is $\left(4, \frac{2\pi}{3}, 3\right)$.

5. Since $r = 2$, the distance from any point to the z-axis is 2. Because θ and z may vary, the surface is a circular cylinder with

radius 2 and axis the z-axis. (See Figure 4.)

Also, $x^2 + y^2 = r^2 = 4$, which we recognize as an equation of this cylinder.

7. Since $r^2 + z^2 = 4$ and $r^2 = x^2 + y^2$, we have $x^2 + y^2 + z^2 = 4$, a sphere centered at the origin with radius 2.

9. (a) Substituting $x^2 + y^2 = r^2$ and $x = r\cos\theta$, the equation $x^2 - x + y^2 + z^2 = 1$ becomes $r^2 - r\cos\theta + z^2 = 1$ or

$z^2 = 1 + r\cos\theta - r^2$.

(b) Substituting $x = r\cos\theta$ and $y = r\sin\theta$, the equation $z = x^2 - y^2$ becomes

$z = (r\cos\theta)^2 - (r\sin\theta)^2 = r^2(\cos^2\theta - \sin^2\theta)$ or $z = r^2\cos 2\theta$.

11.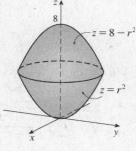

$z = r^2 \iff z = x^2 + y^2$, a circular paraboloid opening upward with vertex the origin,

and $z = 8 - r^2 \iff z = 8 - (x^2 + y^2)$, a circular paraboloid opening downward with

vertex $(0, 0, 8)$. The paraboloids intersect when $r^2 = 8 - r^2 \iff r^2 = 4$. Thus

$r^2 \le z \le 8 - r^2$ describes the solid above the paraboloid $z = x^2 + y^2$ and below the

paraboloid $z = 8 - x^2 - y^2$ for $x^2 + y^2 \le 4$.

13. We can position the cylindrical shell vertically so that its axis coincides with the z-axis and its base lies in the xy-plane. If we

use centimeters as the unit of measurement, then cylindrical coordinates conveniently describe the shell as $6 \le r \le 7$,

$0 \le \theta \le 2\pi, 0 \le z \le 20$.

15.

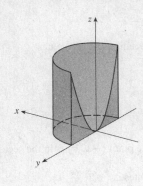

The region of integration is given in cylindrical coordinates by

$$E = \{(r, \theta, z) \mid -\pi/2 \le \theta \le \pi/2, 0 \le r \le 2, 0 \le z \le r^2\}.$$ This

represents the solid region above quadrants I and IV of the xy-plane enclosed

by the circular cylinder $r = 2$, bounded above by the circular paraboloid

$z = r^2$ ($z = x^2 + y^2$), and bounded below by the xy-plane ($z = 0$).

$$\int_{-\pi/2}^{\pi/2} \int_0^2 \int_0^{r^2} r\, dz\, dr\, d\theta = \int_{-\pi/2}^{\pi/2} \int_0^2 \left[rz\right]_{z=0}^{z=r^2} dr\, d\theta = \int_{-\pi/2}^{\pi/2} \int_0^2 r^3\, dr\, d\theta$$

$$= \int_{-\pi/2}^{\pi/2} d\theta \int_0^2 r^3\, dr = \left[\theta\right]_{-\pi/2}^{\pi/2} \left[\tfrac{1}{4}r^4\right]_0^2$$

$$= \pi\,(4 - 0) = 4\pi$$

17. In cylindrical coordinates, E is given by $\{(r, \theta, z) \mid 0 \le \theta \le 2\pi, 0 \le r \le 4, -5 \le z \le 4\}$. So

$$\iiint_E \sqrt{x^2 + y^2}\, dV = \int_0^{2\pi} \int_0^4 \int_{-5}^4 \sqrt{r^2}\, r\, dz\, dr\, d\theta = \int_0^{2\pi} d\theta \int_0^4 r^2\, dr \int_{-5}^4 dz$$

$$= \left[\theta\right]_0^{2\pi} \left[\tfrac{1}{3}r^3\right]_0^4 \left[z\right]_{-5}^4 = (2\pi)\left(\tfrac{64}{3}\right)(9) = 384\pi$$

19. The paraboloid $z = 4 - x^2 - y^2 = 4 - r^2$ intersects the xy-plane in the circle $x^2 + y^2 = 4$ or $r^2 = 4$ $\Rightarrow$ $r = 2$, so in

cylindrical coordinates, E is given by $\left\{(r, \theta, z) \mid 0 \le \theta \le \pi/2, 0 \le r \le 2, 0 \le z \le 4 - r^2\right\}$. Thus

$$\iiint_E (x + y + z)\, dV = \int_0^{\pi/2} \int_0^2 \int_0^{4-r^2} (r\cos\theta + r\sin\theta + z)\, r\, dz\, dr\, d\theta$$

$$= \int_0^{\pi/2} \int_0^2 \left[r^2(\cos\theta + \sin\theta)z + \tfrac{1}{2}rz^2\right]_{z=0}^{z=4-r^2} dr\, d\theta$$

$$= \int_0^{\pi/2} \int_0^2 \left[(4r^2 - r^4)(\cos\theta + \sin\theta) + \tfrac{1}{2}r(4 - r^2)^2\right] dr\, d\theta$$

$$= \int_0^{\pi/2} \left[\left(\tfrac{4}{3}r^3 - \tfrac{1}{5}r^5\right)(\cos\theta + \sin\theta) - \tfrac{1}{12}(4 - r^2)^3\right]_{r=0}^{r=2} d\theta$$

$$= \int_0^{\pi/2} \left[\tfrac{64}{15}(\cos\theta + \sin\theta) + \tfrac{16}{3}\right] d\theta = \left[\tfrac{64}{15}(\sin\theta - \cos\theta) + \tfrac{16}{3}\theta\right]_0^{\pi/2}$$

$$= \tfrac{64}{15}(1 - 0) + \tfrac{16}{3}\cdot\tfrac{\pi}{2} - \tfrac{64}{15}(0 - 1) - 0 = \tfrac{8}{3}\pi + \tfrac{128}{15}$$

21. In cylindrical coordinates, E is bounded by the cylinder $r = 1$, the plane $z = 0$, and the cone $z = 2r$. So

$$E = \{(r, \theta, z) \mid 0 \le \theta \le 2\pi, 0 \le r \le 1, 0 \le z \le 2r\}$$ and

$$\iiint_E x^2\, dV = \int_0^{2\pi} \int_0^1 \int_0^{2r} r^2\cos^2\theta\, r\, dz\, dr\, d\theta = \int_0^{2\pi} \int_0^1 \left[r^3\cos^2\theta\, z\right]_{z=0}^{z=2r} dr\, d\theta = \int_0^{2\pi} \int_0^1 2r^4\cos^2\theta\, dr\, d\theta$$

$$= \int_0^{2\pi} \left[\tfrac{2}{5}r^5\cos^2\theta\right]_{r=0}^{r=1} d\theta = \tfrac{2}{5}\int_0^{2\pi} \cos^2\theta\, d\theta = \tfrac{2}{5}\int_0^{2\pi} \tfrac{1}{2}(1 + \cos 2\theta)\, d\theta = \tfrac{1}{5}\left[\theta + \tfrac{1}{2}\sin 2\theta\right]_0^{2\pi} = \tfrac{2\pi}{5}$$

23. In cylindrical coordinates, E is bounded below by the cone $z = r$ and above by the sphere $r^2 + z^2 = 2$ or $z = \sqrt{2 - r^2}$. The

cone and the sphere intersect when $2r^2 = 2$ $\Rightarrow$ $r = 1$, so $E = \left\{(r, \theta, z) \mid 0 \le \theta \le 2\pi, 0 \le r \le 1, r \le z \le \sqrt{2 - r^2}\right\}$

and the volume is

$$\iiint_E dV = \int_0^{2\pi} \int_0^1 \int_r^{\sqrt{2-r^2}} r\, dz\, dr\, d\theta = \int_0^{2\pi} \int_0^1 \left[rz\right]_{z=r}^{z=\sqrt{2-r^2}} dr\, d\theta = \int_0^{2\pi} \int_0^1 \left(r\sqrt{2 - r^2} - r^2\right) dr\, d\theta$$

$$= \int_0^{2\pi} d\theta \int_0^1 \left(r\sqrt{2 - r^2} - r^2\right) dr = 2\pi \left[-\tfrac{1}{3}(2 - r^2)^{3/2} - \tfrac{1}{3}r^3\right]_0^1$$

$$= 2\pi\left(-\tfrac{1}{3}\right)(1 + 1 - 2^{3/2}) = -\tfrac{2}{3}\pi\left(2 - 2\sqrt{2}\right) = \tfrac{4}{3}\pi\left(\sqrt{2} - 1\right)$$

25. (a) In cylindrical coordinates, E is bounded above by the paraboloid $z = 24 - r^2$ and below by

the cone $z = 2\sqrt{r^2}$ or $z = 2r$ $(r \geq 0)$. The surfaces intersect when

$$24 - r^2 = 2r \quad \Rightarrow \quad r^2 + 2r - 24 = 0 \quad \Rightarrow \quad (r+6)(r-4) = 0 \quad \Rightarrow \quad r = 4, \text{ so}$$

$E = \left\{ (r, \theta, z) \mid 2r \leq z \leq 24 - r^2, 0 \leq r \leq 4, 0 \leq \theta \leq 2\pi \right\}$ and the volume is

$$\iiint_E dV = \int_0^{2\pi} \int_0^4 \int_{2r}^{24-r^2} r \, dz \, dr \, d\theta = \int_0^{2\pi} \int_0^4 r \left(24 - r^2 - 2r \right) dr \, d\theta = \int_0^{2\pi} d\theta \int_0^4 \left(24r - r^3 - 2r^2 \right) dr$$

$$= 2\pi \left[12r^2 - \tfrac{1}{4}r^4 - \tfrac{2}{3}r^3 \right]_0^4 = 2\pi \left(192 - 64 - \tfrac{128}{3} \right) = \tfrac{512}{3}\pi$$

(b) For constant density K, $m = KV = \tfrac{512}{3}\pi K$ from part (a). Since the region is homogeneous and symmetric,

$M_{yz} = M_{xz} = 0$ and

$$M_{xy} = \int_0^{2\pi} \int_0^4 \int_{2r}^{24-r^2} (zK) \, r \, dz \, dr \, d\theta = K \int_0^{2\pi} \int_0^4 r \left[\tfrac{1}{2}z^2 \right]_{z=2r}^{z=24-r^2} dr \, d\theta$$

$$= \tfrac{K}{2} \int_0^{2\pi} \int_0^4 r[(24 - r^2)^2 - 4r^2] \, dr \, d\theta = \tfrac{K}{2} \int_0^{2\pi} d\theta \int_0^4 (576r - 52r^3 + r^5) \, dr$$

$$= \tfrac{K}{2} (2\pi) \left[288r^2 - 13r^4 + \tfrac{1}{6}r^6 \right]_0^4 = \pi K \left(4608 - 3328 + \tfrac{2048}{3} \right) = \tfrac{5888}{3}\pi K$$

Thus $(\overline{x}, \overline{y}, \overline{z}) = \left(\dfrac{M_{yz}}{m}, \dfrac{M_{xz}}{m}, \dfrac{M_{xy}}{m} \right) = \left(0, 0, \dfrac{5888\pi K/3}{512\pi K/3} \right) = \left(0, 0, \tfrac{23}{2} \right)$.

27. The paraboloid $z = 4x^2 + 4y^2$ intersects the plane $z = a$ when $a = 4x^2 + 4y^2$ or $x^2 + y^2 = \tfrac{1}{4}a$. So, in cylindrical

coordinates, $E = \left\{ (r, \theta, z) \mid 0 \leq r \leq \tfrac{1}{2}\sqrt{a}, 0 \leq \theta \leq 2\pi, 4r^2 \leq z \leq a \right\}$. Thus

$$m = \int_0^{2\pi} \int_0^{\sqrt{a}/2} \int_{4r^2}^a Kr \, dz \, dr \, d\theta = K \int_0^{2\pi} \int_0^{\sqrt{a}/2} (ar - 4r^3) \, dr \, d\theta$$

$$= K \int_0^{2\pi} \left[\tfrac{1}{2}ar^2 - r^4 \right]_{r=0}^{r=\sqrt{a}/2} d\theta = K \int_0^{2\pi} \tfrac{1}{16}a^2 \, d\theta = \tfrac{1}{8}a^2 \pi K$$

Since the region is homogeneous and symmetric, $M_{yz} = M_{xz} = 0$ and

$$M_{xy} = \int_0^{2\pi} \int_0^{\sqrt{a}/2} \int_{4r^2}^a Krz \, dz \, dr \, d\theta = K \int_0^{2\pi} \int_0^{\sqrt{a}/2} \left(\tfrac{1}{2}a^2 r - 8r^5 \right) dr \, d\theta$$

$$= K \int_0^{2\pi} \left[\tfrac{1}{4}a^2 r^2 - \tfrac{4}{3}r^6 \right]_{r=0}^{r=\sqrt{a}/2} d\theta = K \int_0^{2\pi} \tfrac{1}{24}a^3 \, d\theta = \tfrac{1}{12}a^3 \pi K$$

Hence $(\overline{x}, \overline{y}, \overline{z}) = \left(0, 0, \tfrac{2}{3}a \right)$.

29. The region of integration is the region above the cone $z = \sqrt{x^2 + y^2}$, or $z = r$, and below the plane $z = 2$. Also, we have

$-2 \leq y \leq 2$ with $-\sqrt{4 - y^2} \leq x \leq \sqrt{4 - y^2}$ which describes a circle of radius 2 in the xy-plane centered at $(0, 0)$. Thus,

$$\int_{-2}^2 \int_{-\sqrt{4-y^2}}^{\sqrt{4-y^2}} \int_{\sqrt{x^2+y^2}}^2 xz \, dz \, dx \, dy = \int_0^{2\pi} \int_0^2 \int_r^2 (r \cos \theta) \, zr \, dz \, dr \, d\theta = \int_0^{2\pi} \int_0^2 \int_r^2 r^2 (\cos \theta) \, z \, dz \, dr \, d\theta$$

$$= \int_0^{2\pi} \int_0^2 r^2 (\cos \theta) \left[\tfrac{1}{2}z^2 \right]_{z=r}^{z=2} dr \, d\theta = \tfrac{1}{2} \int_0^{2\pi} \int_0^2 r^2 (\cos \theta) \left(4 - r^2 \right) dr \, d\theta$$

$$= \tfrac{1}{2} \int_0^{2\pi} \cos \theta \, d\theta \int_0^2 \left(4r^2 - r^4 \right) dr = \tfrac{1}{2} \left[\sin \theta \right]_0^{2\pi} \left[\tfrac{4}{3}r^3 - \tfrac{1}{5}r^5 \right]_0^2 = 0$$

31. (a) The mountain comprises a solid conical region C. The work done in lifting a small volume of material ΔV with density

$g(P)$ to a height $h(P)$ above sea level is $h(P)g(P)\,\Delta V$. Summing over the whole mountain we get

$W = \iiint_C h(P)g(P)\,dV$.

(b) Here C is a solid right circular cone with radius $R = 62{,}000$ ft, height $H = 12{,}400$ ft,

and density $g(P) = 200$ lb/ft^3 at all points P in C. We use cylindrical coordinates:

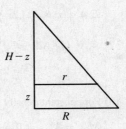

$W = \int_0^{2\pi}\int_0^H\int_0^{R(1-z/H)} z \cdot 200r\,dr\,dz\,d\theta = 2\pi \int_0^H 200z\left[\frac{1}{2}r^2\right]_{r=0}^{r=R(1-z/H)} dz$

$= 400\pi \int_0^H z\,\frac{R^2}{2}\left(1-\frac{z}{H}\right)^2 dz = 200\pi R^2 \int_0^H \left(z - \frac{2z^2}{H} + \frac{z^3}{H^2}\right) dz$

$= 200\pi R^2 \left[\frac{z^2}{2} - \frac{2z^3}{3H} + \frac{z^4}{4H^2}\right]_0^H = 200\pi R^2\left(\frac{H^2}{2} - \frac{2H^2}{3} + \frac{H^2}{4}\right)$

$\frac{r}{R} = \frac{H-z}{H} = 1 - \frac{z}{H}$

$= \frac{50}{3}\pi R^2 H^2 = \frac{50}{3}\pi(62{,}000)^2(12{,}400)^2 \approx 3.1 \times 10^{19}$ ft-lb

15.8 Triple Integrals in Spherical Coordinates

1. (a)

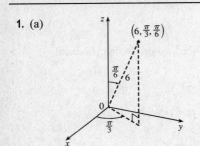

From Equations 1, $x = \rho\sin\phi\cos\theta = 6\sin\frac{\pi}{6}\cos\frac{\pi}{3} = 6\cdot\frac{1}{2}\cdot\frac{1}{2} = \frac{3}{2}$,

$y = \rho\sin\phi\sin\theta = 6\sin\frac{\pi}{6}\sin\frac{\pi}{3} = 6\cdot\frac{1}{2}\cdot\frac{\sqrt{3}}{2} = \frac{3\sqrt{3}}{2}$, and

$z = \rho\cos\phi = 6\cos\frac{\pi}{6} = 6\cdot\frac{\sqrt{3}}{2} = 3\sqrt{3}$, so the point is $\left(\frac{3}{2}, \frac{3\sqrt{3}}{2}, 3\sqrt{3}\right)$ in

rectangular coordinates.

(b)

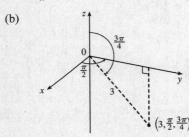

$x = 3\sin\frac{3\pi}{4}\cos\frac{\pi}{2} = 3\cdot\frac{\sqrt{2}}{2}\cdot 0 = 0$,

$y = 3\sin\frac{3\pi}{4}\sin\frac{\pi}{2} = 3\cdot\frac{\sqrt{2}}{2}\cdot 1 = \frac{3\sqrt{2}}{2}$, and

$z = 3\cos\frac{3\pi}{4} = 3\left(-\frac{\sqrt{2}}{2}\right) = -\frac{3\sqrt{2}}{2}$, so the point is $\left(0, \frac{3\sqrt{2}}{2}, -\frac{3\sqrt{2}}{2}\right)$ in

rectangular coordinates.

3. (a) From Equations 1 and 2, $\rho = \sqrt{x^2+y^2+z^2} = \sqrt{0^2+(-2)^2+0^2} = 2$, $\cos\phi = \frac{z}{\rho} = \frac{0}{2} = 0 \;\Rightarrow\; \phi = \frac{\pi}{2}$, and

$\cos\theta = \frac{x}{\rho\sin\phi} = \frac{0}{2\sin(\pi/2)} = 0 \;\Rightarrow\; \theta = \frac{3\pi}{2}$ [since $y < 0$]. Thus spherical coordinates are $\left(2, \frac{3\pi}{2}, \frac{\pi}{2}\right)$.

(b) $\rho = \sqrt{1+1+2} = 2$, $\cos\phi = \frac{z}{\rho} = \frac{-\sqrt{2}}{2} \;\Rightarrow\; \phi = \frac{3\pi}{4}$, and

$\cos\theta = \frac{x}{\rho\sin\phi} = \frac{-1}{2\sin(3\pi/4)} = \frac{-1}{2\left(\sqrt{2}/2\right)} = -\frac{1}{\sqrt{2}} \;\Rightarrow\; \theta = \frac{3\pi}{4}$ [since $y > 0$]. Thus spherical coordinates

are $\left(2, \frac{3\pi}{4}, \frac{3\pi}{4}\right)$.

5. Since $\phi = \frac{\pi}{3}$ but ρ and θ can vary, the surface is the top half of a right circular cone with vertex at the origin and axis the positive z-axis. (See Figure 4.)

7. From Equations 1 we have $z = \rho\cos\phi$, so $\rho\cos\phi = 1$ $\Leftrightarrow$ $z = 1$, and the surface is the horizontal plane $z = 1$.

9. (a) From Equation 2 we have $\rho^2 = x^2 + y^2 + z^2$, so $x^2 + y^2 + z^2 = 9$ $\Leftrightarrow$ $\rho^2 = 9$ $\Rightarrow$ $\rho = 3$ (since $\rho \geq 0$).

(b) From Equations 1 we have $x = \rho\sin\phi\cos\theta$, $y = \rho\sin\phi\sin\theta$, and $z = \rho\cos\phi$, so the equation $x^2 - y^2 - z^2 = 1$

becomes $(\rho\sin\phi\cos\theta)^2 - (\rho\sin\phi\sin\theta)^2 - (\rho\cos\phi)^2 = 1$ $\Leftrightarrow$ $(\rho^2\sin^2\phi)(\cos^2\theta - \sin^2\theta) - \rho^2\cos^2\phi = 1$ $\Leftrightarrow$ $\rho^2(\sin^2\phi\cos 2\theta - \cos^2\phi) = 1$.

11. $\rho \leq 1$ represents the (solid) unit ball. $0 \leq \phi \leq \frac{\pi}{6}$ restricts the solid to that portion on or above the cone $\phi = \frac{\pi}{6}$, and $0 \leq \theta \leq \pi$ further restricts the solid to that portion on or to the right of the xz-plane.

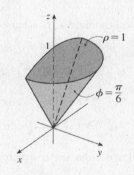

13. $2 \leq \rho \leq 4$ represents the solid region between and including the spheres of radii 2 and 4, centered at the origin. $0 \leq \phi \leq \frac{\pi}{3}$ restricts the solid to that portion on or above the cone $\phi = \frac{\pi}{3}$, and $0 \leq \theta \leq \pi$ further restricts the solid to that portion on or to the right of the xz-plane.

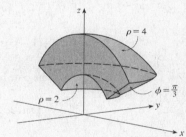

15. $z \geq \sqrt{x^2 + y^2}$ because the solid lies above the cone. Squaring both sides of this inequality gives $z^2 \geq x^2 + y^2$ $\Rightarrow$ $2z^2 \geq x^2 + y^2 + z^2 = \rho^2$ $\Rightarrow$ $z^2 = \rho^2\cos^2\phi \geq \frac{1}{2}\rho^2$ $\Rightarrow$ $\cos^2\phi \geq \frac{1}{2}$. The cone opens upward so that the inequality is $\cos\phi \geq \frac{1}{\sqrt{2}}$, or equivalently $0 \leq \phi \leq \frac{\pi}{4}$. In spherical coordinates the sphere $z = x^2 + y^2 + z^2$ is $\rho\cos\phi = \rho^2$ $\Rightarrow$ $\rho = \cos\phi$. $0 \leq \rho \leq \cos\phi$ because the solid lies below the sphere. The solid can therefore be described as the region in spherical coordinates satisfying $0 \leq \rho \leq \cos\phi$, $0 \leq \phi \leq \frac{\pi}{4}$.

17.

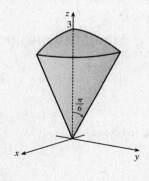

The region of integration is given in spherical coordinates by

$E = \{(\rho, \theta, \phi) \mid 0 \leq \rho \leq 3,\ 0 \leq \theta \leq \pi/2,\ 0 \leq \phi \leq \pi/6\}$. This represents the solid region in the first octant bounded above by the sphere $\rho = 3$ and below by the cone $\phi = \pi/6$.

$$\int_0^{\pi/6}\int_0^{\pi/2}\int_0^3 \rho^2\sin\phi\,d\rho\,d\theta\,d\phi = \int_0^{\pi/6}\sin\phi\,d\phi\int_0^{\pi/2}d\theta\int_0^3\rho^2\,d\rho$$

$$= \left[-\cos\phi\right]_0^{\pi/6}\left[\theta\right]_0^{\pi/2}\left[\tfrac{1}{3}\rho^3\right]_0^3$$

$$= \left(1 - \frac{\sqrt{3}}{2}\right)\left(\frac{\pi}{2}\right)(9) = \frac{9\pi}{4}\left(2 - \sqrt{3}\right)$$

19. The solid E is most conveniently described if we use cylindrical coordinates:

$E = \{(r, \theta, z) \mid 0 \le \theta \le \frac{\pi}{2},\ 0 \le r \le 3,\ 0 \le z \le 2\}$. Then

$\iiint_E f(x, y, z)\, dV = \int_0^{\pi/2} \int_0^3 \int_0^2 f(r\cos\theta, r\sin\theta, z)\, r\, dz\, dr\, d\theta$.

21. In spherical coordinates, B is represented by $\{(\rho, \theta, \phi) \mid 0 \le \rho \le 5, 0 \le \theta \le 2\pi, 0 \le \phi \le \pi\}$. Thus

$$\iiint_B (x^2 + y^2 + z^2)^2\, dV = \int_0^\pi \int_0^{2\pi} \int_0^5 (\rho^2)^2 \rho^2 \sin\phi\, d\rho\, d\theta\, d\phi = \int_0^\pi \sin\phi\, d\phi \int_0^{2\pi} d\theta \int_0^5 \rho^6\, d\rho$$

$$= \left[-\cos\phi\right]_0^\pi \left[\theta\right]_0^{2\pi} \left[\tfrac{1}{7}\rho^7\right]_0^5 = (2)(2\pi)\left(\tfrac{78,125}{7}\right)$$

$$= \tfrac{312,500}{7}\pi \approx 140,249.7$$

23. In spherical coordinates, E is represented by $\{(\rho, \theta, \phi) \mid 2 \le \rho \le 3, 0 \le \theta \le 2\pi, 0 \le \phi \le \pi\}$ and

$x^2 + y^2 = \rho^2 \sin^2\phi \cos^2\theta + \rho^2 \sin^2\phi \sin^2\theta = \rho^2 \sin^2\phi\left(\cos^2\theta + \sin^2\theta\right) = \rho^2 \sin^2\phi$. Thus

$$\iiint_E (x^2 + y^2)\, dV = \int_0^\pi \int_0^{2\pi} \int_2^3 (\rho^2 \sin^2\phi)\, \rho^2 \sin\phi\, d\rho\, d\theta\, d\phi = \int_0^\pi \sin^3\phi\, d\phi \int_0^{2\pi} d\theta \int_2^3 \rho^4\, d\rho$$

$$= \int_0^\pi (1 - \cos^2\phi)\, \sin\phi\, d\phi \left[\theta\right]_0^{2\pi} \left[\tfrac{1}{5}\rho^5\right]_2^3 = \left[-\cos\phi + \tfrac{1}{3}\cos^3\phi\right]_0^\pi (2\pi) \cdot \tfrac{1}{5}(243 - 32)$$

$$= \left(1 - \tfrac{1}{3} + 1 - \tfrac{1}{3}\right)(2\pi)\left(\tfrac{211}{5}\right) = \tfrac{1688\pi}{15}$$

25. In spherical coordinates, E is represented by $\{(\rho, \theta, \phi) \mid 0 \le \rho \le 1, 0 \le \theta \le \frac{\pi}{2}, 0 \le \phi \le \frac{\pi}{2}\}$. Thus

$$\iiint_E x e^{x^2 + y^2 + z^2}\, dV = \int_0^{\pi/2} \int_0^{\pi/2} \int_0^1 (\rho \sin\phi \cos\theta) e^{\rho^2} \rho^2 \sin\phi\, d\rho\, d\theta\, d\phi = \int_0^{\pi/2} \sin^2\phi\, d\phi \int_0^{\pi/2} \cos\theta\, d\theta \int_0^1 \rho^3 e^{\rho^2}\, d\rho$$

$$= \int_0^{\pi/2} \tfrac{1}{2}(1 - \cos 2\phi)\, d\phi \int_0^{\pi/2} \cos\theta\, d\theta \left(\left.\tfrac{1}{2}\rho^2 e^{\rho^2}\right]_0^1 - \int_0^1 \rho e^{\rho^2}\, d\rho\right)$$

$$\left[\text{integrate by parts with } u = \rho^2,\, dv = \rho e^{\rho^2}\, d\rho\right]$$

$$= \left[\tfrac{1}{2}\phi - \tfrac{1}{4}\sin 2\phi\right]_0^{\pi/2} \left[\sin\theta\right]_0^{\pi/2} \left[\tfrac{1}{2}\rho^2 e^{\rho^2} - \tfrac{1}{2}e^{\rho^2}\right]_0^1 = \left(\tfrac{\pi}{4} - 0\right)(1 - 0)\left(0 + \tfrac{1}{2}\right) = \tfrac{\pi}{8}$$

27. The solid region is given by $E = \{(\rho, \theta, \phi) \mid 0 \le \rho \le a, 0 \le \theta \le 2\pi, \frac{\pi}{6} \le \phi \le \frac{\pi}{3}\}$ and its volume is

$$V = \iiint_E dV = \int_{\pi/6}^{\pi/3} \int_0^{2\pi} \int_0^a \rho^2 \sin\phi\, d\rho\, d\theta\, d\phi = \int_{\pi/6}^{\pi/3} \sin\phi\, d\phi \int_0^{2\pi} d\theta \int_0^a \rho^2\, d\rho$$

$$= \left[-\cos\phi\right]_{\pi/6}^{\pi/3} \left[\theta\right]_0^{2\pi} \left[\tfrac{1}{3}\rho^3\right]_0^a = \left(-\tfrac{1}{2} + \tfrac{\sqrt{3}}{2}\right)(2\pi)\left(\tfrac{1}{3}a^3\right) = \tfrac{\sqrt{3}-1}{3}\pi a^3$$

29. (a) Since $\rho = 4\cos\phi$ implies $\rho^2 = 4\rho\cos\phi \iff x^2 + y^2 + z^2 = 4z \iff x^2 + y^2 + (z - 2)^2 = 4$, the equation is that of

a sphere of radius 2 with center at $(0, 0, 2)$. Thus

$$V = \int_0^{2\pi} \int_0^{\pi/3} \int_0^{4\cos\phi} \rho^2 \sin\phi\, d\rho\, d\phi\, d\theta = \int_0^{2\pi} \int_0^{\pi/3} \left[\tfrac{1}{3}\rho^3\right]_{\rho=0}^{\rho=4\cos\phi} \sin\phi\, d\phi\, d\theta = \int_0^{2\pi} \int_0^{\pi/3} \left(\tfrac{64}{3}\cos^3\phi\right) \sin\phi\, d\phi\, d\theta$$

$$= \int_0^{2\pi} \left[-\tfrac{16}{3}\cos^4\phi\right]_{\phi=0}^{\phi=\pi/3} d\theta = \int_0^{2\pi} -\tfrac{16}{3}\left(\tfrac{1}{16} - 1\right) d\theta = 5\theta\Big]_0^{2\pi} = 10\pi$$

(b) By the symmetry of the problem $M_{yz} = M_{xz} = 0$. Then

$$M_{xy} = \int_0^{2\pi} \int_0^{\pi/3} \int_0^{4\cos\phi} \rho^3 \cos\phi \sin\phi\, d\rho\, d\phi\, d\theta = \int_0^{2\pi} \int_0^{\pi/3} \cos\phi \sin\phi\, (64\cos^4\phi)\, d\phi\, d\theta$$

$$= \int_0^{2\pi} 64\left[-\tfrac{1}{6}\cos^6\phi\right]_{\phi=0}^{\phi=\pi/3} d\theta = \int_0^{2\pi} \tfrac{21}{2}\, d\theta = 21\pi$$

Hence $(\overline{x}, \overline{y}, \overline{z}) = (0, 0, 21\pi/(10\pi)) = (0, 0, 2.1)$.

31. (a) By the symmetry of the region, $M_{yz} = 0$ and $M_{xz} = 0$. Assuming constant density K,

$$m = \iiint_E K \, dV = K \iiint_E dV = \tfrac{\pi}{8} K \text{ (from Example 4). Then}$$

$$M_{xy} = \iiint_E z \, K \, dV = K \int_0^{2\pi} \int_0^{\pi/4} \int_0^{\cos\phi} (\rho \cos\phi) \, \rho^2 \sin\phi \, d\rho \, d\phi \, d\theta = K \int_0^{2\pi} \int_0^{\pi/4} \sin\phi \cos\phi \left[\tfrac{1}{4}\rho^4 \right]_{\rho=0}^{\rho=\cos\phi} d\phi \, d\theta$$

$$= \tfrac{1}{4} K \int_0^{2\pi} \int_0^{\pi/4} \sin\phi \cos\phi \left(\cos^4 \phi \right) d\phi \, d\theta = \tfrac{1}{4} K \int_0^{2\pi} d\theta \int_0^{\pi/4} \cos^5 \phi \sin\phi \, d\phi$$

$$= \tfrac{1}{4} K \left[\theta \right]_0^{2\pi} \left[-\tfrac{1}{6} \cos^6 \phi \right]_0^{\pi/4} = \tfrac{1}{4} K (2\pi) \left(-\tfrac{1}{6} \right) \left[\left(\tfrac{\sqrt{2}}{2} \right)^6 - 1 \right] = -\tfrac{\pi}{12} K \left(-\tfrac{7}{8} \right) = \tfrac{7\pi}{96} K$$

Thus the centroid is $(\overline{x}, \overline{y}, \overline{z}) = \left(\dfrac{M_{yz}}{m}, \dfrac{M_{xz}}{m}, \dfrac{M_{xy}}{m} \right) = \left(0, 0, \dfrac{7\pi K/96}{\pi K/8} \right) = \left(0, 0, \tfrac{7}{12} \right)$.

(b) As in Exercise 23, $x^2 + y^2 = \rho^2 \sin^2 \phi$ and

$$I_z = \iiint_E (x^2 + y^2) \, K \, dV = K \int_0^{2\pi} \int_0^{\pi/4} \int_0^{\cos\phi} (\rho^2 \sin^2 \phi) \, \rho^2 \sin\phi \, d\rho \, d\phi \, d\theta = K \int_0^{2\pi} \int_0^{\pi/4} \sin^3 \phi \left[\tfrac{1}{5}\rho^5 \right]_{\rho=0}^{\rho=\cos\phi} d\phi \, d\theta$$

$$= \tfrac{1}{5} K \int_0^{2\pi} \int_0^{\pi/4} \sin^3 \phi \cos^5 \phi \, d\phi \, d\theta = \tfrac{1}{5} K \int_0^{2\pi} d\theta \int_0^{\pi/4} \cos^5 \phi \left(1 - \cos^2 \phi \right) \sin\phi \, d\phi$$

$$= \tfrac{1}{5} K \left[\theta \right]_0^{2\pi} \left[-\tfrac{1}{6} \cos^6 \phi + \tfrac{1}{8} \cos^8 \phi \right]_0^{\pi/4}$$

$$= \tfrac{1}{5} K (2\pi) \left[-\tfrac{1}{6} \left(\tfrac{\sqrt{2}}{2} \right)^6 + \tfrac{1}{8} \left(\tfrac{\sqrt{2}}{2} \right)^8 + \tfrac{1}{6} - \tfrac{1}{8} \right] = \tfrac{2\pi}{5} K \left(\tfrac{11}{384} \right) = \tfrac{11\pi}{960} K$$

33. (a) The density function is $\rho(x, y, z) = K$, a constant, and by the symmetry of the problem $M_{xz} = M_{yz} = 0$. Then

$$M_{xy} = \int_0^{2\pi} \int_0^{\pi/2} \int_0^a K \rho^3 \sin\phi \cos\phi \, d\rho \, d\phi \, d\theta = \tfrac{1}{2} \pi K a^4 \int_0^{\pi/2} \sin\phi \cos\phi \, d\phi = \tfrac{1}{8} \pi K a^4. \text{ But the mass is } K(\text{volume of }$$

the hemisphere) $= \tfrac{2}{3} \pi K a^3$, so the centroid is $\left(0, 0, \tfrac{3}{8} a \right)$.

(b) Place the center of the base at $(0, 0, 0)$; the density function is $\rho(x, y, z) = K$. By symmetry, the moments of inertia about any two such diameters will be equal, so we just need to find I_x:

$$I_x = \int_0^{2\pi} \int_0^{\pi/2} \int_0^a (K \rho^2 \sin\phi) \, \rho^2 (\sin^2 \phi \sin^2 \theta + \cos^2 \phi) \, d\rho \, d\phi \, d\theta$$

$$= K \int_0^{2\pi} \int_0^{\pi/2} (\sin^3 \phi \sin^2 \theta + \sin\phi \cos^2 \phi)(\tfrac{1}{5} a^5) \, d\phi \, d\theta$$

$$= \tfrac{1}{5} K a^5 \int_0^{2\pi} \left[\sin^2 \theta \left(-\cos\phi + \tfrac{1}{3} \cos^3 \phi \right) + \left(-\tfrac{1}{3} \cos^3 \phi \right) \right]_{\phi=0}^{\phi=\pi/2} d\theta = \tfrac{1}{5} K a^5 \int_0^{2\pi} \left[\tfrac{2}{3} \sin^2 \theta + \tfrac{1}{3} \right] d\theta$$

$$= \tfrac{1}{5} K a^5 \left[\tfrac{2}{3} \left(\tfrac{1}{2}\theta - \tfrac{1}{4} \sin 2\theta \right) + \tfrac{1}{3}\theta \right]_0^{2\pi} = \tfrac{1}{5} K a^5 \left[\tfrac{2}{3} (\pi - 0) + \tfrac{1}{3}(2\pi - 0) \right] = \tfrac{4}{15} K a^5 \pi$$

35. In spherical coordinates $z = \sqrt{x^2 + y^2}$ becomes $\phi = \tfrac{\pi}{4}$ (as in Example 4). Then

$$V = \int_0^{2\pi} \int_0^{\pi/4} \int_0^1 \rho^2 \sin\phi \, d\rho \, d\phi \, d\theta = \int_0^{2\pi} d\theta \int_0^{\pi/4} \sin\phi \, d\phi \int_0^1 \rho^2 \, d\rho = 2\pi \left(-\tfrac{\sqrt{2}}{2} + 1 \right) \left(\tfrac{1}{3} \right) = \tfrac{1}{3} \pi (2 - \sqrt{2}),$$

$$M_{xy} = \int_0^{2\pi} \int_0^{\pi/4} \int_0^1 \rho^3 \sin\phi \cos\phi \, d\rho \, d\phi \, d\theta = 2\pi \left[-\tfrac{1}{4} \cos 2\phi \right]_0^{\pi/4} \left(\tfrac{1}{4} \right) = \tfrac{\pi}{8} \text{ and by symmetry } M_{yz} = M_{xz} = 0.$$

Hence $(\overline{x}, \overline{y}, \overline{z}) = \left(0, 0, \dfrac{3}{8(2 - \sqrt{2})} \right)$.

37. (a) If we orient the cylinder so that its axis is the z-axis and its base lies in the xy-plane, then the cylinder is described, in cylindrical coordinates, by $E = \{(r, \theta, z) \mid 0 \le r \le a, \, 0 \le \theta \le 2\pi, \, 0 \le z \le h\}$. Assuming constant density K, the moment of inertia about its axis (the z-axis) is

$$I_z = \iiint_E (x^2 + y^2)\, \rho(x,y,z)\, dV = \int_0^{2\pi} \int_0^a \int_0^h K(r^2)\, r\, dz\, dr\, d\theta = K \int_0^{2\pi} d\theta \int_0^a r^3\, dr \int_0^h dz$$

$$= K\left[\theta\right]_0^{2\pi} \left[\tfrac{1}{4} r^4\right]_0^a \left[z\right]_0^h = K(2\pi)\left(\tfrac{1}{4} a^4\right)(h) = \tfrac{1}{2}\pi K a^4 h$$

(b) By symmetry, the moments of inertia about any two diameters of the base will be equal, and one of the diameters lies on

the x-axis, so we compute:

$$I_x = \iiint_E (y^2 + z^2)\, \rho(x,y,z)\, dV = \int_0^{2\pi} \int_0^a \int_0^h K(r^2 \sin^2 \theta + z^2)\, r\, dz\, dr\, d\theta$$

$$= K \int_0^{2\pi} \int_0^a \int_0^h r^3 \sin^2 \theta\, dz\, dr\, d\theta + K \int_0^{2\pi} \int_0^a \int_0^h rz^2\, dz\, dr\, d\theta$$

$$= K \int_0^{2\pi} \sin^2 \theta\, d\theta \int_0^a r^3\, dr \int_0^h dz + K \int_0^{2\pi} d\theta \int_0^a r\, dr \int_0^h z^2\, dz$$

$$= K\left[\tfrac{1}{2}\theta - \tfrac{1}{4}\sin 2\theta\right]_0^{2\pi} \left[\tfrac{1}{4} r^4\right]_0^a \left[z\right]_0^h + K\left[\theta\right]_0^{2\pi} \left[\tfrac{1}{2} r^2\right]_0^a \left[\tfrac{1}{3} z^3\right]_0^h$$

$$= K(\pi)\left(\tfrac{1}{4} a^4\right)(h) + K(2\pi)\left(\tfrac{1}{2} a^2\right)\left(\tfrac{1}{3} h^3\right) = \tfrac{1}{12}\pi K a^2 h(3a^2 + 4h^2)$$

39. In cylindrical coordinates the paraboloid is given by $z = r^2$ and the plane by $z = 2r \sin \theta$ and the projection of the

intersection onto the xy-plane is the circle $r = 2 \sin \theta$. Then $\iiint_E z\, dV = \int_0^\pi \int_0^{2 \sin \theta} \int_{r^2}^{2r \sin \theta} rz\, dz\, dr\, d\theta = \frac{5\pi}{6}$

[using a CAS].

41. The region E of integration is the region above the cone $z = \sqrt{x^2 + y^2}$ and below the sphere $x^2 + y^2 + z^2 = 2$ in the first

octant. Because E is in the first octant we have $0 \le \theta \le \frac{\pi}{2}$. The cone has equation $\phi = \frac{\pi}{4}$ (as in Example 4), so $0 \le \phi \le \frac{\pi}{4}$,

and $0 \le \rho \le \sqrt{2}$. Then the integral becomes

$$\int_0^{\pi/4} \int_0^{\pi/2} \int_0^{\sqrt{2}} (\rho \sin \phi \cos \theta)(\rho \sin \phi \sin \theta) \rho^2 \sin \phi\, d\rho\, d\theta\, d\phi$$

$$= \int_0^{\pi/4} \sin^3 \phi\, d\phi \int_0^{\pi/2} \sin \theta \cos \theta\, d\theta \int_0^{\sqrt{2}} \rho^4\, d\rho = \left(\int_0^{\pi/4} (1 - \cos^2 \phi) \sin \phi\, d\phi\right) \left[\tfrac{1}{2} \sin^2 \theta\right]_0^{\pi/2} \left[\tfrac{1}{5} \rho^5\right]_0^{\sqrt{2}}$$

$$= \left[\tfrac{1}{3} \cos^3 \phi - \cos \phi\right]_0^{\pi/4} \cdot \tfrac{1}{2} \cdot \tfrac{1}{5} (\sqrt{2})^5 = \left[\tfrac{\sqrt{2}}{12} - \tfrac{\sqrt{2}}{2} - (\tfrac{1}{3} - 1)\right] \cdot \tfrac{2\sqrt{2}}{5} = \tfrac{4\sqrt{2} - 5}{15}$$

43. The region of integration is the solid sphere $x^2 + y^2 + (z - 2)^2 \le 4$ or equivalently

$$\rho^2 \sin^2 \phi + (\rho \cos \phi - 2)^2 = \rho^2 - 4\rho \cos \phi + 4 \le 4 \quad \Rightarrow \quad \rho \le 4 \cos \phi, \text{ so } 0 \le \theta \le 2\pi, 0 \le \phi \le \frac{\pi}{2}, \text{ and }$$

$0 \le \rho \le 4 \cos \phi$. Also $(x^2 + y^2 + z^2)^{3/2} = (\rho^2)^{3/2} = \rho^3$, so the integral becomes

$$\int_0^{\pi/2} \int_0^{2\pi} \int_0^{4 \cos \phi} (\rho^3) \rho^2 \sin \phi\, d\rho\, d\theta\, d\phi = \int_0^{\pi/2} \int_0^{2\pi} \sin \phi \left[\tfrac{1}{6} \rho^6\right]_{\rho=0}^{\rho=4 \cos \phi} d\theta\, d\phi$$

$$= \tfrac{1}{6} \int_0^{\pi/2} \int_0^{2\pi} \sin \phi\, (4096 \cos^6 \phi)\, d\theta\, d\phi$$

$$= \tfrac{1}{6}(4096) \int_0^{\pi/2} \cos^6 \phi \sin \phi\, d\phi \int_0^{2\pi} d\theta = \tfrac{2048}{3} \left[-\tfrac{1}{7} \cos^7 \phi\right]_0^{\pi/2} \left[\theta\right]_0^{2\pi}$$

$$= \tfrac{2048}{3} \left(\tfrac{1}{7}\right)(2\pi) = \tfrac{4096\pi}{21}$$

45. In cylindrical coordinates, the equation of the cylinder is $r = 3$, $0 \le z \le 10$.

The hemisphere is the upper part of the sphere radius 3, center $(0, 0, 10)$, equation

$r^2 + (z - 10)^2 = 3^2$, $z \ge 10$. In Maple, we can use the `coords=cylindrical` option

in a regular `plot3d` command. In Mathematica, we can use `ParametricPlot3D`.

47. If E is the solid enclosed by the surface $\rho = 1 + \frac{1}{5}\sin 6\theta \sin 5\phi$, it can be described in spherical coordinates as

$E = \left\{ (\rho, \theta, \phi) \mid 0 \le \rho \le 1 + \frac{1}{5}\sin 6\theta \sin 5\phi, 0 \le \theta \le 2\pi, 0 \le \phi \le \pi \right\}$. Its volume is given by

$V(E) = \iiint_E dV = \int_0^\pi \int_0^{2\pi} \int_0^{1 + (\sin 6\theta \sin 5\phi)/5} \rho^2 \sin\phi \, d\rho \, d\theta \, d\phi = \frac{136\pi}{99}$ [using a CAS].

49. (a) From the diagram, $z = r \cot\phi_0$ to $z = \sqrt{a^2 - r^2}$, $r = 0$

to $r = a \sin\phi_0$ (or use $a^2 - r^2 = r^2 \cot^2\phi_0$). Thus

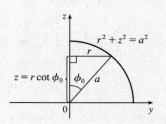

$V = \int_0^{2\pi} \int_0^{a\sin\phi_0} \int_{r\cot\phi_0}^{\sqrt{a^2-r^2}} r \, dz \, dr \, d\theta$

$= 2\pi \int_0^{a\sin\phi_0} \left(r\sqrt{a^2 - r^2} - r^2 \cot\phi_0 \right) dr$

$= \frac{2\pi}{3} \left[-(a^2 - r^2)^{3/2} - r^3 \cot\phi_0 \right]_0^{a\sin\phi_0}$

$= \frac{2\pi}{3} \left[-\left(a^2 - a^2 \sin^2\phi_0\right)^{3/2} - a^3 \sin^3\phi_0 \cot\phi_0 + a^3 \right]$

$= \frac{2}{3}\pi a^3 \left[1 - \left(\cos^3\phi_0 + \sin^2\phi_0 \cos\phi_0\right) \right] = \frac{2}{3}\pi a^3 (1 - \cos\phi_0)$

(b) The wedge in question is the shaded area rotated from $\theta = \theta_1$ to $\theta = \theta_2$.

Letting

V_{ij} = volume of the region bounded by the sphere of radius ρ_i

and the cone with angle ϕ_j ($\theta = \theta_1$ to θ_2)

and letting V be the volume of the wedge, we have

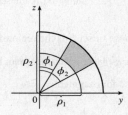

$V = (V_{22} - V_{21}) - (V_{12} - V_{11})$

$= \frac{1}{3}(\theta_2 - \theta_1)\left[\rho_2^3(1 - \cos\phi_2) - \rho_2^3(1 - \cos\phi_1) - \rho_1^3(1 - \cos\phi_2) + \rho_1^3(1 - \cos\phi_1)\right]$

$= \frac{1}{3}(\theta_2 - \theta_1)\left[(\rho_2^3 - \rho_1^3)(1 - \cos\phi_2) - (\rho_2^3 - \rho_1^3)(1 - \cos\phi_1)\right] = \frac{1}{3}(\theta_2 - \theta_1)\left[(\rho_2^3 - \rho_1^3)(\cos\phi_1 - \cos\phi_2)\right]$

Or: Show that $V = \int_{\theta_1}^{\theta_2} \int_{\rho_1\sin\phi_1}^{\rho_2\sin\phi_2} \int_{r\cot\phi_2}^{r\cot\phi_1} r \, dz \, dr \, d\theta$.

(c) By the Mean Value Theorem with $f(\rho) = \rho^3$ there exists some $\tilde{\rho}$ with $\rho_1 \le \tilde{\rho} \le \rho_2$ such that

$f(\rho_2) - f(\rho_1) = f'(\tilde{\rho})(\rho_2 - \rho_1)$ or $\rho_1^3 - \rho_2^3 = 3\tilde{\rho}^2 \Delta\rho$. Similarly there exists ϕ with $\phi_1 \le \tilde{\phi} \le \phi_2$

such that $\cos\phi_2 - \cos\phi_1 = \left(-\sin\tilde{\phi}\right)\Delta\phi$. Substituting into the result from (b) gives

$\Delta V = (\tilde{\rho}^2 \Delta\rho)(\theta_2 - \theta_1)(\sin\tilde{\phi})\Delta\phi = \tilde{\rho}^2 \sin\tilde{\phi}\, \Delta\rho\, \Delta\phi\, \Delta\theta$.

15.9 Change of Variables in Multiple Integrals

1. $x = 2u + v$, $y = 4u - v$.

The Jacobian is $\dfrac{\partial(x, y)}{\partial(u, v)} = \begin{vmatrix} \partial x/\partial u & \partial x/\partial v \\ \partial y/\partial u & \partial y/\partial v \end{vmatrix} = \begin{vmatrix} 2 & 1 \\ 4 & -1 \end{vmatrix} = (2)(-1) - (1)(4) = -6.$

3. $x = s \cos t, \ y = s \sin t.$

$$\frac{\partial(x, y)}{\partial(s, t)} = \begin{vmatrix} \partial x/\partial s & \partial x/\partial t \\ \partial y/\partial s & \partial y/\partial t \end{vmatrix} = \begin{vmatrix} \cos t & -s \sin t \\ \sin t & s \cos t \end{vmatrix} = s \cos^2 t - (-s \sin^2 t) = s(\cos^2 t + \sin^2 t) = s$$

5. $x = uv, \ y = vw, \ z = wu.$

$$\frac{\partial(x, y, z)}{\partial(u, v, w)} = \begin{vmatrix} \partial x/\partial u & \partial x/\partial v & \partial x/\partial w \\ \partial y/\partial u & \partial y/\partial v & \partial y/\partial w \\ \partial z/\partial u & \partial z/\partial v & \partial z/\partial w \end{vmatrix} = \begin{vmatrix} v & u & 0 \\ 0 & w & v \\ w & 0 & u \end{vmatrix} = v \begin{vmatrix} w & v \\ 0 & u \end{vmatrix} - u \begin{vmatrix} 0 & v \\ w & u \end{vmatrix} + 0 \begin{vmatrix} 0 & w \\ w & 0 \end{vmatrix}$$

$$= v(uw - 0) - u(0 - vw) + 0 = uvw + uvw = 2uvw$$

7. The transformation maps the boundary of S to the boundary of the image R, so we first look at side S_1 in the uv-plane. S_1 is

described by $v = 0, 0 \le u \le 3$, so $x = 2u + 3v = 2u$ and $y = u - v = u$. Eliminating u, we have $x = 2y, 0 \le x \le 6$. S_2 is

the line segment $u = 3, 0 \le v \le 2$, so $x = 6 + 3v$ and $y = 3 - v$. Then $v = 3 - y \ \Rightarrow \ x = 6 + 3(3 - y) = 15 - 3y,$

$6 \le x \le 12$. S_3 is the line segment $v = 2, 0 \le u \le 3$, so $x = 2u + 6$ and $y = u - 2$, giving $u = y + 2 \ \Rightarrow \ x = 2y + 10,$

$6 \le x \le 12$. Finally, S_4 is the segment $u = 0, 0 \le v \le 2$, so $x = 3v$ and $y = -v \ \Rightarrow \ x = -3y, 0 \le x \le 6.$

The image of set S is the region R shown
in the xy-plane, a parallelogram bounded
by these four segments.

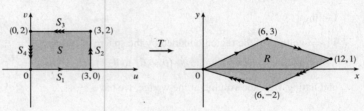

9. S_1 is the line segment $u = v, 0 \le u \le 1$, so $y = v = u$ and $x = u^2 = y^2$. Since $0 \le u \le 1$, the image is the portion of the

parabola $x = y^2, 0 \le y \le 1$. S_2 is the segment $v = 1, 0 \le u \le 1$, thus $y = v = 1$ and $x = u^2$, so $0 \le x \le 1$. The image is

the line segment $y = 1, 0 \le x \le 1$. S_3 is the segment $u = 0, 0 \le v \le 1$, so $x = u^2 = 0$ and $y = v \ \Rightarrow \ 0 \le y \le 1$. The

image is the segment $x = 0, 0 \le y \le 1$. Thus, the image of S is the region R in the first quadrant bounded by the parabola

$x = y^2$, the y-axis, and the line $y = 1$.

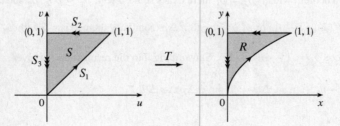

11. R is a parallelogram enclosed by the parallel lines $y = 2x - 1, y = 2x + 1$ and the parallel lines $y = 1 - x, y = 3 - x$. The

first pair of equations can be written as $y - 2x = -1, y - 2x = 1$. If we let $u = y - 2x$ then these lines are mapped to the

vertical lines $u = -1, u = 1$ in the uv-plane. Similarly, the second pair of equations can be written as $x + y = 1, x + y = 3$,

and setting $v = x + y$ maps these lines to the horizontal lines $v = 1, v = 3$ in the uv-plane. Boundary curves are mapped to boundary curves under a transformation, so here the equations $u = y - 2x$, $v = x + y$ define a transformation T^{-1} that maps R in the xy-plane to the square S enclosed by the lines $u = -1$, $u = 1$, $v = 1$, $v = 3$ in the uv-plane. To find the transformation T that maps S to R we solve $u = y - 2x$, $v = x + y$ for x, y: Subtracting the first equation from the second gives $v - u = 3x$ $\Rightarrow$ $x = \frac{1}{3}(v - u)$ and adding twice the second equation to the first gives $u + 2v = 3y$ $\Rightarrow$ $y = \frac{1}{3}(u + 2v)$. Thus one possible transformation T (there are many) is given by $x = \frac{1}{3}(v - u)$, $y = \frac{1}{3}(u + 2v)$.

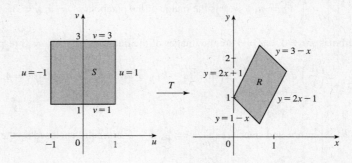

13. R is a portion of an annular region (see the figure) that is easily described in polar coordinates as $R = \{(r, \theta) \mid 1 \le r \le \sqrt{2},\ 0 \le \theta \le \pi/2\}$. If we converted a double integral over R to polar coordinates the resulting region of integration is a rectangle (in the $r\theta$-plane), so we can create a transformation T here by letting u play the role of r and v the role of θ. Thus T is defined by $x = u \cos v$, $y = u \sin v$ and T maps the rectangle $S = \{(u, v) \mid 1 \le u \le \sqrt{2},\ 0 \le v \le \pi/2\}$ in the uv-plane to R in the xy-plane.

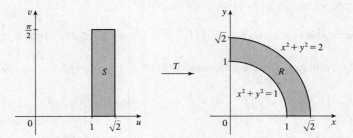

15. $\dfrac{\partial(x, y)}{\partial(u, v)} = \begin{vmatrix} 2 & 1 \\ 1 & 2 \end{vmatrix} = 3$ and $x - 3y = (2u + v) - 3(u + 2v) = -u - 5v$. To find the region S in the uv-plane that corresponds to R we first find the corresponding boundary under the given transformation. The line through $(0, 0)$ and $(2, 1)$ is $y = \frac{1}{2}x$ which is the image of $u + 2v = \frac{1}{2}(2u + v)$ $\Rightarrow$ $v = 0$; the line through $(2, 1)$ and $(1, 2)$ is $x + y = 3$ which is the image of $(2u + v) + (u + 2v) = 3$ $\Rightarrow$ $u + v = 1$; the line through $(0, 0)$ and $(1, 2)$ is $y = 2x$ which is the image of $u + 2v = 2(2u + v)$ $\Rightarrow$ $u = 0$. Thus S is the triangle $0 \le v \le 1 - u$, $0 \le u \le 1$ in the uv-plane and

$$\iint_R (x - 3y)\, dA = \int_0^1 \int_0^{1-u} (-u - 5v)\, |3|\, dv\, du = -3 \int_0^1 \left[uv + \frac{5}{2}v^2 \right]_{v=0}^{v=1-u} du$$

$$= -3 \int_0^1 \left(u - u^2 + \frac{5}{2}(1 - u)^2 \right) du = -3 \left[\frac{1}{2}u^2 - \frac{1}{3}u^3 - \frac{5}{6}(1 - u)^3 \right]_0^1 = -3\left(\frac{1}{2} - \frac{1}{3} + \frac{5}{6} \right) = -3$$

17. $\dfrac{\partial(x,y)}{\partial(u,v)} = \begin{vmatrix} 2 & 0 \\ 0 & 3 \end{vmatrix} = 6$, $x^2 = 4u^2$ and the planar ellipse $9x^2 + 4y^2 \leq 36$ is the image of the disk $u^2 + v^2 \leq 1$. Thus

$$\iint_R x^2\, dA = \iint_{u^2+v^2\leq 1} (4u^2)(6)\, du\, dv = \int_0^{2\pi} \int_0^1 (24r^2 \cos^2 \theta)\, r\, dr\, d\theta = 24 \int_0^{2\pi} \cos^2 \theta\, d\theta \int_0^1 r^3\, dr$$

$$= 24 \left[\tfrac{1}{2}x + \tfrac{1}{4} \sin 2x \right]_0^{2\pi} \left[\tfrac{1}{4} r^4 \right]_0^1 = 24(\pi)\left(\tfrac{1}{4}\right) = 6\pi$$

19. $\dfrac{\partial(x,y)}{\partial(u,v)} = \begin{vmatrix} 1/v & -u/v^2 \\ 0 & 1 \end{vmatrix} = \dfrac{1}{v}$, $xy = u$, $y = x$ is the image of the parabola $v^2 = u$, $y = 3x$ is the image of the parabola

$v^2 = 3u$, and the hyperbolas $xy = 1$, $xy = 3$ are the images of the lines $u = 1$ and $u = 3$ respectively. Thus

$$\iint_R xy\, dA = \int_1^3 \int_{\sqrt{u}}^{\sqrt{3u}} u \left(\dfrac{1}{v}\right) dv\, du = \int_1^3 u\left(\ln \sqrt{3u} - \ln \sqrt{u}\right) du = \int_1^3 u \ln \sqrt{3}\, du = 4 \ln \sqrt{3} = 2 \ln 3.$$

21. (a) $\dfrac{\partial(x,y,z)}{\partial(u,v,w)} = \begin{vmatrix} a & 0 & 0 \\ 0 & b & 0 \\ 0 & 0 & c \end{vmatrix} = abc$ and since $u = \dfrac{x}{a}$, $v = \dfrac{y}{b}$, $w = \dfrac{z}{c}$ the solid enclosed by the ellipsoid is the image of the

ball $u^2 + v^2 + w^2 \leq 1$. So

$$\iiint_E dV = \iiint_{u^2+v^2+w^2 \leq 1} abc\, du\, dv\, dw = (abc)(\text{volume of the ball}) = \tfrac{4}{3}\pi abc$$

(b) If we approximate the surface of the earth by the ellipsoid $\dfrac{x^2}{6378^2} + \dfrac{y^2}{6378^2} + \dfrac{z^2}{6356^2} = 1$, then we can estimate

the volume of the earth by finding the volume of the solid E enclosed by the ellipsoid. From part (a), this is

$$\iiint_E dV = \tfrac{4}{3}\pi(6378)(6378)(6356) \approx 1.083 \times 10^{12} \text{ km}^3.$$

(c) The moment of intertia about the z-axis is $I_z = \iiint_E (x^2 + y^2)\, \rho(x,y,z)\, dV$, where E is the solid enclosed by

$\dfrac{x^2}{a^2} + \dfrac{y^2}{b^2} + \dfrac{z^2}{c^2} = 1$. As in part (a), we use the transformation $x = au$, $y = bv$, $z = cw$, so $\left|\dfrac{\partial(x,y,z)}{\partial(u,v,w)}\right| = abc$ and

$$I_z = \iiint_E (x^2 + y^2)\, k\, dV = \iiint_{u^2+v^2+w^2 \leq 1} k(a^2u^2 + b^2v^2)(abc)\, du\, dv\, dw$$

$$= abck \int_0^\pi \int_0^{2\pi} \int_0^1 (a^2\rho^2 \sin^2\phi \cos^2\theta + b^2\rho^2 \sin^2\phi \sin^2\theta)\, \rho^2 \sin\phi\, d\rho\, d\theta\, d\phi$$

$$= abck \left[a^2 \int_0^\pi \int_0^{2\pi} \int_0^1 (\rho^2 \sin^2\phi \cos^2\theta)\, \rho^2 \sin\phi\, d\rho\, d\theta\, d\phi + b^2 \int_0^\pi \int_0^{2\pi} \int_0^1 (\rho^2 \sin^2\phi \sin^2\theta)\, \rho^2 \sin\phi\, d\rho\, d\theta\, d\phi \right]$$

$$= a^3 bck \int_0^\pi \sin^3\phi\, d\phi \int_0^{2\pi} \cos^2\theta\, d\theta \int_0^1 \rho^4\, d\rho + ab^3 ck \int_0^\pi \sin^3\phi\, d\phi \int_0^{2\pi} \sin^2\theta\, d\theta \int_0^1 \rho^4\, d\rho$$

$$= a^3 bck \left[\tfrac{1}{3}\cos^3\phi - \cos\phi \right]_0^\pi \left[\tfrac{1}{2}\theta + \tfrac{1}{4}\sin 2\theta \right]_0^{2\pi} \left[\tfrac{1}{5}\rho^5 \right]_0^1 + ab^3 ck \left[\tfrac{1}{3}\cos^3\phi - \cos\phi \right]_0^\pi \left[\tfrac{1}{2}\theta - \tfrac{1}{4}\sin 2\theta \right]_0^{2\pi} \left[\tfrac{1}{5}\rho^5 \right]_0^1$$

$$= a^3 bck \left(\tfrac{4}{3}\right)(\pi)\left(\tfrac{1}{5}\right) + ab^3 ck \left(\tfrac{4}{3}\right)(\pi)\left(\tfrac{1}{5}\right) = \tfrac{4}{15}\pi(a^2 + b^2)abck$$

23. Letting $u = x - 2y$ and $v = 3x - y$, we have $x = \tfrac{1}{5}(2v - u)$ and $y = \tfrac{1}{5}(v - 3u)$. Then $\dfrac{\partial(x,y)}{\partial(u,v)} = \begin{vmatrix} -1/5 & 2/5 \\ -3/5 & 1/5 \end{vmatrix} = \dfrac{1}{5}$

and R is the image of the rectangle enclosed by the lines $u = 0$, $u = 4$, $v = 1$, and $v = 8$. Thus

$$\iint_R \dfrac{x - 2y}{3x - y}\, dA = \int_0^4 \int_1^8 \dfrac{u}{v} \left| \dfrac{1}{5} \right| dv\, du = \dfrac{1}{5} \int_0^4 u\, du \int_1^8 \dfrac{1}{v}\, dv = \tfrac{1}{5} \left[\tfrac{1}{2}u^2 \right]_0^4 \left[\ln |v| \right]_1^8 = \tfrac{8}{5} \ln 8.$$

25. Letting $u = y - x$, $v = y + x$, we have $y = \frac{1}{2}(u + v)$, $x = \frac{1}{2}(v - u)$. Then $\dfrac{\partial(x, y)}{\partial(u, v)} = \begin{vmatrix} -1/2 & 1/2 \\ 1/2 & 1/2 \end{vmatrix} = -\dfrac{1}{2}$ and R is the

image of the trapezoidal region with vertices $(-1, 1)$, $(-2, 2)$, $(2, 2)$, and $(1, 1)$. Thus

$$\iint_R \cos\left(\frac{y - x}{y + x}\right) dA = \int_1^2 \int_{-v}^{v} \cos\frac{u}{v} \left| -\frac{1}{2} \right| du\, dv = \frac{1}{2} \int_1^2 \left[v \sin\frac{u}{v} \right]_{u=-v}^{u=v} dv = \frac{1}{2} \int_1^2 2v \sin(1)\, dv = \frac{3}{2} \sin 1$$

27. Let $u = x + y$ and $v = -x + y$. Then $u + v = 2y$ $\Rightarrow$ $y = \frac{1}{2}(u + v)$ and $u - v = 2x$ $\Rightarrow$ $x = \frac{1}{2}(u - v)$.

$\dfrac{\partial(x, y)}{\partial(u, v)} = \begin{vmatrix} 1/2 & -1/2 \\ 1/2 & 1/2 \end{vmatrix} = \dfrac{1}{2}$. Now $|u| = |x + y| \le |x| + |y| \le 1$ $\Rightarrow$ $-1 \le u \le 1$,

and $|v| = |-x + y| \le |x| + |y| \le 1$ $\Rightarrow$ $-1 \le v \le 1$. R is the image of the square

region with vertices $(1, 1)$, $(1, -1)$, $(-1, -1)$, and $(-1, 1)$.

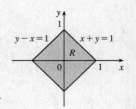

So $\iint_R e^{x+y}\, dA = \frac{1}{2} \int_{-1}^1 \int_{-1}^1 e^u\, du\, dv = \frac{1}{2} \left[e^u \right]_{-1}^1 \left[v \right]_{-1}^1 = e - e^{-1}$.

15 Review

TRUE-FALSE QUIZ

1. This is true by Fubini's Theorem.

3. True by Equation 15.1.11.

5. True. By Equation 15.1.11 we can write $\int_0^1 \int_0^1 f(x)\, f(y)\, dy\, dx = \int_0^1 f(x)\, dx \int_0^1 f(y)\, dy$. But $\int_0^1 f(y)\, dy = \int_0^1 f(x)\, dx$ so

this becomes $\int_0^1 f(x)\, dx \int_0^1 f(x)\, dx = \left[\int_0^1 f(x)\, dx \right]^2$.

7. True: $\iint_D \sqrt{4 - x^2 - y^2}\, dA$ = the volume under the surface $x^2 + y^2 + z^2 = 4$ and above the xy-plane

$= \frac{1}{2}$ (the volume of the sphere $x^2 + y^2 + z^2 = 4$) $= \frac{1}{2} \cdot \frac{4}{3}\pi(2)^3 = \frac{16}{3}\pi$

9. The volume enclosed by the cone $z = \sqrt{x^2 + y^2}$ and the plane $z = 2$ is, in cylindrical coordinates,

$V = \int_0^{2\pi} \int_0^2 \int_r^2 r\, dz\, dr\, d\theta \ne \int_0^{2\pi} \int_0^2 \int_r^2 dz\, dr\, d\theta$, so the assertion is false.

EXERCISES

1. As shown in the contour map, we divide R into 9 equally sized subsquares, each with area $\Delta A = 1$. Then we approximate

$\iint_R f(x, y)\, dA$ by a Riemann sum with $m = n = 3$ and the sample points the upper right corners of each square, so

$$\iint_R f(x, y)\, dA \approx \sum_{i=1}^3 \sum_{j=1}^3 f(x_i, y_j)\, \Delta A$$

$$= \Delta A\, [f(1, 1) + f(1, 2) + f(1, 3) + f(2, 1) + f(2, 2) + f(2, 3) + f(3, 1) + f(3, 2) + f(3, 3)]$$

Using the contour lines to estimate the function values, we have

$$\iint_R f(x, y)\, dA \approx 1[2.7 + 4.7 + 8.0 + 4.7 + 6.7 + 10.0 + 6.7 + 8.6 + 11.9] \approx 64.0$$

3. $\int_1^2 \int_0^2 (y + 2xe^y)\, dx\, dy = \int_1^2 \left[xy + x^2 e^y \right]_{x=0}^{x=2} dy = \int_1^2 (2y + 4e^y)\, dy = \left[y^2 + 4e^y \right]_1^2$

$$= 4 + 4e^2 - 1 - 4e = 4e^2 - 4e + 3$$

5. $\int_0^1 \int_0^x \cos(x^2)\, dy\, dx = \int_0^1 \left[\cos(x^2)y \right]_{y=0}^{y=x} dx = \int_0^1 x\cos(x^2)\, dx = \frac{1}{2}\sin(x^2)\big]_0^1 = \frac{1}{2}\sin 1$

7. $\int_0^\pi \int_0^1 \int_0^{\sqrt{1-y^2}} y\sin x\, dz\, dy\, dx = \int_0^\pi \int_0^1 \left[(y\sin x)z \right]_{z=0}^{z=\sqrt{1-y^2}} dy\, dx = \int_0^\pi \int_0^1 y\sqrt{1-y^2}\sin x\, dy\, dx$

$$= \int_0^\pi \left[-\frac{1}{3}(1-y^2)^{3/2}\sin x \right]_{y=0}^{y=1} dx = \int_0^\pi \frac{1}{3}\sin x\, dx = -\frac{1}{3}\cos x\big]_0^\pi = \frac{2}{3}$$

9. The region R is more easily described by polar coordinates: $R = \{(r, \theta) \mid 2 \le r \le 4, 0 \le \theta \le \pi\}$. Thus

$\iint_R f(x, y)\, dA = \int_0^\pi \int_2^4 f(r\cos\theta, r\sin\theta)\, r\, dr\, d\theta$.

11. $x = r\cos\theta = 2\sqrt{3}\cos\frac{\pi}{3} = 2\sqrt{3}\cdot\frac{1}{2} = \sqrt{3}$, $y = r\sin\theta = 2\sqrt{3}\sin\frac{\pi}{3} = 2\sqrt{3}\cdot\frac{\sqrt{3}}{2} = 3$, $z = 2$, so in rectangular

coordinates the point is $\left(\sqrt{3}, 3, 2\right)$. $\rho = \sqrt{r^2 + z^2} = \sqrt{12+4} = 4$, $\theta = \frac{\pi}{3}$, and $\cos\phi = z/\rho = \frac{1}{2}$, so $\phi = \frac{\pi}{3}$ and spherical

coordinates are $\left(4, \frac{\pi}{3}, \frac{\pi}{3}\right)$.

13. $x = \rho\sin\phi\cos\theta = 8\sin\frac{\pi}{6}\cos\frac{\pi}{4} = 8\cdot\frac{1}{2}\cdot\frac{\sqrt{2}}{2} = 2\sqrt{2}$, $y = \rho\sin\phi\sin\theta = 8\sin\frac{\pi}{6}\sin\frac{\pi}{4} = 2\sqrt{2}$, and

$z = \rho\cos\phi = 8\cos\frac{\pi}{6} = 8\cdot\frac{\sqrt{3}}{2} = 4\sqrt{3}$. Thus rectangular coordinates for the point are $\left(2\sqrt{2}, 2\sqrt{2}, 4\sqrt{3}\right)$.

$r^2 = x^2 + y^2 = 8 + 8 = 16 \quad\Rightarrow\quad r = 4$, $\theta = \frac{\pi}{4}$, and $z = 4\sqrt{3}$, so cylindrical coordinates are $\left(4, \frac{\pi}{4}, 4\sqrt{3}\right)$.

15. (a) $x^2 + y^2 + z^2 = 4$. In cylindrical coordinates, this becomes $r^2 + z^2 = 4$. In spherical coordinates, it becomes $\rho^2 = 4$

or $\rho = 2$.

(b) $x^2 + y^2 = 4$. In cylindrical coordinates: $r^2 = 4$ or $r = 2$. In spherical coordinates: $\rho^2 - z^2 = 4$ or $\rho^2 - \rho^2\cos^2\phi = 4$ or

$\rho^2\sin^2\phi = 4$ or $\rho\sin\phi = 2$.

17.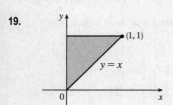

$r = \sin 2\theta$

The region whose area is given by $\int_0^{\pi/2} \int_0^{\sin 2\theta} r\, dr\, d\theta$ is

$\{(r, \theta) \mid 0 \le \theta \le \frac{\pi}{2}, 0 \le r \le \sin 2\theta\}$, which is the region contained in the

loop in the first quadrant of the four-leaved rose $r = \sin 2\theta$.

19.

y↑

(1, 1)

$y = x$

0   x

$\int_0^1 \int_x^1 \cos(y^2)\, dy\, dx = \int_0^1 \int_0^y \cos(y^2)\, dx\, dy$

$$= \int_0^1 \cos(y^2)\left[x \right]_{x=0}^{x=y} dy = \int_0^1 y\cos(y^2)\, dy$$

$$= \left[\frac{1}{2}\sin(y^2) \right]_0^1 = \frac{1}{2}\sin 1$$

21. $\iint_R ye^{xy}\, dA = \int_0^3 \int_0^2 ye^{xy}\, dx\, dy = \int_0^3 \left[e^{xy} \right]_{x=0}^{x=2} dy = \int_0^3 (e^{2y} - 1)\, dy = \left[\frac{1}{2}e^{2y} - y \right]_0^3 = \frac{1}{2}e^6 - 3 - \frac{1}{2} = \frac{1}{2}e^6 - \frac{7}{2}$

23.

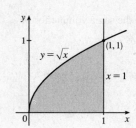

$$\iint_D \frac{y}{1+x^2}\, dA = \int_0^1 \int_0^{\sqrt{x}} \frac{y}{1+x^2}\, dy\, dx = \int_0^1 \frac{1}{1+x^2} \left[\tfrac{1}{2}y^2\right]_{y=0}^{y=\sqrt{x}} dx$$

$$= \tfrac{1}{2} \int_0^1 \frac{x}{1+x^2}\, dx = \left[\tfrac{1}{4}\ln(1+x^2)\right]_0^1 = \tfrac{1}{4}\ln 2$$

25.

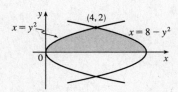

$$\iint_D y\, dA = \int_0^2 \int_{y^2}^{8-y^2} y\, dx\, dy$$

$$= \int_0^2 y\,[x]_{x=y^2}^{x=8-y^2}\, dy = \int_0^2 y(8-y^2-y^2)\, dy$$

$$= \int_0^2 (8y - 2y^3)\, dy = \left[4y^2 - \tfrac{1}{2}y^4\right]_0^2 = 8$$

27.

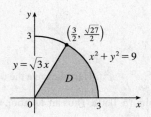

$$\iint_D (x^2+y^2)^{3/2}\, dA = \int_0^{\pi/3} \int_0^3 (r^2)^{3/2} r\, dr\, d\theta$$

$$= \int_0^{\pi/3} d\theta \int_0^3 r^4\, dr = \left[\theta\right]_0^{\pi/3} \left[\tfrac{1}{5}r^5\right]_0^3$$

$$= \frac{\pi}{3}\frac{3^5}{5} = \frac{81\pi}{5}$$

29. $\iiint_E xy\, dV = \int_0^3 \int_0^x \int_0^{x+y} xy\, dz\, dy\, dx = \int_0^3 \int_0^x xy\,[z]_{z=0}^{z=x+y}\, dy\, dx = \int_0^3 \int_0^x xy(x+y)\, dy\, dx$

$= \int_0^3 \int_0^x (x^2y + xy^2)\, dy\, dx = \int_0^3 \left[\tfrac{1}{2}x^2y^2 + \tfrac{1}{3}xy^3\right]_{y=0}^{y=x}\, dx = \int_0^3 \left(\tfrac{1}{2}x^4 + \tfrac{1}{3}x^4\right) dx$

$= \tfrac{5}{6}\int_0^3 x^4\, dx = \left[\tfrac{1}{6}x^5\right]_0^3 = \tfrac{81}{2} = 40.5$

31. $\iiint_E y^2z^2\, dV = \int_{-1}^1 \int_{-\sqrt{1-y^2}}^{\sqrt{1-y^2}} \int_0^{1-y^2-z^2} y^2z^2\, dx\, dz\, dy = \int_{-1}^1 \int_{-\sqrt{1-y^2}}^{\sqrt{1-y^2}} y^2z^2(1-y^2-z^2)\, dz\, dy$

$= \int_0^{2\pi} \int_0^1 (r^2\cos^2\theta)(r^2\sin^2\theta)(1-r^2)\, r\, dr\, d\theta = \int_0^{2\pi} \left(\tfrac{1}{2}\sin 2\theta\right)^2 d\theta \int_0^1 (r^5 - r^7)\, dr$

$= \int_0^{2\pi} \tfrac{1}{4}\left[\tfrac{1}{2}(1-\cos 4\theta)\right] d\theta \int_0^1 (r^5-r^7)\, dr = \tfrac{1}{8}\left[\theta - \tfrac{1}{4}\sin 4\theta\right]_0^{2\pi} \left[\tfrac{1}{6}r^6 - \tfrac{1}{8}r^8\right]_0^1$

$= \tfrac{1}{8}(2\pi)\left(\tfrac{1}{6} - \tfrac{1}{8}\right) = \tfrac{\pi}{4}\cdot\tfrac{1}{24} = \tfrac{\pi}{96}$

33. $\iiint_E yz\, dV = \int_{-2}^2 \int_0^{\sqrt{4-x^2}} \int_0^y yz\, dz\, dy\, dx = \int_{-2}^2 \int_0^{\sqrt{4-x^2}} \left[\tfrac{1}{2}yz^2\right]_{z=0}^{z=y}\, dy\, dx = \tfrac{1}{2}\int_{-2}^2 \int_0^{\sqrt{4-x^2}} y^3\, dy\, dx$

$= \tfrac{1}{2}\int_0^\pi \int_0^2 (r\sin\theta)^3\, r\, dr\, d\theta = \tfrac{1}{2}\int_0^\pi \sin^3\theta\, d\theta \int_0^2 r^4\, dr = \tfrac{1}{2}\int_0^\pi (1-\cos^2\theta)\sin\theta\, d\theta \int_0^2 r^4\, dr$

$= \tfrac{1}{2}\left[-\cos\theta + \tfrac{1}{3}\cos^3\theta\right]_0^\pi \left[\tfrac{1}{5}r^5\right]_0^2 = \tfrac{1}{2}\left(\tfrac{2}{3} + \tfrac{2}{3}\right)\left(\tfrac{32}{5}\right) = \tfrac{64}{15}$

35. $V = \int_0^2 \int_1^4 (x^2 + 4y^2)\, dy\, dx = \int_0^2 \left[x^2y + \tfrac{4}{3}y^3\right]_{y=1}^{y=4}\, dx = \int_0^2 (3x^2 + 84)\, dx = x^3 + 84x\big]_0^2 = 176$

37.

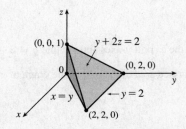

$$V = \int_0^2 \int_0^y \int_0^{(2-y)/2} dz\, dx\, dy = \int_0^2 \int_0^y \left(1 - \tfrac{1}{2}y\right) dx\, dy$$

$$= \int_0^2 \left(y - \tfrac{1}{2}y^2\right) dy = \tfrac{1}{2}y^2 - \tfrac{1}{6}y^3\Big]_0^2 = \tfrac{2}{3}$$

39. Using the wedge above the plane $z = 0$ and below the plane $z = mx$ and noting that we have the same volume for $m < 0$ as

for $m > 0$ (so use $m > 0$), we have

$$V = 2 \int_0^{a/3} \int_0^{\sqrt{a^2 - 9y^2}} mx \, dx \, dy = 2 \int_0^{a/3} \frac{1}{2} m(a^2 - 9y^2) \, dy = m \left[a^2 y - 3y^3 \right]_0^{a/3} = m \left(\frac{1}{3}a^3 - \frac{1}{9}a^3 \right) = \frac{2}{9} m a^3.$$

41. (a) $m = \int_0^1 \int_0^{1-y^2} y \, dx \, dy = \int_0^1 (y - y^3) \, dy = \frac{1}{2} - \frac{1}{4} = \frac{1}{4}$

(b) $M_y = \int_0^1 \int_0^{1-y^2} xy \, dx \, dy = \int_0^1 \frac{1}{2} y (1 - y^2)^2 \, dy = -\frac{1}{12}(1 - y^2)^3 \Big|_0^1 = \frac{1}{12}$,

$M_x = \int_0^1 \int_0^{1-y^2} y^2 \, dx \, dy = \int_0^1 (y^2 - y^4) \, dy = \frac{2}{15}$. Hence $(\overline{x}, \overline{y}) = \left(\frac{1}{3}, \frac{8}{15} \right)$.

(c) $I_x = \int_0^1 \int_0^{1-y^2} y^3 \, dx \, dy = \int_0^1 (y^3 - y^5) \, dy = \frac{1}{12}$,

$I_y = \int_0^1 \int_0^{1-y^2} yx^2 \, dx \, dy = \int_0^1 \frac{1}{3} y (1 - y^2)^3 \, dy = -\frac{1}{24}(1 - y^2)^4 \Big|_0^1 = \frac{1}{24}$,

$\overline{\overline{y}}^2 = I_x / m = \frac{1/12}{1/4} = \frac{1}{3} \Rightarrow \overline{\overline{y}} = \frac{1}{\sqrt{3}}$, and $\overline{\overline{x}}^2 = I_y / m = \frac{1/24}{1/4} = \frac{1}{6} \Rightarrow \overline{\overline{x}} = \frac{1}{\sqrt{6}}$.

43. (a) A right circular cone with axis the z-axis and vertex at the origin has equation $z^2 = c^2(x^2 + y^2)$. Here we have the bottom

frustum, shifted upward h units, and with $c^2 = h^2/a^2$ so that the cone includes the point $(a, 0, 0)$. Thus an equation of the

cone in rectangular coordinates is $z = h - \frac{h}{a} \sqrt{x^2 + y^2}, 0 \le z \le h$. In cylindrical coordinates, the cone is described by

$E = \left\{ (r, \theta, z) \mid 0 \le r \le a, \ 0 \le \theta \le 2\pi, \ 0 \le z \le h \left(1 - \frac{1}{a}r \right) \right\}$, and its volume is $V = \frac{1}{3}\pi a^2 h$. By symmetry

$M_{yz} = M_{xz} = 0$, and

$$M_{xy} = \int_0^{2\pi} \int_0^a \int_0^{h(1-r/a)} z \cdot r \, dz \, dr \, d\theta = \int_0^{2\pi} \int_0^a \left[\frac{1}{2} r z^2 \right]_{z=0}^{z=h(1-r/a)} dr \, d\theta$$

$$= \frac{1}{2} \int_0^{2\pi} \int_0^a rh^2 \left(1 - \frac{r}{a} \right)^2 dr \, d\theta = \frac{1}{2}h^2 \int_0^{2\pi} d\theta \int_0^a \left(r - \frac{2}{a}r^2 + \frac{1}{a^2}r^3 \right) dr$$

$$= \frac{1}{2}h^2 \left[\theta \right]_0^{2\pi} \left[\frac{1}{2}r^2 - \frac{2}{3a}r^3 + \frac{1}{4a^2}r^4 \right]_0^a = \frac{1}{2}h^2 (2\pi) \left(\frac{1}{2}a^2 - \frac{2}{3}a^2 + \frac{1}{4}a^2 \right)$$

$$= \pi h^2 \left(\frac{1}{12}a^2 \right) = \frac{1}{12}\pi a^2 h^2$$

Hence the centroid is $(\overline{x}, \overline{y}, \overline{z}) = \left(0, 0, [\pi a^2 h^2/12]/[\pi a^2 h/3] \right) = \left(0, 0, \frac{1}{4}h \right)$.

(b) The density function is $\rho = \sqrt{x^2 + y^2} = \sqrt{r^2} = r$, so the moment of inertia about the cone's axis (the z-axis) is

$$I_z = \iiint_E (x^2 + y^2) \, \rho(x, y, z) \, dV = \int_0^{2\pi} \int_0^a \int_0^{h(1-r/a)} (r^2)(r) \, r \, dz \, dr \, d\theta$$

$$= \int_0^{2\pi} \int_0^a \left[r^4 z \right]_{z=0}^{z=h(1-r/a)} dr \, d\theta = \int_0^{2\pi} \int_0^a r^4 h \left(1 - \frac{1}{a}r \right) dr \, d\theta$$

$$= h \int_0^{2\pi} d\theta \int_0^a \left(r^4 - \frac{1}{a}r^5 \right) dr = h \left[\theta \right]_0^{2\pi} \left[\frac{1}{5}r^5 - \frac{1}{6a}r^6 \right]_0^a$$

$$= h (2\pi) \left(\frac{1}{5}a^5 - \frac{1}{6}a^5 \right) = \frac{1}{15}\pi a^5 h$$

45. Let D represent the given triangle; then D can be described as the area enclosed by the x- and y-axes and the line $y = 2 - 2x$,

or equivalently $D = \{ (x, y) \mid 0 \le x \le 1, 0 \le y \le 2 - 2x \}$. We want to find the surface area of the part of the graph of

$z = x^2 + y$ that lies over D, so using Equation 15.5.3 we have

$$A(S) = \iint_D \sqrt{1 + \left(\frac{\partial z}{\partial x}\right)^2 + \left(\frac{\partial z}{\partial y}\right)^2}\, dA = \iint_D \sqrt{1 + (2x)^2 + (1)^2}\, dA = \int_0^1 \int_0^{2-2x} \sqrt{2 + 4x^2}\, dy\, dx$$

$$= \int_0^1 \sqrt{2 + 4x^2}\, \big[y \big]_{y=0}^{y=2-2x}\, dx = \int_0^1 (2 - 2x)\sqrt{2 + 4x^2}\, dx = \int_0^1 2\sqrt{2 + 4x^2}\, dx - \int_0^1 2x\sqrt{2 + 4x^2}\, dx$$

Using Formula 21 in the Table of Integrals with $a = \sqrt{2}$, $u = 2x$, and $du = 2\, dx$, we have

$\int 2\sqrt{2 + 4x^2}\, dx = x\sqrt{2 + 4x^2} + \ln\left(2x + \sqrt{2 + 4x^2}\right)$. If we substitute $u = 2 + 4x^2$ in the second integral, then

$du = 8x\, dx$ and $\int 2x\sqrt{2 + 4x^2}\, dx = \frac{1}{4}\int \sqrt{u}\, du = \frac{1}{4} \cdot \frac{2}{3} u^{3/2} = \frac{1}{6}(2 + 4x^2)^{3/2}$. Thus

$$A(S) = \left[x\sqrt{2 + 4x^2} + \ln\left(2x + \sqrt{2 + 4x^2}\right) - \frac{1}{6}(2 + 4x^2)^{3/2} \right]_0^1$$

$$= \sqrt{6} + \ln\left(2 + \sqrt{6}\right) - \frac{1}{6}(6)^{3/2} - \ln\sqrt{2} + \frac{\sqrt{2}}{3} = \ln\frac{2 + \sqrt{6}}{\sqrt{2}} + \frac{\sqrt{2}}{3}$$

$$= \ln\left(\sqrt{2} + \sqrt{3}\right) + \frac{\sqrt{2}}{3} \approx 1.6176$$

47.

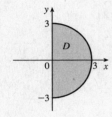

$$\int_0^3 \int_{-\sqrt{9-x^2}}^{\sqrt{9-x^2}} (x^3 + xy^2)\, dy\, dx = \int_0^3 \int_{-\sqrt{9-x^2}}^{\sqrt{9-x^2}} x(x^2 + y^2)\, dy\, dx$$

$$= \int_{-\pi/2}^{\pi/2} \int_0^3 (r\cos\theta)(r^2)\, r\, dr\, d\theta$$

$$= \int_{-\pi/2}^{\pi/2} \cos\theta\, d\theta \int_0^3 r^4\, dr$$

$$= \big[\sin\theta\big]_{-\pi/2}^{\pi/2} \left[\tfrac{1}{5} r^5\right]_0^3 = 2 \cdot \tfrac{1}{5}(243) = \tfrac{486}{5} = 97.2$$

49. From the graph, it appears that $1 - x^2 = e^x$ at $x \approx -0.71$ and at

$x = 0$, with $1 - x^2 > e^x$ on $(-0.71, 0)$. So the desired integral is

$$\iint_D y^2\, dA \approx \int_{-0.71}^0 \int_{e^x}^{1-x^2} y^2\, dy\, dx$$

$$= \tfrac{1}{3} \int_{-0.71}^0 \big[(1 - x^2)^3 - e^{3x}\big]\, dx$$

$$= \tfrac{1}{3}\left[x - x^3 + \tfrac{3}{5} x^5 - \tfrac{1}{7} x^7 - \tfrac{1}{3} e^{3x} \right]_{-0.71}^0 \approx 0.0512$$

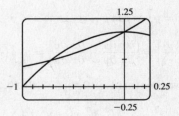

51. (a) $f(x, y)$ is a joint density function, so we know that $\iint_{\mathbb{R}^2} f(x, y)\, dA = 1$. Since $f(x, y) = 0$ outside the rectangle

$[0, 3] \times [0, 2]$, we can say

$$\iint_{\mathbb{R}^2} f(x, y)\, dA = \int_{-\infty}^\infty \int_{-\infty}^\infty f(x, y)\, dy\, dx = \int_0^3 \int_0^2 C(x + y)\, dy\, dx$$

$$= C \int_0^3 \left[xy + \tfrac{1}{2} y^2 \right]_{y=0}^{y=2}\, dx = C \int_0^3 (2x + 2)\, dx = C\big[x^2 + 2x \big]_0^3 = 15C$$

Then $15C = 1 \;\Rightarrow\; C = \tfrac{1}{15}$.

(b) $P(X \leq 2, Y \geq 1) = \int_{-\infty}^2 \int_1^\infty f(x, y)\, dy\, dx = \int_0^2 \int_1^2 \tfrac{1}{15}(x, y)\, dy\, dx = \tfrac{1}{15} \int_0^2 \left[xy + \tfrac{1}{2} y^2 \right]_{y=1}^{y=2}\, dx$

$$= \tfrac{1}{15} \int_0^2 \left(x + \tfrac{3}{2} \right)\, dx = \tfrac{1}{15} \left[\tfrac{1}{2} x^2 + \tfrac{3}{2} x \right]_0^2 = \tfrac{1}{3}$$

(c) $P(X + Y \le 1) = P((X, Y) \in D)$ where D is the triangular region shown in the figure. Thus

$$P(X + Y \le 1) = \iint_D f(x, y)\, dA = \int_0^1 \int_0^{1-x} \tfrac{1}{15}(x + y)\, dy\, dx$$

$$= \tfrac{1}{15} \int_0^1 \left[xy + \tfrac{1}{2} y^2 \right]_{y=0}^{y=1-x} dx$$

$$= \tfrac{1}{15} \int_0^1 \left[x(1 - x) + \tfrac{1}{2}(1 - x)^2 \right] dx$$

$$= \tfrac{1}{30} \int_0^1 (1 - x^2)\, dx = \tfrac{1}{30} \left[x - \tfrac{1}{3} x^3 \right]_0^1 = \tfrac{1}{45}$$

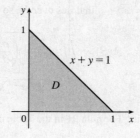

53.

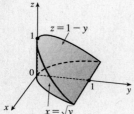

$$\int_{-1}^{1} \int_{x^2}^{1} \int_0^{1-y} f(x, y, z)\, dz\, dy\, dx = \int_0^1 \int_0^{1-z} \int_{-\sqrt{y}}^{\sqrt{y}} f(x, y, z)\, dx\, dy\, dz$$

55. Since $u = x - y$ and $v = x + y$, $x = \tfrac{1}{2}(u + v)$ and $y = \tfrac{1}{2}(v - u)$. Thus $\dfrac{\partial(x, y)}{\partial(u, v)} = \begin{vmatrix} 1/2 & 1/2 \\ -1/2 & 1/2 \end{vmatrix} = \dfrac{1}{2}$.

R is the image under this transformation of the square with vertices $(u, v) = (-2, 2)$, $(0, 2)$, $(0, 4)$, and $(-2, 4)$. So

$$\iint_R \frac{x - y}{x + y}\, dA = \int_2^4 \int_{-2}^0 \frac{u}{v} \left(\frac{1}{2} \right) du\, dv = \frac{1}{2} \int_2^4 \left[\frac{u^2}{2v} \right]_{u=-2}^{u=0} dv = \frac{1}{2} \int_2^4 \left(-\frac{2}{v} \right) dv$$

$$= -\ln v \big]_2^4 = -\ln 4 + \ln 2 = -2\ln 2 + \ln 2 = -\ln 2.$$

57. Let $u = y - x$ and $v = y + x$ so $x = y - u = (v - x) - u \;\Rightarrow\; x = \tfrac{1}{2}(v - u)$ and $y = v - \tfrac{1}{2}(v - u) = \tfrac{1}{2}(v + u)$.

$\left| \dfrac{\partial(x, y)}{\partial(u, v)} \right| = \left| \dfrac{\partial x}{\partial u} \dfrac{\partial y}{\partial v} - \dfrac{\partial x}{\partial v} \dfrac{\partial y}{\partial u} \right| = \left| -\tfrac{1}{2}\left(\tfrac{1}{2}\right) - \tfrac{1}{2}\left(\tfrac{1}{2}\right) \right| = \left| -\tfrac{1}{2} \right| = \tfrac{1}{2}$. R is the image under this transformation of the square

with vertices $(u, v) = (0, 0)$, $(-2, 0)$, $(0, 2)$, and $(-2, 2)$. So

$$\iint_R xy\, dA = \int_0^2 \int_{-2}^0 \frac{v^2 - u^2}{4} \left(\frac{1}{2} \right) du\, dv = \tfrac{1}{8} \int_0^2 \left[v^2 u - \tfrac{1}{3} u^3 \right]_{u=-2}^{u=0} dv = \tfrac{1}{8} \int_0^2 \left(2v^2 - \tfrac{8}{3} \right) dv = \tfrac{1}{8} \left[\tfrac{2}{3} v^3 - \tfrac{8}{3} v \right]_0^2 = 0$$

This result could have been anticipated by symmetry, since the integrand is an odd function of y and R is symmetric about the x-axis.

59. For each r such that D_r lies within the domain, $A(D_r) = \pi r^2$, and by the Mean Value Theorem for Double Integrals there

exists (x_r, y_r) in D_r such that $f(x_r, y_r) = \dfrac{1}{\pi r^2} \iint_{D_r} f(x, y)\, dA$. But $\lim\limits_{r \to 0^+} (x_r, y_r) = (a, b)$,

so $\lim\limits_{r \to 0^+} \dfrac{1}{\pi r^2} \iint_{D_r} f(x, y)\, dA = \lim\limits_{r \to 0^+} f(x_r, y_r) = f(a, b)$ by the continuity of f.

☐ PROBLEMS PLUS

1. Let $R = \bigcup_{i=1}^{5} R_i$, where

$$R_i = \{(x, y) \mid x + y \geq i + 2, x + y < i + 3, 1 \leq x \leq 3, 2 \leq y \leq 5\}.$$

$$\iint_R [\![x + y]\!] \, dA = \sum_{i=1}^{5} \iint_{R_i} [\![x + y]\!] \, dA = \sum_{i=1}^{5} [\![x + y]\!] \iint_{R_i} dA, \text{ since}$$

$[\![x + y]\!] = \text{constant} = i + 2$ for $(x, y) \in R_i$. Therefore

$$\iint_R [\![x + y]\!] \, dA = \sum_{i=1}^{5} (i + 2) \left[A(R_i) \right]$$
$$= 3A(R_1) + 4A(R_2) + 5A(R_3) + 6A(R_4) + 7A(R_5)$$
$$= 3\left(\tfrac{1}{2}\right) + 4\left(\tfrac{3}{2}\right) + 5(2) + 6\left(\tfrac{3}{2}\right) + 7\left(\tfrac{1}{2}\right) = 30$$

3. $f_{\text{ave}} = \dfrac{1}{b - a} \displaystyle\int_a^b f(x) \, dx = \dfrac{1}{1 - 0} \int_0^1 \left[\int_x^1 \cos(t^2) \, dt \right] dx$

$= \int_0^1 \int_x^1 \cos(t^2) \, dt \, dx = \int_0^1 \int_0^t \cos(t^2) \, dx \, dt$ [changing the order of integration]

$= \int_0^1 t \cos(t^2) \, dt = \tfrac{1}{2} \sin(t^2) \big]_0^1 = \tfrac{1}{2} \sin 1$

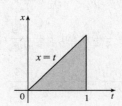

5. Since $|xy| < 1$, except at $(1, 1)$, the formula for the sum of a geometric series gives $\dfrac{1}{1 - xy} = \displaystyle\sum_{n=0}^{\infty} (xy)^n$, so

$$\int_0^1 \int_0^1 \frac{1}{1-xy} \, dx \, dy = \int_0^1 \int_0^1 \sum_{n=0}^{\infty} (xy)^n \, dx \, dy = \sum_{n=0}^{\infty} \int_0^1 \int_0^1 (xy)^n \, dx \, dy = \sum_{n=0}^{\infty} \left[\int_0^1 x^n \, dx \right] \left[\int_0^1 y^n \, dy \right]$$

$$= \sum_{n=0}^{\infty} \frac{1}{n+1} \cdot \frac{1}{n+1} = \sum_{n=0}^{\infty} \frac{1}{(n+1)^2} = \frac{1}{1^2} + \frac{1}{2^2} + \frac{1}{3^2} + \cdots = \sum_{n=1}^{\infty} \frac{1}{n^2}$$

7. **(a)** Since $|xyz| < 1$ except at $(1, 1, 1)$, the formula for the sum of a geometric series gives $\dfrac{1}{1 - xyz} = \displaystyle\sum_{n=0}^{\infty} (xyz)^n$, so

$$\int_0^1 \int_0^1 \int_0^1 \frac{1}{1 - xyz} \, dx \, dy \, dz = \int_0^1 \int_0^1 \int_0^1 \sum_{n=0}^{\infty} (xyz)^n \, dx \, dy \, dz = \sum_{n=0}^{\infty} \int_0^1 \int_0^1 \int_0^1 (xyz)^n \, dx \, dy \, dz$$

$$= \sum_{n=0}^{\infty} \left[\int_0^1 x^n \, dx \right] \left[\int_0^1 y^n \, dy \right] \left[\int_0^1 z^n \, dz \right] = \sum_{n=0}^{\infty} \frac{1}{n+1} \cdot \frac{1}{n+1} \cdot \frac{1}{n+1}$$

$$= \sum_{n=0}^{\infty} \frac{1}{(n+1)^3} = \frac{1}{1^3} + \frac{1}{2^3} + \frac{1}{3^3} + \cdots = \sum_{n=1}^{\infty} \frac{1}{n^3}$$

(b) Since $|-xyz| < 1$, except at $(1, 1, 1)$, the formula for the sum of a geometric series gives $\dfrac{1}{1 + xyz} = \displaystyle\sum_{n=0}^{\infty} (-xyz)^n$, so

$$\int_0^1 \int_0^1 \int_0^1 \frac{1}{1 + xyz} \, dx \, dy \, dz = \int_0^1 \int_0^1 \int_0^1 \sum_{n=0}^{\infty} (-xyz)^n \, dx \, dy \, dz = \sum_{n=0}^{\infty} \int_0^1 \int_0^1 \int_0^1 (-xyz)^n \, dx \, dy \, dz$$

$$= \sum_{n=0}^{\infty} (-1)^n \left[\int_0^1 x^n \, dx \right] \left[\int_0^1 y^n \, dy \right] \left[\int_0^1 z^n \, dz \right] = \sum_{n=0}^{\infty} (-1)^n \frac{1}{n+1} \cdot \frac{1}{n+1} \cdot \frac{1}{n+1}$$

$$= \sum_{n=0}^{\infty} \frac{(-1)^n}{(n+1)^3} = \frac{1}{1^3} - \frac{1}{2^3} + \frac{1}{3^3} - \cdots = \sum_{n=0}^{\infty} \frac{(-1)^{n-1}}{n^3}$$

[continued]

To evaluate this sum, we first write out a few terms: $s = 1 - \dfrac{1}{2^3} + \dfrac{1}{3^3} - \dfrac{1}{4^3} + \dfrac{1}{5^3} - \dfrac{1}{6^3} \approx 0.8998$. Notice that

$a_7 = \dfrac{1}{7^3} < 0.003$. By the Alternating Series Estimation Theorem from Section 11.5, we have $|s - s_6| \le a_7 < 0.003$. This error of 0.003 will not affect the second decimal place, so we have $s \approx 0.90$.

9. (a) $x = r\cos\theta$, $y = r\sin\theta$, $z = z$.   Then $\dfrac{\partial u}{\partial r} = \dfrac{\partial u}{\partial x}\dfrac{\partial x}{\partial r} + \dfrac{\partial u}{\partial y}\dfrac{\partial y}{\partial r} + \dfrac{\partial u}{\partial z}\dfrac{\partial z}{\partial r} = \dfrac{\partial u}{\partial x}\cos\theta + \dfrac{\partial u}{\partial y}\sin\theta$ and

$$\dfrac{\partial^2 u}{\partial r^2} = \cos\theta \left[\dfrac{\partial^2 u}{\partial x^2}\dfrac{\partial x}{\partial r} + \dfrac{\partial^2 u}{\partial y\,\partial x}\dfrac{\partial y}{\partial r} + \dfrac{\partial^2 u}{\partial z\,\partial x}\dfrac{\partial z}{\partial r} \right] + \sin\theta \left[\dfrac{\partial^2 u}{\partial y^2}\dfrac{\partial y}{\partial r} + \dfrac{\partial^2 u}{\partial x\,\partial y}\dfrac{\partial x}{\partial r} + \dfrac{\partial^2 u}{\partial z\,\partial y}\dfrac{\partial z}{\partial r} \right]$$

$$= \dfrac{\partial^2 u}{\partial x^2}\cos^2\theta + \dfrac{\partial^2 u}{\partial y^2}\sin^2\theta + 2\dfrac{\partial^2 u}{\partial y\,\partial x}\cos\theta\,\sin\theta$$

Similarly $\dfrac{\partial u}{\partial \theta} = -\dfrac{\partial u}{\partial x}r\sin\theta + \dfrac{\partial u}{\partial y}r\cos\theta$ and

$$\dfrac{\partial^2 u}{\partial \theta^2} = \dfrac{\partial^2 u}{\partial x^2}r^2\sin^2\theta + \dfrac{\partial^2 u}{\partial y^2}r^2\cos^2\theta - 2\dfrac{\partial^2 u}{\partial y\,\partial x}r^2\sin\theta\,\cos\theta - \dfrac{\partial u}{\partial x}r\cos\theta - \dfrac{\partial u}{\partial y}r\sin\theta. \text{ So}$$

$$\dfrac{\partial^2 u}{\partial r^2} + \dfrac{1}{r}\dfrac{\partial u}{\partial r} + \dfrac{1}{r^2}\dfrac{\partial^2 u}{\partial \theta^2} + \dfrac{\partial^2 u}{\partial z^2} = \dfrac{\partial^2 u}{\partial x^2}\cos^2\theta + \dfrac{\partial^2 u}{\partial y^2}\sin^2\theta + 2\dfrac{\partial^2 u}{\partial y\,\partial x}\cos\theta\,\sin\theta + \dfrac{\partial u}{\partial x}\dfrac{\cos\theta}{r} + \dfrac{\partial u}{\partial y}\dfrac{\sin\theta}{r}$$

$$+ \dfrac{\partial^2 u}{\partial x^2}\sin^2\theta + \dfrac{\partial^2 u}{\partial y^2}\cos^2\theta - 2\dfrac{\partial^2 u}{\partial y\,\partial x}\sin\theta\,\cos\theta$$

$$- \dfrac{\partial u}{\partial x}\dfrac{\cos\theta}{r} - \dfrac{\partial u}{\partial y}\dfrac{\sin\theta}{r} + \dfrac{\partial^2 u}{\partial z^2}$$

$$= \dfrac{\partial^2 u}{\partial x^2} + \dfrac{\partial^2 u}{\partial y^2} + \dfrac{\partial^2 u}{\partial z^2}$$

(b) $x = \rho\sin\phi\cos\theta$, $y = \rho\sin\phi\sin\theta$, $z = \rho\cos\phi$. Then

$$\dfrac{\partial u}{\partial \rho} = \dfrac{\partial u}{\partial x}\dfrac{\partial x}{\partial \rho} + \dfrac{\partial u}{\partial y}\dfrac{\partial y}{\partial \rho} + \dfrac{\partial u}{\partial z}\dfrac{\partial z}{\partial \rho} = \dfrac{\partial u}{\partial x}\sin\phi\,\cos\theta + \dfrac{\partial u}{\partial y}\sin\phi\,\sin\theta + \dfrac{\partial u}{\partial z}\cos\phi, \text{ and}$$

$$\dfrac{\partial^2 u}{\partial \rho^2} = \sin\phi\,\cos\theta \left[\dfrac{\partial^2 u}{\partial x^2}\dfrac{\partial x}{\partial \rho} + \dfrac{\partial^2 u}{\partial y\,\partial x}\dfrac{\partial y}{\partial \rho} + \dfrac{\partial^2 u}{\partial z\,\partial x}\dfrac{\partial z}{\partial \rho} \right]$$

$$+ \sin\phi\,\sin\theta \left[\dfrac{\partial^2 u}{\partial y^2}\dfrac{\partial y}{\partial \rho} + \dfrac{\partial^2 u}{\partial x\,\partial y}\dfrac{\partial x}{\partial \rho} + \dfrac{\partial^2 u}{\partial z\,\partial y}\dfrac{\partial z}{\partial \rho} \right]$$

$$+ \cos\phi \left[\dfrac{\partial^2 u}{\partial z^2}\dfrac{\partial z}{\partial \rho} + \dfrac{\partial^2 u}{\partial x\,\partial z}\dfrac{\partial x}{\partial \rho} + \dfrac{\partial^2 u}{\partial y\,\partial z}\dfrac{\partial y}{\partial \rho} \right]$$

$$= 2\dfrac{\partial^2 u}{\partial y\,\partial x}\sin^2\phi\,\sin\theta\,\cos\theta + 2\dfrac{\partial^2 u}{\partial z\,\partial x}\sin\phi\,\cos\phi\,\cos\theta + 2\dfrac{\partial^2 u}{\partial y\,\partial z}\sin\phi\,\cos\phi\,\sin\theta$$

$$+ \dfrac{\partial^2 u}{\partial x^2}\sin^2\phi\,\cos^2\theta + \dfrac{\partial^2 u}{\partial y^2}\sin^2\phi\,\sin^2\theta + \dfrac{\partial^2 u}{\partial z^2}\cos^2\phi$$

Similarly $\dfrac{\partial u}{\partial \phi} = \dfrac{\partial u}{\partial x}\rho\cos\phi\,\cos\theta + \dfrac{\partial u}{\partial y}\rho\cos\phi\,\sin\theta - \dfrac{\partial u}{\partial z}\rho\sin\phi$, and

$$\frac{\partial^2 u}{\partial \phi^2} = 2 \frac{\partial^2 u}{\partial y\, \partial x} \rho^2 \cos^2 \phi \sin \theta \cos \theta - 2 \frac{\partial^2 u}{\partial x\, \partial z} \rho^2 \sin \phi \cos \phi \cos \theta$$

$$- 2 \frac{\partial^2 u}{\partial y\, \partial z} \rho^2 \sin \phi \cos \phi \sin \theta + \frac{\partial^2 u}{\partial x^2} \rho^2 \cos^2 \phi \cos^2 \theta + \frac{\partial^2 u}{\partial y^2} \rho^2 \cos^2 \phi \sin^2 \theta$$

$$+ \frac{\partial^2 u}{\partial z^2} \rho^2 \sin^2 \phi - \frac{\partial u}{\partial x} \rho \sin \phi \cos \theta - \frac{\partial u}{\partial y} \rho \sin \phi \sin \theta - \frac{\partial u}{\partial z} \rho \cos \phi$$

And $\dfrac{\partial u}{\partial \theta} = -\dfrac{\partial u}{\partial x} \rho \sin \phi \sin \theta + \dfrac{\partial u}{\partial y} \rho \sin \phi \cos \theta$, while

$$\frac{\partial^2 u}{\partial \theta^2} = -2 \frac{\partial^2 u}{\partial y\, \partial x} \rho^2 \sin^2 \phi \cos \theta \sin \theta + \frac{\partial^2 u}{\partial x^2} \rho^2 \sin^2 \phi \sin^2 \theta$$

$$+ \frac{\partial^2 u}{\partial y^2} \rho^2 \sin^2 \phi \cos^2 \theta - \frac{\partial u}{\partial x} \rho \sin \phi \cos \theta - \frac{\partial u}{\partial y} \rho \sin \phi \sin \theta$$

Therefore

$$\frac{\partial^2 u}{\partial \rho^2} + \frac{2}{\rho} \frac{\partial u}{\partial \rho} + \frac{\cot \phi}{\rho^2} \frac{\partial u}{\partial \phi} + \frac{1}{\rho^2} \frac{\partial^2 u}{\partial \phi^2} + \frac{1}{\rho^2 \sin^2 \phi} \frac{\partial^2 u}{\partial \theta^2}$$

$$= \frac{\partial^2 u}{\partial x^2} \left[(\sin^2 \phi \cos^2 \theta) + (\cos^2 \phi \cos^2 \theta) + \sin^2 \theta \right]$$

$$+ \frac{\partial^2 u}{\partial y^2} \left[(\sin^2 \phi \sin^2 \theta) + (\cos^2 \phi \sin^2 \theta) + \cos^2 \theta \right] + \frac{\partial^2 u}{\partial z^2} \left[\cos^2 \phi + \sin^2 \phi \right]$$

$$+ \frac{\partial u}{\partial x} \left[\frac{2 \sin^2 \phi \cos \theta + \cos^2 \phi \cos \theta - \sin^2 \phi \cos \theta - \cos \theta}{\rho \sin \phi} \right]$$

$$+ \frac{\partial u}{\partial y} \left[\frac{2 \sin^2 \phi \sin \theta + \cos^2 \phi \sin \theta - \sin^2 \phi \sin \theta - \sin \theta}{\rho \sin \phi} \right]$$

But $2 \sin^2 \phi \cos \theta + \cos^2 \phi \cos \theta - \sin^2 \phi \cos \theta - \cos \theta = (\sin^2 \phi + \cos^2 \phi - 1) \cos \theta = 0$ and similarly the coefficient of $\partial u / \partial y$ is 0. Also $\sin^2 \phi \cos^2 \theta + \cos^2 \phi \cos^2 \theta + \sin^2 \theta = \cos^2 \theta (\sin^2 \phi + \cos^2 \phi) + \sin^2 \theta = 1$, and similarly the coefficient of $\partial^2 u / \partial y^2$ is 1. So Laplace's Equation in spherical coordinates is as stated.

11. $\int_0^x \int_0^y \int_0^z f(t)\, dt\, dz\, dy = \iiint_E f(t)\, dV$, where

$E = \{(t, z, y) \mid 0 \le t \le z, 0 \le z \le y, 0 \le y \le x\}.$

If we let D be the projection of E on the yt-plane then

$D = \{(y, t) \mid 0 \le t \le x, t \le y \le x\}.$ And we see from the diagram

that $E = \{(t, z, y) \mid t \le z \le y, t \le y \le x, 0 \le t \le x\}.$ So

$$\int_0^x \int_0^y \int_0^z f(t)\, dt\, dz\, dy = \int_0^x \int_t^x \int_t^y f(t)\, dz\, dy\, dt = \int_0^x \left[\int_t^x (y - t)\, f(t)\, dy \right] dt$$

$$= \int_0^x \left[\left(\tfrac{1}{2} y^2 - ty \right) f(t) \right]_{y=t}^{y=x} dt = \int_0^x \left[\tfrac{1}{2} x^2 - tx - \tfrac{1}{2} t^2 + t^2 \right] f(t)\, dt$$

$$= \int_0^x \left[\tfrac{1}{2} x^2 - tx + \tfrac{1}{2} t^2 \right] f(t)\, dt = \int_0^x \left(\tfrac{1}{2} x^2 - 2tx + t^2 \right) f(t)\, dt$$

$$= \tfrac{1}{2} \int_0^x (x - t)^2 f(t)\, dt$$

13. The volume is $V = \iiint_R dV$ where R is the solid region given. From Exercise 15.9.21(a), the transformation $x = au$,

$y = bv$, $z = cw$ maps the unit ball $u^2 + v^2 + w^2 \leq 1$ to the solid ellipsoid

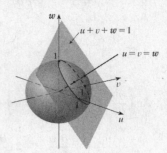

$\dfrac{x^2}{a^2} + \dfrac{y^2}{b^2} + \dfrac{z^2}{c^2} \leq 1$ with $\dfrac{\partial(x, y, z)}{\partial(u, v, w)} = abc$. The same transformation maps the

plane $u + v + w = 1$ to $\dfrac{x}{a} + \dfrac{y}{b} + \dfrac{z}{c} = 1$. Thus the region R in xyz-space

corresponds to the region S in uvw-space consisting of the smaller piece of the

unit ball cut off by the plane $u + v + w = 1$, a "cap of a sphere" (see the figure).

We will need to compute the volume of S, but first consider the general case

where a horizontal plane slices the upper portion of a sphere of radius r to produce

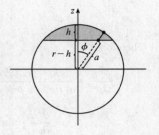

a cap of height h. We use spherical coordinates. From the figure, a line through the

origin at angle ϕ from the z-axis intersects the plane when $\cos \phi = (r - h)/a \;\Rightarrow\;$

$a = (r - h)/\cos \phi$, and the line passes through the outer rim of the cap when

$a = r \;\Rightarrow\; \cos \phi = (r - h)/r \;\Rightarrow\; \phi = \cos^{-1}((r - h)/r)$. Thus the cap

is described by $\{(\rho, \theta, \phi) \mid (r - h)/\cos \phi \leq \rho \leq r, 0 \leq \theta \leq 2\pi, 0 \leq \phi \leq \cos^{-1}((r - h)/r)\}$ and its volume is

$$V = \int_0^{2\pi} \int_0^{\cos^{-1}((r-h)/r)} \int_{(r-h)/\cos \phi}^r \rho^2 \sin \phi \, d\rho \, d\phi \, d\theta$$

$$= \int_0^{2\pi} \int_0^{\cos^{-1}((r-h)/r)} \left[\tfrac{1}{3}\rho^3 \sin \phi\right]_{\rho=(r-h)/\cos \phi}^{\rho=r} d\phi \, d\theta$$

$$= \frac{1}{3} \int_0^{2\pi} \int_0^{\cos^{-1}((r-h)/r)} \left[r^3 \sin \phi - \frac{(r-h)^3}{\cos^3 \phi} \sin \phi\right] d\phi \, d\theta$$

$$= \tfrac{1}{3} \int_0^{2\pi} \left[-r^3 \cos \phi - \tfrac{1}{2}(r-h)^3 \cos^{-2} \phi\right]_{\phi=0}^{\phi=\cos^{-1}((r-h)/r)} d\theta$$

$$= \frac{1}{3} \int_0^{2\pi} \left[-r^3 \left(\frac{r-h}{r}\right) - \frac{1}{2}(r-h)^3 \left(\frac{r-h}{r}\right)^{-2} + r^3 + \frac{1}{2}(r-h)^3\right] d\theta$$

$$= \tfrac{1}{3} \int_0^{2\pi} (\tfrac{3}{2}rh^2 - \tfrac{1}{2}h^3) \, d\theta = \tfrac{1}{3}(\tfrac{3}{2}rh^2 - \tfrac{1}{2}h^3)(2\pi) = \pi h^2(r - \tfrac{1}{3}h)$$

(This volume can also be computed by treating the cap as a solid of revolution and using the single variable disk method;

see Exercise 5.2.49 [ET 6.2.49].)

To determine the height h of the cap cut from the unit ball by the plane

$u + v + w = 1$, note that the line $u = v = w$ passes through the origin with

direction vector $\langle 1, 1, 1 \rangle$ which is perpendicular to the plane. Therefore this line

coincides with a radius of the sphere that passes through the center of the cap and

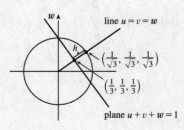

h is measured along this line. The line intersects the plane at $\left(\tfrac{1}{3}, \tfrac{1}{3}, \tfrac{1}{3}\right)$ and the

sphere at $\left(\tfrac{1}{\sqrt{3}}, \tfrac{1}{\sqrt{3}}, \tfrac{1}{\sqrt{3}}\right)$. (See the figure.)

The distance between these points is $h = \sqrt{3\left(\frac{1}{\sqrt{3}} - \frac{1}{3}\right)^2} = \sqrt{3}\left(\frac{1}{\sqrt{3}} - \frac{1}{3}\right) = 1 - \frac{1}{\sqrt{3}}$. Thus the volume of R is

$$V = \iiint_R dV = \iiint_S \left|\frac{\partial(x,y,z)}{\partial(u,v,w)}\right| dV = abc \iiint_S dV = abc\,V(S)$$

$$= abc \cdot \pi h^2\left(r - \tfrac{1}{3}h\right) = abc \cdot \pi\left(1 - \tfrac{1}{\sqrt{3}}\right)^2\left[1 - \tfrac{1}{3}\left(1 - \tfrac{1}{\sqrt{3}}\right)\right]$$

$$= abc\pi\left(\tfrac{4}{3} - \tfrac{2}{\sqrt{3}}\right)\left(\tfrac{2}{3} + \tfrac{1}{3\sqrt{3}}\right) = abc\pi\left(\tfrac{2}{3} - \tfrac{8}{9\sqrt{3}}\right) \approx 0.482\,abc$$

16 □ VECTOR CALCULUS

16.1 Vector Fields

1. $\mathbf{F}(x, y) = 0.3\,\mathbf{i} - 0.4\,\mathbf{j}$

All vectors in this field are identical, with length 0.5 and parallel to $\langle 3, -4 \rangle$.

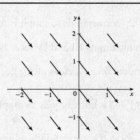

3. $\mathbf{F}(x, y) = -\frac{1}{2}\,\mathbf{i} + (y - x)\,\mathbf{j}$

The length of the vector $-\frac{1}{2}\,\mathbf{i} + (y - x)\,\mathbf{j}$ is

$\sqrt{\frac{1}{4} + (y - x)^2}$. Vectors along the line $y = x$ are

horizontal with length $\frac{1}{2}$.

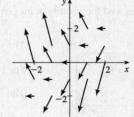

5. $\mathbf{F}(x, y) = \dfrac{y\,\mathbf{i} + x\,\mathbf{j}}{\sqrt{x^2 + y^2}}$

The length of the vector $\dfrac{y\,\mathbf{i} + x\,\mathbf{j}}{\sqrt{x^2 + y^2}}$ is 1.

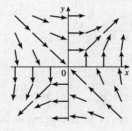

7. $\mathbf{F}(x, y) = \mathbf{i}$

All vectors in this field are identical, with length 1 and pointing in the direction of the positive x-axis.

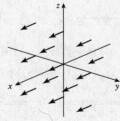

9. $\mathbf{F}(x, y, z) = -y\,\mathbf{i}$

At each point (x, y, z), $\mathbf{F}(x, y, z)$ is a vector of length $|y|$. For $y > 0$, all point in the direction of the negative x-axis, while for $y < 0$, all are in the direction of the positive x-axis. In each plane $y = k$, all the vectors are identical.

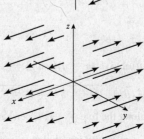

11. $\mathbf{F}(x, y) = \langle x, -y \rangle$ corresponds to graph IV. In the first quadrant all the vectors have positive x-components and negative y-components, in the second quadrant all vectors have negative x- and y-components, in the third quadrant all vectors have negative x-components and positive y-components, and in the fourth quadrant all vectors have positive x- and y-components. In addition, the vectors get shorter as we approach the origin.

13. $\mathbf{F}(x, y) = \langle y, y + 2 \rangle$ corresponds to graph I. As in Exercise 12, all vectors in quadrants I and II have positive x-components while all vectors in quadrants III and IV have negative x-components. Vectors along the line $y = -2$ are horizontal, and the vectors are independent of x (vectors along horizontal lines are identical).

15. $\mathbf{F}(x, y, z) = \mathbf{i} + 2\,\mathbf{j} + 3\,\mathbf{k}$ corresponds to graph IV, since all vectors have identical length and direction.

17. $\mathbf{F}(x, y, z) = x\,\mathbf{i} + y\,\mathbf{j} + 3\,\mathbf{k}$ corresponds to graph III; the projection of each vector onto the xy-plane is $x\,\mathbf{i} + y\,\mathbf{j}$, which points away from the origin, and the vectors point generally upward because their z-components are all 3.

19.

The vector field seems to have very short vectors near the line $y = 2x$.

For $\mathbf{F}(x, y) = \langle 0, 0 \rangle$ we must have $y^2 - 2xy = 0$ and $3xy - 6x^2 = 0$.

The first equation holds if $y = 0$ or $y = 2x$, and the second holds if $x = 0$ or $y = 2x$. So both equations hold [and thus $\mathbf{F}(x, y) = \mathbf{0}$] along the line $y = 2x$.

21. $f(x, y) = y \sin(xy) \quad \Rightarrow$

$$\nabla f(x, y) = f_x(x, y)\,\mathbf{i} + f_y(x, y)\,\mathbf{j} = (y\cos(xy) \cdot y)\,\mathbf{i} + [y \cdot x\cos(xy) + \sin(xy) \cdot 1]\,\mathbf{j}$$
$$= y^2 \cos(xy)\,\mathbf{i} + [xy\cos(xy) + \sin(xy)]\,\mathbf{j}$$

23. $f(x, y, z) = \sqrt{x^2 + y^2 + z^2} \quad \Rightarrow$

$$\nabla f(x, y, z) = f_x(x, y, z)\,\mathbf{i} + f_y(x, y, z)\,\mathbf{j} + f_z(x, y, z)\,\mathbf{k}$$
$$= \tfrac{1}{2}(x^2 + y^2 + z^2)^{-1/2}(2x)\,\mathbf{i} + \tfrac{1}{2}(x^2 + y^2 + z^2)^{-1/2}(2y)\,\mathbf{j} + \tfrac{1}{2}(x^2 + y^2 + z^2)^{-1/2}(2z)\,\mathbf{k}$$
$$= \frac{x}{\sqrt{x^2 + y^2 + z^2}}\,\mathbf{i} + \frac{y}{\sqrt{x^2 + y^2 + z^2}}\,\mathbf{j} + \frac{z}{\sqrt{x^2 + y^2 + z^2}}\,\mathbf{k}$$

25. $f(x, y) = \tfrac{1}{2}(x - y)^2 \quad \Rightarrow$

$\nabla f(x, y) = (x - y)(1)\,\mathbf{i} + (x - y)(-1)\,\mathbf{j} = (x - y)\,\mathbf{i} + (y - x)\,\mathbf{j}$.

The length of $\nabla f(x, y)$ is $\sqrt{(x - y)^2 + (y - x)^2} = \sqrt{2}\,|x - y|$.

The vectors are $\mathbf{0}$ along the line $y = x$. Elsewhere the vectors point away from the line $y = x$ with length that increases as the distance from the line increases.

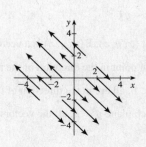

27. We graph $\nabla f(x, y) = \dfrac{2x}{1 + x^2 + 2y^2}\, \mathbf{i} + \dfrac{4y}{1 + x^2 + 2y^2}\, \mathbf{j}$ along with

a contour map of f.

The graph shows that the gradient vectors are perpendicular to the

level curves. Also, the gradient vectors point in the direction in

which f is increasing and are longer where the level curves are closer

together.

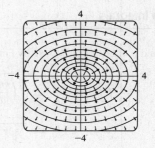

29. $f(x, y) = x^2 + y^2 \;\Rightarrow\; \nabla f(x, y) = 2x\,\mathbf{i} + 2y\,\mathbf{j}$. Thus, each vector $\nabla f(x, y)$ has the same direction and twice the length of

the position vector of the point (x, y), so the vectors all point directly away from the origin and their lengths increase as we

move away from the origin. Hence, ∇f is graph III.

31. $f(x, y) = (x + y)^2 \;\Rightarrow\; \nabla f(x, y) = 2(x + y)\,\mathbf{i} + 2(x + y)\,\mathbf{j}$. The x- and y-components of each vector are equal, so all

vectors are parallel to the line $y = x$. The vectors are $\mathbf{0}$ along the line $y = -x$ and their length increases as the distance from

this line increases. Thus, ∇f is graph II.

33. At $t = 3$ the particle is at $(2, 1)$ so its velocity is $\mathbf{V}(2, 1) = \langle 4, 3 \rangle$. After 0.01 units of time, the particle's change in

location should be approximately $0.01\,\mathbf{V}(2, 1) = 0.01\,\langle 4, 3 \rangle = \langle 0.04, 0.03 \rangle$, so the particle should be approximately at the

point $(2.04, 1.03)$.

35. (a) We sketch the vector field $\mathbf{F}(x, y) = x\,\mathbf{i} - y\,\mathbf{j}$ along with

several approximate flow lines. The flow lines appear to

be hyperbolas with shape similar to the graph of

$y = \pm 1/x$, so we might guess that the flow lines have

equations $y = C/x$.

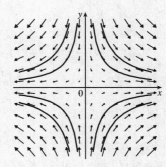

(b) If $x = x(t)$ and $y = y(t)$ are parametric equations of a flow line, then the velocity vector of the flow line at the

point (x, y) is $x'(t)\,\mathbf{i} + y'(t)\,\mathbf{j}$. Since the velocity vectors coincide with the vectors in the vector field, we have

$x'(t)\,\mathbf{i} + y'(t)\,\mathbf{j} = x\,\mathbf{i} - y\,\mathbf{j} \;\Rightarrow\; dx/dt = x,\, dy/dt = -y$. To solve these differential equations, we know

$dx/dt = x \;\Rightarrow\; dx/x = dt \;\Rightarrow\; \ln|x| = t + C \;\Rightarrow\; x = \pm e^{t + C} = Ae^{t}$ for some constant A, and

$dy/dt = -y \;\Rightarrow\; dy/y = -dt \;\Rightarrow\; \ln|y| = -t + K \;\Rightarrow\; y = \pm e^{-t + K} = Be^{-t}$ for some constant B. Therefore

$xy = Ae^{t}Be^{-t} = AB = \text{constant}$. If the flow line passes through $(1, 1)$ then $(1)(1) = \text{constant} = 1 \;\Rightarrow\; xy = 1 \;\Rightarrow$

$y = 1/x,\, x > 0$.

16.2   Line Integrals

1. $x = t^2$ and $y = 2t$, $0 \le t \le 3$, so by Formula 3

$$\int_C y\,ds = \int_0^3 2t \sqrt{\left(\frac{dx}{dt}\right)^2 + \left(\frac{dy}{dt}\right)^2}\,dt = \int_0^3 2t\sqrt{(2t)^2 + (2)^2}\,dt = \int_0^3 2t\sqrt{4t^2 + 4}\,dt$$

$$= \int_0^3 4t\sqrt{t^2 + 1}\,dt = 2 \cdot \tfrac{2}{3}\left(t^2 + 1\right)^{3/2}\Big]_0^3 = \tfrac{4}{3}(10^{3/2} - 1) \ \text{ or } \ \tfrac{4}{3}(10\sqrt{10} - 1)$$

3. Parametric equations for C are $x = 4\cos t$, $y = 4\sin t$, $-\frac{\pi}{2} \le t \le \frac{\pi}{2}$. Then

$$\int_C xy^4\,ds = \int_{-\pi/2}^{\pi/2}(4\cos t)(4\sin t)^4\sqrt{(-4\sin t)^2 + (4\cos t)^2}\,dt = \int_{-\pi/2}^{\pi/2} 4^5\cos t \, \sin^4 t\,\sqrt{16(\sin^2 t + \cos^2 t)}\,dt$$

$$= 4^5\int_{-\pi/2}^{\pi/2}(\sin^4 t\cos t)(4)\,dt = (4)^6\left[\tfrac{1}{5}\sin^5 t\right]_{-\pi/2}^{\pi/2} = 4^6 \cdot \tfrac{2}{5} = 1638.4$$

5. If we choose x as the parameter, parametric equations for C are $x = x$, $y = x^2$ for $0 \le x \le \pi$ and by Equations 7

$$\int_C (x^2 y + \sin x)\,dy = \int_0^\pi \left[x^2(x^2) + \sin x\right] \cdot 2x\,dx = 2\int_0^\pi \left(x^5 + x\sin x\right)dx$$

$$= 2\left[\tfrac{1}{6}x^6 - x\cos x + \sin x\right]_0^\pi \quad \left[\begin{array}{c}\text{where we integrated by parts}\\ \text{in the second term}\end{array}\right]$$

$$= 2\left[\tfrac{1}{6}\pi^6 + \pi + 0 - 0\right] = \tfrac{1}{3}\pi^6 + 2\pi$$

7.

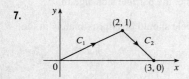

$C = C_1 + C_2$

On C_1: $x = x$, $y = \tfrac{1}{2}x \ \Rightarrow \ dy = \tfrac{1}{2}\,dx$, $0 \le x \le 2$.

On C_2: $x = x$, $y = 3 - x \ \Rightarrow \ dy = -dx$, $2 \le x \le 3$.

Then

$$\int_C (x + 2y)\,dx + x^2\,dy = \int_{C_1}(x + 2y)\,dx + x^2\,dy + \int_{C_2}(x + 2y)\,dx + x^2\,dy$$

$$= \int_0^2 \left[x + 2\left(\tfrac{1}{2}x\right) + x^2\left(\tfrac{1}{2}\right)\right]dx + \int_2^3 \left[x + 2(3 - x) + x^2(-1)\right]dx$$

$$= \int_0^2 \left(2x + \tfrac{1}{2}x^2\right)dx + \int_2^3 \left(6 - x - x^2\right)dx$$

$$= \left[x^2 + \tfrac{1}{6}x^3\right]_0^2 + \left[6x - \tfrac{1}{2}x^2 - \tfrac{1}{3}x^3\right]_2^3 = \tfrac{16}{3} - 0 + \tfrac{9}{2} - \tfrac{22}{3} = \tfrac{5}{2}$$

9. $x = \cos t$, $y = \sin t$, $z = t$, $0 \le t \le \pi/2$. Then by Formula 9,

$$\int_C x^2 y\,ds = \int_0^{\pi/2}(\cos t)^2(\sin t)\sqrt{\left(\frac{dx}{dt}\right)^2 + \left(\frac{dy}{dt}\right)^2 + \left(\frac{dz}{dt}\right)^2}\,dt$$

$$= \int_0^{\pi/2}\cos^2 t\sin t\sqrt{(-\sin t)^2 + (\cos t)^2 + (1)^2}\,dt = \int_0^{\pi/2}\cos^2 t\sin t\sqrt{\sin^2 t + \cos^2 t + 1}\,dt$$

$$= \sqrt{2}\int_0^{\pi/2}\cos^2 t\sin t\,dt = \sqrt{2}\left[-\tfrac{1}{3}\cos^3 t\right]_0^{\pi/2} = \sqrt{2}\left(0 + \tfrac{1}{3}\right) = \tfrac{\sqrt{2}}{3}$$

11. Parametric equations for C are $x = t$, $y = 2t$, $z = 3t$, $0 \le t \le 1$. Then

$$\int_C xe^{yz}\,ds = \int_0^1 te^{(2t)(3t)}\sqrt{1^2 + 2^2 + 3^2}\,dt = \sqrt{14}\int_0^1 te^{6t^2}\,dt = \sqrt{14}\left[\tfrac{1}{12}e^{6t^2}\right]_0^1 = \tfrac{\sqrt{14}}{12}(e^6 - 1).$$

13. $\int_C xye^{yz}\,dy = \int_0^1 (t)(t^2)e^{(t^2)(t^3)} \cdot 2t\,dt = \int_0^1 2t^4 e^{t^5}\,dt = \tfrac{2}{5}e^{t^5}\Big]_0^1 = \tfrac{2}{5}(e^1 - e^0) = \tfrac{2}{5}(e - 1)$

15. Parametric equations for C are $x = 1 + 3t$, $y = t$, $z = 2t$, $0 \le t \le 1$. Then

$$\int_C z^2\, dx + x^2\, dy + y^2\, dz = \int_0^1 (2t)^2 \cdot 3\, dt + (1 + 3t)^2\, dt + t^2 \cdot 2\, dt = \int_0^1 (23t^2 + 6t + 1)\, dt$$

$$= \left[\tfrac{23}{3} t^3 + 3t^2 + t\right]_0^1 = \tfrac{23}{3} + 3 + 1 = \tfrac{35}{3}$$

17. (a) Along the line $x = -3$, the vectors of $\mathbf{F}$ have positive y-components, so since the path goes upward, the integrand $\mathbf{F} \cdot \mathbf{T}$ is always positive. Therefore $\int_{C_1} \mathbf{F} \cdot d\mathbf{r} = \int_{C_1} \mathbf{F} \cdot \mathbf{T}\, ds$ is positive.

(b) All of the (nonzero) field vectors along the circle with radius 3 are pointed in the clockwise direction, that is, opposite the direction to the path. So $\mathbf{F} \cdot \mathbf{T}$ is negative, and therefore $\int_{C_2} \mathbf{F} \cdot d\mathbf{r} = \int_{C_2} \mathbf{F} \cdot \mathbf{T}\, ds$ is negative.

19. $\mathbf{r}(t) = t^3\, \mathbf{i} + t^2\, \mathbf{j}$, so $\mathbf{F}(\mathbf{r}(t)) = (t^3)(t^2)^2\, \mathbf{i} - (t^3)^2\, \mathbf{j} = t^7\, \mathbf{i} - t^6\, \mathbf{j}$ and $\mathbf{r}'(t) = 3t^2\, \mathbf{i} + 2t\, \mathbf{j}$. Then

$$\int_C \mathbf{F} \cdot d\mathbf{r} = \int_0^1 \mathbf{F}(\mathbf{r}(t)) \cdot \mathbf{r}'(t)\, dt = \int_0^1 (t^7 \cdot 3t^2 - t^6 \cdot 2t)\, dt = \int_0^1 (3t^9 - 2t^7)\, dt = \left[\tfrac{3}{10} t^{10} - \tfrac{1}{4} t^8\right]_0^1 = \tfrac{3}{10} - \tfrac{1}{4} = \tfrac{1}{20}.$$

21. $\int_C \mathbf{F} \cdot d\mathbf{r} = \int_0^1 \langle \sin t^3, \cos(-t^2), t^4 \rangle \cdot \langle 3t^2, -2t, 1 \rangle\, dt$

$$= \int_0^1 (3t^2 \sin t^3 - 2t \cos t^2 + t^4)\, dt = \left[-\cos t^3 - \sin t^2 + \tfrac{1}{5} t^5\right]_0^1 = \tfrac{6}{5} - \cos 1 - \sin 1$$

23. $\mathbf{F}(\mathbf{r}(t)) = \sqrt{\sin^2 t + \sin t \cos t}\, \mathbf{i} + \left[(\sin t \cos t)/\sin^2 t\right] \mathbf{j} = \sqrt{\sin^2 t + \sin t \cos t}\, \mathbf{i} + \cot t\, \mathbf{j}$,

$\mathbf{r}'(t) = 2 \sin t \cos t\, \mathbf{i} + (\cos^2 t - \sin^2 t)\, \mathbf{j}$. Then

$$\int_C \mathbf{F} \cdot d\mathbf{r} = \int_{\pi/6}^{\pi/3} \mathbf{F}(\mathbf{r}(t)) \cdot \mathbf{r}'(t)\, dt = \int_{\pi/6}^{\pi/3} \left[2 \sin t \cos t \sqrt{\sin^2 t + \sin t \cos t} + (\cot t)(\cos^2 t - \sin^2 t)\right] dt$$

$$\approx 0.5424$$

25. $x = t^2$, $y = t^3$, $z = \sqrt{t}$ so by Formula 9,

$$\int_C xy \arctan z\, ds = \int_1^2 (t^2)(t^3) \arctan \sqrt{t} \cdot \sqrt{(2t)^2 + (3t^2)^2 + \left[1/(2\sqrt{t})\right]^2}\, dt$$

$$= \int_1^2 t^5 \sqrt{4t^2 + 9t^4 + 1/(4t)} \arctan \sqrt{t}\, dt \approx 94.8231$$

27. We graph $\mathbf{F}(x, y) = (x - y)\, \mathbf{i} + xy\, \mathbf{j}$ and the curve C. We see that most of the vectors starting on C point in roughly the same direction as C, so for these portions of C the tangential component $\mathbf{F} \cdot \mathbf{T}$ is positive. Although some vectors in the third quadrant which start on C point in roughly the opposite direction, and hence give negative tangential components, it seems reasonable that the effect of these portions of C is outweighed by the positive tangential components. Thus, we would expect $\int_C \mathbf{F} \cdot d\mathbf{r} = \int_C \mathbf{F} \cdot \mathbf{T}\, ds$ to be positive.

To verify, we evaluate $\int_C \mathbf{F} \cdot d\mathbf{r}$. The curve C can be represented by $\mathbf{r}(t) = 2 \cos t\, \mathbf{i} + 2 \sin t\, \mathbf{j}$, $0 \le t \le \tfrac{3\pi}{2}$, so $\mathbf{F}(\mathbf{r}(t)) = (2 \cos t - 2 \sin t)\, \mathbf{i} + 4 \cos t \sin t\, \mathbf{j}$ and $\mathbf{r}'(t) = -2 \sin t\, \mathbf{i} + 2 \cos t\, \mathbf{j}$. Then

$$\int_C \mathbf{F} \cdot d\mathbf{r} = \int_0^{3\pi/2} \mathbf{F}(\mathbf{r}(t)) \cdot \mathbf{r}'(t)\, dt$$

$$= \int_0^{3\pi/2} [-2 \sin t (2 \cos t - 2 \sin t) + 2 \cos t (4 \cos t \sin t)]\, dt$$

$$= 4 \int_0^{3\pi/2} (\sin^2 t - \sin t \cos t + 2 \sin t \cos^2 t)\, dt$$

$$= 3\pi + \tfrac{2}{3} \quad \text{[using a CAS]}$$

29. (a) $\int_C \mathbf{F} \cdot d\mathbf{r} = \int_0^1 \left\langle e^{t^2-1}, t^5 \right\rangle \cdot \left\langle 2t, 3t^2 \right\rangle dt = \int_0^1 \left(2te^{t^2-1} + 3t^7 \right) dt = \left[e^{t^2-1} + \frac{3}{8}t^8 \right]_0^1 = \frac{11}{8} - 1/e$

(b) $\mathbf{r}(0) = \mathbf{0}, \quad \mathbf{F}(\mathbf{r}(0)) = \left\langle e^{-1}, 0 \right\rangle;$

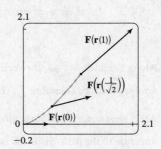

$\mathbf{r}\left(\frac{1}{\sqrt{2}}\right) = \left\langle \frac{1}{2}, \frac{1}{2\sqrt{2}} \right\rangle, \quad \mathbf{F}\left(\mathbf{r}\left(\frac{1}{\sqrt{2}}\right)\right) = \left\langle e^{-1/2}, \frac{1}{4\sqrt{2}} \right\rangle;$

$\mathbf{r}(1) = \langle 1, 1 \rangle, \quad \mathbf{F}(\mathbf{r}(1)) = \langle 1, 1 \rangle.$

In order to generate the graph with Maple, we use the `line` command in

the `plottools` package to define each of the vectors. For example,

```
v1:=line([0,0],[exp(-1),0]):
```

generates the vector from the vector field at the point $(0, 0)$ (but without an arrowhead) and gives it the name `v1`. To show

everything on the same screen, we use the `display` command. In Mathematica, we use `ListPlot` (with the

`PlotJoined -> True` option) to generate the vectors, and then `Show` to show everything on the same screen.

31. $x = e^{-t}\cos 4t, \quad y = e^{-t}\sin 4t, \quad z = e^{-t}, \quad 0 \le t \le 2\pi.$

Then $\dfrac{dx}{dt} = e^{-t}(-\sin 4t)(4) - e^{-t}\cos 4t = -e^{-t}(4\sin 4t + \cos 4t),$

$\dfrac{dy}{dt} = e^{-t}(\cos 4t)(4) - e^{-t}\sin 4t = -e^{-t}(-4\cos 4t + \sin 4t),$ and $\dfrac{dz}{dt} = -e^{-t},$ so

$$\sqrt{\left(\frac{dx}{dt}\right)^2 + \left(\frac{dy}{dt}\right)^2 + \left(\frac{dz}{dt}\right)^2} = \sqrt{(-e^{-t})^2[(4\sin 4t + \cos 4t)^2 + (-4\cos 4t + \sin 4t)^2 + 1]}$$

$$= e^{-t}\sqrt{16(\sin^2 4t + \cos^2 4t) + \sin^2 4t + \cos^2 4t + 1} = 3\sqrt{2}\,e^{-t}$$

Therefore $\int_C x^3 y^2 z \, ds = \int_0^{2\pi} (e^{-t}\cos 4t)^3 (e^{-t}\sin 4t)^2 (e^{-t}) \left(3\sqrt{2}\,e^{-t}\right) dt$

$$= \int_0^{2\pi} 3\sqrt{2}\,e^{-7t}\cos^3 4t \sin^2 4t \, dt = \frac{172,704}{5,632,705}\sqrt{2}\left(1 - e^{-14\pi}\right)$$

33. We use the parametrization $x = 2\cos t, \; y = 2\sin t, \; -\frac{\pi}{2} \le t \le \frac{\pi}{2}.$ Then

$ds = \sqrt{\left(\frac{dx}{dt}\right)^2 + \left(\frac{dy}{dt}\right)^2}\,dt = \sqrt{(-2\sin t)^2 + (2\cos t)^2}\,dt = 2\,dt,$ so $m = \int_C k\,ds = 2k\int_{-\pi/2}^{\pi/2} dt = 2k(\pi),$

$\overline{x} = \frac{1}{2\pi k}\int_C xk\,ds = \frac{1}{2\pi}\int_{-\pi/2}^{\pi/2}(2\cos t)2\,dt = \frac{1}{2\pi}\left[4\sin t\right]_{-\pi/2}^{\pi/2} = \frac{4}{\pi}, \overline{y} = \frac{1}{2\pi k}\int_C yk\,ds = \frac{1}{2\pi}\int_{-\pi/2}^{\pi/2}(2\sin t)2\,dt = 0.$

Hence $(\overline{x}, \overline{y}) = \left(\frac{4}{\pi}, 0\right).$

35. (a) $\overline{x} = \frac{1}{m}\int_C x\rho(x,y,z)\,ds, \; \overline{y} = \frac{1}{m}\int_C y\rho(x,y,z)\,ds, \overline{z} = \frac{1}{m}\int_C z\rho(x,y,z)\,ds$ where $m = \int_C \rho(x,y,z)\,ds.$

(b) $m = \int_C k\,ds = k\int_0^{2\pi}\sqrt{4\sin^2 t + 4\cos^2 t + 9}\,dt = k\sqrt{13}\int_0^{2\pi} dt = 2\pi k\sqrt{13},$

$\overline{x} = \frac{1}{2\pi k\sqrt{13}}\int_0^{2\pi} 2k\sqrt{13}\sin t\,dt = 0, \overline{y} = \frac{1}{2\pi k\sqrt{13}}\int_0^{2\pi} 2k\sqrt{13}\cos t\,dt = 0,$

$\overline{z} = \frac{1}{2\pi k\sqrt{13}}\int_0^{2\pi}\left(k\sqrt{13}\right)(3t)\,dt = \frac{3}{2\pi}\left(2\pi^2\right) = 3\pi.$ Hence $(\overline{x}, \overline{y}, \overline{z}) = (0, 0, 3\pi).$

37. From Example 3, $\rho(x, y) = k(1 - y)$, $x = \cos t$, $y = \sin t$, and $ds = dt$, $0 \le t \le \pi$ $\Rightarrow$

$$I_x = \int_C y^2 \rho(x, y)\, ds = \int_0^\pi \sin^2 t \, [k(1 - \sin t)]\, dt = k \int_0^\pi (\sin^2 t - \sin^3 t)\, dt$$

$$= \tfrac{1}{2} k \int_0^\pi (1 - \cos 2t)\, dt - k \int_0^\pi (1 - \cos^2 t) \sin t\, dt \qquad \left[\begin{array}{l} \text{Let } u = \cos t, \, du = -\sin t\, dt \\ \text{in the second integral} \end{array}\right]$$

$$= k\left[\tfrac{\pi}{2} + \int_1^{-1} (1 - u^2)\, du\right] = k\left(\tfrac{\pi}{2} - \tfrac{4}{3}\right)$$

$$I_y = \int_C x^2 \rho(x, y)\, ds = k \int_0^\pi \cos^2 t \, (1 - \sin t)\, dt = \tfrac{k}{2} \int_0^\pi (1 + \cos 2t)\, dt - k \int_0^\pi \cos^2 t \sin t\, dt$$

$$= k\left(\tfrac{\pi}{2} - \tfrac{2}{3}\right), \text{ using the same substitution as above.}$$

39. $W = \int_C \mathbf{F} \cdot d\mathbf{r} = \int_0^{2\pi} \langle t - \sin t, 3 - \cos t\rangle \cdot \langle 1 - \cos t, \sin t\rangle \, dt$

$$= \int_0^{2\pi} (t - t\cos t - \sin t + \sin t \cos t + 3 \sin t - \sin t \cos t)\, dt$$

$$= \int_0^{2\pi} (t - t\cos t + 2\sin t)\, dt = \left[\tfrac{1}{2} t^2 - (t\sin t + \cos t) - 2\cos t\right]_0^{2\pi} \qquad \left[\begin{array}{l} \text{integrate by parts} \\ \text{in the second term} \end{array}\right]$$

$$= 2\pi^2$$

41. $\mathbf{r}(t) = \langle 2t, t, 1 - t\rangle$, $0 \le t \le 1$.

$$W = \int_C \mathbf{F} \cdot d\mathbf{r} = \int_0^1 \langle 2t - t^2, t - (1 - t)^2, 1 - t - (2t)^2\rangle \cdot \langle 2, 1, -1\rangle \, dt$$

$$= \int_0^1 (4t - 2t^2 + t - 1 + 2t - t^2 - 1 + t + 4t^2)\, dt = \int_0^1 (t^2 + 8t - 2)\, dt = \left[\tfrac{1}{3} t^3 + 4t^2 - 2t\right]_0^1 = \tfrac{7}{3}$$

43. (a) $\mathbf{r}(t) = at^2\, \mathbf{i} + bt^3\, \mathbf{j}$ $\Rightarrow$ $\mathbf{v}(t) = \mathbf{r}'(t) = 2at\, \mathbf{i} + 3bt^2\, \mathbf{j}$ $\Rightarrow$ $\mathbf{a}(t) = \mathbf{v}'(t) = 2a\, \mathbf{i} + 6bt\, \mathbf{j}$, and force is mass times acceleration: $\mathbf{F}(t) = m\, \mathbf{a}(t) = 2ma\, \mathbf{i} + 6mbt\, \mathbf{j}$.

(b) $W = \int_C \mathbf{F} \cdot d\mathbf{r} = \int_0^1 (2ma\, \mathbf{i} + 6mbt\, \mathbf{j}) \cdot (2at\, \mathbf{i} + 3bt^2\, \mathbf{j})\, dt = \int_0^1 (4ma^2 t + 18mb^2 t^3)\, dt$

$$= \left[2ma^2 t^2 + \tfrac{9}{2} mb^2 t^4\right]_0^1 = 2ma^2 + \tfrac{9}{2} mb^2$$

45. The combined weight of the man and the paint is 185 lb, so the force exerted (equal and opposite to that exerted by gravity) is $\mathbf{F} = 185\, \mathbf{k}$. To parametrize the staircase, let $x = 20 \cos t$, $y = 20 \sin t$, $z = \tfrac{90}{6\pi} t = \tfrac{15}{\pi} t$, $0 \le t \le 6\pi$. Then the work done is

$$W = \int_C \mathbf{F} \cdot d\mathbf{r} = \int_0^{6\pi} \langle 0, 0, 185\rangle \cdot \langle -20 \sin t, 20 \cos t, \tfrac{15}{\pi}\rangle \, dt = (185) \tfrac{15}{\pi} \int_0^{6\pi} dt = (185)\left(\tfrac{15}{\pi}\right)(6\pi) \approx 1.67 \times 10^4 \text{ ft-lb}$$

47. (a) $\mathbf{r}(t) = \langle \cos t, \sin t\rangle$, $0 \le t \le 2\pi$, and let $\mathbf{F} = \langle a, b\rangle$. Then

$$W = \int_C \mathbf{F} \cdot d\mathbf{r} = \int_0^{2\pi} \langle a, b\rangle \cdot \langle -\sin t, \cos t\rangle \, dt = \int_0^{2\pi} (-a \sin t + b \cos t)\, dt = \left[a \cos t + b \sin t\right]_0^{2\pi}$$

$$= a + 0 - a + 0 = 0$$

(b) Yes. $\mathbf{F}(x, y) = k\, \mathbf{x} = \langle kx, ky\rangle$ and

$$W = \int_C \mathbf{F} \cdot d\mathbf{r} = \int_0^{2\pi} \langle k \cos t, k \sin t\rangle \cdot \langle -\sin t, \cos t\rangle \, dt = \int_0^{2\pi} (-k \sin t \cos t + k \sin t \cos t)\, dt = \int_0^{2\pi} 0\, dt = 0.$$

49. Let $\mathbf{r}(t) = \langle x(t), y(t), z(t) \rangle$ and $\mathbf{v} = \langle v_1, v_2, v_3 \rangle$. Then

$$\int_C \mathbf{v} \cdot d\mathbf{r} = \int_a^b \langle v_1, v_2, v_3 \rangle \cdot \langle x'(t), y'(t), z'(t) \rangle \, dt = \int_a^b [v_1 \, x'(t) + v_2 \, y'(t) + v_3 \, z'(t)] \, dt$$

$$= [v_1 \, x(t) + v_2 \, y(t) + v_3 \, z(t)]_a^b = [v_1 \, x(b) + v_2 \, y(b) + v_3 \, z(b)] - [v_1 \, x(a) + v_2 \, y(a) + v_3 \, z(a)]$$

$$= v_1 \, [x(b) - x(a)] + v_2 \, [y(b) - y(a)] + v_3 \, [z(b) - z(a)]$$

$$= \langle v_1, v_2, v_3 \rangle \cdot \langle x(b) - x(a), y(b) - y(a), z(b) - z(a) \rangle$$

$$= \langle v_1, v_2, v_3 \rangle \cdot [\langle x(b), y(b), z(b) \rangle - \langle x(a), y(a), z(a) \rangle] = \mathbf{v} \cdot [\mathbf{r}(b) - \mathbf{r}(a)]$$

51. The work done in moving the object is $\int_C \mathbf{F} \cdot d\mathbf{r} = \int_C \mathbf{F} \cdot \mathbf{T} \, ds$. We can approximate this integral by dividing C into

7 segments of equal length $\Delta s = 2$ and approximating $\mathbf{F} \cdot \mathbf{T}$, that is, the tangential component of force, at a point (x_i^*, y_i^*) on

each segment. Since C is composed of straight line segments, $\mathbf{F} \cdot \mathbf{T}$ is the scalar projection of each force vector onto C.

If we choose (x_i^*, y_i^*) to be the point on the segment closest to the origin, then the work done is

$$\int_C \mathbf{F} \cdot \mathbf{T} \, ds \approx \sum_{i=1}^{7} [\mathbf{F}(x_i^*, y_i^*) \cdot \mathbf{T}(x_i^*, y_i^*)] \, \Delta s = [2 + 2 + 2 + 2 + 1 + 1 + 1](2) = 22. \text{ Thus, we estimate the work done to}$$

be approximately 22 J.

16.3 The Fundamental Theorem for Line Integrals

1. C appears to be a smooth curve, and since ∇f is continuous, we know f is differentiable. Then Theorem 2 says that the value

of $\int_C \nabla f \cdot d\mathbf{r}$ is simply the difference of the values of f at the terminal and initial points of C. From the graph, this is

$50 - 10 = 40$.

3. Let $P(x, y) = xy + y^2$ and $Q(x, y) = x^2 + 2xy$. Then $\partial P / \partial y = x + 2y$ and $\partial Q / \partial x = 2x + 2y$. Since $\partial P / \partial y \neq \partial Q / \partial x$,

$\mathbf{F}(x, y) = P\mathbf{i} + Q\mathbf{j}$ is not conservative by Theorem 5.

5. $\dfrac{\partial}{\partial y} \left(y^2 e^{xy} \right) = y^2 \cdot xe^{xy} + 2ye^{xy} = (xy^2 + 2y)e^{xy}$,

$\dfrac{\partial}{\partial x} \left[(1 + xy)e^{xy} \right] = (1 + xy) \cdot ye^{xy} + ye^{xy} = ye^{xy} + xy^2 e^{xy} + ye^{xy} = (xy^2 + 2y)e^{xy}$.

Since these partial derivatives are equal and the domain of $\mathbf{F}$ is $\mathbb{R}^2$ which is open and simply-connected, $\mathbf{F}$ is conservative by

Theorem 6. Thus, there exists a function f such that $\nabla f = \mathbf{F}$, that is, $f_x(x, y) = y^2 e^{xy}$ and $f_y(x, y) = (1 + xy)e^{xy}$. But

$f_x(x, y) = y^2 e^{xy}$ implies $f(x, y) = ye^{xy} + g(y)$ and differentiating both sides of this equation with respect to y gives

$f_y(x, y) = (1 + xy)e^{xy} + g'(y)$. Thus $(1 + xy)e^{xy} = (1 + xy)e^{xy} + g'(y)$ so $g'(y) = 0$ and $g(y) = K$ where K is a

constant. Hence $f(x, y) = ye^{xy} + K$ is a potential function for $\mathbf{F}$.

7. $\partial(ye^x + \sin y)/\partial y = e^x + \cos y = \partial(e^x + x \cos y)/\partial x$ and the domain of $\mathbf{F}$ is $\mathbb{R}^2$. Hence $\mathbf{F}$ is conservative so there

exists a function f such that $\nabla f = \mathbf{F}$. Then $f_x(x, y) = ye^x + \sin y$ implies $f(x, y) = ye^x + x \sin y + g(y)$ and

$f_y(x, y) = e^x + x \cos y + g'(y)$. But $f_y(x, y) = e^x + x \cos y$ so $g(y) = K$ and $f(x, y) = ye^x + x \sin y + K$ is a potential function for $\mathbf{F}$.

9. $\partial(y^2 \cos x + \cos y)/\partial y = 2y \cos x - \sin y = \partial(2y \sin x - x \sin y)/\partial x$ and the domain of $\mathbf{F}$ is $\mathbb{R}^2$ which is open and simply connected. Hence $\mathbf{F}$ is conservative so there exists a function f such that $\nabla f = \mathbf{F}$. Then $f_x(x, y) = y^2 \cos x + \cos y$ implies $f(x, y) = y^2 \sin x + x \cos y + g(y)$ and $f_y(x, y) = 2y \sin x - x \sin y + g'(y)$. But $f_y(x, y) = 2y \sin x - x \sin y$ so $g'(y) = 0 \Rightarrow g(y) = K$ and $f(x, y) = y^2 \sin x + x \cos y + K$ is a potential function for $\mathbf{F}$.

11. (a) $\mathbf{F}$ has continuous first-order partial derivatives and $\dfrac{\partial}{\partial y}(2xy) = 2x = \dfrac{\partial}{\partial x}(x^2)$ on $\mathbb{R}^2$, which is open and simply-connected. Thus, $\mathbf{F}$ is conservative by Theorem 6. Then we know that the line integral of $\mathbf{F}$ is independent of path; in particular, the value of $\int_C \mathbf{F} \cdot d\mathbf{r}$ depends only on the endpoints of C. Since all three curves have the same initial and terminal points, $\int_C \mathbf{F} \cdot d\mathbf{r}$ will have the same value for each curve.

(b) We first find a potential function f, so that $\nabla f = \mathbf{F}$. We know $f_x(x, y) = 2xy$ and $f_y(x, y) = x^2$. Integrating $f_x(x, y)$ with respect to x, we have $f(x, y) = x^2 y + g(y)$. Differentiating both sides with respect to y gives $f_y(x, y) = x^2 + g'(y)$, so we must have $x^2 + g'(y) = x^2 \Rightarrow g'(y) = 0 \Rightarrow g(y) = K$, a constant. Thus $f(x, y) = x^2 y + K$, and we can take $K = 0$. All three curves start at $(1, 2)$ and end at $(3, 2)$, so by Theorem 2, $\int_C \mathbf{F} \cdot d\mathbf{r} = f(3, 2) - f(1, 2) = 18 - 2 = 16$ for each curve.

13. (a) If $\mathbf{F} = \nabla f$ then $f_x(x, y) = x^2 y^3$ and $f_y(x, y) = x^3 y^2$.
$f_x(x, y) = x^2 y^3$ implies $f(x, y) = \frac{1}{3} x^3 y^3 + g(y)$ and $f_y(x, y) = x^3 y^2 + g'(y)$. But $f_y(x, y) = x^3 y^2$ so $g'(y) = 0 \Rightarrow g(y) = K$, a constant. We can take $K = 0$, so $f(x, y) = \frac{1}{3} x^3 y^3$.

(b) C is a smooth curve with initial point $\mathbf{r}(0) = (0, 0)$ and terminal point $\mathbf{r}(1) = (-1, 3)$, so by Theorem 2
$\int_C \mathbf{F} \cdot d\mathbf{r} = \int_C \nabla f \cdot d\mathbf{r} = f(-1, 3) - f(0, 0) = -9 - 0 = -9$.

15. (a) $f_x(x, y, z) = yz$ implies $f(x, y, z) = xyz + g(y, z)$ and so $f_y(x, y, z) = xz + g_y(y, z)$. But $f_y(x, y, z) = xz$ so $g_y(y, z) = 0 \Rightarrow g(y, z) = h(z)$. Thus $f(x, y, z) = xyz + h(z)$ and $f_z(x, y, z) = xy + h'(z)$. But $f_z(x, y, z) = xy + 2z$, so $h'(z) = 2z \Rightarrow h(z) = z^2 + K$. Hence $f(x, y, z) = xyz + z^2$ (taking $K = 0$).

(b) $\int_C \mathbf{F} \cdot d\mathbf{r} = f(4, 6, 3) - f(1, 0, -2) = 81 - 4 = 77$.

17. (a) $f_x(x, y, z) = yze^{xz}$ implies $f(x, y, z) = ye^{xz} + g(y, z)$ and so $f_y(x, y, z) = e^{xz} + g_y(y, z)$. But $f_y(x, y, z) = e^{xz}$ so $g_y(y, z) = 0 \Rightarrow g(y, z) = h(z)$. Thus $f(x, y, z) = ye^{xz} + h(z)$ and $f_z(x, y, z) = xye^{xz} + h'(z)$. But $f_z(x, y, z) = xye^{xz}$, so $h'(z) = 0 \Rightarrow h(z) = K$. Hence $f(x, y, z) = ye^{xz}$ (taking $K = 0$).

(b) $\mathbf{r}(0) = \langle 1, -1, 0 \rangle$, $\mathbf{r}(2) = \langle 5, 3, 0 \rangle$ so $\int_C \mathbf{F} \cdot d\mathbf{r} = f(5, 3, 0) - f(1, -1, 0) = 3e^0 + e^0 = 4$.

19. The functions $2xe^{-y}$ and $2y - x^2e^{-y}$ have continuous first-order derivatives on $\mathbb{R}^2$ and

$\dfrac{\partial}{\partial y}\left(2xe^{-y}\right) = -2xe^{-y} = \dfrac{\partial}{\partial x}\left(2y - x^2e^{-y}\right)$, so $\mathbf{F}(x, y) = 2xe^{-y}\,\mathbf{i} + \left(2y - x^2e^{-y}\right)\mathbf{j}$ is a conservative vector field by

Theorem 6 and hence the line integral is independent of path. Thus a potential function f exists, and $f_x(x, y) = 2xe^{-y}$

implies $f(x, y) = x^2e^{-y} + g(y)$ and $f_y(x, y) = -x^2e^{-y} + g'(y)$. But $f_y(x, y) = 2y - x^2e^{-y}$ so

$g'(y) = 2y \implies g(y) = y^2 + K$. We can take $K = 0$, so $f(x, y) = x^2e^{-y} + y^2$. Then

$\int_C 2xe^{-y}\,dx + (2y - x^2e^{-y})\,dy = f(2, 1) - f(1, 0) = 4e^{-1} + 1 - 1 = 4/e$.

21. If $\mathbf{F}$ is conservative, then $\int_C \mathbf{F} \cdot d\mathbf{r}$ is independent of path. This means that the work done along all piecewise-smooth curves that have the described initial and terminal points is the same. Your reply: It doesn't matter which curve is chosen.

23. $\mathbf{F}(x, y) = x^3\,\mathbf{i} + y^3\,\mathbf{j}$, $W = \int_C \mathbf{F} \cdot d\mathbf{r}$. Since $\partial(x^3)/\partial y = 0 = \partial(y^3)/\partial x$, there exists a function f such that $\nabla f = \mathbf{F}$. In

fact, $f_x(x, y) = x^3 \implies f(x, y) = \frac{1}{4}x^4 + g(y) \implies f_y(x, y) = 0 + g'(y)$. But $f_y(x, y) = y^3$ so

$g'(y) = y^3 \implies g(y) = \frac{1}{4}y^4 + K$. We can take $K = 0 \implies f(x, y) = \frac{1}{4}x^4 + \frac{1}{4}y^4$. Thus

$W = \int_C \mathbf{F} \cdot d\mathbf{r} = f(2, 2) - f(1, 0) = (4 + 4) - \left(\frac{1}{4} + 0\right) = \frac{31}{4}$.

25. We know that if the vector field (call it $\mathbf{F}$) is conservative, then around any closed path C, $\int_C \mathbf{F} \cdot d\mathbf{r} = 0$. But take C to be a circle centered at the origin, oriented counterclockwise. All of the field vectors that start on C are roughly in the direction of motion along C, so the integral around C will be positive. Therefore the field is not conservative.

27.

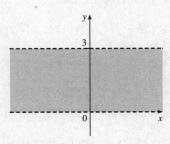

From the graph, it appears that $\mathbf{F}$ is conservative, since around all closed paths, the number and size of the field vectors pointing in directions similar to that of the path seem to be roughly the same as the number and size of the vectors pointing in the opposite direction. To check, we calculate

$\dfrac{\partial}{\partial y}(\sin y) = \cos y = \dfrac{\partial}{\partial x}(1 + x\cos y)$. Thus $\mathbf{F}$ is conservative, by

Theorem 6.

29. Since $\mathbf{F}$ is conservative, there exists a function f such that $\mathbf{F} = \nabla f$, that is, $P = f_x$, $Q = f_y$, and $R = f_z$. Since P,

Q, and R have continuous first order partial derivatives, Clairaut's Theorem says that $\partial P/\partial y = f_{xy} = f_{yx} = \partial Q/\partial x$,

$\partial P/\partial z = f_{xz} = f_{zx} = \partial R/\partial x$, and $\partial Q/\partial z = f_{yz} = f_{zy} = \partial R/\partial y$.

31. $D = \{(x, y) \mid 0 < y < 3\}$ consists of those points between, but not

on, the horizontal lines $y = 0$ and $y = 3$.

(a) Since D does not include any of its boundary points, it is open. More

formally, at any point in D there is a disk centered at that point that

lies entirely in D.

(b) Any two points chosen in D can always be joined by a path that lies

entirely in D, so D is connected. (D consists of just one "piece.")

(c) D is connected and it has no holes, so it's simply-connected. (Every simple closed curve in D encloses only points that are in D.)

33. $D = \{(x, y) \mid 1 \le x^2 + y^2 \le 4,\ y \ge 0\}$ is the semiannular region

in the upper half-plane between circles centered at the origin of radii

1 and 2 (including all boundary points).

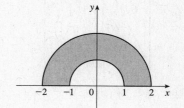

(a) D includes boundary points, so it is not open. [Note that at any

boundary point, $(1, 0)$ for instance, any disk centered there cannot lie

entirely in D.]

(b) The region consists of one piece, so it's connected.

(c) D is connected and has no holes, so it's simply-connected.

35. (a) $P = -\dfrac{y}{x^2 + y^2}$, $\dfrac{\partial P}{\partial y} = \dfrac{y^2 - x^2}{(x^2 + y^2)^2}$ and $Q = \dfrac{x}{x^2 + y^2}$, $\dfrac{\partial Q}{\partial x} = \dfrac{y^2 - x^2}{(x^2 + y^2)^2}$. Thus $\dfrac{\partial P}{\partial y} = \dfrac{\partial Q}{\partial x}$.

(b) C_1: $x = \cos t$, $y = \sin t$, $0 \le t \le \pi$, C_2: $x = \cos t$, $y = \sin t$, $t = 2\pi$ to $t = \pi$. Then

$$\int_{C_1} \mathbf{F} \cdot d\mathbf{r} = \int_0^\pi \frac{(-\sin t)(-\sin t) + (\cos t)(\cos t)}{\cos^2 t + \sin^2 t}\, dt = \int_0^\pi dt = \pi \text{ and } \int_{C_2} \mathbf{F} \cdot d\mathbf{r} = \int_{2\pi}^\pi dt = -\pi$$

Since these aren't equal, the line integral of $\mathbf{F}$ isn't independent of path. (Or notice that $\int_{C_3} \mathbf{F} \cdot d\mathbf{r} = \int_0^{2\pi} dt = 2\pi$ where

C_3 is the circle $x^2 + y^2 = 1$, and apply the contrapositive of Theorem 3.) This doesn't contradict Theorem 6, since the

domain of $\mathbf{F}$, which is $\mathbb{R}^2$ except the origin, isn't simply-connected.

16.4 Green's Theorem

1. (a)

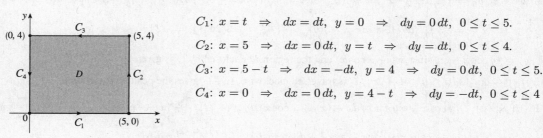

C_1: $x = t$ $\Rightarrow$ $dx = dt$, $y = 0$ $\Rightarrow$ $dy = 0\, dt$, $0 \le t \le 5$.

C_2: $x = 5$ $\Rightarrow$ $dx = 0\, dt$, $y = t$ $\Rightarrow$ $dy = dt$, $0 \le t \le 4$.

C_3: $x = 5 - t$ $\Rightarrow$ $dx = -dt$, $y = 4$ $\Rightarrow$ $dy = 0\, dt$, $0 \le t \le 5$.

C_4: $x = 0$ $\Rightarrow$ $dx = 0\, dt$, $y = 4 - t$ $\Rightarrow$ $dy = -dt$, $0 \le t \le 4$

Thus $\displaystyle\oint_C y^2\, dx + x^2 y\, dy = \oint_{C_1 + C_2 + C_3 + C_4} y^2\, dx + x^2 y\, dy = \int_0^5 0\, dt + \int_0^4 25t\, dt + \int_0^5 (-16 + 0)\, dt + \int_0^4 0\, dt$

$$= 0 + \left[\tfrac{25}{2} t^2\right]_0^4 + \left[-16t\right]_0^5 + 0 = 200 + (-80) = 120$$

(b) Note that C as given in part (a) is a positively oriented, piecewise-smooth, simple closed curve. Then by Green's Theorem,

$$\oint_C y^2\, dx + x^2 y\, dy = \iint_D \left[\tfrac{\partial}{\partial x}(x^2 y) - \tfrac{\partial}{\partial y}(y^2)\right] dA = \int_0^5 \int_0^4 (2xy - 2y)\, dy\, dx = \int_0^5 \left[xy^2 - y^2\right]_{y=0}^{y=4} dx$$

$$= \int_0^5 (16x - 16)\, dx = \left[8x^2 - 16x\right]_0^5 = 200 - 80 = 120$$

3. (a)

C_1: $x = t$ $\Rightarrow$ $dx = dt$, $y = 0$ $\Rightarrow$ $dy = 0\,dt$, $0 \le t \le 1$.

C_2: $x = 1$ $\Rightarrow$ $dx = 0\,dt$, $y = t$ $\Rightarrow$ $dy = dt$, $0 \le t \le 2$.

C_3: $x = 1 - t$ $\Rightarrow$ $dx = -dt$, $y = 2 - 2t$ $\Rightarrow$ $dy = -2\,dt$, $0 \le t \le 1$.

Thus $\oint_C xy\,dx + x^2 y^3\,dy = \oint_{C_1 + C_2 + C_3} xy\,dx + x^2 y^3\,dy$

$$= \int_0^1 0\,dt + \int_0^2 t^3\,dt + \int_0^1 \left[-(1-t)(2-2t) - 2(1-t)^2(2-2t)^3 \right] dt$$

$$= 0 + \left[\tfrac{1}{4}t^4 \right]_0^2 + \int_0^1 \left[-2(1-t)^2 - 16(1-t)^5 \right] dt$$

$$= 4 + \left[\tfrac{2}{3}(1-t)^3 + \tfrac{8}{3}(1-t)^6 \right]_0^1 = 4 + 0 - \tfrac{10}{3} = \tfrac{2}{3}$$

(b) $\oint_C xy\,dx + x^2 y^3\,dy = \iint_D \left[\frac{\partial}{\partial x}(x^2 y^3) - \frac{\partial}{\partial y}(xy) \right] dA = \int_0^1 \int_0^{2x} (2xy^3 - x)\,dy\,dx$

$$= \int_0^1 \left[\tfrac{1}{2}xy^4 - xy \right]_{y=0}^{y=2x} dx = \int_0^1 (8x^5 - 2x^2)\,dx = \tfrac{4}{3} - \tfrac{2}{3} = \tfrac{2}{3}$$

5. The region D enclosed by C is $[0, 3] \times [0, 4]$, so

$$\int_C ye^x\,dx + 2e^x\,dy = \iint_D \left[\frac{\partial}{\partial x}(2e^x) - \frac{\partial}{\partial y}(ye^x) \right] dA = \int_0^3 \int_0^4 (2e^x - e^x)\,dy\,dx$$

$$= \int_0^3 e^x\,dx \int_0^4 dy = \left[e^x \right]_0^3 \left[y \right]_0^4 = (e^3 - e^0)(4 - 0) = 4(e^3 - 1)$$

7. $\int_C \left(y + e^{\sqrt{x}} \right) dx + (2x + \cos y^2)\,dy = \iint_D \left[\frac{\partial}{\partial x}(2x + \cos y^2) - \frac{\partial}{\partial y}\left(y + e^{\sqrt{x}} \right) \right] dA$

$$= \int_0^1 \int_{x^2}^{\sqrt{x}} (2 - 1)\,dy\,dx = \int_0^1 (\sqrt{x} - x^2)\,dx = \left[\tfrac{2}{3}x^{3/2} - \tfrac{1}{3}x^3 \right]_0^1 = \tfrac{1}{3}$$

9. $\int_C y^3\,dx - x^3\,dy = \iint_D \left[\frac{\partial}{\partial x}(-x^3) - \frac{\partial}{\partial y}(y^3) \right] dA = \iint_D (-3x^2 - 3y^2)\,dA = \int_0^{2\pi} \int_0^2 (-3r^2)\,r\,dr\,d\theta$

$$= -3 \int_0^{2\pi} d\theta \int_0^2 r^3\,dr = -3 \left[\theta \right]_0^{2\pi} \left[\tfrac{1}{4}r^4 \right]_0^2 = -3(2\pi)(4) = -24\pi$$

11. $\mathbf{F}(x, y) = \langle y \cos x - xy \sin x, xy + x \cos x \rangle$ and the region D enclosed by C is given by

$\{(x, y) \mid 0 \le x \le 2,\ 0 \le y \le 4 - 2x\}$. C is traversed clockwise, so $-C$ gives the positive orientation.

$\int_C \mathbf{F} \cdot d\mathbf{r} = -\int_{-C}(y \cos x - xy \sin x)\,dx + (xy + x \cos x)\,dy = -\iint_D \left[\frac{\partial}{\partial x}(xy + x \cos x) - \frac{\partial}{\partial y}(y \cos x - xy \sin x) \right] dA$

$$= -\iint_D (y - x \sin x + \cos x - \cos x + x \sin x)\,dA = -\int_0^2 \int_0^{4-2x} y\,dy\,dx$$

$$= -\int_0^2 \left[\tfrac{1}{2}y^2 \right]_{y=0}^{y=4-2x} dx = -\int_0^2 \tfrac{1}{2}(4 - 2x)^2\,dx = -\int_0^2 (8 - 8x + 2x^2)\,dx = -\left[8x - 4x^2 + \tfrac{2}{3}x^3 \right]_0^2$$

$$= -\left(16 - 16 + \tfrac{16}{3} - 0 \right) = -\tfrac{16}{3}$$

13. $\mathbf{F}(x, y) = \langle y - \cos y, x \sin y \rangle$ and the region D enclosed by C is the disk with radius 2 centered at $(3, -4)$.

C is traversed clockwise, so $-C$ gives the positive orientation.

$$\int_C \mathbf{F} \cdot d\mathbf{r} = -\int_{-C}(y - \cos y)\,dx + (x \sin y)\,dy = -\iint_D \left[\frac{\partial}{\partial x}(x \sin y) - \frac{\partial}{\partial y}(y - \cos y) \right] dA$$

$$= -\iint_D (\sin y - 1 - \sin y)\,dA = \iint_D dA = \text{area of } D = \pi(2)^2 = 4\pi$$

15. Here $C = C_1 + C_2$ where

C_1 can be parametrized as $x = t$, $y = 0$, $-\pi/2 \le t \le \pi/2$, and

C_2 is given by $x = -t$, $y = \cos t$, $-\pi/2 \le t \le \pi/2$.

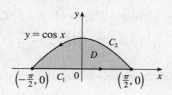

Then the line integral is

$$\oint_{C_1+C_2} x^3 y^4 \, dx + x^5 y^4 \, dy = \int_{-\pi/2}^{\pi/2}(0+0)\,dt + \int_{-\pi/2}^{\pi/2}[(-t)^3(\cos t)^4(-1) + (-t)^5(\cos t)^4(-\sin t)]\,dt$$

$$= 0 + \int_{-\pi/2}^{\pi/2}(t^3\cos^4 t + t^5\cos^4 t \sin t)\,dt = \tfrac{1}{15}\pi^4 - \tfrac{4144}{1125}\pi^2 + \tfrac{7,578,368}{253,125} \approx 0.0779$$

according to a CAS. The double integral is

$$\iint_D \left(\frac{\partial Q}{\partial x} - \frac{\partial P}{\partial y}\right) dA = \int_{-\pi/2}^{\pi/2}\int_0^{\cos x}(5x^4 y^4 - 4x^3 y^3)\,dy\,dx = \tfrac{1}{15}\pi^4 - \tfrac{4144}{1125}\pi^2 + \tfrac{7,578,368}{253,125} \approx 0.0779, \text{ verifying Green's}$$

Theorem in this case.

17. By Green's Theorem, $W = \int_C \mathbf{F} \cdot d\mathbf{r} = \int_C x(x+y)\,dx + xy^2\,dy = \iint_D(y^2 - x)\,dA$ where C is the path described in the question and D is the triangle bounded by C. So

$$W = \int_0^1 \int_0^{1-x}(y^2 - x)\,dy\,dx = \int_0^1 \left[\tfrac{1}{3}y^3 - xy\right]_{y=0}^{y=1-x}\,dx = \int_0^1\left(\tfrac{1}{3}(1-x)^3 - x(1-x)\right)dx$$

$$= \left[-\tfrac{1}{12}(1-x)^4 - \tfrac{1}{2}x^2 + \tfrac{1}{3}x^3\right]_0^1 = \left(-\tfrac{1}{2} + \tfrac{1}{3}\right) - \left(-\tfrac{1}{12}\right) = -\tfrac{1}{12}$$

19. Let C_1 be the arch of the cycloid from $(0,0)$ to $(2\pi, 0)$, which corresponds to $0 \le t \le 2\pi$, and let C_2 be the segment from $(2\pi, 0)$ to $(0,0)$, so C_2 is given by $x = 2\pi - t, y = 0, 0 \le t \le 2\pi$. Then $C = C_1 \cup C_2$ is traversed clockwise, so $-C$ is oriented positively. Thus $-C$ encloses the area under one arch of the cycloid and from (5) we have

$$A = -\oint_{-C} y\,dx = \int_{C_1} y\,dx + \int_{C_2} y\,dx = \int_0^{2\pi}(1-\cos t)(1-\cos t)\,dt + \int_0^{2\pi} 0\,(-dt)$$

$$= \int_0^{2\pi}(1 - 2\cos t + \cos^2 t)\,dt + 0 = \left[t - 2\sin t + \tfrac{1}{2}t + \tfrac{1}{4}\sin 2t\right]_0^{2\pi} = 3\pi$$

21. (a) Using Equation 16.2.8, we write parametric equations of the line segment as $x = (1-t)x_1 + tx_2, y = (1-t)y_1 + ty_2$, $0 \le t \le 1$. Then $dx = (x_2 - x_1)\,dt$ and $dy = (y_2 - y_1)\,dt$, so

$$\int_C x\,dy - y\,dx = \int_0^1 [(1-t)x_1 + tx_2](y_2 - y_1)\,dt + [(1-t)y_1 + ty_2](x_2 - x_1)\,dt$$

$$= \int_0^1 (x_1(y_2 - y_1) - y_1(x_2 - x_1) + t[(y_2 - y_1)(x_2 - x_1) - (x_2 - x_1)(y_2 - y_1)])\,dt$$

$$= \int_0^1 (x_1 y_2 - x_2 y_1)\,dt = x_1 y_2 - x_2 y_1$$

(b) We apply Green's Theorem to the path $C = C_1 \cup C_2 \cup \cdots \cup C_n$, where C_i is the line segment that joins (x_i, y_i) to (x_{i+1}, y_{i+1}) for $i = 1, 2, \ldots, n-1$, and C_n is the line segment that joins (x_n, y_n) to (x_1, y_1). From (5), $\tfrac{1}{2}\int_C x\,dy - y\,dx = \iint_D dA$, where D is the polygon bounded by C. Therefore

$$\text{area of polygon} = A(D) = \iint_D dA = \tfrac{1}{2}\int_C x\,dy - y\,dx$$

$$= \tfrac{1}{2}\left(\int_{C_1} x\,dy - y\,dx + \int_{C_2} x\,dy - y\,dx + \cdots + \int_{C_{n-1}} x\,dy - y\,dx + \int_{C_n} x\,dy - y\,dx\right)$$

To evaluate these integrals we use the formula from (a) to get

$$A(D) = \tfrac{1}{2}[(x_1 y_2 - x_2 y_1) + (x_2 y_3 - x_3 y_2) + \cdots + (x_{n-1} y_n - x_n y_{n-1}) + (x_n y_1 - x_1 y_n)].$$

(c) $A = \frac{1}{2}[(0 \cdot 1 - 2 \cdot 0) + (2 \cdot 3 - 1 \cdot 1) + (1 \cdot 2 - 0 \cdot 3) + (0 \cdot 1 - (-1) \cdot 2) + (-1 \cdot 0 - 0 \cdot 1)]$

$\quad = \frac{1}{2}(0 + 5 + 2 + 2) = \frac{9}{2}$

23. We orient the quarter-circular region as shown in the figure.

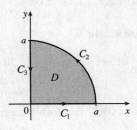

$A = \frac{1}{4}\pi a^2$ so $\overline{x} = \dfrac{1}{\pi a^2/2} \oint_C x^2 \, dy$ and $\overline{y} = -\dfrac{1}{\pi a^2/2} \oint_C y^2 \, dx$.

Here $C = C_1 + C_2 + C_3$ where $C_1\colon x = t, \ y = 0, \ 0 \le t \le a$;

$C_2\colon x = a\cos t, \ y = a\sin t, \ 0 \le t \le \frac{\pi}{2}$; and

$C_3\colon x = 0, y = a - t, 0 \le t \le a$. Then

$\oint_C x^2 \, dy = \int_{C_1} x^2 \, dy + \int_{C_2} x^2 \, dy + \int_{C_3} x^2 \, dy = \int_0^a 0 \, dt + \int_0^{\pi/2} (a\cos t)^2 (a\cos t) \, dt + \int_0^a 0 \, dt$

$\quad = \int_0^{\pi/2} a^3 \cos^3 t \, dt = a^3 \int_0^{\pi/2} (1 - \sin^2 t) \cos t \, dt = a^3 \left[\sin t - \frac{1}{3}\sin^3 t \right]_0^{\pi/2} = \frac{2}{3}a^3$

so $\overline{x} = \dfrac{1}{\pi a^2/2} \oint_C x^2 \, dy = \dfrac{4a}{3\pi}$.

$\oint_C y^2 \, dx = \int_{C_1} y^2 \, dx + \int_{C_2} y^2 \, dx + \int_{C_3} y^2 \, dx = \int_0^a 0 \, dt + \int_0^{\pi/2} (a\sin t)^2 (-a\sin t) \, dt + \int_0^a 0 \, dt$

$\quad = \int_0^{\pi/2} (-a^3 \sin^3 t) \, dt = -a^3 \int_0^{\pi/2} (1 - \cos^2 t) \sin t \, dt = -a^3 \left[\frac{1}{3}\cos^3 t - \cos t \right]_0^{\pi/2} = -\frac{2}{3}a^3$,

so $\overline{y} = -\dfrac{1}{\pi a^2/2} \oint_C y^2 \, dx = \dfrac{4a}{3\pi}$. Thus $(\overline{x}, \overline{y}) = \left(\dfrac{4a}{3\pi}, \dfrac{4a}{3\pi} \right)$.

25. By Green's Theorem, $-\frac{1}{3}\rho \oint_C y^3 \, dx = -\frac{1}{3}\rho \iint_D (-3y^2) \, dA = \iint_D y^2 \rho \, dA = I_x$ and

$\frac{1}{3}\rho \oint_C x^3 \, dy = \frac{1}{3}\rho \iint_D (3x^2) \, dA = \iint_D x^2 \rho \, dA = I_y$.

27. As in Example 5, let C' be a counterclockwise-oriented circle with center the origin and radius a, where a is chosen to

be small enough so that C' lies inside C, and D the region bounded by C and C'. Here

$P = \dfrac{2xy}{(x^2 + y^2)^2} \quad \Rightarrow \quad \dfrac{\partial P}{\partial y} = \dfrac{2x(x^2 + y^2)^2 - 2xy \cdot 2(x^2 + y^2) \cdot 2y}{(x^2 + y^2)^4} = \dfrac{2x^3 - 6xy^2}{(x^2 + y^2)^3}$ and

$Q = \dfrac{y^2 - x^2}{(x^2 + y^2)^2} \quad \Rightarrow \quad \dfrac{\partial Q}{\partial x} = \dfrac{-2x(x^2 + y^2)^2 - (y^2 - x^2) \cdot 2(x^2 + y^2) \cdot 2x}{(x^2 + y^2)^4} = \dfrac{2x^3 - 6xy^2}{(x^2 + y^2)^3}$. Thus, as in the example,

$$\int_C P \, dx + Q \, dy + \int_{-C'} P \, dx + Q \, dy = \iint_D \left(\dfrac{\partial Q}{\partial x} - \dfrac{\partial P}{\partial y} \right) dA = \iint_D 0 \, dA = 0$$

and $\int_C \mathbf{F} \cdot d\mathbf{r} = \int_{C'} \mathbf{F} \cdot d\mathbf{r}$. We parametrize C' as $\mathbf{r}(t) = a\cos t \, \mathbf{i} + a\sin t \, \mathbf{j}, 0 \le t \le 2\pi$. Then

$$\int_C \mathbf{F} \cdot d\mathbf{r} = \int_{C'} \mathbf{F} \cdot d\mathbf{r} = \int_0^{2\pi} \dfrac{2 \, (a\cos t) \, (a\sin t) \, \mathbf{i} + (a^2 \sin^2 t - a^2 \cos^2 t) \, \mathbf{j}}{(a^2 \cos^2 t + a^2 \sin^2 t)^2} \cdot \left(-a\sin t \, \mathbf{i} + a\cos t \, \mathbf{j} \right) dt$$

$$= \dfrac{1}{a} \int_0^{2\pi} (-\cos t \sin^2 t - \cos^3 t) \, dt = \dfrac{1}{a} \int_0^{2\pi} (-\cos t \sin^2 t - \cos t \, (1 - \sin^2 t)) \, dt$$

$$= -\dfrac{1}{a} \int_0^{2\pi} \cos t \, dt = -\dfrac{1}{a} \sin t \Big]_0^{2\pi} = 0$$

29. Since C is a simple closed path which doesn't pass through or enclose the origin, there exists an open region that doesn't

contain the origin but does contain D. Thus $P = -y/(x^2 + y^2)$ and $Q = x/(x^2 + y^2)$ have continuous partial derivatives on

this open region containing D and we can apply Green's Theorem. But by Exercise 16.3.35(a), $\partial P/\partial y = \partial Q/\partial x$, so

$\oint_C \mathbf{F} \cdot d\mathbf{r} = \iint_D 0 \, dA = 0$.

31. Using the first part of (5), we have that $\iint_R dx \, dy = A(R) = \int_{\partial R} x \, dy$. But $x = g(u, v)$, and $dy = \dfrac{\partial h}{\partial u} \, du + \dfrac{\partial h}{\partial v} \, dv$,

and we orient ∂S by taking the positive direction to be that which corresponds, under the mapping, to the positive direction

along ∂R, so

$$\int_{\partial R} x \, dy = \int_{\partial S} g(u, v) \left(\frac{\partial h}{\partial u} \, du + \frac{\partial h}{\partial v} \, dv \right) = \int_{\partial S} g(u, v) \frac{\partial h}{\partial u} \, du + g(u, v) \frac{\partial h}{\partial v} \, dv$$

$$= \pm \iint_S \left[\frac{\partial}{\partial u} \left(g(u, v) \frac{\partial h}{\partial v} \right) - \frac{\partial}{\partial v} \left(g(u, v) \frac{\partial h}{\partial u} \right) \right] dA \quad \text{[using Green's Theorem in the } uv\text{-plane]}$$

$$= \pm \iint_S \left(\frac{\partial g}{\partial u} \frac{\partial h}{\partial v} + g(u, v) \frac{\partial^2 h}{\partial u \, \partial v} - \frac{\partial g}{\partial v} \frac{\partial h}{\partial u} - g(u, v) \frac{\partial^2 h}{\partial v \, \partial u} \right) dA \quad \text{[using the Chain Rule]}$$

$$= \pm \iint_S \left(\frac{\partial x}{\partial u} \frac{\partial y}{\partial v} - \frac{\partial x}{\partial v} \frac{\partial y}{\partial u} \right) dA \quad \text{[by the equality of mixed partials]} \quad = \pm \iint_S \frac{\partial(x, y)}{\partial(u, v)} \, du \, dv$$

The sign is chosen to be positive if the orientation that we gave to ∂S corresponds to the usual positive orientation, and it is

negative otherwise. In either case, since $A(R)$ is positive, the sign chosen must be the same as the sign of $\dfrac{\partial(x, y)}{\partial(u, v)}$.

Therefore $A(R) = \displaystyle\iint_R dx \, dy = \iint_S \left| \frac{\partial(x, y)}{\partial(u, v)} \right| du \, dv$.

16.5   Curl and Divergence

1. (a) $\operatorname{curl} \mathbf{F} = \nabla \times \mathbf{F} = \begin{vmatrix} \mathbf{i} & \mathbf{j} & \mathbf{k} \\ \partial/\partial x & \partial/\partial y & \partial/\partial z \\ xy^2 z^2 & x^2 yz^2 & x^2 y^2 z \end{vmatrix}$

$$= \left[\frac{\partial}{\partial y} (x^2 y^2 z) - \frac{\partial}{\partial z} (x^2 yz^2) \right] \mathbf{i} - \left[\frac{\partial}{\partial x} (x^2 y^2 z) - \frac{\partial}{\partial z} (xy^2 z^2) \right] \mathbf{j} + \left[\frac{\partial}{\partial x} (x^2 yz^2) - \frac{\partial}{\partial y} (xy^2 z^2) \right] \mathbf{k}$$

$$= (2x^2 yz - 2x^2 yz) \mathbf{i} - (2xy^2 z - 2xy^2 z) \mathbf{j} + (2xyz^2 - 2xyz^2) \mathbf{k} = \mathbf{0}$$

(b) $\operatorname{div} \mathbf{F} = \nabla \cdot \mathbf{F} = \dfrac{\partial}{\partial x} (xy^2 z^2) + \dfrac{\partial}{\partial y} (x^2 yz^2) + \dfrac{\partial}{\partial z} (x^2 y^2 z) = y^2 z^2 + x^2 z^2 + x^2 y^2$

3. (a) $\operatorname{curl} \mathbf{F} = \nabla \times \mathbf{F} = \begin{vmatrix} \mathbf{i} & \mathbf{j} & \mathbf{k} \\ \partial/\partial x & \partial/\partial y & \partial/\partial z \\ xye^z & 0 & yze^x \end{vmatrix} = (ze^x - 0) \mathbf{i} - (yze^x - xye^z) \mathbf{j} + (0 - xe^z) \mathbf{k}$

$$= ze^x \mathbf{i} + (xye^z - yze^x) \mathbf{j} - xe^z \mathbf{k}$$

(b) $\operatorname{div} \mathbf{F} = \nabla \cdot \mathbf{F} = \dfrac{\partial}{\partial x} (xye^z) + \dfrac{\partial}{\partial y} (0) + \dfrac{\partial}{\partial z} (yze^x) = ye^z + 0 + ye^x = y(e^z + e^x)$

5. (a) $\text{curl } \mathbf{F} = \nabla \times \mathbf{F} = \begin{vmatrix} \mathbf{i} & \mathbf{j} & \mathbf{k} \\ \partial/\partial x & \partial/\partial y & \partial/\partial z \\ \dfrac{\sqrt{x}}{1+z} & \dfrac{\sqrt{y}}{1+x} & \dfrac{\sqrt{z}}{1+y} \end{vmatrix}$

$\qquad = \left[\sqrt{z}\,(-1)(1+y)^{-2} - 0\right]\mathbf{i} - \left[0 - \sqrt{x}(-1)(1+z)^{-2}\right]\mathbf{j} + \left[\sqrt{y}\,(-1)(1+x)^{-2} - 0\right]\mathbf{k}$

$\qquad = -\dfrac{\sqrt{z}}{(1+y)^2}\,\mathbf{i} - \dfrac{\sqrt{x}}{(1+z)^2}\,\mathbf{j} - \dfrac{\sqrt{y}}{(1+x)^2}\,\mathbf{k}$

(b) $\text{div } \mathbf{F} = \nabla \cdot \mathbf{F} = \dfrac{\partial}{\partial x}\left(\dfrac{\sqrt{x}}{1+z}\right) + \dfrac{\partial}{\partial y}\left(\dfrac{\sqrt{y}}{1+x}\right) + \dfrac{\partial}{\partial z}\left(\dfrac{\sqrt{z}}{1+y}\right)$

$\qquad = \dfrac{1}{2\sqrt{x}\,(1+z)} + \dfrac{1}{2\sqrt{y}\,(1+x)} + \dfrac{1}{2\sqrt{z}\,(1+y)}$

7. (a) $\text{curl } \mathbf{F} = \nabla \times \mathbf{F} = \begin{vmatrix} \mathbf{i} & \mathbf{j} & \mathbf{k} \\ \partial/\partial x & \partial/\partial y & \partial/\partial z \\ e^x \sin y & e^y \sin z & e^z \sin x \end{vmatrix} = (0 - e^y \cos z)\,\mathbf{i} - (e^z \cos x - 0)\,\mathbf{j} + (0 - e^x \cos y)\,\mathbf{k}$

$\qquad = \langle -e^y \cos z, -e^z \cos x, -e^x \cos y \rangle$

(b) $\text{div } \mathbf{F} = \nabla \cdot \mathbf{F} = \dfrac{\partial}{\partial x}\left(e^x \sin y\right) + \dfrac{\partial}{\partial y}\left(e^y \sin z\right) + \dfrac{\partial}{\partial z}\left(e^z \sin x\right) = e^x \sin y + e^y \sin z + e^z \sin x$

9. If the vector field is $\mathbf{F} = P\,\mathbf{i} + Q\,\mathbf{j} + R\,\mathbf{k}$, then we know $R = 0$. In addition, the x-component of each vector of $\mathbf{F}$ is 0, so

$P = 0$, hence $\dfrac{\partial P}{\partial x} = \dfrac{\partial P}{\partial y} = \dfrac{\partial P}{\partial z} = \dfrac{\partial R}{\partial x} = \dfrac{\partial R}{\partial y} = \dfrac{\partial R}{\partial z} = 0$. Q decreases as y increases, so $\dfrac{\partial Q}{\partial y} < 0$, but Q doesn't change

in the x- or z-directions, so $\dfrac{\partial Q}{\partial x} = \dfrac{\partial Q}{\partial z} = 0$.

(a) $\text{div } \mathbf{F} = \dfrac{\partial P}{\partial x} + \dfrac{\partial Q}{\partial y} + \dfrac{\partial R}{\partial z} = 0 + \dfrac{\partial Q}{\partial y} + 0 < 0$

(b) $\text{curl } \mathbf{F} = \left(\dfrac{\partial R}{\partial y} - \dfrac{\partial Q}{\partial z}\right)\mathbf{i} + \left(\dfrac{\partial P}{\partial z} - \dfrac{\partial R}{\partial x}\right)\mathbf{j} + \left(\dfrac{\partial Q}{\partial x} - \dfrac{\partial P}{\partial y}\right)\mathbf{k} = (0 - 0)\,\mathbf{i} + (0 - 0)\,\mathbf{j} + (0 - 0)\,\mathbf{k} = \mathbf{0}$

11. If the vector field is $\mathbf{F} = P\,\mathbf{i} + Q\,\mathbf{j} + R\,\mathbf{k}$, then we know $R = 0$. In addition, the y-component of each vector of $\mathbf{F}$ is 0, so

$Q = 0$, hence $\dfrac{\partial Q}{\partial x} = \dfrac{\partial Q}{\partial y} = \dfrac{\partial Q}{\partial z} = \dfrac{\partial R}{\partial x} = \dfrac{\partial R}{\partial y} = \dfrac{\partial R}{\partial z} = 0$. P increases as y increases, so $\dfrac{\partial P}{\partial y} > 0$, but P doesn't change in

the x- or z-directions, so $\dfrac{\partial P}{\partial x} = \dfrac{\partial P}{\partial z} = 0$.

(a) $\text{div } \mathbf{F} = \dfrac{\partial P}{\partial x} + \dfrac{\partial Q}{\partial y} + \dfrac{\partial R}{\partial z} = 0 + 0 + 0 = 0$

(b) $\text{curl } \mathbf{F} = \left(\dfrac{\partial R}{\partial y} - \dfrac{\partial Q}{\partial z}\right)\mathbf{i} + \left(\dfrac{\partial P}{\partial z} - \dfrac{\partial R}{\partial x}\right)\mathbf{j} + \left(\dfrac{\partial Q}{\partial x} - \dfrac{\partial P}{\partial y}\right)\mathbf{k} = (0 - 0)\,\mathbf{i} + (0 - 0)\,\mathbf{j} + \left(0 - \dfrac{\partial P}{\partial y}\right)\mathbf{k} = -\dfrac{\partial P}{\partial y}\,\mathbf{k}$

Since $\dfrac{\partial P}{\partial y} > 0$, $-\dfrac{\partial P}{\partial y}\,\mathbf{k}$ is a vector pointing in the negative z-direction.

13. $\operatorname{curl}\mathbf{F} = \nabla \times \mathbf{F} = \begin{vmatrix} \mathbf{i} & \mathbf{j} & \mathbf{k} \\ \partial/\partial x & \partial/\partial y & \partial/\partial z \\ y^2 z^3 & 2xyz^3 & 3xy^2 z^2 \end{vmatrix} = (6xyz^2 - 6xyz^2)\,\mathbf{i} - (3y^2 z^2 - 3y^2 z^2)\,\mathbf{j} + (2yz^3 - 2yz^3)\,\mathbf{k} = \mathbf{0}$

and $\mathbf{F}$ is defined on all of $\mathbb{R}^3$ with component functions which have continuous partial derivatives, so by Theorem 4,

$\mathbf{F}$ is conservative. Thus, there exists a function f such that $\mathbf{F} = \nabla f$. Then $f_x(x, y, z) = y^2 z^3$ implies

$f(x, y, z) = xy^2 z^3 + g(y, z)$ and $f_y(x, y, z) = 2xyz^3 + g_y(y, z)$. But $f_y(x, y, z) = 2xyz^3$, so $g(y, z) = h(z)$ and

$f(x, y, z) = xy^2 z^3 + h(z)$. Thus $f_z(x, y, z) = 3xy^2 z^2 + h'(z)$ but $f_z(x, y, z) = 3xy^2 z^2$ so $h(z) = K$, a constant.

Hence a potential function for $\mathbf{F}$ is $f(x, y, z) = xy^2 z^3 + K$.

15. $\operatorname{curl}\mathbf{F} = \nabla \times \mathbf{F} = \begin{vmatrix} \mathbf{i} & \mathbf{j} & \mathbf{k} \\ \partial/\partial x & \partial/\partial y & \partial/\partial z \\ z\cos y & xz\sin y & x\cos y \end{vmatrix}$

$= (-x\sin y - x\sin y)\,\mathbf{i} - (\cos y - \cos y)\,\mathbf{j} + [z\sin y - (-z\sin y)]\,\mathbf{k} = -2x\sin y\,\mathbf{i} + 2z\sin y\,\mathbf{k} \neq \mathbf{0},$

so $\mathbf{F}$ is not conservative.

17. $\operatorname{curl}\mathbf{F} = \nabla \times \mathbf{F} = \begin{vmatrix} \mathbf{i} & \mathbf{j} & \mathbf{k} \\ \partial/\partial x & \partial/\partial y & \partial/\partial z \\ e^{yz} & xze^{yz} & xye^{yz} \end{vmatrix}$

$= [xyze^{yz} + xe^{yz} - (xyze^{yz} + xe^{yz})]\,\mathbf{i} - (ye^{yz} - ye^{yz})\,\mathbf{j} + (ze^{yz} - ze^{yz})\,\mathbf{k} = \mathbf{0}$

$\mathbf{F}$ is defined on all of $\mathbb{R}^3$, and the partial derivatives of the component functions are continuous, so $\mathbf{F}$ is conservative. Thus

there exists a function f such that $\nabla f = \mathbf{F}$. Then $f_x(x, y, z) = e^{yz}$ implies $f(x, y, z) = xe^{yz} + g(y, z) \;\Rightarrow$

$f_y(x, y, z) = xze^{yz} + g_y(y, z)$. But $f_y(x, y, z) = xze^{yz}$, so $g(y, z) = h(z)$ and $f(x, y, z) = xe^{yz} + h(z)$.

Thus $f_z(x, y, z) = xye^{yz} + h'(z)$ but $f_z(x, y, z) = xye^{yz}$ so $h(z) = K$ and a potential function for $\mathbf{F}$ is

$f(x, y, z) = xe^{yz} + K$.

19. No. Assume there is such a $\mathbf{G}$. Then $\operatorname{div}(\operatorname{curl}\mathbf{G}) = \dfrac{\partial}{\partial x}(x\sin y) + \dfrac{\partial}{\partial y}(\cos y) + \dfrac{\partial}{\partial z}(z - xy) = \sin y - \sin y + 1 \neq 0,$

which contradicts Theorem 11.

21. $\operatorname{curl}\mathbf{F} = \begin{vmatrix} \mathbf{i} & \mathbf{j} & \mathbf{k} \\ \partial/\partial x & \partial/\partial y & \partial/\partial z \\ f(x) & g(y) & h(z) \end{vmatrix} = (0 - 0)\,\mathbf{i} + (0 - 0)\,\mathbf{j} + (0 - 0)\,\mathbf{k} = \mathbf{0}.$ Hence $\mathbf{F} = f(x)\,\mathbf{i} + g(y)\,\mathbf{j} + h(z)\,\mathbf{k}$

is irrotational.

For Exercises 23–29, let $\mathbf{F}(x, y, z) = P_1\,\mathbf{i} + Q_1\,\mathbf{j} + R_1\,\mathbf{k}$ and $\mathbf{G}(x, y, z) = P_2\,\mathbf{i} + Q_2\,\mathbf{j} + R_2\,\mathbf{k}$.

23. $\operatorname{div}(\mathbf{F} + \mathbf{G}) = \operatorname{div}\langle P_1 + P_2, Q_1 + Q_2, R_1 + R_2 \rangle = \dfrac{\partial(P_1 + P_2)}{\partial x} + \dfrac{\partial(Q_1 + Q_2)}{\partial y} + \dfrac{\partial(R_1 + R_2)}{\partial z}$

$$= \frac{\partial P_1}{\partial x} + \frac{\partial P_2}{\partial x} + \frac{\partial Q_1}{\partial y} + \frac{\partial Q_2}{\partial y} + \frac{\partial R_1}{\partial z} + \frac{\partial R_2}{\partial z} = \left(\frac{\partial P_1}{\partial x} + \frac{\partial Q_1}{\partial y} + \frac{\partial R_1}{\partial z} \right) + \left(\frac{\partial P_2}{\partial x} + \frac{\partial Q_2}{\partial y} + \frac{\partial R_2}{\partial z} \right)$$

$$= \operatorname{div}\langle P_1, Q_1, R_1 \rangle + \operatorname{div}\langle P_2, Q_2, R_2 \rangle = \operatorname{div}\mathbf{F} + \operatorname{div}\mathbf{G}$$

25. $\operatorname{div}(f\mathbf{F}) = \operatorname{div}(f\langle P_1, Q_1, R_1 \rangle) = \operatorname{div}\langle fP_1, fQ_1, fR_1 \rangle = \dfrac{\partial(fP_1)}{\partial x} + \dfrac{\partial(fQ_1)}{\partial y} + \dfrac{\partial(fR_1)}{\partial z}$

$$= \left(f \frac{\partial P_1}{\partial x} + P_1 \frac{\partial f}{\partial x} \right) + \left(f \frac{\partial Q_1}{\partial y} + Q_1 \frac{\partial f}{\partial y} \right) + \left(f \frac{\partial R_1}{\partial z} + R_1 \frac{\partial f}{\partial z} \right)$$

$$= f\left(\frac{\partial P_1}{\partial x} + \frac{\partial Q_1}{\partial y} + \frac{\partial R_1}{\partial z} \right) + \langle P_1, Q_1, R_1 \rangle \cdot \left\langle \frac{\partial f}{\partial x}, \frac{\partial f}{\partial y}, \frac{\partial f}{\partial z} \right\rangle = f \operatorname{div}\mathbf{F} + \mathbf{F} \cdot \nabla f$$

27. $\operatorname{div}(\mathbf{F} \times \mathbf{G}) = \nabla \cdot (\mathbf{F} \times \mathbf{G}) = \begin{vmatrix} \partial/\partial x & \partial/\partial y & \partial/\partial z \\ P_1 & Q_1 & R_1 \\ P_2 & Q_2 & R_2 \end{vmatrix} = \dfrac{\partial}{\partial x}\begin{vmatrix} Q_1 & R_1 \\ Q_2 & R_2 \end{vmatrix} - \dfrac{\partial}{\partial y}\begin{vmatrix} P_1 & R_1 \\ P_2 & R_2 \end{vmatrix} + \dfrac{\partial}{\partial z}\begin{vmatrix} P_1 & Q_1 \\ P_2 & Q_2 \end{vmatrix}$

$$= \left[Q_1 \frac{\partial R_2}{\partial x} + R_2 \frac{\partial Q_1}{\partial x} - Q_2 \frac{\partial R_1}{\partial x} - R_1 \frac{\partial Q_2}{\partial x} \right] - \left[P_1 \frac{\partial R_2}{\partial y} + R_2 \frac{\partial P_1}{\partial y} - P_2 \frac{\partial R_1}{\partial y} - R_1 \frac{\partial P_2}{\partial y} \right]$$

$$+ \left[P_1 \frac{\partial Q_2}{\partial z} + Q_2 \frac{\partial P_1}{\partial z} - P_2 \frac{\partial Q_1}{\partial z} - Q_1 \frac{\partial P_2}{\partial z} \right]$$

$$= \left[P_2\left(\frac{\partial R_1}{\partial y} - \frac{\partial Q_1}{\partial z} \right) + Q_2\left(\frac{\partial P_1}{\partial z} - \frac{\partial R_1}{\partial x} \right) + R_2\left(\frac{\partial Q_1}{\partial x} - \frac{\partial P_1}{\partial y} \right) \right]$$

$$- \left[P_1\left(\frac{\partial R_2}{\partial y} - \frac{\partial Q_2}{\partial z} \right) + Q_1\left(\frac{\partial P_2}{\partial z} - \frac{\partial R_2}{\partial x} \right) + R_1\left(\frac{\partial Q_2}{\partial x} - \frac{\partial P_2}{\partial y} \right) \right]$$

$$= \mathbf{G} \cdot \operatorname{curl}\mathbf{F} - \mathbf{F} \cdot \operatorname{curl}\mathbf{G}$$

29. $\operatorname{curl}(\operatorname{curl}\mathbf{F}) = \nabla \times (\nabla \times \mathbf{F}) = \begin{vmatrix} \mathbf{i} & \mathbf{j} & \mathbf{k} \\ \partial/\partial x & \partial/\partial y & \partial/\partial z \\ \partial R_1/\partial y - \partial Q_1/\partial z & \partial P_1/\partial z - \partial R_1/\partial x & \partial Q_1/\partial x - \partial P_1/\partial y \end{vmatrix}$

$$= \left(\frac{\partial^2 Q_1}{\partial y\partial x} - \frac{\partial^2 P_1}{\partial y^2} - \frac{\partial^2 P_1}{\partial z^2} + \frac{\partial^2 R_1}{\partial z\partial x} \right)\mathbf{i} + \left(\frac{\partial^2 R_1}{\partial z\partial y} - \frac{\partial^2 Q_1}{\partial z^2} - \frac{\partial^2 Q_1}{\partial x^2} + \frac{\partial^2 P_1}{\partial x\partial y} \right)\mathbf{j}$$

$$+ \left(\frac{\partial^2 P_1}{\partial x\partial z} - \frac{\partial^2 R_1}{\partial x^2} - \frac{\partial^2 R_1}{\partial y^2} + \frac{\partial^2 Q_1}{\partial y\partial z} \right)\mathbf{k}$$

Now let's consider $\operatorname{grad}(\operatorname{div}\mathbf{F}) - \nabla^2\mathbf{F}$ and compare with the above.

(Note that $\nabla^2\mathbf{F}$ is defined on page 1147 [ET 1107].)

$$\text{grad}(\text{div } \mathbf{F}) - \nabla^2 \mathbf{F} = \left[\left(\frac{\partial^2 P_1}{\partial x^2} + \frac{\partial^2 Q_1}{\partial x \partial y} + \frac{\partial^2 R_1}{\partial x \partial z} \right) \mathbf{i} + \left(\frac{\partial^2 P_1}{\partial y \partial x} + \frac{\partial^2 Q_1}{\partial y^2} + \frac{\partial^2 R_1}{\partial y \partial z} \right) \mathbf{j} + \left(\frac{\partial^2 P_1}{\partial z \partial x} + \frac{\partial^2 Q_1}{\partial z \partial y} + \frac{\partial^2 R_1}{\partial z^2} \right) \mathbf{k} \right]$$

$$- \left[\left(\frac{\partial^2 P_1}{\partial x^2} + \frac{\partial^2 P_1}{\partial y^2} + \frac{\partial^2 P_1}{\partial z^2} \right) \mathbf{i} + \left(\frac{\partial^2 Q_1}{\partial x^2} + \frac{\partial^2 Q_1}{\partial y^2} + \frac{\partial^2 Q_1}{\partial z^2} \right) \mathbf{j} \right.$$

$$\left. + \left(\frac{\partial^2 R_1}{\partial x^2} + \frac{\partial^2 R_1}{\partial y^2} + \frac{\partial^2 R_1}{\partial z^2} \right) \mathbf{k} \right]$$

$$= \left(\frac{\partial^2 Q_1}{\partial x \partial y} + \frac{\partial^2 R_1}{\partial x \partial z} - \frac{\partial^2 P_1}{\partial y^2} - \frac{\partial^2 P_1}{\partial z^2} \right) \mathbf{i} + \left(\frac{\partial^2 P_1}{\partial y \partial x} + \frac{\partial^2 R_1}{\partial y \partial z} - \frac{\partial^2 Q_1}{\partial x^2} - \frac{\partial^2 Q_1}{\partial z^2} \right) \mathbf{j}$$

$$+ \left(\frac{\partial^2 P_1}{\partial z \partial x} + \frac{\partial^2 Q_1}{\partial z \partial y} - \frac{\partial^2 R_1}{\partial x^2} - \frac{\partial^2 R_2}{\partial y^2} \right) \mathbf{k}$$

Then applying Clairaut's Theorem to reverse the order of differentiation in the second partial derivatives as needed and comparing, we have curl curl $\mathbf{F} = \text{grad div } \mathbf{F} - \nabla^2 \mathbf{F}$ as desired.

31. (a) $\nabla r = \nabla \sqrt{x^2 + y^2 + z^2} = \dfrac{x}{\sqrt{x^2 + y^2 + z^2}} \mathbf{i} + \dfrac{y}{\sqrt{x^2 + y^2 + z^2}} \mathbf{j} + \dfrac{z}{\sqrt{x^2 + y^2 + z^2}} \mathbf{k} = \dfrac{x\mathbf{i} + y\mathbf{j} + z\mathbf{k}}{\sqrt{x^2 + y^2 + z^2}} = \dfrac{\mathbf{r}}{r}$

(b) $\nabla \times \mathbf{r} = \begin{vmatrix} \mathbf{i} & \mathbf{j} & \mathbf{k} \\ \frac{\partial}{\partial x} & \frac{\partial}{\partial y} & \frac{\partial}{\partial z} \\ x & y & z \end{vmatrix} = \left[\frac{\partial}{\partial y}(z) - \frac{\partial}{\partial z}(y) \right] \mathbf{i} + \left[\frac{\partial}{\partial z}(x) - \frac{\partial}{\partial x}(z) \right] \mathbf{j} + \left[\frac{\partial}{\partial x}(y) - \frac{\partial}{\partial y}(x) \right] \mathbf{k} = \mathbf{0}$

(c) $\nabla \left(\dfrac{1}{r} \right) = \nabla \left(\dfrac{1}{\sqrt{x^2 + y^2 + z^2}} \right)$

$$= -\frac{\frac{1}{2\sqrt{x^2 + y^2 + z^2}}(2x)}{x^2 + y^2 + z^2} \mathbf{i} - \frac{\frac{1}{2\sqrt{x^2 + y^2 + z^2}}(2y)}{x^2 + y^2 + z^2} \mathbf{j} - \frac{\frac{1}{2\sqrt{x^2 + y^2 + z^2}}(2z)}{x^2 + y^2 + z^2} \mathbf{k}$$

$$= -\frac{x\mathbf{i} + y\mathbf{j} + z\mathbf{k}}{(x^2 + y^2 + z^2)^{3/2}} = -\frac{\mathbf{r}}{r^3}$$

(d) $\nabla \ln r = \nabla \ln(x^2 + y^2 + z^2)^{1/2} = \frac{1}{2} \nabla \ln(x^2 + y^2 + z^2)$

$$= \frac{x}{x^2 + y^2 + z^2} \mathbf{i} + \frac{y}{x^2 + y^2 + z^2} \mathbf{j} + \frac{z}{x^2 + y^2 + z^2} \mathbf{k} = \frac{x\mathbf{i} + y\mathbf{j} + z\mathbf{k}}{x^2 + y^2 + z^2} = \frac{\mathbf{r}}{r^2}$$

33. By (13), $\oint_C f(\nabla g) \cdot \mathbf{n} \, ds = \iint_D \text{div}(f\nabla g) \, dA = \iint_D [f \, \text{div}(\nabla g) + \nabla g \cdot \nabla f] \, dA$ by Exercise 25. But $\text{div}(\nabla g) = \nabla^2 g$.

Hence $\iint_D f\nabla^2 g \, dA = \oint_C f(\nabla g) \cdot \mathbf{n} \, ds - \iint_D \nabla g \cdot \nabla f \, dA$.

35. Let $f(x, y) = 1$. Then $\nabla f = \mathbf{0}$ and Green's first identity (see Exercise 33) says

$$\iint_D \nabla^2 g \, dA = \oint_C (\nabla g) \cdot \mathbf{n} \, ds - \iint_D \mathbf{0} \cdot \nabla g \, dA \quad \Rightarrow \quad \iint_D \nabla^2 g \, dA = \oint_C \nabla g \cdot \mathbf{n} \, ds. \text{ But } g \text{ is harmonic on } D, \text{ so}$$

$\nabla^2 g = 0 \quad \Rightarrow \quad \oint_C \nabla g \cdot \mathbf{n} \, ds = 0$ and $\oint_C D_{\mathbf{n}} g \, ds = \oint_C (\nabla g \cdot \mathbf{n}) \, ds = 0$.

37. (a) We know that $\omega = v/d$, and from the diagram $\sin\theta = d/r \Rightarrow v = d\omega = (\sin\theta)r\omega = |\mathbf{w}\times\mathbf{r}|$. But $\mathbf{v}$ is perpendicular

to both $\mathbf{w}$ and $\mathbf{r}$, so that $\mathbf{v} = \mathbf{w}\times\mathbf{r}$.

(b) From (a), $\mathbf{v} = \mathbf{w}\times\mathbf{r} = \begin{vmatrix} \mathbf{i} & \mathbf{j} & \mathbf{k} \\ 0 & 0 & \omega \\ x & y & z \end{vmatrix} = (0\cdot z - \omega y)\,\mathbf{i} + (\omega x - 0\cdot z)\,\mathbf{j} + (0\cdot y - x\cdot 0)\,\mathbf{k} = -\omega y\,\mathbf{i} + \omega x\,\mathbf{j}$

(c) curl $\mathbf{v} = \nabla\times\mathbf{v} = \begin{vmatrix} \mathbf{i} & \mathbf{j} & \mathbf{k} \\ \partial/\partial x & \partial/\partial y & \partial/\partial z \\ -\omega y & \omega x & 0 \end{vmatrix}$

$= \left[\dfrac{\partial}{\partial y}(0) - \dfrac{\partial}{\partial z}(\omega x)\right]\mathbf{i} + \left[\dfrac{\partial}{\partial z}(-\omega y) - \dfrac{\partial}{\partial x}(0)\right]\mathbf{j} + \left[\dfrac{\partial}{\partial x}(\omega x) - \dfrac{\partial}{\partial y}(-\omega y)\right]\mathbf{k}$

$= [\omega - (-\omega)]\,\mathbf{k} = 2\omega\,\mathbf{k} = 2\mathbf{w}$

39. For any continuous function f on $\mathbb{R}^3$, define a vector field $\mathbf{G}(x,y,z) = \langle g(x,y,z), 0, 0\rangle$ where $g(x,y,z) = \int_0^x f(t,y,z)\,dt$.

Then div $\mathbf{G} = \dfrac{\partial}{\partial x}(g(x,y,z)) + \dfrac{\partial}{\partial y}(0) + \dfrac{\partial}{\partial z}(0) = \dfrac{\partial}{\partial x}\int_0^x f(t,y,z)\,dt = f(x,y,z)$ by the Fundamental Theorem of

Calculus. Thus every continuous function f on $\mathbb{R}^3$ is the divergence of some vector field.

16.6 Parametric Surfaces and Their Areas

1. $P(4,-5,1)$ lies on the parametric surface $\mathbf{r}(u,v) = \langle u+v, u-2v, 3+u-v\rangle$ if and only if there are values for u and v

where $u+v = 4$, $u-2v = -5$, and $3+u-v = 1$. From the first equation we have $u = 4-v$ and substituting into the

second equation gives $4-v-2v = -5 \Leftrightarrow v = 3$. Then $u = 1$, and these values satisfy the third equation, so P does lie

on the surface.

$Q(0,4,6)$ lies on $\mathbf{r}(u,v)$ if and only if $u+v = 0$, $u-2v = 4$, and $3+u-v = 6$, but solving the first two equations

simultaneoulsy gives $u = \frac{4}{3}$, $v = -\frac{4}{3}$ and these values do not satisfy the third equation, so Q does not lie on the surface.

3. $\mathbf{r}(u,v) = (u+v)\,\mathbf{i} + (3-v)\,\mathbf{j} + (1+4u+5v)\,\mathbf{k} = \langle 0,3,1\rangle + u\,\langle 1,0,4\rangle + v\,\langle 1,-1,5\rangle$. From Example 3, we recognize

this as a vector equation of a plane through the point $(0,3,1)$ and containing vectors $\mathbf{a} = \langle 1,0,4\rangle$ and $\mathbf{b} = \langle 1,-1,5\rangle$. If we

wish to find a more conventional equation for the plane, a normal vector to the plane is $\mathbf{a}\times\mathbf{b} = \begin{vmatrix} \mathbf{i} & \mathbf{j} & \mathbf{k} \\ 1 & 0 & 4 \\ 1 & -1 & 5 \end{vmatrix} = 4\mathbf{i} - \mathbf{j} - \mathbf{k}$

and an equation of the plane is $4(x-0) - (y-3) - (z-1) = 0$ or $4x - y - z = -4$.

5. $\mathbf{r}(s,t) = \langle s\cos t, s\sin t, s\rangle$, so the corresponding parametric equations for the surface are $x = s\cos t$, $y = s\sin t$, $z = s$.

For any point (x,y,z) on the surface, we have $x^2 + y^2 = s^2\cos^2 t + s^2\sin^2 t = s^2 = z^2$. Since no restrictions are placed on

the parameters, the surface is $z^2 = x^2 + y^2$, which we recognize as a circular cone with axis the z-axis.

7. $\mathbf{r}(u,v) = \langle u^2, v^2, u+v \rangle, \ -1 \leq u \leq 1, \ -1 \leq v \leq 1.$

The surface has parametric equations $x = u^2, \ y = v^2, \ z = u+v, \ -1 \leq u \leq 1, -1 \leq v \leq 1$.

In Maple, the surface can be graphed by entering

```
plot3d([u^2,v^2,u+v],u=-1..1,v=-1..1);
```

In Mathematica we use the `ParametricPlot3D` command.
If we keep u constant at u_0, $x = u_0^2$, a constant, so the
corresponding grid curves must be the curves parallel to the
yz-plane. If v is constant, we have $y = v_0^2$, a constant, so these
grid curves are the curves parallel to the xz-plane.

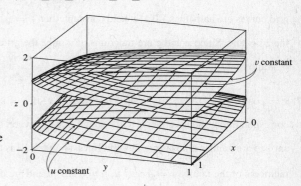

9. $\mathbf{r}(u,v) = \langle u^3, u \sin v, u \cos v \rangle, \ -1 \leq u \leq 1, \ 0 \leq v \leq 2\pi$

The surface has parametric equations $x = u^3, \ y = u \sin v,$
$z = u \cos v, \ -1 \leq u \leq 1, \ 0 \leq v \leq 2\pi$. Note that if $u = u_0$ is
constant then $x = u_0^3$ is constant and $y = u_0 \sin v, \ z = u_0 \cos v$
describe a circle in y, z of radius $|u_0|$, so the corresponding grid
curves are circles parallel to the yz-plane. If $v = v_0$, a constant,

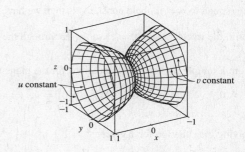

the parametric equations become $x = u^3, \ y = u \sin v_0, \ z = u \cos v_0$. Then $y = (\tan v_0)z$, so these are the grid curves we see
that lie in planes $y = kz$ that pass through the x-axis.

11. $x = \sin v, \ y = \cos u \sin 4v, \ z = \sin 2u \sin 4v, \ 0 \leq u \leq 2\pi, \ -\frac{\pi}{2} \leq v \leq \frac{\pi}{2}.$

Note that if $v = v_0$ is constant, then $x = \sin v_0$ is constant, so the
corresponding grid curves must be parallel to the yz-plane. These
are the vertically oriented grid curves we see, each shaped like a
"figure-eight." When $u = u_0$ is held constant, the parametric
equations become $x = \sin v, \ y = \cos u_0 \sin 4v,$
$z = \sin 2u_0 \sin 4v$. Since z is a constant multiple of y, the
corresponding grid curves are the curves contained in planes
$z = ky$ that pass through the x-axis.

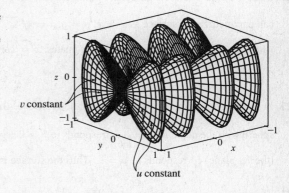

13. $\mathbf{r}(u,v) = u \cos v \, \mathbf{i} + u \sin v \, \mathbf{j} + v \, \mathbf{k}$. The parametric equations for the surface are $x = u \cos v, \ y = u \sin v, \ z = v$. We look at
the grid curves first; if we fix v, then x and y parametrize a straight line in the plane $z = v$ which intersects the z-axis. If u is
held constant, the projection onto the xy-plane is circular; with $z = v$, each grid curve is a helix. The surface is a spiraling
ramp, graph IV.

15. $\mathbf{r}(u, v) = (u^3 - u)\mathbf{i} + v^2\mathbf{j} + u^2\mathbf{k}$. The parametric equations for the surface are $x = u^3 - u$, $y = v^2$, $z = u^2$. If we fix u

then x and z are constant so each corresponding grid curve is contained in a line parallel to the y-axis. (Since $y = v^2 \geq 0$, the

grid curves are half-lines.) If v is held constant, then $y = v^2 = $ constant, so each grid curve is contained in a plane parallel to

the xz-plane. Since x and z are functions of u only, the grid curves all have the same shape. The surface is the cylinder shown

in graph I.

17. $x = \cos^3 u \cos^3 v$, $y = \sin^3 u \cos^3 v$, $z = \sin^3 v$. If $v = v_0$ is held constant then $z = \sin^3 v_0$ is constant, so the

corresponding grid curve lies in a horizontal plane. Several of the graphs exhibit horizontal grid curves, but the curves for this

surface are neither ellipses nor straight lines, so graph III is the only possibility. (In fact, the horizontal grid curves here are

members of the family $x = a \cos^3 u$, $y = a \sin^3 u$ and are called astroids.) The vertical grid curves we see on the surface

correspond to $u = u_0$ held constant, as then we have $x = \cos^3 u_0 \cos^3 v$, $y = \sin^3 u_0 \cos^3 v$ so the corresponding grid curve

lies in the vertical plane $y = (\tan^3 u_0)x$ through the z-axis.

19. From Example 3, parametric equations for the plane through the point $(0, 0, 0)$ that contains the vectors $\mathbf{a} = \langle 1, -1, 0 \rangle$ and

$\mathbf{b} = \langle 0, 1, -1 \rangle$ are $x = 0 + u(1) + v(0) = u$, $y = 0 + u(-1) + v(1) = v - u$, $z = 0 + u(0) + v(-1) = -v$.

21. Solving the equation for x gives $x^2 = 1 + y^2 + \frac{1}{4}z^2 \Rightarrow x = \sqrt{1 + y^2 + \frac{1}{4}z^2}$. (We choose the positive root since we want

the part of the hyperboloid that corresponds to $x \geq 0$.) If we let y and z be the parameters, parametric equations are $y = y$,

$z = z$, $x = \sqrt{1 + y^2 + \frac{1}{4}z^2}$.

23. Since the cone intersects the sphere in the circle $x^2 + y^2 = 2$, $z = \sqrt{2}$ and we want the portion of the sphere above this, we

can parametrize the surface as $x = x$, $y = y$, $z = \sqrt{4 - x^2 - y^2}$ where $x^2 + y^2 \leq 2$.

Alternate solution: Using spherical coordinates, $x = 2\sin\phi\cos\theta$, $y = 2\sin\phi\sin\theta$, $z = 2\cos\phi$ where $0 \leq \phi \leq \frac{\pi}{4}$ and

$0 \leq \theta \leq 2\pi$.

25. In spherical coordinates, parametric equations are $x = 6\sin\phi\cos\theta$, $y = 6\sin\phi\sin\theta$, $z = 6\cos\phi$. The intersection of the

sphere with the plane $z = 3\sqrt{3}$ corresponds to $z = 6\cos\phi = 3\sqrt{3} \Rightarrow \cos\phi = \frac{\sqrt{3}}{2} \Rightarrow \phi = \frac{\pi}{6}$, and the plane $z = 0$

(the xy-plane) corresponds to $\phi = \frac{\pi}{2}$. Thus the surface is described by $\frac{\pi}{6} \leq \phi \leq \frac{\pi}{2}$, $0 \leq \theta \leq 2\pi$.

27. The surface appears to be a portion of a circular cylinder of radius 3 with axis the x-axis. An equation of the cylinder is

$y^2 + z^2 = 9$, and we can impose the restrictions $0 \leq x \leq 5$, $y \leq 0$ to obtain the portion shown. To graph the surface on a

CAS, we can use parametric equations $x = u$, $y = 3\cos v$, $z = 3\sin v$ with the parameter domain $0 \leq u \leq 5$, $\frac{\pi}{2} \leq v \leq \frac{3\pi}{2}$.

Alternatively, we can regard x and z as parameters. Then parametric equations are $x = x$, $z = z$, $y = -\sqrt{9 - z^2}$, where

$0 \leq x \leq 5$ and $-3 \leq z \leq 3$.

29. Using Equations 3, we have the parametrization $x = x$,

$$y = \frac{1}{1+x^2}\cos\theta, \quad z = \frac{1}{1+x^2}\sin\theta, \quad -2 \le x \le 2, \quad 0 \le \theta \le 2\pi.$$

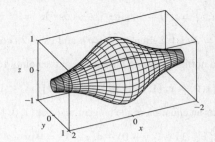

31. (a) Replacing $\cos u$ by $\sin u$ and $\sin u$ by $\cos u$ gives parametric equations

$x = (2 + \sin v)\sin u$, $y = (2 + \sin v)\cos u$, $z = u + \cos v$. From the graph, it

appears that the direction of the spiral is reversed. We can verify this observation by

noting that the projection of the spiral grid curves onto the xy-plane, given by

$x = (2 + \sin v)\sin u$, $y = (2 + \sin v)\cos u$, $z = 0$, draws a circle in the clockwise

direction for each value of v. The original equations, on the other hand, give circular

projections drawn in the counterclockwise direction. The equation for z is identical in

both surfaces, so as z increases, these grid curves spiral up in opposite directions for

the two surfaces.

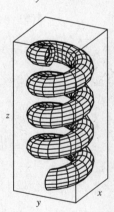

(b) Replacing $\cos u$ by $\cos 2u$ and $\sin u$ by $\sin 2u$ gives parametric equations

$x = (2 + \sin v)\cos 2u$, $y = (2 + \sin v)\sin 2u$, $z = u + \cos v$. From the graph, it

appears that the number of coils in the surface doubles within the same parametric

domain. We can verify this observation by noting that the projection of the spiral grid

curves onto the xy-plane, given by $x = (2 + \sin v)\cos 2u$, $y = (2 + \sin v)\sin 2u$,

$z = 0$ (where v is constant), complete circular revolutions for $0 \le u \le \pi$ while the

original surface requires $0 \le u \le 2\pi$ for a complete revolution. Thus, the new

surface winds around twice as fast as the original surface, and since the equation for z

is identical in both surfaces, we observe twice as many circular coils in the same

z-interval.

33. $\mathbf{r}(u, v) = (u + v)\,\mathbf{i} + 3u^2\,\mathbf{j} + (u - v)\,\mathbf{k}$.

$\mathbf{r}_u = \mathbf{i} + 6u\,\mathbf{j} + \mathbf{k}$ and $\mathbf{r}_v = \mathbf{i} - \mathbf{k}$, so $\mathbf{r}_u \times \mathbf{r}_v = -6u\,\mathbf{i} + 2\,\mathbf{j} - 6u\,\mathbf{k}$. Since the point $(2, 3, 0)$ corresponds to $u = 1$, $v = 1$, a

normal vector to the surface at $(2, 3, 0)$ is $-6\,\mathbf{i} + 2\,\mathbf{j} - 6\,\mathbf{k}$, and an equation of the tangent plane is $-6x + 2y - 6z = -6$ or

$3x - y + 3z = 3$.

35. $\mathbf{r}(u, v) = u\cos v\,\mathbf{i} + u\sin v\,\mathbf{j} + v\,\mathbf{k} \Rightarrow \mathbf{r}\left(1, \frac{\pi}{3}\right) = \left(\frac{1}{2}, \frac{\sqrt{3}}{2}, \frac{\pi}{3}\right)$.

$\mathbf{r}_u = \cos v\,\mathbf{i} + \sin v\,\mathbf{j}$ and $\mathbf{r}_v = -u\sin v\,\mathbf{i} + u\cos v\,\mathbf{j} + \mathbf{k}$, so a normal vector to the surface at the point $\left(\frac{1}{2}, \frac{\sqrt{3}}{2}, \frac{\pi}{3}\right)$ is

$\mathbf{r}_u\left(1, \frac{\pi}{3}\right) \times \mathbf{r}_v\left(1, \frac{\pi}{3}\right) = \left(\frac{1}{2}\mathbf{i} + \frac{\sqrt{3}}{2}\mathbf{j}\right) \times \left(-\frac{\sqrt{3}}{2}\mathbf{i} + \frac{1}{2}\mathbf{j} + \mathbf{k}\right) = \frac{\sqrt{3}}{2}\mathbf{i} - \frac{1}{2}\mathbf{j} + \mathbf{k}$. Thus an equation of the tangent plane at

$\left(\frac{1}{2}, \frac{\sqrt{3}}{2}, \frac{\pi}{3}\right)$ is $\frac{\sqrt{3}}{2}\left(x - \frac{1}{2}\right) - \frac{1}{2}\left(y - \frac{\sqrt{3}}{2}\right) + 1\left(z - \frac{\pi}{3}\right) = 0$ or $\frac{\sqrt{3}}{2}x - \frac{1}{2}y + z = \frac{\pi}{3}$.

37. $\mathbf{r}(u, v) = u^2\,\mathbf{i} + 2u\sin v\,\mathbf{j} + u\cos v\,\mathbf{k} \;\Rightarrow\; \mathbf{r}(1, 0) = (1, 0, 1)$.

$\mathbf{r}_u = 2u\,\mathbf{i} + 2\sin v\,\mathbf{j} + \cos v\,\mathbf{k}$ and $\mathbf{r}_v = 2u\cos v\,\mathbf{j} - u\sin v\,\mathbf{k}$,

so a normal vector to the surface at the point $(1, 0, 1)$ is

$\mathbf{r}_u(1, 0) \times \mathbf{r}_v(1, 0) = (2\,\mathbf{i} + \mathbf{k}) \times (2\,\mathbf{j}) = -2\,\mathbf{i} + 4\,\mathbf{k}$.

Thus an equation of the tangent plane at $(1, 0, 1)$ is

$-2(x - 1) + 0(y - 0) + 4(z - 1) = 0$ or $-x + 2z = 1$.

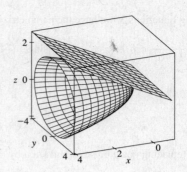

39. The surface S is given by $z = f(x, y) = 6 - 3x - 2y$ which intersects the xy-plane in the line $3x + 2y = 6$, so D is the

triangular region given by $\{(x, y) \mid 0 \le x \le 2,\ 0 \le y \le 3 - \frac{3}{2}x\}$. By Formula 9, the surface area of S is

$$A(S) = \iint_D \sqrt{1 + \left(\frac{\partial z}{\partial x}\right)^2 + \left(\frac{\partial z}{\partial y}\right)^2}\, dA$$
$$= \iint_D \sqrt{1 + (-3)^2 + (-2)^2}\, dA = \sqrt{14} \iint_D dA = \sqrt{14}\, A(D) = \sqrt{14}\left(\tfrac{1}{2}\cdot 2\cdot 3\right) = 3\sqrt{14}.$$

41. Here we can write $z = f(x, y) = \frac{1}{3} - \frac{1}{3}x - \frac{2}{3}y$ and D is the disk $x^2 + y^2 \le 3$, so by Formula 9 the area of the surface is

$$A(S) = \iint_D \sqrt{1 + \left(\frac{\partial z}{\partial x}\right)^2 + \left(\frac{\partial z}{\partial y}\right)^2}\, dA = \iint_D \sqrt{1 + \left(-\tfrac{1}{3}\right)^2 + \left(-\tfrac{2}{3}\right)^2}\, dA = \tfrac{\sqrt{14}}{3} \iint_D dA$$
$$= \tfrac{\sqrt{14}}{3}\, A(D) = \tfrac{\sqrt{14}}{3}\cdot \pi\left(\sqrt{3}\right)^2 = \sqrt{14}\,\pi$$

43. $z = f(x, y) = \frac{2}{3}(x^{3/2} + y^{3/2})$ and $D = \{(x, y) \mid 0 \le x \le 1,\ 0 \le y \le 1\}$. Then $f_x = x^{1/2}$, $f_y = y^{1/2}$ and

$$A(S) = \iint_D \sqrt{1 + (\sqrt{x})^2 + (\sqrt{y})^2}\, dA = \int_0^1 \int_0^1 \sqrt{1 + x + y}\, dy\, dx$$
$$= \int_0^1 \left[\tfrac{2}{3}(x + y + 1)^{3/2}\right]_{y=0}^{y=1} dx = \tfrac{2}{3}\int_0^1 \left[(x + 2)^{3/2} - (x + 1)^{3/2}\right] dx$$
$$= \tfrac{2}{3}\left[\tfrac{2}{5}(x + 2)^{5/2} - \tfrac{2}{5}(x + 1)^{5/2}\right]_0^1 = \tfrac{4}{15}(3^{5/2} - 2^{5/2} - 2^{5/2} + 1) = \tfrac{4}{15}(3^{5/2} - 2^{7/2} + 1)$$

45. $z = f(x, y) = xy$ with $x^2 + y^2 \le 1$, so $f_x = y$, $f_y = x \;\Rightarrow$

$$A(S) = \iint_D \sqrt{1 + y^2 + x^2}\, dA = \int_0^{2\pi}\int_0^1 \sqrt{r^2 + 1}\, r\, dr\, d\theta = \int_0^{2\pi}\left[\tfrac{1}{3}(r^2 + 1)^{3/2}\right]_{r=0}^{r=1} d\theta$$
$$= \int_0^{2\pi} \tfrac{1}{3}(2\sqrt{2} - 1)\, d\theta = \tfrac{2\pi}{3}(2\sqrt{2} - 1)$$

47. A parametric representation of the surface is $x = x$, $y = x^2 + z^2$, $z = z$ with $0 \le x^2 + z^2 \le 16$.

Hence $\mathbf{r}_x \times \mathbf{r}_z = (\mathbf{i} + 2x\,\mathbf{j}) \times (2z\,\mathbf{j} + \mathbf{k}) = 2x\,\mathbf{i} - \mathbf{j} + 2z\,\mathbf{k}$.

Note: In general, if $y = f(x, z)$ then $\mathbf{r}_x \times \mathbf{r}_z = \dfrac{\partial f}{\partial x}\,\mathbf{i} - \mathbf{j} + \dfrac{\partial f}{\partial z}\,\mathbf{k}$, and $A(S) = \iint_D \sqrt{1 + \left(\dfrac{\partial f}{\partial x}\right)^2 + \left(\dfrac{\partial f}{\partial z}\right)^2}\, dA$. Then

$$A(S) = \iint_{0 \le x^2 + z^2 \le 16} \sqrt{1 + 4x^2 + 4z^2}\, dA = \int_0^{2\pi}\int_0^4 \sqrt{1 + 4r^2}\, r\, dr\, d\theta$$
$$= \int_0^{2\pi} d\theta \int_0^4 r\sqrt{1 + 4r^2}\, dr = 2\pi\left[\tfrac{1}{12}(1 + 4r^2)^{3/2}\right]_0^4 = \tfrac{\pi}{6}\left(65^{3/2} - 1\right)$$

49. $\mathbf{r}_u = \langle 2u, v, 0 \rangle$, $\mathbf{r}_v = \langle 0, u, v \rangle$, and $\mathbf{r}_u \times \mathbf{r}_v = \langle v^2, -2uv, 2u^2 \rangle$. Then

$$A(S) = \iint_D |\mathbf{r}_u \times \mathbf{r}_v| \, dA = \int_0^1 \int_0^2 \sqrt{v^4 + 4u^2 v^2 + 4u^4} \, dv \, du = \int_0^1 \int_0^2 \sqrt{(v^2 + 2u^2)^2} \, dv \, du$$

$$= \int_0^1 \int_0^2 (v^2 + 2u^2) \, dv \, du = \int_0^1 \left[\tfrac{1}{3} v^3 + 2u^2 v \right]_{v=0}^{v=2} du = \int_0^1 \left(\tfrac{8}{3} + 4u^2 \right) du = \left[\tfrac{8}{3} u + \tfrac{4}{3} u^3 \right]_0^1 = 4$$

51. From Equation 9 we have $A(S) = \iint_D \sqrt{1 + (f_x)^2 + (f_y)^2} \, dA$. But if $|f_x| \le 1$ and $|f_y| \le 1$ then $0 \le (f_x)^2 \le 1$,

$0 \le (f_y)^2 \le 1 \;\Rightarrow\; 1 \le 1 + (f_x)^2 + (f_y)^2 \le 3 \;\Rightarrow\; 1 \le \sqrt{1 + (f_x)^2 + (f_y)^2} \le \sqrt{3}$. By Property 15.2.11,

$\iint_D 1 \, dA \le \iint_D \sqrt{1 + (f_x)^2 + (f_y)^2} \, dA \le \iint_D \sqrt{3} \, dA \;\Rightarrow\; A(D) \le A(S) \le \sqrt{3} \, A(D) \;\Rightarrow$

$\pi R^2 \le A(S) \le \sqrt{3} \pi R^2$.

53. $z = f(x, y) = \ln(x^2 + y^2 + 2)$ with $x^2 + y^2 \le 1$.

$$A(S) = \iint_D \sqrt{1 + \left(\frac{2x}{x^2 + y^2 + 2} \right)^2 + \left(\frac{2y}{x^2 + y^2 + 2} \right)^2} \, dA = \iint_D \sqrt{1 + \frac{4x^2 + 4y^2}{(x^2 + y^2 + 2)^2}} \, dA$$

$$= \int_0^{2\pi} \int_0^1 \sqrt{1 + \frac{4r^2}{(r^2 + 2)^2}} \, r \, dr \, d\theta = \int_0^{2\pi} d\theta \int_0^1 r \sqrt{\frac{(r^2 + 2)^2 + 4r^2}{(r^2 + 2)^2}} \, dr = 2\pi \int_0^1 \frac{r\sqrt{r^4 + 8r^2 + 4}}{r^2 + 2} \, dr \approx 3.5618$$

55. (a) $A(S) = \iint_D \sqrt{1 + \left(\dfrac{\partial z}{\partial x} \right)^2 + \left(\dfrac{\partial z}{\partial y} \right)^2} \, dA = \int_0^6 \int_0^4 \sqrt{1 + \dfrac{4x^2 + 4y^2}{(1 + x^2 + y^2)^4}} \, dy \, dx$.

Using the Midpoint Rule with $f(x, y) = \sqrt{1 + \dfrac{4x^2 + 4y^2}{(1 + x^2 + y^2)^4}}$, $m = 3$, $n = 2$ we have

$$A(S) \approx \sum_{i=1}^3 \sum_{j=1}^2 f(\overline{x}_i, \overline{y}_j) \, \Delta A = 4 \left[f(1,1) + f(1,3) + f(3,1) + f(3,3) + f(5,1) + f(5,3) \right] \approx 24.2055$$

(b) Using a CAS we have $A(S) = \displaystyle\int_0^6 \int_0^4 \sqrt{1 + \dfrac{4x^2 + 4y^2}{(1 + x^2 + y^2)^4}} \, dy \, dx \approx 24.2476$. This agrees with the estimate in part (a)

to the first decimal place.

57. $z = 1 + 2x + 3y + 4y^2$, so

$$A(S) = \iint_D \sqrt{1 + \left(\frac{\partial z}{\partial x} \right)^2 + \left(\frac{\partial z}{\partial y} \right)^2} \, dA = \int_1^4 \int_0^1 \sqrt{1 + 4 + (3 + 8y)^2} \, dy \, dx = \int_1^4 \int_0^1 \sqrt{14 + 48y + 64y^2} \, dy \, dx.$$

Using a CAS, we have

$\int_1^4 \int_0^1 \sqrt{14 + 48y + 64y^2} \, dy \, dx = \tfrac{45}{8} \sqrt{14} + \tfrac{15}{16} \ln \left(11\sqrt{5} + 3\sqrt{14}\sqrt{5} \right) - \tfrac{15}{16} \ln \left(3\sqrt{5} + \sqrt{14}\sqrt{5} \right)$

or $\tfrac{45}{8} \sqrt{14} + \tfrac{15}{16} \ln \dfrac{11\sqrt{5} + 3\sqrt{70}}{3\sqrt{5} + \sqrt{70}}$.

59. (a) $x = a \sin u \cos v$, $y = b \sin u \sin v$, $z = c \cos u$ $\Rightarrow$

$$\frac{x^2}{a^2} + \frac{y^2}{b^2} + \frac{z^2}{c^2} = (\sin u \cos v)^2 + (\sin u \sin v)^2 + (\cos u)^2$$

$$= \sin^2 u + \cos^2 u = 1$$

and since the ranges of u and v are sufficient to generate the entire graph,

the parametric equations represent an ellipsoid.

(b)

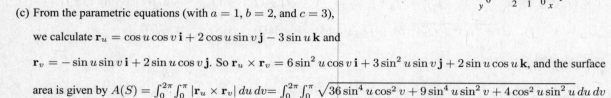

(c) From the parametric equations (with $a = 1$, $b = 2$, and $c = 3$),

we calculate $\mathbf{r}_u = \cos u \cos v\, \mathbf{i} + 2 \cos u \sin v\, \mathbf{j} - 3 \sin u\, \mathbf{k}$ and

$\mathbf{r}_v = -\sin u \sin v\, \mathbf{i} + 2 \sin u \cos v\, \mathbf{j}$. So $\mathbf{r}_u \times \mathbf{r}_v = 6 \sin^2 u \cos v\, \mathbf{i} + 3 \sin^2 u \sin v\, \mathbf{j} + 2 \sin u \cos u\, \mathbf{k}$, and the surface

area is given by $A(S) = \int_0^{2\pi} \int_0^{\pi} |\mathbf{r}_u \times \mathbf{r}_v|\, du\, dv = \int_0^{2\pi} \int_0^{\pi} \sqrt{36 \sin^4 u \cos^2 v + 9 \sin^4 u \sin^2 v + 4 \cos^2 u \sin^2 u}\, du\, dv$

61. To find the region D: $z = x^2 + y^2$ implies $z + z^2 = 4z$ or $z^2 - 3z = 0$. Thus $z = 0$ or $z = 3$ are the planes where the

surfaces intersect. But $x^2 + y^2 + z^2 = 4z$ implies $x^2 + y^2 + (z-2)^2 = 4$, so $z = 3$ intersects the upper hemisphere.

Thus $(z-2)^2 = 4 - x^2 - y^2$ or $z = 2 + \sqrt{4 - x^2 - y^2}$. Therefore D is the region inside the circle $x^2 + y^2 + (3-2)^2 = 4$,

that is, $D = \{(x,y) \mid x^2 + y^2 \le 3\}$.

$$A(S) = \iint_D \sqrt{1 + [(-x)(4 - x^2 - y^2)^{-1/2}]^2 + [(-y)(4 - x^2 - y^2)^{-1/2}]^2}\, dA$$

$$= \int_0^{2\pi} \int_0^{\sqrt{3}} \sqrt{1 + \frac{r^2}{4 - r^2}}\, r\, dr\, d\theta = \int_0^{2\pi} \int_0^{\sqrt{3}} \frac{2r\, dr}{\sqrt{4 - r^2}}\, d\theta = \int_0^{2\pi} \left[-2(4 - r^2)^{1/2}\right]_{r=0}^{r=\sqrt{3}}\, d\theta$$

$$= \int_0^{2\pi} (-2 + 4)\, d\theta = 2\theta \Big]_0^{2\pi} = 4\pi$$

63. Let $A(S_1)$ be the surface area of that portion of the surface which lies above the plane $z = 0$. Then $A(S) = 2A(S_1)$.

Following Example 10, a parametric representation of S_1 is $x = a \sin \phi \cos \theta$, $y = a \sin \phi \sin \theta$,

$z = a \cos \phi$ and $|\mathbf{r}_\phi \times \mathbf{r}_\theta| = a^2 \sin \phi$. For D, $0 \le \phi \le \frac{\pi}{2}$ and for each fixed ϕ, $\left(x - \frac{1}{2}a\right)^2 + y^2 \le \left(\frac{1}{2}a\right)^2$ or

$\left[a \sin \phi \cos \theta - \frac{1}{2}a\right]^2 + a^2 \sin^2 \phi \sin^2 \theta \le (a/2)^2$ implies $a^2 \sin^2 \phi - a^2 \sin \phi \cos \theta \le 0$ or

$\sin \phi\, (\sin \phi - \cos \theta) \le 0$. But $0 \le \phi \le \frac{\pi}{2}$, so $\cos \theta \ge \sin \phi$ or $\sin\left(\frac{\pi}{2} + \theta\right) \ge \sin \phi$ or $\phi - \frac{\pi}{2} \le \theta \le \frac{\pi}{2} - \phi$.

Hence $D = \left\{(\phi, \theta) \mid 0 \le \phi \le \frac{\pi}{2}, \phi - \frac{\pi}{2} \le \theta \le \frac{\pi}{2} - \phi\right\}$. Then

$$A(S_1) = \int_0^{\pi/2} \int_{\phi - (\pi/2)}^{(\pi/2) - \phi} a^2 \sin \phi\, d\theta\, d\phi = a^2 \int_0^{\pi/2} (\pi - 2\phi) \sin \phi\, d\phi$$

$$= a^2 \left[(-\pi \cos \phi) - 2(-\phi \cos \phi + \sin \phi)\right]_0^{\pi/2} = a^2 (\pi - 2)$$

Thus $A(S) = 2a^2(\pi - 2)$.

[continued]

Alternate solution: Working on S_1 we could parametrize the portion of the sphere by $x = x$, $y = y$, $z = \sqrt{a^2 - x^2 - y^2}$.

Then $|\mathbf{r}_x \times \mathbf{r}_y| = \sqrt{1 + \dfrac{x^2}{a^2 - x^2 - y^2} + \dfrac{y^2}{a^2 - x^2 - y^2}} = \dfrac{a}{\sqrt{a^2 - x^2 - y^2}}$ and

$$A(S_1) = \iint\limits_{0 \le (x - (a/2))^2 + y^2 \le (a/2)^2} \frac{a}{\sqrt{a^2 - x^2 - y^2}} \, dA = \int_{-\pi/2}^{\pi/2} \int_0^{a \cos\theta} \frac{a}{\sqrt{a^2 - r^2}} \, r \, dr \, d\theta$$

$$= \int_{-\pi/2}^{\pi/2} -a(a^2 - r^2)^{1/2} \Big]_{r=0}^{r = a\cos\theta} d\theta = \int_{-\pi/2}^{\pi/2} a^2 [1 - (1 - \cos^2\theta)^{1/2}] \, d\theta$$

$$= \int_{-\pi/2}^{\pi/2} a^2 (1 - |\sin\theta|) \, d\theta = 2a^2 \int_0^{\pi/2} (1 - \sin\theta) \, d\theta = 2a^2 \left(\frac{\pi}{2} - 1\right)$$

Thus $A(S) = 4a^2 \left(\frac{\pi}{2} - 1\right) = 2a^2(\pi - 2)$.

Notes:

(1) Perhaps working in spherical coordinates is the most obvious approach here. However, you must be careful in setting up D.

(2) In the alternate solution, you can avoid having to use $|\sin\theta|$ by working in the first octant and then multiplying by 4. However, if you set up S_1 as above and arrived at $A(S_1) = a^2\pi$, you now see your error.

16.7   Surface Integrals

1. The box is a cube where each face has surface area 4. The centers of the faces are $(\pm 1, 0, 0)$, $(0, \pm 1, 0)$, $(0, 0, \pm 1)$. For each face we take the point P_{ij}^* to be the center of the face and $f(x, y, z) = \cos(x + 2y + 3z)$, so by Definition 1,

$$\iint_S f(x, y, z) \, dS \approx [f(1, 0, 0)](4) + [f(-1, 0, 0)](4) + [f(0, 1, 0)](4)$$
$$+ [f(0, -1, 0)](4) + [f(0, 0, 1)](4) + [f(0, 0, -1)](4)$$
$$= 4 \left[\cos 1 + \cos(-1) + \cos 2 + \cos(-2) + \cos 3 + \cos(-3)\right] \approx -6.93$$

3. We can use the xz- and yz-planes to divide H into four patches of equal size, each with surface area equal to $\frac{1}{8}$ the surface area of a sphere with radius $\sqrt{50}$, so $\Delta S = \frac{1}{8}(4)\pi\left(\sqrt{50}\right)^2 = 25\pi$. Then $(\pm 3, \pm 4, 5)$ are sample points in the four patches, and using a Riemann sum as in Definition 1, we have

$$\iint_H f(x, y, z) \, dS \approx f(3, 4, 5)\,\Delta S + f(3, -4, 5)\,\Delta S + f(-3, 4, 5)\,\Delta S + f(-3, -4, 5)\,\Delta S$$
$$= (7 + 8 + 9 + 12)(25\pi) = 900\pi \approx 2827$$

5. $\mathbf{r}(u, v) = (u + v)\,\mathbf{i} + (u - v)\,\mathbf{j} + (1 + 2u + v)\,\mathbf{k}$, $0 \le u \le 2$, $0 \le v \le 1$ and

$\mathbf{r}_u \times \mathbf{r}_v = (\mathbf{i} + \mathbf{j} + 2\,\mathbf{k}) \times (\mathbf{i} - \mathbf{j} + \mathbf{k}) = 3\,\mathbf{i} + \mathbf{j} - 2\,\mathbf{k} \quad \Rightarrow \quad |\mathbf{r}_u \times \mathbf{r}_v| = \sqrt{3^2 + 1^2 + (-2)^2} = \sqrt{14}$. Then by Formula 2,

$$\iint_S (x + y + z) \, dS = \iint_D (u + v + u - v + 1 + 2u + v)\, |\mathbf{r}_u \times \mathbf{r}_v| \, dA = \int_0^1 \int_0^2 (4u + v + 1) \cdot \sqrt{14} \, du \, dv$$
$$= \sqrt{14} \int_0^1 \left[2u^2 + uv + u\right]_{u=0}^{u=2} dv = \sqrt{14} \int_0^1 (2v + 10)\, dv = \sqrt{14} \left[v^2 + 10v\right]_0^1 = 11\sqrt{14}$$

7. $\mathbf{r}(u, v) = \langle u \cos v, u \sin v, v \rangle$, $0 \le u \le 1$, $0 \le v \le \pi$ and

$\mathbf{r}_u \times \mathbf{r}_v = \langle \cos v, \sin v, 0 \rangle \times \langle -u \sin v, u \cos v, 1 \rangle = \langle \sin v, -\cos v, u \rangle \quad \Rightarrow$

$|\mathbf{r}_u \times \mathbf{r}_v| = \sqrt{\sin^2 v + \cos^2 v + u^2} = \sqrt{u^2 + 1}$. Then

$$\iint_S y \, dS = \iint_D (u \sin v) \, |\mathbf{r}_u \times \mathbf{r}_v| \, dA = \int_0^1 \int_0^\pi (u \sin v) \cdot \sqrt{u^2 + 1} \, dv \, du = \int_0^1 u\sqrt{u^2 + 1} \, du \int_0^\pi \sin v \, dv$$

$$= \left[\tfrac{1}{3}(u^2 + 1)^{3/2}\right]_0^1 \, [-\cos v]_0^\pi = \tfrac{1}{3}(2^{3/2} - 1) \cdot 2 = \tfrac{2}{3}(2\sqrt{2} - 1)$$

9. $z = 1 + 2x + 3y$ so $\dfrac{\partial z}{\partial x} = 2$ and $\dfrac{\partial z}{\partial y} = 3$. Then by Formula 4,

$$\iint_S x^2 yz \, dS = \iint_D x^2 yz \sqrt{\left(\frac{\partial z}{\partial x}\right)^2 + \left(\frac{\partial z}{\partial y}\right)^2 + 1} \, dA = \int_0^3 \int_0^2 x^2 y(1 + 2x + 3y)\sqrt{4 + 9 + 1} \, dy \, dx$$

$$= \sqrt{14} \int_0^3 \int_0^2 (x^2 y + 2x^3 y + 3x^2 y^2) \, dy \, dx = \sqrt{14} \int_0^3 \left[\tfrac{1}{2}x^2 y^2 + x^3 y^2 + x^2 y^3\right]_{y=0}^{y=2} dx$$

$$= \sqrt{14} \int_0^3 (10x^2 + 4x^3) \, dx = \sqrt{14}\left[\tfrac{10}{3}x^3 + x^4\right]_0^3 = 171\sqrt{14}$$

11. An equation of the plane through the points $(1, 0, 0)$, $(0, -2, 0)$, and $(0, 0, 4)$ is $4x - 2y + z = 4$, so S is the region in the

plane $z = 4 - 4x + 2y$ over $D = \{(x, y) \mid 0 \le x \le 1,\ 2x - 2 \le y \le 0\}$. Thus by Formula 4,

$$\iint_S x \, dS = \iint_D x \sqrt{(-4)^2 + (2)^2 + 1} \, dA = \sqrt{21} \int_0^1 \int_{2x-2}^0 x \, dy \, dx = \sqrt{21} \int_0^1 [xy]_{y=2x-2}^{y=0} \, dx$$

$$= \sqrt{21} \int_0^1 (-2x^2 + 2x) \, dx = \sqrt{21}\left[-\tfrac{2}{3}x^3 + x^2\right]_0^1 = \sqrt{21}\left(-\tfrac{2}{3} + 1\right) = \tfrac{\sqrt{21}}{3}$$

13. Using y and z as parameters, we have $\mathbf{r}(y, z) = (y^2 + z^2)\mathbf{i} + y\mathbf{j} + z\mathbf{k}$, $y^2 + z^2 \le 1$. Then

$\mathbf{r}_y \times \mathbf{r}_z = (2y\mathbf{i} + \mathbf{j}) \times (2z\mathbf{i} + \mathbf{k}) = \mathbf{i} - 2y\mathbf{j} - 2z\mathbf{k}$ and $|\mathbf{r}_y \times \mathbf{r}_z| = \sqrt{1 + 4y^2 + 4z^2} = \sqrt{1 + 4(y^2 + z^2)}$. Thus

$$\iint_S z^2 \, dS = \iint_{y^2 + z^2 \le 1} z^2 \sqrt{1 + 4(y^2 + z^2)} \, dA = \int_0^{2\pi} \int_0^1 (r\sin\theta)^2 \sqrt{1 + 4r^2}\, r \, dr \, d\theta$$

$$= \int_0^{2\pi} \sin^2\theta \, d\theta \int_0^1 r^3 \sqrt{1 + 4r^2} \, dr \qquad [\text{let } u = 1 + 4r^2 \ \Rightarrow\ r^2 = \tfrac{1}{4}(u - 1) \text{ and } r \, dr = \tfrac{1}{8}du]$$

$$= \left[\tfrac{1}{2}\theta - \tfrac{1}{4}\sin 2\theta\right]_0^{2\pi} \int_1^5 \tfrac{1}{4}(u - 1)\sqrt{u} \cdot \tfrac{1}{8}du = \pi \cdot \tfrac{1}{32}\int_1^5 (u^{3/2} - u^{1/2}) \, du = \tfrac{1}{32}\pi\left[\tfrac{2}{5}u^{5/2} - \tfrac{2}{3}u^{3/2}\right]_1^5$$

$$= \tfrac{1}{32}\pi\left[\tfrac{2}{5}(5)^{5/2} - \tfrac{2}{3}(5)^{3/2} - \tfrac{2}{5} + \tfrac{2}{3}\right] = \tfrac{1}{32}\pi\left(\tfrac{20}{3}\sqrt{5} + \tfrac{4}{15}\right) = \tfrac{1}{120}\pi(25\sqrt{5} + 1)$$

15. Using x and z as parameters, we have $\mathbf{r}(x, z) = x\mathbf{i} + (x^2 + 4z)\mathbf{j} + z\mathbf{k}$, $0 \le x \le 1, 0 \le z \le 1$. Then

$\mathbf{r}_x \times \mathbf{r}_z = (\mathbf{i} + 2x\mathbf{j}) \times (4\mathbf{j} + \mathbf{k}) = 2x\mathbf{i} - \mathbf{j} + 4\mathbf{k}$ and $|\mathbf{r}_x \times \mathbf{r}_z| = \sqrt{4x^2 + 1 + 16} = \sqrt{4x^2 + 17}$. Thus

$$\iint_S x \, dS = \int_0^1 \int_0^1 x\sqrt{4x^2 + 17} \, dz \, dx = \int_0^1 x\sqrt{4x^2 + 17} \, dx = \left[\tfrac{1}{8} \cdot \tfrac{2}{3}(4x^2 + 17)^{3/2}\right]_0^1$$

$$= \tfrac{1}{12}(21^{3/2} - 17^{3/2}) = \tfrac{1}{12}\left(21\sqrt{21} - 17\sqrt{17}\right) = \tfrac{7}{4}\sqrt{21} - \tfrac{17}{12}\sqrt{17}$$

17. Using spherical coordinates to parametrize the sphere we have $\mathbf{r}(\phi, \theta) = 2\sin\phi\cos\theta\,\mathbf{i} + 2\sin\phi\sin\theta\,\mathbf{j} + 2\cos\phi\,\mathbf{k}$ and

$|\mathbf{r}_\phi \times \mathbf{r}_\theta| = 4\sin\phi$ (see Example 16.6.10). Here S is the portion of the sphere corresponding to $0 \le \phi \le \pi/2$, so

$$\iint_S (x^2 z + y^2 z) \, dS = \iint_S (x^2 + y^2)z \, dS = \int_0^{2\pi} \int_0^{\pi/2} (4\sin^2\phi)(2\cos\phi)(4\sin\phi) \, d\phi \, d\theta$$

$$= 32\int_0^{2\pi} d\theta \int_0^{\pi/2} \sin^3\phi\cos\phi \, d\phi = 32\,(2\pi)\left[\tfrac{1}{4}\sin^4\phi\right]_0^{\pi/2} = 16\pi(1 - 0) = 16\pi$$

19. Here S consists of three surfaces: S_1, the lateral surface of the cylinder; S_2, the front formed by the plane $x + y = 5$; and the back, S_3, in the plane $x = 0$.

On S_1: the surface is given by $\mathbf{r}(u, v) = u\,\mathbf{i} + 3\cos v\,\mathbf{j} + 3\sin v\,\mathbf{k}$, $0 \le v \le 2\pi$, and $0 \le x \le 5 - y$ $\Rightarrow$

$0 \le u \le 5 - 3\cos v$. Then $\mathbf{r}_u \times \mathbf{r}_v = -3\cos v\,\mathbf{j} - 3\sin v\,\mathbf{k}$ and $|\mathbf{r}_u \times \mathbf{r}_v| = \sqrt{9\cos^2 v + 9\sin^2 v} = 3$, so

$$\iint_{S_1} xz\, dS = \int_0^{2\pi} \int_0^{5 - 3\cos v} u(3\sin v)(3)\, du\, dv = 9\int_0^{2\pi} \left[\tfrac{1}{2}u^2\right]_{u=0}^{u=5-3\cos v} \sin v\, dv$$

$$= \tfrac{9}{2}\int_0^{2\pi} (5 - 3\cos v)^2 \sin v\, dv = \tfrac{9}{2}\left[\tfrac{1}{9}(5 - 3\cos v)^3\right]_0^{2\pi} = 0.$$

On S_2: $\mathbf{r}(y, z) = (5 - y)\,\mathbf{i} + y\,\mathbf{j} + z\,\mathbf{k}$ and $|\mathbf{r}_y \times \mathbf{r}_z| = |\mathbf{i} + \mathbf{j}| = \sqrt{2}$, where $y^2 + z^2 \le 9$ and

$$\iint_{S_2} xz\, dS = \iint_{y^2 + z^2 \le 9} (5 - y)z\,\sqrt{2}\, dA = \sqrt{2}\int_0^{2\pi}\int_0^3 (5 - r\cos\theta)(r\sin\theta)\, r\, dr\, d\theta$$

$$= \sqrt{2}\int_0^{2\pi}\int_0^3 (5r^2 - r^3\cos\theta)(\sin\theta)\, dr\, d\theta = \sqrt{2}\int_0^{2\pi} \left[\tfrac{5}{3}r^3 - \tfrac{1}{4}r^4\cos\theta\right]_{r=0}^{r=3} \sin\theta\, d\theta$$

$$= \sqrt{2}\int_0^{2\pi} \left(45 - \tfrac{81}{4}\cos\theta\right)\sin\theta\, d\theta = \sqrt{2}\left(\tfrac{4}{81}\right)\cdot\tfrac{1}{2}\left(45 - \tfrac{81}{4}\cos\theta\right)^2 \bigg]_0^{2\pi} = 0$$

On S_3: $x = 0$ so $\iint_{S_3} xz\, dS = 0$. Hence $\iint_S xz\, dS = 0 + 0 + 0 = 0$.

21. From Exercise 5, $\mathbf{r}(u, v) = (u + v)\,\mathbf{i} + (u - v)\,\mathbf{j} + (1 + 2u + v)\,\mathbf{k}$, $0 \le u \le 2$, $0 \le v \le 1$, and $\mathbf{r}_u \times \mathbf{r}_v = 3\mathbf{i} + \mathbf{j} - 2\mathbf{k}$. Then

$$\mathbf{F}(\mathbf{r}(u, v)) = (1 + 2u + v)e^{(u+v)(u-v)}\,\mathbf{i} - 3(1 + 2u + v)e^{(u+v)(u-v)}\,\mathbf{j} + (u + v)(u - v)\,\mathbf{k}$$

$$= (1 + 2u + v)e^{u^2 - v^2}\,\mathbf{i} - 3(1 + 2u + v)e^{u^2 - v^2}\,\mathbf{j} + (u^2 - v^2)\,\mathbf{k}$$

Because the z-component of $\mathbf{r}_u \times \mathbf{r}_v$ is negative we use $-(\mathbf{r}_u \times \mathbf{r}_v)$ in Formula 9 for the upward orientation:

$$\iint_S \mathbf{F} \cdot d\mathbf{S} = \iint_D \mathbf{F} \cdot (-(\mathbf{r}_u \times \mathbf{r}_v))\, dA = \int_0^1 \int_0^2 \left[-3(1 + 2u + v)e^{u^2 - v^2} + 3(1 + 2u + v)e^{u^2 - v^2} + 2(u^2 - v^2)\right] du\, dv$$

$$= \int_0^1 \int_0^2 2(u^2 - v^2)\, du\, dv = 2\int_0^1 \left[\tfrac{1}{3}u^3 - uv^2\right]_{u=0}^{u=2} dv = 2\int_0^1 \left(\tfrac{8}{3} - 2v^2\right) dv$$

$$= 2\left[\tfrac{8}{3}v - \tfrac{2}{3}v^3\right]_0^1 = 2\left(\tfrac{8}{3} - \tfrac{2}{3}\right) = 4$$

23. $\mathbf{F}(x, y, z) = xy\,\mathbf{i} + yz\,\mathbf{j} + zx\,\mathbf{k}$, $z = g(x, y) = 4 - x^2 - y^2$, and D is the square $[0, 1] \times [0, 1]$, so by Equation 10

$$\iint_S \mathbf{F} \cdot d\mathbf{S} = \iint_D [-xy(-2x) - yz(-2y) + zx]\, dA = \int_0^1 \int_0^1 [2x^2 y + 2y^2(4 - x^2 - y^2) + x(4 - x^2 - y^2)]\, dy\, dx$$

$$= \int_0^1 \left[x^2 y^2 + \tfrac{8}{3}y^3 - \tfrac{2}{3}x^2 y^3 - \tfrac{2}{5}y^5 + 4xy - x^3 y - \tfrac{1}{3}xy^3\right]_{y=0}^{y=1} dx$$

$$= \int_0^1 \left(\tfrac{1}{3}x^2 + \tfrac{11}{3}x - x^3 + \tfrac{34}{15}\right) dx = \left[\tfrac{1}{9}x^3 + \tfrac{11}{6}x^2 - \tfrac{1}{4}x^4 + \tfrac{34}{15}x\right]_0^1 = \tfrac{713}{180}$$

25. $\mathbf{F}(x, y, z) = x\,\mathbf{i} + y\,\mathbf{j} + z^2\,\mathbf{k}$, and using spherical coordinates, S is given by $x = \sin\phi\cos\theta$, $y = \sin\phi\sin\theta$, $z = \cos\phi$,

$0 \le \theta \le 2\pi$, $0 \le \phi \le \pi$. $\mathbf{F}(\mathbf{r}(\phi, \theta)) = (\sin\phi\cos\theta)\,\mathbf{i} + (\sin\phi\sin\theta)\,\mathbf{j} + (\cos^2\phi)\,\mathbf{k}$ and, from Example 4,

$\mathbf{r}_\phi \times \mathbf{r}_\theta = \sin^2\phi\cos\theta\,\mathbf{i} + \sin^2\phi\sin\theta\,\mathbf{j} + \sin\phi\cos\phi\,\mathbf{k}$. Thus

$$\mathbf{F}(\mathbf{r}(\phi, \theta)) \cdot (\mathbf{r}_\phi \times \mathbf{r}_\theta) = \sin^3\phi\cos^2\theta + \sin^3\phi\sin^2\theta + \sin\phi\cos^3\phi = \sin^3\phi + \sin\phi\cos^3\phi$$

[continued]

and

$$\iint_S \mathbf{F} \cdot d\mathbf{S} = \iint_D \left[\mathbf{F}(\mathbf{r}(\phi, \theta)) \cdot (\mathbf{r}_\phi \times \mathbf{r}_\theta) \right] dA = \int_0^{2\pi} \int_0^\pi (\sin^3 \phi + \sin \phi \cos^3 \phi) \, d\phi \, d\theta$$

$$= \int_0^{2\pi} d\theta \int_0^\pi (1 - \cos^2 \phi + \cos^3 \phi) \sin \phi \, d\phi = (2\pi) \left[-\cos \phi + \tfrac{1}{3} \cos^3 \phi - \tfrac{1}{4} \cos^4 \phi \right]_0^\pi$$

$$= 2\pi \left(1 - \tfrac{1}{3} - \tfrac{1}{4} + 1 - \tfrac{1}{3} + \tfrac{1}{4} \right) = \tfrac{8}{3}\pi$$

27. Let S_1 be the paraboloid $y = x^2 + z^2$, $0 \le y \le 1$ and S_2 the disk $x^2 + z^2 \le 1$, $y = 1$. Since S is a closed

surface, we use the outward orientation.

On S_1: $\mathbf{F}(\mathbf{r}(x, z)) = (x^2 + z^2)\mathbf{j} - z\mathbf{k}$ and $\mathbf{r}_x \times \mathbf{r}_z = 2x\mathbf{i} - \mathbf{j} + 2z\mathbf{k}$ (since the **j**-component must be negative on S_1). Then

$$\iint_{S_1} \mathbf{F} \cdot d\mathbf{S} = \iint_{x^2 + z^2 \le 1} \left[-(x^2 + z^2) - 2z^2 \right] dA = -\int_0^{2\pi} \int_0^1 (r^2 + 2r^2 \sin^2 \theta) \, r \, dr \, d\theta$$

$$= -\int_0^{2\pi} \int_0^1 r^3 (1 + 2\sin^2 \theta) \, dr \, d\theta = -\int_0^{2\pi} (1 + 1 - \cos 2\theta) \, d\theta \int_0^1 r^3 \, dr$$

$$= -\left[2\theta - \tfrac{1}{2} \sin 2\theta \right]_0^{2\pi} \left[\tfrac{1}{4} r^4 \right]_0^1 = -4\pi \cdot \tfrac{1}{4} = -\pi$$

On S_2: $\mathbf{F}(\mathbf{r}(x, z)) = \mathbf{j} - z\mathbf{k}$ and $\mathbf{r}_z \times \mathbf{r}_x = \mathbf{j}$. Then $\iint_{S_2} \mathbf{F} \cdot d\mathbf{S} = \iint_{x^2 + z^2 \le 1} (1) \, dA = \pi$.

Hence $\iint_S \mathbf{F} \cdot d\mathbf{S} = -\pi + \pi = 0$.

29. Here S consists of the six faces of the cube as labeled in the figure. On S_1:

$\mathbf{F} = \mathbf{i} + 2y\mathbf{j} + 3z\mathbf{k}$, $\mathbf{r}_y \times \mathbf{r}_z = \mathbf{i}$ and $\iint_{S_1} \mathbf{F} \cdot d\mathbf{S} = \int_{-1}^1 \int_{-1}^1 dy \, dz = 4$;

S_2: $\mathbf{F} = x\mathbf{i} + 2\mathbf{j} + 3z\mathbf{k}$, $\mathbf{r}_z \times \mathbf{r}_x = \mathbf{j}$ and $\iint_{S_2} \mathbf{F} \cdot d\mathbf{S} = \int_{-1}^1 \int_{-1}^1 2 \, dx \, dz = 8$;

S_3: $\mathbf{F} = x\mathbf{i} + 2y\mathbf{j} + 3\mathbf{k}$, $\mathbf{r}_x \times \mathbf{r}_y = \mathbf{k}$ and $\iint_{S_3} \mathbf{F} \cdot d\mathbf{S} = \int_{-1}^1 \int_{-1}^1 3 \, dx \, dy = 12$;

S_4: $\mathbf{F} = -\mathbf{i} + 2y\mathbf{j} + 3z\mathbf{k}$, $\mathbf{r}_z \times \mathbf{r}_y = -\mathbf{i}$ and $\iint_{S_4} \mathbf{F} \cdot d\mathbf{S} = 4$;

S_5: $\mathbf{F} = x\mathbf{i} - 2\mathbf{j} + 3z\mathbf{k}$, $\mathbf{r}_x \times \mathbf{r}_z = -\mathbf{j}$ and $\iint_{S_5} \mathbf{F} \cdot d\mathbf{S} = 8$;

S_6: $\mathbf{F} = x\mathbf{i} + 2y\mathbf{j} - 3\mathbf{k}$, $\mathbf{r}_y \times \mathbf{r}_x = -\mathbf{k}$ and $\iint_{S_6} \mathbf{F} \cdot d\mathbf{S} = \int_{-1}^1 \int_{-1}^1 3 \, dx \, dy = 12$.

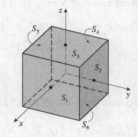

Hence $\iint_S \mathbf{F} \cdot d\mathbf{S} = \sum_{i=1}^6 \iint_{S_i} \mathbf{F} \cdot d\mathbf{S} = 48$.

31. Here S consists of four surfaces: S_1, the top surface (a portion of the circular cylinder $y^2 + z^2 = 1$); S_2, the bottom surface

(a portion of the xy-plane); S_3, the front half-disk in the plane $x = 2$, and S_4, the back half-disk in the plane $x = 0$.

On S_1: The surface is $z = \sqrt{1 - y^2}$ for $0 \le x \le 2$, $-1 \le y \le 1$ with upward orientation, so

$$\iint_{S_1} \mathbf{F} \cdot d\mathbf{S} = \int_0^2 \int_{-1}^1 \left[-x^2(0) - y^2 \left(-\frac{y}{\sqrt{1-y^2}} \right) + z^2 \right] dy \, dx = \int_0^2 \int_{-1}^1 \left(\frac{y^3}{\sqrt{1-y^2}} + 1 - y^2 \right) dy \, dx$$

$$= \int_0^2 \left[-\sqrt{1-y^2} + \tfrac{1}{3}(1-y^2)^{3/2} + y - \tfrac{1}{3}y^3 \right]_{y=-1}^{y=1} dx = \int_0^2 \tfrac{4}{3} \, dx = \tfrac{8}{3}$$

On S_2: The surface is $z = 0$ with downward orientation, so

$$\iint_{S_2} \mathbf{F} \cdot d\mathbf{S} = \int_0^2 \int_{-1}^1 (-z^2) \, dy \, dx = \int_0^2 \int_{-1}^1 (0) \, dy \, dx = 0$$

On S_3: The surface is $x = 2$ for $-1 \le y \le 1$, $0 \le z \le \sqrt{1 - y^2}$, oriented in the positive x-direction. Regarding y and z as parameters, we have $\mathbf{r}_y \times \mathbf{r}_z = \mathbf{i}$ and

$$\iint_{S_3} \mathbf{F} \cdot d\mathbf{S} = \int_{-1}^{1} \int_0^{\sqrt{1-y^2}} x^2 \, dz \, dy = \int_{-1}^{1} \int_0^{\sqrt{1-y^2}} 4 \, dz \, dy = 4A(S_3) = 2\pi$$

On S_4: The surface is $x = 0$ for $-1 \le y \le 1$, $0 \le z \le \sqrt{1 - y^2}$, oriented in the negative x-direction. Regarding y and z as parameters, we use $-(\mathbf{r}_y \times \mathbf{r}_z) = -\mathbf{i}$ and

$$\iint_{S_4} \mathbf{F} \cdot d\mathbf{S} = \int_{-1}^{1} \int_0^{\sqrt{1-y^2}} x^2 \, dz \, dy = \int_{-1}^{1} \int_0^{\sqrt{1-y^2}} (0) \, dz \, dy = 0$$

Thus $\iint_S \mathbf{F} \cdot d\mathbf{S} = \frac{8}{3} + 0 + 2\pi + 0 = 2\pi + \frac{8}{3}$.

33. $z = xe^y \;\Rightarrow\; \partial z/\partial x = e^y$, $\partial z/\partial y = xe^y$, so by Formula 4, a CAS gives

$$\iint_S (x^2 + y^2 + z^2) \, dS = \int_0^1 \int_0^1 (x^2 + y^2 + x^2 e^{2y}) \sqrt{e^{2y} + x^2 e^{2y} + 1} \, dx \, dy \approx 4.5822.$$

35. We use Formula 4 with $z = 3 - 2x^2 - y^2 \;\Rightarrow\; \partial z/\partial x = -4x$, $\partial z/\partial y = -2y$. The boundaries of the region

$3 - 2x^2 - y^2 \ge 0$ are $-\sqrt{\frac{3}{2}} \le x \le \sqrt{\frac{3}{2}}$ and $-\sqrt{3 - 2x^2} \le y \le \sqrt{3 - 2x^2}$, so we use a CAS (with precision reduced to seven or fewer digits; otherwise the calculation may take a long time) to calculate

$$\iint_S x^2 y^2 z^2 \, dS = \int_{-\sqrt{3/2}}^{\sqrt{3/2}} \int_{-\sqrt{3-2x^2}}^{\sqrt{3-2x^2}} x^2 y^2 (3 - 2x^2 - y^2)^2 \sqrt{16x^2 + 4y^2 + 1} \, dy \, dx \approx 3.4895$$

37. If S is given by $y = h(x, z)$, then S is also the level surface $f(x, y, z) = y - h(x, z) = 0$.

$\mathbf{n} = \dfrac{\nabla f(x, y, z)}{|\nabla f(x, y, z)|} = \dfrac{-h_x \mathbf{i} + \mathbf{j} - h_z \mathbf{k}}{\sqrt{h_x^2 + 1 + h_z^2}}$, and $-\mathbf{n}$ is the unit normal that points to the left. Now we proceed as in the derivation of (10), using Formula 4 to evaluate

$$\iint_S \mathbf{F} \cdot d\mathbf{S} = \iint_S \mathbf{F} \cdot \mathbf{n} \, dS = \iint_D (P\mathbf{i} + Q\mathbf{j} + R\mathbf{k}) \dfrac{\dfrac{\partial h}{\partial x} \mathbf{i} - \mathbf{j} + \dfrac{\partial h}{\partial z} \mathbf{k}}{\sqrt{\left(\dfrac{\partial h}{\partial x}\right)^2 + 1 + \left(\dfrac{\partial h}{\partial z}\right)^2}} \sqrt{\left(\dfrac{\partial h}{\partial x}\right)^2 + 1 + \left(\dfrac{\partial h}{\partial z}\right)^2} \, dA$$

where D is the projection of S onto the xz-plane. Therefore $\displaystyle\iint_S \mathbf{F} \cdot d\mathbf{S} = \iint_D \left(P \dfrac{\partial h}{\partial x} - Q + R \dfrac{\partial h}{\partial z} \right) dA$.

39. $m = \iint_S K \, dS = K \cdot 4\pi \left(\frac{1}{2} a^2\right) = 2\pi a^2 K$; by symmetry $M_{xz} = M_{yz} = 0$, and

$$M_{xy} = \iint_S zK \, dS = K \int_0^{2\pi} \int_0^{\pi/2} (a\cos\phi)(a^2 \sin\phi) \, d\phi \, d\theta = 2\pi K a^3 \left[-\frac{1}{4} \cos 2\phi \right]_0^{\pi/2} = \pi K a^3.$$

Hence $(\overline{x}, \overline{y}, \overline{z}) = \left(0, 0, \frac{1}{2} a\right)$.

41. (a) $I_z = \iint_S (x^2 + y^2) \rho(x, y, z) \, dS$

(b) $I_z = \iint_S (x^2 + y^2)\left(10 - \sqrt{x^2 + y^2}\right) dS = \iint\limits_{1 \le x^2 + y^2 \le 16} (x^2 + y^2)\left(10 - \sqrt{x^2 + y^2}\right) \sqrt{2} \, dA$

$= \int_0^{2\pi} \int_1^4 \sqrt{2}\,(10r^3 - r^4) \, dr \, d\theta = 2\sqrt{2}\,\pi \left(\frac{4329}{10}\right) = \frac{4329}{5} \sqrt{2}\,\pi$

43. The rate of flow through the cylinder is the flux $\iint_S \rho \mathbf{v} \cdot \mathbf{n} \, dS = \iint_S \rho \mathbf{v} \cdot d\mathbf{S}$. We use the parametric representation

$\mathbf{r}(u, v) = 2 \cos u \, \mathbf{i} + 2 \sin u \, \mathbf{j} + v \, \mathbf{k}$ for S, where $0 \le u \le 2\pi$, $0 \le v \le 1$, so $\mathbf{r}_u = -2 \sin u \, \mathbf{i} + 2 \cos u \, \mathbf{j}$, $\mathbf{r}_v = \mathbf{k}$, and the

outward orientation is given by $\mathbf{r}_u \times \mathbf{r}_v = 2 \cos u \, \mathbf{i} + 2 \sin u \, \mathbf{j}$. Then

$$\iint_S \rho \mathbf{v} \cdot d\mathbf{S} = \rho \int_0^{2\pi} \int_0^1 \left(v \, \mathbf{i} + 4 \sin^2 u \, \mathbf{j} + 4 \cos^2 u \, \mathbf{k} \right) \cdot \left(2 \cos u \, \mathbf{i} + 2 \sin u \, \mathbf{j} \right) dv \, du$$

$$= \rho \int_0^{2\pi} \int_0^1 \left(2v \cos u + 8 \sin^3 u \right) dv \, du = \rho \int_0^{2\pi} \left(\cos u + 8 \sin^3 u \right) du$$

$$= \rho \left[\sin u + 8 \left(-\tfrac{1}{3} \right) \left(2 + \sin^2 u \right) \cos u \right]_0^{2\pi} = 0 \text{ kg/s}$$

45. S consists of the hemisphere S_1 given by $z = \sqrt{a^2 - x^2 - y^2}$ and the disk S_2 given by $0 \le x^2 + y^2 \le a^2$, $z = 0$.

On S_1: $\mathbf{E} = a \sin \phi \cos \theta \, \mathbf{i} + a \sin \phi \sin \theta \, \mathbf{j} + 2a \cos \phi \, \mathbf{k}$,

$\mathbf{T}_\phi \times \mathbf{T}_\theta = a^2 \sin^2 \phi \cos \theta \, \mathbf{i} + a^2 \sin^2 \phi \sin \theta \, \mathbf{j} + a^2 \sin \phi \cos \phi \, \mathbf{k}$. Thus

$$\iint_{S_1} \mathbf{E} \cdot d\mathbf{S} = \int_0^{2\pi} \int_0^{\pi/2} (a^3 \sin^3 \phi + 2a^3 \sin \phi \cos^2 \phi) \, d\phi \, d\theta$$

$$= \int_0^{2\pi} \int_0^{\pi/2} (a^3 \sin \phi + a^3 \sin \phi \cos^2 \phi) \, d\phi \, d\theta = (2\pi)a^3 \left(1 + \tfrac{1}{3} \right) = \tfrac{8}{3}\pi a^3$$

On S_2: $\mathbf{E} = x \, \mathbf{i} + y \, \mathbf{j}$, and $\mathbf{r}_y \times \mathbf{r}_x = -\mathbf{k}$ so $\iint_{S_2} \mathbf{E} \cdot d\mathbf{S} = 0$. Hence the total charge is $q = \varepsilon_0 \iint_S \mathbf{E} \cdot d\mathbf{S} = \tfrac{8}{3}\pi a^3 \varepsilon_0$.

47. $K \nabla u = 6.5(4y \, \mathbf{j} + 4z \, \mathbf{k})$. S is given by $\mathbf{r}(x, \theta) = x \, \mathbf{i} + \sqrt{6} \cos \theta \, \mathbf{j} + \sqrt{6} \sin \theta \, \mathbf{k}$ and since we want the inward heat flow, we

use $\mathbf{r}_x \times \mathbf{r}_\theta = -\sqrt{6} \cos \theta \, \mathbf{j} - \sqrt{6} \sin \theta \, \mathbf{k}$. Then the rate of heat flow inward is given by

$$\iint_S (-K \nabla u) \cdot d\mathbf{S} = \int_0^{2\pi} \int_0^4 -(6.5)(-24) \, dx \, d\theta = (2\pi)(156)(4) = 1248\pi.$$

49. Let S be a sphere of radius a centered at the origin. Then $|\mathbf{r}| = a$ and $\mathbf{F}(\mathbf{r}) = c\mathbf{r}/|\mathbf{r}|^3 = (c/a^3)(x \, \mathbf{i} + y \, \mathbf{j} + z \, \mathbf{k})$. A

parametric representation for S is $\mathbf{r}(\phi, \theta) = a \sin \phi \cos \theta \, \mathbf{i} + a \sin \phi \sin \theta \, \mathbf{j} + a \cos \phi \, \mathbf{k}$, $0 \le \phi \le \pi$, $0 \le \theta \le 2\pi$. Then

$\mathbf{r}_\phi = a \cos \phi \cos \theta \, \mathbf{i} + a \cos \phi \sin \theta \, \mathbf{j} - a \sin \phi \, \mathbf{k}$, $\mathbf{r}_\theta = -a \sin \phi \sin \theta \, \mathbf{i} + a \sin \phi \cos \theta \, \mathbf{j}$, and the outward orientation is given

by $\mathbf{r}_\phi \times \mathbf{r}_\theta = a^2 \sin^2 \phi \cos \theta \, \mathbf{i} + a^2 \sin^2 \phi \sin \theta \, \mathbf{j} + a^2 \sin \phi \cos \phi \, \mathbf{k}$. The flux of $\mathbf{F}$ across S is

$$\iint_S \mathbf{F} \cdot d\mathbf{S} = \int_0^\pi \int_0^{2\pi} \frac{c}{a^3} \left(a \sin \phi \cos \theta \, \mathbf{i} + a \sin \phi \sin \theta \, \mathbf{j} + a \cos \phi \, \mathbf{k} \right)$$

$$\cdot \left(a^2 \sin^2 \phi \cos \theta \, \mathbf{i} + a^2 \sin^2 \phi \sin \theta \, \mathbf{j} + a^2 \sin \phi \cos \phi \, \mathbf{k} \right) d\theta \, d\phi$$

$$= \frac{c}{a^3} \int_0^\pi \int_0^{2\pi} a^3 \left(\sin^3 \phi + \sin \phi \cos^2 \phi \right) d\theta \, d\phi = c \int_0^\pi \int_0^{2\pi} \sin \phi \, d\theta \, d\phi = 4\pi c$$

Thus the flux does not depend on the radius a.

16.8 Stokes' Theorem

1. Both H and P are oriented piecewise-smooth surfaces that are bounded by the simple, closed, smooth curve $x^2 + y^2 = 4$,

$z = 0$ (which we can take to be oriented positively for both surfaces). Then H and P satisfy the hypotheses of Stokes'

Theorem, so by (3) we know $\iint_H \operatorname{curl} \mathbf{F} \cdot d\mathbf{S} = \int_C \mathbf{F} \cdot d\mathbf{r} = \iint_P \operatorname{curl} \mathbf{F} \cdot d\mathbf{S}$ (where C is the boundary curve).

3. The boundary curve C is the circle $x^2 + z^2 = 16$, $y = 0$ where the hemisphere intersects the xz-plane. The curve should be oriented in the counterclockwise direction when viewed from the right (from the positive y-axis), so a vector equation of C is

$\mathbf{r}(t) = 4\cos(-t)\,\mathbf{i} + 4\sin(-t)\,\mathbf{k} = 4\cos t\,\mathbf{i} - 4\sin t\,\mathbf{k}$, $0 \le t \le 2\pi$. Then $\mathbf{r}'(t) = -4\sin t\,\mathbf{i} - 4\cos t\,\mathbf{k}$ and

$\mathbf{F}(\mathbf{r}(t)) = (-4\sin t)e^0\,\mathbf{i} + (4\cos t)(\cos 0)\,\mathbf{j} + (4\cos t)(-4\sin t)(\sin 0)\,\mathbf{k} = -4\sin t\,\mathbf{i} + 4\cos t\,\mathbf{j}$, and by Stokes' Theorem,

$$\iint_S \operatorname{curl}\mathbf{F}\cdot d\mathbf{S} = \int_C \mathbf{F}\cdot d\mathbf{r} = \int_0^{2\pi} \mathbf{F}(\mathbf{r}(t))\cdot\mathbf{r}'(t)\,dt = \int_0^{2\pi}(-4\sin t\,\mathbf{i} + 4\cos t\,\mathbf{j})\cdot(-4\sin t\,\mathbf{i} - 4\cos t\,\mathbf{k})\,dt$$

$$= \int_0^{2\pi}(16\sin^2 t + 0 + 0)\,dt = 16\left[\tfrac{1}{2}t - \tfrac{1}{4}\sin 2t\right]_0^{2\pi} = 16\pi$$

5. C is the square in the plane $z = -1$. Rather than evaluating a line integral around C we can use Equation 3:

$\iint_{S_1}\operatorname{curl}\mathbf{F}\cdot d\mathbf{S} = \oint_C\mathbf{F}\cdot d\mathbf{r} = \iint_{S_2}\operatorname{curl}\mathbf{F}\cdot d\mathbf{S}$ where S_1 is the original cube without the bottom and S_2 is the bottom face

of the cube. $\operatorname{curl}\mathbf{F} = x^2 z\,\mathbf{i} + (xy - 2xyz)\,\mathbf{j} + (y - xz)\,\mathbf{k}$. For S_2, we choose $\mathbf{n} = \mathbf{k}$ so that C has the same orientation for

both surfaces. Then $\operatorname{curl}\mathbf{F}\cdot\mathbf{n} = y - xz = x + y$ on S_2, where $z = -1$. Thus $\iint_{S_2}\operatorname{curl}\mathbf{F}\cdot d\mathbf{S} = \int_{-1}^1\int_{-1}^1(x + y)\,dx\,dy = 0$

so $\iint_{S_1}\operatorname{curl}\mathbf{F}\cdot d\mathbf{S} = 0$.

7. $\operatorname{curl}\mathbf{F} = -2z\,\mathbf{i} - 2x\,\mathbf{j} - 2y\,\mathbf{k}$ and we take the surface S to be the planar region enclosed by C, so S is the portion of the plane

$x + y + z = 1$ over $D = \{(x,y)\mid 0 \le x \le 1, 0 \le y \le 1 - x\}$. Since C is oriented counterclockwise, we orient S upward.

Using Equation 16.7.10, we have $z = g(x,y) = 1 - x - y$, $P = -2z$, $Q = -2x$, $R = -2y$, and

$$\int_C\mathbf{F}\cdot d\mathbf{r} = \iint_S\operatorname{curl}\mathbf{F}\cdot d\mathbf{S} = \iint_D[-(-2z)(-1) - (-2x)(-1) + (-2y)]\,dA$$

$$= \int_0^1\int_0^{1-x}(-2)\,dy\,dx = -2\int_0^1(1 - x)\,dx = -1$$

9. $\operatorname{curl}\mathbf{F} = -y\,\mathbf{i} - z\,\mathbf{j} - x\,\mathbf{k}$ and we take S to be the part of the paraboloid $z = 1 - x^2 - y^2$ in the first octant. Since C is

oriented counterclockwise (from above), we orient S upward. Then using Equation 16.7.10 with $z = g(x,y) = 1 - x^2 - y^2$

we have

$$\int_C\mathbf{F}\cdot d\mathbf{r} = \iint_S\operatorname{curl}\mathbf{F}\cdot d\mathbf{S} = \iint_D[-(-y)(-2x) - (-z)(-2y) + (-x)]\,dA = \iint_D[-2xy - 2y(1 - x^2 - y^2) - x]\,dA$$

$$= \int_0^{\pi/2}\int_0^1[-2(r\cos\theta)(r\sin\theta) - 2(r\sin\theta)(1 - r^2) - r\cos\theta]\,r\,dr\,d\theta$$

$$= \int_0^{\pi/2}\int_0^1[-2r^3\sin\theta\cos\theta - 2(r^2 - r^4)\sin\theta - r^2\cos\theta]\,dr\,d\theta$$

$$= \int_0^{\pi/2}\left[-\tfrac{1}{2}r^4\sin\theta\cos\theta - 2\left(\tfrac{1}{3}r^3 - \tfrac{1}{5}r^5\right)\sin\theta - \tfrac{1}{3}r^3\cos\theta\right]_{r=0}^{r=1}d\theta$$

$$= \int_0^{\pi/2}\left(-\tfrac{1}{2}\sin\theta\cos\theta - \tfrac{4}{15}\sin\theta - \tfrac{1}{3}\cos\theta\right)d\theta = \left[-\tfrac{1}{4}\sin^2\theta + \tfrac{4}{15}\cos\theta - \tfrac{1}{3}\sin\theta\right]_0^{\pi/2}$$

$$= -\tfrac{1}{4} - \tfrac{4}{15} - \tfrac{1}{3} = -\tfrac{17}{20}$$

11. (a) The curve of intersection is an ellipse in the plane $x + y + z = 1$ with unit normal $\mathbf{n} = \tfrac{1}{\sqrt{3}}(\mathbf{i} + \mathbf{j} + \mathbf{k})$,

$\operatorname{curl}\mathbf{F} = x^2\,\mathbf{j} + y^2\,\mathbf{k}$, and $\operatorname{curl}\mathbf{F}\cdot\mathbf{n} = \tfrac{1}{\sqrt{3}}(x^2 + y^2)$. Then

$$\oint_C\mathbf{F}\cdot d\mathbf{r} = \iint_S\tfrac{1}{\sqrt{3}}(x^2 + y^2)\,dS = \iint_{x^2+y^2\le 9}(x^2 + y^2)\,dx\,dy = \int_0^{2\pi}\int_0^3 r^3\,dr\,d\theta = 2\pi\left(\tfrac{81}{4}\right) = \tfrac{81\pi}{2}$$

(b)

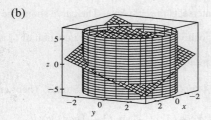

(c) One possible parametrization is $x = 3\cos t$, $y = 3\sin t$,

$$z = 1 - 3\cos t - 3\sin t, 0 \leq t \leq 2\pi.$$

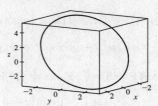

13. The boundary curve C is the circle $x^2 + y^2 = 16$, $z = 4$ oriented in the clockwise direction as viewed from above (since S is

oriented downward). We can parametrize C by $\mathbf{r}(t) = 4\cos t\,\mathbf{i} - 4\sin t\,\mathbf{j} + 4\,\mathbf{k}$, $0 \leq t \leq 2\pi$, and then

$\mathbf{r}'(t) = -4\sin t\,\mathbf{i} - 4\cos t\,\mathbf{j}$. Thus $\mathbf{F}(\mathbf{r}(t)) = 4\sin t\,\mathbf{i} + 4\cos t\,\mathbf{j} - 2\,\mathbf{k}$, $\mathbf{F}(\mathbf{r}(t)) \cdot \mathbf{r}'(t) = -16\sin^2 t - 16\cos^2 t = -16$, and

$$\oint_C \mathbf{F} \cdot d\mathbf{r} = \int_0^{2\pi} \mathbf{F}(\mathbf{r}(t)) \cdot \mathbf{r}'(t)\, dt = \int_0^{2\pi} (-16)\, dt = -16\,(2\pi) = -32\pi$$

Now curl $\mathbf{F} = 2\,\mathbf{k}$, and the projection D of S on the xy-plane is the disk $x^2 + y^2 \leq 16$, so by Equation 16.7.10 with

$z = g(x, y) = \sqrt{x^2 + y^2}$ [and multiplying by -1 for the downward orientation] we have

$$\iint_S \text{curl}\,\mathbf{F} \cdot d\mathbf{S} = -\iint_D (-0 - 0 + 2)\, dA = -2 \cdot A(D) = -2 \cdot \pi(4^2) = -32\pi$$

15. The boundary curve C is the circle $x^2 + z^2 = 1$, $y = 0$ oriented in the counterclockwise direction as viewed from the positive

y-axis. Then C can be described by $\mathbf{r}(t) = \cos t\,\mathbf{i} - \sin t\,\mathbf{k}$, $0 \leq t \leq 2\pi$, and $\mathbf{r}'(t) = -\sin t\,\mathbf{i} - \cos t\,\mathbf{k}$. Thus

$\mathbf{F}(\mathbf{r}(t)) = -\sin t\,\mathbf{j} + \cos t\,\mathbf{k}$, $\mathbf{F}(\mathbf{r}(t)) \cdot \mathbf{r}'(t) = -\cos^2 t$, and $\oint_C \mathbf{F} \cdot d\mathbf{r} = \int_0^{2\pi} (-\cos^2 t)\, dt = \left[-\frac{1}{2}t - \frac{1}{4}\sin 2t\right]_0^{2\pi} = -\pi$.

Now curl $\mathbf{F} = -\mathbf{i} - \mathbf{j} - \mathbf{k}$, and S can be parametrized (see Example 16.6.10) by

$\mathbf{r}(\phi, \theta) = \sin\phi \cos\theta\,\mathbf{i} + \sin\phi \sin\theta\,\mathbf{j} + \cos\phi\,\mathbf{k}$, $0 \leq \theta \leq \pi$, $0 \leq \phi \leq \pi$. Then

$\mathbf{r}_\phi \times \mathbf{r}_\theta = \sin^2\phi \cos\theta\,\mathbf{i} + \sin^2\phi \sin\theta\,\mathbf{j} + \sin\phi \cos\phi\,\mathbf{k}$ and

$$\iint_S \text{curl}\,\mathbf{F} \cdot d\mathbf{S} = \iint_{x^2+z^2 \leq 1} \text{curl}\,\mathbf{F} \cdot (\mathbf{r}_\phi \times \mathbf{r}_\theta)\, dA = \int_0^\pi \int_0^\pi (-\sin^2\phi \cos\theta - \sin^2\phi \sin\theta - \sin\phi \cos\phi)\, d\theta\, d\phi$$

$$= \int_0^\pi (-2\sin^2\phi - \pi\sin\phi \cos\phi)\, d\phi = \left[\frac{1}{2}\sin 2\phi - \phi - \frac{\pi}{2}\sin^2\phi\right]_0^\pi = -\pi$$

17. It is easier to use Stokes' Theorem than to compute the work directly. Let S be the planar region enclosed by the path of the

particle, so S is the portion of the plane $z = \frac{1}{2}y$ for $0 \leq x \leq 1$, $0 \leq y \leq 2$, with upward orientation.

curl $\mathbf{F} = 8y\,\mathbf{i} + 2z\,\mathbf{j} + 2y\,\mathbf{k}$ and

$$\oint_C \mathbf{F} \cdot d\mathbf{r} = \iint_S \text{curl}\,\mathbf{F} \cdot d\mathbf{S} = \iint_D \left[-8y\,(0) - 2z\left(\frac{1}{2}\right) + 2y\right] dA = \int_0^1 \int_0^2 \left(2y - \frac{1}{2}y\right) dy\, dx$$

$$= \int_0^1 \int_0^2 \frac{3}{2}y\, dy\, dx = \int_0^1 \left[\frac{3}{4}y^2\right]_{y=0}^{y=2} dx = \int_0^1 3\, dx = 3$$

19. Assume S is centered at the origin with radius a and let H_1 and H_2 be the upper and lower hemispheres, respectively, of S.

Then $\iint_S \text{curl}\,\mathbf{F} \cdot d\mathbf{S} = \iint_{H_1} \text{curl}\,\mathbf{F} \cdot d\mathbf{S} + \iint_{H_2} \text{curl}\,\mathbf{F} \cdot d\mathbf{S} = \oint_{C_1} \mathbf{F} \cdot d\mathbf{r} + \oint_{C_2} \mathbf{F} \cdot d\mathbf{r}$ by Stokes' Theorem. But C_1 is the

circle $x^2 + y^2 = a^2$ oriented in the counterclockwise direction while C_2 is the same circle oriented in the clockwise direction.

Hence $\oint_{C_2} \mathbf{F} \cdot d\mathbf{r} = -\oint_{C_1} \mathbf{F} \cdot d\mathbf{r}$ so $\iint_S \text{curl}\,\mathbf{F} \cdot d\mathbf{S} = 0$ as desired.

16.9   The Divergence Theorem

1. $\operatorname{div} \mathbf{F} = 3 + x + 2x = 3 + 3x$, so

$\iiint_E \operatorname{div} \mathbf{F} \, dV = \int_0^1 \int_0^1 \int_0^1 (3x + 3) \, dx \, dy \, dz = \frac{9}{2}$ (notice the triple integral is

three times the volume of the cube plus three times $\overline{x}$).

To compute $\iint_S \mathbf{F} \cdot d\mathbf{S}$, on

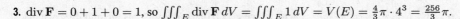

S_1: $\mathbf{n} = \mathbf{i}, \mathbf{F} = 3\mathbf{i} + y\mathbf{j} + 2z\mathbf{k}$, and $\iint_{S_1} \mathbf{F} \cdot d\mathbf{S} = \iint_{S_1} 3 \, dS = 3$;

S_2: $\mathbf{F} = 3x\mathbf{i} + x\mathbf{j} + 2xz\mathbf{k}$, $\mathbf{n} = \mathbf{j}$ and $\iint_{S_2} \mathbf{F} \cdot d\mathbf{S} = \iint_{S_2} x \, dS = \frac{1}{2}$;

S_3: $\mathbf{F} = 3x\mathbf{i} + xy\mathbf{j} + 2x\mathbf{k}$, $\mathbf{n} = \mathbf{k}$ and $\iint_{S_3} \mathbf{F} \cdot d\mathbf{S} = \iint_{S_3} 2x \, dS = 1$;

S_4: $\mathbf{F} = \mathbf{0}$, $\iint_{S_4} \mathbf{F} \cdot d\mathbf{S} = 0$; S_5: $\mathbf{F} = 3x\mathbf{i} + 2x\mathbf{k}$, $\mathbf{n} = -\mathbf{j}$ and $\iint_{S_5} \mathbf{F} \cdot d\mathbf{S} = \iint_{S_5} 0 \, dS = 0$;

S_6: $\mathbf{F} = 3x\mathbf{i} + xy\mathbf{j}$, $\mathbf{n} = -\mathbf{k}$ and $\iint_{S_6} \mathbf{F} \cdot d\mathbf{S} = \iint_{S_6} 0 \, dS = 0$. Thus $\iint_S \mathbf{F} \cdot d\mathbf{S} = \frac{9}{2}$.

3. $\operatorname{div} \mathbf{F} = 0 + 1 + 0 = 1$, so $\iiint_E \operatorname{div} \mathbf{F} \, dV = \iiint_E 1 \, dV = V(E) = \frac{4}{3}\pi \cdot 4^3 = \frac{256}{3}\pi$.

S is a sphere of radius 4 centered at the origin which can be parametrized by $\mathbf{r}(\phi, \theta) = \langle 4\sin\phi\cos\theta, 4\sin\phi\sin\theta, 4\cos\phi \rangle$,

$0 \le \phi \le \pi, 0 \le \theta \le 2\pi$ (similar to Example 16.6.10). Then

$$\mathbf{r}_\phi \times \mathbf{r}_\theta = \langle 4\cos\phi\cos\theta, 4\cos\phi\sin\theta, -4\sin\phi \rangle \times \langle -4\sin\phi\sin\theta, 4\sin\phi\cos\theta, 0 \rangle$$

$$= \langle 16\sin^2\phi\cos\theta, 16\sin^2\phi\sin\theta, 16\cos\phi\sin\phi \rangle$$

and $\mathbf{F}(\mathbf{r}(\phi, \theta)) = \langle 4\cos\phi, 4\sin\phi\sin\theta, 4\sin\phi\cos\theta \rangle$. Thus

$\mathbf{F} \cdot (\mathbf{r}_\phi \times \mathbf{r}_\theta) = 64\cos\phi\sin^2\phi\cos\theta + 64\sin^3\phi\sin^2\theta + 64\cos\phi\sin^2\phi\cos\theta = 128\cos\phi\sin^2\phi\cos\theta + 64\sin^3\phi\sin^2\theta$

and

$$\iint_S \mathbf{F} \cdot d\mathbf{S} = \iint_D \mathbf{F} \cdot (\mathbf{r}_\phi \times \mathbf{r}_\theta) \, dA = \int_0^{2\pi} \int_0^\pi (128\cos\phi\sin^2\phi\cos\theta + 64\sin^3\phi\sin^2\theta) \, d\phi \, d\theta$$

$$= \int_0^{2\pi} \left[\frac{128}{3}\sin^3\phi\cos\theta + 64\left(\frac{1}{3}\cos^3\phi - \cos\phi\right)\sin^2\theta \right]_{\phi=0}^{\phi=\pi} d\theta$$

$$= \int_0^{2\pi} \frac{256}{3}\sin^2\theta \, d\theta = \frac{256}{3}\left[\frac{1}{2}\theta - \frac{1}{4}\sin 2\theta\right]_0^{2\pi} = \frac{256}{3}\pi$$

5. $\operatorname{div} \mathbf{F} = \frac{\partial}{\partial x}(xye^z) + \frac{\partial}{\partial y}(xy^2z^3) + \frac{\partial}{\partial z}(-ye^z) = ye^z + 2xyz^3 - ye^z = 2xyz^3$, so by the Divergence Theorem,

$$\iint_S \mathbf{F} \cdot d\mathbf{S} = \iiint_E \operatorname{div} \mathbf{F} \, dV = \int_0^3 \int_0^2 \int_0^1 2xyz^3 \, dz \, dy \, dx = 2\int_0^3 x \, dx \int_0^2 y \, dy \int_0^1 z^3 \, dz$$

$$= 2\left[\frac{1}{2}x^2\right]_0^3 \left[\frac{1}{2}y^2\right]_0^2 \left[\frac{1}{4}z^4\right]_0^1 = 2\left(\frac{9}{2}\right)(2)\left(\frac{1}{4}\right) = \frac{9}{2}$$

7. $\operatorname{div} \mathbf{F} = 3y^2 + 0 + 3z^2$, so using cylindrical coordinates with $y = r\cos\theta$, $z = r\sin\theta$, $x = x$ we have

$$\iint_S \mathbf{F} \cdot d\mathbf{S} = \iiint_E (3y^2 + 3z^2) \, dV = \int_0^{2\pi} \int_0^1 \int_{-1}^2 (3r^2\cos^2\theta + 3r^2\sin^2\theta) \, r \, dx \, dr \, d\theta$$

$$= 3\int_0^{2\pi} d\theta \int_0^1 r^3 \, dr \int_{-1}^2 dx = 3\left[\theta\right]_0^{2\pi} \left[\frac{1}{4}r^4\right]_0^1 \left[x\right]_{-1}^2 = 3(2\pi)\left(\frac{1}{4}\right)(3) = \frac{9\pi}{2}$$

9. $\operatorname{div} \mathbf{F} = e^y + (-e^y) + 0 = 0$, so by the Divergence Theorem, $\iint_S \mathbf{F} \cdot d\mathbf{S} = \iiint_E 0 \, dV = 0$.

11. $\operatorname{div} \mathbf{F} = 6x^2 + 3y^2 + 3y^2 = 6x^2 + 6y^2$ so

$$\iint_S \mathbf{F} \cdot d\mathbf{S} = \iiint_E 6(x^2 + y^2) \, dV = \int_0^{2\pi} \int_0^1 \int_0^{1-r^2} 6r^2 \cdot r \, dz \, dr \, d\theta = \int_0^{2\pi} \int_0^1 6r^3(1 - r^2) \, dr \, d\theta$$

$$= \int_0^{2\pi} d\theta \int_0^1 (6r^3 - 6r^5) \, dr = \left[\theta\right]_0^{2\pi} \left[\frac{3}{2}r^4 - r^6\right]_0^1 = 2\pi\left(\frac{3}{2} - 1\right) = \pi$$

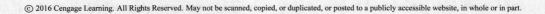

13. $\mathbf{F}(x, y, z) = x\sqrt{x^2 + y^2 + z^2}\,\mathbf{i} + y\sqrt{x^2 + y^2 + z^2}\,\mathbf{j} + z\sqrt{x^2 + y^2 + z^2}\,\mathbf{k}$, so

$$\text{div } \mathbf{F} = x \cdot \tfrac{1}{2}(x^2 + y^2 + z^2)^{-1/2}(2x) + (x^2 + y^2 + z^2)^{1/2} + y \cdot \tfrac{1}{2}(x^2 + y^2 + z^2)^{-1/2}(2y) + (x^2 + y^2 + z^2)^{1/2}$$
$$+ z \cdot \tfrac{1}{2}(x^2 + y^2 + z^2)^{-1/2}(2z) + (x^2 + y^2 + z^2)^{1/2}$$
$$= (x^2 + y^2 + z^2)^{-1/2}\left[x^2 + (x^2 + y^2 + z^2) + y^2 + (x^2 + y^2 + z^2) + z^2 + (x^2 + y^2 + z^2)\right]$$
$$= \frac{4(x^2 + y^2 + z^2)}{\sqrt{x^2 + y^2 + z^2}} = 4\sqrt{x^2 + y^2 + z^2}.$$

Then $\iint_S \mathbf{F} \cdot d\mathbf{S} = \iiint_E 4\sqrt{x^2 + y^2 + z^2}\,dV = \int_0^{\pi/2} \int_0^{2\pi} \int_0^1 4\sqrt{\rho^2} \cdot \rho^2 \sin\phi\,d\rho\,d\theta\,d\phi$

$$= \int_0^{\pi/2} \sin\phi\,d\phi \int_0^{2\pi} d\theta \int_0^1 4\rho^3\,d\rho = [-\cos\phi]_0^{\pi/2}\,[\theta]_0^{2\pi}\,[\rho^4]_0^1 = (1)(2\pi)(1) = 2\pi$$

15. $\iint_S \mathbf{F} \cdot d\mathbf{S} = \iiint_E \sqrt{3 - x^2}\,dV = \int_{-1}^1 \int_{-1}^1 \int_0^{2 - x^4 - y^4} \sqrt{3 - x^2}\,dz\,dy\,dx = \frac{341}{60}\sqrt{2} + \frac{81}{20}\sin^{-1}\left(\frac{\sqrt{3}}{3}\right)$

17. For S_1 we have $\mathbf{n} = -\mathbf{k}$, so $\mathbf{F} \cdot \mathbf{n} = \mathbf{F} \cdot (-\mathbf{k}) = -x^2 z - y^2 = -y^2$ (since $z = 0$ on S_1). So if D is the unit disk, we get

$\iint_{S_1} \mathbf{F} \cdot d\mathbf{S} = \iint_{S_1} \mathbf{F} \cdot \mathbf{n}\,dS = \iint_D (-y^2)\,dA = -\int_0^{2\pi} \int_0^1 r^2 (\sin^2\theta)\,r\,dr\,d\theta = -\frac{1}{4}\pi$. Now since S_2 is closed, we can use

the Divergence Theorem. Since $\text{div } \mathbf{F} = \frac{\partial}{\partial x}(z^2 x) + \frac{\partial}{\partial y}\left(\frac{1}{3}y^3 + \tan z\right) + \frac{\partial}{\partial z}(x^2 z + y^2) = z^2 + y^2 + x^2$, we use spherical

coordinates to get $\iint_{S_2} \mathbf{F} \cdot d\mathbf{S} = \iiint_E \text{div } \mathbf{F}\,dV = \int_0^{2\pi} \int_0^{\pi/2} \int_0^1 \rho^2 \cdot \rho^2 \sin\phi\,d\rho\,d\phi\,d\theta = \frac{2}{5}\pi$. Finally

$\iint_S \mathbf{F} \cdot d\mathbf{S} = \iint_{S_2} \mathbf{F} \cdot d\mathbf{S} - \iint_{S_1} \mathbf{F} \cdot d\mathbf{S} = \frac{2}{5}\pi - \left(-\frac{1}{4}\pi\right) = \frac{13}{20}\pi.$

19. The vectors that end near P_1 are longer than the vectors that start near P_1, so the net flow is inward near P_1 and div $\mathbf{F}(P_1)$ is

negative. The vectors that end near P_2 are shorter than the vectors that start near P_2, so the net flow is outward near P_2 and

div $\mathbf{F}(P_2)$ is positive.

21.

From the graph it appears that for points above the x-axis, vectors starting near a

particular point are longer than vectors ending there, so divergence is positive.

The opposite is true at points below the x-axis, where divergence is negative.

$\mathbf{F}(x, y) = \langle xy, x + y^2 \rangle \implies \text{div } \mathbf{F} = \frac{\partial}{\partial x}(xy) + \frac{\partial}{\partial y}(x + y^2) = y + 2y = 3y.$

Thus div $\mathbf{F} > 0$ for $y > 0$, and div $\mathbf{F} < 0$ for $y < 0$.

23. Since $\dfrac{\mathbf{x}}{|\mathbf{x}|^3} = \dfrac{x\,\mathbf{i} + y\,\mathbf{j} + z\,\mathbf{k}}{(x^2 + y^2 + z^2)^{3/2}}$ and $\dfrac{\partial}{\partial x}\left(\dfrac{x}{(x^2 + y^2 + z^2)^{3/2}}\right) = \dfrac{(x^2 + y^2 + z^2) - 3x^2}{(x^2 + y^2 + z^2)^{5/2}}$ with similar expressions

for $\dfrac{\partial}{\partial y}\left(\dfrac{y}{(x^2 + y^2 + z^2)^{3/2}}\right)$ and $\dfrac{\partial}{\partial z}\left(\dfrac{z}{(x^2 + y^2 + z^2)^{3/2}}\right)$, we have

$\text{div}\left(\dfrac{\mathbf{x}}{|\mathbf{x}|^3}\right) = \dfrac{3(x^2 + y^2 + z^2) - 3(x^2 + y^2 + z^2)}{(x^2 + y^2 + z^2)^{5/2}} = 0$, except at $(0, 0, 0)$ where it is undefined.

25. $\iint_S \mathbf{a} \cdot \mathbf{n}\,dS = \iiint_E \text{div } \mathbf{a}\,dV = 0$ since div $\mathbf{a} = 0$.

27. $\iint_S \text{curl } \mathbf{F} \cdot d\mathbf{S} = \iiint_E \text{div}(\text{curl } \mathbf{F})\, dV = 0$ by Theorem 16.5.11.

29. $\iint_S (f\nabla g) \cdot \mathbf{n}\, dS = \iiint_E \text{div}(f\nabla g)\, dV = \iiint_E (f\nabla^2 g + \nabla g \cdot \nabla f)\, dV$ by Exercise 16.5.25.

31. If $\mathbf{c} = c_1\,\mathbf{i} + c_2\,\mathbf{j} + c_3\,\mathbf{k}$ is an arbitrary constant vector, we define $\mathbf{F} = f\mathbf{c} = fc_1\,\mathbf{i} + fc_2\,\mathbf{j} + fc_3\,\mathbf{k}$. Then

$\text{div } \mathbf{F} = \text{div } f\mathbf{c} = \dfrac{\partial f}{\partial x}\,c_1 + \dfrac{\partial f}{\partial y}\,c_2 + \dfrac{\partial f}{\partial z}\,c_3 = \nabla f \cdot \mathbf{c}$ and the Divergence Theorem says $\iint_S \mathbf{F} \cdot d\mathbf{S} = \iiint_E \text{div } \mathbf{F}\, dV \quad \Rightarrow$

$\iint_S \mathbf{F} \cdot \mathbf{n}\, dS = \iiint_E \nabla f \cdot \mathbf{c}\, dV$. In particular, if $\mathbf{c} = \mathbf{i}$ then $\iint_S f\mathbf{i} \cdot \mathbf{n}\, dS = \iiint_E \nabla f \cdot \mathbf{i}\, dV \quad \Rightarrow$

$\displaystyle\iint_S fn_1\, dS = \iiint_E \dfrac{\partial f}{\partial x}\, dV$ (where $\mathbf{n} = n_1\,\mathbf{i} + n_2\,\mathbf{j} + n_3\,\mathbf{k}$). Similarly, if $\mathbf{c} = \mathbf{j}$ we have $\displaystyle\iint_S fn_2\, dS = \iiint_E \dfrac{\partial f}{\partial y}\, dV$,

and $\mathbf{c} = \mathbf{k}$ gives $\displaystyle\iint_S fn_3\, dS = \iiint_E \dfrac{\partial f}{\partial z}\, dV$. Then

$\displaystyle\iint_S f\mathbf{n}\, dS = \left(\iint_S fn_1\, dS\right)\mathbf{i} + \left(\iint_S fn_2\, dS\right)\mathbf{j} + \left(\iint_S fn_3\, dS\right)\mathbf{k}$

$\displaystyle = \left(\iiint_E \dfrac{\partial f}{\partial x}\, dV\right)\mathbf{i} + \left(\iiint_E \dfrac{\partial f}{\partial y}\, dV\right)\mathbf{j} + \left(\iiint_E \dfrac{\partial f}{\partial z}\, dV\right)\mathbf{k} = \iiint_E \left(\dfrac{\partial f}{\partial x}\,\mathbf{i} + \dfrac{\partial f}{\partial y}\,\mathbf{j} + \dfrac{\partial f}{\partial z}\,\mathbf{k}\right) dV$

$\displaystyle = \iiint_E \nabla f\, dV \quad \text{as desired.}$

16 Review

TRUE-FALSE QUIZ

1. False; div $\mathbf{F}$ is a scalar field.

3. True, by Theorem 16.5.3 and the fact that div $\mathbf{0} = 0$.

5. False. See Exercise 16.3.35. (But the assertion is true if D is simply-connected; see Theorem 16.3.6.)

7. False. For example, $\text{div}(y\,\mathbf{i}) = 0 = \text{div}(x\,\mathbf{j})$ but $y\,\mathbf{i} \neq x\,\mathbf{j}$.

9. True. See Exercise 16.5.24.

11. True. Apply the Divergence Theorem and use the fact that div $\mathbf{F} = 0$.

13. False. By Equations 16.4.5, the area is given by $-\oint_C y\, dx$ or $\oint_C x\, dy$.

EXERCISES

1. (a) Vectors starting on C point in roughly the direction opposite to C, so the tangential component $\mathbf{F} \cdot \mathbf{T}$ is negative.

Thus $\int_C \mathbf{F} \cdot d\mathbf{r} = \int_C \mathbf{F} \cdot \mathbf{T}\, ds$ is negative.

(b) The vectors that end near P are shorter than the vectors that start near P, so the net flow is outward near P and div $\mathbf{F}(P)$ is positive.

3. $\int_C yz\cos x\, ds = \int_0^\pi (3\cos t)(3\sin t)\cos t\, \sqrt{(1)^2 + (-3\sin t)^2 + (3\cos t)^2}\, dt = \int_0^\pi (9\cos^2 t\,\sin t)\sqrt{10}\, dt$

$= 9\sqrt{10}\left(-\tfrac{1}{3}\cos^3 t\right)\Big]_0^\pi = -3\sqrt{10}\,(-2) = 6\sqrt{10}$

5. $\int_C y^3 \, dx + x^2 \, dy = \int_{-1}^{1} \left[y^3(-2y) + (1-y^2)^2 \right] dy = \int_{-1}^{1}(-y^4 - 2y^2 + 1) \, dy$

$$= \left[-\tfrac{1}{5}y^5 - \tfrac{2}{3}y^3 + y \right]_{-1}^{1} = -\tfrac{1}{5} - \tfrac{2}{3} + 1 - \tfrac{1}{5} - \tfrac{2}{3} + 1 = \tfrac{4}{15}$$

7. $C: \; x = 1 + 2t \;\; \Rightarrow \;\; dx = 2 \, dt, \, y = 4t \;\; \Rightarrow \;\; dy = 4 \, dt, \, z = -1 + 3t \;\; \Rightarrow \;\; dz = 3 \, dt, \, 0 \leq t \leq 1.$

$$\int_C xy \, dx + y^2 \, dy + yz \, dz = \int_0^1 [(1+2t)(4t)(2) + (4t)^2(4) + (4t)(-1+3t)(3)] \, dt$$

$$= \int_0^1 (116t^2 - 4t) \, dt = \left[\tfrac{116}{3}t^3 - 2t^2 \right]_0^1 = \tfrac{116}{3} - 2 = \tfrac{110}{3}$$

9. $\mathbf{F}(\mathbf{r}(t)) = e^{-t}\mathbf{i} + t^2(-t)\mathbf{j} + (t^2 + t^3)\mathbf{k}, \, \mathbf{r}'(t) = 2t\,\mathbf{i} + 3t^2\,\mathbf{j} - \mathbf{k}$ and

$$\int_C \mathbf{F} \cdot d\mathbf{r} = \int_0^1 (2te^{-t} - 3t^5 - (t^2 + t^3)) \, dt = \left[-2te^{-t} - 2e^{-t} - \tfrac{1}{2}t^6 - \tfrac{1}{3}t^3 - \tfrac{1}{4}t^4 \right]_0^1 = \tfrac{11}{12} - \tfrac{4}{e}.$$

11. $\frac{\partial}{\partial y}\left[(1+xy)e^{xy} \right] = 2xe^{xy} + x^2ye^{xy} = \frac{\partial}{\partial x}\left[e^y + x^2e^{xy} \right]$ and the domain of $\mathbf{F}$ is $\mathbb{R}^2$, so $\mathbf{F}$ is conservative. Thus there

exists a function f such that $\mathbf{F} = \nabla f$. Then $f_y(x, y) = e^y + x^2e^{xy}$ implies $f(x, y) = e^y + xe^{xy} + g(x)$ and then

$f_x(x, y) = xye^{xy} + e^{xy} + g'(x) = (1 + xy)e^{xy} + g'(x)$. But $f_x(x, y) = (1 + xy)e^{xy}$, so $g'(x) = 0 \;\; \Rightarrow \;\; g(x) = K$.

Thus $f(x, y) = e^y + xe^{xy} + K$ is a potential function for $\mathbf{F}$.

13. Since $\frac{\partial}{\partial y}\left(4x^3y^2 - 2xy^3 \right) = 8x^3y - 6xy^2 = \frac{\partial}{\partial x}\left(2x^4y - 3x^2y^2 + 4y^3 \right)$ and the domain of $\mathbf{F}$ is $\mathbb{R}^2$, $\mathbf{F}$ is conservative.

Furthermore $f(x, y) = x^4y^2 - x^2y^3 + y^4$ is a potential function for $\mathbf{F}$. $t = 0$ corresponds to the point $(0, 1)$ and $t = 1$

corresponds to $(1, 1)$, so $\int_C \mathbf{F} \cdot d\mathbf{r} = f(1, 1) - f(0, 1) = 1 - 1 = 0$.

15. $C_1: \mathbf{r}(t) = t\mathbf{i} + t^2\mathbf{j}, -1 \leq t \leq 1;$

$C_2: \mathbf{r}(t) = -t\mathbf{i} + \mathbf{j}, -1 \leq t \leq 1.$

Then

$$\int_C xy^2 \, dx - x^2y \, dy = \int_{-1}^{1}(t^5 - 2t^5) \, dt + \int_{-1}^{1} t \, dt$$

$$= \left[-\tfrac{1}{6}t^6 \right]_{-1}^{1} + \left[\tfrac{1}{2}t^2 \right]_{-1}^{1} = 0$$

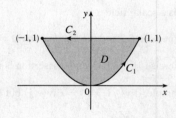

Using Green's Theorem, we have

$$\int_C xy^2 \, dx - x^2y \, dy = \iint_D \left[\frac{\partial}{\partial x}(-x^2y) - \frac{\partial}{\partial y}(xy^2) \right] dA = \iint_D (-2xy - 2xy) \, dA = \int_{-1}^{1} \int_{x^2}^{1} -4xy \, dy \, dx$$

$$= \int_{-1}^{1} \left[-2xy^2 \right]_{y=x^2}^{y=1} dx = \int_{-1}^{1}(2x^5 - 2x) \, dx = \left[\tfrac{1}{3}x^6 - x^2 \right]_{-1}^{1} = 0$$

17. $\int_C x^2y \, dx - xy^2 \, dy = \iint\limits_{x^2+y^2 \leq 4} \left[\frac{\partial}{\partial x}(-xy^2) - \frac{\partial}{\partial y}(x^2y) \right] dA = \iint\limits_{x^2+y^2 \leq 4} (-y^2 - x^2) \, dA = -\int_0^{2\pi} \int_0^2 r^3 \, dr \, d\theta = -8\pi$

19. If we assume there is such a vector field $\mathbf{G}$, then $\operatorname{div}(\operatorname{curl} \mathbf{G}) = 2 + 3z - 2xz$. But $\operatorname{div}(\operatorname{curl} \mathbf{F}) = 0$ for all vector fields $\mathbf{F}$.

Thus such a $\mathbf{G}$ cannot exist.

21. For any piecewise-smooth simple closed plane curve C bounding a region D, we can apply Green's Theorem to

$\mathbf{F}(x, y) = f(x)\,\mathbf{i} + g(y)\,\mathbf{j}$ to get $\int_C f(x) \, dx + g(y) \, dy = \iint_D \left[\frac{\partial}{\partial x}g(y) - \frac{\partial}{\partial y}f(x) \right] dA = \iint_D 0 \, dA = 0.$

23. $\nabla^2 f = 0$ means that $\dfrac{\partial^2 f}{\partial x^2} + \dfrac{\partial^2 f}{\partial y^2} = 0$. Now if $\mathbf{F} = f_y\,\mathbf{i} - f_x\,\mathbf{j}$ and C is any closed path in D, then applying Green's

Theorem, we get

$$\int_C \mathbf{F} \cdot d\mathbf{r} = \int_C f_y\,dx - f_x\,dy = \iint_D \left[\frac{\partial}{\partial x}(-f_x) - \frac{\partial}{\partial y}(f_y) \right] dA$$

$$= -\iint_D (f_{xx} + f_{yy})\,dA = -\iint_D 0\,dA = 0$$

Therefore the line integral is independent of path, by Theorem 16.3.3.

25. $z = f(x,y) = x^2 + 2y$ with $0 \le x \le 1$, $0 \le y \le 2x$. Thus

$$A(S) = \iint_D \sqrt{1 + 4x^2 + 4}\,dA = \int_0^1 \int_0^{2x} \sqrt{5 + 4x^2}\,dy\,dx = \int_0^1 2x\sqrt{5 + 4x^2}\,dx = \tfrac{1}{6}(5 + 4x^2)^{3/2}\Big]_0^1 = \tfrac{1}{6}\left(27 - 5\sqrt{5}\right).$$

27. $z = f(x,y) = x^2 + y^2$ with $0 \le x^2 + y^2 \le 4$ so $\mathbf{r}_x \times \mathbf{r}_y = -2x\,\mathbf{i} - 2y\,\mathbf{j} + \mathbf{k}$. Then

$$\iint_S z\,dS = \iint_{x^2 + y^2 \le 4} (x^2 + y^2)\sqrt{4x^2 + 4y^2 + 1}\,dA$$

$$= \int_0^{2\pi} \int_0^2 r^3\sqrt{1 + 4r^2}\,dr\,d\theta = \tfrac{1}{60}\pi\left(391\sqrt{17} + 1\right)$$

(Substitute $u = 1 + 4r^2$ and use tables.)

29. Since the sphere bounds a simple solid region, the Divergence Theorem applies and

$$\iint_S \mathbf{F} \cdot d\mathbf{S} = \iiint_E \operatorname{div} \mathbf{F}\,dV = \iiint_E (z - 2)\,dV = \iiint_E z\,dV - 2\iiint_E dV$$

$$= 0 \begin{bmatrix} \text{odd function in } z \\ \text{and } E \text{ is symmetric} \end{bmatrix} - 2 \cdot V(E) = -2 \cdot \tfrac{4}{3}\pi(2)^3 = -\tfrac{64}{3}\pi$$

Alternate solution: $\mathbf{F}(\mathbf{r}(\phi, \theta)) = 4\sin\phi\cos\theta\cos\phi\,\mathbf{i} - 4\sin\phi\sin\theta\,\mathbf{j} + 6\sin\phi\cos\theta\,\mathbf{k}$,

$\mathbf{r}_\phi \times \mathbf{r}_\theta = 4\sin^2\phi\cos\theta\,\mathbf{i} + 4\sin^2\phi\sin\theta\,\mathbf{j} + 4\sin\phi\cos\phi\,\mathbf{k}$, and

$\mathbf{F} \cdot (\mathbf{r}_\phi \times \mathbf{r}_\theta) = 16\sin^3\phi\cos^2\theta\cos\phi - 16\sin^3\phi\sin^2\theta + 24\sin^2\phi\cos\phi\cos\theta$. Then

$$\iint_S \mathbf{F} \cdot d\mathbf{S} = \int_0^{2\pi} \int_0^\pi (16\sin^3\phi\cos\phi\cos^2\theta - 16\sin^3\phi\sin^2\theta + 24\sin^2\phi\cos\phi\cos\theta)\,d\phi\,d\theta$$

$$= \int_0^{2\pi} \tfrac{4}{3}(-16\sin^2\theta)\,d\theta = -\tfrac{64}{3}\pi$$

31. Since $\operatorname{curl}\mathbf{F} = \mathbf{0}$, $\iint_S (\operatorname{curl}\mathbf{F}) \cdot d\mathbf{S} = 0$. We parametrize C: $\mathbf{r}(t) = \cos t\,\mathbf{i} + \sin t\,\mathbf{j}$, $0 \le t \le 2\pi$ and

$$\oint_C \mathbf{F} \cdot d\mathbf{r} = \int_0^{2\pi} (-\cos^2 t\,\sin t + \sin^2 t\,\cos t)\,dt = \tfrac{1}{3}\cos^3 t + \tfrac{1}{3}\sin^3 t\Big]_0^{2\pi} = 0.$$

33. The surface is given by $x + y + z = 1$ or $z = 1 - x - y$, $0 \le x \le 1$, $0 \le y \le 1 - x$ and $\mathbf{r}_x \times \mathbf{r}_y = \mathbf{i} + \mathbf{j} + \mathbf{k}$. Then

$$\oint_C \mathbf{F} \cdot d\mathbf{r} = \iint_S \operatorname{curl}\mathbf{F} \cdot d\mathbf{S} = \iint_D (-y\,\mathbf{i} - z\,\mathbf{j} - x\,\mathbf{k}) \cdot (\mathbf{i} + \mathbf{j} + \mathbf{k})\,dA = \iint_D (-1)\,dA = -(\text{area of } D) = -\tfrac{1}{2}.$$

35. $\iiint_E \operatorname{div}\mathbf{F}\,dV = \iiint_{x^2 + y^2 + z^2 \le 1} 3\,dV = 3(\text{volume of sphere}) = 4\pi$. Then

$$\mathbf{F}(\mathbf{r}(\phi, \theta)) \cdot (\mathbf{r}_\phi \times \mathbf{r}_\theta) = \sin^3\phi\cos^2\theta + \sin^3\phi\sin^2\theta + \sin\phi\cos^2\phi = \sin\phi \text{ and}$$

$$\iint_S \mathbf{F} \cdot d\mathbf{S} = \int_0^{2\pi} \int_0^\pi \sin\phi\,d\phi\,d\theta = (2\pi)(2) = 4\pi.$$

37. Because curl $\mathbf{F} = \mathbf{0}$, $\mathbf{F}$ is conservative, so there exists a function f such that $\nabla f = \mathbf{F}$. Then $f_x(x, y, z) = 3x^2yz - 3y$

implies $f(x, y, z) = x^3yz - 3xy + g(y, z)$ $\Rightarrow$ $f_y(x, y, z) = x^3z - 3x + g_y(y, z)$. But $f_y(x, y, z) = x^3z - 3x$, so

$g(y, z) = h(z)$ and $f(x, y, z) = x^3yz - 3xy + h(z)$. Then $f_z(x, y, z) = x^3y + h'(z)$ but $f_z(x, y, z) = x^3y + 2z$,

so $h(z) = z^2 + K$ and a potential function for $\mathbf{F}$ is $f(x, y, z) = x^3yz - 3xy + z^2$. Hence

$\int_C \mathbf{F} \cdot d\mathbf{r} = \int_C \nabla f \cdot d\mathbf{r} = f(0, 3, 0) - f(0, 0, 2) = 0 - 4 = -4$.

39. By the Divergence Theorem, $\iint_S \mathbf{F} \cdot \mathbf{n} \, dS = \iiint_E \operatorname{div} \mathbf{F} \, dV = 3(\text{volume of } E) = 3(8 - 1) = 21$.

41. Let $\mathbf{F} = \mathbf{a} \times \mathbf{r} = \langle a_1, a_2, a_3 \rangle \times \langle x, y, z \rangle = \langle a_2z - a_3y, a_3x - a_1z, a_1y - a_2x \rangle$. Then curl $\mathbf{F} = \langle 2a_1, 2a_2, 2a_3 \rangle = 2\mathbf{a}$,

and $\iint_S 2\mathbf{a} \cdot d\mathbf{S} = \iint_S \operatorname{curl} \mathbf{F} \cdot d\mathbf{S} = \int_C \mathbf{F} \cdot d\mathbf{r} = \int_C (\mathbf{a} \times \mathbf{r}) \cdot d\mathbf{r}$ by Stokes' Theorem.

1. Let S_1 be the portion of $\Omega(S)$ between $S(a)$ and S, and let ∂S_1 be its boundary. Also let S_L be the lateral surface of S_1 [that is, the surface of S_1 except S and $S(a)$]. Applying the Divergence Theorem we have $\iint_{\partial S_1} \frac{\mathbf{r} \cdot \mathbf{n}}{r^3} \, dS = \iiint_{S_1} \nabla \cdot \frac{\mathbf{r}}{r^3} \, dV$.

But

$$\nabla \cdot \frac{\mathbf{r}}{r^3} = \left\langle \frac{\partial}{\partial x}, \frac{\partial}{\partial y}, \frac{\partial}{\partial z} \right\rangle \cdot \left\langle \frac{x}{(x^2 + y^2 + z^2)^{3/2}}, \frac{y}{(x^2 + y^2 + z^2)^{3/2}}, \frac{z}{(x^2 + y^2 + z^2)^{3/2}} \right\rangle$$

$$= \frac{(x^2 + y^2 + z^2 - 3x^2) + (x^2 + y^2 + z^2 - 3y^2) + (x^2 + y^2 + z^2 - 3z^2)}{(x^2 + y^2 + z^2)^{5/2}} = 0$$

$\Rightarrow \iint_{\partial S_1} \frac{\mathbf{r} \cdot \mathbf{n}}{r^3} \, dS = \iiint_{S_1} 0 \, dV = 0$. On the other hand, notice that for the surfaces of ∂S_1 other than $S(a)$ and S,

$\mathbf{r} \cdot \mathbf{n} = 0 \Rightarrow$

$$0 = \iint_{\partial S_1} \frac{\mathbf{r} \cdot \mathbf{n}}{r^3} \, dS = \iint_{S} \frac{\mathbf{r} \cdot \mathbf{n}}{r^3} \, dS + \iint_{S(a)} \frac{\mathbf{r} \cdot \mathbf{n}}{r^3} \, dS + \iint_{S_L} \frac{\mathbf{r} \cdot \mathbf{n}}{r^3} \, dS = \iint_{S} \frac{\mathbf{r} \cdot \mathbf{n}}{r^3} \, dS + \iint_{S(a)} \frac{\mathbf{r} \cdot \mathbf{n}}{r^3} \, dS \Rightarrow$$

$\iint_{S} \frac{\mathbf{r} \cdot \mathbf{n}}{r^3} \, dS = -\iint_{S(a)} \frac{\mathbf{r} \cdot \mathbf{n}}{r^3} \, dS$. Notice that on $S(a)$, $r = a \Rightarrow \mathbf{n} = -\frac{\mathbf{r}}{r} = -\frac{\mathbf{r}}{a}$ and $\mathbf{r} \cdot \mathbf{r} = r^2 = a^2$, so

that $-\iint_{S(a)} \frac{\mathbf{r} \cdot \mathbf{n}}{r^3} \, dS = \iint_{S(a)} \frac{\mathbf{r} \cdot \mathbf{r}}{a^4} \, dS = \iint_{S(a)} \frac{a^2}{a^4} \, dS = \frac{1}{a^2} \iint_{S(a)} dS = \frac{\text{area of } S(a)}{a^2} = |\Omega(S)|$.

Therefore $|\Omega(S)| = \iint_{S} \frac{\mathbf{r} \cdot \mathbf{n}}{r^3} \, dS$.

3. The given line integral $\frac{1}{2} \int_C (bz - cy) \, dx + (cx - az) \, dy + (ay - bx) \, dz$ can be expressed as $\int_C \mathbf{F} \cdot d\mathbf{r}$ if we define the vector field $\mathbf{F}$ by $\mathbf{F}(x, y, z) = P\mathbf{i} + Q\mathbf{j} + R\mathbf{k} = \frac{1}{2}(bz - cy)\mathbf{i} + \frac{1}{2}(cx - az)\mathbf{j} + \frac{1}{2}(ay - bx)\mathbf{k}$. Then define S to be the planar interior of C, so S is an oriented, smooth surface. Stokes' Theorem says $\int_C \mathbf{F} \cdot d\mathbf{r} = \iint_S \text{curl } \mathbf{F} \cdot d\mathbf{S} = \iint_S \text{curl } \mathbf{F} \cdot \mathbf{n} \, dS$. Now

$$\text{curl } \mathbf{F} = \left(\frac{\partial R}{\partial y} - \frac{\partial Q}{\partial z} \right) \mathbf{i} + \left(\frac{\partial P}{\partial z} - \frac{\partial R}{\partial x} \right) \mathbf{j} + \left(\frac{\partial Q}{\partial x} - \frac{\partial P}{\partial y} \right) \mathbf{k}$$

$$= \left(\tfrac{1}{2}a + \tfrac{1}{2}a \right) \mathbf{i} + \left(\tfrac{1}{2}b + \tfrac{1}{2}b \right) \mathbf{j} + \left(\tfrac{1}{2}c + \tfrac{1}{2}c \right) \mathbf{k} = a\mathbf{i} + b\mathbf{j} + c\mathbf{k} = \mathbf{n}$$

so $\text{curl } \mathbf{F} \cdot \mathbf{n} = \mathbf{n} \cdot \mathbf{n} = |\mathbf{n}|^2 = 1$, hence $\iint_S \text{curl } \mathbf{F} \cdot \mathbf{n} \, dS = \iint_S dS$ which is simply the surface area of S. Thus, $\int_C \mathbf{F} \cdot d\mathbf{r} = \frac{1}{2} \int_C (bz - cy) \, dx + (cx - az) \, dy + (ay - bx) \, dz$ is the plane area enclosed by C.

5. $(\mathbf{F} \cdot \nabla) \mathbf{G} = \left(P_1 \frac{\partial}{\partial x} + Q_1 \frac{\partial}{\partial y} + R_1 \frac{\partial}{\partial z} \right) (P_2 \mathbf{i} + Q_2 \mathbf{j} + R_2 \mathbf{k})$

$$= \left(P_1 \frac{\partial P_2}{\partial x} + Q_1 \frac{\partial P_2}{\partial y} + R_1 \frac{\partial P_2}{\partial z} \right) \mathbf{i} + \left(P_1 \frac{\partial Q_2}{\partial x} + Q_1 \frac{\partial Q_2}{\partial y} + R_1 \frac{\partial Q_2}{\partial z} \right) \mathbf{j}$$

$$+ \left(P_1 \frac{\partial R_2}{\partial x} + Q_1 \frac{\partial R_2}{\partial y} + R_1 \frac{\partial R_2}{\partial z} \right) \mathbf{k}$$

$= (\mathbf{F} \cdot \nabla P_2) \mathbf{i} + (\mathbf{F} \cdot \nabla Q_2) \mathbf{j} + (\mathbf{F} \cdot \nabla R_2) \mathbf{k}$.

Similarly, $(\mathbf{G} \cdot \nabla)\mathbf{F} = (\mathbf{G} \cdot \nabla P_1)\mathbf{i} + (\mathbf{G} \cdot \nabla Q_1)\mathbf{j} + (\mathbf{G} \cdot \nabla R_1)\mathbf{k}$. Then

$$\mathbf{F} \times \operatorname{curl}\mathbf{G} = \begin{vmatrix} \mathbf{i} & \mathbf{j} & \mathbf{k} \\ P_1 & Q_1 & R_1 \\ \partial R_2/\partial y - \partial Q_2/\partial z & \partial P_2/\partial z - \partial R_2/\partial x & \partial Q_2/\partial x - \partial P_2/\partial y \end{vmatrix}$$

$$= \left(Q_1\frac{\partial Q_2}{\partial x} - Q_1\frac{\partial P_2}{\partial y} - R_1\frac{\partial P_2}{\partial z} + R_1\frac{\partial R_2}{\partial x} \right)\mathbf{i} + \left(R_1\frac{\partial R_2}{\partial y} - R_1\frac{\partial Q_2}{\partial z} - P_1\frac{\partial Q_2}{\partial x} + P_1\frac{\partial P_2}{\partial y} \right)\mathbf{j}$$

$$+ \left(P_1\frac{\partial P_2}{\partial z} - P_1\frac{\partial R_2}{\partial x} - Q_1\frac{\partial R_2}{\partial y} + Q_1\frac{\partial Q_2}{\partial z} \right)\mathbf{k}$$

and

$$\mathbf{G} \times \operatorname{curl}\mathbf{F} = \left(Q_2\frac{\partial Q_1}{\partial x} - Q_2\frac{\partial P_1}{\partial y} - R_2\frac{\partial P_1}{\partial z} + R_2\frac{\partial R_1}{\partial x} \right)\mathbf{i} + \left(R_2\frac{\partial R_1}{\partial y} - R_2\frac{\partial Q_1}{\partial z} - P_2\frac{\partial Q_1}{\partial x} + P_2\frac{\partial P_1}{\partial y} \right)\mathbf{j}$$

$$+ \left(P_2\frac{\partial P_1}{\partial z} - P_2\frac{\partial R_1}{\partial x} - Q_2\frac{\partial R_1}{\partial y} + Q_2\frac{\partial Q_1}{\partial z} \right)\mathbf{k}.$$

Then

$$(\mathbf{F} \cdot \nabla)\mathbf{G} + \mathbf{F} \times \operatorname{curl}\mathbf{G} = \left(P_1\frac{\partial P_2}{\partial x} + Q_1\frac{\partial Q_2}{\partial x} + R_1\frac{\partial R_2}{\partial x} \right)\mathbf{i} + \left(P_1\frac{\partial P_2}{\partial y} + Q_1\frac{\partial Q_2}{\partial y} + R_1\frac{\partial R_2}{\partial y} \right)\mathbf{j}$$

$$+ \left(P_1\frac{\partial P_2}{\partial z} + Q_1\frac{\partial Q_2}{\partial z} + R_1\frac{\partial R_2}{\partial z} \right)\mathbf{k}$$

and

$$(\mathbf{G} \cdot \nabla)\mathbf{F} + \mathbf{G} \times \operatorname{curl}\mathbf{F} = \left(P_2\frac{\partial P_1}{\partial x} + Q_2\frac{\partial Q_1}{\partial x} + R_2\frac{\partial R_1}{\partial x} \right)\mathbf{i} + \left(P_2\frac{\partial P_1}{\partial y} + Q_2\frac{\partial Q_1}{\partial y} + R_2\frac{\partial R_1}{\partial y} \right)\mathbf{j}$$

$$+ \left(P_2\frac{\partial P_1}{\partial z} + Q_2\frac{\partial Q_1}{\partial z} + R_2\frac{\partial R_1}{\partial z} \right)\mathbf{k}.$$

Hence

$$(\mathbf{F} \cdot \nabla)\mathbf{G} + \mathbf{F} \times \operatorname{curl}\mathbf{G} + (\mathbf{G} \cdot \nabla)\mathbf{F} + \mathbf{G} \times \operatorname{curl}\mathbf{F}$$

$$= \left[\left(P_1\frac{\partial P_2}{\partial x} + P_2\frac{\partial P_1}{\partial x} \right) + \left(Q_1\frac{\partial Q_2}{\partial x} + Q_2\frac{\partial Q_1}{\partial y} \right) + \left(R_1\frac{\partial R_2}{\partial x} + R_2\frac{\partial R_1}{\partial x} \right) \right]\mathbf{i}$$

$$+ \left[\left(P_1\frac{\partial P_2}{\partial y} + P_2\frac{\partial P_1}{\partial y} \right) + \left(Q_1\frac{\partial Q_2}{\partial y} + Q_2\frac{\partial Q_1}{\partial y} \right) + \left(R_1\frac{\partial R_2}{\partial y} + R_2\frac{\partial R_1}{\partial y} \right) \right]\mathbf{j}$$

$$+ \left[\left(P_1\frac{\partial P_2}{\partial z} + P_2\frac{\partial P_1}{\partial z} \right) + \left(Q_1\frac{\partial Q_2}{\partial z} + Q_2\frac{\partial Q_1}{\partial z} \right) + \left(R_1\frac{\partial R_2}{\partial z} + R_2\frac{\partial R_1}{\partial z} \right) \right]\mathbf{k}$$

$$= \nabla(P_1 P_2 + Q_1 Q_2 + R_1 R_2) = \nabla(\mathbf{F} \cdot \mathbf{G}).$$

17 □ SECOND-ORDER DIFFERENTIAL EQUATIONS

17.1 Second-Order Linear Equations

1. The auxiliary equation is $r^2 - r - 6 = 0 \;\Rightarrow\; (r-3)(r+2) = 0 \;\Rightarrow\; r = 3, r = -2$. Then by (8) the general solution is $y = c_1 e^{3x} + c_2 e^{-2x}$.

3. The auxiliary equation is $r^2 + 2 = 0 \;\Rightarrow\; r = \pm\sqrt{2}\,i$. Then by (11) the general solution is
$$y = e^{0x}\left(c_1 \cos\left(\sqrt{2}\,x\right) + c_2 \sin\left(\sqrt{2}\,x\right)\right) = c_1 \cos\left(\sqrt{2}\,x\right) + c_2 \sin\left(\sqrt{2}\,x\right).$$

5. The auxiliary equation is $4r^2 + 4r + 1 = 0 \;\Rightarrow\; (2r+1)^2 = 0 \;\Rightarrow\; r = -\tfrac{1}{2}$. Then by (10), the general solution is
$$y = c_1 e^{-x/2} + c_2 x e^{-x/2}.$$

7. The auxiliary equation is $3r^2 - 4r = r(3r - 4) = 0 \;\Rightarrow\; r = 0, r = \tfrac{4}{3}$, so $y = c_1 e^{0x} + c_2 e^{4x/3} = c_1 + c_2 e^{4x/3}$.

9. The auxiliary equation is $r^2 - 4r + 13 = 0 \;\Rightarrow\; r = \dfrac{4 \pm \sqrt{-36}}{2} = 2 \pm 3i$, so $y = e^{2x}(c_1 \cos 3x + c_2 \sin 3x)$.

11. The auxiliary equation is $2r^2 + 2r - 1 = 0 \;\Rightarrow\; r = \dfrac{-2 \pm \sqrt{12}}{4} = \dfrac{-1 \pm \sqrt{3}}{2}$, so $y = c_1 e^{\left(-1+\sqrt{3}\right)t/2} + c_2 e^{\left(-1-\sqrt{3}\right)t/2}$.

13. The auxiliary equation is $3r^2 + 4r + 3 = 0 \;\Rightarrow\; r = \dfrac{-4 \pm \sqrt{-20}}{6} = -\tfrac{2}{3} \pm \tfrac{\sqrt{5}}{3}i$, so
$$V = e^{-2t/3}\left[c_1 \cos\left(\tfrac{\sqrt{5}}{3}t\right) + c_2 \sin\left(\tfrac{\sqrt{5}}{3}t\right)\right].$$

15. The auxiliary equation is $r^2 + 2r + 2 = 0 \;\Rightarrow\;$

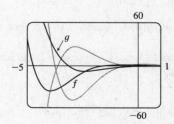

$r = \dfrac{-2 \pm \sqrt{-4}}{2} = -1 \pm i$, so the general solution is

$y = e^{-x}\left(c_1 \cos x + c_2 \sin x\right)$. We graph the basic solutions

$f(x) = e^{-x}\cos x$, $g(x) = e^{-x}\sin x$ as well as

$y = e^{-x}\left(-\cos x - 2\sin x\right)$ and $y = e^{-x}\left(2\cos x + 3\sin x\right)$. All the solutions oscillate with amplitudes that become arbitrarily large as $x \to -\infty$ and the solutions are asymptotic to the x-axis as $x \to \infty$.

17. $r^2 + 3 = 0 \;\Rightarrow\; r = \pm\sqrt{3}\,i$ and the general solution is

$y = e^{0x}\left[c_1 \cos\left(\sqrt{3}\,x\right) + c_2 \sin\left(\sqrt{3}\,x\right)\right] = c_1 \cos\left(\sqrt{3}\,x\right) + c_2 \sin\left(\sqrt{3}\,x\right)$. Then $y(0) = 1 \;\Rightarrow\; c_1 = 1$ and, since

$y' = -\sqrt{3}\,c_1 \sin\left(\sqrt{3}\,x\right) + \sqrt{3}\,c_2 \cos\left(\sqrt{3}\,x\right)$, $y'(0) = 3 \;\Rightarrow\; \sqrt{3}\,c_2 = 3 \;\Rightarrow\; c_2 = \tfrac{3}{\sqrt{3}} = \sqrt{3}$, so the solution to the

initial-value problem is $y = \cos\left(\sqrt{3}\,x\right) + \sqrt{3}\,\sin\left(\sqrt{3}\,x\right)$.

346

19. $9r^2 + 12r + 4 = (3r + 2)^2 = 0 \Rightarrow r = -\frac{2}{3}$ and the general solution is $y = c_1 e^{-2x/3} + c_2 x e^{-2x/3}$. Then $y(0) = 1 \Rightarrow$
$c_1 = 1$ and, since $y' = -\frac{2}{3}c_1 e^{-2x/3} + c_2 \left(1 - \frac{2}{3}x\right) e^{-2x/3}$, $y'(0) = 0 \Rightarrow -\frac{2}{3}c_1 + c_2 = 0$, so $c_2 = \frac{2}{3}$ and the solution to
the initial-value problem is $y = e^{-2x/3} + \frac{2}{3}x e^{-2x/3}$.

21. $r^2 - 6r + 10 = 0 \Rightarrow r = 3 \pm i$ and the general solution is $y = e^{3x}(c_1 \cos x + c_2 \sin x)$. Then $2 = y(0) = c_1$ and
$3 = y'(0) = c_2 + 3c_1 \Rightarrow c_2 = -3$ and the solution to the initial-value problem is $y = e^{3x}(2 \cos x - 3 \sin x)$.

23. $r^2 - r - 12 = (r - 4)(r + 3) = 0 \Rightarrow r = 4, r = -3$ and the general solution is $y = c_1 e^{4x} + c_2 e^{-3x}$. Then
$0 = y(1) = c_1 e^4 + c_2 e^{-3}$ and $1 = y'(1) = 4c_1 e^4 - 3c_2 e^{-3}$ so $c_1 = \frac{1}{7}e^{-4}$, $c_2 = -\frac{1}{7}e^3$ and the solution to the initial-value
problem is $y = \frac{1}{7}e^{-4}e^{4x} - \frac{1}{7}e^3 e^{-3x} = \frac{1}{7}e^{4x-4} - \frac{1}{7}e^{3-3x}$.

25. $r^2 + 16 = 0 \Rightarrow r = \pm 4i$ and the general solution is $y = c_1 \cos 4x + c_2 \sin 4x$. Then $-3 = y(0) = c_1$ and
$2 = y(\pi/8) = c_2$, so the solution of the boundary-value problem is $y = -3 \cos 4x + 2 \sin 4x$.

27. $r^2 + 4r + 4 = (r + 2)^2 = 0 \Rightarrow r = -2$ and the general solution is $y = c_1 e^{-2x} + c_2 x e^{-2x}$. Then $2 = y(0) = c_1$ and
$0 = y(1) = c_1 e^{-2} + c_2 e^{-2}$ so $c_2 = -2$, and the solution of the boundary-value problem is $y = 2e^{-2x} - 2x e^{-2x}$.

29. $r^2 - r = r(r - 1) = 0 \Rightarrow r = 0, r = 1$ and the general solution is $y = c_1 + c_2 e^x$. Then $1 = y(0) = c_1 + c_2$
and $2 = y(1) = c_1 + c_2 e$ so $c_1 = \dfrac{e - 2}{e - 1}$, $c_2 = \dfrac{1}{e - 1}$. The solution of the boundary-value problem is $y = \dfrac{e - 2}{e - 1} + \dfrac{e^x}{e - 1}$.

31. $r^2 + 4r + 20 = 0 \Rightarrow r = -2 \pm 4i$ and the general solution is $y = e^{-2x}(c_1 \cos 4x + c_2 \sin 4x)$. But $1 = y(0) = c_1$ and
$2 = y(\pi) = c_1 e^{-2\pi} \Rightarrow c_1 = 2e^{2\pi}$, so there is no solution.

33. (a) *Case 1* $(\lambda = 0)$: $y'' + \lambda y = 0 \Rightarrow y'' = 0$ which has an auxiliary equation $r^2 = 0 \Rightarrow r = 0 \Rightarrow y = c_1 + c_2 x$
   where $y(0) = 0$ and $y(L) = 0$. Thus, $0 = y(0) = c_1$ and $0 = y(L) = c_2 L \Rightarrow c_1 = c_2 = 0$. Thus $y = 0$.

   Case 2 $(\lambda < 0)$: $y'' + \lambda y = 0$ has auxiliary equation $r^2 = -\lambda \Rightarrow r = \pm\sqrt{-\lambda}$ [distinct and real since $\lambda < 0$] $\Rightarrow$
   $y = c_1 e^{\sqrt{-\lambda}x} + c_2 e^{-\sqrt{-\lambda}x}$ where $y(0) = 0$ and $y(L) = 0$. Thus $0 = y(0) = c_1 + c_2$ (∗) and
   $0 = y(L) = c_1 e^{\sqrt{-\lambda}L} + c_2 e^{-\sqrt{-\lambda}L}$ (†).
   Multiplying (∗) by $e^{\sqrt{-\lambda}L}$ and subtracting (†) gives $c_2\left(e^{\sqrt{-\lambda}L} - e^{-\sqrt{-\lambda}L}\right) = 0 \Rightarrow c_2 = 0$ and thus $c_1 = 0$ from (∗).
   Thus $y = 0$ for the cases $\lambda = 0$ and $\lambda < 0$.

(b) $y'' + \lambda y = 0$ has an auxiliary equation $r^2 + \lambda = 0 \Rightarrow r = \pm i\sqrt{\lambda} \Rightarrow y = c_1 \cos \sqrt{\lambda}x + c_2 \sin \sqrt{\lambda}x$ where
   $y(0) = 0$ and $y(L) = 0$. Thus, $0 = y(0) = c_1$ and $0 = y(L) = c_2 \sin \sqrt{\lambda}L$ since $c_1 = 0$. Since we cannot have a trivial
   solution, $c_2 \neq 0$ and thus $\sin \sqrt{\lambda}L = 0 \Rightarrow \sqrt{\lambda}L = n\pi$ where n is an integer $\Rightarrow \lambda = n^2\pi^2/L^2$ and
   $y = c_2 \sin(n\pi x/L)$ where n is an integer.

35. (a) $r^2 - 2r + 2 = 0 \Rightarrow r = 1 \pm i$ and the general solution is $y = e^x(c_1 \cos x + c_2 \sin x)$. If $y(a) = c$ and $y(b) = d$ then

$e^a(c_1 \cos a + c_2 \sin a) = c \Rightarrow c_1 \cos a + c_2 \sin a = ce^{-a}$ and $e^b(c_1 \cos b + c_2 \sin b) = d \Rightarrow$

$c_1 \cos b + c_2 \sin b = de^{-b}$. This gives a linear system in c_1 and c_2 which has a unique solution if the lines are not parallel.

If the lines are not vertical or horizontal, we have parallel lines if $\cos a = k \cos b$ and $\sin a = k \sin b$ for some nonzero

constant k or $\dfrac{\cos a}{\cos b} = k = \dfrac{\sin a}{\sin b} \Rightarrow \dfrac{\sin a}{\cos a} = \dfrac{\sin b}{\cos b} \Rightarrow \tan a = \tan b \Rightarrow b - a = n\pi$, n any integer. (Note that

none of $\cos a$, $\cos b$, $\sin a$, $\sin b$ are zero.) If the lines are both horizontal then $\cos a = \cos b = 0 \Rightarrow b - a = n\pi$, and

similarly vertical lines means $\sin a = \sin b = 0 \Rightarrow b - a = n\pi$. Thus the system has a unique solution if $b - a \neq n\pi$.

(b) The linear system has no solution if the lines are parallel but not identical. From part (a) the lines are parallel if

$b - a = n\pi$. If the lines are not horizontal, they are identical if $ce^{-a} = kde^{-b} \Rightarrow \dfrac{ce^{-a}}{de^{-b}} = k = \dfrac{\cos a}{\cos b} \Rightarrow$

$\dfrac{c}{d} = e^{a-b} \dfrac{\cos a}{\cos b}$. (If $d = 0$ then $c = 0$ also.) If they are horizontal then $\cos b = 0$, but $k = \dfrac{\sin a}{\sin b}$ also (and $\sin b \neq 0$) so

we require $\dfrac{c}{d} = e^{a-b} \dfrac{\sin a}{\sin b}$. Thus the system has no solution if $b - a = n\pi$ and $\dfrac{c}{d} \neq e^{a-b} \dfrac{\cos a}{\cos b}$ unless $\cos b = 0$, in

which case $\dfrac{c}{d} \neq e^{a-b} \dfrac{\sin a}{\sin b}$.

(c) The linear system has infinitely many solution if the lines are identical (and necessarily parallel). From part (b) this occurs

when $b - a = n\pi$ and $\dfrac{c}{d} = e^{a-b} \dfrac{\cos a}{\cos b}$ unless $\cos b = 0$, in which case $\dfrac{c}{d} = e^{a-b} \dfrac{\sin a}{\sin b}$.

17.2 Nonhomogeneous Linear Equations

1. The auxiliary equation is $r^2 + 2r - 8 = (r - 2)(r + 4) = 0 \Rightarrow r = 2, r = -4$, so the complementary solution is

$y_c(x) = c_1 e^{2x} + c_2 e^{-4x}$. We try the particular solution $y_p(x) = Ax^2 + Bx + C$, so $y_p' = 2Ax + B$ and $y_p'' = 2A$.

Substituting into the differential equation, we have $(2A) + 2(2Ax + B) - 8(Ax^2 + Bx + C) = 1 - 2x^2$ or

$-8Ax^2 + (4A - 8B)x + (2A + 2B - 8C) = -2x^2 + 1$. Comparing coefficients gives $-8A = -2 \Rightarrow$

$A = \frac{1}{4}$, $4A - 8B = 0 \Rightarrow B = \frac{1}{8}$, and $2A + 2B - 8C = 1 \Rightarrow C = -\frac{1}{32}$, so the general solution is

$y(x) = y_c(x) + y_p(x) = c_1 e^{2x} + c_2 e^{-4x} + \frac{1}{4}x^2 + \frac{1}{8}x - \frac{1}{32}$.

3. The auxiliary equation is $9r^2 + 1 = 0$ with roots $r = \pm \frac{1}{3}i$, so the complementary solution is

$y_c(x) = c_1 \cos(x/3) + c_2 \sin(x/3)$. Try the particular solution $y_p(x) = Ae^{2x}$, so $y_p' = 2Ae^{2x}$ and $y_p'' = 4Ae^{2x}$.

Substitution into the differential equation gives $9(4Ae^{2x}) + (Ae^{2x}) = e^{2x}$ or $37Ae^{2x} = e^{2x}$. Thus $37A = 1 \Rightarrow A = \frac{1}{37}$

and the general solution is $y(x) = y_c(x) + y_p(x) = c_1 \cos(x/3) + c_2 \sin(x/3) + \frac{1}{37}e^{2x}$.

5. The auxiliary equation is $r^2 - 4r + 5 = 0$ with roots $r = 2 \pm i$, so the complementary solution is

$y_c(x) = e^{2x}(c_1 \cos x + c_2 \sin x)$. Try $y_p(x) = Ae^{-x}$, so $y_p' = -Ae^{-x}$ and $y_p'' = Ae^{-x}$. Substitution gives

$Ae^{-x} - 4(-Ae^{-x}) + 5(Ae^{-x}) = e^{-x} \Rightarrow 10Ae^{-x} = e^{-x} \Rightarrow A = \frac{1}{10}$. Thus the general solution is

$y(x) = e^{2x}(c_1 \cos x + c_2 \sin x) + \frac{1}{10}e^{-x}$.

7. The auxiliary equation is $r^2 - 2r + 5 = 0$ with roots $r = 1 \pm 2i$, so the complementary solution is

$y_c(x) = e^x(c_1 \cos 2x + c_2 \sin 2x)$. Try the particular solution $y_p(x) = A \cos x + B \sin x$, so $y_p' = -A \sin x + B \cos x$

and $y_p'' = -A \cos x - B \sin x$. Substituting, we have

$(-A \cos x - B \sin x) - 2(-A \sin x + B \cos x) + 5(A \cos x + B \sin x) = \sin x \Rightarrow$

$(4A - 2B) \cos x + (2A + 4B) \sin x = \sin x$. Then $4A - 2B = 0, 2A + 4B = 1 \Rightarrow A = \frac{1}{10}, B = \frac{1}{5}$ and the general

solution is $y(x) = y_c(x) + y_p(x) = e^x(c_1 \cos 2x + c_2 \sin 2x) + \frac{1}{10} \cos x + \frac{1}{5} \sin x$. But $1 = y(0) = c_1 + \frac{1}{10} \Rightarrow c_1 = \frac{9}{10}$

and $1 = y'(0) = 2c_2 + c_1 + \frac{1}{5} \Rightarrow c_2 = -\frac{1}{20}$. Thus the solution to the initial-value problem is

$y(x) = e^x\left(\frac{9}{10} \cos 2x - \frac{1}{20} \sin 2x\right) + \frac{1}{10} \cos x + \frac{1}{5} \sin x$.

9. The auxiliary equation is $r^2 - r = 0$ with roots $r = 0, r = 1$ so the complementary solution is $y_c(x) = c_1 + c_2 e^x$.

Try $y_p(x) = x(Ax + B)e^x$ so that no term in y_p is a solution of the complementary equation. Then

$y_p' = (Ax^2 + (2A + B)x + B)e^x$ and $y_p'' = (Ax^2 + (4A + B)x + (2A + 2B))e^x$. Substitution into the differential equation

gives $(Ax^2 + (4A + B)x + (2A + 2B))e^x - (Ax^2 + (2A + B)x + B)e^x = xe^x \Rightarrow (2Ax + (2A + B))e^x = xe^x \Rightarrow$

$A = \frac{1}{2}, B = -1$. Thus $y_p(x) = \left(\frac{1}{2}x^2 - x\right)e^x$ and the general solution is $y(x) = c_1 + c_2 e^x + \left(\frac{1}{2}x^2 - x\right)e^x$. But

$2 = y(0) = c_1 + c_2$ and $1 = y'(0) = c_2 - 1$, so $c_2 = 2$ and $c_1 = 0$. The solution to the initial-value problem is

$y(x) = 2e^x + \left(\frac{1}{2}x^2 - x\right)e^x = e^x\left(\frac{1}{2}x^2 - x + 2\right)$.

11. The auxiliary equation is $r^2 + 3r + 2 = (r + 1)(r + 2) = 0$, so $r = -1, r = -2$ and $y_c(x) = c_1 e^{-x} + c_2 e^{-2x}$.

Try $y_p = A \cos x + B \sin x \Rightarrow y_p' = -A \sin x + B \cos x, y_p'' = -A \cos x - B \sin x$. Substituting into the differential

equation gives $(-A \cos x - B \sin x) + 3(-A \sin x + B \cos x) + 2(A \cos x + B \sin x) = \cos x$ or

$(A + 3B) \cos x + (-3A + B) \sin x = \cos x$. Then solving the equations

$A + 3B = 1, -3A + B = 0$ gives $A = \frac{1}{10}, B = \frac{3}{10}$ and the general

solution is $y(x) = c_1 e^{-x} + c_2 e^{-2x} + \frac{1}{10} \cos x + \frac{3}{10} \sin x$. The graph

shows y_p and several other solutions. Notice that all solutions are

asymptotic to y_p as $x \to \infty$. Except for y_p, all solutions approach either ∞

or $-\infty$ as $x \to -\infty$.

13. Here $y_c(x) = c_1 e^{2x} + c_2 e^{-x}$, and a trial solution is $y_p(x) = (Ax + B)e^x \cos x + (Cx + D)e^x \sin x$.

15. Here $y_c(x) = c_1 e^{2x} + c_2 e^x$. For $y'' - 3y' + 2y = e^x$ try $y_{p_1}(x) = Axe^x$ (since $y = Ae^x$ is a solution of the complementary equation) and for $y'' - 3y' + 2y = \sin x$ try $y_{p_2}(x) = B \cos x + C \sin x$. Thus a trial solution is

$$y_p(x) = y_{p_1}(x) + y_{p_2}(x) = Axe^x + B \cos x + C \sin x.$$

17. Since $y_c(x) = e^{-x}(c_1 \cos 3x + c_2 \sin 3x)$ we try $y_p(x) = x(Ax^2 + Bx + C)e^{-x} \cos 3x + x(Dx^2 + Ex + F)e^{-x} \sin 3x$

(so that no term of y_p is a solution of the complementary equation).

Note: Solving Equations (7) and (9) in The Method of Variation of Parameters gives

$$u_1' = -\frac{Gy_2}{a\,(y_1 y_2' - y_2 y_1')} \quad \text{and} \quad u_2' = \frac{Gy_1}{a\,(y_1 y_2' - y_2 y_1')}$$

We will use these equations rather than resolving the system in each of the remaining exercises in this section.

19. (a) Here $4r^2 + 1 = 0 \Rightarrow r = \pm\frac{1}{2}i$ and $y_c(x) = c_1 \cos\left(\frac{1}{2}x\right) + c_2 \sin\left(\frac{1}{2}x\right)$. We try a particular solution of the form

$$y_p(x) = A \cos x + B \sin x \Rightarrow y_p' = -A \sin x + B \cos x \text{ and } y_p'' = -A \cos x - B \sin x. \text{ Then the equation}$$

$4y'' + y = \cos x$ becomes $4(-A \cos x - B \sin x) + (A \cos x + B \sin x) = \cos x$ or

$-3A \cos x - 3B \sin x = \cos x \Rightarrow A = -\frac{1}{3}, B = 0$. Thus, $y_p(x) = -\frac{1}{3} \cos x$ and the general solution is

$$y(x) = y_c(x) + y_p(x) = c_1 \cos\left(\frac{1}{2}x\right) + c_2 \sin\left(\frac{1}{2}x\right) - \frac{1}{3} \cos x.$$

(b) From (a) we know that $y_c(x) = c_1 \cos\frac{x}{2} + c_2 \sin\frac{x}{2}$. Setting $y_1 = \cos\frac{x}{2}$, $y_2 = \sin\frac{x}{2}$, we have

$$y_1 y_2' - y_2 y_1' = \frac{1}{2}\cos^2\frac{x}{2} + \frac{1}{2}\sin^2\frac{x}{2} = \frac{1}{2}. \text{ Thus } u_1' = -\frac{\cos x \sin\frac{x}{2}}{4 \cdot \frac{1}{2}} = -\frac{1}{2}\cos\left(2 \cdot \frac{x}{2}\right)\sin\frac{x}{2} = -\frac{1}{2}\left(2\cos^2\frac{x}{2} - 1\right)\sin\frac{x}{2}$$

and $u_2' = \dfrac{\cos x \cos\frac{x}{2}}{4 \cdot \frac{1}{2}} = \frac{1}{2}\cos\left(2 \cdot \frac{x}{2}\right)\cos\frac{x}{2} = \frac{1}{2}\left(1 - 2\sin^2\frac{x}{2}\right)\cos\frac{x}{2}$. Then

$$u_1(x) = \int \left(\frac{1}{2}\sin\frac{x}{2} - \cos^2\frac{x}{2}\sin\frac{x}{2}\right) dx = -\cos\frac{x}{2} + \frac{2}{3}\cos^3\frac{x}{2} \text{ and}$$

$$u_2(x) = \int \left(\frac{1}{2}\cos\frac{x}{2} - \sin^2\frac{x}{2}\cos\frac{x}{2}\right) dx = \sin\frac{x}{2} - \frac{2}{3}\sin^3\frac{x}{2}. \text{ Thus}$$

$$y_p(x) = \left(-\cos\frac{x}{2} + \frac{2}{3}\cos^3\frac{x}{2}\right)\cos\frac{x}{2} + \left(\sin\frac{x}{2} - \frac{2}{3}\sin^3\frac{x}{2}\right)\sin\frac{x}{2} = -\left(\cos^2\frac{x}{2} - \sin^2\frac{x}{2}\right) + \frac{2}{3}\left(\cos^4\frac{x}{2} - \sin^4\frac{x}{2}\right)$$

$$= -\cos\left(2 \cdot \frac{x}{2}\right) + \frac{2}{3}\left(\cos^2\frac{x}{2} + \sin^2\frac{x}{2}\right)\left(\cos^2\frac{x}{2} - \sin^2\frac{x}{2}\right) = -\cos x + \frac{2}{3}\cos x = -\frac{1}{3}\cos x$$

and the general solution is $y(x) = y_c(x) + y_p(x) = c_1 \cos\frac{x}{2} + c_2 \sin\frac{x}{2} - \frac{1}{3}\cos x$.

21. (a) $r^2 - 2r + 1 = (r-1)^2 = 0 \Rightarrow r = 1$, so the complementary solution is $y_c(x) = c_1 e^x + c_2 x e^x$. A particular solution

is of the form $y_p(x) = Ae^{2x}$. Thus $4Ae^{2x} - 4Ae^{2x} + Ae^{2x} = e^{2x} \Rightarrow Ae^{2x} = e^{2x} \Rightarrow A = 1 \Rightarrow y_p(x) = e^{2x}$.

So a general solution is $y(x) = y_c(x) + y_p(x) = c_1 e^x + c_2 x e^x + e^{2x}$.

(b) From (a), $y_c(x) = c_1 e^x + c_2 x e^x$, so set $y_1 = e^x$, $y_2 = xe^x$. Then, $y_1 y_2' - y_2 y_1' = e^{2x}(1+x) - xe^{2x} = e^{2x}$ and so

$u_1' = -xe^x \Rightarrow u_1(x) = -\int xe^x \, dx = -(x-1)e^x$ [by parts] and $u_2' = e^x \Rightarrow u_2(x) = \int e^x \, dx = e^x$. Hence

$y_p(x) = (1-x)e^{2x} + xe^{2x} = e^{2x}$ and the general solution is $y(x) = y_c(x) + y_p(x) = c_1 e^x + c_2 x e^x + e^{2x}$.

23. As in Example 5, $y_c(x) = c_1 \sin x + c_2 \cos x$, so set $y_1 = \sin x$, $y_2 = \cos x$. Then $y_1 y_2' - y_2 y_1' = -\sin^2 x - \cos^2 x = -1$,

so $u_1' = -\dfrac{\sec^2 x \cos x}{-1} = \sec x \quad \Rightarrow \quad u_1(x) = \int \sec x \, dx = \ln(\sec x + \tan x)$ for $0 < x < \frac{\pi}{2}$,

and $u_2' = \dfrac{\sec^2 x \sin x}{-1} = -\sec x \tan x \quad \Rightarrow \quad u_2(x) = -\sec x$. Hence

$y_p(x) = \ln(\sec x + \tan x) \cdot \sin x - \sec x \cdot \cos x = \sin x \ln(\sec x + \tan x) - 1$ and the general solution is

$y(x) = c_1 \sin x + c_2 \cos x + \sin x \ln(\sec x + \tan x) - 1$.

25. $y_1 = e^x$, $y_2 = e^{2x}$ and $y_1 y_2' - y_2 y_1' = e^{3x}$. So $u_1' = \dfrac{-e^{2x}}{(1 + e^{-x})e^{3x}} = -\dfrac{e^{-x}}{1 + e^{-x}}$ and

$u_1(x) = \int -\dfrac{e^{-x}}{1 + e^{-x}} \, dx = \ln(1 + e^{-x})$. $u_2' = \dfrac{e^x}{(1 + e^{-x})e^{3x}} = \dfrac{e^x}{e^{3x} + e^{2x}}$ so

$u_2(x) = \int \dfrac{e^x}{e^{3x} + e^{2x}} \, dx = \ln\left(\dfrac{e^x + 1}{e^x}\right) - e^{-x} = \ln(1 + e^{-x}) - e^{-x}$. Hence

$y_p(x) = e^x \ln(1 + e^{-x}) + e^{2x}[\ln(1 + e^{-x}) - e^{-x}]$ and the general solution is

$y(x) = [c_1 + \ln(1 + e^{-x})]e^x + [c_2 - e^{-x} + \ln(1 + e^{-x})]e^{2x}$.

27. $r^2 - 2r + 1 = (r - 1)^2 = 0 \quad \Rightarrow \quad r = 1$ so $y_c(x) = c_1 e^x + c_2 x e^x$. Thus $y_1 = e^x$, $y_2 = x e^x$ and

$y_1 y_2' - y_2 y_1' = e^x(x + 1)e^x - x e^x e^x = e^{2x}$. So $u_1' = -\dfrac{x e^x \cdot e^x/(1 + x^2)}{e^{2x}} = -\dfrac{x}{1 + x^2} \quad \Rightarrow$

$u_1 = -\int \dfrac{x}{1 + x^2} \, dx = -\dfrac{1}{2} \ln(1 + x^2)$, $u_2' = \dfrac{e^x \cdot e^x/(1 + x^2)}{e^{2x}} = \dfrac{1}{1 + x^2} \quad \Rightarrow \quad u_2 = \int \dfrac{1}{1 + x^2} \, dx = \tan^{-1} x$ and

$y_p(x) = -\frac{1}{2} e^x \ln(1 + x^2) + x e^x \tan^{-1} x$. Hence the general solution is $y(x) = e^x\left[c_1 + c_2 x - \frac{1}{2} \ln(1 + x^2) + x \tan^{-1} x\right]$.

17.3 Applications of Second-Order Differential Equations

1. By Hooke's Law $k(0.25) = 25$ so $k = 100$ is the spring constant and the differential equation is $5x'' + 100x = 0$.

The auxiliary equation is $5r^2 + 100 = 0$ with roots $r = \pm 2\sqrt{5}\,i$, so the general solution to the differential equation is

$x(t) = c_1 \cos(2\sqrt{5}\,t) + c_2 \sin(2\sqrt{5}\,t)$. We are given that $x(0) = 0.35 \quad \Rightarrow \quad c_1 = 0.35$ and $x'(0) = 0 \quad \Rightarrow$

$2\sqrt{5}\,c_2 = 0 \quad \Rightarrow \quad c_2 = 0$, so the position of the mass after t seconds is $x(t) = 0.35 \cos(2\sqrt{5}\,t)$.

3. $k(0.5) = 6$ or $k = 12$ is the spring constant, so the initial-value problem is $2x'' + 14x' + 12x = 0$, $x(0) = 1$, $x'(0) = 0$.

The general solution is $x(t) = c_1 e^{-6t} + c_2 e^{-t}$. But $1 = x(0) = c_1 + c_2$ and $0 = x'(0) = -6c_1 - c_2$. Thus the position is

given by $x(t) = -\frac{1}{5} e^{-6t} + \frac{6}{5} e^{-t}$.

5. For critical damping we need $c^2 - 4mk = 0$ or $m = c^2/(4k) = 14^2/(4 \cdot 12) = \frac{49}{12}$ kg.

7. We are given $m = 1$, $k = 100$, $x(0) = -0.1$ and $x'(0) = 0$. From (3), the differential equation is $\dfrac{d^2x}{dt^2} + c\,\dfrac{dx}{dt} + 100x = 0$

with auxiliary equation $r^2 + cr + 100 = 0$.

If $c = 10$, we have two complex roots $r = -5 \pm 5\sqrt{3}\,i$, so the motion is underdamped and the solution is

$x = e^{-5t}\left[c_1\cos\left(5\sqrt{3}\,t\right) + c_2\sin\left(5\sqrt{3}\,t\right)\right]$. Then $-0.1 = x(0) = c_1$ and $0 = x'(0) = 5\sqrt{3}\,c_2 - 5c_1$ $\;\Rightarrow\;$ $c_2 = -\frac{1}{10\sqrt{3}}$,

so $x = e^{-5t}\left[-0.1\cos\left(5\sqrt{3}\,t\right) - \frac{1}{10\sqrt{3}}\sin\left(5\sqrt{3}\,t\right)\right]$.

If $c = 15$, we again have underdamping since the auxiliary equation has roots $r = -\frac{15}{2} \pm \frac{5\sqrt{7}}{2}i$. The general solution is

$x = e^{-15t/2}\left[c_1\cos\left(\frac{5\sqrt{7}}{2}t\right) + c_2\sin\left(\frac{5\sqrt{7}}{2}t\right)\right]$, so $-0.1 = x(0) = c_1$ and $0 = x'(0) = \frac{5\sqrt{7}}{2}c_2 - \frac{15}{2}c_1$ $\;\Rightarrow\;$ $c_2 = -\frac{3}{10\sqrt{7}}$.

Thus $x = e^{-15t/2}\left[-0.1\cos\left(\frac{5\sqrt{7}}{2}t\right) - \frac{3}{10\sqrt{7}}\sin\left(\frac{5\sqrt{7}}{2}t\right)\right]$.

For $c = 20$, we have equal roots $r_1 = r_2 = -10$, so the oscillation is critically damped and the solution is

$x = (c_1 + c_2t)e^{-10t}$. Then $-0.1 = x(0) = c_1$ and $0 = x'(0) = -10c_1 + c_2$ $\;\Rightarrow\;$ $c_2 = -1$, so $x = (-0.1 - t)e^{-10t}$.

If $c = 25$ the auxiliary equation has roots $r_1 = -5$, $r_2 = -20$, so we have overdamping and the solution is

$x = c_1e^{-5t} + c_2e^{-20t}$. Then $-0.1 = x(0) = c_1 + c_2$ and $0 = x'(0) = -5c_1 - 20c_2$ $\;\Rightarrow\;$ $c_1 = -\frac{2}{15}$ and $c_2 = \frac{1}{30}$,

so $x = -\frac{2}{15}e^{-5t} + \frac{1}{30}e^{-20t}$.

If $c = 30$ we have roots $r = -15 \pm 5\sqrt{5}$, so the motion is

overdamped and the solution is $x = c_1e^{\left(-15 + 5\sqrt{5}\right)t} + c_2e^{\left(-15 - 5\sqrt{5}\right)t}$.

Then $-0.1 = x(0) = c_1 + c_2$ and

$0 = x'(0) = \left(-15 + 5\sqrt{5}\right)c_1 + \left(-15 - 5\sqrt{5}\right)c_2$ $\;\Rightarrow\;$

$c_1 = \frac{-5 - 3\sqrt{5}}{100}$ and $c_2 = \frac{-5 + 3\sqrt{5}}{100}$, so

$x = \left(\frac{-5 - 3\sqrt{5}}{100}\right)e^{\left(-15 + 5\sqrt{5}\right)t} + \left(\frac{-5 + 3\sqrt{5}}{100}\right)e^{\left(-15 - 5\sqrt{5}\right)t}$.

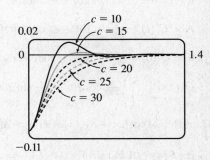

9. The differential equation is $mx'' + kx = F_0\cos\omega_0 t$ and $\omega_0 \neq \omega = \sqrt{k/m}$. Here the auxiliary equation is $mr^2 + k = 0$

with roots $\pm\sqrt{k/m}\,i = \pm\omega i$ so $x_c(t) = c_1\cos\omega t + c_2\sin\omega t$. Since $\omega_0 \neq \omega$, try $x_p(t) = A\cos\omega_0 t + B\sin\omega_0 t$.

Then we need $(m)\left(-\omega_0^2\right)(A\cos\omega_0 t + B\sin\omega_0 t) + k(A\cos\omega_0 t + B\sin\omega_0 t) = F_0\cos\omega_0 t$ or $A\left(k - m\omega_0^2\right) = F_0$ and

$B\left(k - m\omega_0^2\right) = 0$. Hence $B = 0$ and $A = \dfrac{F_0}{k - m\omega_0^2} = \dfrac{F_0}{m(\omega^2 - \omega_0^2)}$ since $\omega^2 = \dfrac{k}{m}$. Thus the motion of the mass is given

by $x(t) = c_1\cos\omega t + c_2\sin\omega t + \dfrac{F_0}{m(\omega^2 - \omega_0^2)}\cos\omega_0 t$.

11. From Equation 6, $x(t) = f(t) + g(t)$ where $f(t) = c_1\cos\omega t + c_2\sin\omega t$ and $g(t) = \dfrac{F_0}{m(\omega^2 - \omega_0^2)}\cos\omega_0 t$. Then f

is periodic, with period $\frac{2\pi}{\omega}$, and if $\omega \neq \omega_0$, g is periodic with period $\frac{2\pi}{\omega_0}$. If $\frac{\omega}{\omega_0}$ is a rational number, then we can say

$\frac{\omega}{\omega_0} = \frac{a}{b} \Rightarrow a = \frac{b\omega}{\omega_0}$ where a and b are non-zero integers. Then

$$x\left(t + a \cdot \tfrac{2\pi}{\omega}\right) = f\left(t + a \cdot \tfrac{2\pi}{\omega}\right) + g\left(t + a \cdot \tfrac{2\pi}{\omega}\right) = f(t) + g\left(t + \tfrac{b\omega}{\omega_0} \cdot \tfrac{2\pi}{\omega}\right) = f(t) + g\left(t + b \cdot \tfrac{2\pi}{\omega_0}\right) = f(t) + g(t) = x(t)$$

so $x(t)$ is periodic.

13. Here the initial-value problem for the charge is $Q'' + 20Q' + 500Q = 12$, $Q(0) = Q'(0) = 0$. Then

$Q_c(t) = e^{-10t}(c_1 \cos 20t + c_2 \sin 20t)$ and try $Q_p(t) = A \Rightarrow 500A = 12$ or $A = \frac{3}{125}$.

The general solution is $Q(t) = e^{-10t}(c_1 \cos 20t + c_2 \sin 20t) + \frac{3}{125}$. But $0 = Q(0) = c_1 + \frac{3}{125}$ and

$Q'(t) = I(t) = e^{-10t}[(-10c_1 + 20c_2)\cos 20t + (-10c_2 - 20c_1)\sin 20t]$ but $0 = Q'(0) = -10c_1 + 20c_2$. Thus the charge

is $Q(t) = -\frac{1}{250}e^{-10t}(6\cos 20t + 3\sin 20t) + \frac{3}{125}$ and the current is $I(t) = e^{-10t}\left(\frac{3}{5}\right)\sin 20t$.

15. As in Exercise 13, $Q_c(t) = e^{-10t}(c_1 \cos 20t + c_2 \sin 20t)$ but $E(t) = 12 \sin 10t$ so try

$Q_p(t) = A \cos 10t + B \sin 10t$. Substituting into the differential equation gives

$(-100A + 200B + 500A)\cos 10t + (-100B - 200A + 500B)\sin 10t = 12 \sin 10t \Rightarrow$

$400A + 200B = 0$ and $400B - 200A = 12$. Thus $A = -\frac{3}{250}$, $B = \frac{3}{125}$ and the general solution is

$Q(t) = e^{-10t}(c_1 \cos 20t + c_2 \sin 20t) - \frac{3}{250}\cos 10t + \frac{3}{125}\sin 10t$. But $0 = Q(0) = c_1 - \frac{3}{250}$ so $c_1 = \frac{3}{250}$.

Also $Q'(t) = \frac{3}{25}\sin 10t + \frac{6}{25}\cos 10t + e^{-10t}[(-10c_1 + 20c_2)\cos 20t + (-10c_2 - 20c_1)\sin 20t]$ and

$0 = Q'(0) = \frac{6}{25} - 10c_1 + 20c_2$ so $c_2 = -\frac{3}{500}$. Hence the charge is given by

$Q(t) = e^{-10t}\left[\frac{3}{250}\cos 20t - \frac{3}{500}\sin 20t\right] - \frac{3}{250}\cos 10t + \frac{3}{125}\sin 10t$.

17. $x(t) = A\cos(\omega t + \delta) \Leftrightarrow x(t) = A[\cos \omega t \cos \delta - \sin \omega t \sin \delta] \Leftrightarrow x(t) = A\left(\frac{c_1}{A}\cos \omega t + \frac{c_2}{A}\sin \omega t\right)$ where

$\cos \delta = c_1/A$ and $\sin \delta = -c_2/A \Leftrightarrow x(t) = c_1 \cos \omega t + c_2 \sin \omega t$. [Note that $\cos^2 \delta + \sin^2 \delta = 1 \Rightarrow c_1^2 + c_2^2 = A^2$.]

17.4 Series Solutions

1. Let $y(x) = \sum\limits_{n=0}^{\infty} c_n x^n$. Then $y'(x) = \sum\limits_{n=1}^{\infty} n c_n x^{n-1}$ and the given equation, $y' - y = 0$, becomes

$\sum\limits_{n=1}^{\infty} n c_n x^{n-1} - \sum\limits_{n=0}^{\infty} c_n x^n = 0$. Replacing n by $n + 1$ in the first sum gives $\sum\limits_{n=0}^{\infty} (n+1)c_{n+1}x^n - \sum\limits_{n=0}^{\infty} c_n x^n = 0$, so

$\sum\limits_{n=0}^{\infty} [(n+1)c_{n+1} - c_n]x^n = 0$. Equating coefficients gives $(n+1)c_{n+1} - c_n = 0$, so the recursion relation is

$c_{n+1} = \frac{c_n}{n+1}$, $n = 0, 1, 2, \dots$. Then $c_1 = c_0$, $c_2 = \frac{1}{2}c_1 = \frac{c_0}{2}$, $c_3 = \frac{1}{3}c_2 = \frac{1}{3} \cdot \frac{1}{2}c_0 = \frac{c_0}{3!}$, $c_4 = \frac{1}{4}c_3 = \frac{c_0}{4!}$, and

in general, $c_n = \frac{c_0}{n!}$. Thus, the solution is $y(x) = \sum\limits_{n=0}^{\infty} c_n x^n = \sum\limits_{n=0}^{\infty} \frac{c_0}{n!}x^n = c_0 \sum\limits_{n=0}^{\infty} \frac{x^n}{n!} = c_0 e^x$.

3. Assuming $y(x) = \sum\limits_{n=0}^{\infty} c_n x^n$, we have $y'(x) = \sum\limits_{n=1}^{\infty} nc_n x^{n-1} = \sum\limits_{n=0}^{\infty} (n+1)c_{n+1} x^n$ and

$-x^2 y = -\sum\limits_{n=0}^{\infty} c_n x^{n+2} = -\sum\limits_{n=2}^{\infty} c_{n-2} x^n$. Hence, the equation $y' = x^2 y$ becomes $\sum\limits_{n=0}^{\infty} (n+1)c_{n+1} x^n - \sum\limits_{n=2}^{\infty} c_{n-2} x^n = 0$

or $c_1 + 2c_2 x + \sum\limits_{n=2}^{\infty} [(n+1)c_{n+1} - c_{n-2}]\, x^n = 0$. Equating coefficients gives $c_1 = c_2 = 0$ and $c_{n+1} = \dfrac{c_{n-2}}{n+1}$

for $n = 2, 3, \ldots$. But $c_1 = 0$, so $c_4 = 0$ and $c_7 = 0$ and in general $c_{3n+1} = 0$. Similarly $c_2 = 0$ so $c_{3n+2} = 0$. Finally

$c_3 = \dfrac{c_0}{3}$, $c_6 = \dfrac{c_3}{6} = \dfrac{c_0}{6 \cdot 3} = \dfrac{c_0}{3^2 \cdot 2!}$, $c_9 = \dfrac{c_6}{9} = \dfrac{c_0}{9 \cdot 6 \cdot 3} = \dfrac{c_0}{3^3 \cdot 3!}, \ldots$, and $c_{3n} = \dfrac{c_0}{3^n \cdot n!}$. Thus, the solution

is $y(x) = \sum\limits_{n=0}^{\infty} c_n x^n = \sum\limits_{n=0}^{\infty} c_{3n} x^{3n} = \sum\limits_{n=0}^{\infty} \dfrac{c_0}{3^n \cdot n!} x^{3n} = c_0 \sum\limits_{n=0}^{\infty} \dfrac{x^{3n}}{3^n n!} = c_0 \sum\limits_{n=0}^{\infty} \dfrac{(x^3/3)^n}{n!} = c_0 e^{x^3/3}$.

5. Let $y(x) = \sum\limits_{n=0}^{\infty} c_n x^n \;\Rightarrow\; y'(x) = \sum\limits_{n=1}^{\infty} nc_n x^{n-1}$ and $y''(x) = \sum\limits_{n=0}^{\infty} (n+2)(n+1)c_{n+2} x^n$. The differential equation

becomes $\sum\limits_{n=0}^{\infty} (n+2)(n+1)c_{n+2} x^n + x \sum\limits_{n=1}^{\infty} nc_n x^{n-1} + \sum\limits_{n=0}^{\infty} c_n x^n = 0$ or $\sum\limits_{n=0}^{\infty} [(n+2)(n+1)c_{n+2} + nc_n + c_n] x^n = 0$

$\left[\text{since } \sum\limits_{n=1}^{\infty} nc_n x^n = \sum\limits_{n=0}^{\infty} nc_n x^n\right]$. Equating coefficients gives $(n+2)(n+1)c_{n+2} + (n+1)c_n = 0$, thus the recursion

relation is $c_{n+2} = \dfrac{-(n+1)c_n}{(n+2)(n+1)} = -\dfrac{c_n}{n+2}$, $n = 0, 1, 2, \ldots$. Then the even

coefficients are given by $c_2 = -\dfrac{c_0}{2}$, $c_4 = -\dfrac{c_2}{4} = \dfrac{c_0}{2 \cdot 4}$, $c_6 = -\dfrac{c_4}{6} = -\dfrac{c_0}{2 \cdot 4 \cdot 6}$, and in general,

$c_{2n} = (-1)^n \dfrac{c_0}{2 \cdot 4 \cdots 2n} = \dfrac{(-1)^n c_0}{2^n\, n!}$. The odd coefficients are $c_3 = -\dfrac{c_1}{3}$, $c_5 = -\dfrac{c_3}{5} = \dfrac{c_1}{3 \cdot 5}$, $c_7 = -\dfrac{c_5}{7} = -\dfrac{c_1}{3 \cdot 5 \cdot 7}$,

and in general, $c_{2n+1} = (-1)^n \dfrac{c_1}{3 \cdot 5 \cdot 7 \cdots (2n+1)} = \dfrac{(-2)^n\, n!\, c_1}{(2n+1)!}$. The solution is

$y(x) = c_0 \sum\limits_{n=0}^{\infty} \dfrac{(-1)^n}{2^n\, n!} x^{2n} + c_1 \sum\limits_{n=0}^{\infty} \dfrac{(-2)^n\, n!}{(2n+1)!} x^{2n+1}$.

7. Let $y(x) = \sum\limits_{n=0}^{\infty} c_n x^n \;\Rightarrow\; y'(x) = \sum\limits_{n=1}^{\infty} nc_n x^{n-1} = \sum\limits_{n=0}^{\infty} (n+1)c_{n+1} x^n$ and $y''(x) = \sum\limits_{n=0}^{\infty} (n+2)(n+1)c_{n+2} x^n$. Then

$(x-1)y''(x) = \sum\limits_{n=0}^{\infty} (n+2)(n+1)c_{n+2} x^{n+1} - \sum\limits_{n=0}^{\infty} (n+2)(n+1)c_{n+2} x^n = \sum\limits_{n=1}^{\infty} n(n+1)c_{n+1} x^n - \sum\limits_{n=0}^{\infty} (n+2)(n+1)c_{n+2} x^n$.

Since $\sum\limits_{n=1}^{\infty} n(n+1)c_{n+1} x^n = \sum\limits_{n=0}^{\infty} n(n+1)c_{n+1} x^n$, the differential equation becomes

$\sum\limits_{n=0}^{\infty} n(n+1)c_{n+1} x^n - \sum\limits_{n=0}^{\infty} (n+2)(n+1)c_{n+2} x^n + \sum\limits_{n=0}^{\infty} (n+1)c_{n+1} x^n = 0 \;\Rightarrow$

$\sum\limits_{n=0}^{\infty} [n(n+1)c_{n+1} - (n+2)(n+1)c_{n+2} + (n+1)c_{n+1}] x^n = 0$ or $\sum\limits_{n=0}^{\infty} [(n+1)^2 c_{n+1} - (n+2)(n+1)c_{n+2}] x^n = 0$.

Equating coefficients gives $(n+1)^2 c_{n+1} - (n+2)(n+1)c_{n+2} = 0$ for $n = 0, 1, 2, \ldots$. Then the recursion relation is

$c_{n+2} = \dfrac{(n+1)^2}{(n+2)(n+1)} c_{n+1} = \dfrac{n+1}{n+2} c_{n+1}$, so given c_0 and c_1, we have $c_2 = \frac{1}{2}c_1$, $c_3 = \frac{2}{3}c_2 = \frac{1}{3}c_1$, $c_4 = \frac{3}{4}c_3 = \frac{1}{4}c_1$, and

in general $c_n = \dfrac{c_1}{n}$, $n = 1, 2, 3, \ldots$. Thus the solution is $y(x) = c_0 + c_1 \sum\limits_{n=1}^{\infty} \dfrac{x^n}{n}$. Note that the solution can be expressed as

$c_0 - c_1 \ln(1-x)$ for $|x| < 1$.

9. Let $y(x) = \sum\limits_{n=0}^{\infty} c_n x^n$. Then $-xy'(x) = -x \sum\limits_{n=1}^{\infty} n c_n x^{n-1} = -\sum\limits_{n=1}^{\infty} n c_n x^n = -\sum\limits_{n=0}^{\infty} n c_n x^n$,

$y''(x) = \sum\limits_{n=0}^{\infty} (n+2)(n+1) c_{n+2} x^n$, and the equation $y'' - xy' - y = 0$ becomes

$\sum\limits_{n=0}^{\infty} [(n+2)(n+1)c_{n+2} - nc_n - c_n] x^n = 0$. Thus, the recursion relation is

$c_{n+2} = \dfrac{nc_n + c_n}{(n+2)(n+1)} = \dfrac{c_n(n+1)}{(n+2)(n+1)} = \dfrac{c_n}{n+2}$ for $n = 0, 1, 2, \ldots$. One of the given conditions is $y(0) = 1$. But

$y(0) = \sum\limits_{n=0}^{\infty} c_n(0)^n = c_0 + 0 + 0 + \cdots = c_0$, so $c_0 = 1$. Hence, $c_2 = \dfrac{c_0}{2} = \dfrac{1}{2}$, $c_4 = \dfrac{c_2}{4} = \dfrac{1}{2 \cdot 4}$, $c_6 = \dfrac{c_4}{6} = \dfrac{1}{2 \cdot 4 \cdot 6}, \ldots$,

$c_{2n} = \dfrac{1}{2^n n!}$. The other given condition is $y'(0) = 0$. But $y'(0) = \sum\limits_{n=1}^{\infty} n c_n(0)^{n-1} = c_1 + 0 + 0 + \cdots = c_1$, so $c_1 = 0$.

By the recursion relation, $c_3 = \dfrac{c_1}{3} = 0$, $c_5 = 0, \ldots, c_{2n+1} = 0$ for $n = 0, 1, 2, \ldots$. Thus, the solution to the initial-value

problem is $y(x) = \sum\limits_{n=0}^{\infty} c_n x^n = \sum\limits_{n=0}^{\infty} c_{2n} x^{2n} = \sum\limits_{n=0}^{\infty} \dfrac{x^{2n}}{2^n n!} = \sum\limits_{n=0}^{\infty} \dfrac{(x^2/2)^n}{n!} = e^{x^2/2}$.

11. Assuming that $y(x) = \sum\limits_{n=0}^{\infty} c_n x^n$, we have $xy = x \sum\limits_{n=0}^{\infty} c_n x^n = \sum\limits_{n=0}^{\infty} c_n x^{n+1}$, $x^2 y' = x^2 \sum\limits_{n=1}^{\infty} n c_n x^{n-1} = \sum\limits_{n=0}^{\infty} n c_n x^{n+1}$,

$y''(x) = \sum\limits_{n=2}^{\infty} n(n-1) c_n x^{n-2} = \sum\limits_{n=-1}^{\infty} (n+3)(n+2) c_{n+3} x^{n+1}$ [replace n with $n+3$]

$\qquad = 2c_2 + \sum\limits_{n=0}^{\infty} (n+3)(n+2) c_{n+3} x^{n+1}$,

and the equation $y'' + x^2 y' + xy = 0$ becomes $2c_2 + \sum\limits_{n=0}^{\infty} [(n+3)(n+2)c_{n+3} + nc_n + c_n] x^{n+1} = 0$. So $c_2 = 0$ and the

recursion relation is $c_{n+3} = \dfrac{-nc_n - c_n}{(n+3)(n+2)} = -\dfrac{(n+1)c_n}{(n+3)(n+2)}$, $n = 0, 1, 2, \ldots$. But $c_0 = y(0) = 0 = c_2$ and by the

recursion relation, $c_{3n} = c_{3n+2} = 0$ for $n = 0, 1, 2, \ldots$. Also, $c_1 = y'(0) = 1$, so $c_4 = -\dfrac{2c_1}{4 \cdot 3} = -\dfrac{2}{4 \cdot 3}$,

$c_7 = -\dfrac{5c_4}{7 \cdot 6} = (-1)^2 \dfrac{2 \cdot 5}{7 \cdot 6 \cdot 4 \cdot 3} = (-1)^2 \dfrac{2^2 5^2}{7!}, \ldots, c_{3n+1} = (-1)^n \dfrac{2^2 5^2 \cdots (3n-1)^2}{(3n+1)!}$. Thus, the solution is

$y(x) = \sum\limits_{n=0}^{\infty} c_n x^n = x + \sum\limits_{n=1}^{\infty} \left[(-1)^n \dfrac{2^2 5^2 \cdots (3n-1)^2 x^{3n+1}}{(3n+1)!} \right]$.

17 Review

TRUE-FALSE QUIZ

1. True. See Theorem 17.1.3.

3. True. $\cosh x$ and $\sinh x$ are linearly independent solutions of this linear homogeneous equation.

EXERCISES

1. The auxiliary equation is $4r^2 - 1 = 0 \Rightarrow (2r+1)(2r-1) = 0 \Rightarrow r = \pm\frac{1}{2}$. Then the general solution

is $y = c_1 e^{x/2} + c_2 e^{-x/2}$.

3. The auxiliary equation is $r^2 + 3 = 0 \Rightarrow r = \pm\sqrt{3}\,i$. Then the general solution is $y = c_1 \cos\left(\sqrt{3}\,x\right) + c_2 \sin\left(\sqrt{3}\,x\right)$.

5. $r^2 - 4r + 5 = 0 \Rightarrow r = 2 \pm i$, so $y_c(x) = e^{2x}(c_1 \cos x + c_2 \sin x)$. Try $y_p(x) = Ae^{2x} \Rightarrow y_p' = 2Ae^{2x}$

and $y_p'' = 4Ae^{2x}$. Substitution into the differential equation gives $4Ae^{2x} - 8Ae^{2x} + 5Ae^{2x} = e^{2x} \Rightarrow A = 1$ and

the general solution is $y(x) = e^{2x}(c_1 \cos x + c_2 \sin x) + e^{2x}$.

7. $r^2 - 2r + 1 = 0 \Rightarrow r = 1$ and $y_c(x) = c_1 e^x + c_2 x e^x$. Try $y_p(x) = (Ax + B)\cos x + (Cx + D)\sin x \Rightarrow$

$y_p' = (C - Ax - B)\sin x + (A + Cx + D)\cos x$ and $y_p'' = (2C - B - Ax)\cos x + (-2A - D - Cx)\sin x$. Substitution

gives $(-2Cx + 2C - 2A - 2D)\cos x + (2Ax - 2A + 2B - 2C)\sin x = x\cos x \Rightarrow A = 0, B = C = D = -\frac{1}{2}$.

The general solution is $y(x) = c_1 e^x + c_2 x e^x - \frac{1}{2}\cos x - \frac{1}{2}(x+1)\sin x$.

9. $r^2 - r - 6 = 0 \Rightarrow r = -2, r = 3$ and $y_c(x) = c_1 e^{-2x} + c_2 e^{3x}$. For $y'' - y' - 6y = 1$, try $y_{p_1}(x) = A$. Then

$y_{p_1}'(x) = y_{p_1}''(x) = 0$ and substitution into the differential equation gives $A = -\frac{1}{6}$. For $y'' - y' - 6y = e^{-2x}$ try

$y_{p_2}(x) = Bxe^{-2x}$ [since $y = Be^{-2x}$ satisfies the complementary equation]. Then $y_{p_2}' = (B - 2Bx)e^{-2x}$ and

$y_{p_2}'' = (4Bx - 4B)e^{-2x}$, and substitution gives $-5Be^{-2x} = e^{-2x} \Rightarrow B = -\frac{1}{5}$. The general solution then is

$y(x) = c_1 e^{-2x} + c_2 e^{3x} + y_{p_1}(x) + y_{p_2}(x) = c_1 e^{-2x} + c_2 e^{3x} - \frac{1}{6} - \frac{1}{5}xe^{-2x}$.

11. The auxiliary equation is $r^2 + 6r = 0$ and the general solution is $y(x) = c_1 + c_2 e^{-6x} = k_1 + k_2 e^{-6(x-1)}$. But

$3 = y(1) = k_1 + k_2$ and $12 = y'(1) = -6k_2$. Thus $k_2 = -2, k_1 = 5$ and the solution is $y(x) = 5 - 2e^{-6(x-1)}$.

13. The auxiliary equation is $r^2 - 5r + 4 = 0$ and the general solution is $y(x) = c_1 e^x + c_2 e^{4x}$. But $0 = y(0) = c_1 + c_2$

and $1 = y'(0) = c_1 + 4c_2$, so the solution is $y(x) = \frac{1}{3}(e^{4x} - e^x)$.

15. $r^2 + 4r + 29 = 0 \Rightarrow r = -2 \pm 5i$ and the general solution is $y = e^{-2x}(c_1 \cos 5x + c_2 \sin 5x)$. But $1 = y(0) = c_1$ and

$-1 = y(\pi) = -c_1 e^{-2\pi} \Rightarrow c_1 = e^{2\pi}$, so there is no solution.

17. Let $y(x) = \sum\limits_{n=0}^{\infty} c_n x^n$. Then $y''(x) = \sum\limits_{n=0}^{\infty} n(n-1)c_n x^{n-2} = \sum\limits_{n=0}^{\infty} (n+2)(n+1)c_{n+2}x^n$ and the differential equation

becomes $\sum\limits_{n=0}^{\infty} [(n+2)(n+1)c_{n+2} + (n+1)c_n]x^n = 0$. Thus the recursion relation is $c_{n+2} = -c_n/(n+2)$

for $n = 0, 1, 2, \ldots$. But $c_0 = y(0) = 0$, so $c_{2n} = 0$ for $n = 0, 1, 2, \ldots$. Also $c_1 = y'(0) = 1$, so $c_3 = -\dfrac{1}{3}$, $c_5 = \dfrac{(-1)^2}{3 \cdot 5}$,

$c_7 = \dfrac{(-1)^3}{3 \cdot 5 \cdot 7} = \dfrac{(-1)^3 2^3 3!}{7!}, \ldots, c_{2n+1} = \dfrac{(-1)^n 2^n n!}{(2n+1)!}$ for $n = 0, 1, 2, \ldots$. Thus the solution to the initial-value problem

is $y(x) = \sum\limits_{n=0}^{\infty} c_n x^n = \sum\limits_{n=0}^{\infty} \dfrac{(-1)^n 2^n n!}{(2n+1)!} x^{2n+1}$.

19. Here the initial-value problem is $2Q'' + 40Q' + 400Q = 12$, $Q(0) = 0.01$, $Q'(0) = 0$. Then

$Q_c(t) = e^{-10t}(c_1 \cos 10t + c_2 \sin 10t)$ and we try $Q_p(t) = A$. Thus the general solution is

$Q(t) = e^{-10t}(c_1 \cos 10t + c_2 \sin 10t) + \frac{3}{100}$. But $0.01 = Q'(0) = c_1 + 0.03$ and $0 = Q''(0) = -10c_1 + 10c_2$,

so $c_1 = -0.02 = c_2$. Hence the charge is given by $Q(t) = -0.02e^{-10t}(\cos 10t + \sin 10t) + 0.03$.

21. (a) Since we are assuming that the earth is a solid sphere of uniform density, we can calculate the density ρ as follows:

$\rho = \dfrac{\text{mass of earth}}{\text{volume of earth}} = \dfrac{M}{\frac{4}{3}\pi R^3}$. If V_r is the volume of the portion of the earth which lies within a distance r of the

center, then $V_r = \frac{4}{3}\pi r^3$ and $M_r = \rho V_r = \dfrac{Mr^3}{R^3}$. Thus $F_r = -\dfrac{GM_r m}{r^2} = -\dfrac{GMm}{R^3}r$.

(b) The particle is acted upon by a varying gravitational force during its motion. By Newton's Second Law of Motion,

$m\dfrac{d^2y}{dt^2} = F_y = -\dfrac{GMm}{R^3}y$, so $y''(t) = -k^2 y(t)$ where $k^2 = \dfrac{GM}{R^3}$. At the surface, $-mg = F_R = -\dfrac{GMm}{R^2}$, so

$g = \dfrac{GM}{R^2}$. Therefore $k^2 = \dfrac{g}{R}$.

(c) The differential equation $y'' + k^2 y = 0$ has auxiliary equation $r^2 + k^2 = 0$. (This is the r of Section 17.1,

not the r measuring distance from the earth's center.) The roots of the auxiliary equation are $\pm ik$, so by (11) in

Section 17.1, the general solution of our differential equation for t is $y(t) = c_1 \cos kt + c_2 \sin kt$. It follows that

$y'(t) = -c_1 k \sin kt + c_2 k \cos kt$. Now $y(0) = R$ and $y'(0) = 0$, so $c_1 = R$ and $c_2 k = 0$. Thus $y(t) = R \cos kt$ and

$y'(t) = -kR \sin kt$. This is simple harmonic motion (see Section 17.3) with amplitude R, frequency k, and phase angle 0.

The period is $T = 2\pi/k$. $R \approx 3960$ mi $= 3960 \cdot 5280$ ft and $g = 32$ ft/s^2, so $k = \sqrt{g/R} \approx 1.24 \times 10^{-3}$ s^{-1} and

$T = 2\pi/k \approx 5079$ s ≈ 85 min.

(d) $y(t) = 0 \Leftrightarrow \cos kt = 0 \Leftrightarrow kt = \frac{\pi}{2} + \pi n$ for some integer $n \Rightarrow y'(t) = -kR \sin\left(\frac{\pi}{2} + \pi n\right) = \pm kR$. Thus the

particle passes through the center of the earth with speed $kR \approx 4.899$ mi/s $\approx 17,600$ mi/h.

☐ APPENDIX

Complex Numbers

1. $(5 - 6i) + (3 + 2i) = (5 + 3) + (-6 + 2)i = 8 + (-4)i = 8 - 4i$

3. $(2 + 5i)(4 - i) = 2(4) + 2(-i) + (5i)(4) + (5i)(-i) = 8 - 2i + 20i - 5i^2 = 8 + 18i - 5(-1)$

$$= 8 + 18i + 5 = 13 + 18i$$

5. $\overline{12 + 7i} = 12 - 7i$

7. $\dfrac{1 + 4i}{3 + 2i} = \dfrac{1 + 4i}{3 + 2i} \cdot \dfrac{3 - 2i}{3 - 2i} = \dfrac{3 - 2i + 12i - 8(-1)}{3^2 + 2^2} = \dfrac{11 + 10i}{13} = \dfrac{11}{13} + \dfrac{10}{13}i$

9. $\dfrac{1}{1 + i} = \dfrac{1}{1 + i} \cdot \dfrac{1 - i}{1 - i} = \dfrac{1 - i}{1 - (-1)} = \dfrac{1 - i}{2} = \dfrac{1}{2} - \dfrac{1}{2}i$

11. $i^3 = i^2 \cdot i = (-1)i = -i$

13. $\sqrt{-25} = \sqrt{25}\, i = 5i$

15. $\overline{12 - 5i} = 12 + 15i$ and $|12 - 15i| = \sqrt{12^2 + (-5)^2} = \sqrt{144 + 25} = \sqrt{169} = 13$

17. $\overline{-4i} = \overline{0 - 4i} = 0 + 4i = 4i$ and $|-4i| = \sqrt{0^2 + (-4)^2} = \sqrt{16} = 4$

19. $4x^2 + 9 = 0 \;\Leftrightarrow\; 4x^2 = -9 \;\Leftrightarrow\; x^2 = -\frac{9}{4} \;\Leftrightarrow\; x = \pm\sqrt{-\frac{9}{4}} = \pm\sqrt{\frac{9}{4}}\, i = \pm\frac{3}{2}i.$

21. By the quadratic formula, $x^2 + 2x + 5 = 0 \;\Leftrightarrow\; x = \dfrac{-2 \pm \sqrt{2^2 - 4(1)(5)}}{2(1)} = \dfrac{-2 \pm \sqrt{-16}}{2} = \dfrac{-2 \pm 4i}{2} = -1 \pm 2i.$

23. By the quadratic formula, $z^2 + z + 2 = 0 \;\Leftrightarrow\; z = \dfrac{-1 \pm \sqrt{1^2 - 4(1)(2)}}{2(1)} = \dfrac{-1 \pm \sqrt{-7}}{2} = -\dfrac{1}{2} \pm \dfrac{\sqrt{7}}{2}i.$

25. For $z = -3 + 3i$, $r = \sqrt{(-3)^2 + 3^2} = 3\sqrt{2}$ and $\tan\theta = \frac{3}{-3} = -1 \;\Rightarrow\; \theta = \frac{3\pi}{4}$ (since z lies in the second quadrant).

   Therefore, $-3 + 3i = 3\sqrt{2}\left(\cos\frac{3\pi}{4} + i\sin\frac{3\pi}{4}\right)$.

27. For $z = 3 + 4i$, $r = \sqrt{3^2 + 4^2} = 5$ and $\tan\theta = \frac{4}{3} \;\Rightarrow\; \theta = \tan^{-1}\left(\frac{4}{3}\right)$ (since z lies in the first quadrant). Therefore,

   $3 + 4i = 5\left[\cos\left(\tan^{-1}\frac{4}{3}\right) + i\sin\left(\tan^{-1}\frac{4}{3}\right)\right]$.

29. For $z = \sqrt{3} + i$, $r = \sqrt{\left(\sqrt{3}\right)^2 + 1^2} = 2$ and $\tan\theta = \frac{1}{\sqrt{3}} \;\Rightarrow\; \theta = \frac{\pi}{6} \;\Rightarrow\; z = 2\left(\cos\frac{\pi}{6} + i\sin\frac{\pi}{6}\right)$.

   For $w = 1 + \sqrt{3}\, i$, $r = 2$ and $\tan\theta = \sqrt{3} \;\Rightarrow\; \theta = \frac{\pi}{3} \;\Rightarrow\; w = 2\left(\cos\frac{\pi}{3} + i\sin\frac{\pi}{3}\right)$.

   Therefore, $zw = 2 \cdot 2\left[\cos\left(\frac{\pi}{6} + \frac{\pi}{3}\right) + i\sin\left(\frac{\pi}{6} + \frac{\pi}{3}\right)\right] = 4\left(\cos\frac{\pi}{2} + i\sin\frac{\pi}{2}\right)$,

   $z/w = \frac{2}{2}\left[\cos\left(\frac{\pi}{6} - \frac{\pi}{3}\right) + i\sin\left(\frac{\pi}{6} - \frac{\pi}{3}\right)\right] = \cos\left(-\frac{\pi}{6}\right) + i\sin\left(-\frac{\pi}{6}\right)$, and $1 = 1 + 0i = 1(\cos 0 + i\sin 0) \;\Rightarrow\;$

$1/z = \frac{1}{2}\left[\cos\left(0 - \frac{\pi}{6}\right) + i\sin\left(0 - \frac{\pi}{6}\right)\right] = \frac{1}{2}\left[\cos\left(-\frac{\pi}{6}\right) + i\sin\left(-\frac{\pi}{6}\right)\right]$. For $1/z$, we could also use the formula that precedes

Example 5 to obtain $1/z = \frac{1}{2}\left(\cos\frac{\pi}{6} - i\sin\frac{\pi}{6}\right)$.

31. For $z = 2\sqrt{3} - 2i$, $r = \sqrt{\left(2\sqrt{3}\right)^2 + (-2)^2} = 4$ and $\tan\theta = \frac{-2}{2\sqrt{3}} = -\frac{1}{\sqrt{3}} \;\Rightarrow\; \theta = -\frac{\pi}{6} \;\Rightarrow$

$z = 4\left[\cos\left(-\frac{\pi}{6}\right) + i\sin\left(-\frac{\pi}{6}\right)\right]$. For $w = -1 + i$, $r = \sqrt{2}$, $\tan\theta = \frac{1}{-1} = -1 \;\Rightarrow\; \theta = \frac{3\pi}{4} \;\Rightarrow$

$w = \sqrt{2}\left(\cos\frac{3\pi}{4} + i\sin\frac{3\pi}{4}\right)$. Therefore, $zw = 4\sqrt{2}\left[\cos\left(-\frac{\pi}{6} + \frac{3\pi}{4}\right) + i\sin\left(-\frac{\pi}{6} + \frac{3\pi}{4}\right)\right] = 4\sqrt{2}\left(\cos\frac{7\pi}{12} + i\sin\frac{7\pi}{12}\right)$,

$z/w = \frac{4}{\sqrt{2}}\left[\cos\left(-\frac{\pi}{6} - \frac{3\pi}{4}\right) + i\sin\left(-\frac{\pi}{6} - \frac{3\pi}{4}\right)\right] = \frac{4}{\sqrt{2}}\left[\cos\left(-\frac{11\pi}{12}\right) + i\sin\left(-\frac{11\pi}{12}\right)\right] = 2\sqrt{2}\left(\cos\frac{13\pi}{12} + i\sin\frac{13\pi}{12}\right)$, and

$1/z = \frac{1}{4}\left[\cos\left(-\frac{\pi}{6}\right) - i\sin\left(-\frac{\pi}{6}\right)\right] = \frac{1}{4}\left(\cos\frac{\pi}{6} + i\sin\frac{\pi}{6}\right)$.

33. For $z = 1 + i$, $r = \sqrt{2}$ and $\tan\theta = \frac{1}{1} = 1 \;\Rightarrow\; \theta = \frac{\pi}{4} \;\Rightarrow\; z = \sqrt{2}\left(\cos\frac{\pi}{4} + i\sin\frac{\pi}{4}\right)$. So by De Moivre's Theorem,

$$(1+i)^{20} = \left[\sqrt{2}\left(\cos\frac{\pi}{4} + i\sin\frac{\pi}{4}\right)\right]^{20} = (2^{1/2})^{20}\left(\cos\frac{20\cdot\pi}{4} + i\sin\frac{20\cdot\pi}{4}\right) = 2^{10}(\cos 5\pi + i\sin 5\pi)$$

$$= 2^{10}[-1 + i(0)] = -2^{10} = -1024$$

35. For $z = 2\sqrt{3} + 2i$, $r = \sqrt{\left(2\sqrt{3}\right)^2 + 2^2} = \sqrt{16} = 4$ and $\tan\theta = \frac{2}{2\sqrt{3}} = \frac{1}{\sqrt{3}} \;\Rightarrow\; \theta = \frac{\pi}{6} \;\Rightarrow\; z = 4\left(\cos\frac{\pi}{6} + i\sin\frac{\pi}{6}\right)$.

So by De Moivre's Theorem,

$$\left(2\sqrt{3} + 2i\right)^5 = \left[4\left(\cos\frac{\pi}{6} + i\sin\frac{\pi}{6}\right)\right]^5 = 4^5\left(\cos\frac{5\pi}{6} + i\sin\frac{5\pi}{6}\right) = 1024\left[-\frac{\sqrt{3}}{2} + \frac{1}{2}i\right] = -512\sqrt{3} + 512i.$$

37. $1 = 1 + 0i = 1(\cos 0 + i\sin 0)$. Using Equation 3 with $r = 1$, $n = 8$, and $\theta = 0$, we have

$w_k = 1^{1/8}\left[\cos\left(\frac{0 + 2k\pi}{8}\right) + i\sin\left(\frac{0 + 2k\pi}{8}\right)\right] = \cos\frac{k\pi}{4} + i\sin\frac{k\pi}{4}$, where $k = 0, 1, 2, \ldots, 7$.

$w_0 = 1(\cos 0 + i\sin 0) = 1$, $w_1 = 1\left(\cos\frac{\pi}{4} + i\sin\frac{\pi}{4}\right) = \frac{1}{\sqrt{2}} + \frac{1}{\sqrt{2}}i$,

$w_2 = 1\left(\cos\frac{\pi}{2} + i\sin\frac{\pi}{2}\right) = i$, $w_3 = 1\left(\cos\frac{3\pi}{4} + i\sin\frac{3\pi}{4}\right) = -\frac{1}{\sqrt{2}} + \frac{1}{\sqrt{2}}i$,

$w_4 = 1(\cos\pi + i\sin\pi) = -1$, $w_5 = 1\left(\cos\frac{5\pi}{4} + i\sin\frac{5\pi}{4}\right) = -\frac{1}{\sqrt{2}} - \frac{1}{\sqrt{2}}i$,

$w_6 = 1\left(\cos\frac{3\pi}{2} + i\sin\frac{3\pi}{2}\right) = -i$, $w_7 = 1\left(\cos\frac{7\pi}{4} + i\sin\frac{7\pi}{4}\right) = \frac{1}{\sqrt{2}} - \frac{1}{\sqrt{2}}i$

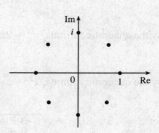

39. $i = 0 + i = 1\left(\cos\frac{\pi}{2} + i\sin\frac{\pi}{2}\right)$. Using Equation 3 with $r = 1$, $n = 3$, and $\theta = \frac{\pi}{2}$, we have

$w_k = 1^{1/3}\left[\cos\left(\frac{\frac{\pi}{2} + 2k\pi}{3}\right) + i\sin\left(\frac{\frac{\pi}{2} + 2k\pi}{3}\right)\right]$, where $k = 0, 1, 2$.

$w_0 = \left(\cos\frac{\pi}{6} + i\sin\frac{\pi}{6}\right) = \frac{\sqrt{3}}{2} + \frac{1}{2}i$

$w_1 = \left(\cos\frac{5\pi}{6} + i\sin\frac{5\pi}{6}\right) = -\frac{\sqrt{3}}{2} + \frac{1}{2}i$

$w_2 = \left(\cos\frac{9\pi}{6} + i\sin\frac{9\pi}{6}\right) = -i$

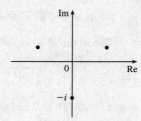

41. Using Euler's formula (6) with $y = \frac{\pi}{2}$, we have $e^{i\pi/2} = \cos\frac{\pi}{2} + i\sin\frac{\pi}{2} = 0 + 1i = i$.

43. Using Euler's formula (6) with $y = \dfrac{\pi}{3}$, we have $e^{i\pi/3} = \cos\dfrac{\pi}{3} + i\sin\dfrac{\pi}{3} = \dfrac{1}{2} + \dfrac{\sqrt{3}}{2}i$.

45. Using Equation 7 with $x = 2$ and $y = \pi$, we have $e^{2+i\pi} = e^2 e^{i\pi} = e^2(\cos\pi + i\sin\pi) = e^2(-1 + 0) = -e^2$.

47. Take $r = 1$ and $n = 3$ in De Moivre's Theorem to get

$$[1(\cos\theta + i\sin\theta)]^3 = 1^3(\cos 3\theta + i\sin 3\theta)$$

$$(\cos\theta + i\sin\theta)^3 = \cos 3\theta + i\sin 3\theta$$

$$\cos^3\theta + 3(\cos^2\theta)(i\sin\theta) + 3(\cos\theta)(i\sin\theta)^2 + (i\sin\theta)^3 = \cos 3\theta + i\sin 3\theta$$

$$\cos^3\theta + (3\cos^2\theta\,\sin\theta)i - 3\cos\theta\,\sin^2\theta - (\sin^3\theta)i = \cos 3\theta + i\sin 3\theta$$

$$(\cos^3\theta - 3\sin^2\theta\,\cos\theta) + (3\sin\theta\,\cos^2\theta - \sin^3\theta)i = \cos 3\theta + i\sin 3\theta$$

Equating real and imaginary parts gives $\cos 3\theta = \cos^3\theta - 3\sin^2\theta\,\cos\theta$ and $\sin 3\theta = 3\sin\theta\,\cos^2\theta - \sin^3\theta$.

49. $F(x) = e^{rx} = e^{(a+bi)x} = e^{ax+bxi} = e^{ax}(\cos bx + i\sin bx) = e^{ax}\cos bx + i(e^{ax}\sin bx)$ $\Rightarrow$

$$F'(x) = (e^{ax}\cos bx)' + i(e^{ax}\sin bx)'$$

$$= (ae^{ax}\cos bx - be^{ax}\sin bx) + i(ae^{ax}\sin bx + be^{ax}\cos bx)$$

$$= a[e^{ax}(\cos bx + i\sin bx)] + b[e^{ax}(-\sin bx + i\cos bx)]$$

$$= ae^{rx} + b[e^{ax}(i^2\sin bx + i\cos bx)]$$

$$= ae^{rx} + bi[e^{ax}(\cos bx + i\sin bx)] = ae^{rx} + bie^{rx} = (a + bi)e^{rx} = re^{rx}$$